Ionel Haiduc, Frank T. Edelmann

Supramolecular Organometallic Chemistry

Ionel Haiduc,
Frank T. Edelmann

Supramolecular Organometallic Chemistry

Weinheim · New York · Chichester
Brisbane · Singapore · Toronto

Prof. Dr. Ionel Haiduc
Facultatea de Chimie
Universitatea Babes-Bolyai
RO-3400 Cluj-Napoca
Roumania

Prof. Dr. Frank T. Edelmann
Chemisches Institut der Universität Magdeburg
Universitätsplatz 21
D-39106 Magdeburg
Germany

This book was carefully produced. Nevertheless, authors and publisher do not warrant the information contained therein to be free of errors. Readers are advised to keep in mind that statements, data, illustrations, procedural details or other items may inadvertently be inaccurate.

Library of Congress Card No.: applied for

A catalogue record for this book is available from the British Library.

Deutsche Bibliothek Cataloguing-in-Publication Data:
Haiduc, Ionel:
Supramolecular organometallic chemistry / Ionel Haiduc ; Frank T. Edelmann. –
Weinheim ; New York ; Chichester ; Brisbane ; Singapore ; Toronto ; Wiley-VCH, 1999
ISBN 3-527-29533-X

© WILEY-VCH Verlag GmbH, D-69469 Weinheim (Federal Republic of Germany). 1999

Printed on acid-free and chlorine-free paper.

All rights reserved (including those of translation in other languages). No part of this book may be reproduced in any form – by photoprinting, microfilm, or any other means – nor transmitted or translated into machine language without written permission from the publishers. Registered names, trademarks, etc. used in this book, even when not specifically marked as such, are not to be considered unprotected by law.
Composition: Asco Typesetters, Hong Kong.
Printing: betz-druck, gmbh, D-64291 Darmstadt.
Bookbinding: J. Schäffer GmbH 810. KG, D-67269 Grünstadt.
Printed in the Federal Republic of Germany.

Foreword

Supramolecular chemistry has become a major field of chemistry. Its basic concepts have penetrated vast areas covering the whole domain of chemistry and its interfaces with biology, physics, and materials science. They have provided novel grids through which to read, novel points of view from which to consider, novel perspectives towards which to project the different established domains of chemistry. This is true of organic as well as inorganic, physical and analytical, and biological and medicinal chemistry. It is therefore inevitable and highly desirable that organometallic chemistry, an area of great breadth, strong impact, and rapid development be also analyzed and characterized in terms of its relationship to supramolecular chemistry. Indeed by the variety of bonding types and of constituents it presents, organometallic chemistry entertains an especially rich set of connections with supramolecular chemistry.

This book by Ionel Haiduc and Frank Edelmann fills the need. It is particularly welcome for its timeliness and even more for its intrinsic qualities. Indeed, it provides a broad coverage, a thorough description, and incisive analysis of organometallic compounds and bonding features in terms of the concepts of, and relationships with, supramolecular chemistry.

This book is, of course, a must for organometallic chemists and supramolecular chemists alike, but it is also of great value for all other types of chemist who at one time or another encounter these two domains.

The authors are to be warmly congratulated for their excellent job and for the service they render to the chemical community, and for carrying the flame of supramolecular chemistry into yet another domain, thus setting the stage of supramolecular organometallic chemistry.

Strasbourg, December 1998   *Jean-Marie Lehn*

Author's Preface

This book covers a topic at the interface between two major directions of current research – organometallic and supramolecular chemistry.

Supramolecular chemistry is a young branch of science, growing with teenage enthusiasm and approaching maturity. By dealing with intermolecular forces and interactions and using *molecules as building blocks* for supramolecular architectures, supramolecular chemistry covers very broad fields. On the other hand, organometallic chemistry, now a mature science, offers a wide range of interesting and even unusual molecules to be used as building blocks in more complex, supramolecular structures.

When their studies went *beyond the molecule*, quite often organometallic chemists were surprised to find unexpected intermolecular associations, i.e. self-assembly and self-organization. Thus, they stepped (sometimes unconsciously) into the fertile field of supramolecular chemistry.

In this book we have tried to apply the basic concepts and principles of supramolecular science to organometallic chemistry and to collect as many as possible examples of supramolecular interactions and structures of organometallics. Although we have tried to be comprehensive, we realize that an exhaustive treatment is impossible and we know that many examples might have escaped our attention. We apologize to the authors of such uncited work and would be grateful to anyone who would mention these to us. Most overlooked examples probably come from transition metal chemistry, a very vast area which would require a second volume for adequate coverage. We hope, however, that by accepting the basic concepts and principles of supramolecular chemistry, introduced in the first chapter of the book, the reader will be able to recognize examples of supramolecular interactions and structures not included in these pages. If so, the main purpose of this book was achieved. The book should be regarded only as illustrative rather than exhaustive.

In writing this book we have enjoyed the encouragement of Professor Jean-Marie Lehn, the father of supramolecular chemistry and we are extremely honored and grateful to him for writing the Foreword, which underscores the task assumed by us.

The book was initiated, and in the most part written, during a Humboldt Fellowship awarded to one of us (IH) and spent at the Universität Magdeburg. We are grateful to the Humboldt Foundation for making this possible. For the use of library facilities we also thank the Anorganisch-Chemisches Institut, Universität

Göttingen (Professor H. W. Roesky) and the Institut für Anorganische Chemie, Universität Braunschweig (Professor R. Schmutzler and Dr. I. Neda). One of us (IH) is also thankful to the Babes-Bolyai University and to the Romanian Academy for leave of absence, and to the Universität Magdeburg for hospitality. Thanks are also addressed to the NATO Science Division for supporting the participation of IH in NATO Advanced Research Workshops on Supramolecular Chemistry and for a Collaborative Research Grant. We also acknowledge support of our activity in supramolecular chemistry by the National Council for Academic Research (Bucharest), the Romanian Ministry of Research and Technology, and the Deutsche Forschungsgemeinschaft, through research grants to IH and FTE, respectively.

Any critical observations and suggestions will be accepted with pleasure, with the hope that they will be used in a future, improved, edition.

December 1998

Ionel Haiduc
Frank T. Edelmann

Table of Contents

1 Basic Concepts and Principles   1
  1.1 Definitions   1
       1.1.1 Supramolecular chemistry   1
       1.1.2 Molecular recognition and host–guest interactions   3
       1.1.3 Self-assembly and self-organization   4
  1.2 Intermolecular Bond Types in Organometallic Supramolecular Systems   9
       1.2.1 Dative bonding (electron-pair donor–acceptor bonding or Lewis acid–base interactions)   9
       1.2.2 The secondary bond concept   14
       1.2.3 Hydrogen-bond interactions   18
       1.2.4 Ionic interactions   20
       1.2.5 π-Bonding interactions   22

2 Molecular Recognition and Host–Guest Interactions   27
  2.1 Organometallic Receptors and their Host–Guest Complexes   27
       2.1.1 Organomercury macrocyclic receptors   27
       2.1.2 Organocopper and organosilver potential receptors   29
       2.1.3 Organocyclosiloxane receptors   29
       2.1.4 Organotin macrocyclic receptors   35
       2.1.5 Ferrocene-containing coronands and cryptands   37
            2.1.5.1 Ferrocene polyoxa coronands   37
            2.1.5.2 Ferrocene polyaza–oxa coronands and cryptands   40
            2.1.5.3 Ferrocene polyaza coronands and cryptands   45
            2.1.5.4 Ferrocene polyoxa-thia coronands   48
            2.1.5.5 Ferrocene polythia coronands   49
            2.1.5.6 Non-cyclic ferrocene receptors   50
       2.1.6 Cobaltocenium receptors   52
       2.1.7 Other metallocene receptors   56
       2.1.8 Calixarene receptors modified with organometallic groups   58
       2.1.9 Organometallic cyclotriveratrylene receptors   64

2.2	Organometallic Guests		65
	2.2.1 Organometallic guests in inorganic hosts		65
		2.2.1.1 Organometallics in zeolites and mesoporous silica	65
		2.2.1.2 Organometallic compounds in layered chalcogenides, oxohalides, and oxides	68
	2.2.2 Organometallic guests in organic hosts		70
		2.2.2.1 Organometallic complexes of crown ethers, cryptands and related receptors	70
		2.2.2.2 Organometallic molecules in cyclodextrin receptors	76
		2.2.2.3 Organometallics in the thiourea lattice host	78

3 Supramolecular Self-Assembly by Dative Bonding (Electron-Pair Donor–Acceptor or Lewis Acid–Base Interactions) — 95

3.1	Group 12 Metals – Zinc, Cadmium, Mercury		95
	3.1.1 Self-assembly of organozinc compounds		95
		3.1.1.1 Organozinc halides	95
		3.1.1.2 Organozinc–oxygen compounds	96
		3.1.1.3 Organozinc–sulfur compounds	100
		3.1.1.4 Organozinc–selenium compounds	102
		3.1.1.5 Organozinc–nitrogen compounds	103
	3.1.2 Self-assembly of organocadmium compounds		104
	3.1.3 Self-assembly of organomercury compounds		106
3.2	Group 13 Metals – Aluminum, Gallium, Indium, Thallium		107
	3.2.1 Self-assembly of organoaluminum compounds		108
		3.2.1.1 Organoaluminum halides	108
		3.2.1.2 Organoaluminum–oxygen derivatives	109
		3.2.1.3 Organoaluminum–sulfur derivatives	114
		3.2.1.4 Organoaluminum–selenium and tellurium derivatives	116
		3.2.1.5 Organoaluminum–nitrogen compounds	116
		3.2.1.6 Organoaluminum–phosphorus and organoaluminum–arsenic compounds	122
	3.2.2 Self-assembly of organogallium compounds		123
		3.2.2.1 Organogallium halides	123
		3.2.2.2 Organogallium–oxygen compounds	124
		3.2.2.3 Organogallium–sulfur compounds	126
		3.2.2.4 Organogallium–selenium and –tellurium compounds	127
		3.2.2.5 Organogallium–nitrogen compounds	128
		3.2.2.6 Organogallium–phosphorus, –arsenic and –antimony compounds	131
	3.2.3 Self-assembly of organoindium compounds		133
		3.2.3.1 Organoindium–halogen derivatives	134
		3.2.3.2 Organoindium–oxygen derivatives	135
		3.2.3.3 Organoindium–sulfur and –selenium compounds	136
		3.2.3.4 Organoindium–nitrogen derivatives	137

			3.2.3.5	Organoindium–phosphorus and –arsenic derivatives	139

	3.2.4	Self-assembly of organothallium compounds	141
3.3	Group 14 Metals – Tin and Lead		146
	3.3.1	Self-assembly of organotin compounds	146
		3.3.1.1 Organotin halides	146
		3.3.1.2 Dimeric organodistannoxanes $[XR_2SnOSnR_2X]_2$ and related compounds	148
		3.3.1.3 Organotin hydroxides, alkoxides, and related compounds	150
		3.3.1.4 Organotin carboxylates	156
		3.3.1.5 Monoorganotin oxo-carboxylates	168
		3.3.1.6 Organotin derivatives of inorganic acids	171

4 Supramolecular Self-Assembly by Formation of Secondary Bonds — 195

- 4.1 Homoatomic Interactions — 195
- 4.2 Group 12 Metals – Zinc, Cadmium, Mercury — 202
  - 4.2.1 Self-assembly of organomercury compounds — 202
    - 4.2.1.1 Organomercury halides — 202
    - 4.2.1.2 Organomercury–oxygen compounds — 205
    - 4.2.1.3 Organomercury–sulfur compounds — 209
    - 4.2.1.4 Organomercury–nitrogen compounds — 212
- 4.3 Group 13 Metals – Gallium, Indium, Thallium — 213
  - 4.3.1 Self-assembly of organogallium compounds — 213
  - 4.3.2 Self-assembly of organoindium compounds — 214
    - 4.3.2.1 Organoindium halides — 214
    - 4.3.2.2 Organoindium–oxygen derivatives — 216
    - 4.3.2.3 Organoindium–nitrogen derivatives — 217
  - 4.3.3 Self-assembly of organothallium compounds — 218
    - 4.3.3.1 Organothallium halides — 218
    - 4.3.3.2 Organothallium–sulfur compounds — 220
    - 4.3.3.3 Organothallium–nitrogen compounds — 224
- 4.4 Group 14 – Tin and Lead — 227
  - 4.4.1 Self-assembly of organotin compounds — 227
    - 4.4.1.1 Organotin halides — 227
    - 4.4.1.2 Organotin–oxygen compounds — 234
    - 4.4.1.3 Organotin–sulfur compounds — 244
    - 4.4.1.4 Organotin–nitrogen compounds — 249
  - 4.4.2 Self-assembly in organolead compounds — 259
- 4.5 Group 15 – Arsenic, Antimony, Bismuth — 265
  - 4.5.1 Self-assembly in organoarsenic compounds — 265
  - 4.5.2 Self-assembly in organoantimony compounds — 268
    - 4.5.2.1 Organoantimony halides — 268
    - 4.5.2.2 Organoantimony–oxygen compounds — 274
    - 4.5.2.3 Organoantimony–sulfur compounds — 276
    - 4.5.2.4 Organoantimony–nitrogen compounds — 282

		4.5.3	Self-assembly in organobismuth compounds	282
			4.5.3.1 Organobismuth–halogen compounds	282
			4.5.3.2 Organobismuth–oxygen compounds	287
			4.5.3.3 Organobismuth–sulfur compounds	288
			4.5.3.4 Organobismuth–nitrogen compounds	290
	4.6	Group 16 – Selenium and Tellurium		290
		4.6.1	Self-assembly in organoselenium compounds	290
		4.6.2	Self-assembly in organotellurium compounds	293
			4.6.2.1 Organotellurium–halogen compounds	293
			4.6.2.2 Organotellurium–oxygen compounds	298
			4.6.2.3 Organotellurium–sulfur compounds	299
			4.6.2.4 Organotellurium–nitrogen compounds	301
5	**Supramolecular Self-Assembly by Hydrogen-Bond Interactions**			**319**
	5.1	Introduction		319
	5.2	Group 14 Elements – Si, Ge, Sn, Pb		319
		5.2.1	Organosilanols and organosiloxanols	319
		5.2.2	Organogermanium hydroxides	327
		5.2.3	Organotin compounds	328
	5.3	Group 15 Elements – As, Sb, Bi		333
		5.3.1	Organoarsenic compounds	333
		5.3.2	Organoantimony compounds	335
	5.4	Transition Metals (Organometallic Crystal Engineering)		337
		5.4.1	Peripheral hydrogen-bonding	338
			5.4.1.1 O–H···O hydrogen bonds involving carboxyl groups	338
			5.4.1.2 O–H···O hydrogen bonds involving hydroxyl groups	342
			5.4.1.3 N–H···O hydrogen bonds involving amido groups	352
			5.4.1.4 C–H···O hydrogen bonds involving carbonyl ligands	352
		5.4.2	Hydrogen bonds involving transition metal atoms	357
			5.4.2.1 M–H···O≡C intermolecular hydrogen bonds	358
		5.4.3	Hydrogen-bonds assisted by ionic interactions	359
6	**Supramolecular Self-Assembly Caused by Ionic Interactions**			**371**
	6.1	Alkali Metals		371
		6.1.1	Hydrocarbyls	371
			6.1.1.1 Base-free alkali metal hydrocarbyls	373
			6.1.1.2 Solvated alkali metal hydrocarbyls	377
		6.1.2	Amides and related compounds	386
			6.1.2.1 Amides and related nitrogen derivatives	387
			6.1.2.2 Phosphides and arsenides	398
		6.1.3	Alkoxides and their higher homologs	402
	6.2	Alkaline Earth Metals		415

7	**Supramolecular Self-Assembly as a Result of π-Interactions**		425
	7.1	Introduction	425
	7.2	Main Group Elements	425
		7.2.1 π-Interactions with acetylenes and allyl ligands	425
		7.2.2 Cyclopentadienyl, indenyl, and fluorenyl complexes	426
		7.2.3 Five-membered heterocycles	441
		7.2.4 π-Interactions with arene rings	441
	7.3	Transition Elements	452
		7.3.1 Olefin complexes	452
		7.3.2 Cyclopentadienyl complexes	452
		7.3.3 Arene complexes	454
	7.4	*f*-Elements	455
		7.4.1 Acetylide complexes	456
		7.4.2 Cyclopentadienyl complexes	456
		7.4.3 Five-membered heterocycles	461
		7.4.4 Arene complexes	462
Subject Index			469

1 Basic Concepts and Principles

1.1 Definitions

1.1.1 Supramolecular chemistry

Because the basic concepts and principles of supramolecular organometallic chemistry are not necessarily familiar to every organometallic chemist, the basic concepts and principles of supramolecular chemistry and its specific vocabulary, which created a new language in chemistry, will be briefly introduced here. Chemists are, in general, preoccupied with the molecular structure of their compounds, i.e. with the nature and reciprocal position or arrangement of the atoms and the nature of bonding forces which hold together the atoms in a molecule. Only relatively recently have chemists become intensively preoccupied with intermolecular forces, acting mainly in condensed phases, and with the structures of the systems which might result from such intermolecular interaction. The subject became particularly fashionable after J. M. Lehn [1], D. J. Cram [2], and C. J. Pedersen [3] received the Nobel Prize for their work originating from the study of crown ethers, further extended over the so-called cryptands, carcerands, and other species capable of trapping ions or molecules in their cavities [4]. New ideas about molecular recognition and '*host–guest*' chemistry [2a] added new dimensions to the field of intermolecular interactions. Thus, Supramolecular Chemistry was born. Since every subject has a history, it should be mentioned that a book about intermolecular interactions was published in German, in 1927 by P. Pfeiffer [5] and the term '*Übermoleküle*' was used to describe systems formed by hydrogen-bond association of carboxylic acids with formation of dimeric species [6].

It seems that the term '*supramolecular chemistry*' was first used in 1978 by J. M. Lehn with the statement: "*Just as there is a field of molecular chemistry based on the covalent bond, there is a field of supramolecular chemistry, the chemistry of molecular assemblies and of the intermolecular bond*" [7]. Today supramolecular chemistry is defined as "*the chemistry of molecular assemblies and of the intermolecular bond*". It is the "*chemistry beyond the molecule* and deals with *organized entities of higher complexity that result from the association of two or more chemical species held together by intermolecular forces*" [1, 8]. This definition is very generous and covers a very broad area of chemical phenomena and structures and extends to biological molecules, coordination compounds, and new materials. The new field is attracting an enormous interest, and to get an impression of the broad area it covers and about its fuzzy borders a recent eleven-volume treatise [9] is recommended for consultation.

By operating with covalent bonds, both organic and inorganic chemistry have created a large diversity of molecules, with remarkable compositions and structures. The power of organic chemistry and its ability to create almost any designed organic molecule is illustrated by the synthesis of most fascinating natural compounds, but also of structures which have never before existed. Inorganic chemistry in turn, followed with the synthesis of carbon-free organic-like compounds, such as inorganic heterocycles [10–12] and polymers [13]. Cooperation between organic and inorganic chemistry resulted in wonderful coordination compounds (with organic ligands coordinated to metal atoms or ions through intermediate donor atoms such as oxygen, nitrogen, sulfur, phosphorus, etc.), some identified as important biological molecules [14]. The direct attachment of organic groups to metal and semimetal atoms has, furthermore, led to the development of a great diversity of no less fascinating organometallic compounds, many of which are biologically active [15]. To be defined as *organometallic* a chemical compound should contain a direct metal–carbon bond, M–C, where M is an element more electropositive than carbon, which results in a $M^{\delta+}-C^{\delta-}$ bond polarization [16]. According to this definition, derivatives of non-metals such as boron, silicon, germanium, arsenic, and antimony are also referred to as organometallic. There is also a great diversity of compounds containing organic groups attached to a metal via oxygen, nitrogen or sulfur, e.g. metal alkoxides, amides, thiolates; sometimes these are described as *metal–organic*.

This book will deal with supramolecular aspects of organometallic chemistry, but not exclusively with authentic organometallic compounds. Metal–organic derivatives will also be considered occasionally, especially in the discussion of ionic supramolecular aggregates of alkali and alkaline earth metals, e.g. amides. This area will not, however, be comprehensively covered. Thus, supramolecular associations, i.e. self-assembly and self-organization, host–guest chemistry and related aspects of organometallic chemistry will be presented.

As currently defined, supramolecular chemistry deals with two major aspects:

i) the organization of molecular units into **supramolecular assemblies** *or systems*, also called *supramolecular arrays*, i.e. "*polymolecular entities that result from the spontaneous* **association of a large undefined number of components**" and
ii) **supermolecules**, i.e. "*well-defined discrete oligomolecular species that result from the intermolecular* **association of a few components**" [1, 8, 9]. In the solid state the supermolecules and supramolecular arrays can associate with one another to yield gigantic *macroscopic conglomerates*, i.e. supramolecular structures of higher order [17].

According to the definition of supramolecular chemistry cited above, the smallest *supermolecule* can be a dimer; at the other extreme we find very large, polymeric aggregates containing an infinite number of building units held together by non-covalent intermolecular forces, i.e. the *supramolecular array*. In both cases, the spatial arrangement of the building units and the nature of forces which hold them together define the supramolecular architecture and determine their properties. The types of non-covalent interactions between the building blocks of a supramolecular system are: metal-ion coordination (dative bonds), electrostatic forces, hydrogen-bonding, donor–acceptor interactions, even van der Waals interactions, etc [8]. Another type, i.e. secondary bonds or semibonding interactions will also be considered in this book. Compared with classical coordination chemistry, in which a central (metal) atom (ion) is surrounded (coordinated) by ligands, supramolecular chemistry deals with a broader area of species, consisting of a *receptor* (e.g. a host molecule) and a *substrate* (e.g. a guest atom, ion or molecule). Thus, supramolecular chemistry became a '*generalized coordination chemistry*' [18].

Supermolecules include host–guest complexes [19], but also molecular species such as large ring systems, catenanes, rotaxanes [20], etc. In highly branched macromolecules, called dendrimers or arboroles [21, 21], the building blocks are connected through covalent bonds, yet they are sometimes described as supermolecules. These will not be covered in this book.

1.1.2 Molecular recognition and host–guest interactions

The formation of a supramolecular architecture through intermolecular interactions requires so-called *molecular recognition*. This implies the presence of a built-in *information* consisting of a *geometric (steric) and energetic (electronic) double complementarity* between the interacting species. The information required for intermolecular recognition is stored in the *binding sites* of the interacting molecules. The binding sites are characterized by steric or geometric properties (shape, size, number, position) and energetic or electronic properties (charge, polarity, polarizability, presence of electron pairs or vacant sites, etc.). A large contact area between the interacting species and the presence of multiple interaction sites might result in a strong overall binding, although individual non-covalent interactions acting in the process of supramolecular organization might be much weaker than covalent bonds. Only when such organization is achieved through dative or Lewis acid–base interactions does the strength of each individual connection (bond) approach that of the covalent bond.

Molecular recognition is well illustrated by the complexation between crown ethers and alkali metal ions [22]. Thus, the size and the shape of the cavity of the polyether macrocycle favors binding of a spherical, e.g. alkali metal, cation. Although for alkali metals the M···O interaction forces are weak, the cumulative effect of several binding sites leads to strong complexation. Examples from organometallic chemistry are also known and will be discussed in the appropriate place. For illustration we mention here only the potassium complex, **1**, of a permethylcycloheptasiloxane, an organosilicon analog of a typical crown ether complex, **2**.

A remarkable process is the recognition of anions [22a] by macrocycles containing ammonium sites, **3**. Similar performance can be achieved with macrocycles which contain metal atoms as binding sites, e.g. organotin macrocycles, **4**.

In the examples cited the macrocyclic ligands play the role of *molecular receptors*. Molecular receptors are defined by J. M. Lehn as "*organic structures held by covalent bonds, that are able to bind selectively ionic or molecular substrates (or both) by means of various intermolecular interactions, leading to an assembly of two or more species, a supramolecule*". The molecular receptor is a host and the bonded substrate is a guest. In architectures with macrocyclic receptors the host is an *endo* receptor, i.e. the non-covalent binding interactions are convergently oriented towards the centre of a concave receptor. *exo*-Receptors are also possible when the orientation of the binding interactions is directed outwards of the molecular structure of the receptor. We will show in Chapter 2 how organometallic receptors can perform the same function of binding cationic or anionic substrates.

An important role play some inorganic hosts, and a broad diversity of fascinating supramolecular architectures has been developed on this basis [23, 24]. The incorporation of organometallic guests into inorganic hosts will be dealt with briefly in Section 2.2.1.

1.1.3 Self-assembly and self-organization

Self-assembly is a process best described as *a spontaneous association of molecules under equilibrium conditions into stable aggregates held together by non-covalent*

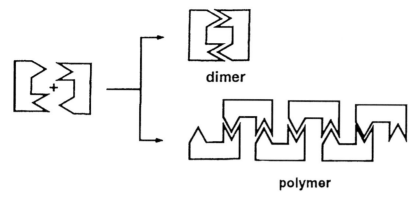

Fig. 1.1. Schematic illustration of geometric self-complementarity leading to formation of dimeric supermolecules and polymeric supramolecular arrays.

bonds [8, 25]. Self-assembly can occur in solution or in the process of crystallization. In crystallization the supramolecular aggregates exist only in solid state and dissolution can lead to partial or complete dissociation into molecular components.

The selective processes which determine the growth of some nuclei into a crystal lattice are little understood at the molecular level and some recent static and dynamic calculations have been performed on small organic molecules in liquids, solutions, and crystals aimed at throwing some light upon these processes [26]. The results might be also relevant to organometallic species.

When the product of self-assembly is a structurally well defined aggregate the term *self-organization* is used to describe the intermolecular association. Such phenomena are frequent in complex biological systems, but many other molecules can self-organize into very interesting supramolecular structures.

Classical coordination compounds are formed by interactions between acceptor species (Lewis acids) and donor species (Lewis bases). When potential acceptor and donor sites are present simultaneously in the same molecule these can be described as *self-complementary* and their reciprocal recognition can lead to self-assembly (or self-organization). This often occurs with organometallic compounds. For self-assembly to occur an additional steric fitness or geometric complementarity is required, as shown schematically in Figure 1.1.

To illustrate these ideas we present the example of self-complementary organomercury dithiophosph(in)ate molecules, $RHgS(S)PR'_2$ (where R = alkyl, aryl, R' = alkyl, aryl or alkoxy group). Although the valences of all component atoms are satisfied in these molecules, the mercury atom retains some chemical affinity as a potentially weak acceptor and sulfur sites in the same molecule display weak donor properties. This built-in information is used by the molecules when they self-assemble into dimeric supermolecules, **5**, or polymeric supramolecular arrays, **6**, through weak intermolecular forces called secondary bonds (see Section 1.2.2 and Chapter 4). In these supramolecular structures the bridging dithiophosph(in)ato ligands connect the organometallic moieties RHg into a self-organized structure.

Self-assembly processes are very frequent in organometallic compounds when organometallic moieties R_nM are connected through 'anionic' groups. Thus, for example, monofunctional $R_3M^{IV}X$ derivatives can form supramolecular aggregates, as either linear zigzag or helical chains, with functional groups X, such as halogens, hydroxo, alkoxo, thiolato (to give monoatomic bridges, as in **7**), cyano (to give diatomic bridges, as in **8**) or carboxylato, phosph(in)ato or dithiophosph(in)ato (to give triatomic bridges, as in **9**).

The arrows represent here dative bonds (see Section 1.2.1. and Chapter 3) but secondary bonds (soft–soft interactions) can also play a similar role (see Section 1.2.2. and Chapter 4). The latter are characterized by interatomic distances larger than covalent bond lengths but shorter than van der Waals distances.

Increased functionality of the organometallic moiety will lead to more intricate supramolecular assemblies. A suggestive organometallic example is provided by tetra(*p*-cyanophenyl)silane, $Si(C_6H_4CN\text{-}p)_4$, which forms a three-dimensional self-organized supramolecular network of interpenetrating double layers by coordination with silver ions [27] and a layered supramolecular structure with $TiCl_4$ [28], both imposed by the directionality of tetrahedral silicon centers. Supramolecular assemblies are defined as "*extended polymolecular arrays presenting a more or less well-defined microscopic organization*" [8]. Self-assembled supermolecules can occur in the liquid state, but weak intermolecular forces do not always survive in solution (quite often because of the interactions with the solvent, if the latter is a more or less good donor). In non-polar solvents, e.g. hydrocarbons, the polar organometallic molecules containing electronegative functional groups, or those containing functional groups able to form hydrogen-bonds (e.g. organosilanols such as R_3SiOH), will have a tendency to self-assemble. The polar groups will interact to build an internal core made up of the 'inorganic parts' of the molecules assembled into a suprastructure, wrapped into a lipophilic external jacket pointing towards the solvent. This process of self-assembly is similar to the formation of micelles or bi-layers at the interface between two non-miscible solvents, one polar and one non-polar, e.g. water and hydrocarbons. In our case the process of self-assembly occurs at the molecular level.

The self-assembly and the self-organization is more evident in molecular solids. Solid state compounds can be divided into two groups:

i) *non-molecular solids*, made up of atoms or ions, connected into a three-dimensional network, in which electrostatic and dative bonds in ionic or nearly ionic crystals (e.g. halides, oxides, sulfides, nitrides, etc.) and covalent bonds in atomic crystals (e.g. diamond) determine the crystal structure in such a way that no discrete groups of atoms or ions can be identified as building blocks; and

ii) *molecular solids*, made up of discrete molecules connected by various types of intermolecular force. In general the molecules are arranged in the crystal according to the rules of compact packing, *i.e.* they tend to fill the space as completely as possible. Of course, the mode of packing depends very much on the shape of packed molecules. For example, homoleptic organometallic compounds such as $Sn(C_6H_5)_4$ form molecular solids.

A new situation can occur when the organometallic molecules interact through some *directional intermolecular forces* able to induce self-assembly and self-organization. The dimeric, trimeric, tetrameric, or other oligomeric supermolecules and polymeric supramolecular aggregates thus formed will play the role of building blocks in the crystal. As a consequence, in supramolecular compounds one can distinguish three levels of structural organization in the solid state [29]:

i) a *primary structure*, at the molecular level, based upon the covalent bonds which connect the atoms in the molecule;

ii) a *secondary structure*, consisting of supramolecular entities resulted from intermolecular interactions, e.g. dimeric, trimeric, oligomeric or polymeric supramolecular aggregates, as building blocks of the crystal; and

iii) a *tertiary structure*, i.e. the crystal packing of the secondary or supramolecular entities. Numerous organic and organometallic compounds have only the primary structure (molecular) and the tertiary structure (molecule packing) and form molecular crystals without any intermediate organization of molecules into a secondary structure. One must distinguish between common molecular crystals, based upon simple (more or less) compact packing of molecular entities, lacking the intermediate level of structural organization, and supramolecular crystals formed after self-assembly and self-organization of molecules into definite polymolecular entities which act as building blocks of the crystal.

The molecules which play the role of building blocks in a self-assembled, ordered supramolecular structure are called *tectons* [17, 30]. A *tecton* is defined as *"any molecule whose interactions are dominated by particular associative forces that induce self-assembly of an organized network with specific architectural or functional features"* [30]. The term *synthon* has also been suggested, as describing *"structural units within supramolecules which can be formed and/or assembled by known or conceivable synthetic operations involving intermolecular interactions"* [31]. The use of the term *synthon* with a different meaning in organic chemistry leaves the term *tecton* as more acceptable.

The design and construction of multicomponent supermolecules or supramolecular arrays utilizing non-covalent bonding of tectons is called *supramolecular synthesis* [17]. This is a new extension of synthetic chemistry, which so far used mostly covalent bonds to build molecules. Now the molecules (as tectons) become the building blocks of chemical structures of a higher order, i.e. supramolecular architectures. The supramolecular synthesis successfully exploits hydrogen-bonding, and other types of non-covalent interaction, in building supramolecular systems [17, 31, 32].

Another aspect of synthetic supramolecular chemistry is the *supramolecular assistance to molecular synthesis*, in which intermolecular non-covalent interactions are used to bring molecular components in appropriate position and orientation in the intermediate steps of the synthesis of a molecular compound [17]. In both these aspects the self-assembly is the main actor [32–34].

Self-assembly is an important process in natural systems and its better understanding in model systems has focused considerable attention in recent years with synthetic chemistry producing fascinating novel supramolecular structures [32, 34–36]. Most studies of self-assembly have so far focused on purely organic and/or metal coordination systems. A contribution from the organometallic chemistry can add a new dimension to this field.

An interesting study dealing with the packing of 2-benzimidazole molecules into self-assembled organic tape-like (ribbon) supramolecular arrays in the solid state [37] set the stage for the understanding of such architectures. The intermolecular forces used were hydrogen-bonds. Numerous organometallic tape-like systems can also be cited, and we will mention here as an illustration only the example of the dimethyltin hydroxo derivative of benzene-1,2-disulfonimide (HL), [Me$_2$Sn(μ-OH)$_2$L]$_2$ [38]. In this compound there is a self-assembly into dimeric units via hydroxo bridges, followed by a self-organization of the dimers into ribbons (or tapes) via hydrogen-bonds, and then packing of the tapes in the crystal architecture, with some participation of Sn–C–H$\cdots$O bonds (Figure 1.2). Other examples can be found in Chapter 4 (organometallic ribbons based upon secondary interactions) and Chapter 5 (organometallic ribbons based upon hydrogen-bonds).

Some authors describe any molecular crystal as a supramolecular architecture [39]. This is based upon the idea that intermolecular interactions of any nature (in-

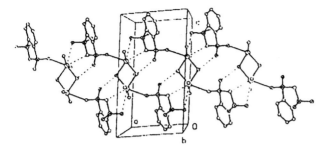

Fig. 1.2. Supramolecular structure of the dimethyltin hydroxo derivative of benzene-1,2-disulfonimide, showing successive levels of self-assembly. (Reproduced from ref. [38])

cluding van der Waals forces) can be considered as a driving force for the formation of supramolecular structures. Perhaps such an approach is too general and allows for too little distinction between simple crystal packing and real supramolecular organization.

The non-covalent forces which hold together the molecules in a crystal can be of various types, and will be discussed in Section 2. These can be dative bonds (electron-pair donor–acceptor or Lewis acid–base interactions), secondary bonds (or soft–soft interactions), hydrogen-bonds, ionic (electrostatic) interactions and π-bonding interactions. They cover a broad range of bond strengths but all are weaker than covalent bonds. In some crystals a large degree of order other than that expected from just compact packing is observed because of even weaker interactions, such as van der Waals forces. These will not be considered in this book, but the packing of molecules containing several phenyl groups, which involves interactions between the aromatic rings (based upon CH$\cdots\pi$-interactions) is covered briefly. These lead to so-called 'multiple phenyl embraces' and result in spectacular modes of supramolecular organization in the crystals. For further details the reader is referred to some significant papers on the subject [40–44].

1.2 Intermolecular Bond Types in Organometallic Supramolecular Systems

1.2.1 Dative bonding (electron-pair donor–acceptor bonding or Lewis acid–base interactions)

This type of bonding is familiar to every chemist and does not need much discussion. It is known under various names and involves sharing of a lone electron pair from an atom called *donor*, with an empty orbital of another atom, called *acceptor*. The whole of coordination chemistry is based upon this type of interaction.

Like the covalent bond, the donor–acceptor bond is a two-electron interaction; the only difference is that the covalent bond is formed by electron pairing, i.e. each of the two atoms contributes one electron to the formation of the electron pair, whereas in the dative bond the electron pair comes from a single atom. There is a general tendency to consider that there is no difference when a given pair of atoms, A and B, are bonded by electron pairing through covalent bonds (A–B), or by electron pair donation through dative bonds (A ← B), because in both cases the atoms are connected by two electrons. It has, however, been elegantly demonstrated by Haaland [45] that distinction between the two types is necessary. Thus, if two isoelectronic molecules, H_3C–CH_3 and $H_3N \rightarrow BH_3$ are considered as prototypes for covalent and dative bonds respectively, significant differences can be distinguished between the two types. These differences reside in:

i) the nature of fragments formed when the central bonds are broken;
ii) the nature of the bond rupture process; and
iii) the magnitude of the bond cleavage enthalpy.

The rupture of the C–C bond in ethane proceeds homolytically and the neutral species produced are CH_3 free radicals; an energy of 89.8 kcal mol^{-1} is spent in the process. On the other hand, B ← N bond rupture proceeds heterolytically, the neutral species formed being diamagnetic molecules NH_3 and BH_3; here the dissociation enthalpy is significantly smaller, only 31.1 kcal mol^{-1}. It can be seen that the process requires only approximately one third of the energy needed to cleave a normal covalent bond. Consequently, Haaland [45] defines as *normal covalent* a bond in a neutral diamagnetic molecule if the minimum energy rupture proceeds homolytically to yield neutral, radical (paramagnetic) species, and as *dative* a bond which ruptures heterolytically to yield neutral diamagnetic species.

When *normal covalent*, **10**, and *dative*, **11**, bonds between the same atoms (e.g. the boron-nitrogen pair) are compared, it is found that the dative bonds are longer than normal covalent bonds. In addition, the dative bonds are more sensitive to inductive effects of the substituents. Thus, the boron–nitrogen interatomic distances in dative bonds are 1.658(2) Å in $H_3N \rightarrow BH_3$, 1.656(2) Å in $Me_3N \rightarrow BH_3$, 1.70(1) Å in $Me_3N \rightarrow BMe_3$ and 1.65(6) Å in $Me_3N \rightarrow BCl_3$, whereas for a normal (single covalent) B–N bond the interatomic distance is estimated to be 1.58 Å. Boron–nitrogen interatomic distances can be even shorter because of π-bonding contributions, e.g. in H_2BNH_2 (1.391(2) Å) [46] or in graphite-like, hexagonal boron nitride $(BN)_x$ 1.446 Å [47]. The strength of a normal covalent B–N bond is esti-

covalent
10

dative
11

mated at 91 kcal mol^{-1}, which is three times the strength of a dative N → B bond. The strength of a dative bond seldom exceeds half the strength of a normal covalent bond between the same pair of atoms.

The distinction between normal covalent and dative bonds is underscored by interatomic distances when both types are present in the same compound, as shown in **12** and **13**, below [45]:

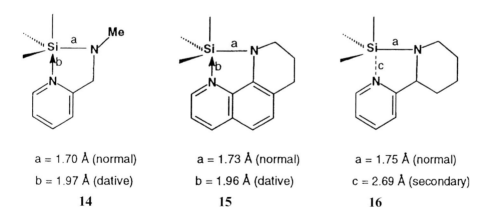

normal	a = 1.97 Å
dative	b = 2.13 Å

12

normal	a = 1.89 Å
dative	b = 2.08 Å

13

Large variations in the lengths of dative bonds can be observed, and some high values can be regarded as 'secondary bonds' as discussed in Chapter 4. Thus, silicon–nitrogen interatomic distances cover a range from ca. 1.70 Å (normal covalent bond) to 1.97 Å (dative bond) and 2.69 Å (secondary bond) in the three structurally related compounds **14–16** [45].

a = 1.70 Å (normal)
b = 1.97 Å (dative)

14

a = 1.73 Å (normal)
b = 1.96 Å (dative)

15

a = 1.75 Å (normal)
c = 2.69 Å (secondary)

16

These considerations should be born in mind when interpreting the self-assembly of numerous molecules which contain both *donor and acceptor binding sites*, with formation of dimeric, trimeric, or tetrameric supermolecules or polymeric supramolecular arrays. That some chemical classes of compound can exist both as monomeric species and as oligomeric or polymeric self-assembled systems (depending on the steric bulkiness of the substituents and other factors), and the facile interconversion between monomers and associated species (because of the low dissociation energy of the dative bonds), are arguments in favor of a supramolecular chemistry approach in their description.

Dative bonds are involved in the self-assembly of numerous organometallic derivatives of the type R_nM-X where X is a functional group with at least one lone electron pair. X may be halogen (F, Cl, Br, I), hydroxy (OH), alkoxy (OR'), (un)-substituted amido (NR'R''), phosphido (PR'R''), arsenido (AsR'R''), thiolato (SR'), or selenolato (SeR') groups, etc. A good example of dative bond supramolecular self-assembly is Group 13 diorgano–metal(III) functional derivatives of the type $R_2M-ER'_2$, **17**. Depending on the nature of organic groups, these compounds may associate into cyclic oligomeric structures (dimers, trimers, tetramers, or larger rings), with four-coordinate metal, the functional group acting as a bridge. In these cyclic structures the M–(μ-E) interatomic distances are frequently equalized (by resonance or delocalization), although sometimes small differences are observed which could define the distinction between a dative and a covalent bond. It could be argued that oligomers of the type $[R_2M(\mu\text{-}E)]_n$ are simply derivatives of n-membered ring systems M_nE_n. The existence of monomeric forms and cyclic oligomeric forms within the same class of compound suggests, however, that these should be regarded as products of self-assembly and self-organization, i.e. supramolecular systems, rather than authentic heterocyclic compounds. To illustrate this, Group 13–15 organometallic functional derivatives can be cited. Examples of structurally characterized supermolecules are the dimers $[R_2M-ER'_2]_2$, **18**, and the trimers $[R_2M-ER'_2]_3$, **19**, with M = Al, Ga, In, Tl, ER'_2 = NRR', PRR', AsRR', SbRR'.

Distinction should be made between the self-assembled dimeric or trimeric species cited above and authentic heterocycles in which the metal(III) is three-coordinate, e.g. the four-membered ring derivatives $R_2M_2E_2R'_2$, **20**, and the six-membered ring derivatives $R_3M_3E_3R'_3$, **21**.

For Group 13 elements these heterocycles are themselves coordinatively unsaturated, unless there is internal (*endo*cyclic) electron delocalization. Otherwise, the acceptor (metal) and donor (E) sites will favor the supramolecular self-assembly

1.2 Intermolecular Bond Types in Organometallic Supramolecular Systems 13

M = Al, Ga, In, Tl
X = N, P, As, Sb

20   **21**

with formation of tetranuclear cubane-type species [RMER']₄, **22**, prismane-type hexanuclear species [RMER']₆, **23**, or other cage structures.

22   **23**

These cages can be regarded as dimers, **24**, or trimers, **25**, of heterocyclic R₂M₂E₂R'₂ compounds, as shown schematically below (the R and R' groups are omitted for clarity). Alternatively, the hexagonal prism can also be regarded as a dimer, **26**, which results from the stacking of two six-membered heterocyclic compounds R₃M₃E₃R'₃.

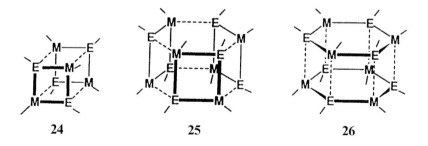

24   **25**   **26**

The frequently observed equalization of the interatomic distances in the cages does not enable distinction between covalent and dative bonds, and this approach may look somewhat artificial to the synthetic chemist or crystallographer. But, from the view point of supramolecular chemistry we believe this formalism to be a way of distinguishing between common molecular compounds and self-assembled supermolecules.

If the organic groups at either the metallic site or in the functional groups are bulky, the self-assembly might be sterically hindered and the compounds remain as molecular monomers, with three-coordinate metal. Examples are again numerous. In all the monomeric tricoordinate metal(III) derivatives and in the authentic heterocyclic trimers the M–X bond lengths are shorter than in four-coordinate dative bond self-assembled dimers and trimers. Thus, in the trimer $(2,4,6-Ph_3C_6H_2GaPCy)_3$ the Ga–P bond lengths (2.295 to 2.34 Å) are shorter than in four-coordinate dimeric and trimeric supermolecules (ca 2.41–2.45 Å). The short bonds are normal covalent single bonds (perhaps with some π-bond character), whereas the longer bonds observed in the four-coordinate dimers and trimers are the result of mixing between single covalent M–X and dative and M ← X (gallium–phosphorus) bonds. More information is available elsewhere [45].

The use of dative bonds to build supramolecular architectures is not limited to the self-assembly of cyclic and cage supermolecules, as illustrated above. Subsequent sections of this book will present numerous other examples, e.g. the self-assembly of triorganotin carboxylates, oxocarboxylates, or phosphinates into intricate supramolecular arrays (Section 3.3).

Dative bonds play an essential role in the supramolecular chemistry of transition metals and numerous examples of self-assembled supermolecules can be cited, including basically all-bridged dinuclear, oligonuclear, or polynuclear coordination compounds. We mention here only a few recent examples of transition-metal organometallic self-assembly – these include organometallic fluorides [48, 48a] and the formation of transition-metal metallacycles, e.g. squares, boxes, hexagons, and polygons made up from tectons containing M–C bonds, interconnected by coordination [49–53, 53a,b]. Finally, the use of organometallic compounds, such as organosilicon-bipyridine molecules, as *exo*-receptor spacers to build supramolecular arrays through coordination [54] illustrates another possibility.

Transition metal supramolecular chemistry based upon dative (electron-pair donor–acceptor) bonds is so extensive that its coverage would require another volume. This topic will not, therefore, be covered in this book in any more detail. Numerous examples can be identified in the various chapters of the treatise *Comprehensive Organometallic Chemistry II*.

1.2.2 The secondary bond concept

The concept of the '*secondary bond*' was introduced by Alcock [55, 55a] to describe *interactions characterized by interatomic distances longer than single covalent bonds but shorter than van der Waals interatomic distances*. In crystal structure determinations interatomic distances longer than the sum of covalent radii but shorter than the sum of van der Waals radii are quite frequently observed. They represent interactions weaker than covalent or dative bonds, but are strong enough to influence the coordination geometry of the atoms involved and to hold together pairs of atoms, either to intramolecularly close a ring or to lead to intermolecular association.

1.2 Intermolecular Bond Types in Organometallic Supramolecular Systems 15

Secondary interactions can be *intramolecular* or *intermolecular*. An unusually long intramolecular silicon–nitrogen interatomic distance was cited as example in the previous section. In this section we will be concerned mainly with *inter*molecular secondary bonds, which might serve to connect individual molecules (tectons) into self-assembled (and self-organized) supermolecules or supramolecular arrays. Sometimes intra- and intermolecular secondary bonds may occur simultaneously in the same supramolecular structure. Usually secondary bonds are not strong enough to survive in solution, especially in coordinating solvents, but they can have spectacular effects in the building of a crystal, by creating the secondary structure which defines the supramolecular systems in the solid state.

Two examples (from our own work) illustrate *intramolecular secondary interactions*. Thus, in triphenyltellurium(IV) tetraphenyldithioimidodiphosphinate, $Ph_3Te^+[SPh_2PNPPh_2S]^-$, **27**, the tellurium–sulfur interatomic distances, 3.264 and 3.451 Å, are significantly longer than the sum of the covalent radii (2.443 Å) but significantly shorter than the sum of the van der Waals radii (3.86 Å) [56]. In bis(diphenyldithiophosphinato)ditellurium(I), $Te_2(S_2PPh_2)_2$, **28**, there are two types of tellurium–sulfur interaction – normal covalent, i.e. primary bonds (Te–S 2.471 and 2.493 Å) and secondary bonds (Te···S 2.989 and 3.066 Å) [57].

27   **28**

Cyclic systems in which ring closure occurs via secondary bonds were called '*quasi-cyclic*' systems [29, 58].

An example of *intermolecular secondary bonds* for the same tellurium–sulfur pair, leading to a regular supramolecular self-assembly, **29**, is provided by phenyltellurium(II) diphenyldithiophosphinate, $PhTe-S(S)PPh_2$, in which intermolecular, secondary Te···S bonds (3.383 Å) are observed; the intramolecular primary Te–S bonds are shorter (2.406 Å) [59, 60] and fall within the range of single, covalent tellurium–sulfur bonds [61].

Interatomic distances between a given pair of elements can vary substantially, from covalent bond lengths and dative (or donor–acceptor) bond lengths to secondary bond interatomic distances. Bond lengths vary to some extent with the coordination number of the central atom, and there is no clear-cut distinction between these three types of bond. Thus, for tin–oxygen bonds, there is a broad range of observed interatomic distances. Covalent Sn–O bonds are shortest in four-

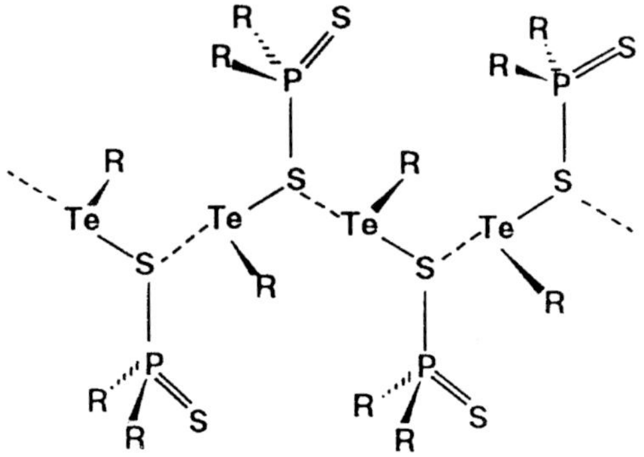

29

coordinate (tetrahedral) tin–oxygen compounds, as observed in sterically hindered monomeric organotin hydroxides, e.g. $(2,4,6\text{-Me}_3\text{C}_6\text{H}_2)_3\text{Sn–OH}$ (1.999 Å) [62], cyclotristannoxanes $(\text{R}_2\text{SnO})_3$ with R = Bu^t (1.965 Å) [63], R = amyl (1.952–1.978 Å), R = 2,6-Et$_2$C$_6$H$_3$ and 2,4,6-Me$_3$C$_6$H$_2$ (1.958 Å) [64], or R$_2$ = (Me)[C(SiMe$_3$)$_3$] (1.931–1.968 Å) [65], linear hexaorganodistannoxanes R$_3$Sn–O–SnR$_3$ with R = benzyl (1.919 Å) [66] or Bu^t (1.936 Å) [67], and bent Ph$_3$Sn–O–SnPh$_3$ (1.952–1.958 Å) [68]. The tin–oxygen bond becomes longer when the coordination number of tin is five (usually distorted trigonal bipyramidal geometry) or six (distorted octahedral geometry). Thus, in polymeric [R$_3$Sn–OR′]$_n$ two sets of Sn–O bonds are observed, short and long; the one which is slightly longer can be described as dative, e.g. in [Me$_2$BzSn(μ-OH)]$_n$ (2.17 and 2.29 Å) [69], [Ph$_3$Sn(μ-OH)]$_n$ (2.197 and 2.255 Å) [70], and [Me$_3$Sn(μ-OMe)]$_n$ (2.202 and 2.256 Å) [71]. In dimeric [XR$_2$Sn(μ-OH)]$_2$ again long and short tin–oxygen distances are observed, for R = Bu^t, X = F (2.194 and 2.012 Å), X = Cl (2.238 and 2.036 Å), and X = Br (2.225.7 and 2.048 Å) [72].

All these interatomic distances correspond to two-electron bonds, i.e. covalent and dative. When much longer tin–oxygen interatomic distances are observed, associated with short distances, as in dimeric dioxastannolanes, **30**, (dioxastannacyclopentanes, R = Bu^n, with Sn–O 1.98 and Sn$\cdots$O 2.50 Å) [73] or dioxastannacyclohexanes, **31** (R = Bu, Sn–O 2.05 Å, Sn$\cdots$O 2.57 Å) [74], the longer distances should be considered as secondary bonds.

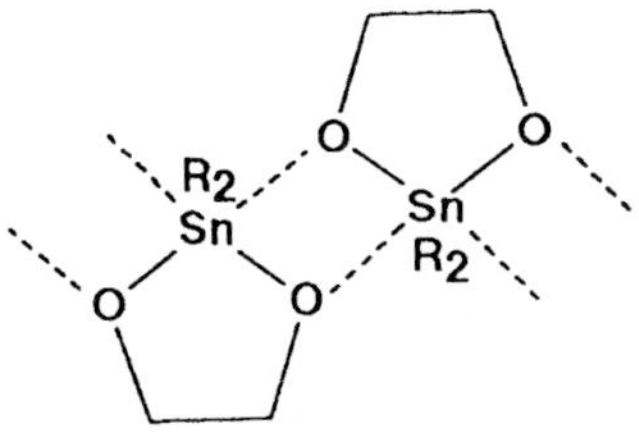

30 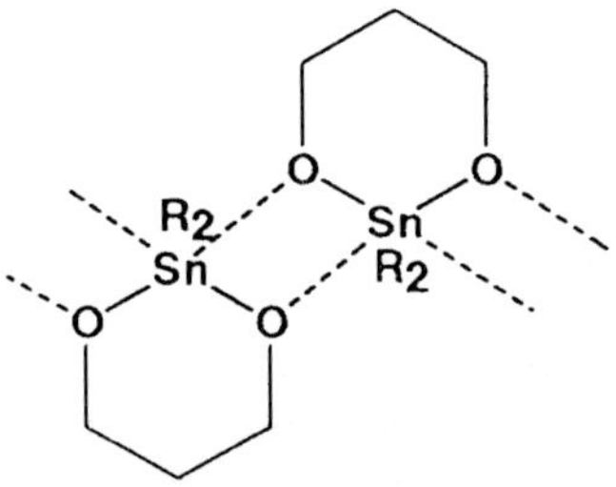   **31**

1.2 Intermolecular Bond Types in Organometallic Supramolecular Systems

In some organotin carboxylates, Sn···O intermolecular distances as long as 3.37 Å are observed (e.g. in tributyltin pyridine-2,6-dicarboxylate) [75]. Such secondary bonds are rare in tin–oxygen chemistry, because of the strong donor properties of oxygen toward tin, but with softer donors such as sulfur secondary bonds (Sn···S) are quite frequent. These will be discussed in Chapter 4. We have discussed the tin–oxygen example here to illustrate the broad range of interatomic distances, which can be encountered for a pair of two different elements, when normal covalent, dative, and secondary bonds are possible. The situation is similar for many pairs of elements, and can be the force driving supramolecular self-assembly.

Some formal similarity between secondary bonding and hydrogen-bonding should be mentioned. Typically, secondary bonding interaction occurs as a basically linear X–A···Y system, in which A–X is a normal covalent bond and A···Y is the secondary bond, **32**. Thus, like the hydrogen-bond, this system is linear and non-symmetric, although exceptions can also exist in both cases (more information is available elsewhere [75a]):

X——A········Y   X——H········Y

normal   secondary   *normal   hydrogen-*
covalent   *covalent   bond*

e.g. X = Hg, Tl, Sn, Sb, Te   X = F, O, N
A = S, Se, etc.

32

The explanation advanced by Alcock [55] suggests that the secondary bond is formed either by donation from the lone pair on X into a s^* orbital of the A–Y bond, or (alternatively and equivalently) as an asymmetric three-centre system, with three s symmetry atomic orbitals on A, X, and Y, combined to form three molecular orbitals: one filled bonding molecular orbital located between A and X, one filled non-bonding or weakly bonding orbital located between A and Y and one empty antibonding orbital [76]. It seems that such secondary interactions are associated with a high electron density at the acceptor binding site (e.g. because of the presence of one or more lone pairs of electrons). The strength of the secondary bonds is comparable with that of hydrogen-bonds.

Secondary bonds are particularly important and frequently occur in compounds of heavier main group elements [29], and many examples of supramolecular self-assembly through secondary bonds are known. Chapter 4 deals extensively with self-assembly and self-organization through secondary bonding.

The secondary bonds can, in fact, be regarded as a particular case of donor–acceptor bonds. A survey of the available data suggests that stronger acceptors, e.g. metals of the upper rows of the periodic table, associated with stronger donors such as fluorine, oxygen, and nitrogen (generally known as hard acids and bases) tend to form stronger dative bonds of the type discussed in the previous section. The metals and non-metals situated in the bottom rows of the periodic table, especially when associated with weaker donors (also known as soft acids and bases) tend to form weaker donor–acceptor bonds and 'secondary bonds' (also described as soft–soft interactions, closed-shell interaction, semibonding interactions, non-bonding interactions, etc.). A detailed discussion of this type of bonding has recently been published [77].

A particular case is that of d^{10}–d^{10} secondary interactions, first identified in gold compounds and described as 'aurophilic' attraction [78–80], later generalized as 'metallophilic' attraction [81]. This leads to supramolecular self-assembly in many gold compounds, especially when associated with three-center Au$\cdots$C$\cdots$Au bonds, and can be also found in other d^{10} metal compounds, such as silver, gold, and thallium complexes. Such supermolecules are illustrated by the quasi-cyclic pentamers Au$_5$(Mes)$_5$ and Cu$_5$(Mes)$_5$, **33**, and the tetramer Ag$_4$(Mes)$_4$, **34** [82].

The dependence of the degree of aggregation of RCu compounds on various factors demonstrate that the formation of (RCu)$_n$ polynuclear species is a process of supramolecular self-assembly rather than a molecular synthesis [83].

The supramolecular assemblies formed through secondary bonds are discussed in some detail in Chapter 4 with numerous examples. It will be seen that this type of interaction is mostly characteristic of main group metals and semimetals.

1.2.3 Hydrogen-bond interactions

Hydrogen-bonds are intermolecular interactions which occur in compounds in which hydrogen is bonded to an electronegative element, such as fluorine, oxygen, or nitrogen. The hydrogen-bonds are typically of the type X–H$\cdots$Y, where both X and Y (identical or different) are electronegative atoms. The pair consists of a normal covalent bond X–H, associated with a weaker interaction H$\cdots$Y, the hydrogen-bond proper. The interatomic distance between Y and H is longer than a covalent bond but shorter than the expected van der Waals distance. The strength of hydrogen-bonds is of the order of 5–10 kcal mol^{-1}, but both stronger (e.g. with participation of F$^-$ anions, in inorganic systems) and weaker hydrogen-bonds (with participation of sulfur, chlorine, or other non-metallic elements) have also been reported.

Hydrogen-bonding is an important feature of organic chemistry and leads to dimerization of carboxylic acids, association of functional derivatives (alcohols, amines, amides, aminoalcohols, and amino acids, etc.) and other types of intermolecular association. Hydrogen-bonding plays an important role in biological systems (remember the structure of DNA!) and is a well studied phenomenon. In organic crystals hydrogen-bond association leads to formation of very diverse supramolecular structures and can be manipulated to build pre-established crystal architectures [84]. This is called *crystal engineering*. Actually, it is considered that any non-covalent interactions that are selective, directional, and strongly attractive can induce self-assembly of predictable supramolecular architectures [30]. Thus, the term *crystal engineering* has a broader meaning and has been defined as "*the planning of a crystal structure from its building blocks, i.e. molecules or ions chosen on the basis of their size, shape and extramolecular bonding capacities*" [84a]. So far this is best achieved by rational manipulation of hydrogen-bonding capabilities of molecules, but it is not limited to this type of interaction.

It can be expected that organometallic compounds containing functional groups such as OH, COOH, CONH$_2$, to name only a few, will undergo intermolecular self-assembly by hydrogen-bonding, leading to supramolecular architectures. Selected but illustrative examples will be discussed in Chapter 5, with emphasis upon organosilanols and functional organotin derivatives. A more recent development was the recognition of C–H$\cdots$O weak hydrogen-bonding, which plays a significant role in the supramolecular self-organization of metal carbonyl organometallic compounds of transition metals [85–87, 87a]. Negatively charged O–H$\cdots$O$^-$ and charge-assisted C–H$^{\delta+}\cdots$O$^{\delta-}$ hydrogen-bonds are particularly strong and enable the design of organometallic supramolecular crystals with pre-established structures [87b].

Transition metal organometallics display a broader diversity of hydrogen-bonds, in addition to the traditional types known in organic chemistry. Thus, X–H$\cdots\pi$ (where X = O, N, C) hydrogen-bonds to electron-rich π-ligands (alkynes, arenes, cyclopentadienyls) [87c] and M–H$\cdots$H–C intermolecular interactions [87d] play an important role.

The hydrogen-bond has been explained as an electrostatic dipole–dipole interaction:

$$X^{\delta-}-H^{\delta+}\cdots Y^{\delta-}$$

but more recent explanations describe it as a donor–acceptor type of interaction [88]. The hydrogen-bond is asymmetric, linear, and rarely branched. It can be readily detected in crystal structures from the interatomic distances measured.

The simplest organic supramolecular aggregates are the dimers of carboxylic acids, **35**, and the cyclic tetramers of alcohols, **36**. Of the organic compounds amides have a great propensity to hydrogen-bond formation; amides also form dimers, **37**.

$$\text{35} \qquad \text{36} \qquad \text{37}$$

These and other structural motifs, e.g. some based upon phosphinic groups, will be encountered often in supramolecular self-assembly of organometallic compounds containing appropriate functional groups [89–91].

Two examples of hydrogen-bond interactions in complementary amides, leading to cyclic supermolecules or to polymeric supramolecular aggregates, are illustrated by the pair barbituric acid and 2,4,6-triaminopyrimidine (Figure 1.3). Similar self-complementary molecules, able to self-organize through hydrogen-bonds into supramolecular architectures, are also possible among organometallic compounds; these will be illustrated in Chapter 5.

1.2.4 Ionic interactions

Ionic interactions leading to supramolecular organometallic structures can often be expected, e.g.:

i) in compounds of alkali metals and alkaline earths, in which the nature of the metal–carbon bond can be described as essentially ionic;
ii) in alkali metal and alkaline earth amides, alkoxides, thiolates, and other organic derivatives not containing direct metal–carbon bonds but with chemical behavior similar to that of organic compounds (e.g. sensitivity to oxygen, water, and other reagents containing mobile hydrogen), and
iii) in organometallic compounds containing functional groups able to displace a proton by an electropositive metal and to form cation–anion ion pairs (e.g. metal organosilanolates). Such species are seldom mononuclear and frequently occur as associated species. i.e. supermolecules of definite size or supramolecular aggregates (polymeric structures) in which the building units are hold together by essentially electrostatic forces. Thus, for example, ring, cubane, ladder, and other supramolecular structures (Figure 1.4) can be found in organonitrogen–lithium compounds [92, 93].

Organometallic cations, e.g. η^5-$C_5Me_5Ru^+$ fragments, can be incorporated into tectons of particular shapes and geometries of localized positive charges, which in

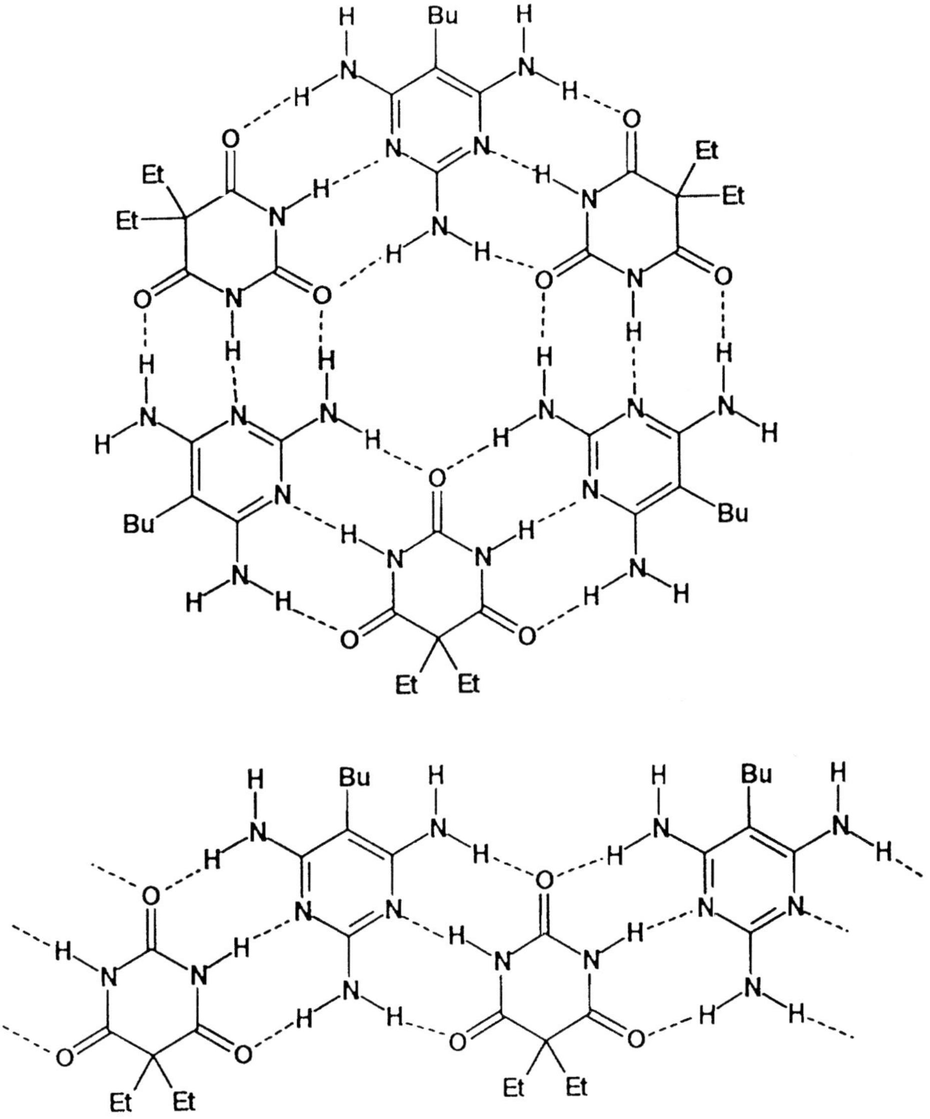

Fig. 1.3. Cyclic and polymeric supramolecular aggregates formed by hydrogen-bond self-assembly of barbituric acid and 2,4,6-triaminopyrimidine derivatives.

turn can serve for the crystal engineering of solid-state materials with pre-established supramolecular structures. Thus, linear, triangular, tetrahedral, or octahedral polycations were prepared by attaching the η^5-C$_5$Me$_5$Ru$^+$ moiety to paracyclophanes, triptycene, E(C$_6$H$_5$)$_4$ molecules (E = C, Si, Ge), or C$_6$(OC$_6$H$_4$OMe-p)$_6$ derivatives. By combination of the resulting cations with appropriately chosen anions, crystals of predetermined structures have been built in a rational manner [94, 95].

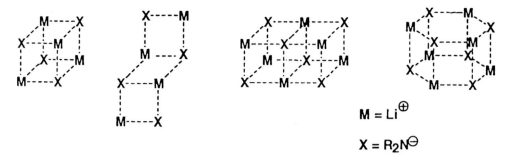

Fig. 1.4. Polynuclear cages found in supermolecules formed through ionic interactions.

In this book all these types of compounds will be considered, even though some, e.g. alkali metal amides and alkoxides, might not contain direct metal–carbon bonds. This is line with other books of organometallic chemistry (e.g. the multi-volume treatise *Comprehensive Organometallic Chemistry*) which includes such compounds.

Ionic supramolecular self-assembly will be discussed in Chapter 6, but some supramolecular systems based upon ionic interactions will be discussed earlier, e.g. the organocyclosiloxanolates, which form sandwich compounds by intercalating transition metal ions between two macrocyclic rings (held together by ionic interactions) and acting as *endo* receptors which concurrently have crown-ether-type complexing properties (as *exo* receptors) (see Section 2.1.2).

1.2.5 π-Bonding interactions

The formation of multiple-decker polymetallic π-complexes, with metal atoms situated on both sides of cyclopentadienyl rings, suggests that oligomers or polymers of high nuclearity can also be synthesized. Indeed, in recent years such multisandwich complexes, with both cyclopentadienyl [96–98], indenyl [99], and carborane rings [100] bridging several metal atoms, as shown in **38**, have been reported. Such compounds can be regarded as supramolecular assemblies held together by π-bonds.

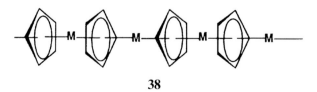

38

π-Bonding aggregates are formed by alkali metals [101] and transition metals. In the former, ionic interactions between metal cations M^+ and cyclopentadienyl anions $C_5H_5^-$ play an important role, but the stacking indicates that π-bonding should also be considered. Obviously, in transition metal complexes this type of bonding is the main factor in the formation of the supramolecular arrays. It is difficult to separate the two types of compounds, ionic and π-bonded, but alkali metal and alkaline earth derivatives will be described in the section dealing with ionic forces.

Transition metal supramolecular arrays can be constructed with the aid of π-bonding interactions by using various [2n]cyclophanes and polymetallic clusters. This area is only in its infancy and its potential is illustrated by the formation of a supramolecular hexagonal two-dimensional network by using metal centers linked through the [2.2.2]paracyclophane ligand [102].

References

[1] J. M. Lehn, *Angew. Chem.* **1988**, *100*, 91; *Angew. Chem. Int. Ed. Engl.* **1988**, *27*, 89.
[2] D. J. Cram, *Angew. Chem.* **1988**, *100*, 1041; *Angew. Chem. Int. Ed. Engl.* **1988**, *27*, 1009.
[2a] D. J. Cram and J. M. Cram, *Science* **1974**, *183*, 803.
[3] C. J. Pedersen, *Angew. Chem.* **1988**, *100*, 1053; *Angew. Chem. Int. Ed. Engl.* **1988**, *27*, 1053.
[4] J. M. Lehn, *Struct. Bonding* **1973**, *6*, 1.
[5] P. Pfeiffer, *Organische Molekülverbindungen*, F. Enke Verlag, Stuttgart, **1927**.
[6] K. L. Wolf, H. Frahm and H. Harms, *Z. Phys. Chem.* **1937**, *B36*, 237; K. L. Wolf, and R. Wolf, *Angew. Chem.* **1949**, *61*, 191.
[7] J. M. Lehn, *Pure Appl. Chem.* **1978**, *50*, 871.
[8] J. M. Lehn, *Supramolecular Chemistry. Concepts and Perspectives*, VCH, Weinheim, **1995**.
[9] J. M. Lehn, J. L. Atwood, J. E. D. Davies, D. D. MacNicol and F. Vögtle (Editors), *Comprehensive Supramolecular Chemistry*, Vols.1–11, Pergamon Press, Oxford, **1996**.
[10] I. Haiduc, *The Chemistry of Inorganic Ring Systems*, Wiley–Interscience, London, **1970**.
[11] I. Haiduc and D. B. Sowerby (Eds.), *The Chemistry of Inorganic Homo- and Heterocycles*, Academic Press, London, New York, **1987**.
[12] D. A. Armitage, *Inorganic Rings and Cages*, E. Arnold, London, **1972**.
[13] H. R. Allcock, *Heteroatom Ring Systems and Polymers*, Academic Press, New York, London, **1967**.
[14] G. L. Eichhorn (Editor), *Inorganic Biochemistry*, Elsevier, New York, **1973**.
[15] J. S. Thayer, *Organometallic Compounds and Living Organisms*, Academic Press, New York, **1984**.
[16] I. Haiduc and J. J. Zuckerman, *Basic Organometallic Chemistry*, Walter de Gruyter, Berlin, New York, **1985**.
[17] M. C. T. Fyfe and J. F. Stoddart, *Acc. Chem. Res.* **1997**, *30*, 393.
[18] J. M. Lehn, in vol. *Perspectives in Coordination Chemistry*, Edited by A. F. Williams, C. Floriani and A. E. Mehrbach, Verlag Helvetica Chimica Acta, Basel, and VCH, Weinheim, **1992**, p. 447.
[19] B. Dietrich, P. Viout and J. M. Lehn, *Macrocyclic Chemistry. Aspects of Organic and Inorganic Supramolecular Chemistry*, VCH, Weinheim, **1993**.
[20] J. A. Semlyen, *Large Ring Molecules*, J. Wiley, Chichester, **1997**
[21] G. R. Newcome, C. N. Morefield and F. Vögtle, *Dendritic Molecules*, VCH, Weinheim, **1996**.
[21a] H. Frey, C. Lach and K. Lorenz, *Adv. Mater.* **1998**, *10*, 279.
[21b] C. Gorman, *Adv. Mater.* **1998**, *10*, 295.
[21c] J. P. Majoral and A. M. Caminade, *Chem. Rev.* **1999**, *99*, 845.
[21d] M. Fischer and F. Vögtle, *Angew. Chem.* **1999**, *111*, 935; *Angew. Chem. Int. Ed.* **1999**, *38*, 884.

[21e] M. A. Hearshaw and J. R. Moss, *Chem. Commun.* **1999**, 1.
[21f] G. R. Newkome, E. He and C. N. Moorefield, *Chem. Rev.* **1999**, *99*, 1689.
[21g] E. C. Constable, O. Eich and C. E. Mousecraft, *Dalton Trans.* **1999**, 1363.
[21h] G. E. Oosterom, R. J. van Haaren, J. N. M. Reek, P. C. J. Kamer and P. W. N. M. van Leeuwen, *Chem. Commun.* **1999**, 1363.
[22] G. Gokel, *Crown Ethers and Cryptands*, The Royal Society of Chemistry, Cambridge, **1991**.
[22a] A. Bianchi, K. Bowmann-James and E. Garcia-Espana (Editors), *Supramolecular Chemistry of Anions*, Wiley-VCH, New York, Weinheim, **1998**.
[23] A. Müller, H. Reuter and S. Dillinger, *Angew. Chem.* **1995**, *107*, 2505; *Angew. Chem.* Int. Ed. Engl. **1995**, *34*, 2328.
[24] J. M. Lehn, J. L. Atwood, J. E. D. Davies, D. D. MacNicol and F. Vögtle (Editors), *Comprehensive Supramolecular Chemistry. Volume 7. Solid State Supramolecular Chemistry: Two- and Three-dimensional Inorganic Networks*, G. Alberti and T. Bein, Volume Editors, Pergamon Press, Oxford, **1996**.
[25] B. König, *European Chem. Chronicle* **1998**, *3*, 17.
[26] A. Gavezotti and G. Filippini, *Chem. Commun.* **1998**, 287.
[27] F. Q. Liu and T. D. Tiley, *Inorg. Chem.* **1997**, *36*, 5090.
[28] F. Q. Liu and T. D. Tiley, *Chem. Commun.* **1998**, 103.
[29] I. Haiduc, *Coord. Chem. Rev.* **1997**, *158*, 325.
[30] S. Simard, D. Su and J. D. Wuest, *J. Am. Chem. Soc.* **1991**, *113*, 4696.
[30a] D. Braga, F. Grepioni and G. R. Desiraju, *Chem. Rev.* **1998**, *98*, 1375.
[30b] D. Braga and F. Grepioni, *J. Chem. Soc., Dalton Trans.* **1999**, 1.
[31] G. R. Desiraju, *Angew. Chem.* **1995**, *107*, 2541; *Angew. Chem. Int. Ed. Engl.* **1995**, *34*, 2311.
[32] G. M. Whitesides, E. E. Simanek, J. P. Mathias, C. T. Seto, D. N. Chin, M. Maamen and D. M. Gordon, *Acc. Chem. Res.* **1995**, *28*, 37.
[33] D. S. Lawrence, T. Jiang and M. Levett, *Chem. Rev.* **1995**, *95*, 2229.
[34] D. Philp and J. F. Stoddart, *Angew. Chem. Int. Ed. Engl.* **1996**, *35*, 1155.
[34a] V. Balzani, M. Gomez-Lopez and J. F. Stoddart, *Acc. Chem. Res.* **1998**, *31*, 405.
[35] G. Tsoucaris, J. L. Atwood and J. Lipkowski (Editors), *Crystallography of Supramolecular Compounds* (NATO ASI Series, Series C: Mathematical and Physical Sciences, Volume 480), Kluwer Academic Publishers, Dordrecht, **1996**.
[36] A. Müller, A. Dress and F. Vögtle (Editors), *From Simplicity to Complexity in Chemistry and Beyond*, Vieweg Verlag, Braunschweig, **1996**.
[37] K. E. Schwiebert, D. N. Chin, J. C. MacDonald and G. M. Whitesides, *J. Am. Chem. Soc.* **1996**, *118*, 4018.
[38] A. Wirth, D. Henschel, A. Blaschette and P. G. Jones, *Z. Anorg. Allg. Chem.* **1997**, *623*, 587.
[39] G. Desiraju (Editor), *The Crystal as a Supramolecular Entity*, J. Wiley, Chichester, **1996**.
[40] I. G. Dance and M. L. Scudder, *J. Chem. Soc., Chem. Commun.* **1995**, 1039.
[41] I. G. Dance and M. L. Scudder, *J. Chem. Soc., Dalton Trans.* **1996**, 3755.
[42] I. G. Dance and M. L. Scudder, *Chem. Eur. J.* **1996**, *2*, 481.
[43] C. Hasselgren, P. A. W. Dean, M. L. Scudder, D. C. Craig and I. G. Dance, *J. Chem. Soc., Dalton Trans.* **1997**, 2019.
[44] M. Scudder and I. G. Dance, *J. Chem. Soc., Dalton Trans.* **1998**, 329.
[45] A. Haaland, *Angew. Chem.* **1989**, *101*, 1017; *Angew. Chem. Int. Ed. Engl.* **1989**, *28*, 992.
[46] P. Paetzold, *Adv. Inorg. Chem. Radiochem.* **1987**, *31*, 123.
[47] M. Sugie, H. Takeo and C. Matsumura, *J. Mol. Spectr.* **1987**, *123*, 286.
[48] E. F. Murphy, R. Murugavel and H. W. Roesky, *Chem. Revs.* **1997**, *97*, 3425.
[48a] H. W. Roesky and I. Haiduc, *J. Chem. Soc., Dalton Trans.* **1999**, 2249.
[49] H. Plenio, *Angew. Chem.* **1997**, *109*, 358; *Angew. Chem. Int. Ed. Engl.* **1997**, *36*, 348.
[50] P. J. Stang and B. Olenyuk, *Acc. Chem. Res.* **1997**, *30*, 502.
[51] P. J. Stang, *Chem. Eur. J.* **1998**, *4*, 19.
[52] J. R. Hall, S. J. Loeb, G. K. Shimizu and G. P. A. Yap, *Angew. Chem.* **1998**, *110*, 130; *Angew. Chem. Int. Ed. Engl.* **1998**, *37*, 121.
[53] R. Anwander, *Angew. Chem. Int. Ed. Engl.* **1998**, *37*, 599.
[53a] K. K. Klausmeyer, T. B. Rauchfuss and S. R. Wilson, *Angew. Chem.* **1998**, *110*, 1808; *Angew. Chem. Int. Ed. Engl.* **1998**, *37*, 1694.

[53b] F. A. Cotton, L. M. Daniels, C. Lin and C. A. Murillo, *Chem. Commun.* **1999**, 841.
[54] C. Kaes, M. W. Hosseini, A. De Cian and J. Fischer, *Chem. Commun.* **1997**, 2229.
[55] N. W. Alcock, *Adv. Inorg. Chem. Radiochem.* **1972**, *15*, 1.
[55a] N. W. Alcock and R. M. Countryman, *J. Chem. Soc. Dalton Trans.* **1977**, 217.
[56] A. Silvestru, I. Haiduc, R. Toscano and H. J. Breunig, *Polyhedron*, **1995**, *14*, 2047.
[57] M. G. Newton, R. B. King, I. Haiduc and A. Silvestru, *Inorg. Chem.* **1993**, *32*, 3795.
[58] I. Haiduc, *Phosphorus, Sulfur and Silicon*, **1994**, *93/94*, 345.
[59] A. Silvestru, I. Haiduc, K. H. Ebert and H. J. Breunig, *Inorg. Chem.* **1994**, *33*, 1253.
[60] A. Silvestru, I. Haiduc, K. H. Ebert, H. J. Breunig and D. B. Sowerby, *J. Organomet. Chem.* **1994**, *482*, 253.
[61] I. Haiduc, R. B. King and M. G. Newton, *Chem. Rev.* **1994**, *94*, 301.
[62] H. Reuter and H. Puff, *J. Organomet. Chem.* **1989**, *379*, 223.
[63] H. Puff, W. Schuh, R. Sievers, W. Wald and R. Zimmer, *J. Organomet. Chem.* **1984**, *260*, 271.
[64] U. Weber, N. Pauls, W. Winter and H. B. Stegmann, *Z. Naturforsch.* **1982**, *37b*, 1316.
[65] V. K. Belsky, N. N. Zemlyansky, I. V. Borisova, N. D. Kolosova and I. P. Beletskaya, *J. Organomet. Chem.* **1983**, *254*, 189.
[66] C. Glidewell and D. C. Liles, *Acta Cryst.* **1979**, *B35*, 1689.
[67] S. Kerschl, B. Wrackmayer, D. Männig, H. Nöth and R. Staudigl, *Z. Naturforsch.* **1987**, *42b*, 387.
[68] C. Glidewell and D. C. Liles, *Acta Cryst.* **1978**, *B34*, 1693.
[69] U. Wannagat, V. Damrath, V. Huch, M. Veith and U. Harder, *J. Organomet. Chem.* **1993**, *443*, 153.
[70] C. Glidewell and D. C. Liles, *Acta Cryst.* **1978**, *B34*, 129.
[71] A. M. Domingos and G. M. Sheldrick, *Acta Cryst.* **1974**, B30, 519.
[72] H. Puff, H. Hevendehl, K. Hofer, H. Reuter and W. Schuh, *J. Organomet. Chem.* **1985**, *287*, 163.
[73] A. G. Davies, A. J. Price, H. M. Dawes, and M. B. Hursthouse, *J. Chem. Soc. Dalton Trans.* **1986**, 297.
[74] J. C. Pommier, F. Mendes, J. Valade and J. Housty, *J. Organomet. Chem.* **1973**, *55*, C 19.
[75] S. W. Ng, V. G. Kumar Das and E. R. T. Tiekink, *J. Organomet. Chem.* **1991**, *403*, 111.
[75a] G. A. Landrum and R. Hoffmann, *Angew. Chem.* **1998**, *110*, 1989; *Angew. Chem. Int. Ed. Engl.* **1998**, *37*, 1887.
[76] N. W. Alcock, *Bonding and Structure. Structural Principles in Inorganic and Organic Chemistry*, Ellis Horwood, New York, London, **1993**, p. 195.
[77] P. Pyykkö, *Chem. Rev.* **1997**, *97*, 597.
[78] F. Scherbaum, A. Grohmann, B. Huber, C. Kruger, H. Schmidbaur, *Angew. Chem.* **1988**, *100*, 1602; *Angew. Chem. Int. Ed. Engl.* **1988**, *27*, 1544.
[79] H. Schmidbaur, *Gold Bull.* **1990**, *23*, 11.
[80] H. Schmidbaur, *Chem. Soc. Rev.* **1995**, *24*, 383.
[81] P. Pyykkö, J. Li and R. Runeberg, *Chem. Phys. Lett.* **1990**, *23*, 11.
[82] E. M. Meyer, S. Gambarotta, C. Floriani, A. Chiesi-Villa and C. Guastini, *Organometallics* **1989**, *8*, 1067.
[83] A. Gerold, J. T. B. H. Jastrzebski, C. H. P. Kronenburg, N. Krause and G. van Koten, *Angew. Chem.* **1997**, *109*, 778; *Angew. Chem. Int. Ed. Engl.* **1997**, *36*, 755.
[84] G. R. Desiraju, *Crystal Engineering. The Design of Organic Solids*, Elsevier, Amsterdam, **1989**.
[84a] D. Braga and F. Grepioni, *Comments Inorg. Chem.* **1997**, *19*, 185.
[85] G. R. Desiraju, *Acc. Chem. Res.* **1991**, *24*, 290.
[86] D. Braga and F. Grepioni, *Acc. Chem. Res.* **1994**, *27*, 51.
[87] D. Braga, F. Grepioni and G. R. Desiraju, *J. Organomet. Chem.* **1997**, *548*, 33.
[87a] D. Braga, F. Grepioni and G. R. Desiraju, *Chem. Rev.* **1998**, *98*, 1375.
[87b] D. Braga, A. Angeloni, E. Tagliavini and F. Grepioni, *J. Chem. Soc. Dalton Trans.* **1998**, 1961.
[87c] D. Braga, F. Grepioni and E. Tedesco, *Organometallics* **1998**, *17*, 2669.
[87d] D. Braga, P. De Leonardis, F. Grepioni, E. Tedesco and M. J. Calhorda, *Inorg. Chem.* **1998**, *37*, 3337.

[88] N. W. Alcock, *Bonding and Structure. Structural Principles in Inorganic and Organic Chemistry*, Ellis Horwood, New York, p. 191.
[89] J. M. Lehn, J. L. Atwood, J. E. D. Davies, D. D. MacNicol, F. Vögtle (Editors), *Comprehensive Supramolecular Chemistry. Volume 6. Solid State Supramolecular Chemistry: Crystal Engineering*, D. D. MacNicol, F. Toda and R. Bishop, Pergamon, Oxford, **1996**.
[90] J. Yang, J. L. Marendez, A. Zafar, S. J. Geib and A. D. Hamilton, in vol. *Supramolecular Stereochemistry*, Edited by J. S. Siegel, Kluwer Academic Publishers, Dordrecht, **1995**, 141.
[91] F. H. Allen, P. R. Raithby, G. P. Shields and R. Taylor, *Chem. Commun.* **1998**, 1043.
[92] K. Gregory, P. von R. Schleyer and R. Snaith, *Adv. Inorg. Chem.* **1991**, *37*, 48.
[93] A. M. Sapse and P. von R. Schleyer (Editors), *Lithium Chemistry. A Theoretical and Experimental Overview*, J. Wiley & Sons, Inc., New York, **1995**.
[94] P. J. Fagan, M. D. Ward, and J. C. Calabrese, *J. Am. Chem. Soc.* **1989**, *111*, 1698.
[95] M. Ward, P. J. Fagan, J. C. Calabrese and D. C. Johnson, *J. Am. Chem. Soc.* **1989**, *111*, 1719.
[96] P. Bergerat, J. Blümel, M. Fritz, J. Hiermeier, P. Hudeczek, O. Khan and F. H. Köhler, *Angew. Chem. Int. Ed. Engl.* **1992**, *31*, 1258.
[97] H. Atzkern, J. Hiermeier, B. Kanellakopoulos, F. H. Köhler, G. Müller and O. Steigelmann, *J. Chem. Soc. Chem. Commun.* **1991**, 997.
[98] M. Fritz, J. Hiermeier, N. Hertkorn, F. H. Köhler, G. Müller, G. Reber and O. Steigelmann, *Chem. Ber.* **1991**, *124*, 1531.
[99] B. Oelckers, I. Chavez, J. M. Manriquez and E. Roman, *Organometallics* **1993**, *12*, 3396.
[100] X. Meng, S. Sabat and R. N. Grimes, *J. Am. Chem. Soc.* **1993**, *115*, 6143.
[101] R. E. Dinnebier, U. Behrens and F. Olbrich, *Organometallics* **1997**, *16*, 3855.
[102] B. F. G. Johnson, C. M. Martin and P. Schooler, *Chem. Commun.* **1998**, 1239.

2 Molecular Recognition and Host–Guest Interactions

2.1 Organometallic Receptors and their Host–Guest Complexes

2.1.1 Organomercury macrocyclic receptors

Macrocyclic organomercury compounds have been known for some time, but their ability to bind anionic guests in their cavity was demonstrated relatively recently. This molecular recognition process is somewhat similar to the formation of the crown ether complexes, except that the binding sites of the macrocycle are occupied by metal atoms, able to function as Lewis acids, and therefore to bind halogen and other Lewis base anions. This approach is only in its beginnings and the recent discovery of mercuracarboranes as anion binding cyclic receptors gave a new impetus to research in the field.

One of the earliest known organomercury macrocycles was the trimer of *ortho*-phenylene mercury, $(o\text{-}C_6H_4Hg)_3$, known from X-ray diffraction studies to have a planar cyclic structure, **1** [1, 2]. It has recently been found, with the aid of ^{199}Hg NMR spectroscopy, that this compound forms complexes with halide anions (Cl^-, Br^- and I^-) in dichloromethane solution, but the complexes could not be isolated in solid state [3]. The fluorinated analog $(o\text{-}C_6F_4Hg)_3$, **2**, is a much better electron acceptor and forms stable complexes with the same halide anions; these could be isolated as solid crystalline compounds and their crystal structure has been determined by X-ray diffraction [4, 5].

The perfluorinated organomercury compound forms both 1:1 and 3:2 complexes, which can be isolated with Br$^-$ and I$^-$ guests. The 1:1 complex has a polynuclear structure, in which the Br$^-$ ions are coordinated to six mercury atoms, and are sandwiched between two macrocyclic trimercury molecules, to form irregular stacking with unequal Hg$\cdots$Br$^-$ interatomic distances (3.07–3.391 Å, and one longer 3.61 Å). A chloride complex [(o-C$_6$F$_4$Hg)$_3$Cl$_2$]$^{2-}$, probably having a triple decker structure, has also been obtained [3].

The same fluorinated mercuracycle links thiocyanate anions (in acetone solution) and the tetra-n-butylammonium salt of [(o-C$_6$F$_4$Hg)$_3$SCN]$^-$ has been isolated and structurally characterized [6]. The crystal structure contains a helical chain of alternating (o-C$_6$F$_4$Hg)$_3$ molecules and SCN$^-$ ions. The stacks are held by sulfur atoms at Hg$\cdots$S distances in the range 3.06–3.36 Å, with one longer at 3.87 Å. Thus, both host–guest complexation and supramolecular self-assembly if observed with these compounds.

Another macrocyclic organomercury receptor for halide anions is the pentamer [HgC(CF$_3$)$_2$]$_5$, **3**, which contains a nearly planar Hg$_5$C$_5$ ring [7]. It reacts with tetraphenylphosphonium chloride to form a 1:2 complex, {[HgC(CF$_3$)$_2$]$_5$Cl$_2$}$^{2-}$, with each chlorine coordinated to all five mercury atoms (ave Hg$\cdots$Cl 3.284 Å above and 3.221 Å below the ring), the macrocycle being stacked between the two halide anions [8].

3

The nature of chemical bonds between the ring metal atoms and the incorporated anions has been analyzed theoretically by the MNDO method [9].

Tri- and tetranuclear mercuracarbaboranes were recently prepared and were found to bind halide and other anions by forming host–guest complexes such as **4** and **5** [10–19]. The subject has recently been covered by an excellent review [20].

The several interconvertible modes of interaction between a tetramercuracarbaborane receptor and iodide guests, are shown schematically in Figure 2.1.

The mercuracarbaboranes are versatile anion receptors and interesting new results are expected in this field.

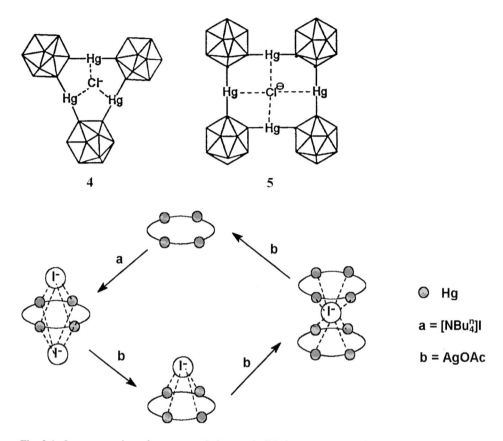

Fig. 2.1. Interconversion of mercuracarbaborane iodide host–guest complexes.

2.1.2 Organocopper and organosilver potential receptors

Organocopper and -silver macrocycles are also known and it would be interesting to check their molecular recognition of anions, if any. Thus, the pentameric (μ-mesityl)copper(I) derivative [21] and the tetrameric (μ-mesityl)silver(I) derivative [22] would be interesting candidates as receptors in host–guest chemistry.

2.1.3 Organocyclosiloxane receptors

Large cyclosiloxane rings are well known and have been reviewed [23]; examples include $(Me_2SiO)_n$ with $n = 5, 6, 7 \cdots$ (in addition to normal size rings with $n = 3, 4$) and their structure is reminiscent of that of crown ethers in that an alternation of oxygen atoms with silicon atoms may be taken as an analogy with the alternation of $-CH_2CH_2-$ groups with oxygen atoms in the organic crown ethers. The donor properties of oxygen atoms differ from those in crown ethers and are expected to be less pronounced because of possible delocalization of some electron density into the $3d$-orbitals of silicon. The tendency of macrocyclic siloxanes to complex with alkali metal cations has, however, been observed in solvent extraction studies. Thus, large-ring polydimethylsiloxanes, $(Me_2SiO)_n$ with $n = 5$–9 and 12 have selective ion-binding properties towards alkali and alkaline earth metal ions. Large cations, such as K^+, Cs^+, Ca^{2+}, Sr^{2+} are preferentially extracted by the cyclododecasiloxane, $(Me_2SiO)_{12}$ and the occurrence of host–guest complexation is confirmed by stability-constant measurements. The estimated diameter of the cavity in the macrocyclic siloxane (ca 3.5 Å) is of the same order of magnitude as the cavity in 18-crown-6 polyether [24, 25].

Alkali metal ion host–guest complexation by organocyclosiloxanes became certitude with the serendipitous isolation and then X-ray crystal structure determination of a potassium complex of the 14-membered ring tetradecamethylcycloheptasiloxane obtained as $[K(Me_2SiO)_7]^+[InH(CH_2Bu^t)_3]^-$, in an accidental reaction of $K^+[InH(CH_2Bu^t)_3]^-$ with stopcock silicone grease [26]. The same cationic host – guest complex, **6**, has been found in $[K(SiMe_2O)_7][K\{C(SiMe_3)_2(SiMe_2Vi)\}_2]$ also obtained by a similar lucky accident [27].

6

The cycloheptasiloxane ring is basically planar and the cation guest is coplanar with the ring. The $K^+ \cdots O$ distances are 2.86–2.99 Å (average 2.93 Å).

A beautiful family of host–guest assemblies consists of sandwich complexes made of two cyclosiloxane rings, incorporating several metal ions. All contain the 12-membered cyclohexasiloxane ring, bearing an organic substituent and a silanolate Si–O$^-$ function at each silicon site. It is formed when a mono-organo-substituted siloxanolate reacts with metal cations, under strongly basic conditions – the parent cyclohexasiloxanolate hexanegative anion is unknown. Two views of this macrocyclic anion are shown in **7**, one to indicate the *cis* positions of all the silanolate functions, **7a**, the other to illustrate the crown structure of the hexasiloxane ring, **7b**. It will be seen in the compounds to be discussed that the silanolate functions bind the metal ions *endo*-hedrally between the siloxane rings, whereas the ring oxygen atoms bind *exo*-hedrally some alkali metal ions, just like the crown ethers.

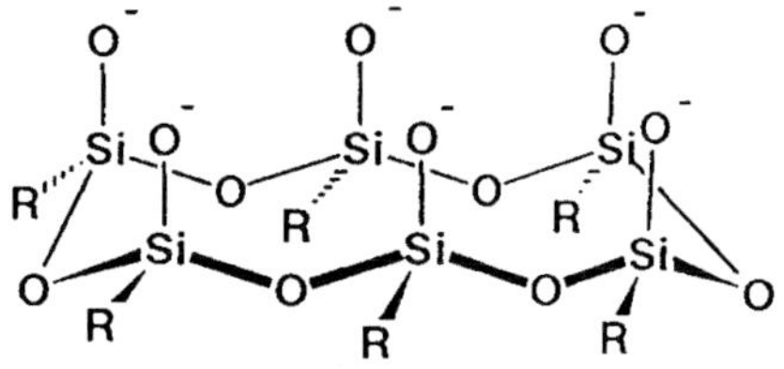

7a 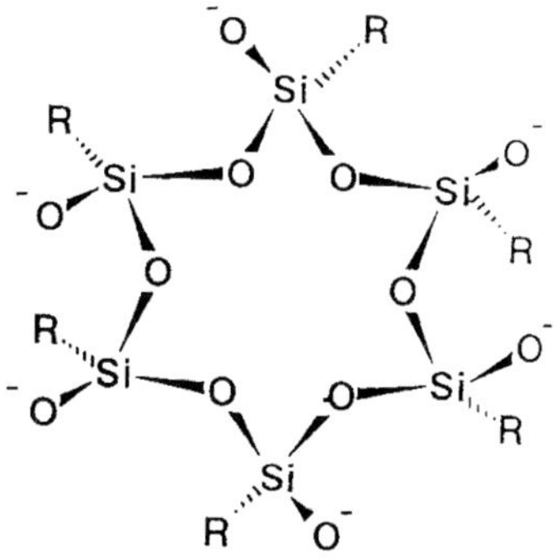 **7b**

A whole series of host–guest complexes of this macrocyclic anion has been prepared and examined by X-ray diffraction [28, 29]. One of the first examples is the nickel complex Na$^+$[(PhSiO$_2$)$_6$Ni$_6$(Cl)(O$_2$SiPh)$_6$]$^-$·14Bu^{n}OH, obtained via ring expansion from the cyclotrisiloxanolate Na$_3$[PhSiO(O$^-$)]$_3$ with nickel chloride in alcoholic solution. The structure of this supramolecular architecture, established by X-ray diffraction [30, 31], has several features worthy of comment. All six silanolate functions are oriented *cis* axial, thus enabling metal ion complexation, as shown in **8**. In these supramolecular architectures the organocyclosiloxanolate acts as both *endo* and *exo* receptor. The intermolecular forces involved are *endo*-hedral ionic interactions and *exo*-hedral ion–dipole interactions.

In the structure diagram of **8** the phenyl groups and the solvating alcohol molecules (externally coordinated to nickel) are not shown for clarity. The structure contains six nickel atoms sandwiched between two Si$_6$O$_6$ macrocycles bonded

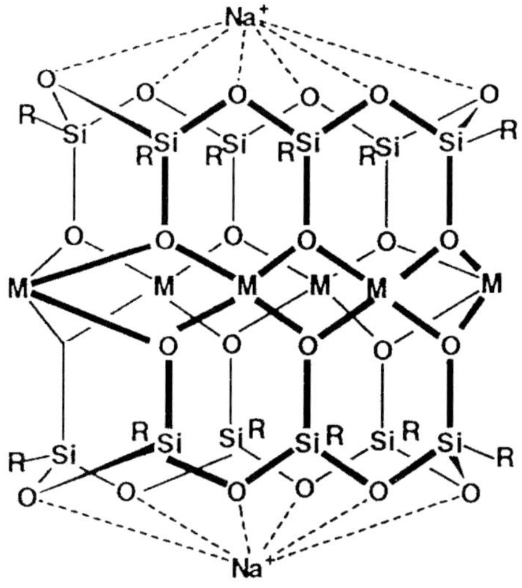

8

through basically ionic bonds to the silanolate functions. The six nickel atoms form a tiara-shaped Ni_6O_{12} macrocyclic unit, which in turn encapsulates a chloride ion coordinated by six metal ions. Each nickel ion is surrounded by four oxygens (all being *cis*-silanolate functions), by the central chlorine ion and by a solvent (alcohol) molecule (not shown) coordinated *exo*-hedrally. The intercalation of the $L_6Ni_6O_{12}Cl$ core thus formed between the two 12-membered rings leads to the sandwich structure shown in **8**.

This type of coordination stabilizes a crown conformation of the cyclohexasiloxane ring. There are, in addition, two sodium cations attached at the top and the bottom of the sandwich in a crown ether-type complexation mode, with each sodium ion bonded to six oxygens of the rings, **9**, and further solvated by alcohol molecules (not shown). The macrocyclic siloxane acts as both *endo* and *exo* receptors and the formation of the supramolecular architecture involves both ionic self-assembly and crown-ether-type ion recognition.

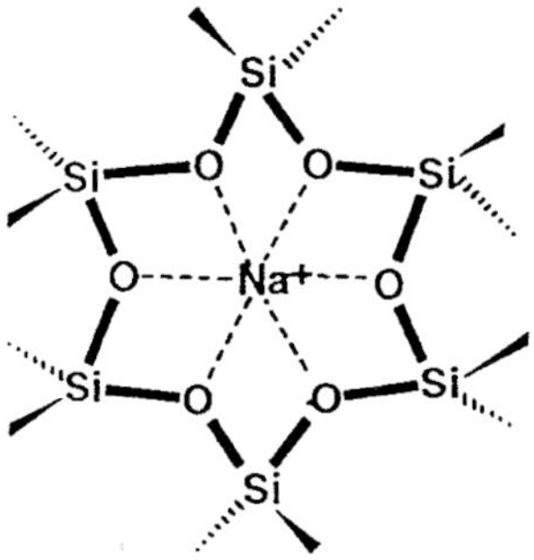

9

An additional layer of nickel atoms can be inserted between the two cyclohexasiloxane macrocycles, in the previous structure leading to the composition $[(PhSiO_2)_6Ni_8O_2(O_2SiPh)_6] \cdot 14Bu^nOH \cdot 10H_2O \cdot 2acetone$ [32], but nickel can be partially replaced by sodium within the sandwich, to give the composition $Na_2[(PhSiO_2)Na_4Ni_4(OH)_2(O_2SiPh)_6] \cdot 16Bu^nOH$ [33] (see also [33a]). Manganese $Na[(PhSiO_2)_6Mn_6(Cl)(O_2SiPh)_6] \cdot solv.$, and cobalt $Na^+[(PhSiO_2)_6Co_6(Cl)(O_2SiPh)_6]^-$, compounds with similar structures have also been investigated [8].

A related copper(II) compound $K_4[(PhSiO_2)_6Cu_6(O_2SiPh)_6] \cdot 6EtOH$, has a similar core, but the potassium cations are attached to only four *endo*-cyclic oxygens [34, 35]. In the compound $[(PhSiO_2)_6Cu(PhSiO_2)_6 \cdot 6DMF] \cdot 4DMF$ only dimethylformamide molecules are externally coordinated to copper atoms [36].

Analogous compounds with other organic groups at silicon have been also obtained, with similar structures and only minor differences. Thus, in the ethyl derivative $K_2[(EtSiO_2)K_2Cu_4(O_2SiEt)_6] \cdot 4Bu^nOH$ two copper ions have been replaced by two potassium ions, externally coordinated to four silanolate functions, whereas the other two potassium ions are coordinated in a crown-ether fashion to six oxygens of the rings, as shown in **10** [37].

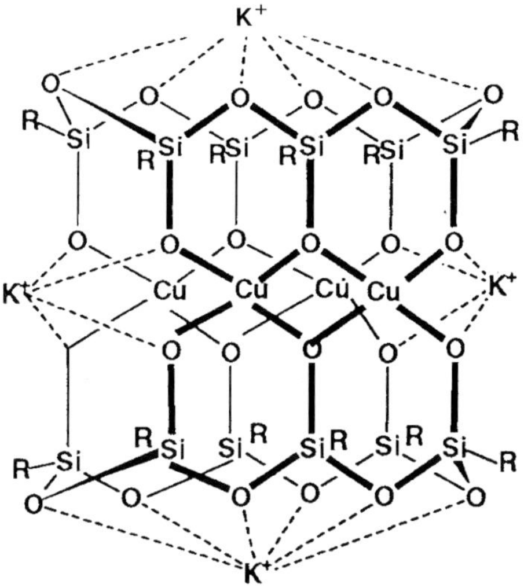

2.1 Organometallic Receptors and their Host–Guest Complexes

10

Under slightly different preparative conditions, two other macrocyclic siloxanolate host–guest copper phenyl and vinyl complexes, $Na_4[(PhSiO_2)_{12}Cu_4]\cdot 8Bu^nOH$ and $K_4[(ViSiO_2)_{12}Cu_4]\cdot 6Bu^nOH$ (Vi = vinyl), have been obtained from $(RSiO_{1.5})_x$, $CuCl_2$, and NaOH in n-butanol. These have a totally different structure, containing a 24-membered cyclosiloxane ring, folded in a saddle conformation, **11**, imposed by the need to accommodate two planar Cu_2O_2 central units [11].

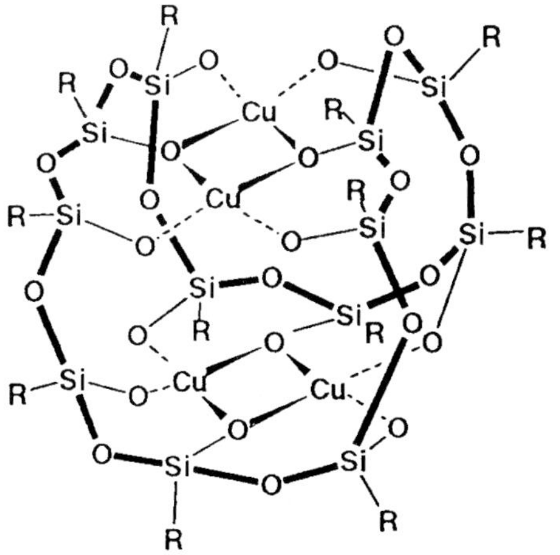

11

The formation of these architectures is not yet fully understood. A self-assembly process occurs spontaneously in solution, and it seems that the metal ions present act as templates at one stage of the process. This is suggested by the formation of the 24-membered macrocyclic siloxane just cited, and is further supported by the fact that cations with a significantly larger ionic radius, such as lanthanides, determine the formation of 16-membered cyclooctasiloxane rings, coordinated basically in a similar mode, to form sandwich complexes of the type shown in **12**. Thus, a vinyl derivative $K_5[(ViSiO_2)_8La_4(OH)(O_2SiVi)_8]\cdot 5Bu^nOH\cdot 2H_2O$ [38] and phenyl derivatives $Na_6[(PhSiO_2)_8M_4(O)(O_2SiPh)_8]\cdot 10EtOH\cdot 8H_2O$ with M = Nd and Gd [39, 40], containing 16-membered cyclooctasiloxane rings, are formed under conditions rather similar to those in which smaller divalent cations produced compounds derived from cyclohexasiloxane rings.

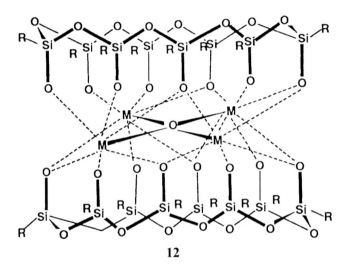

12

In this case, the supramolecular structure contains a central M_4O unit, and the lanthanide cations do not behave as *endo* receptors and are not externally solvated by additional ligands. The encapsulation of the M_4O fragment and its connectivity to cyclooctasiloxanolate ring is shown schematically in **13**.

A completely different composition, $Na_6[(PhSiO_{1.5})_{22}Co_3O_6]\cdot 7H_2O$, has been obtained by reaction of $Na_3[PhSiO(O^-)]_3\cdot 3H_2O$, with $(PhSiO_{1.5})_8$, NaOH, and $CoCl_2$. This retains a major fragment of the parent cubane-like octaphenylsilsesquioxane and contains a $[(PhSiO_{1.5})_{22}Co_3O_6]^{6-}$ framework, built from three open-edge silsequioxane cubes, each having a corner occupied by a cobalt atom which replaced a silicon. These cubes are connected by a $PhSi(O_{0.5})_3$ moiety in a complex polycyclic structure, which also incorporates three Na^+ cations as guest; the other three Na^+ cations are *exo*-hedrally coordinated to pairs of oxygen atoms of the siloxane skeleton and are further solvated by water molecules [41].

Beyond their attraction as fascinating examples of inorgano–organometallic supramolecular constructions, these metallosiloxane sandwich host–guest complexes are of great interest as catalyst precursors [42–44], and molecular magnets [45, 46], and their properties along these lines are intensively investigated. They also

[Structure 13, M = La, Nd, Gd, Dy]

13

represent models for the attachment of catalytically active metal ions to silica surfaces [47].

The encapsulated metal can be eliminated by trimethylsilylation, with preservation of the cyclosiloxane skeleton. This provides a unique opportunity for preparing stereoregular macrocyclic siloxanes [47a].

An electrochemical study has shown that in DMF most of the cyclosiloxanolate sandwich complexes are electrolytes and the conductivity is mainly because of the free Na^+ cation, the anion contribution being negligible. Cyclic voltammetry experiments suggest a high degree of delocalization of the negative charge over the complex core of the anion [48].

Other types of metallosiloxane guests for alkali metals are presumably possible, as suggested by the unusual structure of an indium–siloxane cage which incorporates sodium cations [48a].

It seems that molecular recognition and incorporation of alkali metal ions is not limited to cyclosiloxanes or heterometallosiloxanes. The recently reported host–guest complexation of sodium by cyclic organogallium phosphonates in $[Na_4(\mu_2\text{-}OH_2)_2(THF)_2][(Me_2GaO_3PBu^t)_2]_2 \cdot 2THF$ and potassium in $[K(THF)_6]$-$[K_5(THF)_2\{(Me_2GaO_3PBu^t)_2\}_3]$ opens a new chapter in inclusion chemistry [48b]. Sandwich complexes with four lithium cation guests intercalated between two $Ga_3O_6P_3$ macrocycles in a crown-ether fashion, $Li_4[(MeGa)_6(\mu_3\text{-}O)_2(O_3PBu^t)_6]\cdot 4THF$, have also been reported [48c]. The Ga–O–P macrocycles are weaker complexing agents than the organic crown ethers and can be delithiated with 12-crown-4 to leave cubane cage molecules void of cationic guests [48d].

Crown ether complexation of lithium by the eight-membered ring tetramethylcyclotetraalumoxane $(MeAlO)_4$ has been revealed by single-crystal X-ray diffraction studies [48e].

2.1.4 Organotin macrocyclic receptors

Tin-containing macrocycles **14–16** use the capacity of the metal atom acceptor to capture halide anions in the cavity, acting like 'inverse crown ethers' [49–51]. Thus, tri- and tetrastanna macrocycles, prepared as methyl derivatives (R=Me) can be chlorodemethylated, and the products (R=Cl) act as Lewis acids and form Cl⁻ complexes. A crystal structure determination of $(Cl_2SnCH_2CH_2CH_2)_3Cl^-$ established that the chloride anion guest is bridging two tin atoms [52].

Organotin cryptands, able to encapsulate Cl⁻ anions, were similarly prepared, with $n = 8$, 10, and 12. X-Ray crystal structure analyses of $\{ClSn[(CH_2)_8]_3SnCl\}Cl^-$ and $\{ClSn[(CH_2)_8]_3SnCl\}F^-$ showed that the Cl⁻ anion is trapped in the middle of the bicyclic cryptand and forms a stronger bond to tin (2.610 Å) and a weaker bond (Sn···Cl 3.888 Å). The fluoride anion is more symmetrically placed (Sn–F 2.12 and 2.28 Å) [53]. The complexing capacity of such macrocycles towards Cl⁻ and Br⁻ has been confirmed by electrochemical transport experiments [54]. In the monocyclic species **17** the Cl⁻ oscillates between the two tin atoms. In the bicyclic species **18** only one Cl⁻ ion per host is incorporated; obviously, there is no room for two. Negligible binding occurs for $n = 6$, the strongest is observed for $n = 8$ and 10, and weaker for $n = 12$, in agreement with the steric complementarity between the anionic guest and the host cavity [55].

Monocyclic $Cl_2Sn[(CH_2)_n]_2SnCl_2$ ($n = 8$, 10, and 12) bind two Cl⁻ anions rapidly but are poorly selective and there is little dependence on ring size [56]. The anion-binding capacity of tricyclic organotin cryptands depends on the tin–tin distance in the host [57, 58].

Strong anion binding capacity (towards F⁻ and Cl⁻) is also observed for non-cyclic methylene bridged, tri- and tetratin hosts, $X(Ph_2SnCH_2)_nSnPh_2X$ ($n = 2$ or 3, X = F and Cl) which form $[X(Ph_2SnCH_2)_nSnPh_2X](X^-)_{n-1}$ anionic complexes [59].

2.1.5 Ferrocene-containing coronands and cryptands

Several receptors containing metallocene building units and serving multiple purposes have been designed and synthesized. It has been found that the metallocene-based receptors are redox-active species and can bear positive charges (and thus serve as anion receptors) [60]. They can also contain functional groups (such as amido and carboxylic) able to form hydrogen-bonds (and thus to enhance the binding possibilities of anions or to afford binding of neutral molecules). The presence of nitrogen or sulfur in such receptors can facilitate metal cation binding. Receptors incorporating metallocene units into a coronand, cryptand, or cavitand structure might, therefore, perform multiple functions. If the guest species changes the physical properties of the receptor in a detectable way, the resulting supramolecular system can serve as a sensor. Great structural variety is possible and there is already a large volume of literature on the subject. Here we will not be able to cover it exhaustively, and some published reviews [61–65] are recommended for more detailed discussion of various aspects of their chemistry, especially their electrochemical behavior and other properties. On the basis of redox-active metallocene-based receptors the new discipline of supramolecular electrochemistry has emerged [66].

Ferrocene-containing macrocycles (coronands) were prepared as early as 1979–1980 [67–69], but intensive research in the area is of more recent date.

2.1.5.1 Ferrocene polyoxa coronands

There are several structural variations in the building of ferrocene-containing polyoxa coronands, depending on whether the crown ether is bonded directly to the ferrocene moiety through oxygen, or there are spacers, which can be short (e.g. methylene) or larger groups. Thus, the first type is illustrated by compounds **19** ($n = 2, 3, 4,$ and 5) [70–73]. In other coronands, the ferrocene unit is separated by the oxygen of the macrocycle through methylene groups (compounds **20–23**) [74]. In all these receptors the ferrocene unit is incorporated in the macrocycles.

n = 1 - 4   n = 2 - 5   n = 2 - 5
   19   **20**   **21**

Other types include crown ether (coronands) modified by ferrocene units attached as side groups. These are illustrated by compounds in which the ferrocene is sepa-

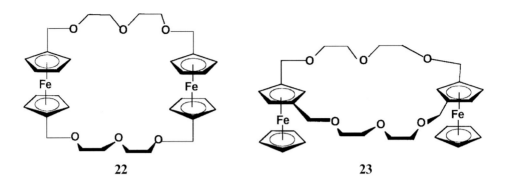

rated from the macrocycles by spacers such as aromatic, **24** [75], alkenylaryl, **25** and **26** [76], ketimino and aminoaryl, **27–29**, and amidoaryl, **30**, moieties [77–79].

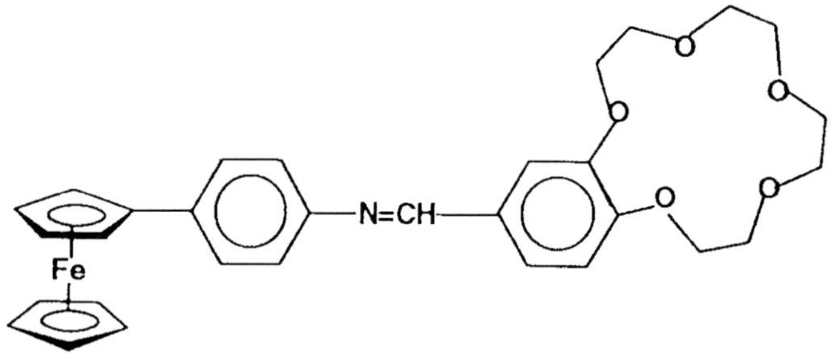

27

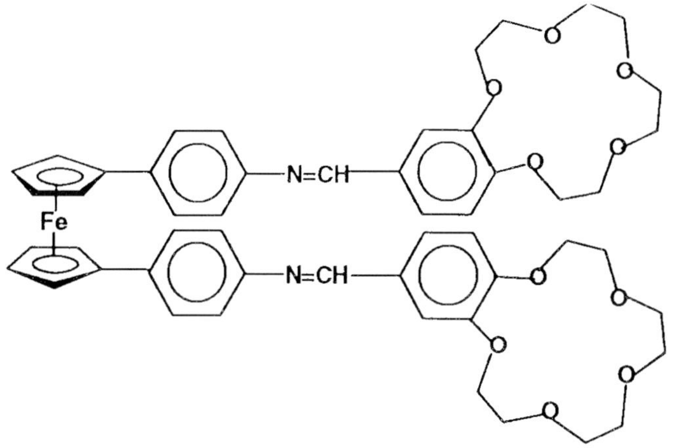

28

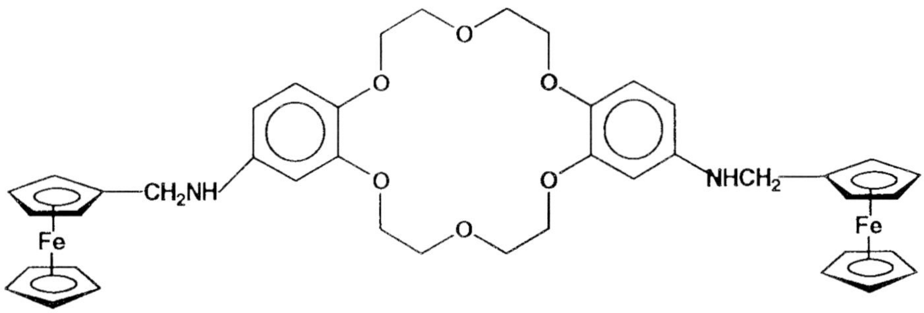

29

The ferrocene coronands **31** undergo reversible one-electron oxidation [72] and the redox potential is usually sensitive to metal-ion complexation by the crown ether part of the molecule. Such complexes have been prepared in reactions with thiocyanates, MSCN (M = Li, Na, K) [71]. The macrocyclic ligands extract thal-

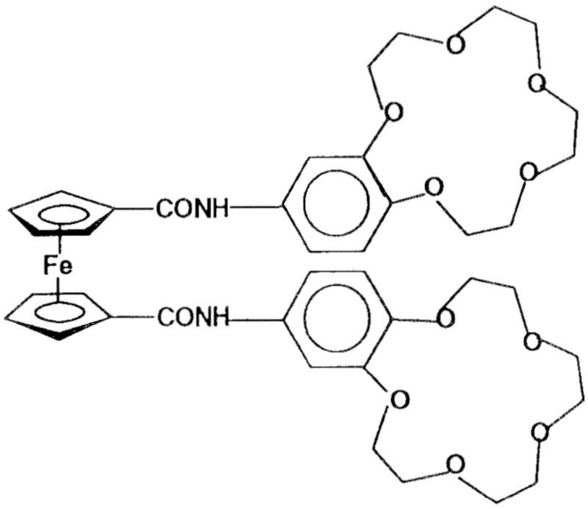

30

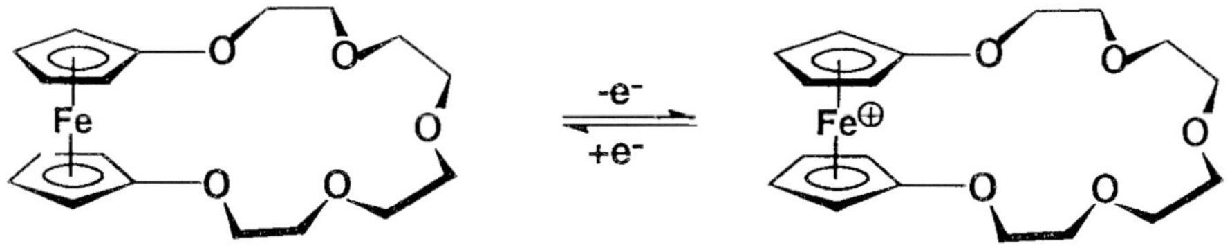

31a ⇌ (-e⁻ / +e⁻) **31b**

lium and alkali metals in the order $Tl^+ > Rb^+ > K^+ > Cs^+ > Na^+$ [71, 74] and can transport sodium ions across membranes [73].

The aryl- and arylalkenyl ferroceno-coronands also form 1:1 complexes with Na^+ and 2:1 complexes with K^+ and are redox-responsive. Thus, coordination of K^+, Na^+, and Mg^{2+} shifts the ferrocene oxidation half-wave potential to more positive values [75, 76].

Selective complexation of alkali metals by metalloceno-coronands has also been observed; some have strong selectivity for potassium relative to both Na^+ and Cs^+ [77].

2.1.5.2 Ferrocene polyaza–oxa coronands and cryptands

A broad variety of aza–oxa coronands is known with ferrocene units bonded as side groups to the macrocycle in various modes. The ferrocene units can be connected by hydrocarbon bridges, **32–35**, or by functional groups, **36–38** [80–84].

These ligands form different host–guest complexes, **39**, and the redox potential of the ferrocene–ferricinium couple is sensitive to coronand complexation by alkali metals [84a].

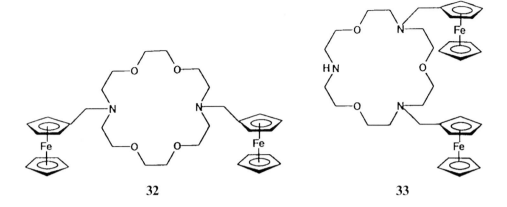

32

33

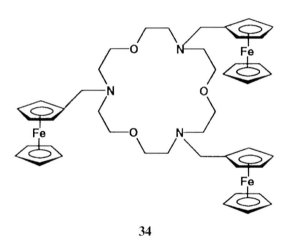

34

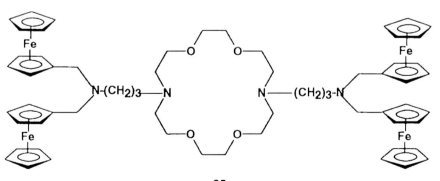

35

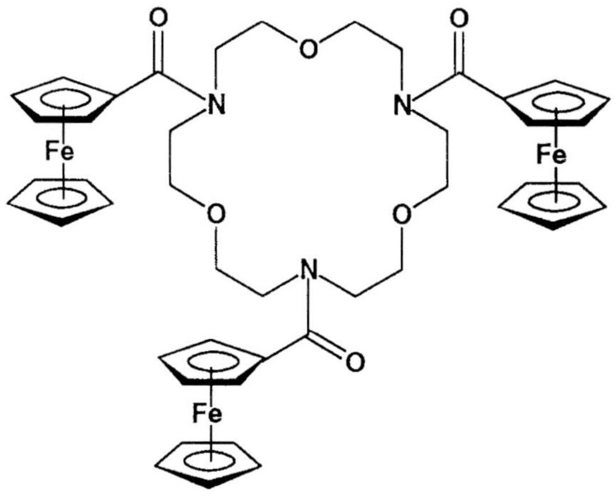

36

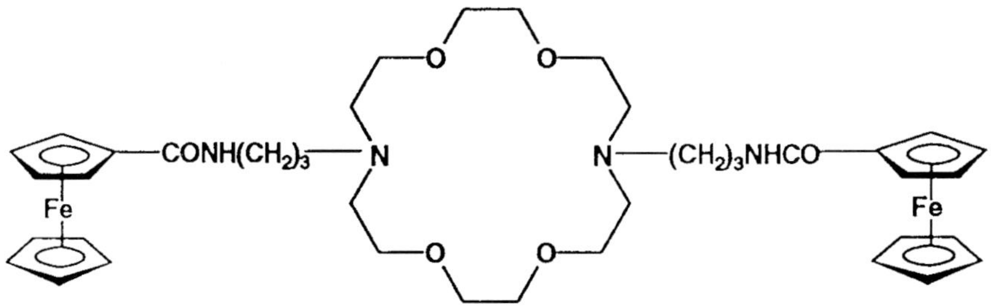

37

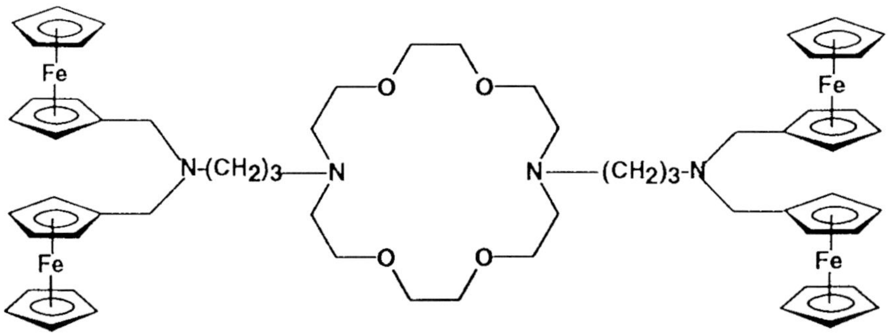

37a

Selective electrochemical recognition of mercury in water can be achieved with a ferrocene-functionalized 12-azacrown-4 derivative [84b].

Simultaneous cation–anion complexation, **40**, is possible with appropriately constructed receptors.

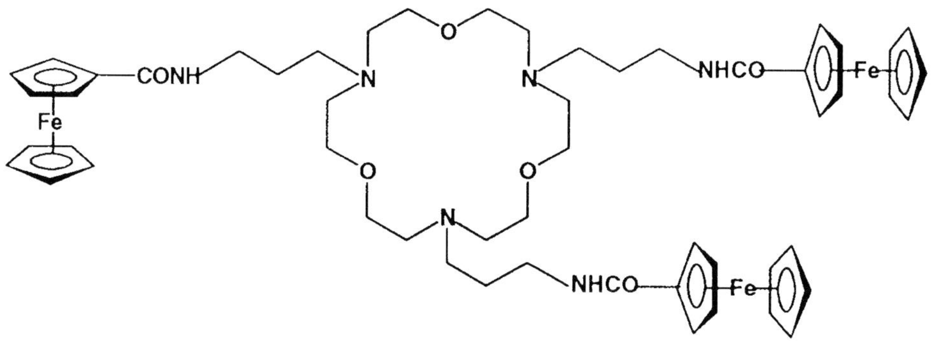

2.1 *Organometallic Receptors and their Host–Guest Complexes*   43

38

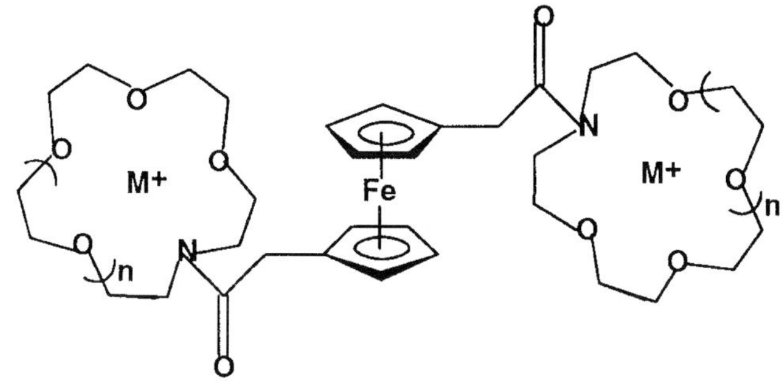

n = 1, M⁺ = Na⁺
n = 2, M⁺ = Na⁺, K⁺

39

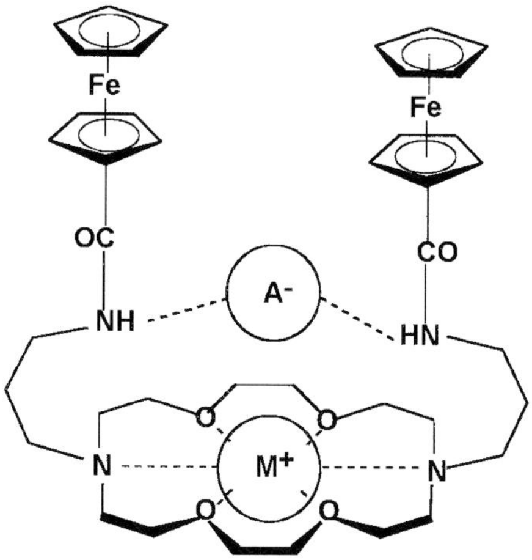

40

A tris-ferrocenyl coronand recognizes the ammonium cation in the presence of potassium; X-ray diffraction analysis of the ammonium complex has revealed a host–guest relationship [81].

A host–guest complex, **41**, of silver with a bis-ferrocene cryptand has been thoroughly investigated, revealing strong evidence of an interaction between Ag^+ and the ferrocene unit, in addition to silver–nitrogen (and some silver–oxygen) interactions with the macrocycles [85].

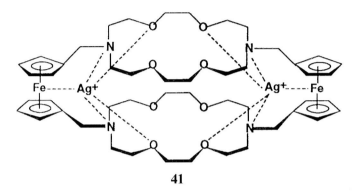

41

A nitrogen atom can be incorporated in the macrocycle with a pyridine moiety, **42**, to modify the chemical behavior of the receptor. Such receptors ($n = 0$) have poor complexing capacity towards alkali metal ions but excellent complexation properties towards Ag^+, In^{3+} and Zr^{4+} cations [86].

$n = 0, 1, 2, 3$

42

Several ferrocene aza–oxa cryptands incorporating a ferrocene unit in the main body of the receptor have been synthesized and their complexing properties have been thoroughly investigated [84a, 87–93]. The ferrocene unit can be connected directly to an oxa–aza macrocycle to close a cryptand structure, as shown in **43**, or can be part of a polyether arm of the tricyclic cryptand, **44**, thus creating a larger cavity:

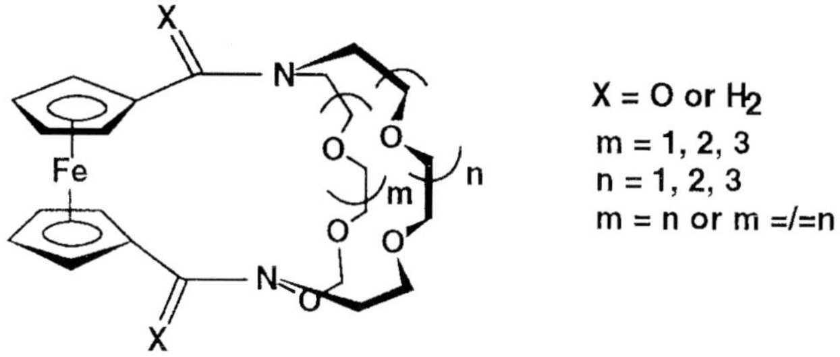

X = O or H₂
m = 1, 2, 3
n = 1, 2, 3
m = n or m =/=n

43

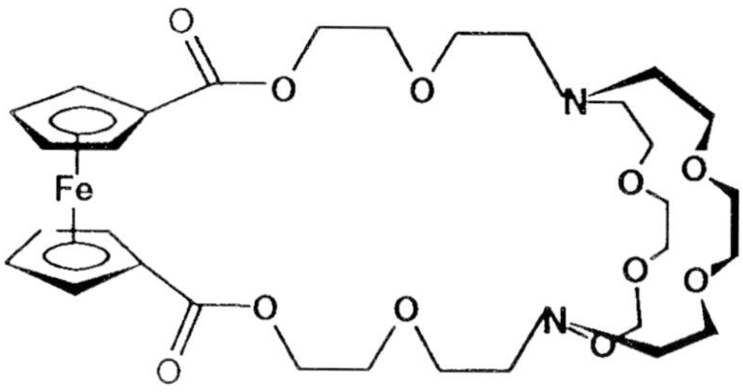

44

Obviously, many structural variations are possible in the design of ferrocene oxa–aza cryptands. Some of the oxa–azaferrocene cryptands form alkali metal and lanthanide complexes [90]. FAB mass spectrometry experiments have shown that the cryptands have strong selectivity for the potassium cation compared with Li^+, Na^+, or Cs^+ [94]. In these complexes the macrocycle functions as a host, but in Mg^{2+} complexes the cation is coordinated by the amide carbonyl groups [95]. In the lanthanide complexes the metallocene moiety acts as an efficient center for radiationless deactivation of the lanthanide excited state [96].

Because of their structures the ferrocene aza–oxa cryptands, **45**, can simultaneously bind anions by interaction with the ferrocene CO–NH group (with hydrogen-bonding participation) and alkali metal ions (in the crown ether moiety) [92].

Novel redox-active basket-shaped ferrocene-containing receptors with diphenylglycoluril fragments can bind cations in the crown rings and neutral organic guests in the cavity [93].

2.1.5.3 Ferrocene polyaza coronands and cryptands

Polyaza ferrocene coronands with three, four, and five nitrogen atoms in the macrocycle, **46** [97, 98], have been prepared and investigated. They form nickel(II) host–guest complexes, and reversible oxidation of the ferrocene moiety can be

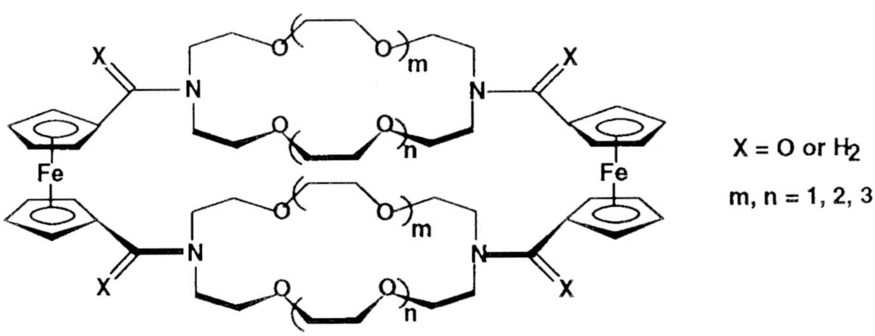

46   2 *Molecular Recognition and Host–Guest Interactions*

X = O or H$_2$

m, n = 1, 2, 3

45

modulated by changing the solvent medium. The new receptors can complex with and electrochemically recognize transition metal cations (Ni^{2+}, Cu^{2+}, Zn^{2+}) and the HPO$_4^-$ anion in water [98]. A cathodic shift of the ferrocene–ferrocenium redox couple in the tetraaza-ferrocene macrocycle can occur as a result of the binding of the HPO$_4^{2-}$ and ATP^{4-} anions; the compound is thus sensitive to their presence.

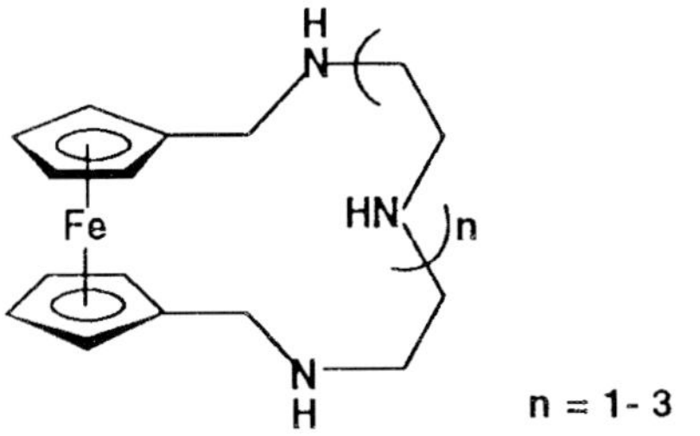

n = 1-3

46

A more sophisticated structure incorporating a 1,1-bipyridyl fragment, **47**, can coordinate as an *exo* receptor through its external coordination sites; it is remarkable because of its high selectivity for Cl$^-$ relative to H$_2$PO$_4^-$ [99].

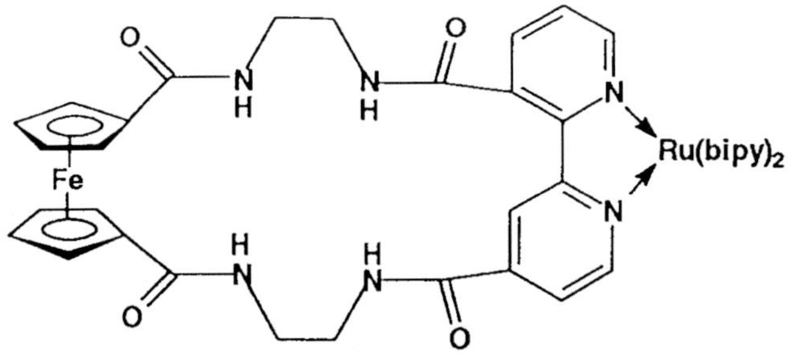

47

A novel redox-active cryptand, **48**, containing two bipyridyl and two ferrocene units incorporated in a macrocycle has been recently described [100, 100a].

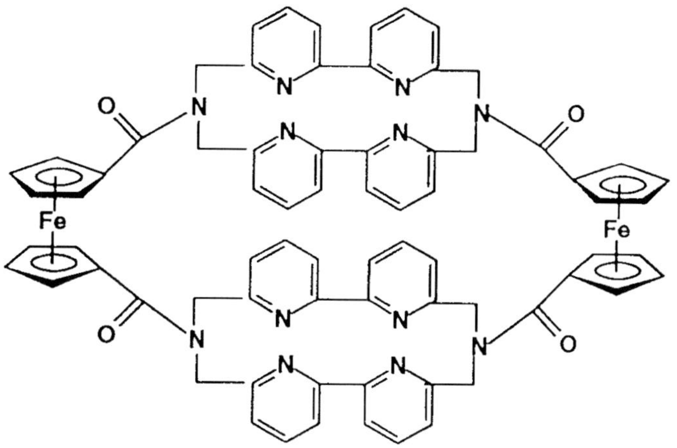

48

A remarkable cryptand, incorporating three ferrocene moieties connected by two tripodal tetramine fragments, **49**, has been reported [101].

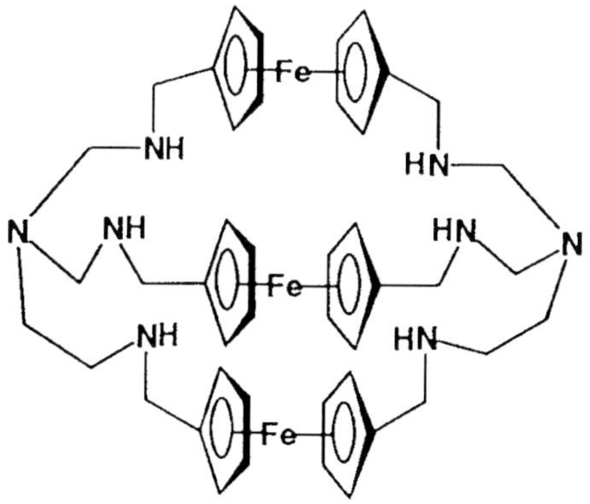

49

A less traditional nitrogen-containing ferrocene coronand is the phosphorus–nitrogen macrocycle **50**, which binds anions by hydrogen-bond interactions and is selective for $H_2PO_4^-$ against HSO_4^- and Cl^- anions in dichloromethane solution [102].

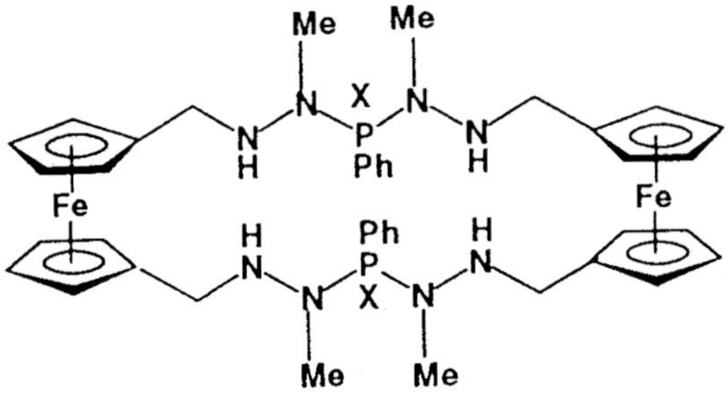

50

The tendency of ferrocene coronands containing NH groups to participate in hydrogen-bonding enhances their anion-binding capacities.

2.1.5.4 Ferrocene polyoxa-thia coronands

The sulfur-containing ferrocene coronands were among the first reported [103] and several types, e.g. **51–54**, are now known [104–107]. Various other related compounds have been characterized and can be found in the references cited.

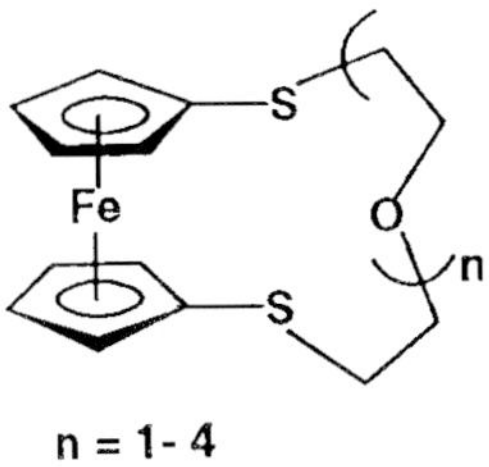

n = 1- 4

51

n = 1- 4

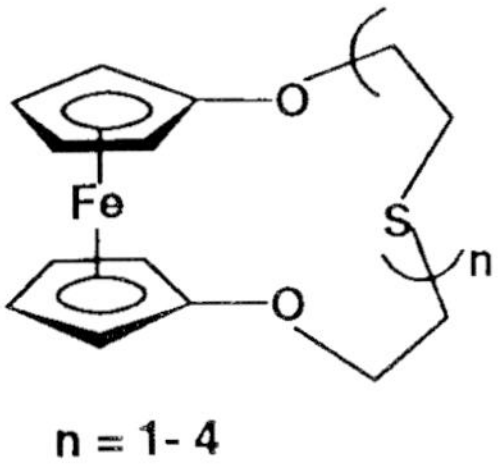

52

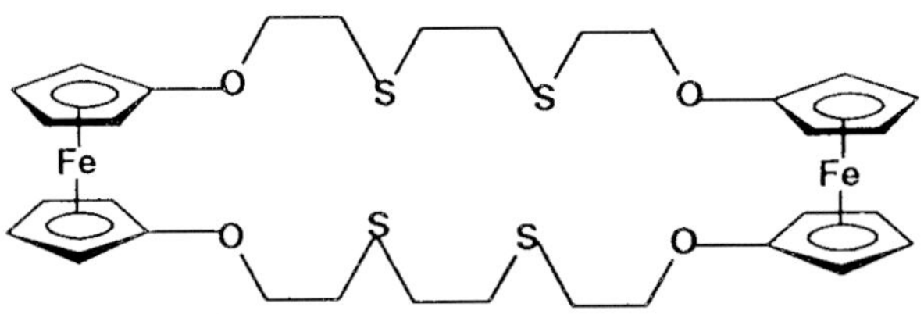

53

These coronands are efficient complexing agents for silver(I), **55**, but not for alkali metals. They can serve as extracting agents for Ag^+, Tl^+, Cd^{2+}, Co^{2+}, Cu^{2+}, Ni^{2+}, and Hg^{2+} and they complex selectively with Hg^{2+} and Cu^{2+}.

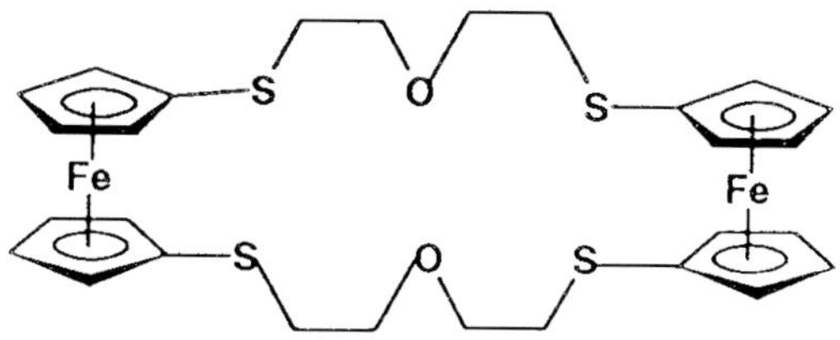

54

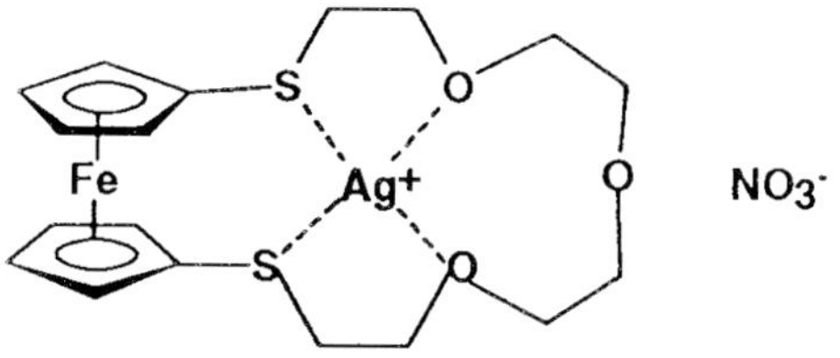

55

A tetrathiafulvalene unit has been incorporated into the macrocycle **56** to modify its electrochemical (redox) properties [108].

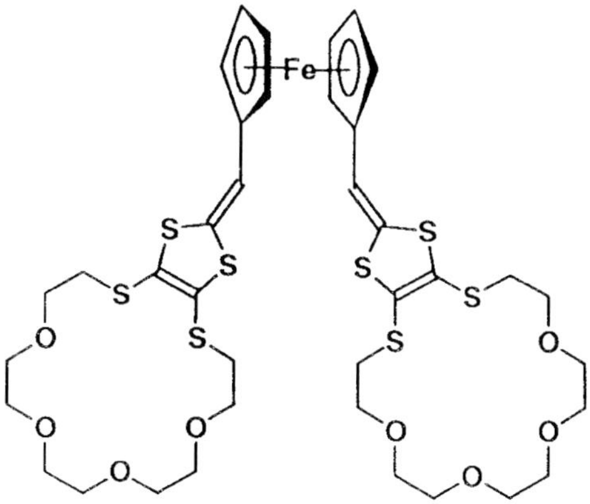

56

2.1.5.5 Ferrocene polythia coronands

Sulfur-containing ferrocene macrocyclic receptors are known containing from three to six sulfur atoms and one or two ferrocene moieties [109–112]. Examples include **57–59**.

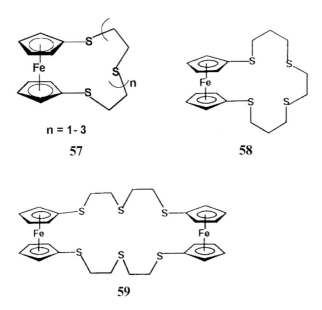

n = 1- 3
57   58

59

The sulfur-containing receptors form host–guest complexes with soft metal cations such as Cu^+, Ag^+, Hg^{2+}, and Pd^{2+}; their stability depends on the size of macrocycle and in some cases spectral data suggest an interaction between the iron atom and the metal coordinated by the macrocycle, as shown in **60**.

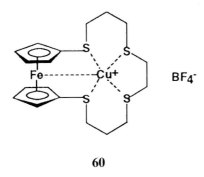

60

Nitrogen can be incorporated in the macrocycle to give aza-thia hosts, **61** and **62** [113, 114].

2.1.5.6 Non-cyclic ferrocene receptors

A series of ferrocene derivatives with amine substituents behave as anion or neutral molecule receptors and can recognize various anions, a property facilitated by their

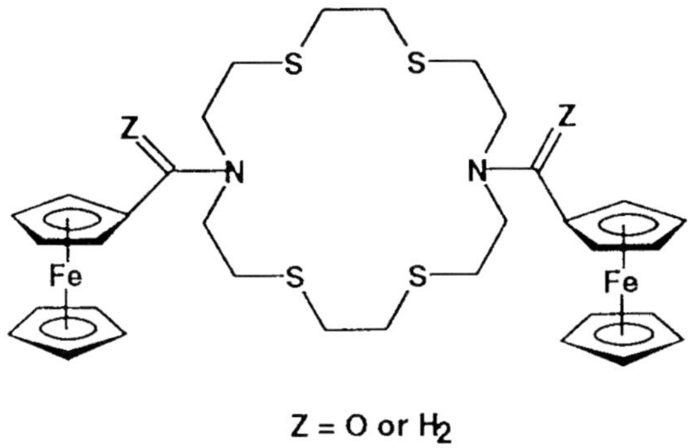

Z = O or H₂

61

n = 1- 3

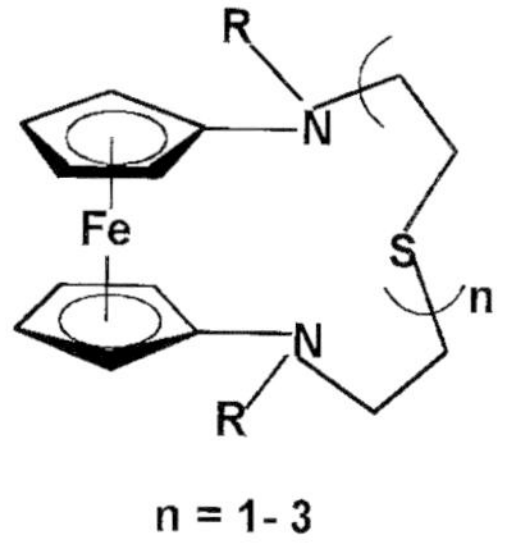

62

capacity to form N–H···X⁻ hydrogen-bonds. Receptors of this type are ferrocenecarboxylic acid amides, **63** [115], including pyridyl amides, **64** and **65**, and polyamino derivatives, **66** and **67** [116–118].

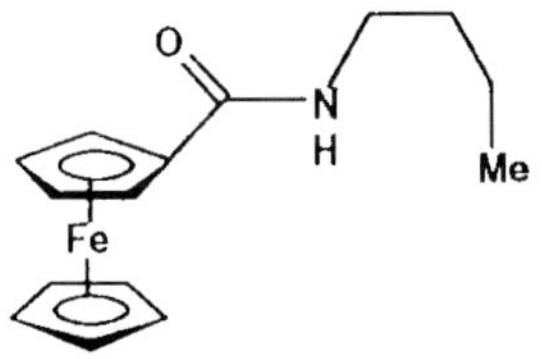

63

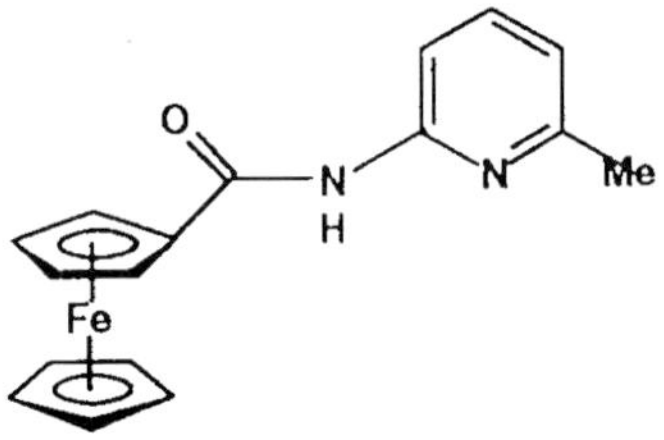

64

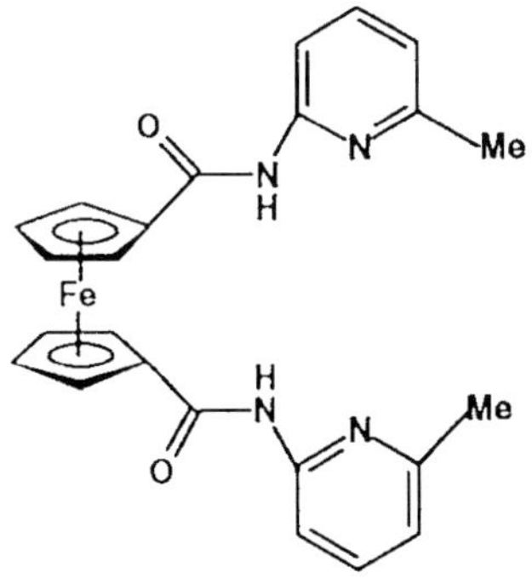

65

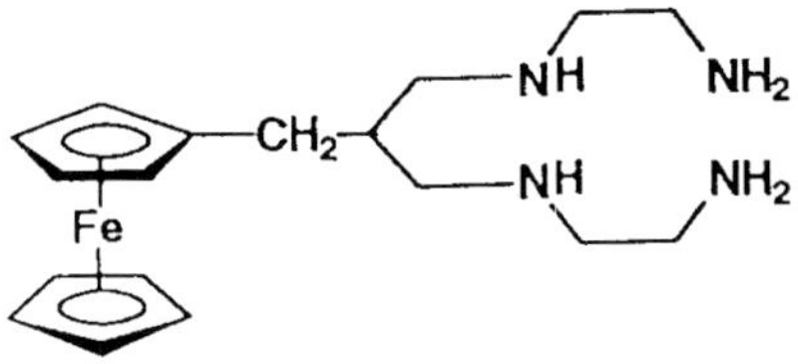

66

The tripodal receptor containing three ferrocene units and four nitrogen atoms, **68**, selectively forms a host–guest complex with the $H_2PO_4^-$ anion and electrochemically recognizes it in the presence of Cl^- and HSO_4^- anions [118]. The ferrocenes containing bipyridyl side chains ('pyridyl podands') are good receptors for Cu^+ and Ag^+ cations [119].

A new class of ferrocene (also cobaltocenium) receptors which sense anions both spectrally and electrochemically, is based upon attachment of metallocene moieties to a porphyrin skeleton. The porphyrin fragment can be simultaneously complexed by zinc [120].

2.1.6 Cobaltocenium receptors

A great variety of molecular hosts, mostly devised as anion receptors, contain cobaltocenium moieties. These can be combined with polyoxo macrocyclic ethers to give redox responsive coronands, **69**, with electrochemical sensitivity to anion complexation [121].

2.1 Organometallic Receptors and their Host–Guest Complexes

The cobaltocenium fragment is involved in anion recognition and complexation, whereas the crown ether moiety can interact with alkali metal ions. The result is that the anion recognition properties can be switched on and off through the absence or presence of potassium cation simultaneously bound by the crown.

The receptor containing two crown ether macrocycles forms a sandwich 1:1 complex with potassium, **70**, and a 1:2 complex with sodium, **71**, because of the different recognition properties of the crown towards the two cations [121].

70   **71**

Another variety is the ester type of receptors, containing carboxycobaltocenium units esterified with various diphenols, **72** and **73**, which introduce aromatic spacers [122]. Such compounds accept Br^- guests and shift the cobaltocenium reduction wave towards more negative potentials. In this instance anion recognition and fixation seem to be based simply on electrostatic interactions.

72   **73**

Polyaza–oxa macrocycles can also be attached to cobaltocenium moieties, and function as both cation and anion receptors [123, 124].

Cooperative enhancement of anion complexation is promoted by sodium cation complexation of the aza–oxa crown, combined with the amide-linked cobaltocenium, **74**. The result is that the compound acts as a host to both the cation and anion of sodium chloride.

Comparison of cyclic, **75**, and non-cyclic hosts revealed that the former is much more effective in chloride ion complexation [125]. Ditopic macrocyclic hosts containing two cobaltocenium units, **76**, can be yet more efficient.

Several acyclic cobaltocenium aminoarylamides have been investigated [126–128]; they have strong selectivity for $H_2PO_4^-$ over halide anions.

The tripod ligand **77** forms 1:1 host–guest complexes with Cl^-, Br^-, and NO_3^- and a 2:1 complex with F^-.

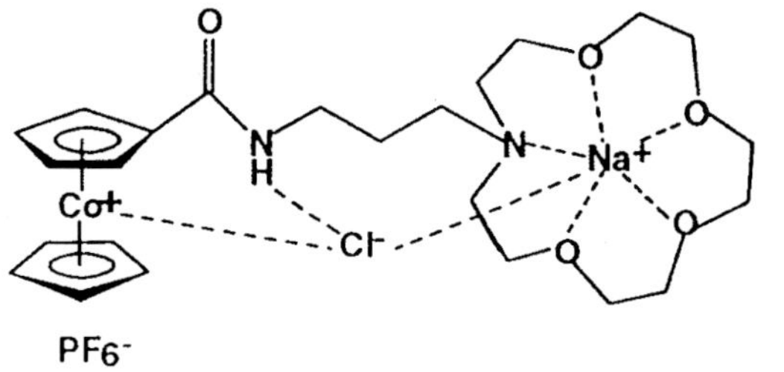

54   2 *Molecular Recognition and Host–Guest Interactions*

74

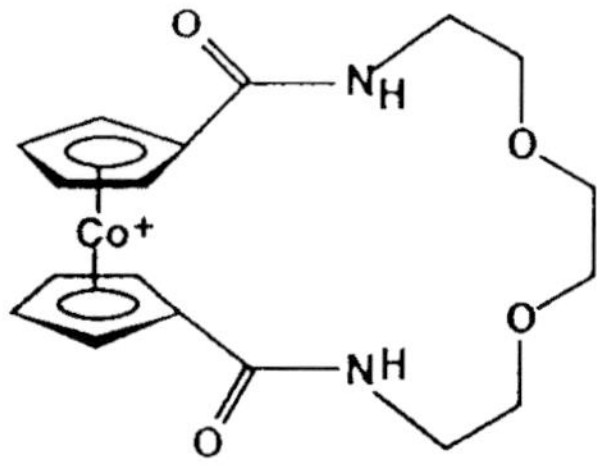

75

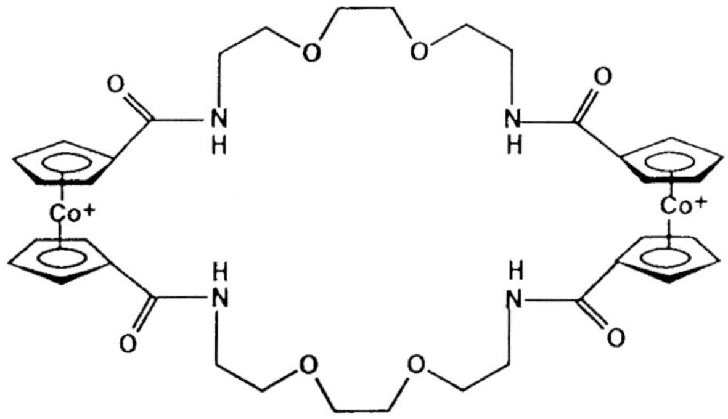

76

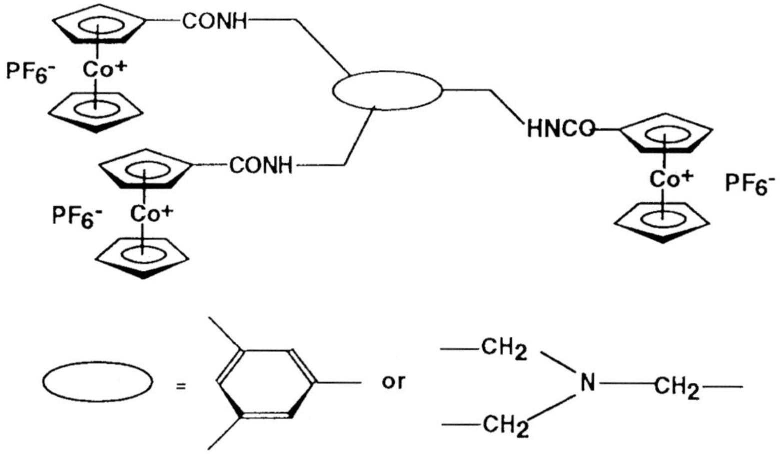

77

In anion fixation by cobaltocenium aminoaryl derivatives **78** the positive charge on the cobalt and the hydrogen-bond capacity of amino substituents both play a role.

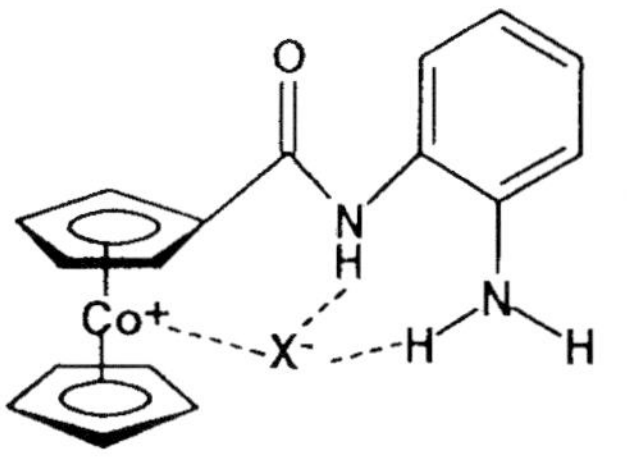

78

The crystal structure of the Br⁻ host–guest complex has been determined and supports the conclusion of anion interaction with both cobalt and amino groups [128].

Inclusion of pyridinium moieties in cobaltocenium receptors, **79** and **80**, was found to enhance their anion-binding capacity (compared with analogous pyridyl complexes). All the receptors shown recognize the X⁻ (halide), HSO_4^- and $H_2PO_4^-$ anions and form 1:1 complexes [129].

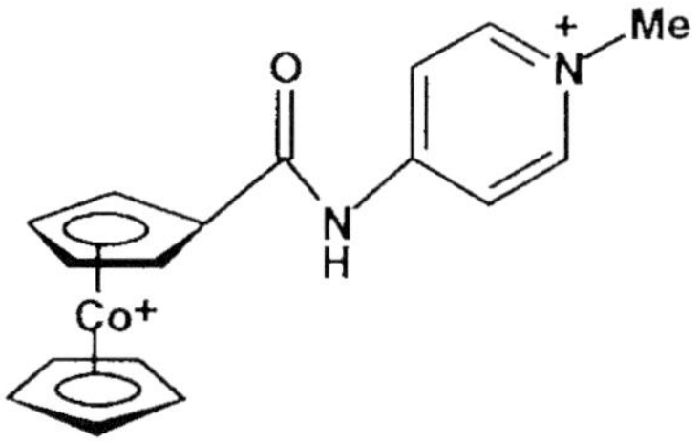

79 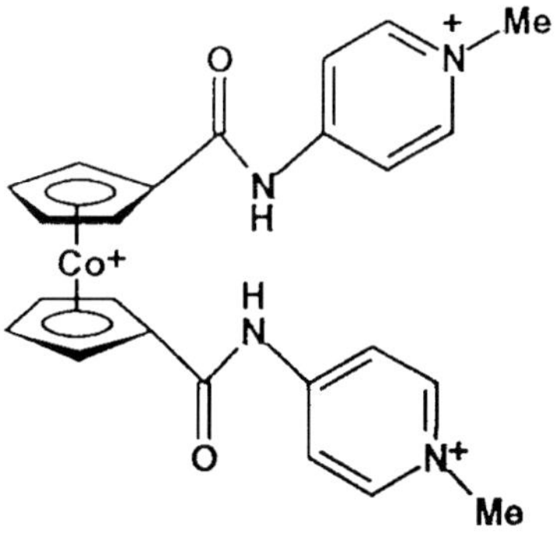   **80**

Among ditopic cobaltocenium receptors, **81**, those containing alkyl spacers form 1:1 host–guest complexes with halide anions, whereas those containing larger, aromatic spacers form 2:1 complexes. The selectivity and complexing capacity of the receptors decreases with increasing distance between the cobaltocenium units [130, 131].

With a chiral cobaltocenium receptor chiral selectivity and recognition towards camphor-10-sulfonate was observed [132, 133] raising hopes for future use of such systems in the separations of optically active compounds. These species also have receptor properties towards Cl⁻, Br⁻, acetate, and tosylate anions.

A structural modification of organocobalt-containing receptors was achieved by introduction of cyclopentadienylcobalt moieties attached to dithiolene-crown ethers, **82** [134]. The electrochemical behavior of this new type of macrocyclic host has been investigated by cyclic voltammetry.

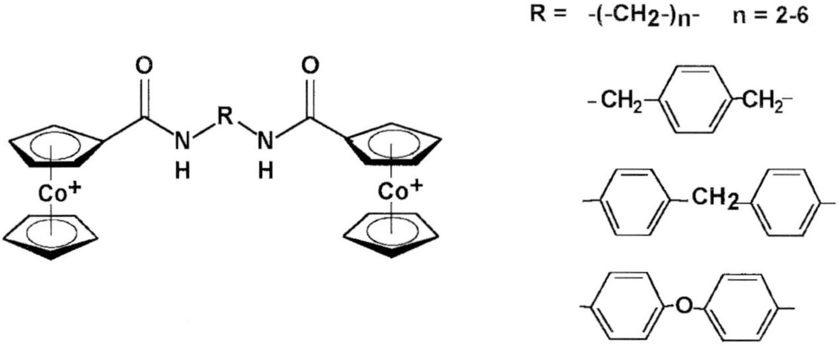

81

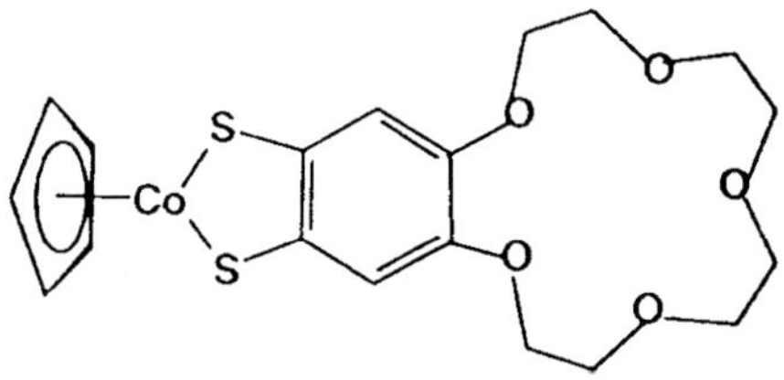

R = -(-CH$_2$-)$_n$- n = 2-6

82

2.1.7 Other metallocene receptors

Large differences in composition are, in principle, possible by changing the metal center in metallocenes combined with macrocyclic ethers or cryptands, but these were mostly limited to air- and water-stable compounds, which could be considered for further cation and anion sensing devices. So far ruthenocene, titanocene, molybdocene, and tungstocene moieties associated with crown ether-type macrocycles have been described.

Ruthenium-containing compounds are, in general, structurally similar to their iron analogs. Thus, coronands containing two polyoxa- and polyaza–oxa macrocycles, and from three to six polar atoms in each ring (oxygen and/or sulfur) have been synthesized and investigated [135–137].

The sulfur-containing receptors have little affinity for hard cations but can form extractable host–guest complexes with soft cations such as Ag$^+$, Hg^{2+}, Tl$^+$, Pd^{2+}, and Pt^{2+}. Silver is found to interact strongly with sulfur [138, 139]. It was even possible to isolate 1:1 host–guest silver and mercury complexes; in a large sulfur-containing macrocycle a linear HgCl$_2$ molecule is hosted in the cavity, but in the 1:2 complex a second mercury atom is externally attached to the ruthenium atom [140–142]. Palladium and platinum are also selectively complexed by polythia-coronands [143–148]. Some X-ray diffraction studies on palladium [147, 148] and platinum [143, 145] complexes established that the coordination of the metal occurs by interaction with the sulfur sites.

A series of receptors containing two polyoxa- [149, 150] and aza-polyoxa [84a, 151–153] macrocycles connected by a ruthenocene unit, have been prepared and their behavior towards alkali metal cations investigated. In the 1:1 host–guest complexes formed the cation is held between the macrocycles and clear selectivity toward potassium is observed.

Polyaza–oxa cryptands with one or two ruthenocene units have been reported [154, 155]. They form Mg^{2+} host guest complexes.

Lanthanide, e.g. dysprosium complexes obtained with the same type of receptor are of interest because the incorporated metallocene unit can act as a radiationless deactivation center of the photochemically excited state [156–158].

Recently, a triruthenocene receptor, **83**, derived from 1,4,7-triazacyclononane, which incorporates a proton in the cavity, has been reported [159].

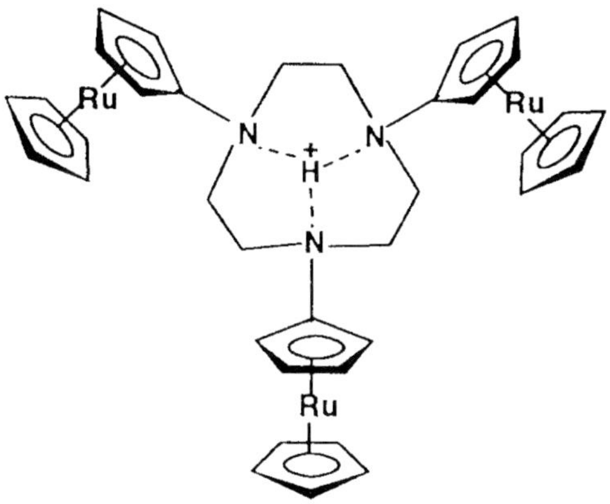

83

New redox active receptors $Ru_6C(CO)_{14}(\eta^6$-benzo-15-crown-5) and $Ru_6C(CO)_{14}$-$(\eta^6$-benzo-18-crown-6) have been prepared by attaching ruthenium clusters to the arene fragment of the benzo-crown ethers [159a].

Bis(cyclopentadienyl)titanium)dithiolene-polyoxa macrocycles, **84**, have been described and their electrochemical properties investigated by cyclic voltammetry [160].

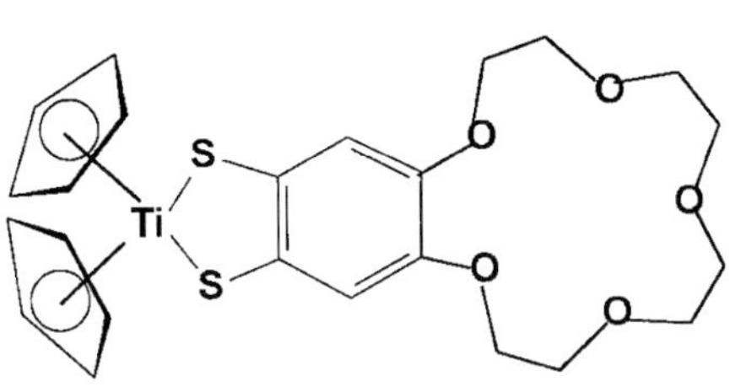

84

Molybdocene and tungstocene building units have been attached to crown-ether-type metallocycles **85** via sulfur [161].

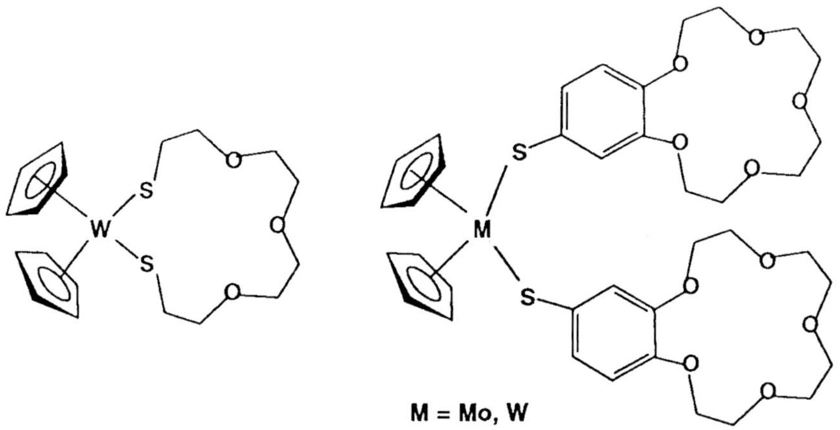

M = Mo, W

85

The compounds with two polyoxo macrocycles have reversible one-electron electrochemical oxidation–reduction cycles and the half-wave potential is quite sensitive to the presence of alkali metal ions. The thio-oxo coronand also undergoes reversible one electron redox reactions, but is insensitive to alkali metal cations.

Although not metallocene-type compounds, crown ethers containing metal carbonyl centers are mentioned here; they are reviewed elsewhere [161a,b].

2.1.8 Calixarene receptors modified with organometallic groups

The calixarenes are important molecular receptors and play a major role in host–guest supramolecular chemistry [162, 163, 163a]. Their conic shape favors host–guest complexation of various hydrophobic molecules and the functionalized 'rim' enables interaction with polar molecules. Modification of calixarenes by organometallic moieties can be achieved:

i) by π-complexing one or more of the aromatic nuclei with transition metal organometallic fragments;
ii) by functionalization of phenolic groups by attaching organometallic moieties;
iii) by coordinating phenolic oxygens directly to a metal atom; and
iv) by introducing organometallic groups as substituents on the aromatic rings or on the methylene bridges connecting the aromatic rings.

Only selected examples will be cited here. A more comprehensive review on metal-containing calixarenes is available [164].

A modification of calixarenes by *π-complexation of the aromatic rings*, e.g. **86** and **87** was performed by attaching chromium tricarbonyl, $Cr(CO)_3$ groups [165–169], η^6-arene-ruthenium groups, $Ru(\eta^6$-arene) [170], and η^5-(pentamethylcyclopentadienyl)-iridium and -rhodium groups, $M(\eta^5$-$C_5Me_5)$, M = Ir, Rh [171, 172]. The η^6-arene-ruthenium and (η^5-C_5Me_5)iridium derivatives can be obtained with two or four organometallic groups and are water-soluble.

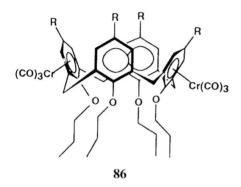

86

87

M = Rh, Ir

The X-ray crystal structure determination of the tetrametallic derivative [{Ru(p-cymene)}$_4(\eta^6 : \eta^6 : \eta^6 : \eta^6$-calix[4]arene-2H][BF$_4$]$_6$ confirmed the complexation of each aromatic group; the molecule incorporates a BF_4^- anion host in the cavity. The pentamethylcyclopentadienyl-iridium complex [(η^5-C_5Me_5)Ir]$_2(\eta^6 : \eta^6$-p-tert-butyl-calix[4]arene][BF$_4$]$_3 \cdot$MeNO$_2 \cdot$Et$_2$O crystallizes incorporating a diethyl ether molecule as a guest [171, 172].

Silver π-complexes with the Ag^+ ion trapped as a guest in a calix[4]arene cone have been reported and structurally characterized by X-ray diffraction [173–175].

The π-complexation of the aromatic ring with transition metal organometallic moieties induces major changes of properties such as solubility, acidity, and host–guest activity.

Calixarenes have also been modified by *functionalization of phenolic groups* by attaching organometallic moieties. Thus, esterification with ferrocene-, ruthenocene-, and cobaltocene-carboxylic acids gave a broad variety of compounds; some are illustrated by **88** and **89** [176–178, 178a].

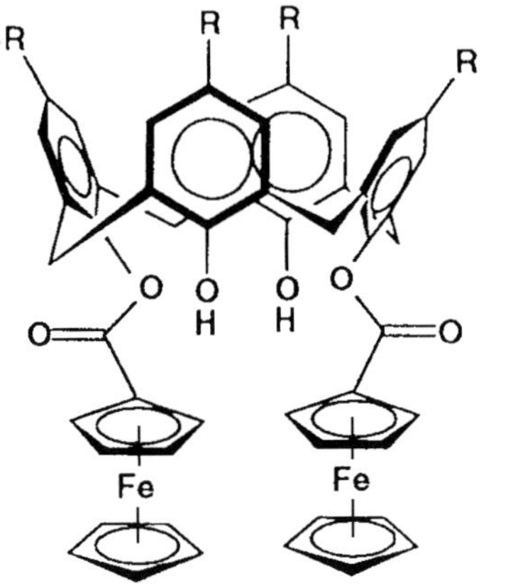

88

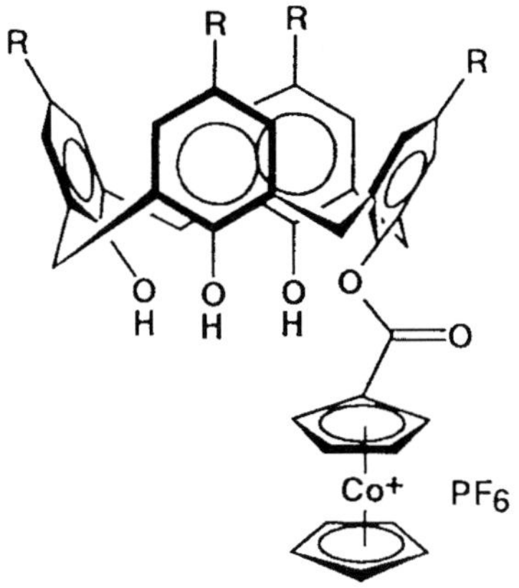

89

With ferrocene- and ruthenocene dicarboxylic acids two calix[4]arenes were connected to give **90** [179].

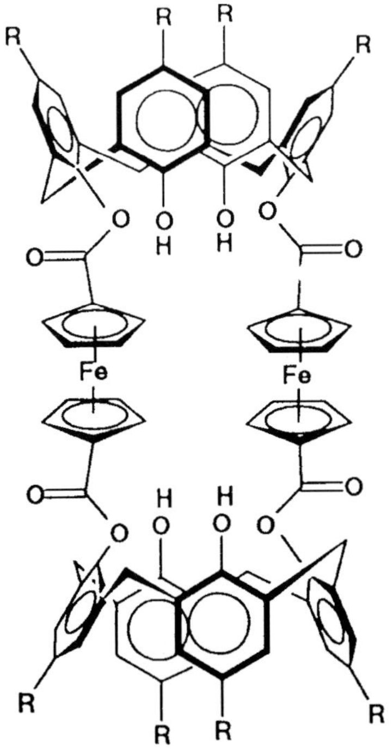

90

Solvent-dependent anion selectivity was observed for a receptor consisting of two calixarene units bridged by a ferrocenyl moiety [179a].

Two of the phenolic groups in *tert*-butyl-calix[4]arene were condensed with a ferrocene carboxylic diamide bridge, to give a receptor, **91**, which electrochemically recognizes the $H_2PO_4^-$ anion in the presence of excess HSO_4^- and Cl^- [180]. Many similar derivatives have also been reported [181, 182].

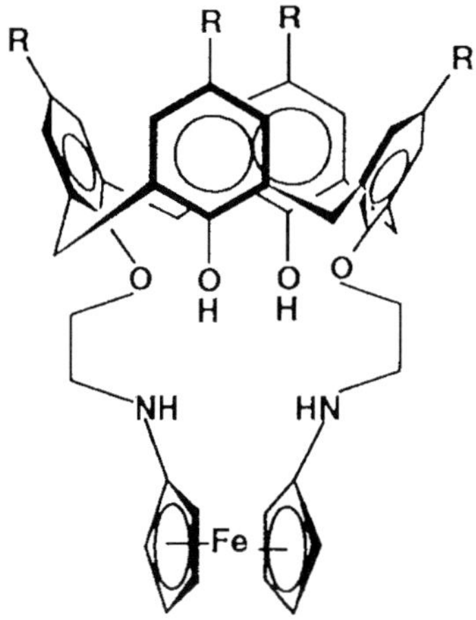

91

Cobaltocenium moieties have been attached to the calixarene skeleton by functionalizing the upper rim (opposite to phenolic functions) and such a receptor **92** can recognize dicarboxylate anions such as adipate [183, 184]. Functionalization of the upper rim and insertion of organopalladium building sites has also been reported [184a].

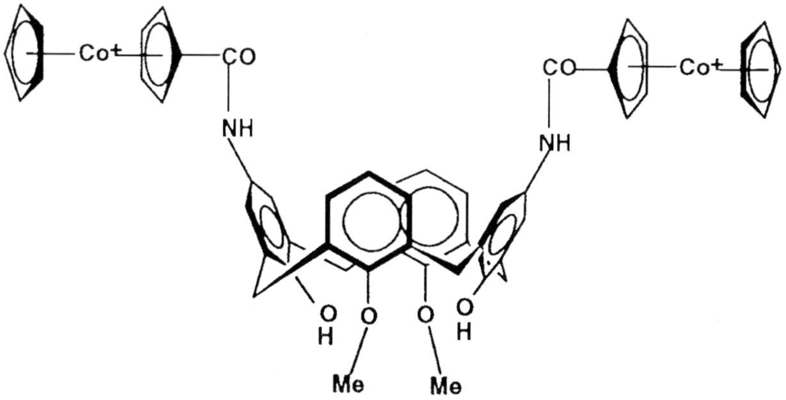

92a

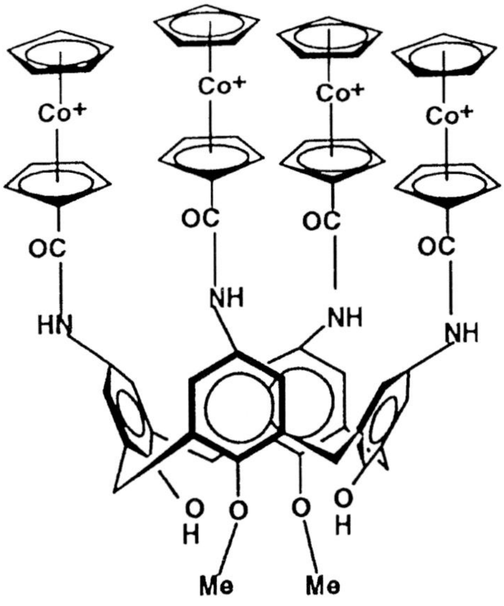

92b

Phenolic oxygens can be directly connected to metal atoms and thus the calixarene becomes a ligand. Examples, **93**, are provided by organozirconium coordination centers, $Zr(CH_2Ph)$, $ZrMe_2$, $ZrMeCl$, and $Zr(\eta^4\text{-}C_4H_6)$ [185, 185a]. All four oxygens of a calix[4]arene are coordinated to the metal.

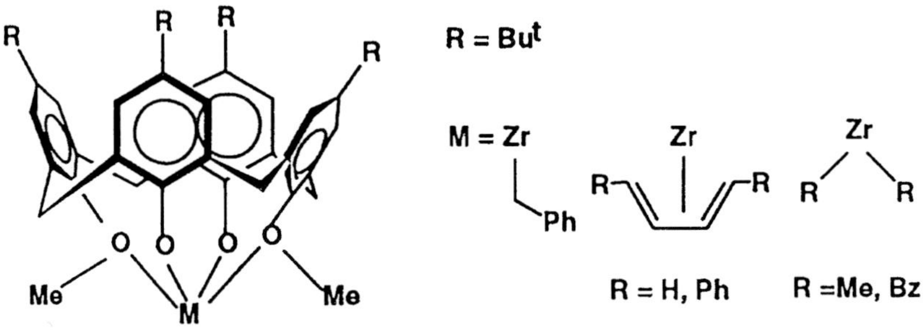

93

Similar cyclopentadienylniobium and -tantalum complexes, **94**, with $M(\eta^5\text{-}C_5R_5)$ groups (M = Nb, Ta, R = H or Me) coordinated to oxygen have been synthesized and structurally characterized [186]. An iron–carbene complex with the metal attached to the phenolic oxygen has also been reported [186a].

Organoaluminum *p-tert*-butylcalix[4]arene complexes are formed in reactions

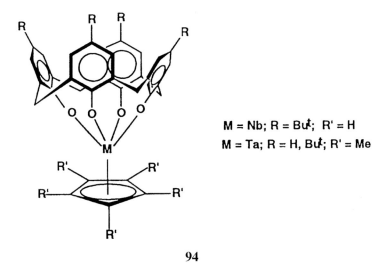

M = Nb; R = Bu^t; R' = H
M = Ta; R = H, Bu^t; R' = Me

94

with trimethylaluminum; some contain two [AlMe(NCMe)] groups attached to a pair of oxygen atoms each. The calix[4]arene adopts a flattened partial-cone conformation [187]. Derivatives with four aluminum atoms have also been reported [187a].

The functionalization of phenolic groups with trimethylsilyl moieties is a similar type of modification and produces useful precursors for further transformation [188, 188a, 189]. Thus, silylated calixarenes form subvalent germanium and tin complexes [189a].

Resorcinarenes were modified by introducing $-SiR_2-$ bridges [R = Me, Et, and $R_2 = -(CH_2)_5-$] between the oxygens of adjacent aromatic groups. The resulting receptors have narrower cavities and can host slim molecules only, e.g. carbon disulfide, propyne, and dioxygen [189b,c]. Calix[4]arene-modified oligosiloxanes behave as a quasi-immobilized neutral carrier for silicone rubber membrane sodium-ion-selective electrodes [189d].

Another type of modified calixarene has been prepared by attaching the ferrocene moieties to the bridge connecting the benzene rings in the calix[4]arenes [190] or both to the bridge and the aromatic rings, **95** [191, 192].

Finally, calixarenes can be modified by replacing the $-CH_2-$ connecting groups with $-SiMe_2-$ groups. Thus, silacalix[4]arenes [192a] and similar macrocyclic silacompounds [192b–e] have been described.

The simultaneous combination of the calix[4]arene skeleton with ferrocene and crown ether moieties gave a multiple functional redox-active receptor capable of hosting alkali metal cations in the crown and hydrophobic molecules in the calixarene fragment [193, 194].

All these examples illustrate the great versatility of calixarenes as supports for a variety of other functions able to participate in host–guest complexing and cation and anion sensing.

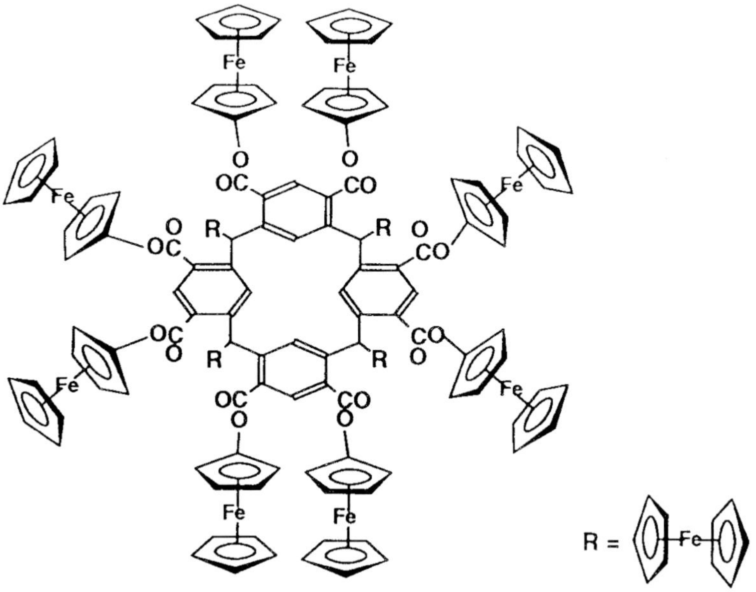

95

R = (ferrocenyl-based group)

2.1.9 Organometallic cyclotriveratrylene receptors

Cyclotriveratrylene is a trimeric, bowl-shaped molecular skeleton, known to form solid-state inclusion compounds. A variety of cyclotriveratrylene derivatives, e.g. **96** and **97**, resulted from π-complexation of the aromatic nuclei with one, two, or three Ru(η^6-arene) (arene = C_6H_6, p-MeC$_6$H$_4$Pr^i) and Ir(η^5-C$_5$Me$_5$) organometallic moieties [195–197]. Some of their structures were established by X-ray diffraction.

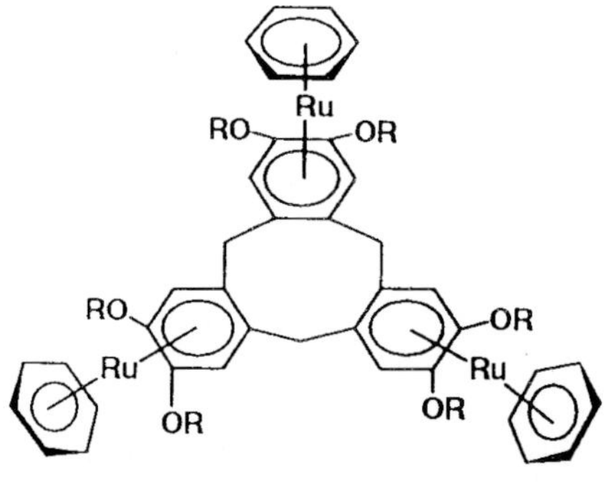

96 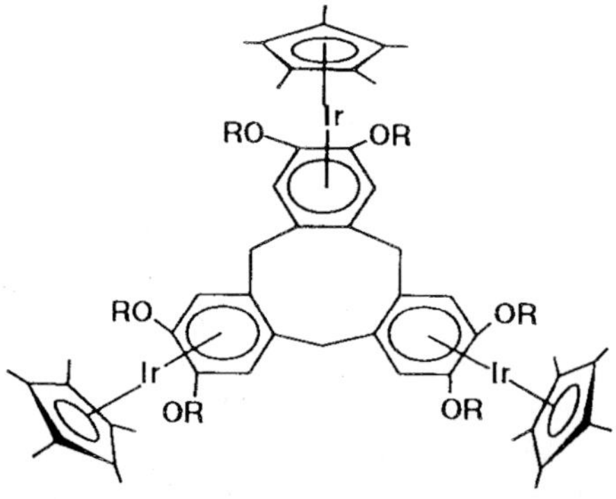   **97**

All the new receptors have interesting host–guest complexing properties, and novel specific activity for large tetrahedral cations. The dimetalated ruthenium derivative [{Ru(η^6-MeC$_6$H$_4$Pr^i-p)}(η^6:η^6-CTV)]$^{4+}$ (CTV = cyclotriveratrylene) selectively extracts ^{99}TcO$_4^-$ and ReO$_4^-$ from aqueous solutions in the presence of a large excess of Cl$^-$, CF$_3$SO$_3^-$, NO$_3^-$, SO$_4^{2-}$, even ClO$_4^-$ (to some extent), which promises interesting applications.

Silacyclotriveratrylenes with one, two or three silicon atoms replacing the methylene bridges (as –SiMe$_2$– groups) have also been synthesized [198] but at present the field seems to lie dormant. Cyclotriveratrylenes can act as guests for metallocenes. Thus, the metallocenium cation [(C$_5$H$_5$)Fe(C$_6$H$_6$)]$^+$ as a guest in cyclotriveratrylene gives rise to a two-dimensional structure as a result of π-stacking interactions between the electron-rich host and electron-deficient guest [199].

2.2 Organometallic Guests

2.2.1 Organometallic guests in inorganic hosts

2.2.1.1 Organometallics in zeolites and mesoporous silica

The importance of zeolites as catalyst supports and of many organometallic compounds (metal carbonyls in particular) as active catalysts or catalyst precursors explains the interest in the host–guest complexes of the two types of compounds. Because the subject has been reviewed [200–204, 204a] a brief presentation only will be made here. Zeolites are silicates with a regular, rigid tridimensional framework containing cavities (pores) and/or channels of variable sizes and shapes. As such, these materials are selective hosts for specific guests, normally organic molecules. The interior of zeolite pores is surface-covered with oxygen atoms bearing negative charges or present as Si–OH groups, and this structural peculiarity determines their chemical (acidic) properties. The guests can be anchored by hydrogen-bonds such as C–H$\cdots$O or X$\cdots$H–O, through M$^+$ $^-$O–Si electrostatic interactions, or perhaps through dipole–dipole interactions between the positive pole of the guest and the negative poles of the oxygen atoms of the silicate network, to form supramolecular inclusion compounds.

Both main group and transition metal organometallic compounds have been incorporated as guests in zeolite hosts.

Main group organometallics incorporated in zeolites include some Group 12 and 13 metal alkyls as precursors for semiconducting materials. Thus, dimethylzinc, ZnMe$_2$, and dimethylcadmium, CdMe$_2$, react with the acidic sites of the pores to anchor RZn and RCd moieties to the internal surface; further reaction with H$_2$S or H$_2$Se produces intrazeolite nanoclusters of metal chalcogenides [205, 206]. It has been shown that host–guest fixation of dibenzylmercury in a zeolite changes its pattern of reactivity towards arenes [207].

Ethylene polymerization catalysts can be made by incorporating methylalumoxane or trimethylaluminum, $AlMe_3$, into zeolites, together with $(\eta^5-C_5H_5)_2MCl_2$ (M = Ti, Zr) [208].

Trimethylgallium, $GaMe_3$, reacts with the surface Si–OH groups of the pores to form $Me_{3-x}Ga$ moieties attached to the cavity surface, which further react with excess phosphine to produce intrazeolite GaP on heating [209].

Transition metal organometallic guests in zeolite hosts have received considerably more attention than their main group counterparts.

Group 6 metal (chromium, molybdenum, tungsten) carbonyls trapped in zeolites have been found to undergo partial thermal decomposition to produce $M(CO)_3$ species which react with phosphines, polyolefins, and arenes to form the expected complexes, e.g. $(\eta^6-C_6H_6)Cr(CO)_3$ [210].

Decomposition of $Cr(CO)_6$ in a zeolite host at 200 °C produced materials with catalytic activity in the hydrogenation and polymerization of olefins [211]. The $Cr(CO)_4$ species formed in zeolites also has catalytic activity in the hydrogenation of butadiene [212]. Other Group 6 metal carbonyls were incorporated in zeolites and the properties of the host–guest complexes have been investigated in some detail [213–219].

The photooxidation of $Mo(CO)_6$ guest served for the production of intrazeolite Mo(IV) and Mo(VI) oxides [220]. Similarly, the photooxidation of $W(CO)_6$ encapsulated in faujasite zeolites was used for the preparation of non-stoichiometric tungsten oxides WO_{3-x}, which can be further readily oxidized or reduced [220a,b]. It has been shown by use of EXAFS that $W(CO)_6$ maintains its molecular integrity at room temperature in a zeolite host, but that photooxidation with molecular oxygen and thermal decomposition at 300–400 °C results in the formation of various intrazeolite tungsten oxides [221, 222].

The intrazeolite fixation and chemical behavior of bimetallic carbonyl complexes, such as $(CO)_5MGeCl_2 \cdot THF$ (M = Mo, W) [223] and $\eta^5-C_5H_5Mo(CO)_3SnMe_3$ [224] has also been investigated.

Dibenzenechromium cations, $Cr(\eta^6-C_6H_6)_2^+$ incorporated in zeolites have well resolved ESR spectra [225]. The redox chemistry and other properties of intrazeolite chromium (also iron and cobalt) metallocenes $M(\eta^5-C_5H_5)_2$ (M = Cr, Fe, Co) have been reported [226].

Among Group 7 metals (Mn, Tc Re) several manganese and rhenium compounds were studied. When $(CO)_5MnSnMe_3$ is anchored in zeolite pores and heated to 150 °C the guest decomposes with loss of carbon monoxide and Mn–Sn bond cleavage occurs and metal clusters are formed at higher temperatures [227]. Interaction of the metal carbonyl groups in manganese and rhenium carbonyl $Re_2(CO)_{10}$ [227, 228] with sodium ions of the zeolite host was observed. A rhenium carbonyl hydride intermediate in the decomposition of the carbonyl, leading finally to rhenium metal particles, has been suggested and high selectivity in some hydrogenation reactions has been reported [229]. The intrazeolite complex $(CO)_5RePt(CN)_4$ served as precursor for a benzene hydrogenation catalyst [230]. Encapsulation of methyltrioxorhenium, $MeReO_3$, a remarkable and versatile catalyst [230a,b] into zeolites produces a tunable olefin metathesis catalyst [231].

Group 8 metals (Fe, Ru, Os) have enjoyed more attention. Numerous studies have dealt with the behavior of mono- and polynuclear iron carbonyl guests in zeolite frameworks [216, 232–242], and with the incorporation of cycloolefin complexes Fe(COT)(CO)$_3$ [243] or cyclopentadienyliron carbonyl [η^5-C$_5$H$_5$Fe(CO)$_2$]$_2$ [244]. The metal carbonyls retain their molecular integrity at room temperature but heating causes decarbonylation with formation of anchored Fe(CO)$_4$ [232–234, 238]. Catalytically active, highly dispersed iron particles are formed on photolysis of the iron carbonyl guests [235]. These materials are important because of their high catalytic activity and selectivity in the Fischer–Tropsch hydrogenation of carbon monoxide [237, 242, 245, 246]. Such studies have been extended to ruthenium carbonyls [247–250] and osmium carbonyls [251–254].

Ferrocene has also been incorporated in zeolite hosts [244, 255] and hydrocarbon loss on heating and oxidation to ferricinium cation were observed. Similarly, substituted ruthenocenes were also investigated [256, 257]. The ring opening polymerization of a [1]silaferrocenophane within the channels of mesoporous silica (MCM-41) gave precursors of magnetic iron nanostructures [257a,b].

Carbon monoxide hydrogenation catalysts have been obtained by thermal decomposition of intrazeolite iron bis(toluene) complexes [258].

Incorporation of anionic ruthenium cluster carbonylates [Ru$_6$C(CO)$_{16}$]$^{2-}$ or [Ru$_{10}$H$_2$(CO)$_{25}$]$^{2-}$ interspersed with [Ph$_3$PNPPh$_3$]$^+$ cations into MCM-41 mesoporous silica produced well ordered tightly packed arrays. Gentle thermolysis of these materials produced catalytically active ruthenium particles [258a]. Bimetallic nanoparticle catalysts were produced by thermolysis of [AsPh$_4$]$^+$-[Ag$_3$Ru$_{10}$C$_2$(CO)$_{28}$Cl]$^{2-}$ in MCM-41 silica [258b].

Group 9 metals (Co, Rh, Ir) have been the subject of much interest because of their known catalytic properties. Dicobalt octacarbonyl, Co$_2$(CO)$_8$, can be trapped in zeolites and undergoes various transformations leading to polynuclear Co$_4$(CO)$_{12}$, anionic [Co(CO)$_4$]$^-$ and other species [259–263]. Thermal decarbonylation of η^5-C$_5$H$_5$Co(CO)$_2$ [264] and (CO)$_4$CoSnMe$_3$ [265, 266] in zeolites has also been investigated.

Ethylene–cobalt(II) complexes in zeolites have been reported [267].

Cobaltocene cations, Co(η^5-C$_5$H$_5$)$_2^+$, can serve as templates for the synthesis of molecular sieves as metallocenium inclusion compounds [268, 269].

The formation and reactivity of catalytically important intra-zeolite rhodium carbonyls have been investigated in some detail [270–276]. Such materials are of interest as catalysts in hydroformylation [277, 278] and water-gas shift [279] reactions.

Interestingly, the formation of an intrazeolite dinitrogen complex [Rh(CO)(N$_2$)] was reported in the presence of dihydrogen and dinitrogen [280].

Other zeolite host–guest systems include Rh(CO)Cl(PPh$_3$)$_2$ [281, 282]. Chiral complexes of rhodium as zeolite guests are good enantioselective hydrogenation catalysts [283].

Cyclopentadienylmetal carbonyls, η^5-C$_5$H$_5$M(CO)$_2$ and η^5-C$_5$Me$_5$M(CO)$_2$ (M = Rh, Ir) have also been incorporated into zeolites and their interaction with the cage has been investigated. Decarbonylation occurs on heating [284, 285].

Mixed rhodium–iridium clusters incorporated in zeolites have been formed by thermal decomposition of the appropriate metal carbonyls [286].

The η^3-allyl rhodium complex, Rh(η^3-C$_3$H$_5$)$_3$, can be incorporated into an acidic zeolite with the formation and loss of propene, and several species formed in subsequent reactions with dihydrogen, hydrogen chloride, carbon monoxide, and phosphines, have been investigated [270, 287]. The same intrazeolite triallyl complex has been studied as an alkene hydrogenation catalyst [288].

Some studies have dealt with Group 10 metals (Ni, Pd, Pt). The trapping and chemical behavior of nickel tetracarbonyl, Ni(CO)$_4$, in zeolites has been investigated [289, 290] again showing preservation of molecular integrity at room temperature and decarbonylation on heating.

Hydrocracking catalysts were obtained by hydrogenation of an intrazeolite butadiene–nickel(0) complex at 500 °C [291].

Allyl(cyclopentadienyl)palladium, η^5-C$_5$H$_5$Pd(η^3-C$_3$H$_5$), deposited onto zeolites and the thermally decomposed produced incorporated palladium clusters [292].

Platinum carbonyls stable under normal conditions are practically unknown, but their formation in zeolites has been detected [293, 294].

The Group 11 metals (Cu, Ag, Au) have been almost ignored; copper(I) and gold(I) carbonyls are known only as guests in zeolite hosts [295, 296] and silver carbonyls are unknown.

2.2.1.2 Organometallic compounds in layered chalcogenides, oxohalides, and oxides

Intercalation compounds contain a rigid bidimensional crystalline lattice (matrix) which can incorporate molecular or ionic species between the layers to form a peculiar type of host–guest complex [297–299]. Intercalation of metallocenes into various inorganic hosts lattices has been the subject of much attention in recent years [300–304].

Intercalation of organometallic compounds modifies the physical properties of the host and has been used to produce new materials with potential uses as heterogeneous catalysts or in electrochemical applications [305]. The intercalation is reversible and the integrity of the host lattice is preserved.

Early studies reported the intercalation of cobaltocene, Co(η^5-C$_5$H$_5$)$_2$, and chromocene, Cr(η^5-C$_5$H$_5$)$_2$, guests into a series of layered metal disulfide and diselenide hosts, including ZrS$_2$, NbS$_2$, TaS$_2$, SnS$_2$, TiSe$_2$, HfSe$_2$, NbSe$_2$, and TaSe$_2$ [306]. Extensive investigations of intercalation of various metallocenes and related compounds into layered inorganic materials followed.

Transition metal and tin disulfides and diselenides are important hosts for organometallic compounds. They readily accept metallocenes, especially those of low oxidation potential, e.g. cobaltocene, and charge transfer from the guest to the host occurs, i.e. the metallocene is (partially) oxidized. Examples include intercalation compounds of cobaltocene into ZrS$_2$ [306], TaS$_2$ [306–311], TaSe$_2$, TiS$_2$, TiSe$_2$, NbSe$_2$, HfSe$_2$ [306], VSe$_2$ [312], SnS$_2$, SnSe$_2$, and SnS$_x$Se$_{2-x}$ [306, 313–325], substituted cobaltocenes into ZrS$_2$ [312], substituted cobaltocenes into TaS$_2$ [312], chromocene into ZrS$_2$ [326, 327], HfSe$_2$, NbSe$_2$, TaS$_2$, and TaSe$_2$ [306], and (η^5-cyclopentadienyl)chromium(η^6-benzene) [326], (η^5-cyclopentadienyl)chromium(η^7-

cycoheptatrienyl) [312, 326], bis(benzene)chromium, bis(benzene)molybdenum, bis(toluene)molybdenum, bis(mesitylene)molybdenum [326, 328], and (η^5-cyclopentadienyl)titanium(η^8-cyclooctatetraene) into ZrS$_2$ [326].

The change of electric properties and resulting mixed valence as a result of charge transfer from host to guest have been carefully investigated in SnS$_{2-x}$Se$_x$(CoCp$_2$)$_{0.33}$ [321].

Intercalates of bis(benzene)molybdenum, (η^5-cyclopentadienyl)titanium(η^8-cyclooctatetraene) and (η^5-methylcyclopentadienyl)tungsten(η^7-cycloheptatrienyl) into SnS$_2$ and SnSe$_2$ have also been investigated in detail [314, 320].

An important finding was the discovery that intercalation of metallocenes into TaS$_2$ produce a superconducting material with $T_c = 3.2$ K [329] and similar effects were observed by intercalation of cobaltocene into semiconducting SnSe$_2$ (producing a material with $T_c = 8.3$ K) [316, 319].

A subject of considerable debate in the literature was the orientation of the metallocene molecules in the space between the inorganic layers. Broad-band ^{1}H NMR and other physical measurements suggested that cobaltocene adopts a parallel orientation between TaS$_2$ layers [307, 308]; other work [312] has indicated perpendicular orientation, but solid-state ^{2}H NMR indicated that both parallel and perpendicular orientation occur at room temperature, either separate or as a mixture [309, 312]. A single-crystal study of [SnS$_2$(CoCp$_2$)$_{0.29}$] supported these findings [313, 325].

Another important family of inorganic hosts include lamellar metal–phosphorus sulfides MPS$_3$ (or better M$_2$P$_2$S$_6$) with M = Mn, Fe, Ni, Zn, Cd. These can incorporate metallocenes or metallocenium cations. Thus cobaltocene and bis(benzene)chromium have been incorporated into MnPS$_3$ [330–334], FePS$_3$ and NiPS$_3$ [335], ZnPS$_3$ [336], and CdPS$_3$ [313, 317, 337–341]. The most intensively investigated subject, by far, is the intercalation of cobaltocene into various metal–phosphorus sulfides [330, 338, 339, 342–346]. Unlike metal disulfides, the metal–phosphorus sulfides (e.g. CdPS$_3$) can also incorporate ferrocene [347].

Metallocenium cations, such as Co(C$_5$H$_5$)$_2^+$ and Cr(C$_6$H$_6$)$_2^+$ in MnPS$_3$ [348], FePS$_3$ [349], NiPS$_3$ [350], ZnPS$_3$ [330], and CdPS$_3$ [349], and [Fe(C$_5$H$_5$)(C$_6$H$_6$)]$^+$ in MnPS$_3$ and ZnPS$_3$ [330] have also been investigated.

Some interesting results on these intercalation compounds have been reported. Thus, investigation of the magnetic properties of Mn$_{0.83}$PS$_3$(CoCp$_2$)$_{0.34}$(H$_2$O)$_{0.3}$ revealed dramatic changes of the magnetic behavior upon intercalation – the antiferromagnetic host is converted into ferro- and ferrimagnetic materials [351]. Diffraction, infrared and ^{2}H NMR solid-state spectroscopy [346] and neutron scattering [352] studies indicate parallel orientation of cobaltocene in this intercalation complex.

Intercalation into lamellar metal oxides and oxide chlorides has also received some attention [353, 354] in particular because of the capacity to intercalate ferrocene. Thus, FeOCl intercalation complexes with ferrocene [355–357] and substituted ferrocenes [358, 359], chromocene [355], cobaltocene [355, 360], ruthenocene [358], and intercalation host–guest complexes of cobaltocene into TiOCl and VOCl [361] have been described and investigated in some detail. Intercalation of some organotin compounds into FeOCl has also been reported [362].

Parallel orientation between the layers is observed for the ferrocene molecules

[355, 361] and the electrical properties of these materials have been investigated by means of Mössbauer, diffraction, and conductivity measurements [363, 364].

Other inorganic hosts for metallocenes reported include metal oxides and phosphates. Thus, vanadium pentoxide, V_2O_5 can incorporate ferrocene [365] (in neutral or cationic form) and molybdenum trioxide, MoO_3, incorporates substituted ferrocenes [366] and cobaltocene [312]. Ferrocene and substituted ferrocenes [367, 368, 368a], rhodium carbonyl chloride, $Rh_2(CO)_4Cl_2$ [368b], and cobaltocene [366, 369] have also been incorporated as guests into oxovanadium phosphates, α-$VOPO_4 \cdot H_2O \cdot EtOH$, and $VOPO_4 \cdot 2H_2O$. Host–guest complexes of intercalation ferrocene and cobaltocene with oxovanadium phosphate α-$VOPO_4$ and zirconium hydrogen phosphate, α-$Zr(HPO_4) \cdot 2H_2O$ [369a, 370] and dioxouranium phosphate, UO_2HPO_4-$2H_2O$ [370] and tin hydrogen phosphate, α-$SnHPO_4 \cdot H_2O$ [367] have also been reported. Ferrocenium and cobaltocenium cations have also been incorporated into $V_2O_5 \cdot nH_2O$ gels [365, 371].

The cubane cluster organometallic sulfides $(\eta^5\text{-}C_5H_4Me)_4M_4S_4$ (M = Mo, Fe) have been intercalated into metal oxide hosts, MoO_3 or $FeOCl$, to produce bronzes which are electrically conducting along the crystal axis but are non-conductive in the interlayer direction [372].

Novel bidimensional magnetic materials $[Fe(C_5Me_5)_2][M^{II}M^{III}(ox)_3]$ with M^{II} = Mn, Fe, Co, Ni, Cu, Zn, M^{III} = Cr, Fe, ox = oxalate, were obtained by intercalating decamethylferrocenium cations in layered bimetallic oxalates. Their structure has been established by X-ray diffraction and the magnetic properties of the new materials have been investigated over a broad range of temperatures [372a].

2.2.2 Organometallic guests in organic hosts

2.2.2.1 Organometallic complexes of crown ethers, cryptands and related receptors

The crown ethers and related macrocyclic molecules can act as hosts for organometallic guests and form complexes similar to those of naked metal ions. The situation can, however, be more complex because the crown ethers can behave both as *endo* and *exo* receptors toward organometallic coordination centers. In the latter circumstance the crown ether acts as a simple multidentate ligand and molecular recognition is not necessarily part of the process.

Macrocyclic complexes of organometallics have been reviewed in the treatise *Comprehensive Supramolecular Chemistry* [373, 374] and elsewhere [375, 376] and the subject will be surveyed only briefly here.

There are two types of macrocyclic organometallic complex:

i) host–guest complexes in which the macrocycle donor atoms interact directly with the metal atom [373]; and
ii) second-sphere complexes, in which the macrocycle heteroatoms interact (e.g. through hydrogen-bonds) with the ligands of the organometallic compound [374].

2.2 Organometallic Guests

Both types will be covered in this section.

Most of the organometallic host–guest complexes of crown ethers and cryptands contain main group organometallic compounds.

Among Group 2 metals only magnesium has been reported to form organometallic crown ether complexes. Diorganomagnesium compounds form threaded host–guest complexes with crown ethers and this type of interaction is demonstrated by X-ray crystal structure determinations. Thus, linear diethylmagnesium is incorporated into 18-crown-6 with the six oxygens lying in a plane around the metal, as shown in **98a**, at Mg···O distances of 2.767, 2.778, and 2.792 Å [377]. On the other hand, bent diphenylmagnesium molecules (C–Mg–C 163.8°) in 1,3-xylyl-18-crown-5 form a less symmetrical threaded host–guest complex, **98b**, with two short (2.204 and 2.222 Å), two longer (2.516 and 2.520 Å) Mg···O bonds and a very long (4.038 Å) non-interacting magnesium-oxygen distance [378].

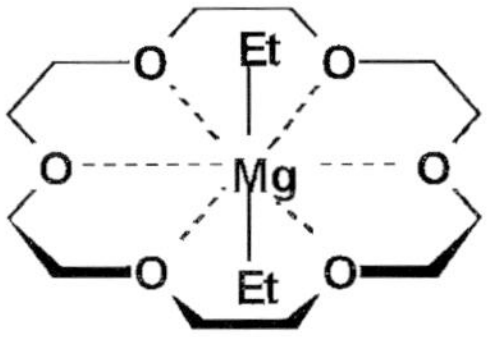

98a 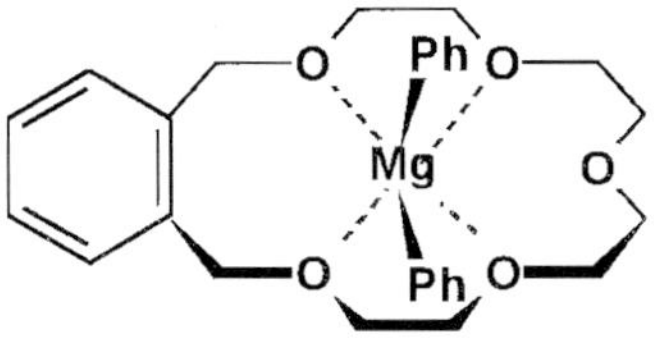   **98b**

Other threaded crown ether complexes are $Mg(C_6H_4CH_3\text{-}p)_2 \cdot$crown (where crown = 18-crown-6 and 15-crown-5) [379] and the first organometallic catenane, **99**, obtained via the intermediate step of an *exo*-hedral coordination complex [380].

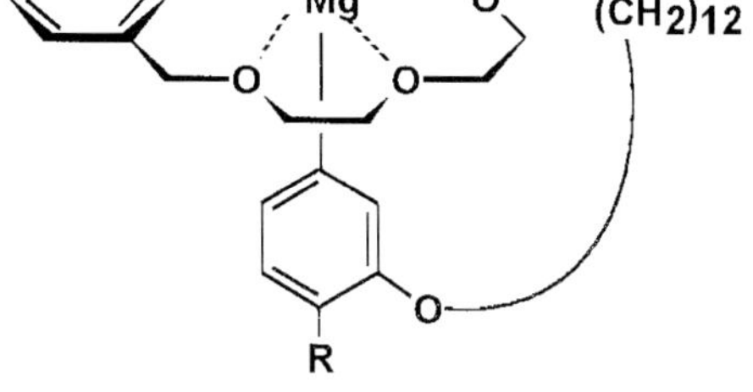

99

Other types of crown ether and cryptand complex, **100a** and **100b**, include RMg^+ encapsulated cations, compensated by outer magnesate MgR_3^- or $Mg_2R_6^{2-}$ anions, which are formed in reactions between diorganomagnesium derivatives and macrocyclic ligands (crown ethers and cryptands) [379, 381, 382]. Thus, dimethylmagnesium forms $MeMg^+$ complexes with 15-crown-5 and 18-crown-6. An X-ray diffraction study showed that [MeMg(15-crown-5)]$^+$ cations are associated with polymeric $[Mg_2Me_5^-]_n$ chains in the crystal [383]. Solutions containing similar species have been intensively investigated by NMR spectroscopy [379].

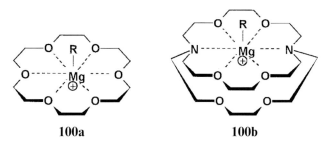

100a **100b**

Diethylmagnesium forms $[EtMg(2.2.1\text{-crypt})]_2^+[Mg_2Et_6]^{2-}$ with the metal connected to three oxygen atoms (Mg···O 2.089, 2.124 and 2.180 Å) and two nitrogens (Mg···N 2.408 and 2.551 Å) with an apical ethyl group in a pentagonal pyramidal coordination [381].

Dineopentylmagnesium with 2.1.1-cryptand forms a complex containing [NpMg(2.1.1-crypt)]$^+$ and $[MgNp_3]^-$ (Np = neopentyl) in which the metal has pentagonal bipyramidal coordination, with three equatorial Mg···O 2.189, 2.234, 2.34 Å and one axial Mg···O 2.349 Å; also equatorial Mg···N (2.398 and 2.434 Å), and an axial position occupied by a neopentyl group [381]. Similar complexes are obtained from Grignard reagents with crown ethers [384–386]. Spectroscopic (mainly ^{1}H NMR) evidence suggests that three possible types of complex are in equilibrium [378–380, 387].

A third type of crown ether magnesium complex includes compounds containing an Mg–C bond to a carbon atom of the macrocycle, formed as internally complexed Grignard reagents **101** [375, 387, 388]. The crystal structure determination of **101** (X=Br) shows that the metal coordination is distorted pentagonal pyramidal, with bromine in apical position (Mg–Br 2.517 Å) and two normal Mg–O distances (2.12 and 2.13 Å) and two large interatomic distances (2.33 and 2.49 Å) [388].

101

Another structurally characterized diorganomagnesium internal complex has a structure similar to that of the bromo derivative [389, 390].

Some Group 12 metal organometallic host–guest complexes with crown ethers are also known. Thus, diethylzinc forms threaded complexes with 18-crown-6 (Zn···O 2.837, 2.890, and 2.873 Å) [377, 379, 379a] and diphenylzinc forms non-threaded complexes with 1,3-xylyl-18-crown-5 and other macrocycles [391]. Internally coordinated organozinc derivatives with an intra-ring Zn–C bond have been reported with phenylzinc moieties [392]. Complexes of the type [RZn(macrocycle)]$^+$[ZnR$_3$]$^-$ are formed with nitrogen-containing macrocycles and 2.1.1- and 2.2.1-cryptands [393].

Organomercury crown ether derivatives have been structurally characterized as threaded host–guest complexes of bis(trifluoromethyl)mercury with dibenzocrown ethers [394, 395]. Other threaded diorganomercury host–guest complexes of crown ethers have been studied in solution [396–398]. Internally complexed organomercury compounds (with Hg–Me and other groups) and 1,3-xylyl-18-crown-5, e.g. **102**, are also known [399].

Much attention has been paid to crown ether and related complexes of Group 13 metals, by using both crown ethers and thia- and aza-macrocycles [400–402]. Both *endo-* and *exo-*cyclic coordination of aluminum has been established by X-ray diffraction studies. Thus, dicyclohexano-18-crown-6 coordinates as an *exo* receptor to two AlMe$_3$ molecules in [(AlMe$_3$)dicyh-18-crown-6)]. A similarly coordinated trimethylgallium complex is also formed [403], but 18-azacrown-6 acts as an *endo* and *exo* receptor simultaneously, coordinating an AlMe$_2^+$ group in the cavity and AlMe$_3$ molecul externally to nitrogen **103** [404]. A diethylaluminum complex [(AlEt$_2$)diaza-18-crown-6)] has also been described [405].

102   **103**

R = Me

Trimethylaluminum deprotonates the NH sites of the aza macrocycles, to form organometallic host–guest complexes with AlMe$_2$ groups coordinated in the cavity, in addition to externally coordinated AlMe$_3$ molecules, with ligands such as dibenzo-diaza-15-crown-4 and dibenzo-triaza-17-crown-5 [406]. Several other crown ether complexes with dimethyl- and trimethylaluminum moieties and 12-crown-4 [407], 15-crown-5 [408, 409], 18-crown-6 [409–411], and dibenzo-18-crown-6 [408] have been described. In [Me$_2$Al(15-crown-5)]$^+$[Me$_2$AlCl$_2$]$^-$ the linear AlMe$_2^+$ cation is threaded into the macrocycle (C–Al–C 178°, Al···O 2.13–2.26 Å). In [Me$_2$Al(18-crown-6)]$^+$[Me$_2$Al$_2$]$^-$ the AlMe$_2^+$ cation is bent (C–Al–C 140.6°) and only three Al–O bonds are present (Al···O 1.929, 2.181, and 2.435 Å), the other aluminum–oxygen distances being larger (3.093, 3.460, and 3.800 Å) and suggesting non-bonding [409]. Only compounds with anionic, non-coordinated organoaluminum groups, [AlCl$_2$(crown)]$^+$[EtAlCl$_3$]$^-$, have been obtained from EtAlCl$_2$ with 12-crown-4, benzo-15-crown-5, and 18-crown-6 [412, 413]. Solution NMR studies also suggest threaded structures for R$_2$Al(15-crown-5) and RAlX(15-crown-5) complexes [414].

Organoaluminum groups can be attached to phenolic oxygens, in the cavity of 1,3-xylyl-17-crown-5 macrocyclic ligands [415].

The trimethylaluminum complexes of tetraaza macrocycles contain AlMe$_2$ groups and externally coordinated AlMe$_3$ molecules [416–419]; tetrathia macrocycles [420, 421] are perhaps better described as simple coordination compounds of polydentate ligands, rather than host–guest complexes. This is suggested by the short Al–N distances (ca 2.0 Å) and the external coordination of the metal.

Several organogallium complexes of crown ethers and cryptands have been investigated. With 18-diazacrown-6 trimethylgallium forms a complex, **104**, with one GaMe$_2^+$ cation located in the cavity and two GaMe$_3$ molecules coordinated externally to nitrogen [422]; some unusual inclusion complexes containing (Me$_2$GaOH)$_3$ and (Me$_2$GaOH)$_4$ attached through hydrogen-bonds to the crown ethers were, however, obtained with dibenzodiaza-15-crown-4 and dibenzodiaza-17-crown-5 [406]. Tetraaza-12-crown-4 macrocycles also were reported to form trimethylgallium complexes [416, 423].

104

R = Me

Apparently no organoindium crown ether complexes are known.

Organothallium host–guest complexes of various crown ethers have been reported and most are threaded structures. These include dimethylthallium complexes with dibenzo-18-crown-6 (linear Me–Tl–Me cation with C–Tl–C 178° and Tl···O 2.69–2.82 Å) [424, 425], substituted 18-crown-6 ethers [426, 427] and dicyclohexyl-18-crown-6 [428, 429].

In Group 14 few organometallic crown ether complexes have been investigated. An internally coordinated trimethyltin derivative of 1,3-xylyl-17-benzo-crown-5 has been reported as being obtained via an organocalcium complex [430]. Group 15 organometallic compounds have also received little attention, but a 15-crown-5 complex of phenylantimony dibromide has been structurally characterized [431].

Surprisingly, few transition organometallic compounds seem to form crown ether complexes. A complex containing a cyclopentadienyltitanium moiety attached to 18-crown-6 has been structurally characterized. There are three Ti–O short distances (2.09, 2.14, and 2.15 Å), suggesting dative bonding, and the macrocycle behaves mostly as an *exo* receptor [432].

The sulfur-containing macrocycle 18-thiacrown-6 forms a pentamethylcyclopentadienylrhodium complex, **105**, with two (η^5-C$_5$Me$_5$)Rh groups externally coordinated to the ring acting as an *exo* receptor [433], i.e. more like a traditional polydentate ligand.

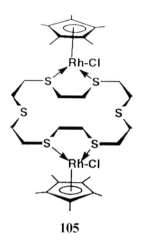

105

Second sphere complexes are sometimes obtained from organometallics and crown ethers. In these cases the crown ethers do not interact directly with the metal center of the organometallic compound, but rather with the ligands coordinated to the metal. The interaction usually involves hydrogen-bonding.

An example is the complex, **106**, of Me$_2$SnCl$_2$·H$_2$O with 18-crown-6. The compound dimerizes through secondary Sn···Cl bonds (3.311 and 3.050 Å) and the water molecules coordinated to tin are hydrogen-bonded to two crown ether oxygen atoms [433a]. Similar behavior is observed with MePhSnCl$_2$·H$_2$O and 18-crown-6 [433b]. With 15-crown-5 mononuclear hydrogen-bond complexes of hydrated Me$_2$SnCl$_2$ are formed [433c,d].

In another example, ammonia molecules coordinated to a rhenium center of a half-sandwich complex, **107**, are hydrogen-bonded to 18-crown-6 [434].

Other second sphere complexes reported include complexes of (η^5-C$_5$H$_5$)Fe(CO)$_2$-(NH$_3$), W(CO)$_5$(NH$_3$) [435], and of [Rh(COD)(NH$_3$)$_2$]$^+$ with 18-crown-6 and dibenzocrown ethers [436, 437] and with cryptands [438].

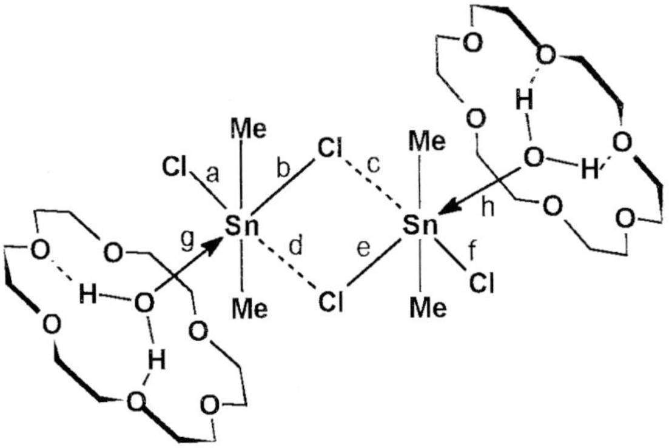

Sn-Cl

a = 2.419 Å
b = 2.524 Å
c = 3.311 Å
d = 3.050 Å
e = 2.536 Å
f = 2.393 Å

Sn-O

g = 2.360 Å
h = 2.313 Å

106

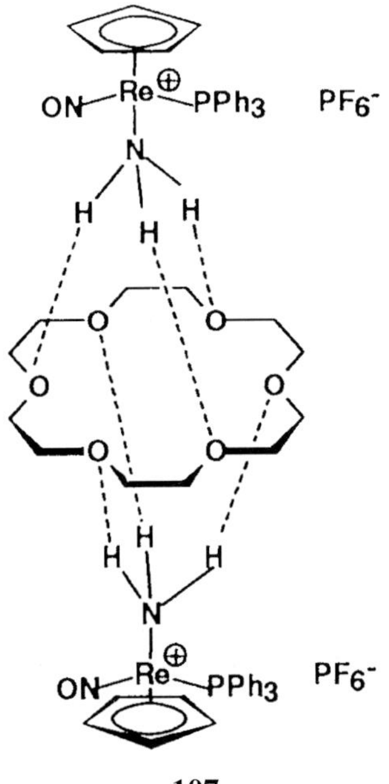

107

2.2.2.2 Organometallic molecules in cyclodextrin receptors

Cyclodextrins are probably the best studied organic hosts for organometallic molecules. Cyclodextrins are cyclic oligosaccharides containing six, seven, or eight α-D-glucopyranose groups (referred to as α-, β-, and γ-cyclodextrin or CD, respectively) which can encapsulate various organic, organometallic, or even inorganic hosts into their cavities [439]. The host–guest interactions involve non-covalent bonds. These inclusion (or host–guest) complexes are considered as enzyme models in biomimetic chemistry [440–442a].

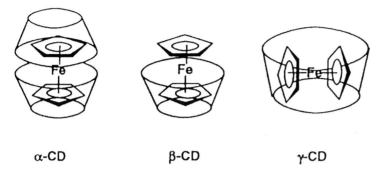

Fig. 2.2. Possible orientations of ferrocene guests in α-, β-, and γ-CD hosts.

The host–guest chemistry of organometallic cyclodextrin complexes has been reviewed [443, 443a] and will be surveyed only briefly here.

The cavities of cyclodextrins can well accommodate the cyclopentadienyl and benzene rings and as a result many metallocenes and half-sandwich π-complexes have been incorporated as guests.

Ferrocene was the first organometallic guest incorporated and numerous spectroscopic and electrochemical studies have been performed on ferrocene, substituted ferrocene, and related metallocene (e.g. cobaltocene) inclusion complexes [444–469]. Half-sandwich cyclopentadienyl- and benzene-metal carbonyl complexes have also been studied quite extensively [470–479] as have η^3-allyl metal (palladium) complexes [480], diene metal (rhodium) complexes [481–484], acetylene cobalt carbonyl cluster complexes [485], and complexes with metal carbonyls, e.g. $Fe(CO)_5$, $Mn_2(CO)_{10}$, and $CoNO(CO)_3$ [485a].

Ferrocene forms 1:2 host–guest complexes with α-CD and 1:1 host–guest complexes with β- and γ-CD and can occupy 'vertical' or 'horizontal' positions (Figure 2.2) in agreement with the size of the cavity [486].

X-Ray crystal structure investigations of the ferrocene–α-CD 1:2 complex [487] and of the $[(\eta^5\text{-}C_5H_5)Fe(\eta^6\text{-}C_6H_6)]^+$ complex [488] have established that the metallocenes are tilted relative to the six-fold axis of the cyclodextrin (by 42° for ferrocene and 33° for the sandwich cation).

Dinuclear half-sandwich iron complexes are believed to encapsulate each cyclopentadienyl group in a separate cyclodextrin host (Figure 2.3) [479].

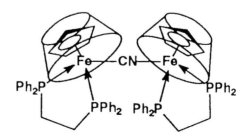

Fig. 2.3. Probable structure of dinuclear iron–cyclodextrin host–guest complexes.

The geometric (size and shape) complementarity between the cyclodextrin host and the organometallic guest determines a well manifested selectivity in the formation of inclusion complexes and can be used for the separation of ferrocene from dimethylferrocene (only the latter forms a complex with β-CD) [471], and of (benzene)-chromium tricarbonyl, $(\eta^6\text{-}C_6H_6)Cr(CO)_3$ from (hexamethylbenzene)-chromium tricarbonyl, $(\eta^6\text{-}C_6Me_6)Cr(CO)_3$ (only the latter forms a complex with γ-CD) [471]. Cyclodextrin host–guest complexation also affords the resolution of hydroxyethylferrocene enantiomers [489].

The encapsulation of organometallic molecules into cyclodextrin hosts significantly changes their chemical and electrochemical properties. Thus, the rate of hydrolysis of ferrocene acrylic acid ester is increased by six orders of magnitude [447] and the oxidation of encapsulated ferrocene requires a higher anodic potential than free ferrocene [453, 490, 491]. Cyclodextrine-cobaltocenium and ferrocene dendrimers were recently reported [491a,b]

2.2.2.3 Organometallics in the thiourea lattice host

Thiourea forms a layered lattice able to act as host for some organometallic guest molecules. The ferrocene–thiourea inclusion compound has been investigated by spectroscopic methods (NMR, Mössbauer) and it has been shown that the thiourea host structure is preserved [492–502]. Other inclusion compounds of thiourea with organometallics such as $(\eta^4\text{-}1,3\text{-cyclohexadiene})Mn(CO)_3$, $(\eta^6\text{-}C_6H_6)Cr(CO)_3$, $(\eta^4\text{-}1,3\text{-cyclohexadiene})Fe(CO)_3$, and $\eta^4\text{-}C(CH_2)_3Fe(CO)_3$ have been reported [503, 504].

References

[1] D. S. Brown, A. G. Massey and D. A. Wickens, *Acta Cryst.* **1978**, *B34*, 1695.
[2] D. S. Brown, A. G. Massey and D. A. Wickens, *Inorg. Chim. Acta*, **1980**, *44*, L 193.
[3] V. B. Shur, I. A. Tikhonova, P. V. Petrovskii and M. E. Vol'pin, *Metalloorg. Khim. USSR*, **1989**, *2*, 1435.
[4] V. B. Shur, I. A. Tikhonova, A. I. Yanovsky, Yu. T. Struchkov, P. V. Petrovskii, S. Yu. Panov, G. G. Furin and M. E. Vol'pin, *Izv. Akad. Nauk SSSR, Ser. Khim.* **1991**, 1491.
[5] V. B. Shur, I. A. Tikhonova, A. I. Yanovsky, Yu. T. Struchkov, P. V. Petrovskii, S. Yu. Panov, G. G. Furin and M. E. Vol'pin, *J. Organomet. Chem.* **1991**, *418*, C 29; *Doklady Akad. Nauk SSSR* **1991**, *321*, 1002.
[6] I. A. Tikhonova, F. M. Dolgushin, A. I. Yanovsky, Yu. T. Struchkov, A. N. Gavrilova, L. N. Saitkulova, E. S. Shubina, L. M. Epstein, G. G. Furin and V. B. Shur, *J. Organomet. Chem.* **1996**, *528*, 271.
[7] M. Yu. Antipin, Yu. T. Struchkov, A. Yu. Volkovskii and E. M. Rokhlin, *Bull. Acad. Sci. USSR*, **1983**, *32*, 410.
[8] V. B. Shur, I. A. Tikhonova, F. M. Dolgushin, A. I. Yanovsky, Yu. T. Struchkov, A. Yu. Volkonsky, E. V. Solodova, S. Yu. Panov, P. V. Petrovskii and M. E. Vol'pin, *J. Organomet. Chem.* **1993**, *443*, C 19.
[9] A. L. Chistyakov, I. V. Stankevich, N. P. Gambarayan, Yu. T. Struchkov, A. I. Yanovsky, I. A. Tikhonova and V. B. Shur, *J. Organomet. Chem.* **1997**, *536/537*, 413.

[10] X. Yang, C. B. Knobler, Z. Zheng and M. F. Hawthorne, *Angew. Chem.* **1991**, *103*, 1519; *Angew. Chem. Int. Ed. Engl.* **1991**, *30*, 1507.
[11] X. Yang, S. E. Johnson, S. I. Khan and M. F. Hawthorne, *Angew. Chem.* **1992**, *104*, 886; *Angew. Chem. Int. Ed. Engl.* **1992**, *31*, 893.
[12] X. Yang, C. B. Knobler and M. F. Hawthorne, *J. Am. Chem. Soc.* **1992**, *114*, 380.
[13] X. Yang, C. B. Knobler and M. F. Hawthorne, *J. Am. Chem. Soc.* **1993**, *115*, 193.
[14] X. Yang, C. B. Knobler and M. F. Hawthorne, *J. Am. Chem. Soc.* **1993**, *115*, 5320.
[15] X. Yang, C. B. Knobler, Z. Zheng and M. F. Hawthorne, *J. Am. Chem. Soc.* **1994**, *116*, 7142.
[16] A. A. Zinn, Z. Zheng, C. B. Knobler and M. F. Hawthorne, *J. Am. Chem. Soc.* **1996**, *118*, 70.
[17] Z. Zheng, C. B. Knobler, M. D. Mortimer, G. Kong and M. F. Hawthorne, *Inorg. Chem.* **1996**, *35*, 1235.
[18] R. J. Blanch, M. Williams, G. D. Fallon, M. G. Gardiner, R. Kaddour and C. L. Raston, *Angew. Chem. Int. Ed. Engl.* **1997**, *36*, 504.
[19] X. Yang, C. B. Knobler and M. F. Hawthorne, *J. Am. Chem. Soc.* **1993**, *115*, 4904.
[20] M. F. Hawthorne and Z. Zheng, *Acc. Chem. Res.* **1997**, *30*, 267.
[21] S. Gambarotta, C. Floriani, A. Chiesi-Villa and C. Guastini, *J. Chem. Soc. Chem. Comm.*, **1983**, 1156.
[22] S. Gambarotta, C. Floriani, A. Chiesi-Villa and C. Guastini, *J. Chem. Soc. Chem. Comm.*, **1983**, 1087.
[23] I. Haiduc and R. B. King, in vol. *Large Ring Molecules*, Edited by J. A. Semlyen, John Wiley & Sons, Chichester, **1996**, p. 525.
[24] C. J. Olliff, G. R. Pickering and K. J. Pitt, *J. Inorg. Nucl. Chem.* **1980**, *42*, 288.
[25] C. J. Olliff, G. R. Pickering and K. J. Pitt, *J. Inorg. Nucl. Chem.* **1980**, *42*, 1201.
[26] M. R. Churchill, C. H. Lake, S. H. L. Chao and O. T. Beachley, Jr., *J. Chem. Soc. Chem. Commun.* **1993**, 1577.
[27] C. Eaborn, P. P. Hitchcock, K. Izod and J. D. Smith, *Angew. Chem. Int. Ed. Engl.* **1995**, *34*, 2679.
[28] Yu. T. Struchkov and S. V. Lindeman, *J. Organomet. Chem.* **1995**, *488*, 9.
[29] O. I. Shchegolikhina, I. Blagodashkikh and A. A. Zhdanov, in vol. *Tailor-made Silicon-Oxygen Compounds. From Molecules to Materials*, Edited by R. Corriu and P. Jutzi, Vieweg Verlag, Braunschweig, **1996**, p. 177.
[30] O. I. Shchegolikhina, A. A. Zhadov, V. A. Igonin, Yu. E. Ovchinnikov, V. E. Shklover and Yu. T. Struchkov, *Metalloorg. Khim. (Russ.)* **1991**, *4*, 74; *Organomet. Chem. USSR (Engl.)*, **1991**, *4*, 39.
[31] V. A. Igonin, O. I. Shchegolikhina, S. V. Lindeman, M. M. Levitsky, Yu. T. Struchkov and A. A. Zhdanov, *J. Organomet. Chem.* **1992**, *423*, 351.
[32] M. M. Levitsky, A. A. Zhdanov, V. A. Igonin, Yu. E. Ovchinnikov, V. E. Shklover and Yu. T. Struchkov, *J. Organomet. Chem.* **1991**, *401*, 199.
[33] V. A. Igonin, S. V. Lindeman, K. A. Potekhin, V. E. Shklover, Yu. T. Struchkov, O. I. Shchegolikhina, A. A. Zhdanov and I. V. Razumovskaya, *Metalloorg. Khim. (Russ.)* **1991**, *4*, 790; Organomet. Chem. USSR (Engl) **1991**, *4*, 383.
[33a] A. Cornia, A. C. Fabretti, G. Gavioli, C. Zucchi, M. Pizzotti, A. Viz-Orosz, O. I. Shchegolikhina, Yu. A. Pozdniakova and G. Pályi, *J. Cluster Sci.* **1998**, *9*, 295.
[34] V. A. Igonin, S. V. Lindeman, Yu. T. Struchkov, O. I. Shchegolikhina, A. A. Zhdanov, Yu. A. Molodtsova and I. V. Razumovskaya, *Metalloorg. Khim. (Russ.)* **1991**, *4*, 1355; *Organomet. Chem. USSR (Engl.)* **1991**, *4*, 672.
[35] A. Cornia, A. C. Fabretti, D. Gatteschi, G. Pályi, E. Rentschler, O. I. Shchegolikhina and A. A. Zhdanov, *Inorg. Chem.* **1995**, *34*, 5383.
[36] S. V. Lindeman, O. I. Shchegolikhina, Y. A. Molodtsova and A. A. Zhdanov, *Acta Cryst.* **1997**, *C 53*, 305.
[37] V. A. Igonin, S. V. Lindeman, Yu. T. Struchkov, Yu. A. Molodtsova, O. I. Shchegolikhina and A. A. Zhdanov, *Izv. Akad. Nauk Ser. Khim.* **1993**, 752; *Russ. Chem. Bull (Engl.)* **1993**, *42*, 718.
[38] V. A. Igonin, S. V. Lindeman, Yu. T. Struchkov, O. I. Shchegolikhina, Yu. A. Molodtsova,

Yu. A. Pozdnyakova and A. A. Zhdanov, *Izv. Akad. Nauk, Ser. Khim.* **1993**, 184; *Russ. Chem. Bull. (Engl.)* **1993**, *42*, 168.

[39] V. A. Igonin, S. V. Lindeman, Yu. T. Struchkov, Yu. A. Molodtsova, Yu. A. Pozdyakova, O. I. Shchegolikhina and A. A. Zhdanov, *Izv. Akad. Nauk Ser. Khim.* **1993**, 193; *Russ. Chem. Bull. (Engl.)* **1993**, *42*, 176.

[40] O. I. Shchegolikhina, Yu. A. Pozdnyakova, S. V. Lindeman, A. A. Zhdanov, R. Psaro, R. Ugo, G. Gavioli, R. Battistuzzi, M. Borsari, T. Rüffer, C. Zucchi and G. Palyi, *J. Organomet. Chem.* **1996**, *514*, 29.

[41] Yu. E. Ovchinnikov, V. E. Shklover, Yu. T. Struchkov, M. M. Levitsky and A. A. Zhdanov, *J. Organomet. Chem.* **1988**, *347*, 253; Yu. E. Ovchinnikov, A. A. Zhdanov, M. M. Levitsky, V. E. Shklover and Yu. T. Struchkov, *Izv. Akad. Nauk SSSR, Ser. Khim.* **1986**, 1206.

[42] V. I. Kuznetsov, G. L. Elizarova, L. G. Matvienko, I. G. Lantyukhova, V. N. Kolomiichuk, A. A. Zhdanov, and O. I. Shchegolikhina, *J. Organomet. Chem.* **1994**, *475*, 65.

[43] G. Gavioli, M. Borsari, C. Zucchi, G. Layi, R. Psaro, R. Ugo, O. I. Shchegolikhina and A. A. Zhdanov, *J. Organomet. Chem.* **1994**, *467*, 165.

[44] V. L. Kuznetsov and A. N. Usoltseva, in vol. *Tailor-made Silicon-Oxygen Compounds. From Molecules to Materials*, Edited by R. Corriu and P. Jutzi, Vieweg Verlag, Braunschweig, **1996**, p. 193.

[45] E. Rentschler, D. Gatteschi, A. Cornia, A. C. Fabretti, A. L. Barra, O. I. Shchegolikhina and A. A. Zhdanov, *Inorg. Chem.* **1996**, *35*, 4427.

[46] A. Cornia, A. C. Fabretti, D. Gatteschi, G. Pályi, E. Rentschler, O. I. Schegolikhina and A. A. Zhdanov, *Inorg. Chem.* **1995**, *34*, 5383.

[47] R. Murugavel, A. Voigt, M. G. Walawalkar and H. W. Roesky, *Chem. Rev.* **1996**, *96*, 2205.

[47a] O. I. Shchegolikhina, V. A. Igonin, Yu. A. Molodtsova, Yu. A. Pozdniakova, A. A. Zhdanov, T. V. Strelkova and S. V. Lindeman, *J. Organomet. Chem.* **1998**, *562*, 141.

[48] M. Borsari, G. Gavioli, C. Zucchi, G. Palyi, R. Psaro, R. Ugo, O. I. Shchegolikhina and A. A. Zhdanov, *Inorg. Chim. Acta* **1997**, *258*, 139.

[48a] A. Voigt, M. G. Walawalkar, R. Murugavel, H. W. Roesky, E. Parizini and P. Lubini, *Angew. Chem.* **1997**, *109*, 2313; *Angew. Chem. Int. Ed. Engl.* **1997**, *36*, 2203.

[48b] M. G. Walawalkar, R. Murugavel, H. W. Roesky, I. Uson and R. Kraetzner, *Inorg. Chem.* **1998**, *37*, 473.

[48c] M. G. Walawalkar, R. Murugavel, A. Voigt, H. W. Roesky and H. G. Schmidt, *J. Am. Chem. Soc.* **1997**, *119*, 4656.

[48d] M. G. Walawalkar, R. Murugavel, H. W. Roesky and H. G. Schmidt, *Inorg. Chem.* **1997**, *36*, 4202.

[48e] J. Storre, C. Schnitter, H. W. Roesky, H.-G. Schmidt, M. Noltemeyer, R. Fleischer and D. Stalke, *J. Am. Chem. Soc.* **1997**, *119*, 7505.

[49] Y. Azuma and M. Newcomb, *Organometallics* **1984**, *3*, 9.

[50] J. H. Horner and M. Newcomb, *Organometallics* **1991**, *10*, 1732.

[51] A. G. Davies, M. W. Tse, J. D. Kennedy, W. McFarlane, G. S. Pyne, M. F. C. Ladd and D. C. Povey, *J. Chem. Soc. Perkin Trans.* II, **1981**, 369.

[52] K. Jurkschat, H. G. Kuivila, S. Liu and J. A. Zubieta, *Organometallics*, **1989**, *8*, 2755.

[53] M. Newcomb, J. H. Horner, M. T. Blanda and P. J. Squattrito, *J. Am. Chem. Soc.* **1989**, *111*, 6294.

[54] K. Jurkschat, A. Rühleman and A. Tzschach, *J. Organomet. Chem.* **1990**, *381*, C 53.

[55] M. Newcomb, J. H. Horner and M. T. Blanda, *J. Am. Chem. Soc.* **1987**, *109*, 7878.

[56] M. Newcomb, A. M. Madonik, M. T. Blanda and J. K. Judice, *Organometallics*, **1987**, *6*, 145.

[57] M. T. Blanda and M. Newcomb, *Tetrahedorn Letters*, **1989**, *30*, 3501.

[58] J. H. Horner, P. J. Squattrito, N. McGuire, J. P. Reibenspies and M. Newcomb, *Organometallics*, **1991**, *10*, 1741.

[59] R. Altmann, K. Jurkschat, M. Schurmann, D. Dakternieks and A. Duthie, *Organometallics* **1997**, *16*, 5716.

[60] P. D. Beer, *Chem. Commun.* **1996**, 689.

[60a] P. D. Beer, *Chem. Commun.* **1998**, 71.

[60b] J. E. Kingston, L. Ashford, P. D. Beer and M. G. B. Drew, *J. Chem. Soc., Dalton Trans.* **1999**, 251.
[61] P. D. Beer, *Advan. Inorg. Chem.* **1992**, *39*, 79.
[62] P. D. Beer and D. K. Smith, *Progr. Inorg. Chem.* **1997**, *46*, 1.
[63] F. P. Schmidtchen and M. Berger, *Chem. Rev.* **1997**, *97*, 1623.
[64] P. D. Beer, *Acc. Chem. Res.* **1998**, *31*, 71.
[65] C. D. Hall, in vol. *Ferrocenes*, Edited by A. Togni and T. Hayashi, VCH Weinheim, **1995**, pag. 279.
[66] P. L. Boulas, M. Gomez-Kaifer and L. Echegoyen, *Angew. Chem.* **1998**, *110*, 226.
[67] G. Oepen and F. Vögtle, *Liebigs Ann. Chem.* **1979**, 1094.
[68] A. P. Bell and C. D. Hall, *J. Chem. Soc. Chem. Commun.* **1980**, 163.
[69] B. Czech and A. Ratayczak, *Polish J. Chem.* **1980**, *54*, 57.
[70] J. F. Biernat and T. Wilchewski, *Tetrahedron* **1980**, *36*, 2521.
[71] S. Akabori, Y. Habata, Y. Sakamoto, M. Sato and S. Ebine, *Bull. Chem. Soc. Japan*, **1982**, *56*, 537.
[72] T. Saji, *Chem. Lett*, **1986**, 275.
[73] T. Saji and I. Kinoshita, *J. Chem. Soc. Chem. Commun.* **1986**, 716.
[74] T. Izumi, K. Saitou, S. Matsunaga and A. Kasahara, *Bull. Chem. Soc. Japan*, **1986**, *59*, 2425.
[75] P. D. Beer, H. Sikanyika, C. Blackburn, J. F. McAleer and M. G. B. Drew, *J. Chem. Soc. Dalton Trans.* **1990**, 3295.
[76] P. D. Beer, C. Blackburn, J. F. McAleer and H. Sikanyika, *Inorg. Chem.* **1990**, *29*, 378.
[77] P. D. Beer, *J. Chem. Soc. Chem. Commun.* **1985**, 1115.
[78] P. D. Beer, H. Sikanyika, A. M. Z. Slawin and D. J. Williams, *Polyhedron* **1989**, *8*, 879.
[79] P. D. Beer and S. E. Stokes, *Polyhedron*, **1995**, *19*, 2631.
[80] P. D. Beer and D. B. Crowe, *J. Organomet. Chem.* **1989**, *375*, C35.
[81] P. D. Beer, D. C. Crowe, M. I. Ogden, M. G. B. Drew and B. Main, *J. Chem. Soc. Dalton Trans.* **1993**, 2107.
[82] J. C. Medina, C. Li, S. Bott, J. L Atwood and G. W. Gokel, *J. Am. Chem. Soc.* **1991**, *113*, 366.
[83] C. D. Hall, I. P. Danks and N. W. Sharpe, *J. Organomet. Chem.* **1990**, *390*, 227.
[84] H. Plenio and C. Aberle, *Organometallics*, **1997**, *16*, 5950.
[84a] G. C. Dol, P. C. J. Kamer, F. Hartl, P. W. N. M. van Leuwen and R. J. M. Nolte, *J. Chem. Soc. Dalton Trans.* **1998**, 2083.
[84b] J. M. Lloris, R. Martinez-Manez, T. Parlo, J. Soto and M. E. Paddilla-Tosta, *Chem. Commun.* **1998**, 837.
[85] J. C. Medina, T. T. Goodnow, M. T. Rojas, J. L Atwood, B. C. Lynn, A. E. Kaifer and G. W. Gokel, *J. Am. Chem. Soc.* **1992**, *114*, 10583.
[86] T. Izumi, S. Murakami, and A. Kasahara, *Bull. Chem. Soc. Japan*, **1988**, *61*, 3565.
[87] P. J. Hammond, P. D. Beer and C. D. Hall, *J. Chem. Soc. Chem. Comm.* **1983**, 1161.
[88] P. J. Hammond, A. P. Bell and C. D. Hall, *J. Chem. Soc. Perkin Trans.* I, **1983**, 707.
[89] P. D. Beer, J. Elliot, P. J. Hammond, C. Dudman and C. D. Hall, *J. Organomet. Chem.* **1984**, *263*, C37.
[90] C. D. Hall, N. W. Sharpe, I. P. Danks and Y. P. Sung, *J. Chem. Soc. Chem. Commun.* **1989**, 419.
[91] C. D. Hall and N. W. Sharpe, *Organometallics* **1990**, *9*, 952.
[92] P. D. Beer, Z. Chen and M. I. Ogden, *J. Chem. Soc. Faraday Trans.* **1995**, *91*, 295.
[93] M. C. Grossel, D. G. Hamilton, J. I. Fuller and E. Millan-Barios, *J. Chem. Soc. Dalton Trans.* **1997**, 3471.
[94] P. D. Beer, C. G. Crane, A. D. Keefe and A. R. Whyman, *J. Organomet. Chem.* **1986**, *314*, C9.
[95] C. D. Hall and N. W. Sharpe, *J. Organomet. Chem.* **1991**, *405*, 365.
[96] C. D. Hall and N. W. Sharpe, *J. Photochem. Photobiol. A. Chem.* **1991**, *56*, 255.
[97] G. De Santis, L. Fabrizzi, M. Lichelli, C. Mangano, L. Pallavicini and A. Poggi, *Inorg. Chem.* **1993**, *32*, 854.
[98] P. D. Beer, Z. Chen, M. G. B. Drew, J. Kingston, M. Ogden and P. Spencer, *J. Chem. Soc. Chem. Commun.* **1993**, 1046.

[99] P. D. Beer, and F. Szemes, *J. Chem. Soc. Chem. Comm.* **1995**, 2245.
[100] C. D. Hall, T. K. U. Truong and S. C. Nyburg, *J. Organomet. Chem.* **1997**, *547*, 281.
[100a] C. D. Hall, T. K. U. Truong, J. H. R. Tucker and J. W. Steed, *Chem. Commun.* **1997**, 2195.
[101] P. D. Beer, O. Kocian, R. J. Mortimer and P. Spencer, *J. Chem. Soc. Chem. Commun.* **1992**, 602.
[102] B. Delavaux-Nicot, Y. Guari, B. Douziech and R. Mathieu, *J. Chem. Soc. Chem. Commun.* **1995**, 585.
[103] B. Czech and A. Ratajczak, *Polish J. Chem.* **1980**, *54*, 57.
[104] M. Sato, M. Kubo, S. Ebine and S. Akabori, *Tetrahedron Lett.* **1981**, *22*, 185.
[105] S. Akabori, S. Shibahara, Y. Habata and M. Sato, *Bull. Chem. Soc. Japan*, **1984**, *57*, 63.
[106] M. Sato, M. Kubo, S. Ebine and S. Akabori, *Bull. Chem. Soc. Japan*, **1984**, *57*, 421.
[107] I. Bernal, E. Raabe, G. M. Reisner, R. A. Bartsch, R. A. Holwerda, B. P. Czech and Z. Huang, *Organometallics*, **1988**, *7*, 247.
[108] T. Jorgensen, T. K. Hansen and J. Becher, *Chem. Soc. Rev.* **1994**, 41.
[109] A. Ratajczak and B. Czech, *Polish. J. Chem.* **1980**, *54*, 767.
[110] M. Sato, S. Tanaka, S. Ebine and S. Akabori, *Bull. Chem. Soc. Japan*, **1984**, *57*, 1929.
[111] H. Sato, S. Tanaka, S. Akabori and Y. Habata, *Bull. Chem. Soc. Japan*, **1986**, *59*, 1515.
[112] M. Sato, K. Suzuki and S. Akabori, *Bull. Chem. Soc. Japan*, **1986**, *59*, 3611.
[113] P. D. Beer, J. E. Nation and S. L. Brown, *J. Organomet. Chem.* **1989**, *377*, C23.
[114] P. D. Beer, J. E. Nation, S. L. W. McWhinnie, M. E. Harman, M. B. Hursthouse, M. I. Ogden and A. H. White, *J. Chem. Soc. Dalton Trans.* **1991**, 2485.
[115] P. D. Beer, C. A. P. Dickson, N. C. Fletcher, A. J. Goulden, A. Grieve, J. Hodacova and T. Wear, *J. Chem. Soc. Chem. Commun.* **1993**, 828.
[116] J. D. Carr, L. Lambert, D. E. Hibbs, M. B. Hurthouse, K. M. Abdul Malik and J. H. R. Tucker, *Chem. Commun.* **1997**, 1649.
[117] G. De Santis, L. Fabrizzi, M. Lichelli, P. Pallavicini and A. Perotti, *J. Chem. Soc. Dalton Trans.* **1992**, 3283.
[118] P. D. Beer, Z. Chen, A. J. Goulden, A. Graydon, S. E. Stokes and T. Wear, *J. Chem. Soc. Chem. Commun.* **1993**, 1834.
[119] N. Sachsinger and C. D. Hall, *J. Organomet. Chem.* **1997**, *531*, 61.
[120] P. D. Beer, M. G. B. Drew and R. Jgessar, *J. Chem. Soc. Chem. Commun.* **1995**, 1187.
[121] P. D. Beer and S. E. Stokes, *Polyhedron* **1995**, *19*, 2631.
[122] P. D. Beer and A. D. Keefe, *J. Organomet. Chem.* **1989**, *375*, C40.
[123] P. D. Beer and S. E. Stokes, *Polyhedron*, **1995**, *14*, 873.
[124] P. D. Beer and A. R. Graydon, *J. Organomet. Chem.* **1994**, *466*, 241.
[125] P. D. Beer, M. G. B. Drew, J. Hodacova and S. E. Stokes, *J. Chem. Soc. Dalton Trans.* **1995**, 3447.
[126] P. D. Beer, D. Hesek, J. Hodacova and S. E. Stokes, *J. Chem. Soc. Chem. Commun.* **1992**, 270.
[127] P. D. Beer, C. Hazlewood, D. Hesek, J. Hodacova and S. E. Stokes, *J. Chem. Soc. Dalton Trans.* **1993**, 1327.
[128] P. D. Beer, M. G. B. Drew, A. R. Graydon, D. K. Smith and S. E. Stokes, *J. Chem. Soc. Dalton Trans.* **1995**, 403.
[129] P. D. Beer and S. E. Stokes, *Polyhedron* **1995**, *14*, 873.
[130] P. D. Beer, M. G. B. Drew, D. Hesek, J. Kingston, D. K. Smith and S. E. Stokes, *Organometallics*, **1995**, *14*, 3288.
[131] P. D. Beer, M. G. B. Drew, C. Hazlewood, D. Hesek, J. Hodacova and S. E. Stokes, *J. Chem. Soc. Chem. Commun.* **1993**, 229.
[132] M. Uno, N. Komatsuzaki, K. Shirai and S. Takahashi, *J. Organomet. Chem.* **1993**, *462*, 343.
[133] N. Komatsuzaki, M. Uno, K. Shirai, Y. Takai, T. Tanaka, M. Sawada and S. Takahashi, *Bull. Chem. Soc. Japan*, **1996**, *69*, 17.
[134] N. D. Lowe and C. D. Garner, *J. Chem. Soc. Dalton Trans.* **1993**, 2197.
[135] S. Akabori, S. Sato, K. Kawazoe, C. Tamura, M. Sato and Y. Habata, *Chem. Lett.* **1987**, 1783.
[136] L. Ogierman, A. Palka, B. Chech and A. Rataiczak, *J. Chromatogr.* **1983**, *268*, 144.
[137] T. Izumi, S. Murakami and A. Kasahara, *Bull. Chem. Soc. Japan*, **1988**, *61*, 3565.

[138] Y. Habata, S. Akabori and M. Sato, *Bull. Chem. Soc. Japan*, **1985**, *58*, 3540.
[139] S. Akabori, Y. Habata and M. Sato, *Chem. Lett.* **1985**, 1063.
[140] S. Akabori, H. Munegumi, Y. Habata, S. Sato, K. Kawazoe, C. Tamura and M. Sato, *Bull. Chem. Soc. Japan*, **1985**, *58*, 2185.
[141] S. Akabori, Y. Habata, H. Munegumi and M. Sato, *Tetrahedron Lett.* **1984**, *25*, 1991.
[142] S. Sato, Y. Habata, M. Sato and S. Akabori, *Bull. Chem. Soc. Japan* **1989**, *62*, 3963.
[143] S. Akabori, S. Sato, T. Tokuda, Y. Habata, K. Kawazoe, C. Tamura and M. Sato, *Bull. Chem. Soc. Japan*, **1986**, *59*, 3189.
[144] S. Akabori, H. Munegumi, S. Sato and M. Sato, *J. Organomet. Chem.* **1984**, *272*, C 54.
[145] S. Akabori, S. Sato, T. Tokuda, Y. Habata, K. Kawazoe, C. Tamura and M. Sato, *Chem. Lett.* **1986**, 121.
[146] S. Akabori, S. Sato, Y. Habata, K. Kawazoe, C. Tamura and M. Sato, *Chem Lett.*, **1987**, 787.
[147] S. Akabori, S. Sato, K. Kawazoe, M. Sato and Y. Habata, *Bull. Soc. Chem. Japan*, **1988**, *61*, 1695.
[148] S. Akabori, Y. Habata, S. Sato, K. Kawazoe, C. Tamura and M. Sato, *Acta Cryst.* **1986**, *42C*, 682.
[149] P. D. Beer, *J. Chem. Soc. Chem. Commun.* **1985**, 1115.
[150] P. D. Beer, H. Sikanyika, A. M. Z. Slawin and D. J. Williams, *Polyhedron*, **1989**, *8*, 879.
[151] P. D. Beer and A. D. Keefe, *J. Organomet. Chem.* **1986**, *306*, C 10.
[152] P. D. Beer, A. D. Keefe, H. Sikanyika, C. Blackburn, and J. M. Mc Aleer, *J. Chem. Soc. Dalton Trans.* **1990**, 3289.
[153] P. D. Beer, C. G. Crane, A. D. Keefe and A. R. Whyman, *J. Organomet. Chem.* **1986**, *314*, C 9.
[154] P. D. Beer, J. Elliot, P. J. Hammond, C. Dudman and C. D. Hall, *J. Organomet. Chem.* **1984**, *263*, C 37.
[155] C. D. Hall, A. W. Parkins, S. C. Nyburg and N. W. Sharpe, *J. Organomet. Chem.* **1991**, *407*, 107.
[156] C. D. Hall and N. W. Sharpe, *Organometallics*, **1990**, *9*, 952.
[157] C. D. Hall and N. W. Sharpe, *J. Organomet. Chem.* **1991**, *405*, 365.
[158] C. D. Hall and N. W. Sharpe, *J. Photochem. Photobiol.*, **1991**, *A 56*, 255.
[159] Y. G. Yin, K. K. Cheung and W. T. Wong, *Polyhedron*, **1997**, *16*, 2889.
[159a] D. S. Shephard, B. F. G. Johnson, J. Matters and S. Parsons, *J. Chem. Soc. Dalton Trans.* **1998**, 2289.
[160] N. D. Lowe and C. D. Garner, *J. Chem. Soc. Dalton Trans.* **1993**, 2197.
[161] E. Fu, M. L. H. Green, V. J. Lowe and S. R. Marder, *J. Organomet. Chem.* **1988**, *341*, C 39.
[161a] G. M. Gray, *Comments Inorg. Chem.* **1995**, *17*, 95.
[161b] E. Lindner, M. F. Gunther, H. A. Mayer, R. Fawzi and M. Steinmann, *Chem. Ber.* **1997**, *130*, 1815.
[162] C. D. Gutsche, *Calixarenes*, Royal Society of Chemistry, Cambridge, **1989**.
[163] J. Vincenes and V. Bohmer (Editors) *Calixarenes: A Versatile Class of Macrocyclic Compounds*, Kluwer, Dordrecht, **1991**.
[163a] R. Ungaro and A. Pochini, in J. M. Lehn, J. L. Atwood, J. E. D. Davies, D. D. MacNicol and F. Vögtle, Editors, *Comprehensive Supramolecular Chemistry, Volume 2. Molecular Recognition: Receptors for Molecular Guests*, F. Vögtle, Volume Editor, Pergamon Press, Oxford, **1996**, p. 103.
[164] C. Wieser, C. B. Dieleman and D. Matt, *Coord. Chem. Revs.* **1997**, *165*, 93.
[165] H. Iki, T. Kikuchi and S. Shinkai, *J. Chem. Soc., Perkin Trans.* I, **1992**, 669.
[166] H. Iki, T. Kikuchi and S. Shinkai, *J. Chem. Soc. Perkin Trans.* I, **1993**, 205.
[167] H. Iki, T. Kikuchi, H. Tsuzuki and S. Shinkai, *Chem. Lett.* **1993**, 1735.
[168] T. Kikuchi, H. Iki, H. Tsuzuki and S. Shinkai, *Supramol. Chem.* **1993**, *1*, 103.
[169] H. Iki, T. Kikuchi, H. Tsuzuki and S. Shinkai, *J. Incl. Phenom.* **1994**, *19*, 227.
[170] J. W. Steed, R. K. Juneja and J. L. Atwood, *Angew. Chem.* **1994**, *106*, 2571; *Angew. Chem. Int. Ed. Engl.* **1994**, *33*, 2456.
[171] J. W. Steed, R. K. Juneja, R. S. Burkhalter and J. L Atwood, *J. Chem. Soc. Chem. Commun.* **1994**, 2205.

[172] J. W. Steed, C. P. Johnson, R. K. Juneja, R. S. Burkhalter and J. L. Atwood, *Supramol. Chem.* **1996**, *6*, 235.
[173] A. Ikeda, H. Tsuzuki and S. Shinkai, *J. Chem. Soc., Perkin Trans.* II, **1994**, 2073.
[174] A. Ikeda and S. Shinkai, *J. Am. Chem. Soc.* **1994**, *116*, 3102.
[175] W. Xu, R. J. Puddephatt, K. W. Muir and A. A. Torabi, *Organometallics* **1994**, *13*, 3054.
[176] P. D. Beer, A. D. Keefe and M. G. B. Drew, *J. Organomet. Chem.* **1988**, *353*, C10.
[177] P. D. Beer, A. D. Keefe and M. G. B. Drew, *J. Organomet. Chem.* **1989**, *378*, 437.
[178] P. D. Beer, *Chem. Commun.* **1996**, 689.
[178a] P. A. Gale, Z. Chen, M. G. B. Drew, J. A. Heath and P. D. Beer, *Polyhedron* **1998**, *17*, 405.
[179] P. D. Beer, A. D. Keefe, A. M. Z. Slawin and D. J. Williams, *J. Chem. Soc. Dalton Trans.* **1990**, 3675.
[179a] P. D. Beer and M. Shate, *Chem. Commun.* **1997**, 2377.
[180] P. D. Beer, Z. Chen, A. J. Goulden, A. Graydon, S. E. Strokes and T. Wear, *J. Chem. Soc. Chem. Commun.* **1993**, 1834.
[181] P. D. Beer and A. D. Keefe, *J. Incl. Phenom.* **1987**, *5*, 499.
[182] P. D. Beer, Z. Chen, M. G. B. Drew and P. A. Gale, *J. Chem. Soc. Chem. Commun.* **1995**, 1851.
[183] P. D. Beer, M. G. B. Drew, C. Hazlewood, D. Hesek, J. Hodacova and S. E. Stokes, *J. Chem. Soc. Chem. Commun.* **1993**, 229.
[184] P. D. Beer, D. Hesek, J. E. Kingston, D. K. Smith, S. E. Stokes and M. G. B. Drew, *Organometallics*, **1995**, *14*, 3288.
[184a] B. R. Cameron, S. J. Loeb and G. P. A. Yap, *Inorg. Chem.* **1997**, *36*, 5498.
[185] L. Giannini, E. Solari, A. Zanotti-Gerosa, C. Floriani, A. Chiesi-Villa and C. Rizzoli, *Angew. Chem. Int. Ed. Engl.* **1996**, *35*, 85.
[185a] A. Caselli, L. Giannini, E. Solari, C. Floriani, N. Re, A. Chiesi-Villa and C. Rizzoli, *Organometallics* **1997**, *16*, 5457.
[186] J. A. Acho, L. H. Doerrer and S. J. Lippard, *Inorg. Chem.* **1995**, *34*, 2542.
[186a] M. Giusti, E. Solari, L. Giannini, C. Floriani, A. Chiesi-Villa and C. Rizzoli, *Organometallics*, **1997**, *16*, 5610.
[186b] B. Castellayno, E. Solari, C. Floriani, N. Re, A. Chiesi-Villa and C. Rizzoli, *Organometallics* **1998**, *17*, 2328.
[187] V. C. Gibson, C. Redshaw, W. Clegg and M. R. J. Elsegood, *Polyhedron* **1997**, *16*, 4385.
[187a] J. L. Atwood, M. Gardiner, C. Jones, C. L. Raston, B. W. Selton and A. H. White, *J. Chem. Soc. Chem. Commun.* **1996**, 2487.
[187b] M. G. Gardiner, S. M. Lawrence, C. L. Raston, B. W. Skelton and A. H. White, *J. Chem. Soc. Chem. Commun.* **1996**, 2491.
[188] I. Neda, A. Vollbrecht, J. Grunenberg and R. Schmutzler, *Heteroatom Chem.*, **1998**, *9*, 553.
[188a] I. Neda, T. Kankorat and R. Schmutsler, *Main Group Chem. News* **1998**, *6*, 4, and references cited therein.
[189] S. Shang, D. V. Khasnis, J. M. Burton, C. J. Santini, M. Fan, A. C. Small and M. Lattman, *Organometallics*, **1994**, *13*, 5157.
[189a] T. Hascall, A. L. Rheingold, I. Guzei and G. Parkin, *Chem. Commun.* **1998**, 101.
[189b] D. J. Cram, K. D. Stewart, I. Goldberg, K. N. Trueblood, *J. Am. Chem. Soc.* **1985**, *107*, 2574.
[189c] I. Goldberg, *J. Incl. Phenom.* **1986**, *4*, 191.
[189d] K. Kimura, Y. Tsujimura, M. Yokoyama and T. Maeda, *Bull. Chem. Soc. Japan* **1998**, *71*, 657.
[190] P. D. Beer, M. G. B. Drew, A. Ibbotson and E. L. Tite, *J. Chem. Soc.* **1988**, 1498.
[191] P. D. Beer and E. L Tite, *Tetrahedron Lett.* **1988**, 2349.
[192] P. D. Beer, E. L. Tite, M. G. B. Drew and A. Ibbotson, *J. Chem. Soc. Dalton Trans.* **1990**, 2543.
[192a] B. König, M. Rodel, P. Bubenitschek, P. G. Jones and I. Thondorf, *J. Org. Chem.* **1995**, *60*, 7406.
[192b] B. König, M. Rodel, P. Bubenitschek and P. G. Jones, *Angew. Chem.* **1995**, *107*, 752; *Angew. Chem. Int. Ed. Engl.*, **1995**, *34*, 661.
[192c] M. Fan, H. Zhang and M. Lattman, *Organometallics* **1996**, *15*, 5216.

[192d] S. S. H. Mao, F. Q. Liu and T. D. Tilley, *J. Am. Chem. Soc.* **1998**, *120*, 1193.
[192e] M. Yoshida, M. Goto and F. Nakanishi, *Organometallics* **1999**, *18*, 1465.
[193] P. D. Beer, E. L. Tite and A. Ibbotson, *J. Chem. Soc. Chem. Commun.* **1989**, 1874.
[194] P. D. Beer, E. L. Tite and A. Ibbotson, *J. Chem. Soc. Dalton Trans.* **1991**, 1691.
[195] K. T. Holman, M. M. Halihan, J. W. Steed, S. S. Jurisson and J. L. Atwood, *J. Am. Chem. Soc.* **1995**, *117*, 7848.
[196] K. T. Holman, M. M. Halihan, S. S. Jurisson, J. L. Atwood, R. S. Burkhalter, A. R. Mitchell and J. W. Steed, *J. Am. Chem. Soc.* **1996**, *118*, 9567.
[197] J. W. Steed, R. K. Juneja and J. L. Atwood, *Angew. Chem.* **1994**, *196*, 2571.
[197a] K. S. B. Hancock and J. W. Steed, *Chem. Commun.* **1998**, 1409.
[198] H. Sakurai, Y. Eriyama, A. Hosomi, Y. Nakadaira and C. Kabuto, *Chem. Lett.* **1984**, 595.
[199] K. T. Holman, J. W. Steed and J. L. Atwood, *Angew. Chem. Int. Ed. Engl.* **1997**, *36*, 1736.
[199a] K. T. Holman, G. W. Orr, J. W. Steed and J. L. Atwood, *Chem. Commun.* **1998**, 2109.
[200] T. Bein, in J. M. Lehn, J. L. Atwood, J. E. D. Davies, D. D. MacNicol, and F. Vögtle, Editors, *Comprehensive Supramolecular Chemistry, Volume 7, Solid State Supramolecular Chemistry: Two- and Three-dimensional Inorganic Networks*, G. Alberti and T. Bein, Volume Editors, Pergamon Press, Oxford, **1996**, p. 579.
[201] D. C. Bailey and S. H. Langer, *Chem. Rev.* **1981**, *81*, 109.
[202] G. A. Ozin and C. Gil, *Chem. Rev.* **1989**, *89*, 1749.
[203] W. H. M. Sachtler and Z. Zhang, *Adv. Catal.* **1993**, *39*, 129.
[204] K. J. Balkus, Jr. and A. G. Gabrielov, *J. Incl. Phenom.* **1995**, *21*, 159.
[204a] G. A. Ozin, A. Kuperman and A. Stein, *Angew. Chem./Adv. Mater.* **1989**, *101*, 353; *Angew. Chem. Int. Ed. Engl./Adv. Mater.* **1989**, *28*, 359.
[205] G. A. Ozin, M. R. Steele and A. J. Holmes, *Chem. Mater.* **1994**, *6*, 999.
[206] M. R. Steele, P. M. Macdonald and G. A. Ozin, *J. Am. Chem. Soc.* **1993**, *115*, 7285.
[207] H. A. Girbas and R. A. Jacobson, *J. Chem. Soc. Perkin Trans. II*, **1985**, 2041. 167.
[208] S. I. Woo, Y. S. Ko and T. K. Han, *Macromol. Rapid Commun.* **1995**, *16*, 484.
[209] J. E. MacDougall, H. Eckert, G. D. Stucky, N. Herron, Y. Wang, K. Möller, T. Bein and D. Cos, *J. Am. Chem. Soc.* **1989**, *111*, 8006.
[210] S. Ozkar, G. A. Ozin, K. Moller and T. Bein, *J. Am. Chem. Soc.* **1990**, *112*, 9575.
[211] J. K. Wang, S. Namba and T. Yoshima, *J. Mol. Catal.* **1989**, *53*, 155.
[212] Y. Okamoto, Y. Inui, H. Onimatsu and T. Imanaka, *J. Phys. Chem.* **1991**, *95*, 4596.
[213] S. Abdo and R. F. Howe, *J. Phys. Chem.* **1983**, *87*, 1713.
[214] S. Abdo and R. F. Howe, *J. Phys. Chem.* **1983**, *87*, 1722.
[215] M. B. Ward, K. Mizuno and J. H. Lunsford, *J. Mol. Catal.* **1984**, *27*, 1.
[216] S. L. T. Anderson and R. F. Howe, *J. Phys. Chem.* **1989**, *93*, 4913.
[217] Y. Okamoto, T. Imanaka, K. Asakura and Y. Iwasawa, *J. Phys. Chem.* **1991**, 95.
[218] H. O. Pastore, G. A. Ozin and A. J. Poe, *J. Am. Chem. Soc.* **1993**, *115*, 1215.
[219] C. Bremard and M. Le Maire, *J. Mol. Struct.* **1995**, *349*, 49. 3700.
[220] S. Ozkar, G. A. Ozin and R. A. Prokopowicz, *Chem. Mater.* **1992**, *4*, 1380.
[220a] G. A. Ozin, *Adv. Mater.* **1992**, *4*, 612.
[220b] G. A. Ozin and S. Ozkar, *Adv. Mater.* **1992**, *4*, 11.
[221] K. Moller, T. Bein, S. Ozkar and G. A. Ozin, *J. Phys. Chem.* **1991**, *95*, 5276.
[222] G. A. Ozin, R. A. Prokopowicz and S. Ozkar, *J. Am. Chem. Soc.* **1992**, *114*, 8953.
[223] A. Borvornwattananont, K. Moller and T. Bein, *J. Phys. Chem.* **1992**, *96*, 6713.
[224] C. Huber, K. Moller and T. Bein, *J. Chem. Soc. Chem. Commun.* **1994**, 2619.
[225] D. J. Kippenberger, E. F. Vasant and J. H. Lunsford, *J. Catal.* **1974**, *35*, 447.
[226] G. A. Ozin and J. Godber, *J. Phys. Chem.* **1989**, *93*, 878.
[227] A. Borvornwattananont and T. Bein, *J. Phys. Chem.* **1992**, *96*, 9447.
[228] A. Zecchina, S. Bordiga, C. Otero-Arean and E. Escalona-Platero, *J. Mol. Catal.* **1991**, *70*, 43.
[229] C. Dossi, J. Schaefer and W. M. H. Sachtler, *J. Mol. Catal.* **1989**, *52*, 193.
[230] E. Fritsch, J. Heidrich, K. Pohlborn and W. Beck, *J. Organomet. Chem.* **1992**, *441*, 203.
[230a] W. A. Hermann, *Angew. Chem. Int. Ed. Engl.* **1988**, *27*, 1297.

[230b] W. A. Hermann, P. Kiprof, K. Rypdal, J. Tremmel, R. Blom, R. Alberto, J. Behm, R. W. Albach, H. Boch, B. Solouki, J. Mink, D. Lichtenberger and N. E. Gruhn, *J. Am. Chem. Soc.* **1991**, *113*, 6527.
[231] T. Bein, C. Huber, K. Moller, C. G. Wu and L. Xu, *Chem. Mater.* **1997**, *9*, 2252.
[231a] P. E. Riley and K. Seff, *J. Am. Chem. Soc.* **1973**, *95*, 8180.
[232] D. Ballivet-Tkatchenko and G. Coudurier, *Inorg. Chem.* **1979**, *18*, 558.
[233] T. Bein and P. A. Jacobs, *J. Chem. Soc. Faraday Trans. I*, **1983**, *79*, 1819.
[234] T. Bein and P. A. Jacobs, *J. Chem. Soc. Faraday Trans. I*, **1984**, *80*, 1391.
[235] S. L. Suib, A. Kostapapas, K. C. McMahon, J. C. Baxter and A. M. Winiecki, *Inorg. Chem.* **1985**, *24*, 858.
[236] T. Bein, F. Schmidt and P. A. Jacobs, *Zeolites*, **1985**, *5*, 240.
[237] T. Bein, G. Schmieser and P. A. Jacobs, *J. Phys. Chem.* **1986**, *90*, 4851.
[238] C. Bowers and P. K. Dutta, *J. Phys. Chem.* **1989**, *93*, 2596.
[239] P. K. Dutta and M. Borja, *Zeolites*, **1992**, *12*, 142.
[240] J. Pires, M. Brotas de Carvalho, F. R. Ribeiro and E. G. Derouane, *Microporous Mater.* **1995**, *3*, 573.
[241] L. Reven and E. Oldfield, *Inorg. Chem.* **1992**, *31*, 243.
[242] M. Iwamoto, S. Nakamura, H. Kusano and S. Kagawa, *J. Phys. Chem.* **1986**, *90*, 5244.
[243] A. Bovornwattananont, K. Moller and T. Bein, *J. Phys. Chem.* **1989**, *93*, 4205.
[244] K. Moller, A. Bovornwattananont and T. Bein, *J. Phys. Chem.* **1989**, *93*, 4562.
[245] D. Ballivet-Tkatchenko, N. D. Chau, H. Mozzanega, M. C. Roux and I. Tkatchenko, *ACS Symp. Ser.* **1981**, *152*, 187.
[246] T. A. Mitsudo, A. Ishihara and Y. Watanabe, *J. Mol. Catal.* **1987**, *40*, 119.
[247] J. J. Verdonk, R. A. Schoonheydt and P. A. Jacobs, *J. Phys. Chem.* **1983**, *87*, 683.
[248] W. R. Hastings, C. J. Cameron, M. J. Thomas and M. C. Baird, *Inorg. Chem.* **1988**, *27*, 3024.
[249] J. J. Bergmeister and B. E. Hanson, *Inorg. Chem.* **1990**, *29*, 4055.
[250] U. Kiizi, T. Venalainen, T. A. Pakkanen and O. Krause, *J. Mol. Catal.* **1991**, *64*, 163.
[251] M. Lenarda, R. Ganzerla, M. Graziani and R. Spogliarich, *J. Organomet. Chem.* **1985**, *290*, 213.
[252] M. Lenarda, J. Kaspar, R. Ganzerla, A. Trovarelli and M. Graziani, *J. Catal.* **1988**, *112*, 1.
[253] S. D. Maloney, P. L. Zhou, M. J. Kelley and B. C. Gates, *J. Phys. Chem.* **1991**, *95*, 5409.
[254] P. L. Zhou, S. D. Maloney and B. C. Gates, *J. Catal.* **1991**, *129*, 315.
[255] P. K. Dutta and M. A Thomson, *Chem. Phys. Lett.* **1986**, *131*, 435.
[256] G. Lemay, S. Kaliaguine, A. Adnot, S. Nahar, D. Kozak and J. Monnier, *Can. J. Chem.* **1986**, *64*, 1943.
[257] K. J. Balkus, Jr., and K. Nowinska, *Microporous Mater.* **1995**, *3*, 665.
[257a] M. J. MacLahlan, P. Aroca, N. Coombs, I. Manners and G. A. Ozin, *Advan. Mater.* **1998**, *10*, 144.
[257b] B. F. G. Johnson, S. A. Raynor, D. S. Shephard, T. Mashmeyer, J. M. Thomas, G. Sankar, S. Bromley, R. Oldroyd, L. Gladden and M. D. Mantle, *Chem. Commun.* **1999**, 1167.
[258] L. F. Nazar, G. A. Ozin, F. Hugues, J. Godber and D. Rancourt, *J. Mol. Catal.* **1983**, *21*, 313.
[258a] W. Zhou, J. M. Thomas, D. S. Shephard, B. F. G. Johnson, D. Ozkaya, T. Masachmeyer, R. G. Bell and Q. Ge, *Science* **1998**, *280*, 705.
[258b] D. S. Shephard, T. Maschmeyer, B. F. G. Johnson, J. M. Thomas, G. Sankar, D. Ozkaya, W. Zhou, R. D. Oldroyd and R. G. Bell, *Angew. Chem.* **1997**, *109*, 2337; *Angew. Chem. Int. Ed. Engl.* **1997**, *36*, 2242.
[259] R. L. Schneider, R. F. Howe and K. L. Watters, *Inorg. Chem.* **1984**, *23*, 4600.
[260] R. P. Zerger, K. C. McMahon, M. D. Seltzer, R. G. Michel and S. L. Suib, *J. Catal.* **1986**, *99*, 498.
[261] B. A. Morrow, M. I. Baraton, Y. Lijour and J. L. Roustan, *Spectrochim. Acta* **1987**, *43A*, 1583.
[262] T. P. Newcomb, P. G. Gopal and K. L. Waters, *Inorg. Chem.* **1987**, *26*, 809.
[263] L. Alvila, T. A. Pakkanen, T. T. Pakkanen and O. Krause, *J. Mol. Catal.* **1992**, *75*, 333.
[264] X. Li, G. A. Ozin and S. Ozkar, *J. Phys. Chem.* **1991**, *95*, 4463.

[265] C. Huber, K. Moller and T. Bein, *J. Phys. Chem.* **1994**, *98*, 12067.
[266] C. Huber, K. Moller, S. B. Ogunwumi and T. Bein, *J. Phys. Chem.* **1994**, *98*, 13651.
[267] P. E. Riley, K. B. Kunz and K. Seff, *J. Am. Chem. Soc.* **1975**, *97*, 537.
[268] K. J. Balkus, Jr. and S. Shepelev, *Microporous Mater.* **1993**, *1*, 383.
[269] K. J. Balkus, Jr., A. G. Gabrielov and S. Shepelev, *Microporous Mater.* **1995**, *3*, 489.
[270] D. F. Taylor, B. E. Hanson and M. E. Davis, *Inorg. Chim. Acta* **1987**, *128*, 55.
[271] M. E. Davis, J. Schnitzer, J. A. Rossin, D. F. Taylor and B. E. Hanson, *J. Mol. Catal.* **1987**, *39*, 243.
[272] H. Miessner, I. Burkhardt, D. Gutschik, A. Zecchina, C. Morterra and G. Spoto, *J. Chem. Soc. Faraday Trans.* **1990**, *86*, 2321.
[273] L. F. Rao, A. Fukuoka, N. Kosugi, H. Kuroda and M. Ichikawa, *J. Phys. Chem.* **1990**, *94*, 5317.
[274] L. Basini, R. Patrini, A. Aragno and B. C. Gates, *J. Mol. Catal.* **1991**, *70*, 29.
[275] T. J. Lee and B. C. Gates, *Catal. Lett.* **1991**, *8*, 15.
[276] T. T. Wong, A. Y. Stakheev and W. M. H. Sachtler, *J. Phys. Chem.* **1992**, *96*, 7733.
[277] A. Fukuoka, L. F. Rao, N. Kosugi, H. Kuroda and M. Ichikawa, *Appl. Catal.* **1989**, *50*, 295.
[278] N. Takahashi, A. Mijin, H. Suematsu, S. Shinohara and H. Matsuoka, *J. Catal.* **1989**, *117*, 348.
[279] J. M. Basset, A. Theolier, D. Commereuc and Y. Chauvin, *J. Organomet. Chem.* **1985**, *279*, 147.
[280] H. Miessner, *J. Chem. Soc. Chem. Commun.* **1994**, 927.
[281] A. Rahman, A. Adnot, G. Lemay, S. Kaliaguine and G. Jean, *Appl. Catal.* **1989**, *50*, 131.
[282] C. V. Rode, S. P. Gupta, R. V. Chaudhari, C. D. Pirozhkov and A. L. Lapidus, *J. Mol. Catal.* **1994**, *91*, 195.
[283] A. Corma, M. Iglesias, C. del Pino and F. Sanchez, *J. Organomet. Chem.* **1992**, *431*, 233.
[284] G. A. Ozin, D. M. Haddleton and C. J. Gil, *J. Phys. Chem.* **1989**, *93*, 6710.
[285] L. Crowfoot, G. A. Ozin and S. Ozkar, *J. Am. Chem. Soc.* **1991**, *113*, 2033.
[286] M. Ichikawa, L. F. Rao, T. Kimura and A. Fukuoka, *J. Mol. Catal.* **1990**, *62*, 15.
[287] T. N. Huang and J. Schwartz, *J. Am. Chem. Soc.* **1982**, *104*, 5244.
[288] D. R. Corbin, W. C. Seidel, L. Abrams, N. Herron, G. D. Stucky and C. A. Tolman, *Inorg. Chem.* **1985**, *24*, 1800.
[289] D. Olivier, M. Richard and M. Che, *Chem. Phys. Lett.* **1978**, *60*, 77.
[290] T. Bein, S. J. McLain, D. R. Corbin, R. D. Farlee, K. Moller, G. D. Stucky, G. Woolery and D. Sayers, *J. Am. Chem. Soc.* **1988**, *110*, 1801.
[291] K. I. Abad-Zade, M. I. Rustamov, V. M. Akhmedov and V. G. Mardanov, *Azerb. Neft. Khozyaistvo*, 1984, *11*, 48; *Chem. Abs.* **1985**, *102*, 134568.
[292] C. Dossi, R. Psaro, R. Ugo, Z. C. Zhang and W. M. H. Sachtler, *J. Catal.* **1994**, *149*, 92.
[293] J. R. Chang, Z. Xu, S. K. Purnell and B. C. Gates, *J. Mol. Catal.* **1993**, *80*, 49.
[294] H. Bischoff, N. I. Jaeger, G. Schulz-Ekloff and L. Kubelkova, *J. Mol. Catal.* **1993**, *80*, 95.
[295] S. Qiu, R. Ohnishi and M. Ichikawa, *J. Phys. Chem.* **1994**, *98*, 2719.
[296] V. Y. Borovkov and H. G. Karge, *J. Chem. Soc. Faraday Trans.* **1995**, *91*, 2035.
[297] M. S. Whittingham and A. J. Jacobson (Editors), *Intercalation Chemistry*, Academic Press, New York, **1982**.
[298] W. Müller-Warmuth and R. Schöllhorn (Editors), *Progress in Intercalation Research*, Kluwer, Dordrecht, **1994**.
[299] J. Rouxel in *Advances in the Synthesis and Reactivity of Solids*, T. E. Mallouk, Editor, JAI Press, London, **1994**, Vol. 2, p. 27.
[300] D. O'Hare in *Inorganic Materials*, D. W. Bruce and D. O'Hare (Editors), Wiley, New York, **1992**, p. 165.
[301] A. J. Jacobson, in vol. *Solid State Chemistry*, A. K. Cheetham and P. Day (Editors), Clarendon Press, Oxford, **1992**, p. 183.
[302] D. O'Hare and J. S. O. Evans, *Comments Inorg. Chem.* **1993**, *14*, 155.
[303] J. Rouxel in J. M. Lehn, J. L. Atwood, J. E. D. Davies, D. D. Mac Nicol and F. Vögtle (Editors), *Comprehensive Supramolecular Chemistry. Volume 7. Solid State Chemistry: Two and Three-dimensional Inorganic Networks*, G. Alberti and T. Bein, Volume Editors, Pergamon Press, Oxford, **1996**, p. 93.

[304] R. Schöllhorn, *Chem. Mater.* **1996**, *8*, 1747.
[305] M. S. Whittingham and L. B. Eberl, in *Intercalated Layered Materials*, Edited by E. Levy, D. Reidel Publ. Co., Dordrecht, **1979**, p. 533.
[306] M. B. Dines, *Science*, **1975**, *188*, 1210.
[306a] J. S. O. Evans, S. J. Price, H. V. Wong and D. O'Hare, *J. Am. Chem. Soc.* **1998**, *120*, 10837.
[307] B. G. Silbernagel, *Chem. Phys. Lett.* **1975**, *34*, 298.
[308] L. F. Nazar and A. J. Jacobson, *J. Chem. Soc. Chem. Commun.* **1986**, 570.
[309] S. J. Heyes, N. J. Clayden, C. M. Dobson, M. L. H. Green and P. J. Wiseman, *J. Chem. Soc. Chem. Commun.* **1987**, 1560.
[310] J. S. O. Evans and D. O'Hare, *Chem. Mater.* **1995**, *7*, 1668.
[311] B. Silbernagel, M. B. Dines, F. R. Gamble, L. A. Gebhard and M. S. Whittingham, *J. Chem. Phys.* **1976**, *65*, 1906.
[312] R. P. Clement, W. B. Davies, K. A. Ford, M. L. H. Green and A. J. Jacobson, *Inorg. Chem.* **1978**, *17*, 2754.
[313] D. O'Hare, W. Jaegerman, D. L. Williamson, F. S. Ohuchi and B. A. Parkinson, *Inorg. Chem.* **1988**, *27*, 1537.
[314] H. V. Wong, R. Millet, J. S. O. Evans, S. Barlw and D. O'Hare, *Chem. Mater.* **1995**, *7*, 210.
[315] L. Benes, J. Votinski, P. Lostak, J. Kalousova and J. Klikorka, *Phys. Status Solidi* **1985**, *89*, K1.
[316] D. O'Hare, H. V. Wong, S. Hazel and J. W. Hodby, *Adv. Mater.*, **1992**, *4*, 658.
[317] D. A. Cleary and D. R. Baer, *Chem. Mater.* **1992**, *4*, 112.
[318] H. V. Wong, J. S. O. Evans, S. Barlow and D. O'Hare, *J. Chem. Soc. Chem. Commun.* **1993**, 1589.
[319] C. A. Formstone, E. T. FitzGerald, D. O'Hare, P. A. Cox, M. Kurmoo, J. W. Hodby, D. Lillicrap and M. Goss-Custard, *J. Chem. Soc. Chem. Commun.* **1990**, 501.
[320] H. V. Wong, J. S. O. Evans, S. Barlow, S. J. Mason and D. O Hare, *Inorg. Chem.* **1994**, *33*, 5515.
[321] D. O'Hare, *Chem. Soc. Rev.* **1992**, *21*, 121.
[322] C. A. Formstone, M. Kurmoo, E. T. Fitzgerald, P. A. Cox, D. O'Hare, *J. Mater. Chem.* **1991**, *1*, 51.
[323] C. P. Grey, J. S. O. Evans, D. O'Hare and S. J. Heyes, *J. Chem. Soc. Chem. Commun.* **1991**, 1381.
[324] C. A. Formstone, E. T. FitzGerlad, P. A. Cox and D. O'Hare, *Inorg. Chem.* **1990**, *29*, 3860.
[325] D. O'Hare, J. S. O. Evans, C. K. Prout and P. J. Wiseman, *Angew. Chem. Int. Ed. Engl.* **1991**, *30*, 1156.
[326] W. B. Davies, M. L. H. Green and A. J. Jacobsen, *J. Chem. Soc. Chem. Commun.* **1976**, 781.
[327] K. Chatakondu, M. L. H. Green, M. E. Thompson and K. S. Suslik, *J. Chem. Soc. Chem. Commun.* **1987**, 900.
[328] J. O. Evans and D. O'Hare, *Adv. Mater.* **1995**, *7*, 163.
[329] D. J. Williams, *Angew. Chem. Int. Ed. Engl.* **1984**, *23*, 690.
[330] R. Clement and M. L. H. Green, *J. Chem. Soc. Dalton Trans.* **1979**, 1566.
[331] R. Clement, I. J. Girard and I. Morgenstern-Badarau, *Inorg. Chem.* **1980**, *19*, 2852.
[332] A. H. Reis, Jr., V. S. Hagley and S. W. Peterson, *J. Am. Chem. Soc.* **1977**, *99*, 4184.
[333] A. H. Reis, Jr. and S. W. Peterson, *Ann. N.Y. Acad. Sci.* **1978**, *313*, 560.
[334] A. Michalowicz and R. Clement, *J. Incl. Phenom.* **1986**, *4*, 265.
[335] R. Clement, O. Garnier and Y. Mathey, *Nouv. J. Chim.* **1982**, *6*, 13.
[336] C. Sourisseau, J. P. Forgerit, and Y. Mathey, *J. Phys. Chem. Solids*, **1983**, *44*, 119.
[337] J. P. Audiere, R. Clement, Y. Mathey and C. Mazieres, *Physica B*, **1980**, *99*, 133.
[338] D. O'Hare, J. S. O. Evans, P. A. Turner, S. M. Stephens, J. Heyes and J. Greenwood, *J. Mater. Chem.* **1995**, *5*, 1383.
[339] Y. Mathey, R. Clement, C. Sourisseau and G. Lucazeau, *Inorg. Chem.* **1980**, *19*, 2773.
[340] D. G. Clerq and D. A. Cleary, *Chem. Mater.* **1992**, *4*, 1344.
[341] S. J. Mason, S. J. Heyes and D. O'Hare, *J. Chem. Soc. Chem. Commun.* **1995**, 1657.
[342] C. Sourisseau, J. P. Forgerit and J. Mathey, *J. Solid State Chem.* **1983**, *49*, 134.
[343] K. Kim, D. J. Little and D. A. Cleary, *J. Phys. Chem.* **1990**, *94*, 3205.

[344] K. Kim and D. A. Cleary, *J. Phys. Chem.* **1990**, *94*, 3816.
[345] G. T. Long and D. A. Cleary, *J. Solid State Chem.* **1990**, *87*, 77.
[346] D. A. Cleary and A. H. Francis, *J. Phys. Chem.* **1985**, 89.
[347] B. Bal, S. Ganguli and M. Bhattacharya, *Physica B&C*, **1985**, 133.
[348] R. Clement, *J. Chem. Soc. Chem. Commun.* **1980**, 647.
[349] R. Clement, O. Garnier and J. Jegoudez, *Inorg. Chem.* **1986**, *125*, 1404.
[350] M. Doeuff, C. Cartier and R. Clement, *J. Chem. Soc. Chem. Commun.* **1988**, 629.
[351] R. Clement, J. P. Audiere and J. P. Renaud, *Rev. Chim. Miner.* **1982**, *19*, 560.
[352] C. Sourisseau, Y. Mathey and C. Poizignon, *Chem. Phys.* **1982**, *71*, 257.
[353] M. L. H. Green, J. Qin and D. O'Hare, *J. Organomet. Chem.* **1988**, *358*, 375.
[354] M. L. H. Green, J. Qin, D. O'Hare, H. E. Bunting, M. E. Thompson, S. R. Marder and K. Chatakondu, *Pure Appl. Chem.* **1989**, *51*, 817.
[355] H. Schaefer-Stahl and R. Abele, *Angew. Chem. Int. Ed. Engl.* **1980**, *19*, 477.
[356] H. Schaefer-Stahl and R. Abele, *Z. Anorg. Allg. Chem.* **1980**, *465*, 147.
[357] H. Schaefer-Stahl and R. Abele, *Mat. Res. Bull.* **1980**, *15*, 1157.
[358] H. Schaefer-Stahl, *Synth. Metals* **1981**, *4*, 65.
[359] H. Schaefer-Stahl, *Inorg. Nucl. Chem. Lett.* **1980**, *16*, 271.
[360] T. R. Halbert and J. C. Scanlon, *Mat. Res. Bull.* **1979**, *14*, 415.
[361] T. R. Halbert, D. C. Johnston, L. E. McCandlish, A. H. Thompson, J. C. Scanlon and J. A. Dumesic, *Physica B*, **1980**, *99*, 128.
[362] J. E. Phililips and R. H. Herber, *Inorg. Chem.* **1986**, *25*, 3081.
[363] P. Palvadeau, L. Coic and J. Rouxel, *Mater. Res. Bull.* **1981**, *16*, 1055.
[364] G. Villeneuve, P. Dordor, P. Palvadeau and J. P. Venien, *Mater. Res. Bull.* **1982**, *17*, 1407.
[365] P. Aldebert, N. Baffier, J. J. Legendre and J. Livage, *Rev. Chim. Miner.* **1982**, *19*, 485.
[366] K. Chatakondu, G. Villeneuve, L. Fournes and H. Smith, *Mat. Res. Bull.* **1992**, *27*, 357.
[367] E. Rodriguez-Castellon, A. Jimenez-Lopez, M. Martinez-Lara and L. Moreno-Real, *J. Incl. Phenom.* **1987**, *5*, 335.
[368] G. E. Matsubayashi, S. Ohta and S. Okuno, *Inorg. Chim. Acta* **1991**, *184*, 47.
[368a] S. Okuno and G. E. Matsubayashi, *J. Chem. Soc. Dalton Trans.* **1992**, 2441.
[368b] A. Datta, S. Bhaduri, R. Y. Kelkar and H. I. Khwaja, *J. Phys. Chem.* **1994**, *98*, 11811.
[369] G. Matsubayashi and S. Ohta, Chem. Lett. 1990, 787.
[369a] S. Okuno and G. Matsubayashi, *Chem. Lett.* **1993**, 799.
[370] W. Johnson, *J. Chem. Soc. Chem. Commun.* **1980**, 263.
[371] P. Aldebert and V. Paul-Boncour, *Mat. Res. Bull.* **1983**, *18*, 1263.
[372] K. Chatakondu, M. L. H. Green, J. Qin, M. E. Thompson and P. J. Wiseman, *J. Chem. Soc. Chem. Commun.* **1988**, 223.
[372a] M. Clemente-Leon, E. Coronado, J. R. Galan-Mascaros and C. J. Gomez-Garcia, *Chem. Commun.* **1997**, 1727.
[373] H. G. Richey, Jr., in J. M. Lehn, J. L. Atwood, J. E. D. Davies, D. D. MacNicol and F. Vögtle, Editors, *Comprehensive Supramolecular Chemistry. Volume 1. Molecular Recognition: Receptors for cationic Guests*, G. W. Gokel, Volume Editor, Pergamon Press, Oxford, **1996**, Chapter 21, p. 755.
[374] R. D. Rogers, *ibidem*, p. 315.
[375] P. R. Markies, T. Nomoto, O. S. Akkerman, F. Bickelhaupt, W. J. Smeets and A. L. Spek, *Adv. Organomet. Chem.* **1991**, *32*, 147.
[376] G. J. M. Gruter, G. P. M. van Klink, O. S. Akkerman and F. Bickelhaupt, *Chem. Rev.* **1995**, *95*, 2405.
[377] A. D. Pajerski, G. L. Bergstresser, M. Parvez and H. G. Richey, Jr., *J. Am. Chem. Soc.* **1988**, *110*, 4844.
[378] P. R. Markies, T. Nomoto, O. S. Akkerman, F. Bickelhaupt, W. J. J. Smeets and A. L. Spek, *J. Am. Chem. Soc.* **1988**, *110*, 4845.
[379] H. G. Richey, Jr. and D. M. Kushlan, *J. Am. Chem. Soc.* **1987**, *109*, 2510.
[379a] J. E. Chubb and H. G. Richey, Jr., *Organometallics* **1998**, *17*, 3208.
[380] G. J. M. Gruter, F. J. J. de Kanter, P. R. Markies, T. Nomoto, O. S. Akkerman and F. Bickelhaupt, *J. Am. Chem. Soc.* **1993**, *115*, 12179.
[381] E. P. Squiller, R. R. Whittle and H. G. Richey, Jr., *J. Am. Chem. Soc.* **1985**, *107*, 432.

[382] E. P. Squiller, R. R. Whittle and H. G. Richey, Jr., *Organometallics*, **1985**, *4*, 1154.
[383] A. D. Pajerski, M. Parvez and H. G. Richey, Jr., *J. Am. Chem. Soc.* **1988**, *110*, 2660.
[384] A. V. Bogatski, T. K. Chumachenko, N. G. Lukánenko, L. N. Lyamtseva and I. A. Starovoit, *Doklady Akad. Nauk SSSR*, **1980**, *251*, 113; *Doklady-Chemistry (Engl. transl.)* **1980**, *251*, 105.
[385] R. M. Fabicon, A. D. Pajerski and H. G. Richey, Jr., *J. Am. Chem. Soc.* **1993**, *115*, 9333.
[386] H. G. Richey, Jr. and B. A. King, *J. Am. Chem. Soc.* **1982**, *104*, 4672.
[387] F. Bickelhaupt, *Acta Chem. Scand.* **1992**, *46*, 409.
[388] P. R. Markies, O. S. Akkerman, F. Bickelhaupt, W. J. J. Smeets and A. L. Spek, *J. Am. Chem. Soc.* **1988**, *110*, 4284.
[389] P. R. Markies, T. Nomoto, G. Schat, O. S. Akkerman, F. Bickelhaupt, W. J. J. Smeets and A. L. Spek, *Angew. Chem.* **1988**, *100*, 1143; *Angew. Chem. Int. Ed. Engl.* **1988**, *27*, 1084.
[390] P. R. Markies, T. Nomoto, G. Schat, O. S. Akkerman, F. Bickelhaupt, W. J. J. Smeets and A. L. Spek, *Organometallics*, **1991**, *10*, 2535.
[391] P. R. Markies, G. Schat, O. S. Akkerman, F. Bickelhaupt, W. J. J. Smeets and A. L. Spek, *Organometallics*, **1991**, *10*, 3538.
[392] G. J. M. Gruter, O. S. Akkerman, F. Bickelhaupt, W. J. J. Smeets and A. L. Speek, *Rec. Trav. Chim. Pays-Bas* **1993**, *112*, 425.
[393] R. M. Fabicon, A. D. Pajerski and H. G. Richey, Jr., *J. Am. Chem. Soc.* **1991**, *113*, 6680.
[394] K. Onan, J. Rebek, Jr., T. Costello and L. Marshall, *J. Am. Chem. Soc.* **1983**, *105*, 6759.
[395] J. Rebek, Jr., T. Costello, L. Marshall, R. Whattley, R. C. Gadwood and K. Onan, *J. Am. Chem. Soc.* **1985**, *107*, 7481.
[396] J. Rebek, Jr., and L. Marshall, *J. Am. Chem. Soc.* **1983**, *105*, 6668.
[397] J. Rebek, Jr., S. V. Luis and L. Marshall, *J. Am. Chem. Soc.* **1986**, *108*, 5011.
[398] S. V. Luis, M. I. Burguete and R. L. Salvador, *J. Incl. Phenom.* **1991**, *10*, 341.
[399] P. R. Markies, A. Villena, O. S. Akkerman, F. Bickelhaupt, W. J. J. Smeets and A. L. Spek, *J. Organomet. Chem.* **1993**, *463*, 7.
[400] J. L. Atwood, in *Inclusion Compounds*, J. L. Atwood, J. E. D. Davies and D. D. MacNicol, Editors, Academic Press, London, **1984**, Vol. 1, Chapter 5.
[401] G. H. Robinson, *Coord. Chem. Rev.* **1992**, *112*, 227.
[402] G. H. Robinson, Editor, *Coordination Compounds of Aluminum*, VCH New York, **1993**, Chapter 2.
[403] G. H. Robinson, W. E. Hunter, S. G. Bott and J. L. Atwood, *J. Organomet. Chem.* **1987**, *326*, 9.
[404] A. D. Pajerski, T. P. Cleary, M. Parvez, G. W. Gokel and H. G. Richey, Jr., *Organometallics* **1992**, *11*, 1400.
[405] M. F. Self, W. T. Pennington, J. A. Laske and G. H. Robinson, *Organometallics* **1991**, *10*, 36.
[406] Q. Zhao, H. Sun, W. Chen, C. Duan, Y. Liu, Y. Pan and X. You, *Organometallics* **1998**, *17*, 156.
[407] G. H. Robinson, S. G. Bott, H. Elgamal, W. E. Hunter and J. L. Atwood, *J. Incl. Phenom.* **1985**, *3*, 65.
[408] J. L. Atwood, D. C. Hrncir, R. Shakir, M. S. Dalton, R. D. Priester and R. D. Rogers, *Organometallics* **1982**, *1*, 1021.
[409] S. G. Bott, A. Alvanipour, S. D. Morley, D. A. Atwood, C. M. Means, A. W. Coleman and J. L. Atwood, *Angew. Chem.* **1987**, *99*, 476; *Angew. Chem. Int. Ed. Engl.* **1987**, *26*, 485.
[410] J. L. Atwood, R. D. Prieste, R. D. Rogers and L. G. Canada, *J. Incl. Phenom.* **1983**, *1*, 61.
[411] H. Zhang, C. M. Means and J. L. Atwood, *J. Cryst. Spec. Res.* **1985**, *15*, 445.
[412] J. L. Atwood, H. Elgamal, G. H. Robinson, S. G. Bott, J. A. Weeks and W. E. Hunter, *J. Incl. Phenom.* **1984**, *2*, 367.
[413] S. G. Bott, H. Elgamal and J. L. Atwood, *J. Am. Chem. Soc.* **1985**, *107*, 1796.
[414] H. G. Richey, Jr., and G. L. Bergstresser, *Organometallics* **1988**, *7*, 1459.
[415] M. T. Reetz, B. M. Johnson and K. Harms, *Tetrahedron Lett.* **1994**, *35*, 2525.
[416] G. H. Robinson, W. T. Pennington, B. Lee, M. F. Self and D. C. Hrncir, *Inorg. Chem.* **1991**, *30*, 809.
[417] G. H. Robinson, H. Zhang and J. L Atwood, *J. Organomet. Chem.* **1987**, *331*, 153.

[418] F. Moize, W. T. Pennington, G. H. Robinson and S. A. Sangokoyu, *Acta Cryst.* **1990**, *46C*, 1110.
[419] G. H. Robinson and S. A. Sangokoya, *J. Am. Chem. Soc.* **1987**, *109*, 6852.
[420] G. H. Robinson and S. A. Sangokoya, *J. Am. Chem. Soc.* **1988**, *110*, 1494.
[421] G. H. Robinson, H. Zhang and J. L. Atwood, *Organometallics* **1987**, *6*, 887.
[422] B. Lee, W. T. Pennington and G. H. Robinson, *Organometallics* **1990**, *9*, 1709.
[423] B. Lee, W. T. Pennington, G. H. Robinson and R. D. Rogers, *J. Organomet. Chem.* **1990**, *396*, 269.
[424] K. Henrick, R. W. Mathews, B. L. Podejma and P. A. Tasker, *J. Chem. Soc. Chem. Commun.* **1982**, 118.
[425] J. Crowder, K. Henrick, R. W. Mathews and B. L. Podejma, *J. Chem. Res. (S)* **1983**, 82.
[426] K. Kobiro, S. Takada, Y. Odaira and Y. Kawasaki, *J. Chem. Soc. Dalton Trans.* **1986**, 1767.
[427] K. Kobiro, Y. Odaira, Y. Kawasaki, Y. Kai and N. Kasai, *J. Chem. Soc. Dalton Trans.* **1986**, 2613.
[428] D. L. Hughes and M. R. Truter, *J. Chem. Soc. Chem. Commun.* **1982**, 727.
[429] D. L. Hughes and M. R. Truter, *Acta Cryst.* **1983**, *39B*, 329.
[430] P. R. Markies, Thesis, Free Univ. Amsterdam, **1990**, cited by G. J. M. Gruter *et al*, *Chem. Rev.* **1995**, *95*, 2405.
[431] M. Schafer, J. Pebler and K. Dehnicke, *Z. Anorg. Allg. Chem.* **1992**, *611*, 149.
[432] A. Alvanipour, J. L. Atwood, S. G. Bott, P. C. Junk, V. H. Kynast and H. Prinz, *J. Chem. Soc. Dalton Trans.* **1998**, 1223.
[433] M. N. Bell, A. J. Blake, M. Schroder and T. A. Stephenson, *J. Chem. Soc. Dalton Trans.* **1986**, 471.
[433a] M. M. Amini, A. L. Rheingold, R. W. Taylor and J. J. Zuckerman, *J. Am. Chem. Soc.* **1984**, *106*, 7289.
[433b] M. M. Amini, J. J. Zuckerman, A. L. Rheingold and S. W. Ng, *Z. Kristallogr.* **1994**, *209*, 682.
[433c] M. M. Amini, J. J. Zuckerman, A. L. Rheingold and S. W. Ng, *Z. Kristallogr.* **1994**, *209*, 613.
[433d] G. P. A. Yap, M. M. Amini, S. W. Ng, A. E. Counterman and A. L. Rheingold, *Main Group Chem.* **1996**, *1*, 359.
[434] A. J. Amoroso, A. M. Arif and J. A. Gladysz, *Organometallics* **1997**, *16*, 6032.
[435] H. M. Colquhoun and J. F. Stoddart, *J. Chem. Soc. Chem. Commun.* **1981**, 612.
[436] H. M. Colquhoun, S. M. Doughty, J. F. Stoddart and D. J. Williams, *Angew. Chem. Int. Ed. Engl.* **1984**, *23*, 325.
[437] H. M. Colquhoun, S. M. Doughty, J. F. Stoddart, A. M. Z. Slawin and D. J. Williams, *J. Chem. Soc. Dalton Trans.* **1986**, 1639.
[438] D. R. Alston, A. M. Z. Slawin, J. F. Stoddart, D. J. Williams and R. Zarzycki, *Angew. Chem. Int. Ed. Engl.* **1987**, *26*, 692.
[439] J. F. Stoddart and R. Zarzycki, *Rec. Trav. Chim. Pays-Bas*, **1988**, *107*, 515.
[440] M. L. Bender, *Cyclodextrin Chemistry*, Springer, New York, **1978**.
[440a] J. Szeitli, *Cyclodextrins and their Inclusion Complexes*, Akadémiai Kiadó, Budapest, **1982**.
[441] S. Li and W. C. Purdy, *Chem. Rev.* **1992**, *92*, 1457.
[442] G. Wenz, *Angew. Chem.* **1994**, *106*, 851; *Angew. Chem. Int. Ed. Engl.* **1994**, *33*, 803.
[442a] J. Szeitli, *Chem. Rev.* **1998**, *98*, 1743.
[443] E. Fenyvesi, L. Szente, N. R. Russell and M. McNamara, in J. M. Lehn, J. L. Atwood, J. E. D. Davies, D. D. MacNicol and F. Vögtle, Editors, *Comprehensive Supramolecular Chemistry. Volume 3. Cyclodextrins*, J. Szeitli and T. Osa, Volume Editors, Pergamon Press, Oxford, **1996**, Chapter 3, p. 305.
[443a] S. A. Nepogodiev and J. F. Stoddart, *Chem. Rev.* **1998**, *98*, 1959.
[444] B. Sigel and R. Breslow, *J. Am. Chem. Soc.* **1975**, *97*, 6869.
[445] M. F. Czarniecki and R. Breslow, *J. Am. Chem. Soc.* **1978**, *100*, 7771.
[446] R. Breslow, M. F. Czarniecki, J. Emert and H. Hamaguchi, *J. Am. Chem. Soc.* **1980**, *102*, 762.
[447] G. L. Trainor and R. Breslow, *J. Am. Chem. Soc.* **1981**, *103*, 154.

[448] R. Breslow, G. L. Trainor and A. Ueno, *J. Am. Chem. Soc.* **1983**, *105*, 2739.
[449] W. J. LeNoble, S. Srivastava, R. Breslow and G. L. Trainor, *J. Am. Chem. Soc.* **1983**, *105*, 2745.
[450] A. Harada and T. Takahashi, *J. Chem. Soc. Chem. Commun.* **1984**, 645.
[451] A. Harada and T. Takahashi, *Chem. Lett.* **1984**, 2089.
[452] A. Harada and T. Takahashi, *J. Incl. Phenom.* **1984**, *2*, 791.
[453] T. Matsue, D. H. Evans, T. Osa and N. Kobayashi, *J. Am. Chem. Soc.* **1985**, *107*, 3411.
[454] A. Ueno, F. Moriwaki, T. Osa, F. Hamada and K. Murai, *Tetrahedron Lett.* **1985**, *26*, 899.
[455] N. Kobayashi and T. Osa, *Chem. Lett.* **1986**, 421.
[456] Y. Maeda, N. Ogawa and Y. Takashima, *J. Chem. Soc. Dalton Trans.* **1987**, 627.
[457] N. J. Claydon, C. M. Dobson, S. J. Heyes and P. J. Wiseman, *J. Incl. Phenom.* **1987**, *5*, 65.
[458] A. Harada, Y. Hu, S. Yamamoto and S. Takahashi, *J. Chem. Soc. Dalton Trans.* **1988**, 729.
[458a] V. I. Sokolov, *Metalloorg. Khim. (Russ.)* **1988**, *1*, 25.
[459] H. J. Thiem, M. Brandl and R. Breslow, *J. Am. Chem. Soc.* **1988**, *110*, 8612.
[459a] V. I. Sokolov, V. L. Bondareva and I. F. Golovaneva, *J. Organomet. Chem.* **1988**, *358*, 401.
[460] A. D. Ryabov, E. M. Tyapochkin, E. D. Varfolomeev and A. A. Karyakin, *J. Electroanal. Chem.* **1990**, *24*, 257.
[461] B. Klingert and G. Rihs, *J. Chem. Soc. Dalton Trans.* **1991**, 2749.
[462] V. V. Strelets, L. A. Mamedjarova, M. N. Nefedova, N. I. Pynograeva, V. I. Sokolov, L. Pospizil and J. Hanzlik, *J. Electroanal. Chem.* **1991**, *310*, 179.
[462a] A. D. Ryabov, *Angew. Chem.* **1991**, *103*, 945; *Angew. Chem. Int. Ed. Engl.* **1991**, *30*, 931.
[463] B. Klingert and G. Rihs, *J. Incl. Phenom.* **1991**, *10*, 255.
[464] R. Isnin, C. Salam, and A. E. Kaifer, *J. Org. Chem.* **1991**, *56*, 35.
[465] S. McCormack, N. R. Russel and J. F. Cassidy, *Electrochim. Acta* **1992**, *37*, 1939.
[466] R. M. G. Roberts and J. F. Warmsley, *J. Organomet. Chem.* **1991**, *405*, 357.
[467] J. A. Imonigie and D. H. Macaetney, *Inorg. Chim. Acta* **1994**, *225*, 51 A.
[468] L. A. Godinez, S. Patel, C. M. Criss and A. E. Kaifer, *J. Phys. Chem.* **1995**, *99*, 17449.
[469] Y. Wang, S. Mendoza and A. E. Kaifer, *Inorg. Chem.* **1998**, *37*, 317.
[470] A. Harada, K. Saeki and S. Takahashi, *Chem. Lett.* **1985**, 1157.
[471] A. Harada, K. Saeki and S. Takahashi, *Organometallics* **1989**, *8*, 730.
[471a] S. Aime, H. C. Canuto, R. Gobetto and F. Napolitano, *Chem. Commun.* **1999**, 281.
[472] D. E. Eaton, A. G. Anderson, W. Tamm and Y. Wang, *J. Am. Chem. Soc.* **1987**, *109*, 1886.
[473] M. Shimada, A. Harada and S. Takahashi, *J. Chem. Soc. Chem. Commun.* **1991**, 263.
[474] G. Meister, H. Steoecker-Evans and G. Süss-Fink, *J. Organomet. Chem.* **1993**, *453*, 249.
[475] L. X. Song, Q. J. Meng and X. Z. You, *Chinese Chem. Lett.* **1994**, 1047.
[476] L. X. Song, Q. J. Meng and X. Z. You, *Chinese Chem. J.* **1995**, *13*, 311.
[477] L. X. Song, Q. J. Meng and X. Z. You, *J. Organomet. Chem.* **1995**, *498*, C1.
[477a] L. X. Song, Q. J. Meng and X. Z. You, *Syn. React. Inorg. Metal-org. Chem.* **1995**, *25*, 671.
[478] C. Diaz and A. Aranciba, *Bol. Soc. Chile. Chim.* **1996**, *41*, 291.
[479] C. Diaz and A. Aranciba, *J. Incl. Phenom.* **1998**, *30*, 127.
[480] A. Harada, M. Takeukchi and S. Takahashi, *Chem. Lett.* **1986**, 1893.
[481] D. R. Alston, A. M. Z. Slawin, J. F. Stoddart and D. J. Williams, *Angew. Chem.* **1985**, *97*, 771; *Angew. Chem. Int. Ed. Engl.* **1985**, *24*, 786.
[482] A. Harada and S. Takahashi, *J. Chem. Soc. Chem. Commun.* **1986**, 1229.
[483] P. R. Ashton, J. F. Stoddart and R. Zarzycki, *Tetrahedron Lett.* **1988**, *29*, 2103.
[484] J. F. Stoddart and R. Zarzycki, *Rec. Trav. Chim. Pays-Bas* **1988**, *107*, 515.
[485] M. Shimada, A. Harada and S. Takahashi, *J. Organomet. Chem.* **1992**, *428*, 199.
[485a] M. Shimada, Y. Morimoto and S. Takahashi, *J. Organomet. Chem.* **1993**, *443*, C8.
[486] F. Menger and M. J. Sherrod, *J. Am. Chem. Soc.* **1988**, *110*, 8606.
[487] Y. Odaki, K. Hirotsu, T. Higuchi, A. Harada and S. Takahashi, *J. Chem. Soc. Perkin Trans. I* **1990**, 1230.
[488] B. Klingert and G. Rihs, *Organometallics* **1990**, *9*, 1136.
[489] A. Harada, K. Saeki and S. Takahashi, *J. Incl. Phenom.* **1987**, *5*, 601.
[490] T. Matsue, T. Kato, U. Akiba and T. Osa, *Chem. Lett.* **1986**, 843.
[491] T. Matsue, M. Suda, I. Uchita, T. Kato, U. Akiba and T. Osa, *J. Electroanal. Chem.* **1987**, *234*, 163.

[491a] B. Gonzalez, C. M. Casado, B. Alonso, I. Cuadrado, M. Morán, Y. Wang and A. E. Kaifer, *Chem. Commun.* **1998**, 2569.
[491b] A. E. Kaifer, *Acc. Chem. Res.* **1999**, *32*, 62.
[492] R. Clement, R. Claudes and C. Mezieres, *J. Chem. Soc. Chem. Commun.* **1974**, 654.
[493] R. E. Brazak and A. D. Barrone, *Chem. Lett.* **1975**, 75.
[494] A. N. Nesmeyanov, G. B. Shulpin and M. G. Rybinskaya, *Doklady Akad. Nauk SSSR* **1975**, *221*, 624.
[495] E. Hough and D. G. Nicholson, *J. Chem. Soc. Dalton Trans.* **1978**, 15.
[496] T. C. Gibb, *J. Phys. C, Solid State Phys.* **1976**, *9*, 2627.
[497] R. Clement, M. Gourdji and L. Guibe, *Chem. Phys. Lett.* **1980**, *72*, 466.
[498] M. Sorai, K. Ogasahara and H. Suga, *Mol. Cryst. Liq. Cryst.* **1981**, *73*, 231.
[499] T. Nakai, T. Terao, F. Imashiro and A. Saika, *Chem. Phys. Lett.* **1986**, *132*, 554.
[500] M. Sorai, and Y. Shiomi, *Thermochim. Acta* **1986**, *109*, 29.
[501] M. D. Lowery, R. J. Wittebort, M. Sorai and D. N. Hendrickson, *J. Am. Chem. Soc.* **1990**, *112*, 4214.
[502] S. T. Heyes, N. J. Clayden and C. M. Dobson, *J. Phys. Chem.* **1991**, *95*, 1547.
[503] W. Tam, D. F. Eaton, J. C. Calabrese, I. D. Williams, Y. Wang and A. G. Anderson, *Chem. Mater.* **1989**, *1*, 128.
[504] A. G. Anderson, J. C. Calabrese, W. Tam and I. D. Williams, *Chem. Phys. Lett.* **1987**, *134*, 392.

3 Supramolecular Self-Assembly by Dative Bonding (Electron-Pair Donor–Acceptor or Lewis Acid–Base Interactions)

3.1 Group 12 Metals – Zinc, Cadmium, Mercury

Among the Group 12 metals, zinc and cadmium compounds have the tendency to self-organize in supramolecular structures through intermolecular dative bonds. In these compounds the metal sites are strong Lewis acids and compounds with halogen, oxygen, nitrogen, and sulfur donor sites in the molecule (Lewis basic sites) are self-complementary and undergo intermolecular association, leading to an increased metal coordination number. Mercury is a 'soft' metal and tends to form weaker and longer donor–acceptor intermolecular bonds, which can be in most cases regarded as secondary bonds, in the sense discussed in Section 1.2.2. In this chapter, therefore, we will deal only with zinc and cadmium, leaving mercury for Chapter 4, where some borderline cases will be discussed together with authentic secondary bond self-assembled structures.

3.1.1 Self-assembly of organozinc compounds

3.1.1.1 Organozinc halides

According to molecular weight determinations, unsolvated organozinc halides, e.g. EtZnCl and EtZnBr, are associated in organic solution, with the formation of tetrameric species [1]. Little is known about the nature of these associates.

In the solid state, apparently only one structure has been elucidated by X-ray diffraction. Thus, ethylzinc iodide, EtZnI, was found to contain polymeric chains, **1**, in which each zinc atom forms two long (2.91 Å) and one short (2.64 Å) zinc–iodine bonds [2]. This leads to the formation of Zn_3I_3 quasi-rings, which are interconnected in long chains.

A dodecameric macrocyclic compound $[ClZnCHMePEt_2N(SiMe_3)]_{12}$ consisting of six units, shown in **2**, is formed by zinc–halogen bridging (Zn–Cl 2.395 Å) [3].

3.1.1.2 Organozinc–oxygen compounds

Unsolvated organozinc compounds of the type RZnOR' are associated in solution and the formation of trimeric aggregates has been established by cryoscopic measurements for EtZnOCHPh$_2$ and [RZnOC$_6$H$_4$NMe$_2$]$_3$ in benzene solution [4, 5]. The structures of these trimers are unknown (although presumably cyclic) and X-ray crystal structure determinations of a significant number of solid state compounds reveal a much more complex reality, with dimers, trimers, and tetramers being identified.

Several dimeric organozinc alkoxides containing a four-membered Zn$_2$O$_2$ planar ring have been described. Thus, ethylzinc (2,6-*tert*-butyl)phenolate, EtZnO(2,6-Bu^{t}C$_6$H$_3$) contains a planar four-membered ring with alternating Zn–O bond lengths (1.970 and 1.990 Å) [6]. Not surprisingly, other organozinc alkoxides with bulky substituents, such as Me$_3$SiCH$_2$ZnOC$_6$H$_3$Pr^i$_2$ (Zn–O 1.944 and 1.975 Å) and Me$_3$SiCH$_2$ZnOC$_6$H$_2$Bu^t$_3$ (Zn–O 1.958 and 2.021 Å) are also cyclic dimers in the solid state [7]. In all these compounds the metal atom and the oxygens are tri-coordinate planar. Surprisingly, a trinuclear zinc alkoxide was also found to contain four-membered zinc–oxygen rings, rather than an expected six-membered ring. Thus, the compound [Zn$_3${O(2,6-Pr^i$_2$C$_6$H$_3$)}$_4$(CH$_2$SiMe$_3$) $_2$] contains a central four-coordinate (tetrahedral) zinc atom bridged by aryloxy groups to two terminal tricoordinate zinc atoms [7]. The organozinc hydroxide, $(Me_2PhSi)_3CZnOH$, is also dimeric, with OH bridges between zinc atoms, and contains a planar Zn_2O_2 ring [8].

Four-coordinate zinc is also present in the structurally characterized anion $[Et_2Zn(\mu-OBu^t)_2ZnEt_2]^{2-}$ (potassium salt), **3**. The Zn–O bonds are basically identical (2.091 and 2.097 Å), the rhomboid Zn_2O_2 ring is planar (O–Zn–O 79.6° and Zn–O–Zn 100.4°) and two potassium cations are situated above and below the ring plane [9].

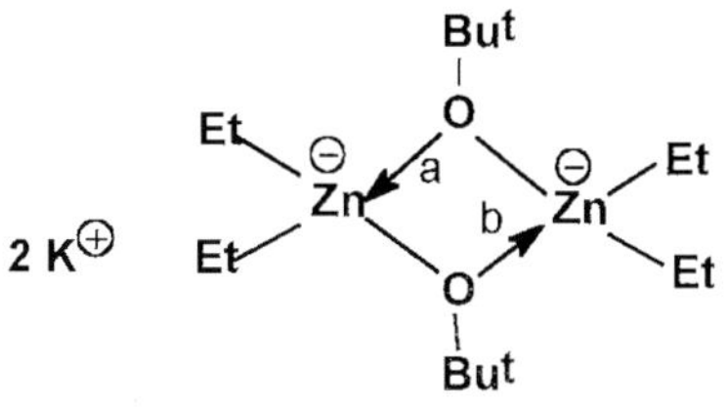

a = 2.097 Å
b = 2.091 Å

3

Derivatives of aminoalcohols, in which chelate rings are formed by nitrogen donation, are also dimers formed by O → Zn donation, and several examples, e.g. **4**, are known [10–12].

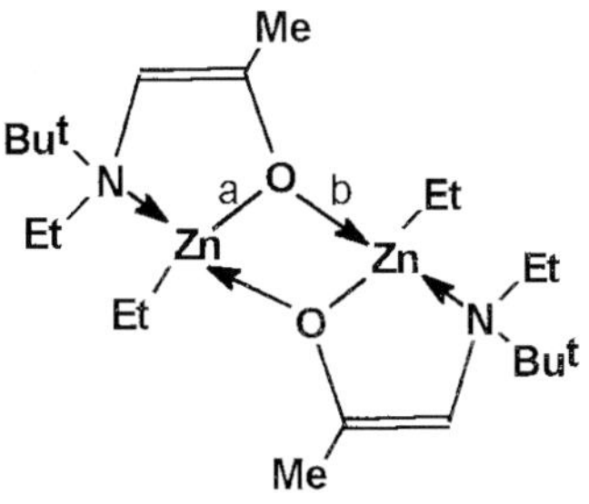

a = 2.02 Å
b = 2.12 Å

4

The ethylzinc derivative of 2-(1-methyliminoethyl)phenol is also a dimer with non-equivalent Z–O bonds (Zn–O 1.965 and 2.064 Å) [13].

The structure of bis[phenyl(acetylacetonato)zinc][bis(acetylacetonato)zinc], $2PhZn(acac)\cdot Zn(acac)_2$, **5**, also contains a Zn_2O_2 ring [14, 15] as does the ethylzinc derivative of pivaloylacetone [16].

Six-membered zinc–oxygen rings are rare – only one example has been authenticated. Pyrazolylboratozinc hydroxide, $[\{\eta^2-H_2B(3-Bu^tPz)_2\}Zn(\mu-OH)\}]_3$, is a cyclic trimer, **6**, with four-coordinate zinc atoms and alternating zinc–oxygen bond lengths (1.90 and 1.97 Å) [17].

Tetranuclear zinc–oxygen aggregates seem to be preferred and they occur in various structural variants. The cubane structure **7** has been identified in the solid

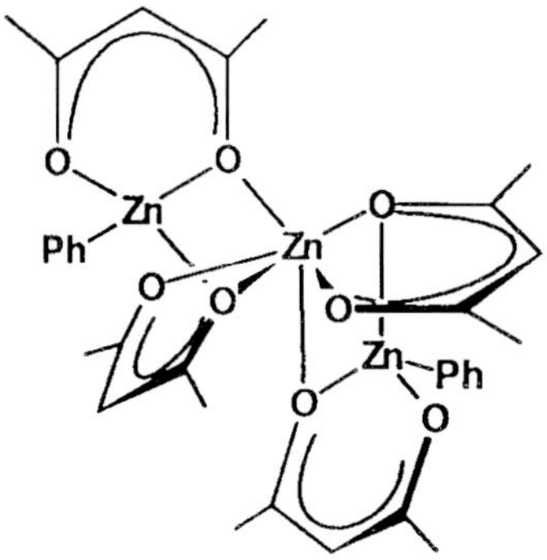

5

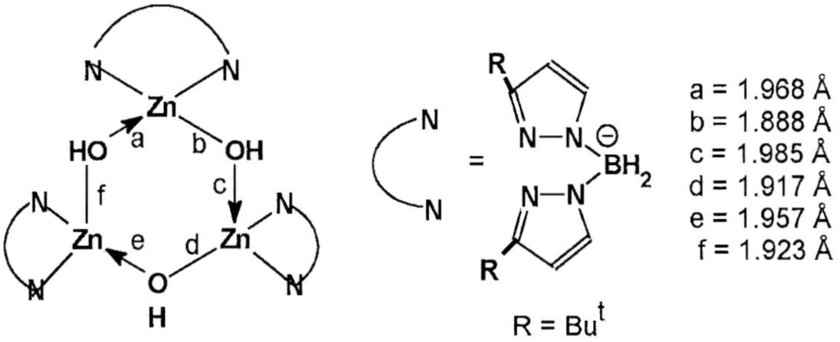

6

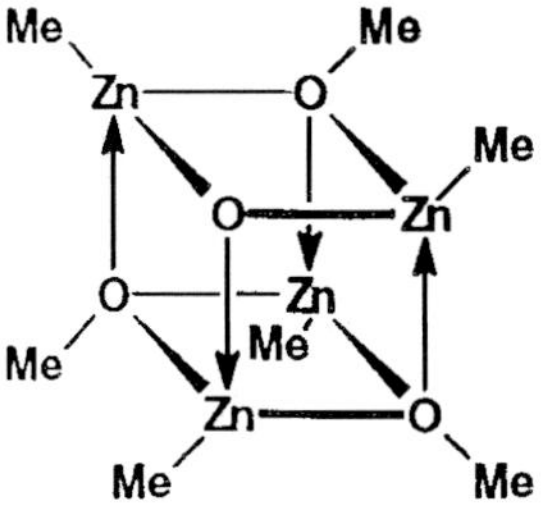

a = 1.968 Å
b = 1.888 Å
c = 1.985 Å
d = 1.917 Å
e = 1.957 Å
f = 1.923 Å

R = Bu^t

7

state structures of methylzinc alkoxides [MeZnOMe]$_4$ [18], [MeZnOBu^t]$_4$ [19], and [MeZnOSiMe$_3$]$_4$ [20]. Molecular weight determinations in benzene solution also indicate tetrameric aggregation for [MeZnOR]$_4$ (R = Me, Ph, and Bu^t) [21]. A similar structure has also been reported for [Me$_3$SiCH$_2$ZnOAd]$_4$ (Ad = 1-adamantyl), which is formed despite the bulky substituents [22]. Dicubane structures, **8**, containing a six-coordinate zinc atom and six four-coordinate zinc atoms were found in Me$_6$Zn$_7$(OMe)$_8$ [23, 24] and [(EtZnOMe)$_6$Zn(OMe)$_2$] [25].

Another type of tetrameric aggregate, containing two six-coordinate and two

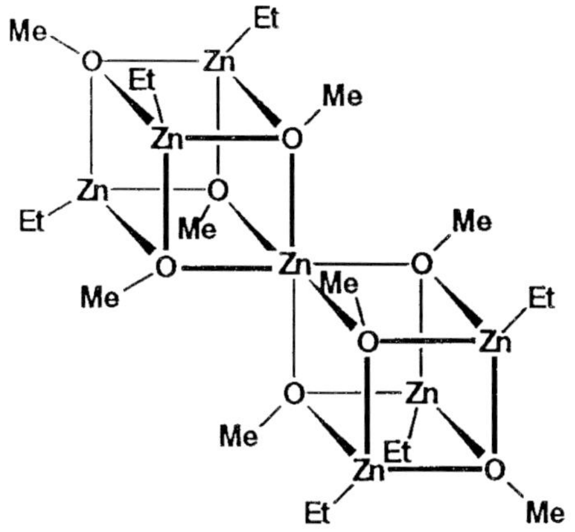

8

four-coordinate zinc atoms, **9**, is formed with pivaloylacetone (= Pac) and RZnOR. Thus, the phenyl derivative [PhZnOPh·Zn(Pac)$_2$]$_2$ contains a pentacyclic zinc–oxygen core, held together by OPh and diketonato bridges [26]. In benzene solution the tetranuclear species is in equilibrium with a dinuclear species formed by dissociation of the tetranuclear compound. A similar structure was found for [EtZnOMe·Zn(Pac)$_2$]$_2$ [16]. This tetranuclear structure can be described as consisting of two incomplete cubane units sharing one face.

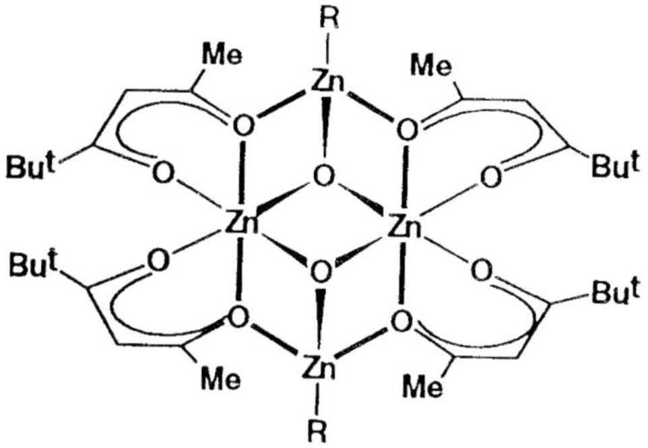

9

A monocyclic eight-membered ring system, **10**, representing another variant of tetranuclear aggregates, has been identified in an aminovinyl alcohol derivative of composition [EtZnOC(OMe)=CHNBu^{t}Me]$_4$. All zinc atoms are four-coordinate and a tendency of Zn–O bond length alternation is observed in the eight-membered ring [27].

In hydrocarbon solution the tetramer **11a** is in equilibrium with the corresponding dimer **11b**, indicating facile dissociation.

a = 2.076 Å
b = 2.030 Å
c = 2.068 Å
d = 2.034 Å
e = 2.061 Å
f = 2.037 Å
g = 2.056 Å
h = 2.028 Å

10

11a ⇌ 2 **11b**

Finally, two other modes of supramolecular aggregation leading to tetranuclear species was found in some organozinc carbamates [Me$_2$Zn$_4$(O$_2$CNEt$_2$)$_6$] and [Me$_4$Zn$_4$(O$_2$CNEt$_2$)$_4$], involving bidentate bridging by the anion [28, 29].

The remarkable structural diversity of supramolecular aggregates of organozinc alkoxides and related compounds is possible because of the variations in the coordination numbers and geometry of the metal; it is favored by the relatively strong donor–acceptor bonds formed between oxygen and zinc. Most of the supramolecular aggregates are preserved in hydrocarbon solution (e.g. benzene), when there is no competition from the solvent for the vacant coordination sites of the metal.

3.1.1.3 Organozinc–sulfur compounds

Organozinc thiolates, dithiocarbamates, and possibly other sulfur derivatives, are frequently associated through additional Zn–S bonds, although the Zn–S bond is less polar. Supramolecular aggregates of various degrees of association, including dimers, trimers, tetramers, pentamers, and octamers, have been reported.

Dimeric association, **12**, has been established for the compound Me$_3$SiCH$_2$ZnCPh$_3$; it results in the formation of a planar rhomboidal ring, with slightly differing Zn–S bond lengths (2.381 and 2.416 Å). Other trimethylsilylmethyl derivatives are cyclic trimers, **13**, e.g. [Me$_3$SiCH$_2$ZnS(2,4,6-Pr$^i_3$C$_6$H$_2$)]$_3$ and [Me$_3$SiCH$_2$ZnS(2,4,6-Bu$^t_3$C$_6$H$_2$)]$_3$, both containing planar Zn$_3$S$_3$ six-membered rings with a slight tendency of bond-length alternation in the ring [22]. It is worth emphasizing that in the dimers the coordination geometry at sulfur is trigonal pyramidal, whereas in the trimers it is trigonal planar; this led to some speculation about quasiaromatic delocalization in the six-membered rings.

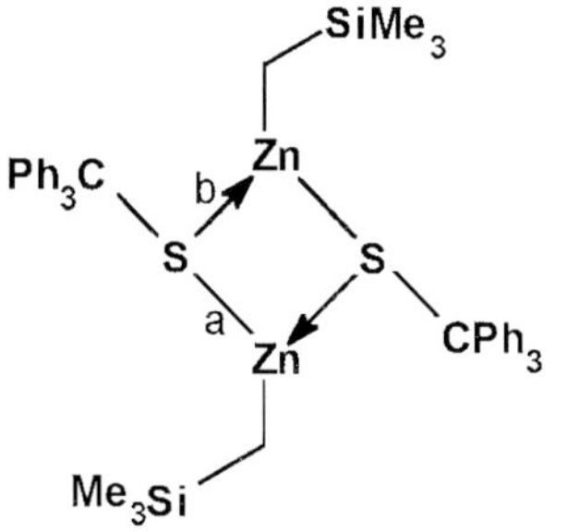

a = 2.381 Å
b = 2.416 Å

12

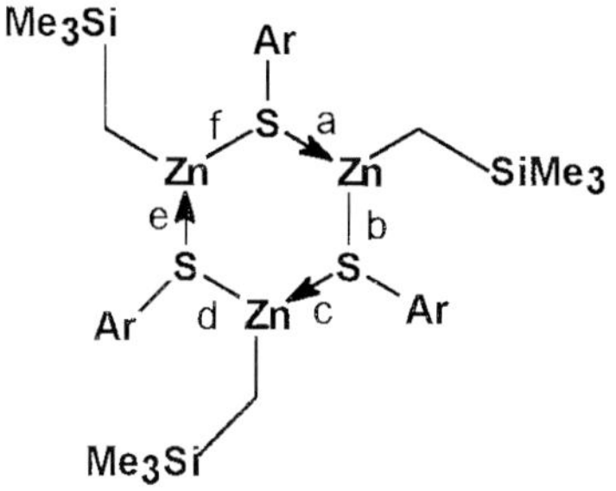

Ar = 2,6-Pr$^i_2$C$_6$H$_3$

a = 2.317 Å
b = 2.289 Å
c = 2.324 Å
d = 2.294 Å
e = 2.327 Å
f = 2.298 Å

Ar = 2,4,6-Bu$^t_3$C$_6$H$_2$

a = 2.356 Å
b = 2.311 Å
c = 2.325 Å
d = 2.316 Å
e = 2.372 Å
f = 2.291 Å

13

Chelated aminothiol derivatives dimerize via Zn–S dative bonds, as shown for the compound [MeZnSC$_6$H$_4$CHMeNMe$_2$-2]$_2$, **14**, which contains an almost planar Zn$_2$S$_2$ ring, with pyramidal coordination at sulfur and differing Zn–S bond lengths (2.390 and 2.453 Å) alternating in the ring [30].

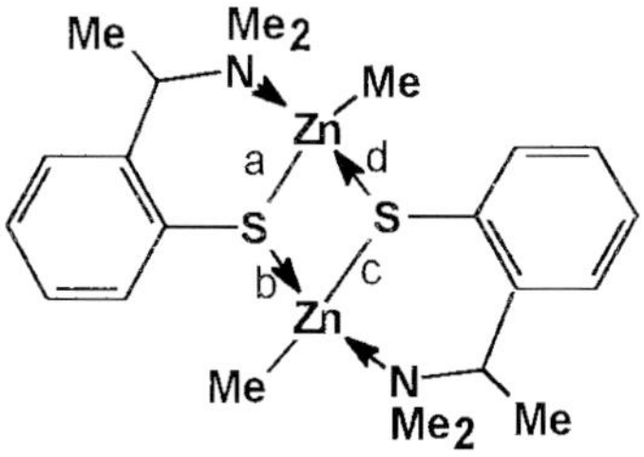

a = 2.390 Å
b = 2.420 Å
c = 2.389 Å
d = 2.453 Å

14

Methylzinc diethyldithiocarbamate, MeZnS$_2$CNEt$_2$, a useful material for zinc sulfide deposition, is a chelate dimer, **15** [31, 32].

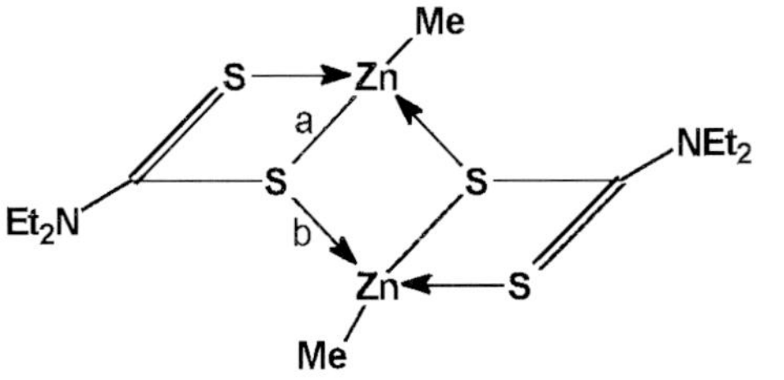

15

a = 2.501 Å
b = 2.512 Å

Some methylzinc thiolates with bulky groups on the sulfur form self-assembled aggregates with cage structures and four-coordinate zinc. Thus, the *tert*-butyl derivative is a pentamer, [MeZnSBu^t]$_5$, **16**, (in benzene solution) [33] and the isopropyl derivative is an octamer, [MeZnSPr^i]$_8$, **17** [34]. According to cryoscopic measurements [4] the latter is hexameric in benzene solution, indicating facile interconversion between variously aggregated (supramolecular) forms.

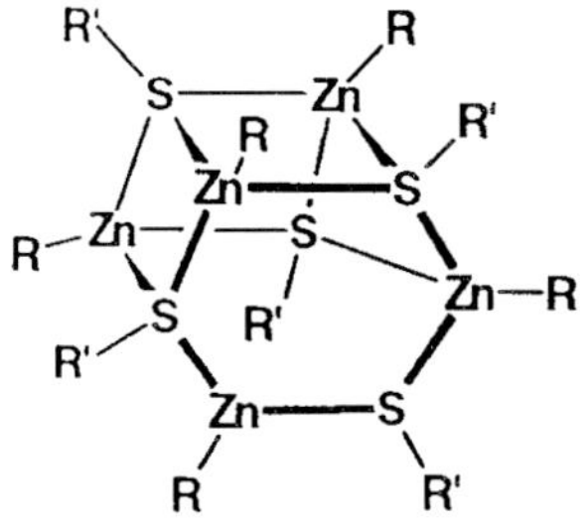

16 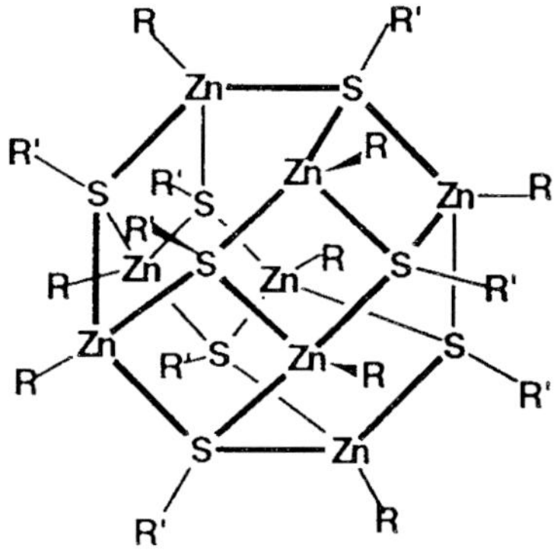 **17**

The examples cited show that zinc–sulfur self-assembly promises an interesting structural chemistry if a broader range of compounds can be investigated.

3.1.1.4 Organozinc–selenium compounds

As expected, the behavior of organozinc–selenium compounds is similar to that of the sulfur analogs. Dimeric self-assembled aggregates, **18**, were reported for organozinc diselenocarbamates, [RZnSe$_2$CNR$'_2$]$_2$, R = Me, Et, R' = Et [35] and R = Me$_3$CCH$_2$, R' = Et [36].

A trimeric compound, [Me$_3$SiCH$_2$ZnSe(2,4,6-Bu$^t_3$C$_6$H$_2$)]$_3$, **19**, contains a twisted six-membered ring and trigonal pyramidal coordination of selenium (differing from the sulfur analog which is planar) [37].

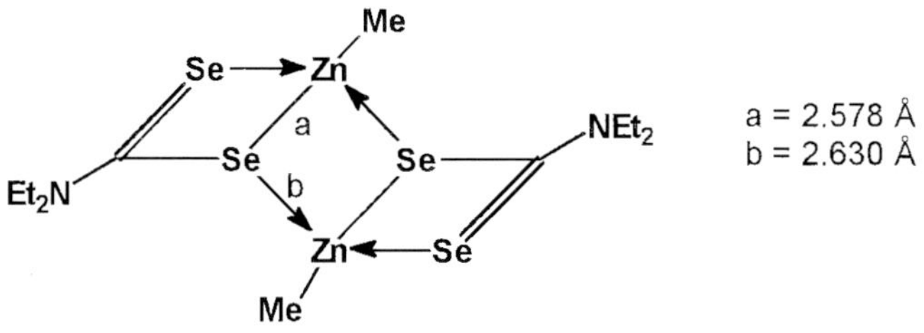
Me
Se
Zn
a
NEt2
Et2N
Se
Se
b
Zn
Se
Me
a = 2.578 Å
b = 2.630 Å

18

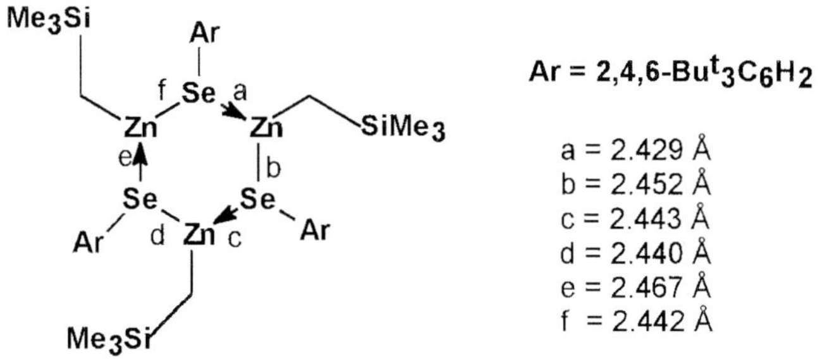
Me3Si
Ar
f
Se
a
Zn
Zn
SiMe3
e
b
Se
Se
Ar
d
Zn
c
Ar
Me3Si
Ar = 2,4,6-But3C6H2
a = 2.429 Å
b = 2.452 Å
c = 2.443 Å
d = 2.440 Å
e = 2.467 Å
f = 2.442 Å

19

3.1.1.5 Organozinc–nitrogen compounds

Self-assembly via zinc–nitrogen bonds occurs in numerous compounds containing this pair. Dimeric $[MeZnNPh_2]_2$, **20** [38], and $[EtZnNPh_2]_2$ [39] have been known for a long time.

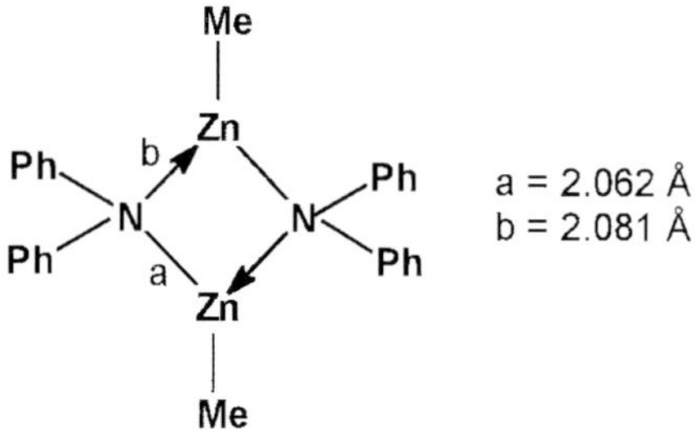
Me
Zn
b
Ph
Ph
N
N
Ph
Ph
a
Zn
Me
a = 2.062 Å
b = 2.081 Å

20

More recently chelate derivatives of iminoalcohols were found to be dimeric with planar Zn_2N_2 four-membered rings; $[EtZnOC(NEt_2)CH_2NBu^t]_2$, **21**, has been investigated by X-ray diffraction [40].

Trimeric $[EtZnNHC_{10}H_7{\cdot}THF]_3$, **22**, derived from 1-naphthylamine and containing four-coordinate zinc by addition of THF, came as a surprise, in view of the

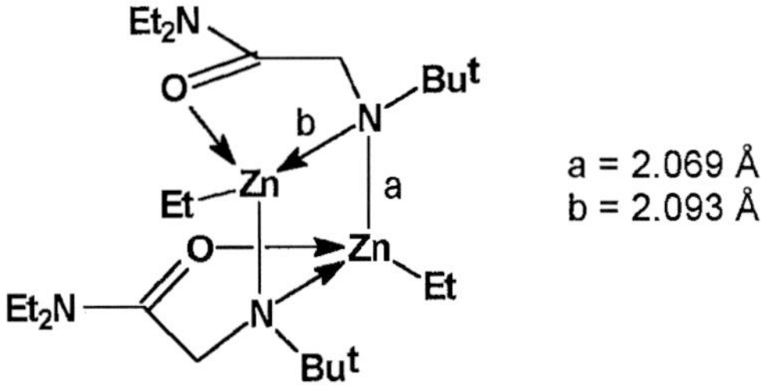

104   3 *Supramolecular Self-Assembly by Dative Bonding*

$$a = 2.069 \text{ Å}$$
$$b = 2.093 \text{ Å}$$

21

bulky substituent [41]. The six-membered Zn_3N_3 ring is non-planar (chair-shaped) and the H atoms on the nitrogen are all axial.

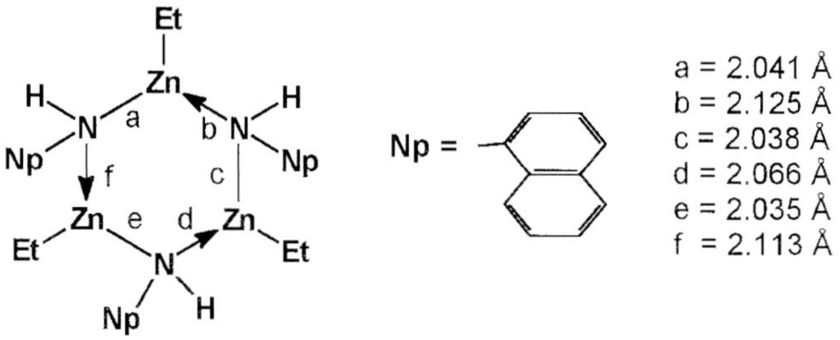

$$a = 2.041 \text{ Å}$$
$$b = 2.125 \text{ Å}$$
$$c = 2.038 \text{ Å}$$
$$d = 2.066 \text{ Å}$$
$$e = 2.035 \text{ Å}$$
$$f = 2.113 \text{ Å}$$

22

Several tetranuclear aggregates have been reported. For example, a tetranuclear zinc–nitrogen cage is present in $Et_2Zn_4(NPhOOCMe)_6 \cdot 2C_6H_6$ [42].

Mixed zinc–nitrogen and zinc–oxygen self-assembled species are also known. Thus, methylzinc acetoximate, which is tetrameric in benzene solution, maintains this aggregation state in the crystal and the molecular structure is based upon a cage made up of five-membered Zn_2O_2N rings [43].

3.1.2 Self-assembly of organocadmium compounds

Dimethylcadmium forms an 1:1 adduct with dioxane which is used in metal–organic chemical vapor deposition of cadmium sulfide. This material is a self-organized monodimensional array, **23**, formed via O → Cd bonds (2.88 and 2.75 Å). The coordination geometry around cadmium is rather curious: the methyl groups are arranged linearly (C–Cd–C 173.0°), other angles being as follows: O–Cd–O 114.2°, C–Cd–O 88.8° and 93.5°. Some association between dimethylcadmium and dioxane is also maintained in benzene solution [44].

Little information is available about the structure of organocadmium halides. They are poorly soluble in organic solvents, suggesting supramolecular association in the solid state. Some organocadmium halides with bulky substituents were found

3.1 Group 12 Metals – Zinc, Cadmium, Mercury

23

to be more soluble and crystals could be grown for investigation by single-crystal X-ray diffraction. Tris(trimethylsilyl)methylcadmium chloride, (Me$_3$Si)$_3$CCdCl is a tetramer, based on a cubane Cd$_4$Cl$_4$ skeleton in the solid state (av. Cd–Cl 2.626 Å); mass spectrometric measurements show that some association is maintained even in the vapor phase, dimeric species being detected. The corresponding bromide is a dimer [(Me$_3$Si)$_3$CCdBr]$_2$ in the solid state (Cd–Br 2.605 and 2.667 Å). A halogen-bridged lithium–dicadmate trinuclear anion, **24**, [{(Me$_3$Si)$_3$CCd}$_2$LiCl$_4$·2THF]$^-$ was also identified by X-ray diffraction (Cd–Cl 2.526 Å, 2.576 and 2.582 Å, Li–Cl 2.40 Å) [45].

a = 2.526 Å
b = 2.576 Å
c = 2.582 Å

24

In the solid state pentafluorophenylcadmium hydroxide, C$_6$F$_5$CdOH, is a tetramer, **25**, based on a Cd$_4$O$_4$ cubane skeleton (Cd–O 2.204–2.248 Å), reminiscent of the structures of organozinc alkoxides [46]. An organocadmium–sulfur cubane supermolecule, [C$_6$F$_5$CdSCPPh$_3$]$_4$, and related self-assembled anionic species have also been reported [47].

Dimeric [(σ-C$_5$Me$_5$)CdN(SiMe$_3$)$_2$]$_2$ contains a nearly planar, asymmetric Cd$_2$N$_2$ ring, **26**, (Cd–N 2.257 and 2.338 Å) with trigonal planar coordination of the cadmium atoms [48].

Dimethylcadmium reacts with gallium tris(di-*tert*-butyl phosphide), Ga(PBu^t$_2$)$_3$, to give a trimeric derivative [MeCdPBu^t$_2$]$_3$; crystal structure determination revealed cyclic self-assembly, **27**, with trigonally planar coordinated cadmium and a tendency of bond length alternation in the six-membered ring (Cd–P 2.568 and 2.586 Å) [49]. This compound serves as a precursor for cadmium phosphide (Cd$_3$P$_2$) nanoparticles [49a].

The scarce structural data available for functional organocadmium derivatives suggest that supramolecular self-assembly can be expected for many such compounds, with some probable similarity to organozinc compounds.

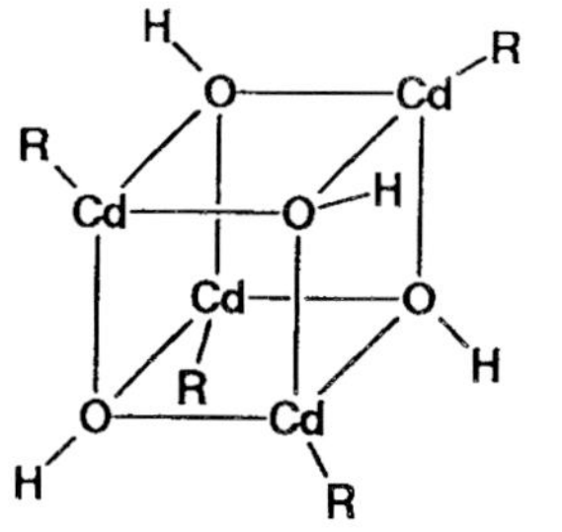

25

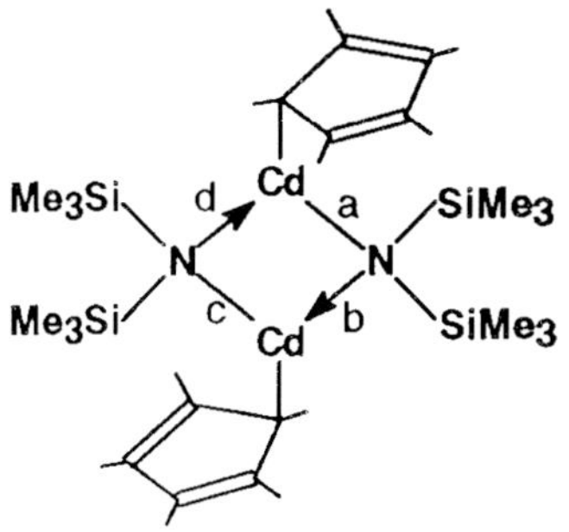

R = C₆F₅

26

a = 2.257 Å
b = 2.338 Å
c = 2.254 Å
d = 2.323 Å

27

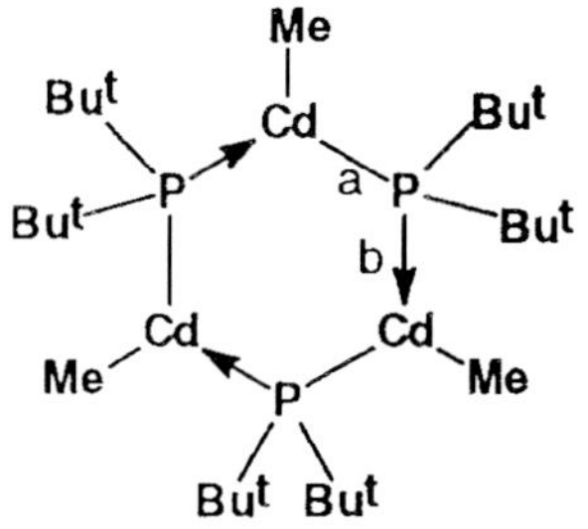

a = 2.568 Å
b = 2.586 Å

3.1.3 Self-assembly of organomercury compounds

Self-assembly of organomercury compounds by dative bonding is difficult to distinguish, on the basis of bond lengths, from self-assembly by secondary bonding. Significantly longer interatomic distances than expected for covalent bonds are observed but small differences between the values of 'long' bonds cannot be used to distinguish between primary and secondary bonds; the latter will, therefore, be interpreted as dative.

For the mercury–iodine pair a covalent bond length of 2.73 Å is expected and observed terminal, i.e. clearly covalent, Hg–I bonds in this range were found for two compounds which also contain longer iodine bridges. Thus in *trans*-di-μ-iodobis-(triphenylphosphoniumcyclopentadienylide) dimercury(II) [50], **28**, and *trans*-bis-(3-dimethylsulfoniumcyclopentadienylide)-di-μ-iodo-diiododimercury(II) [51], **29**, the bridging Hg···I distances are close to 2.9 Å and should be regarded as dative.

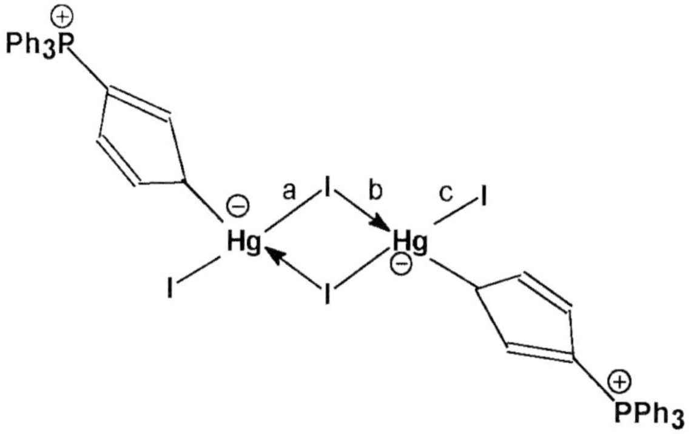

a = 2.937 Å
b = 2.982 Å
c = 2.681 Å

28

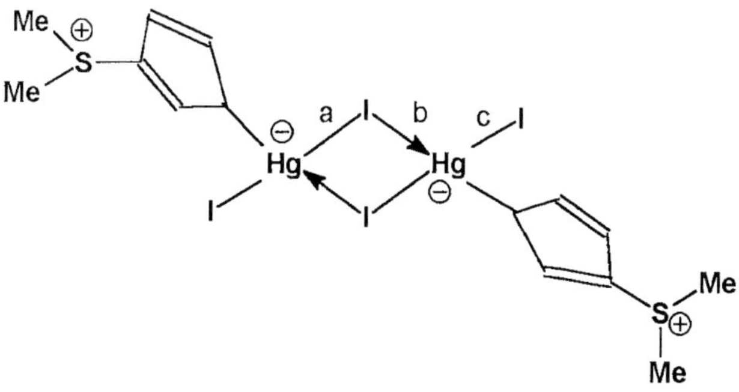

a = 2.986 Å
b = 3.031 Å
c = 2.706 Å

29

The general tendency of organomercury compounds is to form secondary-bond self-assembled supermolecules or supramolecular arrays through secondary bonds; the subject will be treated in Section 4.2.1.

3.2 Group 13 Metals – Aluminum, Gallium, Indium, Thallium

Metals of this group all form self-organized supramolecular structures via intermolecular dative bonds. This tendency is strong for the elements in the upper part

108   3 Supramolecular Self-Assembly by Dative Bonding

of the group and is probably the only mode of supramolecular self-assembly in Group 13 organometallic fluorides [52]. The lowest element in the group, thallium, is a 'soft' metal which has a predominant tendency to form weaker dative bonds which can be best regarded as secondary bonds. Only in some thallium–oxygen compounds are the intermolecular bonds stronger and can be regarded as dative.

3.2.1 Self-assembly of organoaluminum compounds

Organoaluminum compounds have a great propensity toward supramolecular association, because of the vacant valence orbital able to accept electron pairs from almost any functional group containing atoms of Group 15–17 elements. Numerous examples have been given in a recent review on the crystal structures of organoaluminum compounds [53] and in books dealing with this metal [54, 55].

3.2.1.1 Organoaluminum halides

The lone pairs of halogens in both di- and mono-organoaluminum halides, R_2AlX and $RAlX_2$, enable the supramolecular self-assembly of these species, both in solution and in the solid state. The expected covalent bond lengths are: Al–F 1.80 Å, Al–Cl 2.20 Å, Al–Br 2.38 Å, and Al–I 2.58 Å. Slightly longer bonds can be interpreted as dative and this corresponds with observed values in many self-organized supramolecular organoaluminum halides.

Dimethylaluminum halides, Me_2AlX (X = Cl, Br, I) are dimers, **30**, as are Et_2AlCl, Ph_2AlCl, $EtAlCl_2$, and $PhAlCl_2$, **31**, whereas the fluorides R_2AlF (R = Me, Et, Pr^n, Bu^i, Bu^t) are trimers or tetramers in solution (e.g. **32**) all associated by halogen bridging based upon dative bonding [56–60]. Association in the solid state has been confirmed by X-ray diffraction analysis of [Me$_2$AlCl]$_2$ [61], [MeAlCl$_2$]$_2$ (*trans* isomer) [62], and of [Bu$^t_2$Al(μ-Cl)]$_2$ (Al–Cl 2.316 and 2.324 Å) [63], [Mes(Cl)Al(μ-Cl)]$_2$ (Mes = 2,4,6-Bu^{t}C$_6$H$_2$) (Al–Cl 2.315 and 2.346 Å) [64], [MeAl(OC$_6$H$_2$Bu$^t_2$-2,6-Me-4)(μ-Cl)]$_2$ (Al–Cl 2.277 and 2.291 Å) [64a], and [(Me$_3$Si)$_2$CHAlCl(μ-Cl)]$_2$ (Al–Cl 2.068 and 2286 Å) [64b].

Even organoaluminum halides containing bulky organic groups tend to self-assemble. Thus, X-ray determination of the crystal structure of the bromide, (Trip)$_2$AlBr, where Trip = 2,4,6-Pr$^i_3$C$_6$H$_2$, established that the compound is a dimer (Al–Br 2.500 and 2.475 Å, compared with 2.311 Å in monomeric TripAlBr$_2$·Et$_2$O and the sum of Al–Br covalent radii, 2.38 Å) [65]. The pentamethylcyclopentadienyl aluminum chlorides are also dimers, e.g. [(η^5-C$_5$Me$_5$)(R)AlCl]$_2$ (R = Me, Bu^i) with Al$_2$Cl$_2$ cores [66, 67], [(η^5-C$_5$Me$_5$)AlCl(μ-Cl)]$_2$ (Al–Cl 2.149 and 2.340 Å), and [(η^5-C$_5$Me$_4$Et)AlCl(μ-Cl)]$_2$ (Al–Cl 2.148 and 2.346 Å) [67a].

More complex mixed metal fluorides of the type {[(η^5-C$_5$Me$_5$)MMeF][Me$_2$AlF$_2$]}$_2$, with M = Zr and Hf have been prepared and structurally characterized [68]. These can be regarded as supermolecules self-assembled via fluorine bridges, **33**.

33

A unique structure, self-assembled around a six-coordinate oxygen octahedrally surrounded by six aluminum atoms, interconnected by fluorine bridges was found in the compounds [RR′NAlF$_2$]$_4$[AlF$_2$(THF)$_2$]$_2$O with R/R′ = 2,6-Me$_2$C$_6$H$_3$/SiMe$_2$Bu^t and R/R′ = 2,6-Pr$^i_2$C$_6$H$_2$/SiMe$_3$ [69].

3.2.1.2 Organoaluminum–oxygen derivatives

Although organoaluminum–oxygen derivatives (alumoxanes) are extremely important compounds in Ziegler–Natta polymerization of olefins and molecular precursors for ceramic materials [70, 71], their structure was considered an enigma as recently as 1994 [72] although considerable experimental data indicating association (self-assembly) in solution and in the solid state has been accumulated [73, 74]. Recent years have seen major clarification, and a clearer picture is now emerging.

The sum of the covalent bond radii for the Al–O pair is 1.84 Å and the values observed should be compared with this value. Indeed, observed covalent Al–O bonds are close to it and slightly longer bonds can be regarded as dative.

Organoaluminum alkoxide and phenolate molecules self-assemble via strong Al–O dative bonds to form dimeric, **34**, and trimeric, **35**, supermolecules. Thus, [R_2AlOR']$_2$ with R = Me, R′ = Bu^t, SiMe$_3$, SiPh$_3$, and R = Et, R′ = Et, Bu^t, Ph, CH$_2$C$_6$H$_2$Bu$^t_3$, etc. [75–78] are examples of dimers and [R_2AlOMe]$_3$ (R = Me, Et) are trimers [79], according to molecular weight determinations in solution. Even dialumoxanes such as Et$_2$AlOAlEt$_2$ are associated in solution, and molecular weight determinations indicate tetrameric self-assembly; these are supported by ^{1}H, ^{27}Al and ^{17}O NMR investigations which suggest the presence of tetracoordinate aluminum and tricoordinate oxygen [80].

The structures suggested on the basis of solution studies remained only speculative, however, until X-ray single-crystal structure determinations were performed on some key compounds, including [Bu$^t_2$Al(μ-OH)]$_3$, [$R_2AlOAlR_2$]$_2$ (R = Bu^t), and others [81]. Structurally characterized dimers include planar [Me$_2$AlOSiMe$_3$]$_2$ [82] and [Me$_2$AlOC$_6$F$_5$]$_2$ (with unsymmetric Al–O 1.880 and 1.911 Å), which contain four-membered Al$_2$O$_2$ rings [83]. Cyclic dimers also include [Bu$^t_2$Al(μ-OPr^n)]$_2$ (Al–O 1.843 and 1.855 Å) and [Bu$^t_2$AlO(CH$_2$)$_3$NMe$_2$]$_2$ (Al–O 1.848 and 1.862 Å) [84], the latter demonstrating that oxygen is a more powerful donor than nitrogen towards aluminum. X-ray diffraction has established that K[Me$_{16}$Al$_7$O$_6$]·C$_6$H$_6$, **35a**, is a self-assembled tricyclic anion containing seven four-coordinate aluminum atoms [84a].

Another cyclic dimer is the dialumoxane anion [Me$_3$AlOAlMe$_2$]$_2^{2-}$, **36**, (Al–O 1.78 and 1.80 Å); this has been structurally characterized as tetramethylammonium salt [85].

The 1,3-dihydroxodialumoxane (HO)RAlOAlR(OH) [R = C(SiMe$_3$)$_2$] self-assembles into dimeric supermolecules with an adamantane-like structure [85a].

Trialumadioxane [Mes*(Et)AlOAlEt$_2$(Mes*AlO)]$_2$ (Mes* = 2,4,6-Bu^{t}C$_6$H$_2$), **37**, has a ladder-like supramolecular structure [86] of a type frequently observed in organotin–oxygen compounds.

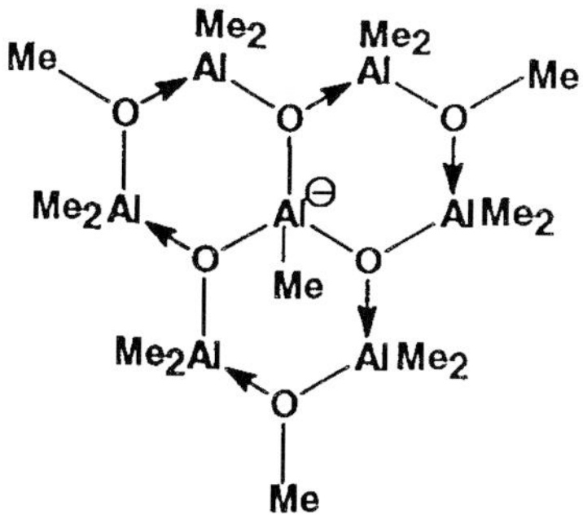

35a

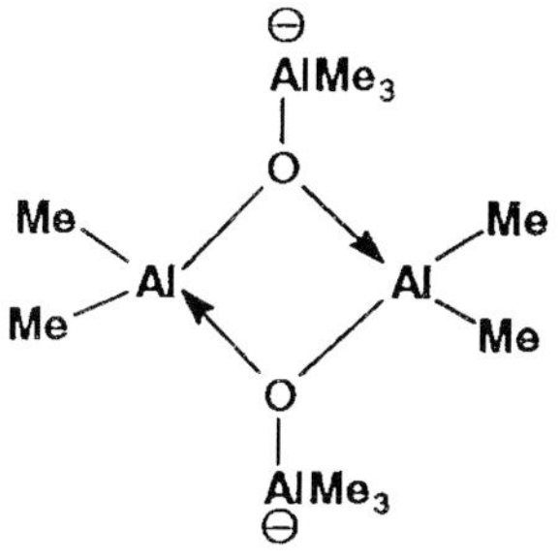

36

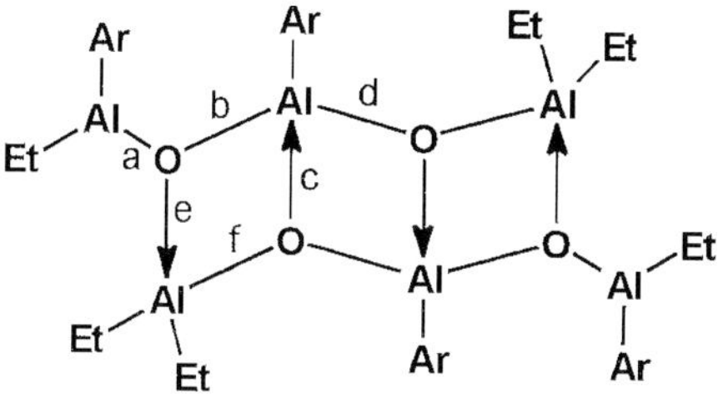

Ar = 2,4,6-Bu$^t_3$C$_6$H$_2$

a = 1.762 Å
b = 1.850 Å
c = 1.853 Å
d = 1.819 Å
e = 1.901 Å
f = 1.811 Å

37

The notable asymmetry of the four-membered Al$_2$O$_2$ rings is preserved in five-coordinate alumoxane derivatives, [R$_2$Al(OC$_6$H$_4$-2-XMe)]$_2$, **38** (Table 3.1), in which additional (but weaker) coordination from a donor group (OMe or SMe) attached to the aromatic substituent is observed [87]. ^{1}H and ^{13}C NMR studies show that the dimeric structures are preserved in solution.

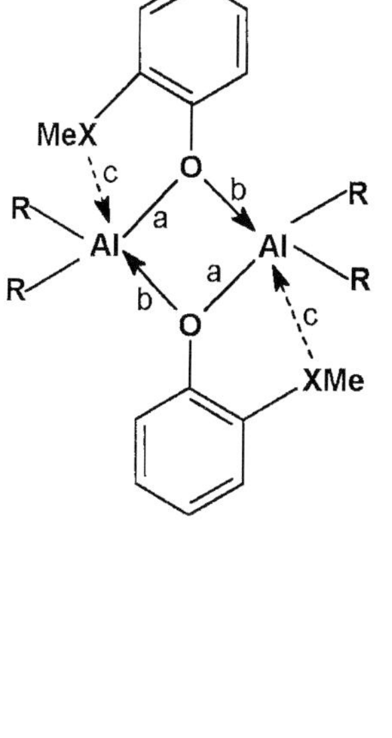

a,b,c values in Table 3.1

38

112 3 Supramolecular Self-Assembly by Dative Bonding

Table 3.1. Interatomic distances (Å) in dimeric compounds of type **38**.

Compound	a	b	c
[Me$_2$Al(OC$_6$H$_4$-2-SMe)]$_2$	1.865	1.963	2.734 (Al–S)
	1.856	1.980	2.714
	1.865	1.946	2.767
[Bu$^i_2$Al(OC$_6$H$_4$-2-SMe)]$_2$	1.870	1.966	2.778 (Al–S)
[Et$_2$Al(OC$_6$H$_4$-2-OMe)]$_2$	1.859	1.952	2.249 (Al–O)
[Bu$^i_2$Al(OC$_6$H$_4$-2-OMe)]$_2$	1.861	1.950	2.267 (Al–O)

In dimeric [Me(I)AlOCH$_2$CH$_2$OMe]$_2$, **39**, three different values were observed for the aluminum–oxygen interatomic distances (Al–O 1.826, 1.875, and 2.026 Å) [88].

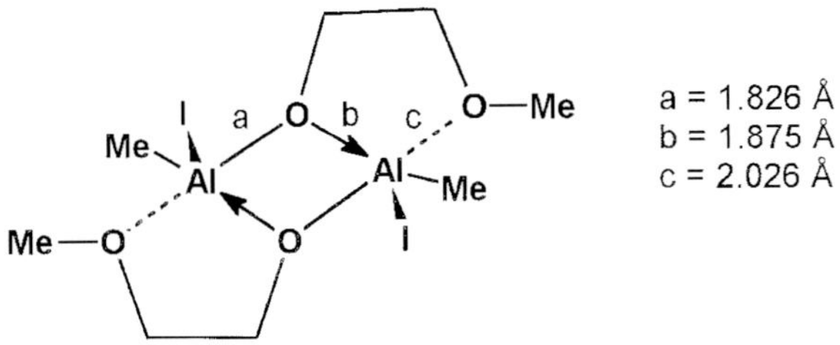

39

The sum of Al–O covalent radii is 1.839 Å; both normal covalent and dative Al–O bonds observed in the cyclic supermolecules are all longer.

Dimeric association has been observed in some more complex compounds such as various dialkylaluminum chelates derived from hydroxy carbonyl ligands (e.g. lactate, tropolonate, 2-acetyl-4-chlorophenolate) [89] or in Al$_4$Me$_6$(μ_3-O)$_2$(dpa) [dpa = deprotonated di-2-pyridylamine] [90] or in the cyclic chelate aminoalkoxy derivatives **40a** (Al–O 1.847 and 1.938 Å) and **40b** (Al–O 1.843 and 1.927 Å) [91].

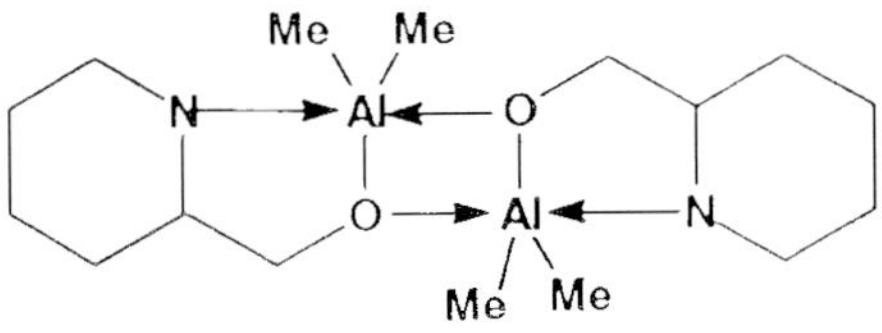

40a 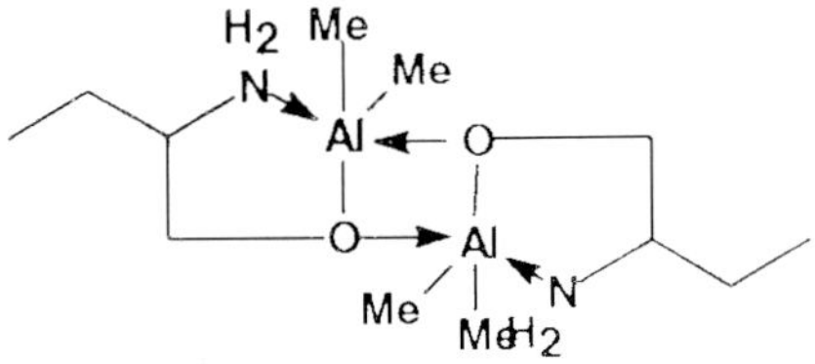   **40b**

A rich structural chemistry of self-assembled supermolecules is provided by organoaluminum hydroxides. NMR studies have shown that hydrolysis of aluminum trialkyls, AlR$_3$, produces dialkylaluminum hydroxides and aluminoxanes, which

3.2 Group 13 Metals – Aluminum, Gallium, Indium, Thallium

41 **42**

R = Bu^t

form 'stable associates' in solution. The structures suggested were only speculative [92]. The first compounds isolated in the solid state, and structurally characterized by X-ray diffraction, were obtained by hydrolysis of AlBu$^t_3$ under various conditions. They included trimeric [Bu$^t_2$Al(μ-OH)]$_3$, dimeric [Bu$^t_2$Al(μ-OAlBu$^t_2$)]$_2$, [Bu$^t_2$Al(Py)]$_2$(μ-O), basket [Bu$^t_7$Al$_5$(μ_3-O)$_3$(μ-OH)$_2$], [Bu$^t_8$Al$_6$(μ_3-O)$_4$(μ-OH)$_2$], and cage-like [Bu^{t}AlO]$_n$ ($n = 6, 7, 8$, and 9), e.g. **41** and **42** [81, 93]. The *tert*-butylaluminum derivative of butyrolactone [Al$_6$Bu$^t_6$(μ^3-O)$_4$(μ^3-O$_2$CCH$_2$CHMeO)$_2$] also contains an Al$_6$O$_6$ cage [94].

With bulky organic substituents self-organized organoaluminum hydroxides, such as dimeric [Mes$_2$Al(μ-OH)]$_2$ [95], [{(Me$_3$Si)$_2$CH}$_2$AlOH]$_3$ [96], [RMeAl(μ-OH)]$_2$·2THF, and the adamantane-like cage [R$_4$Al$_4$(μ-O)$_2$(μ-OH)$_4$], where R = C(SiMe$_3$)$_3$, have been obtained [97]. Dimeric and trimeric di-*tert*-butylaluminum hydroxides, [Bu$^t_2$Al OH]$_n$ ($n = 2$ and 3) have also been structurally characterized [81]. Several diorganoaluminum hydroxides [R$_2$AlOH]$_n$ with R = Me, Et, Bu^i are known [92] and all are expected to be self-assembled oligomeric supermolecules, but further investigation of their structure is necessary.

Self-assembled cage structures have been established for *tert*-butyl alumoxanes of various degrees of association formed by hydrolysis of AlBu$^t_3$ [81]. Thus, compounds such as Bu$^t_7$Al$_4$O$_2$(OH), Bu$^t_7$Al$_5$O$_3$(OH)$_2$, Bu$^t_8$Al$_6$O$_4$(OH)$_2$ [93], and Bu$^t_6$Al$_6$O$_4$(OH)$_4$ have been structurally characterized [98].

Theoretical calculations on Al–O cages throw some light upon the formation of such superstructures [99].

Organoaluminum derivatives of other oxygen-containing ligands (e.g. carboxylates, phosphinates, etc.) are expected to self-assemble in a similar manner via Al–O dative bonds. Indeed, several diorganoaluminum carboxylates, Bu$^t_2$AlOOCR (R = Bu^t, Ph, CH$_2$Ph, CHPh$_2$, and CH$_2$OCH$_2$CH$_2$OCH$_3$) have been structurally investigated and found to be dimers with bridging carboxylato groups [100]. The diethylaluminum derivative of salicylic acid is also a self-assembled dimer, **43** [101], as is the dimethylaluminum derivative of methyl salicylate [101a].

More complex supramolecular structures, containing oxo, hydroxo, and carboxylato bridges, and self-assembled from four-, six- and eight-membered Al$_n$O$_n$ rings ($n = 2, 3$, or 4) have been found in Al$_5$Bu$^t_5$(μ^3-O)$_3$(μ-OH)$_2$(μ-OOCPh)$_2$ and in

$Al_6Bu^t{}_6(\mu^3\text{-}O)_4(\mu\text{-}OH)_2(\mu\text{-}O_2CCl_3)_2$ [102]. These recent ground-breaking results suggest that a great structural diversity, comparable with that found in organotin chemistry, can be expected in the area of organoaluminum carboxylates.

It has been found by single-crystal X-ray diffraction [103] that dimethylaluminum diphenylphosphinate, **44**, is dimeric in the solid state, in agreement with early molecular-weight measurements in solution [104].

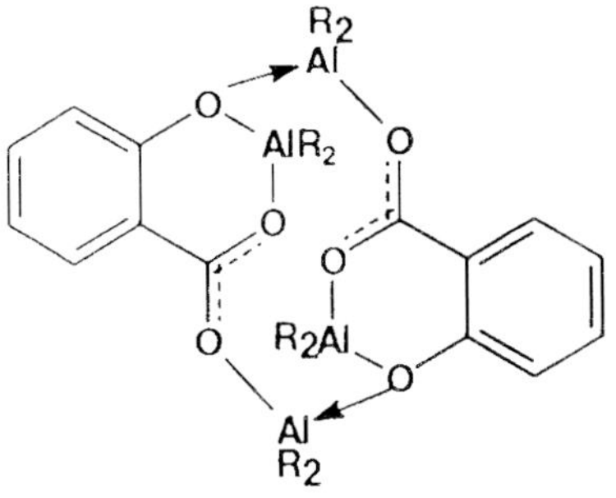

a = 1.781 Å
b = 1.794 Å

43

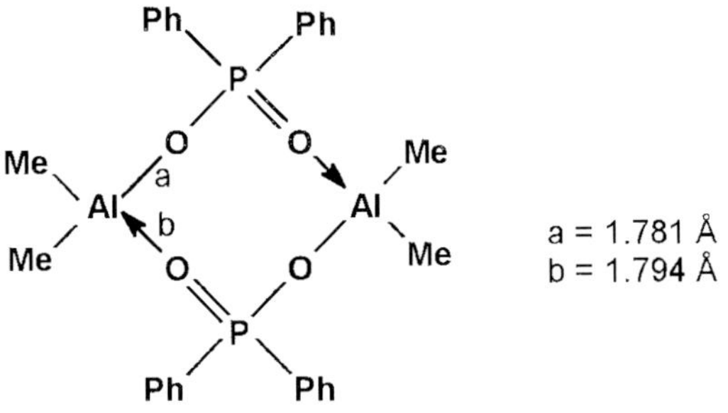

44

Monoorganoaluminum derivatives have two acceptor binding sites on the metal and form more complex supramolecular aggregates. Thus, the phosphonate $Bu^iAlO_3PBu^t$, based upon an Al–O–P–O cage, is a cubane-type self-organized tetramer (Al–O 1.764 and 1.757 Å) [105] whereas $MeAlO_3PBu^t$ forms a self-assembled cubane tetramer, **45a**, and a prismane hexamer, **45b** [106, 107]. The arsonate $[Bu^tAlO_3AsPh]_4$ also contains a cubic cage [108] (see also [108a,b]).

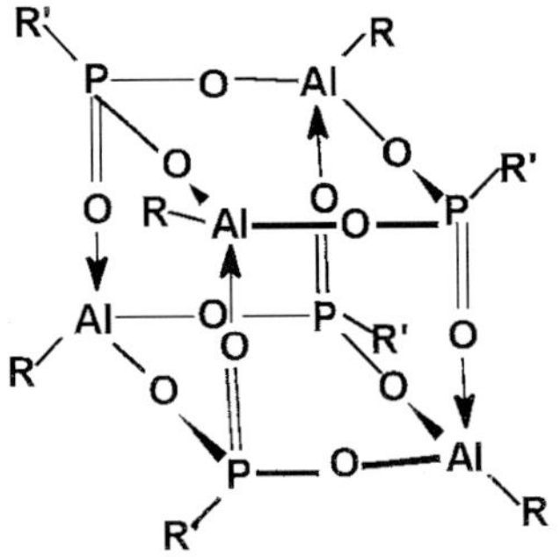

45a

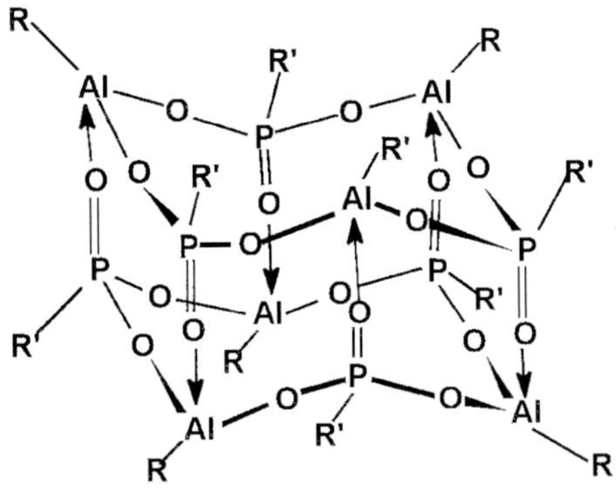

45b

3.2.1.3 Organoaluminum–sulfur derivatives

Organoaluminum thiolates undergo self-assembly with the formation of dimeric, **46**, trimeric, **47**, and tetrameric, **48**, supermolecules and polymeric supramolecular

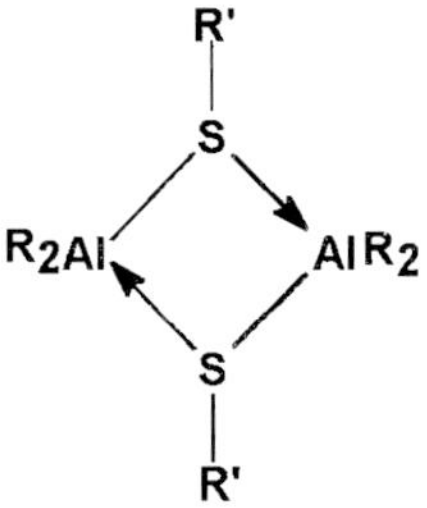

3.2 Group 13 Metals – Aluminum, Gallium, Indium, Thallium

46 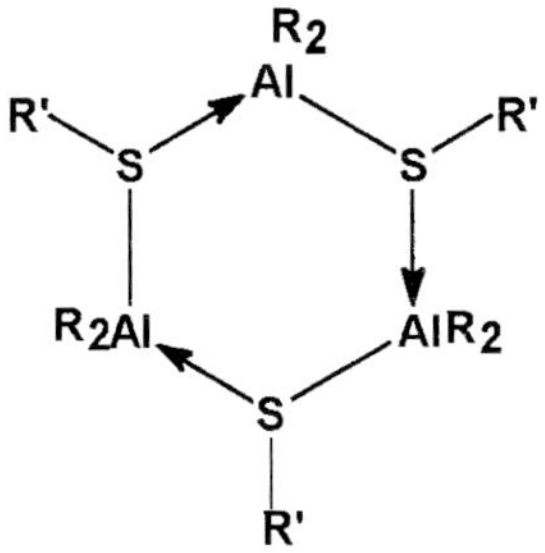 **47**

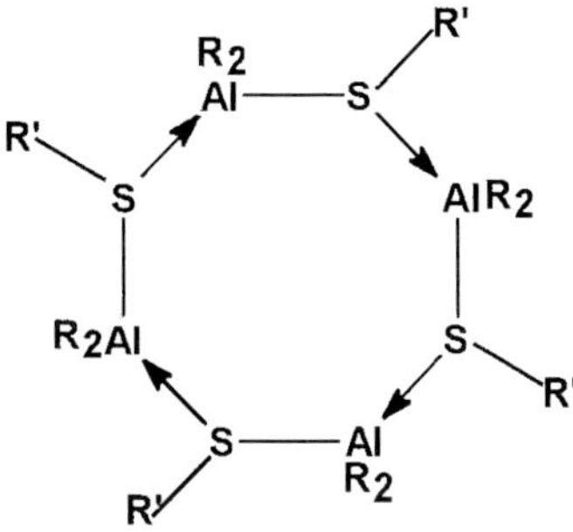

48

arrays. Thus, X-ray diffraction studies showed that Me$_2$AlSMe forms a polymeric, chain-like assembly in the solid state (2.345 and 2.352 Å) [109], whereas in the gas phase it forms a dimer, the molecular structure of which has been established by electron diffraction [110]. Other dimeric species include [Me$_2$AlSC$_6$F$_5$]$_2$ [111], [Mes$_2$AlSBz]$_2$ (Al–S 2.358 and 2.405 Å), [Me$_2$AlSSiPh$_3$]$_2$ (Al–S 2.355–2.372 Å), and [Mes$_2$AlSPh]$_2$ (Al–S 2.371–2.416 Å) [112]. Trimeric aggregates (with chair or twisted-boat conformation) include [Me$_2$AlSAr]$_3$, where Ar = 2-Pr^{i}C$_6$H$_4$ (Al–S 2.356–2.380 Å), 2-Bu^{t}C$_6$H$_4$ (Al–S 2.336–2.367 Å), and 2-Me$_3$SiC$_6$H$_4$ (Al–S 2.351–2.369 Å) and [Bu$^i{}_2$AlSC$_6$H$_2$Pr$^i{}_3$-2,4,6]$_3$ (Al–S 2.36–2.38 Å). Even a tetramer, [Me$_2$AlS(C$_6$H$_3$Me$_2$-2,6)]$_4$, has been investigated [112]. All have been structurally characterized by X-ray diffraction.

The sum of the covalent radii for the aluminum–sulfur pair is 2.23 Å and the observed interatomic distances in the cited compounds fall in the range 2.35–2.40 Å (see the review [113]). In monomeric Al[S(2,4,6-Bu$^t{}_3$C$_6$H$_2$)]$_3$, in which bulky organic groups prevent self-assembly, the Al–S interatomic distances are in the range 2.177–2.191 Å [114], suggesting a small amount of π-delocalization. Supramolecular self-assembly persists in solution, and the dimer–trimer equilibria have been investigated by NMR spectroscopy.

$$2[\text{Me}_2\text{AlSMe}]_3 \leftrightarrow 3[\text{Me}_2\text{AlSMe}]_2$$

116 3 Supramolecular Self-Assembly by Dative Bonding

In solutions of [Me$_2$AlSMe]$_n$ the trimer predominates, and *syn–anti* isomers also participate [113]. Aluminum–sulfur (and –selenium) cubane [(Me$_2$EtC)AlE]$_4$ (E = S, Se) and prismane [(Me$_2$EtC)AlS]$_6$ supermolecules have been also structurally characterized [114a].

3.2.1.4 Organoaluminum–selenium and –tellurium derivatives

Organoaluminum selenolates have been less investigated. A dimer [Mes$_2$AlSeMe]$_2$, **49**, has been structurally characterized [115]; it contains a planar Al$_2$Se$_2$ ring with Al–Se bond lengths (2.519 Å) longer than the sum of the covalent radii (2.38 Å) because of dative bond contribution. In the tellurium derivative [Bu$^t_2$AlTeBu^t]$_2$, **50**, the Al–Te bond length is 2.732 Å, compared with the sum of the covalent radii of 2.57 Å [116]. In solutions of [Me$_2$AlSeMe]$_n$ a dimer–trimer equilibrium, with the dimer predominating, has been observed [113].

The self-assembly of organoaluminum selenolates and tellurolates, by dative bonding, into supermolecules is remarkable in view of the relatively weak donor properties of selenium and tellurium (soft bases) towards aluminum (hard acid).

3.2.1.5 Organoaluminum–nitrogen compounds

The presence of both donor and acceptor binding sites in organoaluminum amides makes them ideally suited for supramolecular self-assembly by dative bonding. Indeed, aminoalanes can occur either as monomeric molecules, **51**, or as cyclic dimeric, **52**, or trimeric, **53**, supermolecules [R$_2$AlNR$'$R$''$]$_n$ ($n = 1, 2$, or 3), depending mostly on the steric properties of the organic groups.

These compounds, and their gallium and indium analogs, serve as precursors for Group 13 nitride materials (AlN, GaN, InN) [117]. Because aluminum–nitrogen compounds have been reviewed [118–121], selected recent examples only will be discussed here. It is worth mentioning that recent mass spectral measurements show that the dimeric aggregates are preserved even in the gas phase, with partial disso-

3.2 Group 13 Metals – Aluminum, Gallium, Indium, Thallium

51 **52** **53**

ciation into monomers [122]. Curiously, for the dimers [Me$_2$AlN(SiMe$_3$)(2,6-Pr$^i_2$C$_6$H$_3$)]$_2$ and [Me$_2$AlN(2,6-Me$_2$C$_6$H$_3$)(SiMe$_2$C$_6$HMe$_3$-2,4,6)]$_2$ the dimerization occurs via tricenter Al(μ-Me)$_2$Al bridges rather than Al–N donor–acceptor bonds [123, 124], probably because of the extreme bulkiness of the substituents on the nitrogen, which make this donor site unavailable. The less congested compounds are all aluminum–nitrogen cyclic dimeric supermolecules based upon Al$_2$N$_2$ rings, e.g. [Me$_2$AlNPhSiMe$_3$]$_2$ (Al–N 1.991–2.024 Å) and [Me$_2$AlNHSiMe$_3$]$_2$ (Al–N 1.991–2.024) [124, 125].

The influence of organic groups on the extent of association of aminoalanes [R$_2$AlNR'R"]$_n$ is illustrated by the list given in Table 3.2 [118].

It is obvious that small groups favor the formation of trimeric associates and bulky groups facilitate the formation of dimers.

The Al$_2$N$_2$ ring in the dimers can be planar or folded, depending upon the nature of organic groups [122]. The Al$_3$N$_3$ ring in the trimer can have the skew-boat (in [Me$_2$AlNH$_2$]$_3$), planar (in [Bu$^t_2$AlNH$_2$]$_3$) [126], or chair (in cis-[Me$_2$AlNHMe]$_3$) [127] conformation.

It seems that in the monomers some intramolecular donation occurs from nitrogen to aluminum, leading to an increase in the bond order, e.g. in Mes$_2$AlN(SiMe$_3$)$_2$ [128] (1.813 Å) and Bu$^t_2$AlN(Mes)$_2$ (1.823 Å), but is not significant in Bu$^t_2$AlN(SiPh$_3$)$_2$ (1.880 Å) [129], as demonstrated by comparison with the sum of the covalent radii of aluminum and nitrogen, 1.882 Å. The Al–N interatomic distances in dimers and trimers (1.92–1.98 Å) are longer, because of contributions from the dative bonds. The Al–N bond lengths in the dimers and trimers are intermediate between the bond lengths of normal covalent bonds (cited above for the monomers) and a purely dative bond measured in Me$_3$Al ← NH$_2$Bu^t (2.027 Å) [128], which is the longest observed.

In many dimeric and trimeric supermolecules there is a certain amount of asymmetry, in the form of alternating long and short bond lengths, which supports the idea that these species should be regarded as self-assembled supermolecules rather than authentic heterocycles. Thus, an alternation of Al–N bond lengths in the six-membered rings is obvious, e.g. in skew-boat [Me$_2$AlNH$_2$]$_3$ (1.927, 1.942, 1.933, 1.921, 1.938, and 1.940 Å) and in planar [Bu$^t_2$AlNH$_2$]$_3$ (1.918 and 1.985 Å) [126]. Non-equivalent Al–N bonds, but not clear bond length alternation, have also been observed in [Me$_2$AlNHPh]$_3$ (Al–N 1.955–1.988 Å) [130].

Table 3.2. The influence of organic substituents upon the association of aminoalanes $[R_2AlNR'R'']_n$.

R	R'	R''	n
Mes	SiMe$_3$	SiMe$_3$	1
H	H	Bu^t	2
H	Et	Et	2
Me	H	Pr^i, Bu^t, 1-Ad*	2
Me	H	o-tol, 2,6-Me$_2$C$_6$H$_3$, 2,6-Pr^i$_2$C$_6$H$_3$	2
Me	H	SiMe$_3$, SiEt$_3$, SiPh$_3$	2
Me	Me	Me, Ph	2
Me, Et	Ph	Ph	2
Et	Me, Et	Me, Et	2
Me$_3$SiCH$_2$	H	Me	2
Bu^i	H	Ph, 2,6-Pr^i$_2$C$_6$H$_3$	2
SiMe$_3$	H	Ph	2
Ph	Me	Me	2
Cl, Br, I	Me	Me	2
Cl	H	Pr^i, Bu^t	2
Me	H	H	3
H	Me, Pr^i	Me, Pr^i	3
Me	H	Me, Et	3
Et	H	H, Me	3
Bu^t	H	H	3

* 1-Adamantyl.

The thermolysis of some aluminum–nitrogen dimers occurs with the formation of self-assembled supermolecules (rings and cages) based upon the Al$_2$N$_2$ ring, which seems to be readily formed [131].

The hydrazido derivatives are also dimers. Thus, [Me$_2$Al(μ-NHNMe$_2$)]$_2$ (Al–N 1.953 and 1.958 Å) [132] and [Bu^t$_2$Al(μ-NHNHBu^t)]$_2$ (Al–N 1.997 and 2.001 Å) [133] contain Al$_2$N$_2$ four-membered rings rather than six-membered Al$_2$N$_4$ rings. The nitrogen bonded directly to aluminum seems to be more basic than the terminal nitrogen.

According to vibrational spectroscopic evidence the azides [Me$_2$AlN$_3$]$_3$ [134, 135], [Et$_2$AlN$_3$]$_3$ [136], and [EtClAlN$_3$]$_3$ [137] also contain six-membered Al$_3$N$_3$ rings. A trimeric azide, [{(Me$_3$Si)$_2$CH}$_2$AlN$_3$]$_3$, investigated by single-crystal X-ray diffraction, was found to contain a twelve-membered macrocycle formed through N–Al dative bonds (Al–N 1.904–1.944 Å) [138]. Related trimers are the nitrilimines [R$_2$AlNNCPR$_2$]$_3$ [139]. The organolaluminum azides seem to be excellent precursors for AlN semiconducting materials [139a].

Organoiminoalanes, RAl=NR', cannot be isolated in monomeric state. The Al=N double bond has a strong tendency to polymerize (no such monomer isolated so far), but this rarely results in heterocyclic alazanes containing tricoordinate aluminum. It happens only with very bulky organic substituents; an example is the borazine analog [MeAlN(2,6-Pr$^i_2$C$_6$H$_3$)]$_3$, **54**, which contains a planar Al$_3$N$_3$ heterocycle with rather short Al–N interatomic distances (1.782 Å) [140, 141].

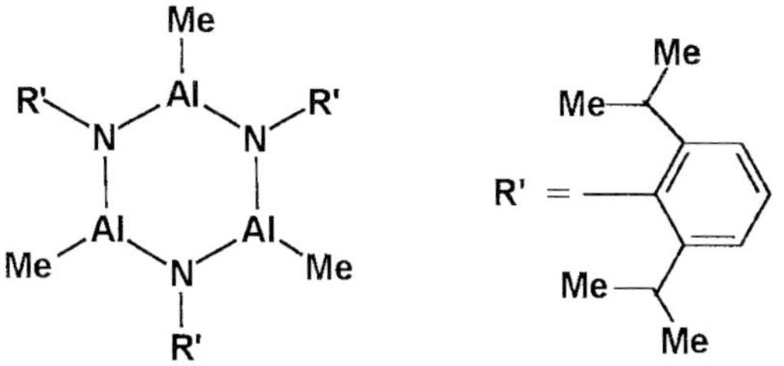

54

All other iminoalanes known are self-assembled polycyclic cages, **55–57**, containing four-coordinate aluminum and four-coordinate nitrogen, thus making full use of the s and p orbitals in bonding.

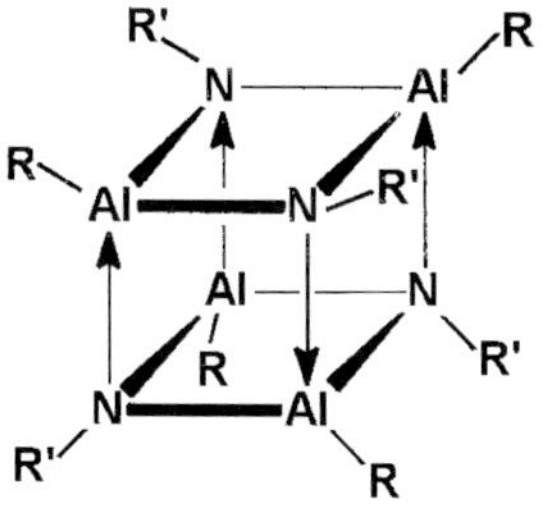

55 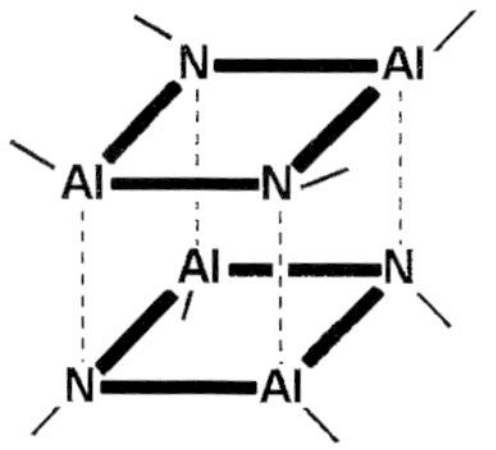   **56**

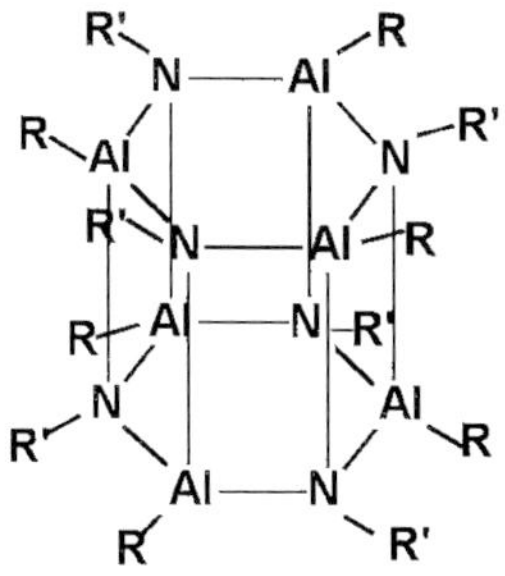

57 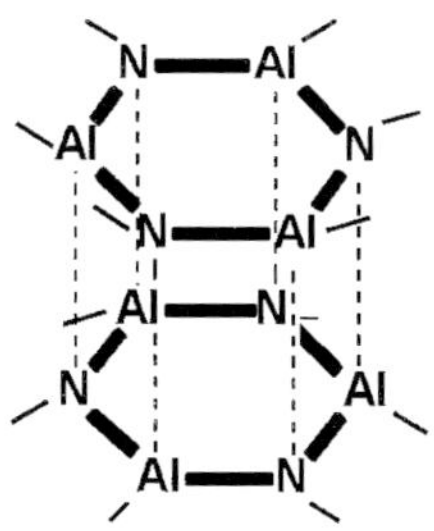   **57a** 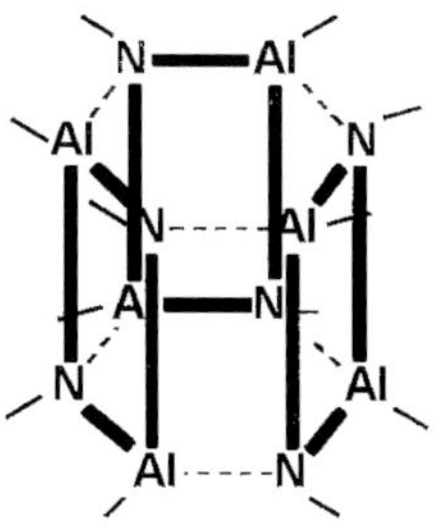   **57b**

The Al–N bond lengths in these cages are longer than in the heterocycles; bond equalization sometimes occurs but different Al–N bond lengths are also observed

Table 3.3. Aluminum–nitrogen interatomic distances in Al–N cage supermolecules.

Compound and ref.	Al–N in Al_3N_3 ring (Å)	Al–N in Al_2N_2 ring (Å)
$(HAlNPr^i)_6$ [145]	1.898	1.956
$(HAlNPr^n)_6$ [146]	1.890	1.959
$(HAlNCHMePh)_6$ [147]	1.890	1.981
$(MeAlNPr^i)_6$ [148]	1.917	1.964

occasionally. Among these cage supermolecules the cubanes $[RAlNR']_4$ are common – e.g. $(HAlNPr^i)_4$, $(MeAlNPr^i)_4$ [142], $(MeAlNMes)_4$ [141], $(PhAlNPh)_4$ [143], and $[MeAlNC_6F_5]_4$ [143a] – but prismane hexamers $[RAlNR']_6$, e.g. $(HAlNPr^n)_6$, $(HAlNPr^i)_6$, $(MeAlNPr^i)_6$, $(HAlNCHMePh)_6$, and $(ClAlNPr^i)_6$, and more complex structures are also known [121]. The cubane can be regarded as a pair of stacked four-membered cyclodialazanes, **56**, and the hexameric prismane as a stack of two six-membered cycloalazanes, **57a**, or as an assembly of three cyclodialazanes, **57b**. Because the Al–N bonds in the six-membered rings of the prismane are shorter than the transversal Al–N bonds (in the four-membered Al_2N_2 rings, connecting the two Al_3N_3 rings), describing the prismane as a self-assembled dimer seems justified. Theoretical calculations on Al_2N_2 and Al_4N_4 cubes explain the electronic details of their formation [144].

The Al–N interatomic distances in some hexameric cage supermolecules are given in Table 3.3.

Heptameric $(MeAlNMe)_7$, **58** [149], and octameric $(HAlNPr^n)_8$, **59** [150], are also known, as are the adamantane-like amide–imide $[\{AlCl(NMe_2)_2\}_2NMe]_2$, **60** (with Al–N 1.92 Å to four-coordinate nitrogen and 1.79 Å to three-coordinate nitrogen) [151], pentanuclear $[(H_2AlNHPr^i)_3(HAlNPr^i)_2]$, **61** [152], $[(H_2AlNHPr^i)_2-(HAlNPr^i)_2[HAlNCHMeCH_2NMe_2]$, hexanuclear $[HAlN(CH_2)_3NMe_2]_6$ [153], and octanuclear $[(Me_2AlNHMe)_2(MeAlNMe)_6]$, **62** [154]. The last three are mixed compounds self-assembled from amide $R_2AlNR'R''$ and imide $RAlNR'$ units. Other aluminum–nitrogen cage supermolecules include organofluorinated derivatives $(MeAlNR_F)_{4,6}$ (R = C_6H_4F-p) [155]. A tricyclic ladder consisting of three Al_2N_2 fused rings has also been reported [155a].

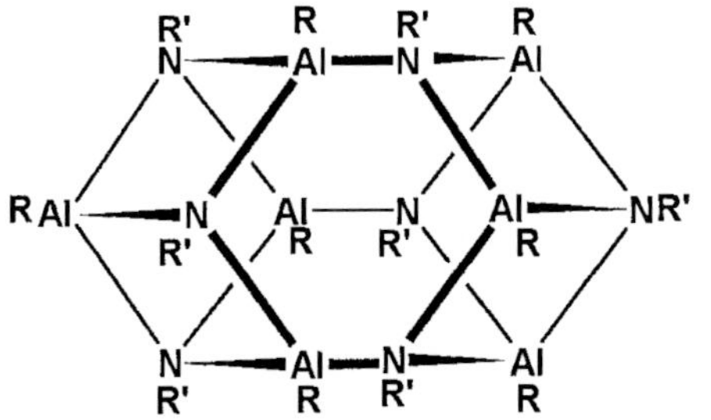

58

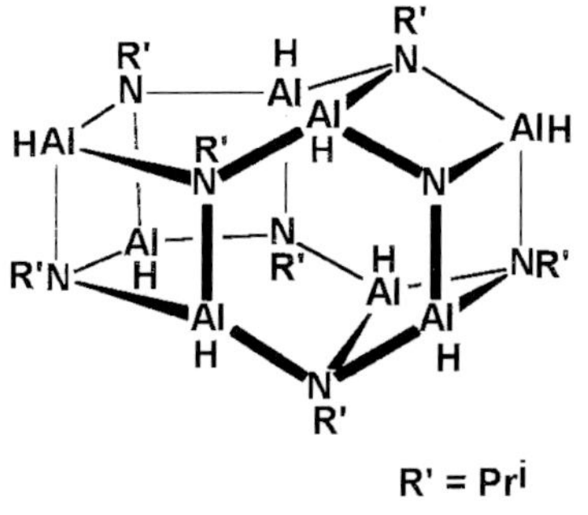

59

R' = Pr^i

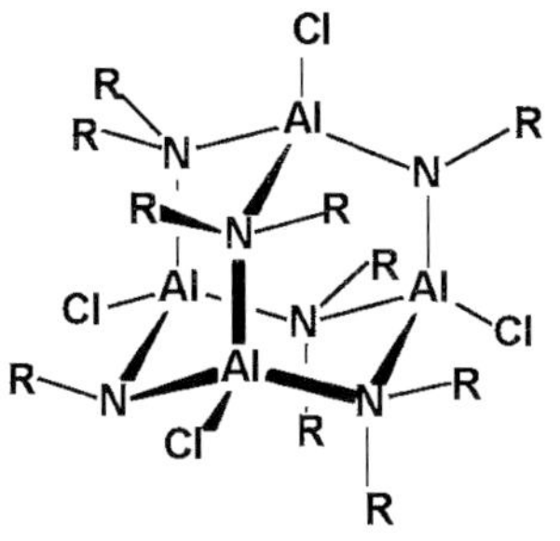

3.2 Group 13 Metals – Aluminum, Gallium, Indium, Thallium

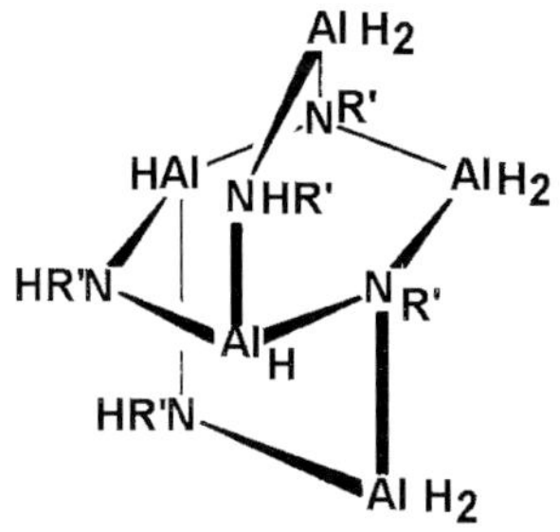

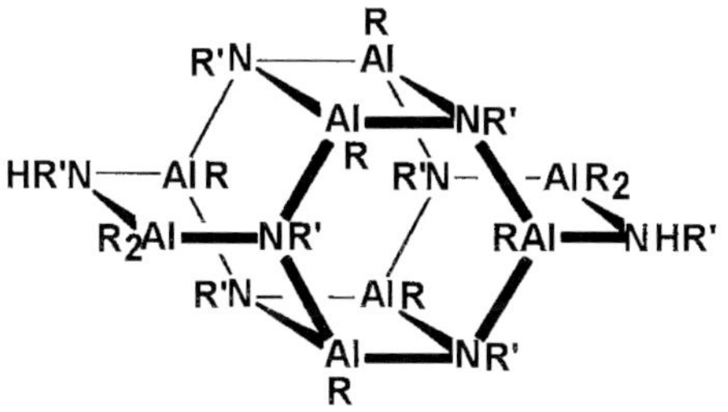

60 R = Me

61 R' = Pr^i

62 R = R' = Me

The number of known cage aluminum–nitrogen supermolecules is much larger and the interested reader is referred to the review by Cucinella [121].

The significance of bond length variations as a result of dative bond contributions is underscored by compounds containing different types of bond in the same molecule. The tendency of the Al–N bond to promote self-assembly is manifest even when this is incorporated into a heterocycle. Thus, the cycloalumadisilatriazane $(Me_2Si)_2(ClAl)(NMe)_3$ dimerizes by Al–N dative bonding. The resulting dimer, **63**, contains normal covalent Al–N bonds (c), 'pure' dative Al–N bonds (b) and covalent bonds with dative bond contribution (a), each different in length [156].

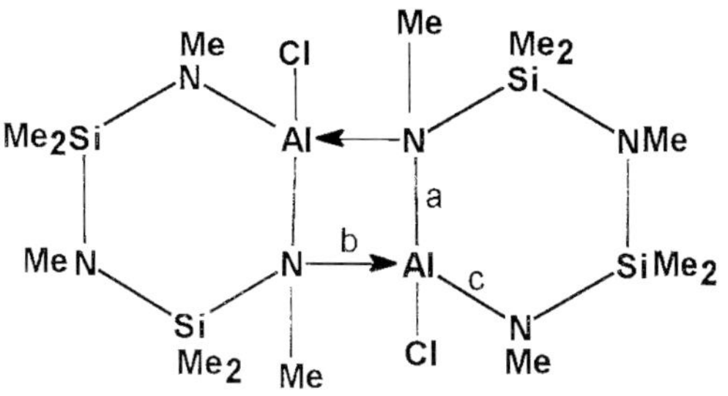

a = 1.941 Å
b = 1.952 Å
c = 1.803 Å

63

3.2.1.6 Organoaluminum–phosphorus and organoaluminum–arsenic compounds

Phosphinoalanes, $R_2AlPR'R''$, also contain both donor and acceptor binding sites and can undergo supramolecular self-assembly by dative bonding [157]. Dimers $[R_2AlPR'R'']_2$ [e.g. **64**, R = Me, R' = Ph] [158] or $[Me_2AlP(SiMe_3)_2]_2$ [159] and trimers $[R_2AlPR'_2]_3$ [e.g. **65**, R = Me, R' = Me, Et] [160] have been identified by molecular-weight determination; some were structurally characterized by X-ray diffraction, e.g. $[(Me_3Si)_2AlPPh_2]_2$ and $[(Me_3Si)_2AlP(SiMe_3)]_2$, in which the Al–P bond lengths (2.46 Å) [161] are longer than the sum of the covalent radii (2.268 Å) indicating dative bond contribution.

Similar aluminum–arsenic dimeric, **66**, and trimeric, **67**, $[R_2AlAsR'R'']_n$ ($n = 2$ and 3) supermolecules have been reported. Thus, $[Me_2AlAsMe_2]_n$ is polymeric in solid state and trimeric in solution [160].

Phosphorus–aluminum cages are illustrated by the cubane $[Bu^iAlPSiPh_3]_4$, **68** [162].

In $[Et_2AlAsBu^t_2]_2$ the Al–As interatomic distances are 2.571 and 2.562 Å [163], longer than the sum of the covalent radii (2.376 Å). The crystal structure of $[Et_2AlAs(SiMe_3)_2]$ has been determined (Al–As 2.539 and 2.531 Å) [164]. In the trimers $[Me_2AlAsPh_2]_3 \cdot 2C_7H_8$ and $[Me_2AlAsPh(CH_2SiMe_3)]_3$ (chair conformation) the Al–As interatomic distances are in the range 2.50–2.54 Å [165].

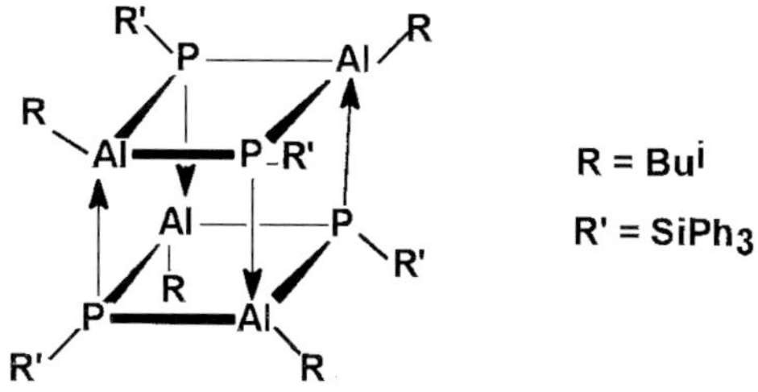

68

R = Bu^i
R' = SiPh$_3$

3.2.2 Self-assembly of organogallium compounds

3.2.2.1 Organogallium halides

Organogallium halides self-assemble as dimers, **69** [166], in solution and in the solid state. Examples include [{(Me$_3$SiCH$_2$)$_2$As}$_2$GaBr]$_2$ [167] and [(Bu^t$_3$C$_6$H$_2$)(PHBu^t)-GaCl]$_2$ [168]. Few crystal-structure determinations have been reported. [(C$_5$Me$_5$)$_2$GaCl]$_2$ contains bridging chlorine with Ga–Cl interatomic distances of 2.432 and 2.499 Å. In the dimer [(C$_5$Me$_5$)Ga(μ-Cl)Cl]$_2$ the bridging Ga–Cl bond lengths (2.352 and 2.373 Å) are longer than terminal Ga–Cl bond lengths (2.124 Å) [169]. The sum of covalent bond radii for the Ga–Cl pair is 2.203 Å. The dimers dissociate easily in solution and only monomers of (C$_5$Me$_5$)$_2$GaCl are detected cryoscopically in benzene. The mesityl derivative is dimeric, [Mes$_2$GaCl]$_2$, in benzene solution and probably in the solid state; the dichloro derivative is a monodimensional supramolecular array [MesGaCl$_2$]$_n$, **70**, with bridging Ga–Cl (2.363–2.369 Å) and terminal Ga–Cl (2.150 Å) [170].

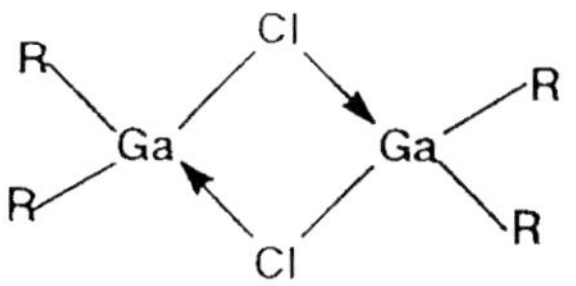

69   **70**   R = mesityl

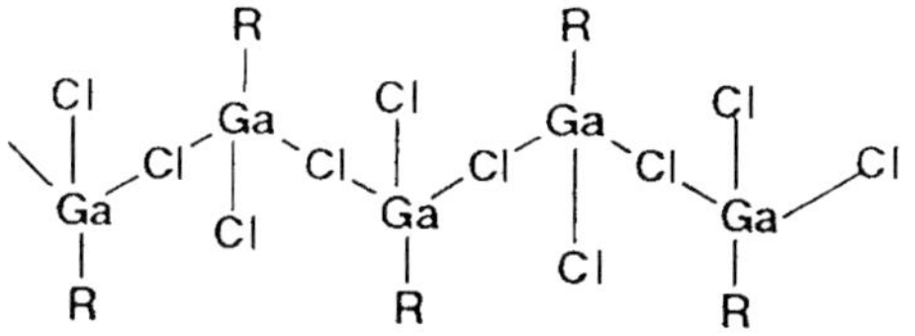

In [Bu^{t}GaCl$_2$]$_2$ the interatomic distances are Ga–Cl (bridging) 2.343 and 2.352 Å and Ga–Cl (terminal) 2.136 Å; in [Cy$_2$GaCl]$_2$ Ga–Cl (bridging) is 2.390 Å [171]. Another determination for the di-*tert*-butyl trimer [Bu^{t}ClGa(μ-Cl)]$_3$ gave Ga–Cl 2.338 and 2.346 Å (bridging Cl) and 2.151 Å (terminal Cl) [172].

Curiously, the tetrameric salts [Cs{R$_3$GaF}]$_4$, are not formed via F–Ga dative bonds; instead they contain a Cs$_4$F$_4$ central cubane assembly, with GaR$_3$ groups attached to the fluorine corners of the cube [173]. This is in contrast with the dimeric association of [Mes$_2$GaF]$_2$ through Ga–F bridges (Ga–F 1.947 Å, compared with the sum of covalent atomic radii 1.87 Å) [174]. A complex supramolecular architecture is present in the crystal of [Cs(THF)$_{0.5}$(Me$_2$GaF$_2$)]$_x$ [175].

Another dimer, [Bu$^t_2$GaI]$_2$, contains iodine bridges (Ga–I 2.755 Å, compared with the sum of the covalent radii 2.645 Å) in a planar four-membered ring [176].

For a number of organogallium halides self-assembly involves longer, secondary gallium–halogen bonds; these are discussed in Section 4.3.1.

3.2.2.2 Organogallium–oxygen compounds

Many organogallium–oxygen compounds are associated, self-assembled supermolecules. Their structure and chemistry have been reviewed [177–179].

Dimethylgallium hydroxide has been reported to be a trimer [Me$_2$GaOH]$_3$, **71**, on the basis of molecular-weight determinations in solution [180], but was found to be a tetramer, **72**, by X-ray diffraction in the solid state, with Ga–O 1.94–1.99 Å [181]. A trimer [Bu$^t_2$GaOH]$_3$ has, however, been characterized and contains a planar Ga$_3$O$_3$ ring (Ga–O 1.957 Å) [182–184].

Interestingly, (Me$_2$GaOH)$_3$ and (Me$_2$GaOH)$_4$ rings sandwiched between two oxa-aza crown ethers, attached through hydrogen-bonds to form unprecedented supramolecular structures, were formed from trimethylgallium in reactions with crown ethers [185]. These compounds combine host–guest interactions with hydrogen-bond and dative bond self-assembly, illustrating an intricate possibility of building supramolecular architectures [185a].

Some new organogallium–oxygen supermolecules formed by self-assembly include cyclic [RMeGa(μ-OH)]$_3$, the adamantane-like cage [R$_4$Ga$_4$(μ-O)$_2$(μ-OH)$_4$] where R = C(SiMe$_3$)$_3$ [186], hexanuclear [Mes$_6$Ga$_6$(μ^3-O)$_4$(μ^3-OH)$_4$·4THF]·6THF [187], and dodecanuclear [Bu$^t_{12}$Ga(μ^3-O)$_8$(μ-OH)$_4$] [98].

The sum of the covalent radii for the Ga–O pair is 1.909 Å, which compares well with the value found in monomeric Bu$^t_2$GaOPh$_3$ (1.831 Å) [188] and Bu$^t_2$GaOC$_6$H$_2$Bu$^t_3$-2,4,6 [189], which contain a 'normal' covalent Ga–O bond, but is considerably shorter than the dative Ga–O bonds.

3.2 Group 13 Metals – Aluminum, Gallium, Indium, Thallium

Most diorganogallium alkoxides are self-assembled dimers [R$_2$GaOR']$_2$, **73**, e.g. R = Me, R' = Me, Pr^n, CHEt$_2$, Bu, CH$_2$Bu^t, n-C$_6$H$_{13}$, Ph, p-Bu^{t}C$_6$H$_4$, p-ClC$_6$H$_4$ [190–192, 192a,b], OC$_6$H$_4$COOMe-2 [101a], and R = σ-C$_5$Me$_5$, R' = Et [193]. Similarly, dimethylgallium aminoalkoxides dimerize via Ga–O bonds [194].

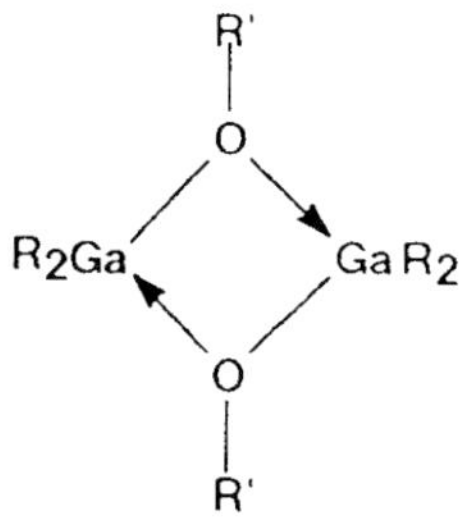

73

The peroxide [Bu$^t_2$GaOOBu^t]$_2$ also contains a Ga$_2$O$_2$ ring (rather than a possible Ga$_2$O$_4$ ring) with Ga–O 2.005 Å [172]. Dimethylgallium dimethylaminoethanolate, Me$_2$GaOCH$_2$CH$_2$NMe$_2$, dimerizes by Ga–O dative bonding (1.913 and 2.078 Å) rather than by nitrogen donation [195].

Cage molecules formed by self-assembly include structurally characterized organogallium oxides, e.g. [MesGaO]$_9$ [196].

Organogallium carboxylates are also expected to self-assemble, but few compounds have been structurally characterized. The acetate MeGa(O$_2$CCH$_3$)$_2$ forms an intricate polymeric bidimensional supramolecular array by acetate bridging and contains five-coordinate trigonal-bipyramidal gallium. The gallium–oxygen interatomic distances are 2.153 Å (apical), 1.943 Å (equatorial, bridge), and 1.873 Å (equatorial, terminal) [197].

Organogallium phosphinates, arsinates, and sulfinates are also self-assembled cyclic dimers [104]. [Me$_2$GaO$_2$PPh$_2$]$_2$ [198], [Et$_2$GaO$_2$PPh$_2$]$_2$ [199], [Bu$^t_2$GaO$_2$PPh$_2$]$_2$ [200], and [(Bu$^t_2$Ga)$_2$(O$_3$PPh)]$_2$ [200a] were structurally characterized and contain a chair-shaped eight-membered ring, **74**, with Ga–O distances

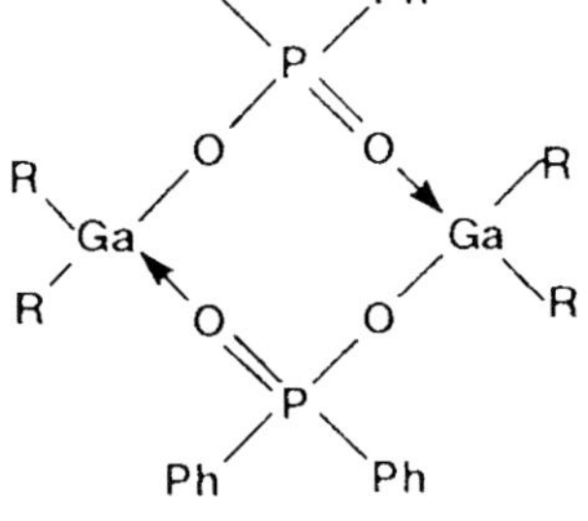

74

Table 3.4. Bond lengths in gallium–sulfur dimers.

Compound and ref.	Ga–S bond length (Å)
[Me$_2$GaS(cyclo-C$_5$H$_9$)]$_2$ [204]	Ave 2.388
[Me$_2$GaSC$_6$F$_5$]$_2$ [205]	2.436–2.460
[Bu$^t_2$GaSH]$_2$ [206]	2.421, 2.444
[Ph$_2$GaSEt]$_2$ [207, 208]	2.373, 2.384
[Ph$_2$GaS(cyclo-C$_5$H$_9$)]$_2$ [204]	Ave 2.388
[Ph$_2$GaSSiMe$_3$]$_2$ [204]	Ave 2.383
[Ph$_2$GaSSn(cyclo-C$_6$H$_{11}$)$_3$]$_2$ [209]	Ave 2.342

in the range 1.950 to 1.969 Å. The dimeric aggregate persists even in the vapor phase, as demonstrated by the mass spectrum.

Monoorganogallium derivatives form more complex supramolecular aggregates, e.g. [MeGaO$_3$PBu^t]$_4$ and Li$_4$[(MeGa)$_6$(μ_3-O)$_2$(Bu^{t}PO$_3$)$_6$]·4THF [201].

Self-assembled phosphonic acid derivatives have been reported, including dimeric [Bu$^t_2$Ga{μ-O$_2$PPh(OH)}]$_2$ and cubane [Bu^{t}Ga(μ^3-O$_3$PPh)]$_4$ [202]. The phosphonates can form ionic host–guest complexes incorporating alkali metal guests, e.g. sodium in [Na$_4$(μ^2-OH$_2$)$_2$(THF)$_2$][(Me$_2$GaO$_3$PBu^t)$_2$]$_2$·2THF and potassium in the salt [K(THF)$_6$][K$_5$(THF)$_2${(Me$_2$GaO$_3$PBu^t)$_2$}$_3$] [203] (see also ref. [203a]).

3.2.2.3 Organogallium–sulfur compounds

Molecular-weight determination revealed dimeric self-assembly of diorganogallium thiolates [R$_2$GaSR']$_2$, **75**, (R = Me, R' = Me, Ph, p-ClC$_6$H$_4$) [190]. Several structurally characterized solid-state dimers are known. Some Ga–S interatomic distances in associated organogallium thiolates are collected in Table 3.4 [204–209].

R$_2$Ga⟨S(R')⟩GaR$_2$ with bridging S–R' groups

75

These values can be compared with the sum of the covalent radii for the Ga–S pair, which is 2.302 Å. A tetramer [Me$_2$GaSC$_6$H$_3$Me$_2$-2,6]$_4$ (Ga–S 2.392–2.408 Å) has also been reported [112].

Monoorganogallium sulfides are associated as clusters, **76–78**; a tetramer [Bu^{t}GaS]$_4$ (Ga–S 2.365 Å) [116, 206, 210, 211], a hexamer [Bu^{t}GaS]$_6$ (Ga–S 2.316–2.392 Å), and a heptamer [Bu^{t}GaS]$_7$ (Ga–S 2.282–2.372 Å) have been structurally characterized and several interconversions have been described [212]. They have cubane, prismane (with non-planar six-membered rings) and double incomplete cubane structures, respectively. An octamer [Bu^{t}GaS]$_8$ has also been described but not analyzed by single-crystal X-ray diffraction [212, 213]. Other structurally characterized cubane supramolecules include [(Me$_2$Et)CGaE]$_4$ (E = S, Se, Te) [114a].

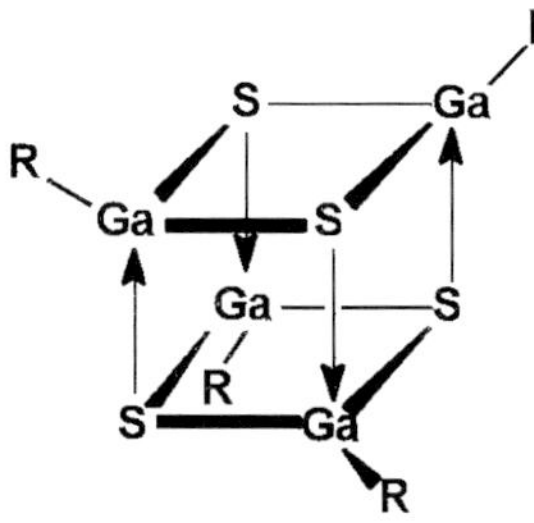

76 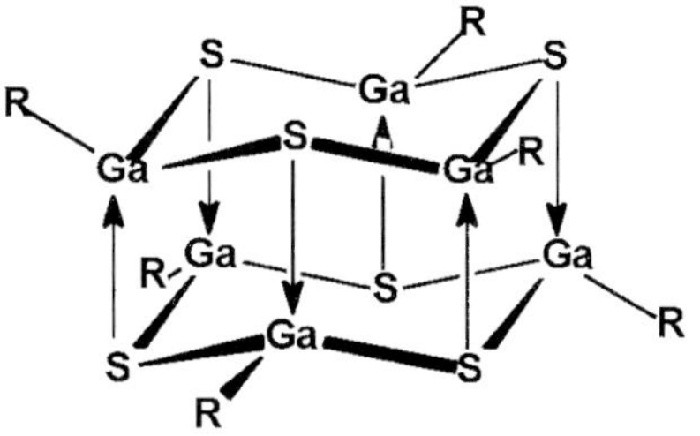   **77**

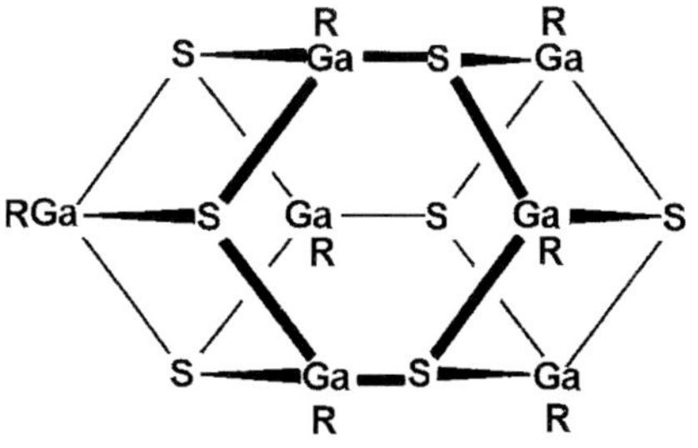

78

It is worth mentioning that in [Bu^{t}GaS]$_6$ the Ga–S interatomic distances in the six-membered rings (2.316–2.323 Å) are shorter than those connecting them (2.379–2.392 Å) in agreement with the view of this compound as a dimer of two authentic six-membered Ga$_3$S$_3$ heterocycles. The facile interconversions of these cages [212] support their description as self-assembled supermolecules built from RGaS units (Scheme 3.1).

3.2.2.4 Organogallium–selenium and –tellurium compounds

Like their sulfur analogs, diorganogallium selenols undergo self-assembly in solution with formation of dimeric supermolecules, e.g. [R$_2$GaSeR′]$_2$, (e.g. **79**) R = Me,

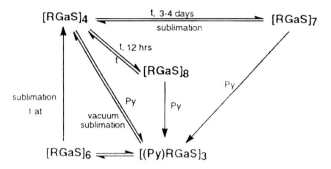

Scheme 3.1. Interconversions of gallium–sulfur cage supermolecules.

R' = Me) [190], [Bu$^t_2$GaSeBu^t]$_2$, [Ph$_2$GaSeMe]$_2$ (Ga–Se 2.501 Å) [214], and [(Me$_3$SiCH$_2$)$_2$GaSeCH$_2$SiMe$_3$)]$_2$ (Ga–Se 2.529–2.539 Å [215]. A cyclic trimer [Pr^i(Br)GaSeEt]$_3$ [216] and a silylated dimer [Et$_2$GaSeSiEt$_3$]$_2$ [217] have also been reported. The sum of the Ga–Se covalent radii is 2.453 Å.

$$\begin{array}{c} R' \\ | \\ Se \\ R_2Ga \quad GaR_2 \\ Se \\ | \\ R' \end{array}$$

79

Monoorganogallium selenides form cubane tetrameric supermolecules, e.g. [Bu^{t}GaSe]$_4$ [212] and [σ-C$_5$Me$_5$GaSe]$_4$ [218]. These can serve as precursors for CVD deposition of Ga$_2$Se$_3$ [218]. Tellurium analogs [Bu$^t_2$GaTeBu^t]$_2$ and [Bu^{t}GaTe]$_4$ are also known [213].

The tellurium derivatives [(Bu^{t}CH$_2$)$_2$GaTePh]$_2$ (Ga–Te 2.7435–2.7623 Å) and [Np$_2$GaTePh]$_2$ (Np = Bu^{t}CH$_2$; Ga–Te 2.755 Å) [219] have been structurally characterized; the observed bond lengths are longer than the sum of the covalent radii (2.641 Å). An experimental value for a covalent gallium–tellurium bond length observed in [(Me$_3$Si)$_2$CH]$_2$GaTeSi(SiMe$_3$)$_3$ is 2.535 Å, obviously shorter than the dative bond [220].

3.2.2.5 Organogallium–nitrogen compounds

Dimeric self-assembly, **80**, in solution of diorganogallium amides, e.g. [R$_2$GaNR'R'']$_2$ with R = Me, R', R'' = H, Me, Ph, 1,6-Pr$^i_2$C$_6$H$_3$ and N(CH$_2$Ph)$_2$

Table 3.5. Interatomic distances in gallium–nitrogen dimers.

Compound and ref.	Ga–N bond length (Å)
[Me$_2$GaNHBu^t]$_2$ [226]	2.024, 2.022
[Me$_2$GaNHPh]$_2$ [222]	2.039, 2.036
[Me$_2$GaNH(2,6-Pr$^i{}_2$C$_6$H$_3$)]$_2$ [222]	2.026, 2.024
[MeGa(μ-NHC$_6$H$_2$Me$_2$-4,6-CH$_2$-2)]$_2$ [222]	2.016, 2.029
[MeGa(μ-NHC$_6$H$_3$Pr^i-6-CHMeCH$_2$-2)]$_2$ [222]	2.045, 2.023
[Me$_2$GaN(CH$_2$Ph)$_2$]$_2$ [223]	2.045, 2.042
[Me$_2$GaN(CH$_2$CH$_2$)$_2$NMe]$_2$ [223]	2.042, 2.022
[Me$_2$GaNPh(SiMe$_3$)]$_2$ [124]	2.071, 2.072
[Me(Cl)GaNHSiMe$_3$]$_2$ [227]	2.012
[(PhMe$_2$CCH$_2$)$_2$GaNHPr]$_2$ [228]	2.013, 2.029
[Bu$^t{}_2$GaNHPh]$_2$ [229]	2.103, 2.018

[219] was found by molecular-weight determination [158, 221–223]. Trimeric supermolecules, **81**, include [Me$_2$GaNH$_2$]$_3$ [184], [Me$_2$GaNHMe]$_3$ [224], and [Et$_2$GaN$_3$]$_3$ [225], among others. Some derivatives have been characterized by X-ray diffraction; their bond lengths are given in Table 3.5 [226–229].

80

81

The trimer [Me$_2$GaNH$_2$]$_3$ contains a planar six-membered Ga$_3$N$_3$ ring with Ga–N 2.017 Å. Compare this value with the sum of the covalent radii (1.952 Å) and with the purely dative Ga–N bond lengths observed in Bu$^t{}_3$Ga·NH$_2$Ph (2.246 Å) [229], Me$_3$Ga·NHCy$_2$ (2.151 Å), and Me$_3$Ga·NH[*cyclo*-(CHMe)(CH$_2$)$_3$CHMe] (2.199 Å) [230].

Di-*tert*-butylgallium amide is a self-assembled cyclic trimer [Bu$^t{}_2$GaNH$_2$]$_3$ (Ga–N 2.017 Å) [184]. Cubane-type cages including [MeGaNC$_6$F$_5$]$_4$ [231], (MesGa)$_3$(GaNHC$_6$F$_5$)(μ^3-NC$_6$F$_5$)$_4$ [143a], and a hexaprismane [MeGaNC$_6$F$_5$]$_6$·7THF [155] have been structurally characterized.

The isolation and structural characterization of the polycyclic cage [(Me$_2$GaNHMe)$_2$(MeGaNMe)$_6$], **82**, suggests that it might be an analogy between aluminum and gallium–nitrogen self-assembled supermolecules [154].

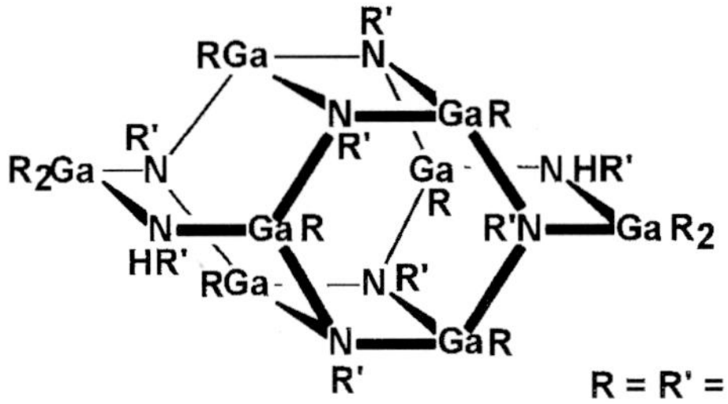

82

R = R' = Me

A unique self-assembled polycyclic cage based upon five-membered Ga_2N_3 rings, of composition [MeGa(μ-NHNPh)]$_4$, **83**, has been prepared in a reaction between GaMe$_3$ and PhNHNH$_2$ and structurally characterized [232].

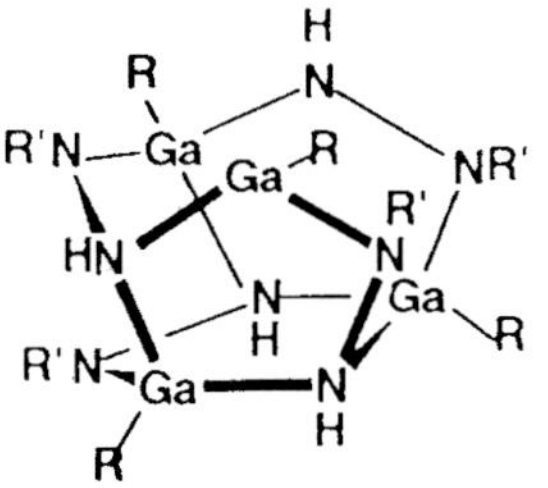

R = Me
R' = Ph

83

In the solid state dimethylgallium azide [Me$_2$GaN$_3$]$_x$ is a supramolecular array, **84**, consisting of spiral chains (Ga–N 2.051 and 2.039 Å), but in solution cryoscopic measurements indicate trimeric aggregation [233].

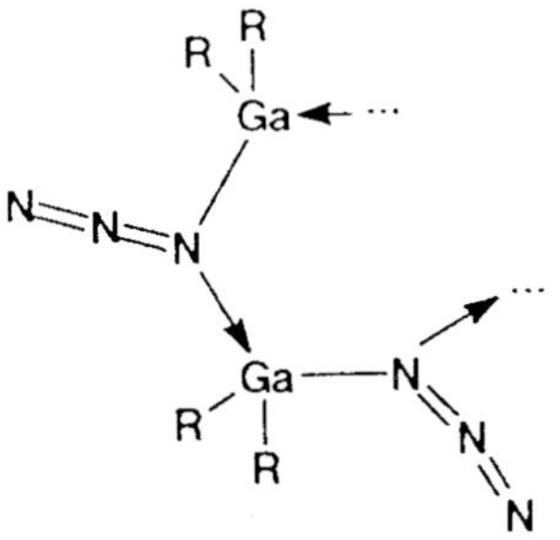

84

Diorganogallium derivatives of azoles (e.g. imidazole, benzimidazole, pyrazole, *s*-triazole, and benzotriazole) were found associated in solution [234]. This was confirmed by the isolation in the solid state of crystalline dimers which have been structurally characterized by single-crystal X-ray diffraction. Thus, the pyrazolyl, 3-methylpyrazolyl, indazolyl [235], and 3,5-dimethylpyrazolyl [236] derivatives, are all dimers in which two gallium atoms are bridged by azole rings by dative Ga–N bonds, as shown in **85**.

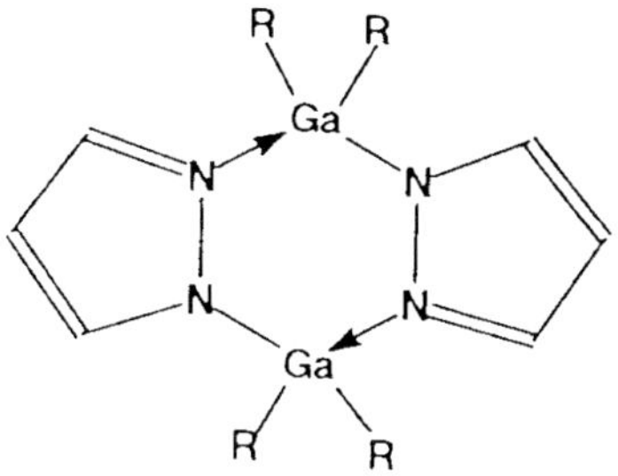

85

3.2.2.6 Organogallium–phosphorus, –arsenic and –antimony compounds

Extensive studies have been performed on self-assembled dimeric and trimeric organogallium phosphides, **86** and **87**, and arsenides, **88** and **89**, in relation to their use as precursors for gallium phosphide and gallium arsenide semiconductors [237–240]. Their association in solution was observed some time ago as a result of molecular weight determinations in solution [158, 160] and many of the compounds have been structurally characterized. As already shown, the degree of association depends upon the steric bulkiness of the organic groups.

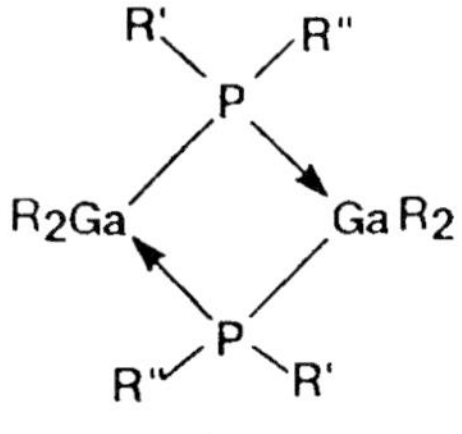

86 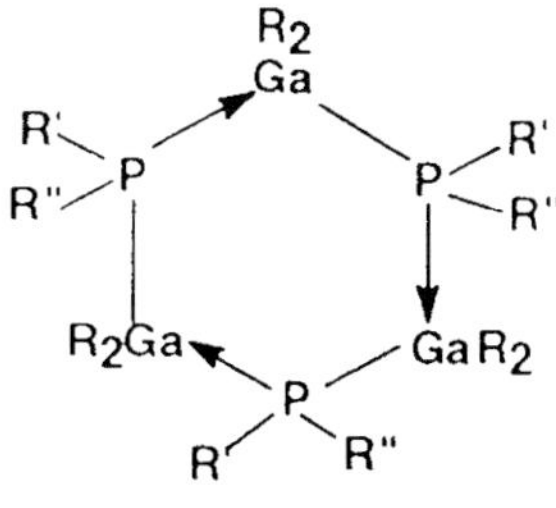   **87**

A list of dimeric and trimeric supermolecules is given in Table 3.6, with their Ga–P bond lengths [241–251].

The sum of the covalent bond radii are 2.338 Å for Ga–P and 2.446 Å for Ga–As. These values can be compared with the Ga–P interatomic distance in monomeric $Ph_3Ga \leftarrow P(SiMe_3)_3$ (2.539 Å) and $Ph_2ClGa \leftarrow P(SiMe_3)_3$ (2.459 Å) which contain 'pure' dative P → Ga bonds [252], with monomeric $(Me_3Si)_3$-

132   3 Supramolecular Self-Assembly by Dative Bonding

88 **89**

Table 3.6. Interatomic distances in some gallium–phosphorus dimeric and trimeric supermolecules.

Compound and ref.	Ga–P bond length (Å)
[Pr$^i_2$GaPPr$^i_2$]$_2$ [241]	2.475, 2.465
	2.453, 2.477
[Bu$^n_2$GaPBu$^t_2$]$_2$ [242]	2.477, 2.476, 2.468, 2.483
[Bu$^t_2$GaPHBu^t]$_2$ [243]	2.459
[Bu$^t_2$GaPH(*cyclo*-C$_5$H$_9$)]$_2$ [244]	2.451
[(Bu^{t}CH$_2$)$_2$GaPPh$_2$]$_2$ [245]	2.482, 2.512, 2.488, 2.479
[(Bu^t)(Me$_3$SiCC)GaPEt$_2$]$_2$ [246]	2.399, 2.415
[Me$_2$GaPPh$_2$]$_3$ [247]	2.416–2.445
[Me$_2$GaPMePh]$_3$ [248]	2.413–2.407
[Me$_2$GaPPr$^i_2$]$_3$ [249]	2.419–2.457
[Bu$^t_2$GaPH$_2$]$_3$ [250]	2.439
[(Bu^{t}CH$_2$)ClGaPPh$_2$]$_3$ [251]	2.414–2.433

As → Ga(Cl)(CH$_2$Bu^t)$_2$ (2.626 Å) [253] which contain dative Ga ← As bonds, and with monomeric Bu$^t_2$GaAsBu$^t_2$ (2.466 Å) [254] and (C$_5$Me$_5$)$_2$GaAs(SiMe$_3$)$_2$ (2.433 Å) [255], containing normal covalent As–Ga bonds. In structurally characterized [ClPr^{i}GaAsBu$^t_2$]$_2$ the Ga–As bonds are basically identical (2.529 and 2.520 Å) [256].

Data relating to gallium–arsenic dimeric supermolecules are collected in Table 3.7.

Facile conversion of [Me$_2$GaPPr$^i_2$]$_3$ into the corresponding dimer has been observed on vacuum sublimation, and the dimeric association persists in the gas phase, as demonstrated by mass spectrometry [249].

For [R(Cl)GaAs(CH$_2$SiMe$_3$)$_2$]$_n$ (R = Me, Ph) the dimer–trimer equilibrium is detected spectroscopically in solution [261]. Cryoscopic and ^{1}H and ^{31}P NMR studies suggest that the degree of association of R$_2$M–ER$'_2$ compounds in solution is controlled by thermodynamic factors such as solvation energies and entropic effects rather than kinetic factors [245, 262].

A cubane supermolecule [Bu^{t}GaPCy]$_4$ has also been prepared and structurally characterized [162].

Table 3.7. Interatomic distances in gallium–arsenic supermolecules.

Compound and ref.	Ga–As bond lengths (Å)
[Me$_2$GaAsBu$^t_2$]$_2$ [257]	2.541, 2.558
[Me$_2$GaAs(CH$_2$Bu^t)$_2$]$_2$ [258]	2.529, 2.532
[(Me$_3$SiCH$_2$)$_2$GaAs(SiMe$_3$)$_2$]$_2$ [259]	2.562, 2.571
[Bu$^t_2$GaAs(CH$_2$SiMe$_3$)$_2$]$_2$ [260]	2.630
[(Bu^{t}CH$_2$)$_2$GaAs(SiMe$_3$)$_2$]$_2$ [253]	2.584, 2.590
[Ph$_2$GaAs(CH$_2$SiMe$_3$)$_2$]$_2$ [260]	2.518, 2.530

Mixed halide–phosphide and arsenide bridged dimers **90** and **91** are formed by association of diorganogallium halides with diorganogallium phosphides.

90   **91**

The compound [Ph$_2$Ga{μ-P(SiMe$_3$)$_2$}GaPh$_2$(μ-Cl)] has been structurally characterized [252] and several organoarsenido analogs have also been reported [259, 263–265].

Similar gallium–antimony self-assembled compounds have also been reported. These include dimeric [Bu$^t_2$GaSb(SiMe$_3$)$_2$]$_2$, the mixed bridge dinuclear compound [Bu$^t_2$GaSb(SiMe$_3$)$_2$GaBu$^t_2$Cl] [266] and polymeric [(Me$_3$SiCH$_2$)$_2$GaSb(SiMe$_3$)$_2$]$_x$ [267]. In the cyclic dimer [Bu$^t_2$GaSb(SiMe$_3$)$_2$]$_2$ the Ga–Sb bonds (2.7684 and 2.7648 Å) are, as expected, shorter than the purely dative bonds in Bu$^t_3$Ga ← Sb(SiMe$_3$)$_3$ (3.027 Å), but comparable with the value observed in Et$_3$Ga ← Sb(SiMe$_3$)$_3$ (2.846 Å). A trimeric supermolecule based upon Ga–Sb bonds is known only in a compound lacking Ga–C bonds, namely [Cl$_2$GaSbBu$^t_2$]$_3$ [268].

3.2.3 Self-assembly of organoindium compounds

With increasing atom size and soft-acid character of the metals on descending in Group 13, higher coordination numbers and secondary bond associations become favored and this is reflected in the structure of organoindium derivatives. In addition to a few compounds which can be described as supermolecules formed by dative bonding, there are many examples of supramolecular self-assembly via secondary bonds. The dative bond supermolecules are discussed in this section and supramolecular self-assembly via secondary bonds is treated in Section 4.3.2.

3.2.3.1 Organoindium–halogen derivatives

The self-organization of organoindium halides splendidly illustrates the two alternatives of self-assembly, i.e. via dative bonds and secondary bonds. Molecular-weight determinations in solution often reveal dimeric association of organoindium halides. These will not be discussed here, and only compounds structurally characterized in the solid state by X-ray diffraction will be treated. An interesting feature of organoindium halides is the tendency to self-organize into dimeric structures by dative bonding, followed by further association into supramolecular arrays via secondary bonds (see Section 4.3.2.1.). The exact nature of supramolecular organization is determined by the steric properties of the organic groups. Isolated dimeric supermolecules are limited to compounds with very bulky organic groups. Thus, in the solid state dimesitylindium fluoride is a trimer $[Mes_2In(\mu-F)]_3$ (In–F 2.095–2.140 Å) [269] and dimesitylindium chloride is a dimer $[Mes_2In(\mu-Cl)]_2$ (In–Cl 2.584 Å) [270]. The In–Cl bond length compares well with the sum of the covalent radii (2.433 Å) and with the In–Cl bond length (2.540 Å) measured in the anion $[Mes_3InCl]^-$, in which the metal–halogen bond can probably be considered as purely dative. Another example is the structure of $[\{(Me_3Si)_2CH\}(Pr^i)In(\mu-Cl)]_2$ which contains discrete dimers, **92**, in the unit cell, with In–Cl 2.575 and 2.586 Å [271], but even in $[(Me_3SiCH_2)(Bu^tCH_2)In(\mu-Cl)]_2$ with In–Cl distances of 2.572 and 2.654 Å the dimeric assemblies are further associated through secondary bonds (In···Cl 3.528 Å) into a ladder-like supramolecular structure (see Section 4.3.2.1.) [272].

92

Dimesitylindium iodide is also a dimer, $[Mes_2InI]_2$, with In–I 2.900 and 2.988 Å, with significant asymmetry in the four-membered In_2I_2 ring (In–I 2.900 and 2.988 Å) [273]. The sum of the In–I covalent radii is 2.805 Å.

Di-*tert*-butylindium chloride forms a chain-like polymeric supramolecular array, **93**, with a saw-tooth pattern and In–Cl distances of 2.523 and 2.58 Å [274].

R = Bu^t
a = 2.523 Å
b = 2.58 Å

93

Interestingly, indium–halogen bonding can serve as 'cement' for supramolecular constructions in some unlikely cases, such as the polymeric structure of di-isopropylindium tetrafluoroborate, $[Pr^i_2In(BF_4)\cdot 2THF]_x$, containing anion bridges in a chain, **94** [275]. The In–F dative bond lengths (2.591 and 2.646 Å) are significantly longer than the sum of the covalent radii (2.029 Å) and might suggest ionic interactions between BF_4^- anions and $[R_2In(THF)_2]^+$ cations.

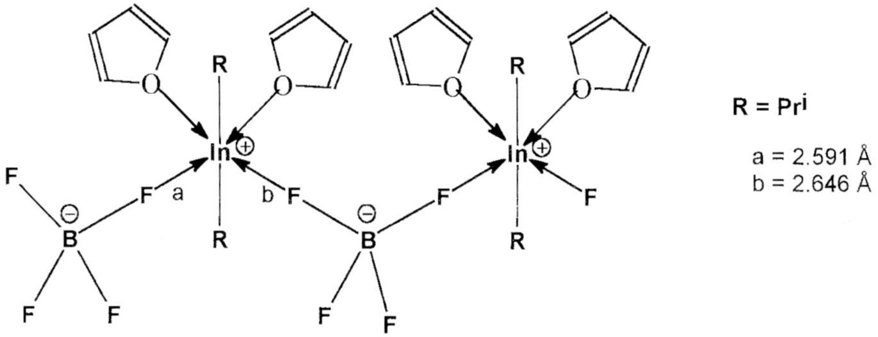

$R = Pr^i$

a = 2.591 Å
b = 2.646 Å

94

3.2.3.2 Organoindium–oxygen derivatives

There are similarities between indium and gallium chemistry. Diorganoindium derivatives containing an indium–oxygen bond tend to self-organize into supermolecules, mostly dimeric, **95**, both in solution and in the solid state [276]. Some compounds have been structurally characterized. Thus, in the dimeric alkoxide $[Bu^t_2InOEt]_2$ the In–O interatomic distances in the In_2O_2 ring are 2.147 and 2.165 Å [277], compared with the sum of the covalent radii, 2.069 Å. A peroxide $[Bu^t_2InOOBu^t]_2$ also forms In_2O_2 rings, with In–O 2.191 Å [278].

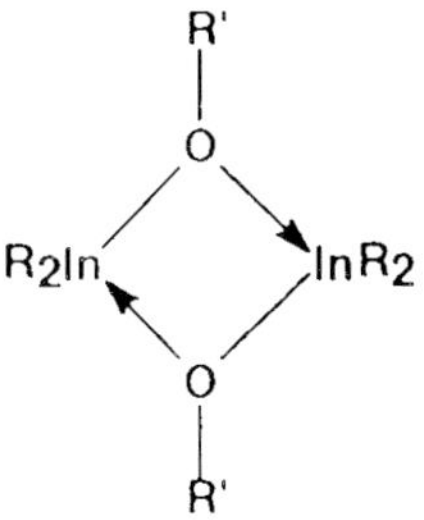

95

Other dimers are the dimethylindium derivatives of salicylaldehyde, **96** [279] and methyl salicylate [101a], and the methylindium derivative of bis(N-methylimidazol-2-yl)(pyridon-2-yl)methanol [280]. An electrospray mass spectrometry study also revealed the existence of dimeric species in solutions of several indium–oxygen compounds [281].

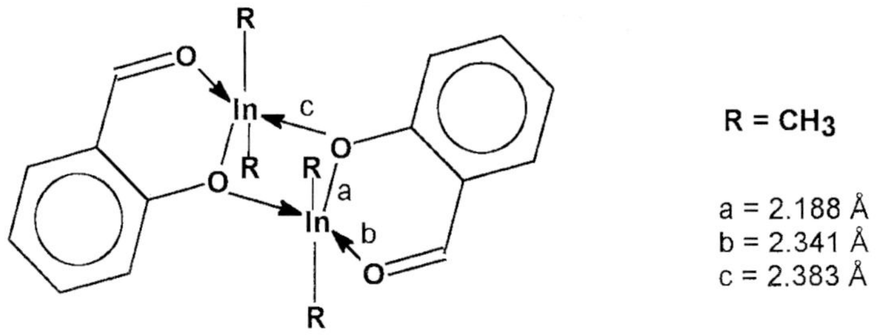

R = CH₃

a = 2.188 Å
b = 2.341 Å
c = 2.383 Å

96

A tetranuclear self-organized structure, **97**, containing an oxygen-centered adamantane-like aggregate $In_4O(\mu\text{-OH})_6$ has been identified in the oxo-hydroxide [{(Me₃Si)₃C}₄In₄O(OH)₆] (In–O 2.12 and 2.17 Å) [282] and a cubane structure skeleton is present in [{(Me₃Si)₃C}₄In₄O₄] [283].

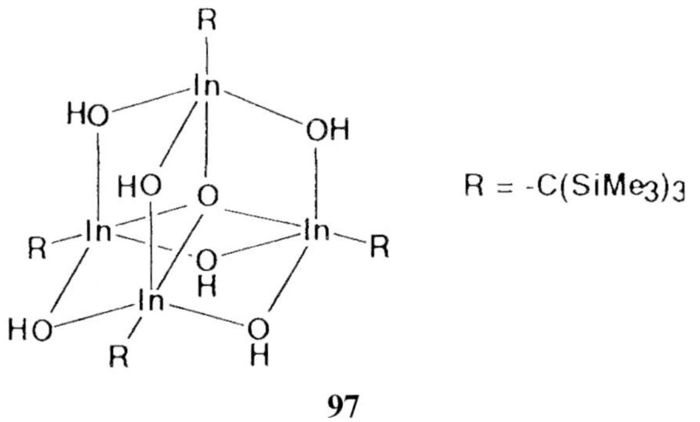

R = -C(SiMe₃)₃

97

Some supramolecular assemblies are formed via mixed bonds. One example is the dimesylamide derivative, [(Me₃SiCH₂)₂InN(SO₂Me)₂]₂, **98**, associated through both In–N (2.331 and 2.348 Å) and In–O bonds (2.302 and 2.386 Å) [284].

3.2.3.3 Organoindium–sulfur and –selenium compounds

This class can be illustrated by the dimers [(Bu^{t}CH₂)₂InSCH₂Bu^t]₂ (In–S 2.618 and 2.623 Å), [(Bu^{t}CH₂)₂InSeCH₂Bu^t]₂ (In–Se 2.705 and 2.719 Å) [215], [(Me₃CCH₂)₂InSePh]₂, and polymeric [Me(PhSe)In(μ-SePh)]$_x$ [285]. The compounds have been reviewed [113]. Some interatomic distances for other indium–sulfur and indium–selenium supermolecules are collected in Table 3.8.

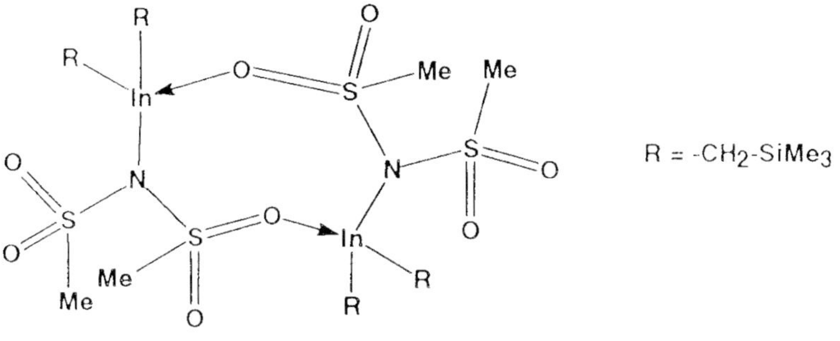

98

Table 3.8. Interatomic distances in indium–sulfur supermolecules.

Compound and ref.	In–S bond length (Å)
[Bu^{t}CH$_2$InSCH$_2$Bu^t]$_2$ [286]	2.618, 2.623
[Mes$_2$InSBu^t]$_2$ [287]	2.615–2.622
[Mes$_2$InSAmyl^t]$_2$ [287]	2.586–2.598
[Mes$_2$InSSiPh$_3$]$_2$ [287]	2.597–2.644
[Me$_2$InSSiPh$_3$]$_3$ [287]	2.589–2.644
[Mes$_2$InSSn(cyclo-C$_6$H$_{11}$)$_3$]$_2$ [288]	2.551
[Bu^{t}CH$_2$InSeCH$_2$Bu^t]$_2$ [286]	2.705, 2.719
[Bu^{t}CH$_2$InSePh]$_2$ [289]	2.745, 2.756
[Mes$_2$InSePh]$_2$ [113]	Ave 2.372
[Mes$_2$InSeMes]$_2$ [113]	Ave 2.715
[MeIn(SePh)$_2$]$_x$ [113]	2.541, 2.682
[(Bu^{t}CH$_2$)$_2$InSePh]$_2$ [285]	2.735, 2.745
	2.738, 2.756
[Mes$_2$In TePh]$_2$ [290]	Ave 2.915
[Mes$_2$In TePr^n]$_2$ [290]	Ave 2.909

3.2.3.4 Organoindium–nitrogen derivatives

A relatively large number of diorganoindium amides, R$_2$InNR'R', have been synthesized and several have been structurally characterized by X-ray diffraction. Practically all are self-organized cyclic dimers, **99**, containing an In$_2$N$_2$ ring.

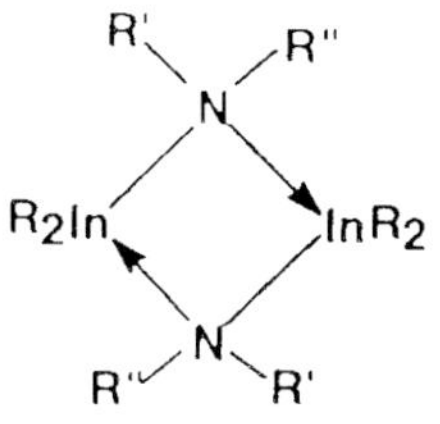

99

Table 3.9. Interatomic distances in indium–nitrogen supermolecules.

Compound and ref.	In–N bond length (Å)
[Me$_2$InNMe$_2$]$_2$ [292]	2.247, 2.225
[Me$_2$InNEt$_2$]$_2$ [293, 294]	2.234, 2.236
[Me$_2$InNPr$^i_2$]$_2$ [293]	2.242, 2.284
[Me$_2$InN(SiMe$_3$)$_2$]$_2$ [293]	2.304, 2.305
[Me$_2$InN(GeMe$_3$)$_2$]$_2$ [295]	2.259, 2.262
[Me$_2$InN(CH$_2$CH$_2$)$_2$NMe]$_2$ [296]	2.230, 2.235
[Me$_2$InNMePh]$_2$ [297]	2.280, 2.284
[Pr^i(Cl)InNHBu^t]$_2$ [298]	2.183, 2.187
[Pr$^i_2$InNHBu^t]$_2$ [299]	2.234
[Me$_2$InN=CMe$_2$]$_2$ [300]	2.194, 2.195
[Mes(Cl)InN(SiMe$_3$)$_2$]$_2$ [301]	2.018, 2.022
[Me$_2$InCN]$_x$ [302]	2.260

Structural data are collected in Table 3.9. The measured In–N interatomic distances observed are in the range 2.02–2.30 Å, usually larger than the sum of the covalent radii (2.112 Å). 'Purely' dative bonds in the amine adducts Me$_3$In·NHMe(CH$_2$)$_2$NHMe·InMe$_3$ (2.369 and 2.393 Å) and Me$_3$In·NHCMe$_2$(CH$_2$)$_3$-CMe$_2$ (2.502 Å) [291] are larger than the values cited. This indicates that in the supramolecular assemblies formed via In–N bonds, the bonds are a mixture of covalent and dative bonding.

The association is preserved in solution, even occasionally in the gas phase, as demonstrated by numerous NMR, mass spectrometric, and molecular weight determinations performed not only on the compounds characterized structurally, and cited here, but also on numerous others [293, 298, 303–305]. Indium–nitrogen cubane supermolecules, e.g. [MeInNBu^t]$_4$ [306], [(THF)MeInNC$_6$H$_4$F-p]$_4$ [155], and [MeInNC$_6$F$_5$]$_4$ [231a] have also been described.

Dimethylindium cyanide, [Me$_2$InCN]$_x$, has a supramolecular spiral chain structure, **100**, associated by In–N dative bonding (2.260 Å) [302].

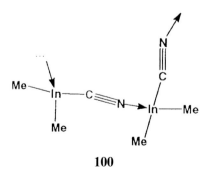

100

Linear arrays formed by nitrogen bridging of indium atoms have been reported

3.2 Group 13 Metals – Aluminum, Gallium, Indium, Thallium 139

in $[In\{(CH_2)_3NMe_2)\}_2(\mu-N_3)]_x$ (In–N 2.536 Å) and $[(CF_3SO_3)In\{(CH_2)_3NMe_2)\}_2]_x$ (In–N 2.564 Å). These compounds can serve as precursors for indium nitride [307].

A less common self-organized structure, **101**, is that of dimeric $[Me_2Si(NBu^t)_2InMe]_2$, which contains an In_2N_2 ring, as a result of dative bond association (In–N 2.277 and 2.256 Å) of two SiN_2In chelate rings [307].

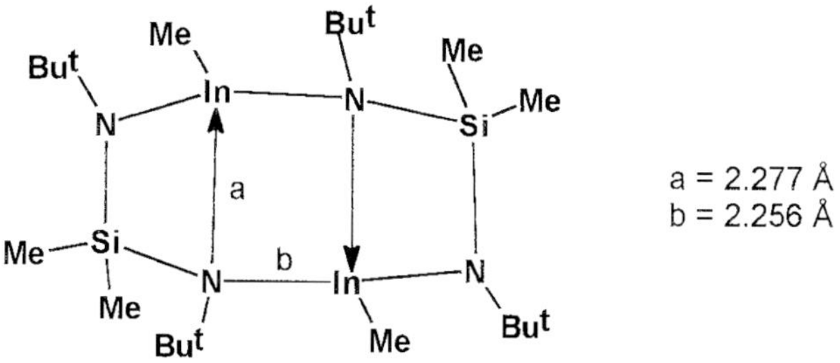

101

a = 2.277 Å
b = 2.256 Å

Also interesting is the supramolecular association of $Et_2InS_2CNMe(CH_2)_3NMe_2$ via In–N (2.66 Å) bonds to form polymeric chains, **102** [309].

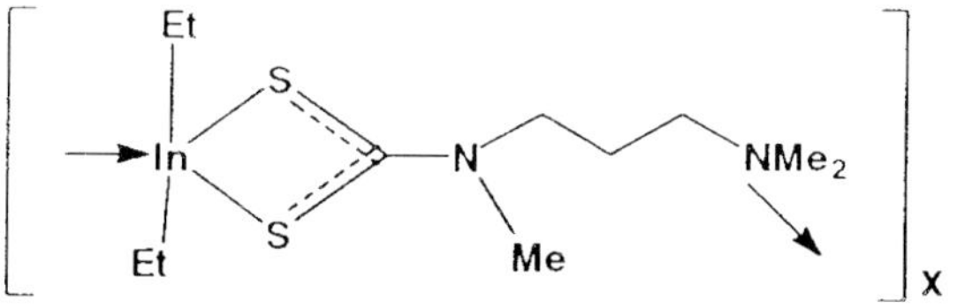

102

This type of self-assembly can probably be expected for many other organoindium derivatives of polyfunctional ligands.

3.2.3.5 Organoindium–phosphorus and –arsenic derivatives

Solution studies indicated the association of $[Me_2InPR_2]$ in dimeric and trimeric forms [310]. Solid-state structure determinations confirmed the existence of dimeric and trimeric $[R_2InPR'R'']_n$ ($n = 2$ and 3) supermolecules, **103** and **104**, formed by dative In–P bonding [311]. Some interatomic distances are collected in Table 3.10.

The In–P interatomic distances are in the range 2.61–2.69 Å, i.e. longer than the sum of the covalent In–P radii (2.498 Å) and close to the values measured for dative In–P bonds in molecular adducts $Me_3In\cdot PMe_3$ (2.683 Å) and $Me_3In\cdot PPh_2\text{-}CH_2CH_2PPh_2\cdot InMe_3$ (2.755 Å) [323]. In the tetrameric cubane-like compound $[Pr^iInPSiMe_3]_4$ (2.582–2.603 Å) [322], **105**, and $[MesInPMes]_4$ (In–P 2.582–2.623 Å) [318] the In–P bonds are somewhat shorter. Other organoindium phosphides, e.g. Et_2InPEt_2 and Me_2InPR_2 (R = Me, Et) are trimeric in solution but not char-

Table 3.10. Interatomic distances in indium–phosphorus supermolecules.

Compound and ref.	In–P bond length (Å)
[Me$_2$InPBu$^t_2$]$_2$ [305]	2.637, 2.656
[Et$_2$InPBu$^t_2$]$_2$ [312]	2.635
[(σ-C$_5$Me$_5$)(Cl)InP(SiMe$_3$)$_2$]$_2$ [313]	2.648, 2.594
[(Me$_3$SiCH$_2$)$_2$InPPh$_2$]$_2$ [314] A	2.664, 2.643
B	2.659, 2.632
[(Bu$^t_2$CH$_2$)$_2$InPEt$_2$]$_2$ [315]	2.623, 2.641
[(Bu$^t_2$CH$_2$)$_2$InPHCy]$_2$ [315]	2.613, 2.659
[(Me$_3$SiCH$_2$)$_2$InP(SiMe$_3$)$_2$]$_2$ [316]	2.656, 2.654
[(Me$_3$SiCH$_2$)(Cl)InP(SiMe$_3$)$_2$]$_2$ [317]	2.591, 2.595
[(Me$_3$SiCH$_2$)MeInP(SiMe$_3$)$_2$]$_2$ [317]	2.632, 2.638
[Ph$_2$InP(SiMe$_3$)$_2$]$_2$ [318]	2.612
[Me$_2$InPPh$_2$]$_3$ [319]	2.593, 2.633
[Pr$^i_2$InPPh$_2$]$_3$ [320]	2.635, 2.653
	2.650, 2.664 (two independent molecules)
[(Bu^{t}CH$_2$)$_2$InPPh$_2$]$_3$ [321]	2.699–2.677
[(PhCH$_2$)$_2$InPPh$_2$]$_3$ [320]	2.594–2.638
[Pr^{i}InPSiMe$_3$]$_4$ [322]	2.599, 2.595, 2.603
	2.590, 2.582, 2.583

Table 3.11. Interatomic distances in indium–arsenic supermolecules.

Compound and ref.	In–As bond length (Å)
[Me$_2$InAsMe$_2$]$_2$ [327]	
[(Me$_3$SiCH$_2$)$_2$InAs(SiMe$_3$)$_2$]$_2$ [328]	2.733, 2.722
[Ph$_2$InAs(SiMe$_3$)$_2$]$_2$ [329]	2.689, 2.683
[(PhCH$_2$)$_2$InAsBu$^t_2$]$_2$ [330]	2.716, 2.712
[Me$_2$InAsMe$_2$]$_3$ [331] planar	2.673–2.679
puckered	2.657–2.688

acterized in the solid state [324]. Compounds of this type are promising precursors for indium phosphide III–V semiconductors [325, 326]. Interatomic distances in indium–arsenic supermolecules are listed in Table 3.11.

3.2.4 Self-assembly of organothallium compounds

Few, mostly thallium–oxygen, compounds are mentioned here, because most organothallium derivatives self-assemble via weaker bonds which can be properly regarded as secondary bonds; in some borderline cases differentiation is difficult.

Supramolecular self-organization in the solid state has been observed (by X-ray diffraction) for several thallium–oxygen compounds. The secondary Tl···O bonds in these compounds are rather strong and it is debatable whether these should be regarded as electron-pair donor–acceptor bonds or as secondary bonds.

Bis(pentafluorophenyl)thallium(III) hydroxide, $(C_6F_5)_2TlOH$, forms a ladder architecture, **106**, involving Tl_2O_2 quasi-cyclic steps, with longer bridging Tl···O distances (2.51 and 2.69 Å) than the primary Tl–O(H) bonds (2.23 Å) [332]. The coordination geometry around thallium is very distorted trigonal bipyramidal, with oxygen atoms in two axial and one in equatorial positions. This leads to an uncommon angle between the aromatic groups (C–Tl–C 138.5°); R–Tl–R groups are usually almost linear.

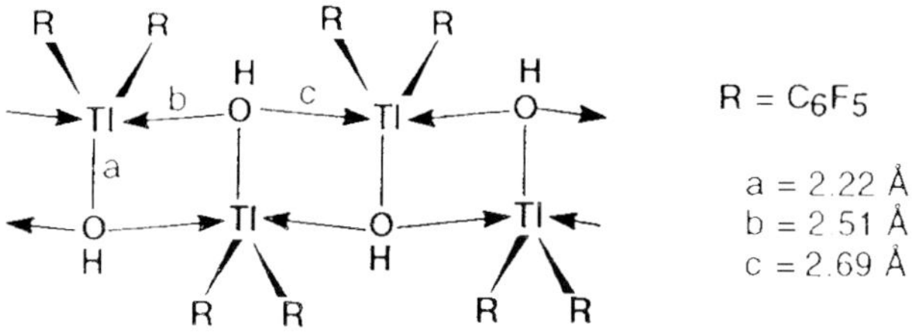

106

Two dimethylthallium(III) phenoxides, Me_2TlOPh and $Me_2TlOC_6H_4Cl$-2, **107** and **108**, are dimeric, with basically identical Tl–O distances in the four-membered ring formed by self-assembly [333]. These are longer than the Tl–O distance expected for a covalent bond, however. Molecular weight determinations in benzene show that the association is maintained in solutions of R_2TlOAr derivatives [334]

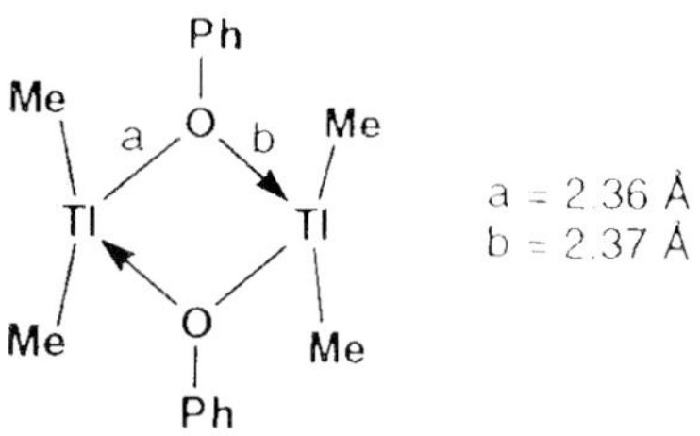

107 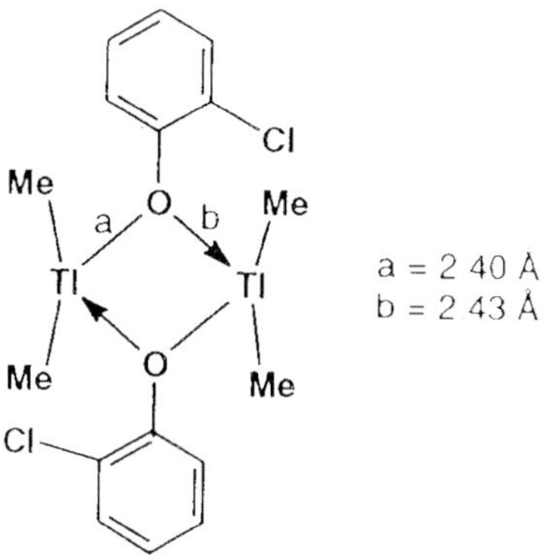   **108**

and results from ^{1}H and ^{205,203}Tl NMR studies also are indicative of the presence of dimers in solutions of R_2TlOEt [335].

Aryloxide compounds of Tl(I) without Tl–C bonds, also dimerize. Thus, 2,4,6-tris(trifluoromethyl)phenoxide, TlOC$_6$H$_2$(CF$_3$)$_3$, is a dimer, **109**, with identical but longer Tl–O bonds in a four-membered Tl$_2$O$_2$ ring [336].

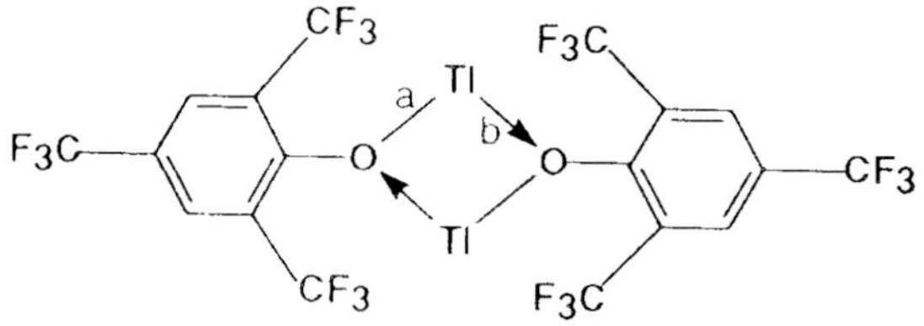

a = 2.461 Å
b = 2.469 Å

109

Polymeric self-assembled structures via Tl···O association have been identified in dimethylthallium(III) acetate, Me$_2$TlOOCMe, **110**, dimethylthallium(III) tropolonate, **111**, dimethylthallium(III) acetylacetonate, **112** [337], and the diethylthallium(III) salicylaldehyde derivative **113** [338].

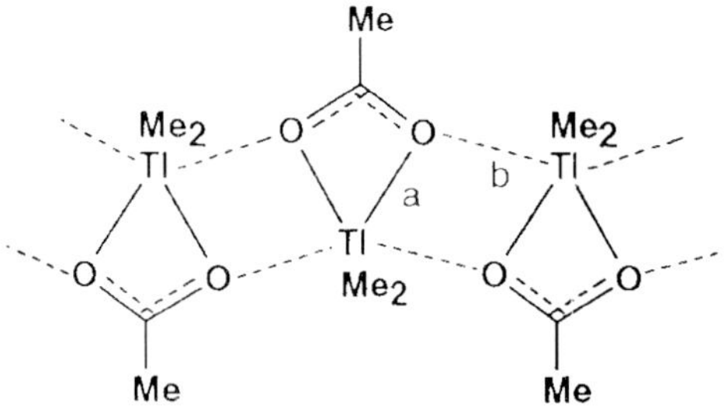

a = 2.61 Å
b = 2.67 Å

110

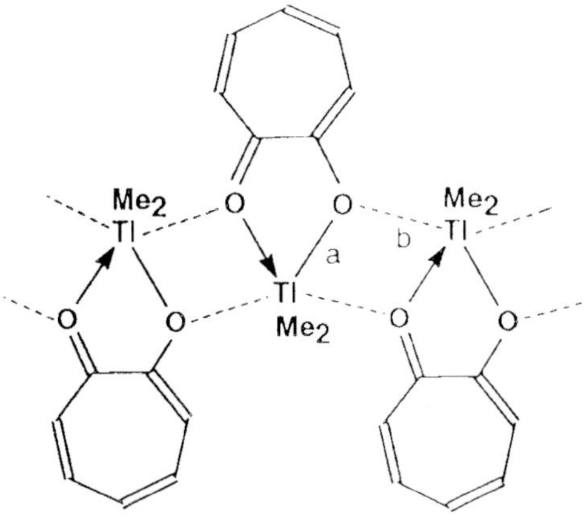

a = 2.47 Å
b = 2.74 Å

111

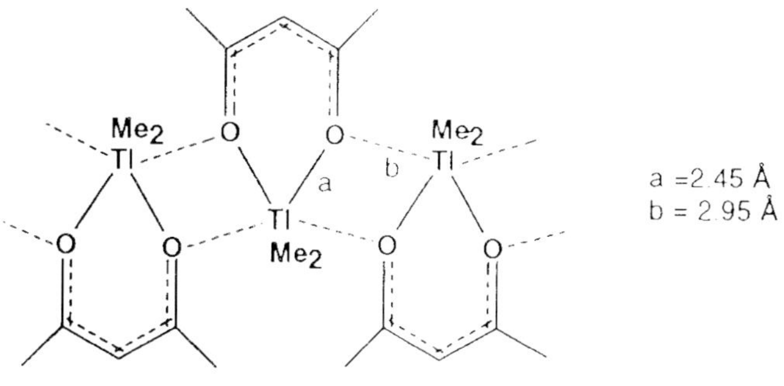

112

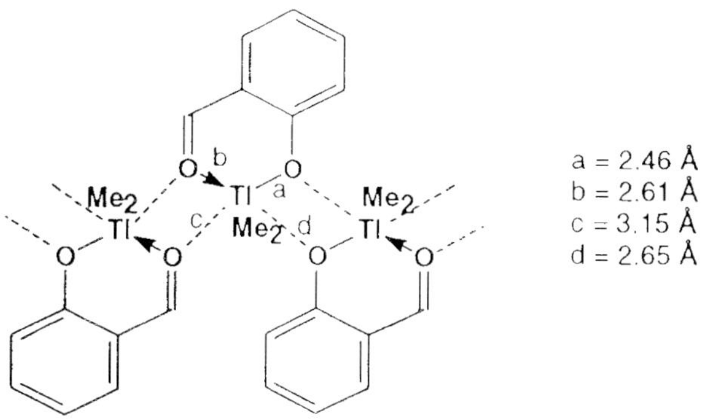

a = 2.45 Å
b = 2.95 Å

113

a = 2.46 Å
b = 2.61 Å
c = 3.15 Å
d = 2.65 Å

In all these compounds the supramolecular chain structure is based upon double bridging by Tl ← O bonds forming four-membered quasi-rings, alternating with the chelate rings formed by the ligands. The Tl···O bonds leading to self-organization are significantly longer than the intramolecular Tl–O bonds determined in the molecular tectons. Diphenylthallium tropolonate is a dimer (Tl–O 2.421 and 2.612 Å) [338a].

Bis(pentafluorophenyl)thallium(III) pentafluorobenzoate triphenylphosphine oxide adduct, $(C_6F_5)_2TlO_2CC_6F_5 \cdot OPPh_3$, **114**, associates as dimeric species via Tl···O bonds (2.789 Å) longer than the Tl–O bonds within the molecular unit (2.531 Å) [339]. Obviously, the coordination of the triphenylphosphine oxide donor prevents further extension of the assembly and limits it to a dinuclear pair.

The tendency of Tl···O bond formation, leading to self-organized supramolecular structures, is obvious in several instances of bridge formation between molecules which otherwise would contain four-coordinate thallium(III). Thus, two molecules of dimethylthallium(III) tryptophanate (with intramolecular Tl–O bonds of 2.468 Å and long Tl···N interactions of 3.434 Å) dimerize via longer Tl···O secondary bonds (Tl···O 2.593 Å) to form a central Tl_2O_2 four-membered moiety, **115** [340].

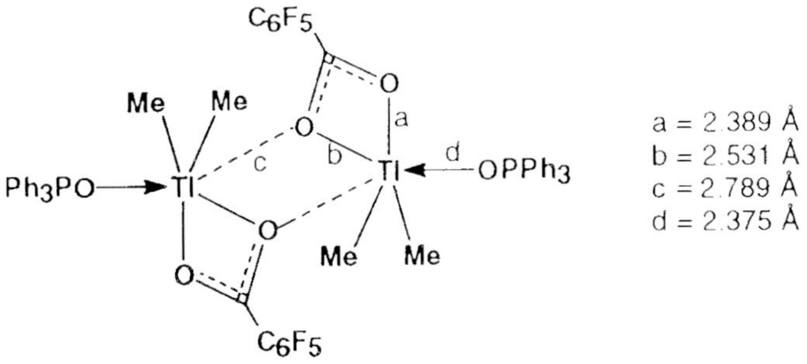

144   *3 Supramolecular Self-Assembly by Dative Bonding*

a = 2.389 Å
b = 2.531 Å
c = 2.789 Å
d = 2.375 Å

114

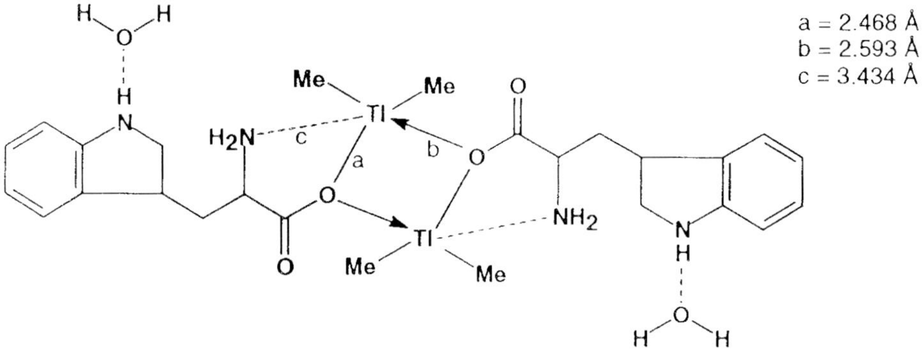

a = 2.468 Å
b = 2.593 Å
c = 3.434 Å

115

Secondary Tl···O bonds can connect polymeric chains of organothallium derivatives into ribbons extended along one of the crystal axes. Thus, dimethylthallium(III) rodaninate, **116** [341], and dimethylthallium(III) (L-phenylalaninate), **117** [342], which are themselves polymers based upon thallium–oxygen bonds from the ligand to the metal, are interconnected in double chains by weaker (and longer)

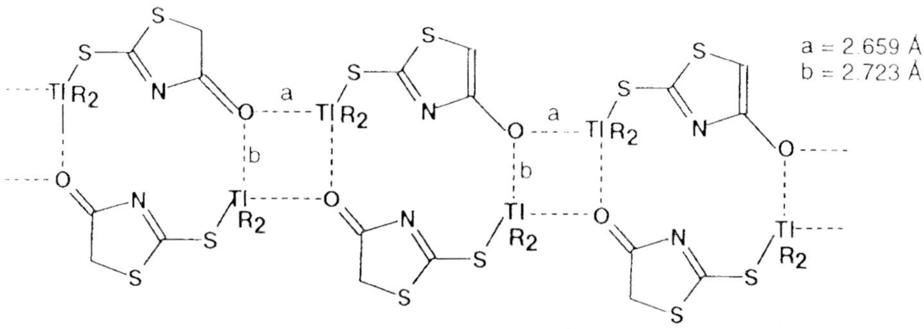

a = 2.659 Å
b = 2.723 Å

116

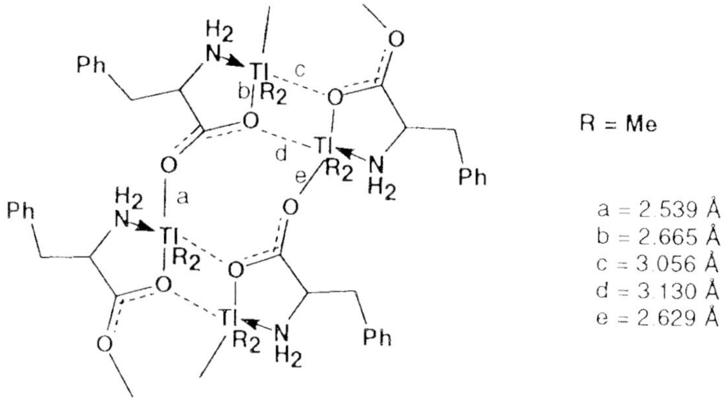

3.2 Group 13 Metals – Aluminum, Gallium, Indium, Thallium

R = Me

a = 2.539 Å
b = 2.665 Å
c = 3.056 Å
d = 3.130 Å
e = 2.629 Å

117

Tl···O secondary bonds, again with the formation of the same Tl$_2$O$_2$ motif encountered in the previous structures.

The apparent strong affinity of thallium for oxygen and its tendency to increase its coordination number at any price, sometimes results in unusual structures in which the donor oxygen atoms belong to such unlikely ligands as perchlorate or mesylamide. Thus, the dimethylthallium(III) perchlorate 1,10-dimethylphenanthroline complex, Me$_2$TlClO$_4$·Me$_2$Phen, **118**, forms helical chains by perchlorate bridging [343] and bis(trimethylsilylmethyl)thallium(III) dimesylamide, (Me$_3$SiCH$_2$)$_2$TlN(SO$_2$Me)$_2$, self-organizes into bidimensional layers, **119** [344].

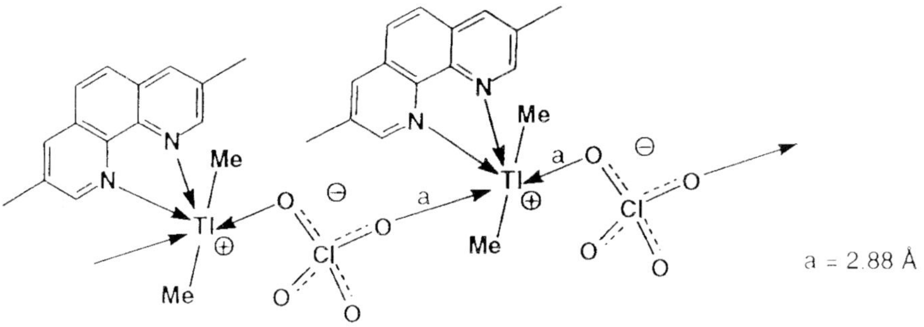

a = 2.88 Å

118

All these examples demonstrate the versatility of using a heavy metal, in a low oxidation state and with the tendency to achieve large coordination numbers, as a coordination center in tectons able to self-assemble into supramolecular structures. Manipulation of these possibilities might be successfully exploited in the construction of new supramolecular architectures.

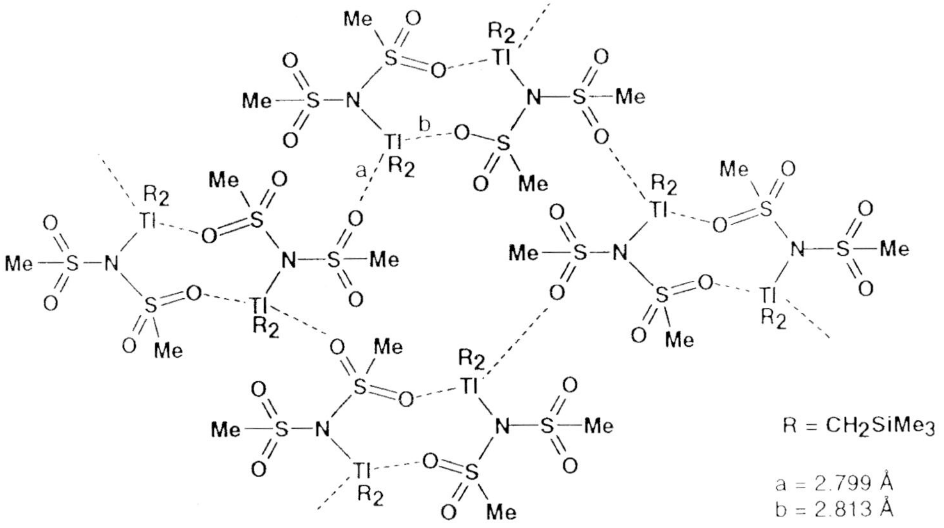

119

R = CH$_2$SiMe$_3$

a = 2.799 Å
b = 2.813 Å

3.3 Group 14 Metals – Tin and Lead

In Group 14 organotin and -lead compounds undergo supramolecular self-assembly in solution and self-organization in the solid state by dative bonding when the donor is a very electronegative element, e.g. fluorine or oxygen. In almost all other circumstances self-assembly is determined by secondary bonding and will be discussed in the next chapter. Some borderline cases are encountered in organotin–oxygen chemistry, e.g. with cyclic compounds such as stannolanes in which a broad interval of Sn–O interatomic distances covers the whole area from dative to secondary bonds. To avoid a discontinuity in presentation in this book, all these will be dealt with in the next chapter, within the framework of secondary bond self-assembly.

The ability of organotin compounds to self-assemble, with an increase of the coordination number of tin from four to five and six, has been recognized quite early, and the resulting compounds were termed 'autocomplexes' [345].

3.3.1 Self-assembly of organotin compounds

3.3.1.1 Organotin halides

Among the organotin halides, as suggested by the intermolecular bond distances in associated structures, only fluorides seem to contain dative bonds. The chlorides

Table 3.12. Triorganotin fluorides.

Compound and ref.	Form of association	Sn–F bond length (Å)	Sn←F bond length (Å)	Sn–F–Sn bond angle (°)
SnMe$_3$F [349]	Polymer	2.1	2.2–2.6	Non–Linear
SnMes$_3$F [350]	Monomer	1.961	–	–
SnPh$_3$F [351]	Polymer	2.146	2.146	Linear
Sn(CH$_2$SiMe$_3$)$_3$F [352]	Polymer	n.a.	–	–
SnBu$^n_3$F [353]	Polymer	n.a.	–	–
SnMe$_2$[C(SiMe$_2$Ph)$_3$]F [354]	Monomer	1.965	–	–
SnPh$_2$[C(SiMe$_3$)$_3$]F [354]	Monomer	1.965	–	–
SnCy$_3$F [348]	Polymer	2.051	2.303	Linear

and other halides are best described as self-assembled by secondary bonding interactions; these will be discussed in Section 4.3.1.

Triorganotin fluorides form polymeric chains in the solid state. The tin atom is five-coordinate and has trigonal bipyramidal geometry, with the halogen atoms in axial positions.

Some triorganotin fluorides have been found to be monomeric, **120**, and the bond length of 1.961 Å in trimesityltin fluoride has been suggested as a standard value for the single covalent Sn–F bond [346].

Except for the unusually high value initially reported for tricyclohexyltin fluoride (Sn–F 2.45 Å) [347], in other monomeric triorganotin fluorides the observed Sn–F interatomic distances are similar (see Table 3.12). It was recently demonstrated that this figure is wrong, that the correct value is 2.051 Å, and that the structure is asymmetric, with a second, dative Sn ← F intermolecular bond of 2.303 Å [348]. In trimethyltin fluoride a bent chain structure was reported in an early determination of crystal structure [349]. The Sn–F bond lengths, however, differ very little.

Triphenyltin fluoride has been found to have a symmetrical linear structure, **121** [351], the two Sn–F distances being identical (2.146 Å). The solid-state ^{119}Sn NMR spectrum shows equal coupling of tin with the two fluorine atoms (J_{SnF} 1500 Hz) [355], in agreement with a symmetrical structure. Similarly, for Me$_3$SnF and Bu$^i_3$SnF solid-state ^{119}Sn NMR spectra indicate five-coordinate tin and equal coupling [355]. Thus formula **121** describes the compound as a coordination polymer formed by dative (donor–acceptor) bonds. Some other triorganotin fluorides have also been found to be polymeric (see Table 3.12), and all are associated by dative bonding.

Some dinuclear, fluoride-bridged triorganotin complexes of the type [XR$_3$SnFSnR$_3$X]$^-$, **122**, with R = Me, but not with R = Ph, are formed in solution and have been detected with the aid of ^{119}Sn and ^{19}F NMR [356]. Such compounds can be regarded as fragments of the parent associated fluoride discussed above.

$$X-\underset{R}{\overset{R\quad R}{Sn}}\leftarrow F \rightarrow \underset{R\quad R}{\overset{R}{Sn}}-X \quad \ominus$$

122

Dimethyltin difluoride is a bidimensional infinite sheet with octahedrally coordinated tin and identical Sn–F donor–acceptor bonds (2.12 Å) [357]. It is, therefore, atypical of this class of compound. An unusual organodifluorostannate(IV), [NEt$_4$][Sn$_2$Me$_4$F$_5$], obtained from [NEt$_4$]F and Me$_2$SnF$_2$ in dichloroethane or acetonitrile, has a chain structure which can be formally derived from a linear polymeric [Me$_2$SnF$_3$]$^-$ anion; this contains a main chain of alternating tin and fluorine atoms, and each metal atom is coordinated by four equatorial fluorines. Additional Me$_2$SnF$_2$ molecules are associated with the main chain by fluorine bridges. Two terminal fluorine atoms complete the octahedral coordination around tin atoms.

In the main chain the Me–Sn–Me fragments are linear (C–Sn–C 180°) but the octahedral coordination at lateral tin atoms is distorted (C–Sn–C 167.0°). The Sn–F interatomic distances are non-equivalent; bridging bonds are significantly longer (Sn–F 2.272 Å, 2.115 Å, and 2.147 Å) than terminal bonds (Sn–F 2.026 Å) which explains the facile loss of tetrahedral Me$_2$SnF$_2$ units (as a precipitate) when the compound is dissolved in organic solvents [358]. Perhaps these longer distances correspond to secondary tin–fluorine bonds.

3.3.1.2 Dimeric organodistannoxanes [XR$_2$SnOSnR$_2$X]$_2$ and related compounds

The functional derivatives of distannoxanes, XR$_2$SnOSnR$_2$X, obtained as partial hydrolysis products of R$_2$SnCl$_2$ [359, 360] are, almost without exception, dimeric aggregates, formed via Sn ← O dative bonds, and as such can be regarded as supramolecular species. The dimers contain unsymmetrical, rhombus-shaped Sn$_2$O$_2$ four-membered rings, in which two sets of tin–oxygen bonds can be distinguished – a pair of short (ca 2.0 Å) Sn–O bonds, corresponding to the sum of the covalent radii of tin and oxygen (2.049 Å) and a pair of slightly longer (by ca. 20–25%), i.e. in the range 2.2–2.4 Å, which can be regarded as dative Sn ← O bonds. The general structure of this type of compounds is represented by **123**. Such structures have been demonstrated by crystal-structure determinations for the dichlorides [ClMe$_2$SnOSnMe$_2$Cl]$_2$ [361–363], [ClPr$^i_2$SnOSnPr$^i_2$Cl]$_2$ [364], [ClBu$^n_2$SnOSnBu$^n_2$Cl]$_2$ [365], hydroxo-chlorides [ClPr$^i_2$SnOSnPr$^i_2$(OH)]$_2$, and [Cl(Me$_3$SiCH$_2$)$_2$SnOSn(CH$_2$SiMe$_3$)$_2$(OH)]$_2$ [366], [ClPh$_2$SnOSnPh$_2$(OH)]·2DMF

(DMF = dimethylformamide) [367], and [ClPh$_2$SnOSnPh$_2$(OH)] [368]. Occasionally additional intramolecular coordination from the terminal substituent to exocyclic tin atoms transforms the dimer into a 'ladder'-type structure, **124**.

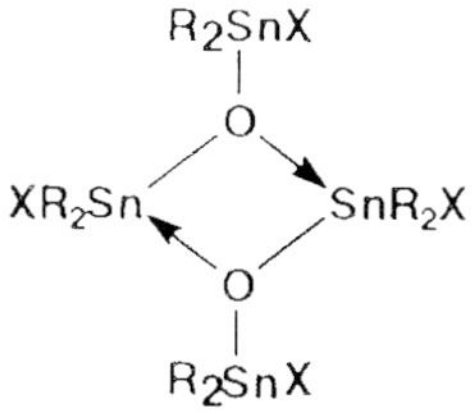

123

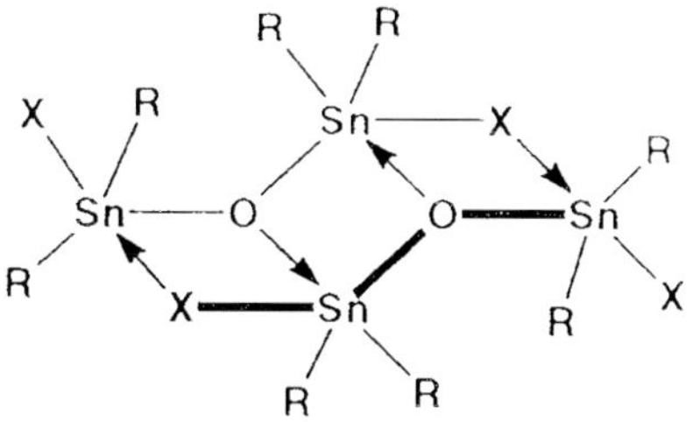

124

The simplest derivative, bis(chlorodimethyltin)oxide (or 1,3-dichlorotetramethyldistannoxane), (ClMe$_2$Sn)$_2$O, forms centrosymmetric dimers held together in the crystal by chlorine bridges. The dimeric unit skeleton is formed via O → Sn bonds only slightly longer (2.115 Å) than covalent Sn–O bonds (2.054 Å). Additional intra-unit Sn–Cl secondary bonds (2.710 and 3.348 Å) and large inter-unit Sn···Cl secondary interactions (3.411 and 3.3458 Å) significantly longer than Sn–Cl covalent bonds (Sn–Cl 2.438 Å) are also observed [369]. The *n*-butyl derivative (ClBu$^n_2$Sn)$_2$O is also dimeric in the solid state but the dimers are discrete units and are not further associated. In the dimer intramolecular chlorine bridges complete a ladder structure, **124** [365]. The fluoro derivative (FBu$^t_2$Sn)$_2$O has been structurally characterized only recently [370].

A series of dimers involving coordination from both oxygen and chlorine to tin, [ClR$_2$SnOSnR′$_2$Cl]$_2$ with R, R′ = Me, Pr^i, Ph, Bu^t, **125**, have been structurally characterized and their transformation in solution involving various self-assembling reactions has been investigated [371].

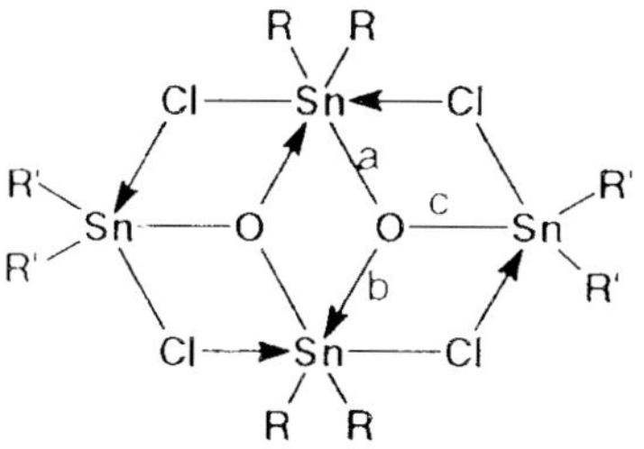

R = Me; R′ = Bu^t	R = Bu^n; R′ = Bu^t
a = 2.074 Å	a = 2.080 Å
b = 2.104 Å	b = 2.075 Å
c = 2.028 Å	c = 2.034 Å

125

A ladder supramolecular structure, **126**, of three fused Sn$_2$O$_2$ rings is present in the dimer [(EtO)Me$_2$SnOSnMe$_2$I]$_2$ [373].

R = Me

a = 2.131 Å
b = 2.178 Å
c = 2.171 Å
d = 2.014 Å
e = 2.131 Å

126

Similar dimers are [(2-MeOC$_6$H$_4$O)Bu$_2$SnOSnBu$_2$(OC$_6$H$_4$OMe-2)]$_2$ [373] and [(EtO)Me$_2$SnOSnMe$_2$(S$_2$COPr^i$_2$)]$_2$ [374].

The dimers survive in solution in a wide range of solvents, as has been demonstrated by molecular-weight determinations [375–377] and supported by ^{119}Sn NMR and Mössbauer spectroscopy [378].

A particular example is the thiocyanato derivative [(SCN)$_2$Me$_2$SnOSnMe$_2$·(SCN)]$_2$. In the crystal the dimers are connected by Sn···S secondary bonds (3.33 Å) to form infinite chains [379].

Beautiful supramolecular aggregates are formed by self-assembly of the reaction products of trimethylene-bridged chlorodi- and tristannanes and related tetrastannanes with cyclic trimeric di-*tert*-butyltin oxide (Bu^t$_2$SnO)$_3$. Thus, distannanes [Me$_3$SiCH$_2$(Cl$_2$)SnCH$_2$]$_2$CH$_2$ produced an octatin aggregate **127** [380] whereas tri- and tetratin chains, **128** and **129**, resulted in more complex structures [381]. NMR experiments have shown that the solid-state structure is retained in solution [380].

α,ω-Difunctional tristannadioxanes, X(R$_2$SnO)$_2$SnR$_2$X, are little known. The structurally characterized azido derivative (X = N$_3$) was found to dimerize into a supramolecular ladder structure via four interchain tin–oxygen bonds (2.045–2.068 Å) [382]. Carboxylates (X = OOCR') are discussed in Section 3.3.1.4.

Related dimeric supramolecular aggregates can be obtained with other terminal functional groups, e.g. β-diketonates, in the distannoxanes; an example is the (benzoyltrifluoroacetonato)(trifluoroacetato)tetramethyldistannoxane dimer [383]. This chemistry has not been pursued further.

3.3.1.3 Organotin hydroxides, alkoxides, and related compounds

Organotin hydroxides are frequently associated by strong tin–oxygen bonds, as reflected in interatomic distances just slightly longer than the covalent bonds. The sum of covalent bond radii for tin and oxygen is 2.049 Å. Because the observed tin–oxygen distances cover a very broad range of values, it will be arbitrarily assumed that those which are shorter than 2.5 Å are dative (donor–acceptor or Lewis acid–base interaction bonds) whereas those longer than 2.5 Å correspond to Sn···O secondary bonds, as discussed in Section 4.4.1.2.

The structure of the simplest organotin hydroxide, trimethyltin hydroxide, Me$_3$SnOH, has not been fully characterized although it has been subjected to an

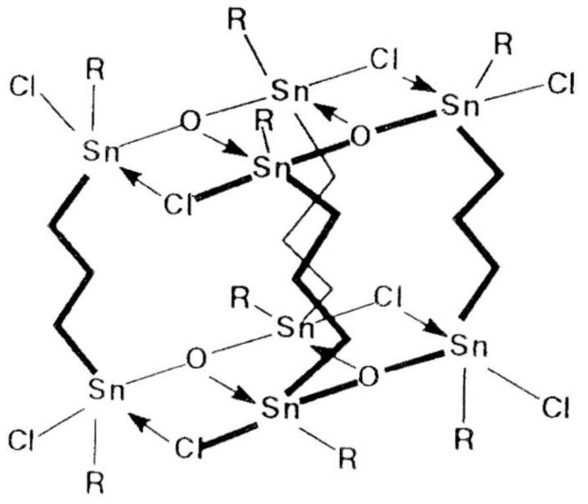

3.3 Group 14 Metals – Tin and Lead   151

127

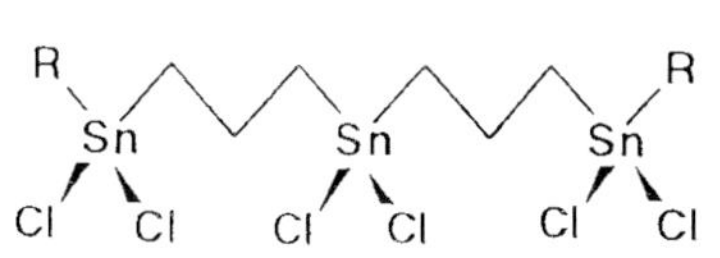

128 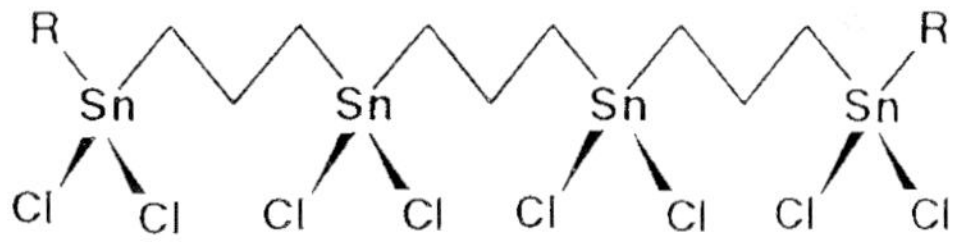   **129**

X-ray diffraction analysis, and is described as polymeric [384]. An indication of the association of trimethyltin moieties via hydroxo groups is given by the structure of the adduct 2Me$_3$SnOH·Me$_3$SnI, which contains a trimetallic cationic species, **130** [386].

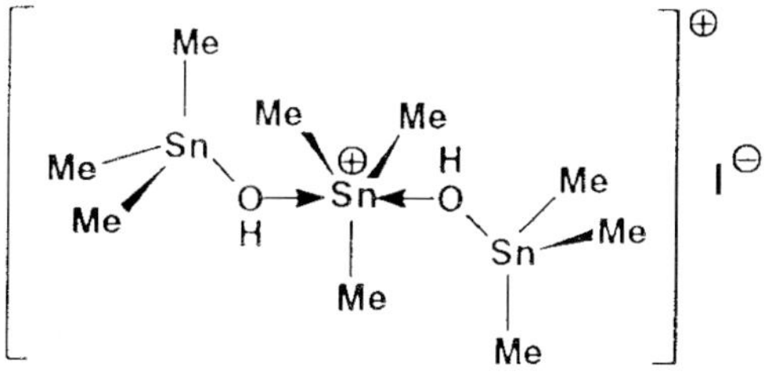

130

Triethyltin hydroxide [Et$_3$SnOH]$_x$ has been fully characterized structurally and contains trigonal-bipyramidal tin with OH groups in axial positions (Sn–O 2.156 and 2.244 Å) forming zigzag chains bent at oxygen (O–Sn–O 177.9°, Sn–O–Sn 145.5°) [387].

A clear supramolecular self-organized structure, **131**, associated through hydroxo bridges (Sn–O 2.197 Å, Sn ← O 2.255 Å), has been determined for triphenyltin hydroxide, Ph$_3$SnOH [388]; benzyldimethyltin hydroxide has a similar zigzag chain structure (Sn–O 2.17 Å, Sn ← O 2.29 Å). Cryoscopic measurements show that the compound is dimeric in benzene solution [389].

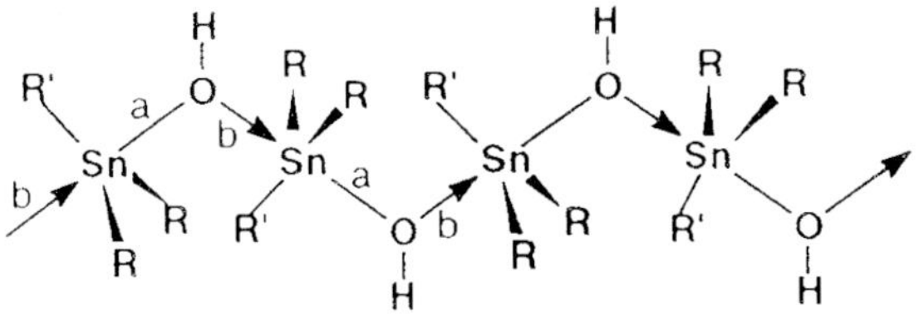

R = R' = Ph
a = 2.197 Å
b = 2.25 Å

R = Me, R' = PhCH$_2$
a = 2.17 Å
b = 2.29 Å

131

Other triorganotin hydroxides, containing bulky substituents, e.g. trineophyltin hydroxide [390] and trimesityltin hydroxide (Sn–O 1.999 Å) [391], are monomeric, non-associated.

Some mixed functional triorganotin derivatives have both hydroxo and other groups bridging. The mixed cyanate–hydroxide, Me$_3$SnNCO·Me$_3$SnOH contains alternating cyanato and hydroxo bridges in the chain, connecting planar Me$_3$Sn units, the tin atom becoming five-coordinate trigonal-pyramidal [392]. Further association occurs via hydrogen-bonding between the chains.

Another example is the supramolecular compound [Me$_3$Sn(μ-OH)SnMe$_3$-{μ-N(SO$_2$Me)$_2$}]$_x$, with hydroxo and bis(methylsulfonamido) bridges, **132** [393], and related dimers [Me$_2$Sn(μ-OH){N(SO$_2$Me)$_2$}]$_2$ [394] and [Me$_2$Sn(μ-OH)-{N(SO$_2$)$_2$C$_6$H$_4$}]$_2$ [394a]. Some adducts such as [Me$_2$Sn(μ-OH){N(SO$_2$Me)$_2$}(2L)]$_2$ with L = OP(NMe$_2$)$_3$ and PPh$_3$, **133** [395], and [Me$_2$Sn(μ-OH){N(SO$_2$Me)$_2$}-(phen)]$_2$ [394a] are also known.

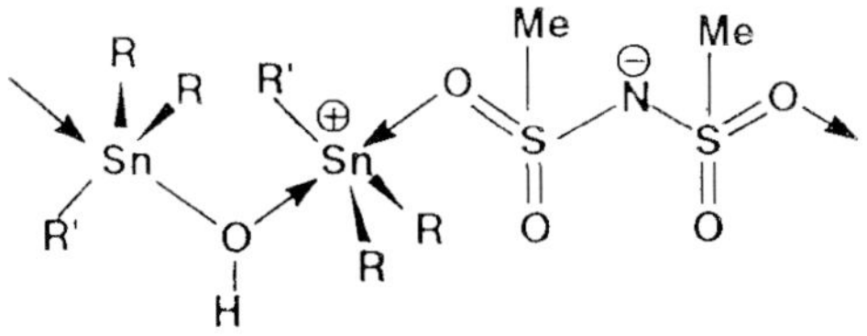

132

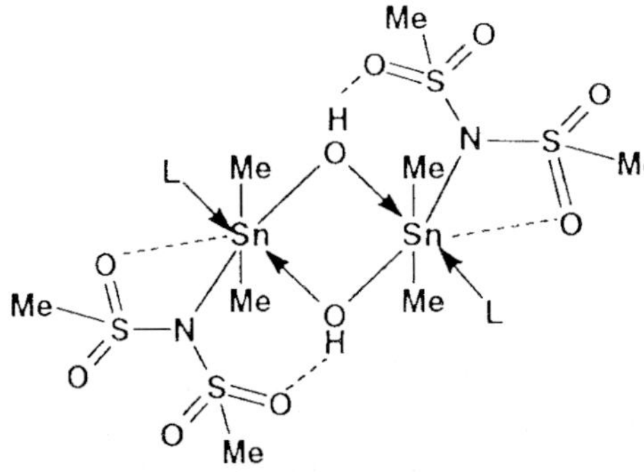

133

Diorganotin halide hydroxides [Bu$^t_2$Sn(OH)X]$_2$ (X = F, Cl, Br) are also dimers, **134**, associated by hydroxo bridges and held together in the crystal by hydrogen-bonding [396]. Tin is five-coordinate.

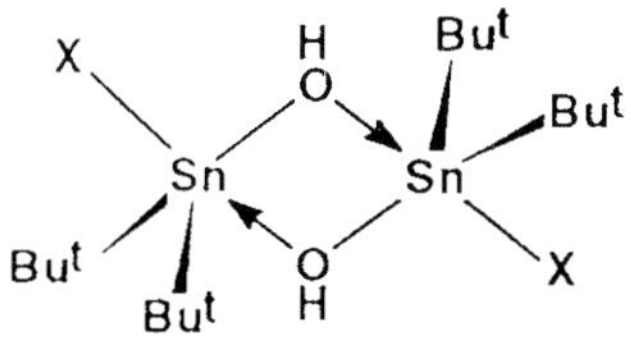

	X = F	Cl	Br
Sn–O	2.012	2.036	2.048 Å
Sn←O	2.119	2.237	2.257 Å

134

A similar structure (X = OH) has been observed for the dimethylaminomethyl ferrocenyl derivative [(C$_5$H$_5$FeC$_5$H$_3$-2-CH$_2$NMe$_2$)$_2$Sn(OH)$_2$]$_2$·2CHCl$_3$ [396a].

In 1-chloro-3-hydroxodistannoxanes the ladder structure is closed by hydroxo bridges, as found in the dimers [ClPr$^i_2$SnOSnPr$^i_2$OH]$_2$ and [Cl(Me$_3$SiCH$_2$)$_2$SnOSn(CH$_2$SiMe$_3$)$_2$OH]$_2$ [366]. Similar structures were observed in [ClPh$_2$SnOSnPh$_2$OH]$_2$ [397] and [HOR$_2$SnOSnR$_2$OH]$_2$ (R = alkyl) and in the aryloxy derivative [{(2-2-MeOC$_6$H$_4$O)Bu$^n_2$Sn}$_2$O]$_2$. The latter contains three sets of tin–oxygen bonds which can de described as normal covalent (Sn–O 2.011 and 2.083 Å), dative (O → Sn 2.150 and 2.198 Å) and secondary (Sn···O 2.792, 2.893, and 3.236 Å), illustrating the broad range of possible tin–oxygen interactions [398].

Polycyclic supermolecules containing boron, **135** [399], and silicon, **136** [400, 401], heterocycles fused to bicyclic Sn$_3$O$_3$ systems have been reported (R = Bu^t).

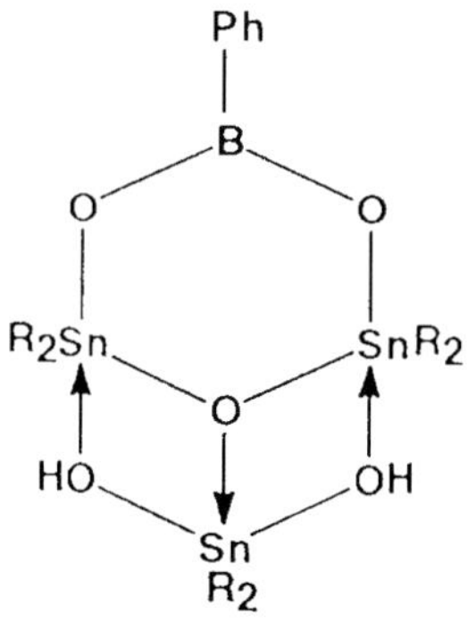

135 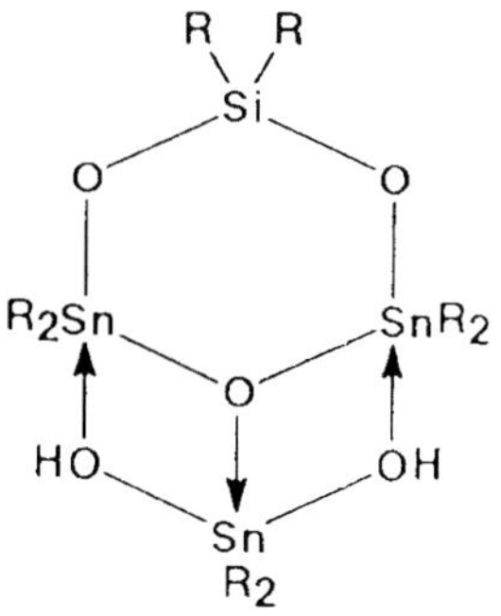   **136**

A trigonal prismatic self-assembled cluster [{Sn(OCMe$_2$C$_6$H$_4$)$_2$}$_3$Li$_2$O] incorporating two lithium cationic guests in an unprecedented Sn$_3$O$_7$ cage, has recently been reported and is one of the few known examples of heterometallic oxotin supermolecules [402].

Diorganotin carboxylates usually have complex polynuclear structures and are discussed in Section 3.3.1.4. Occasionally, hydroxo bridged dimers are formed; the *tert*-butyl acetate [Bu$^t{}_2$Sn(OH)(OAc)]$_2$ belongs to this class [403]. The perchlorate [Bu$_2$Sn(μ-OH)(ClO$_4$)]$_2$ has a similar structure [403a].

The dimeric itaconic acid chloro-hydroxo ester tin complex, [(MeOOCCH$_2$-(COOMe)CH$_2$Sn(μ-OH)Cl]$_2$·2dioxane also contains a dinuclear hydroxo-bridged nucleus [404]. The dioxane molecules are hydrogen-bonded to the OH bridges.

Monoorganotin hydroxo derivatives can also be associated. Thus, dimeric ethyltin hydroxide dichloride, **137**, [EtSn(μ-OH)Cl$_2$]$_2$ (Sn–O 2.067 Å, Sn ← O 2.152 Å) [405] and butyltin hydroxide dichloride, **138**, [BuSn(μ-OH)Cl$_2$]$_2$ [406] are illustrative examples. Tin is six-coordinate.

137   **138**

Other hydroxo-bridged dimers, [R$_2$Sn(μ-OH)X]$_2$, where X is an inorganic or organophosphorus anion, are discussed in Section 3.3.1.6.

A similar structure has been reported for [Bu^{n}Sn(μ-OH)(H$_2$O)Cl$_2$]$_2$ [406].

Careful hydrolysis of Pr^{i}SnCl$_3$ enables isolation of the crystalline compound (Pr^{i}Sn)$_9$O$_8$(OH)$_6$Cl$_5$·6DMSO, which is formed after several weeks storage of a DMSO solution of the Pr^{i}Sn(OH)$_2$Cl·3/4H$_2$O, the primary hydrolysis product. Crystal structure determination reveals a self-organized cage structure built up of five-coordinate (trigonal bipyramidal) and six-coordinate (octahedral) tin atoms linked by bridging hydroxo groups and trigonal oxygen atoms. Four-membered Sn$_2$O$_2$ and six-membered Sn$_3$O$_3$ rings can be distinguished in the structure [407].

Another cage made up from tin and oxygen with participation of primary and dative tin–oxygen bonds, formed by self-assembly, was found in the cation [(Pr^{i}Sn)$_{12}$O$_{14}$(OH)$_6$]$^{2+}$ identified in the compounds [(Pr^{i}Sn)$_{12}$O$_{14}$(OH)$_6$]Cl$_2$·L with L = 3H$_2$O, 2DMF, or 4H$_2$O, 4DMPU (DMPU = 1,3-dimethyltetrahydro-2(1H)-pyrimidinone), all structurally characterized by single-crystal X-ray diffraction. The cage contains five-coordinate (square pyramidal) and six-coordinate (octahedral) tin; six-membered Sn$_3$O$_3$ rings (half-chair conformation) fused to four-membered Sn$_2$O$_2$ rings are part of the framework. The Sn–O bond lengths vary between the limits known for primary and dative bonds [407]. A detailed solid-state ^{119}Sn NMR study of the compound correlated the spectrometric and diffractometric data [408]. The incomplete hydrolysis of butyltin trichloride leads to [(BuSn)$_{12}$O$_{14}$(OH)$_6$]-Cl$_2$·2H$_2$O, which has been structurally characterized [408a]. Full hydrolysis leads to [(BuSn)$_{12}$O$_{14}$(OH)$_6$](OH)$_2$, which can be formulated as [(BuSn)$_{12}$(μ_3-O)$_{14}$(μ-OH)$_6$]-(OH)$_2$ on the basis of a single-crystal X-ray analysis [408b–d].

3.3 Group 14 Metals – Tin and Lead   155

The formation of [(RSn)$_{12}$(μ_3-O)$_{14}$(μ-OH)$_6$](OH)$_2$ by hydrolysis of trialkynyltin derivatives RSn(CCR')$_3$ is somewhat surprising and demonstrates the high tendency of such self-assembly [409]. A related cation has been identified in the salt [(Pr^{i}Sn)$_{12}$O$_4$(OH)$_{24}$][Ag$_7$I$_{11}$]·NaCl·H$_2$O·10DMSO. In this compound the tin–oxygen cage acts as a host for the sodium cation encapsulated as a guest in the center of the polynuclear cation formed by self-assembly. The DMSO molecules are held by the cation via SnOH···O(S) hydrogen-bonds [410]. Solution and solid-state NMR investigations of the structure of self-assembled [(BuSn)$_{12}$O$_{14}$(OH)$_6$](O$_2$PPh$_2$)$_2$ throw some light upon the behavior of such species [411].

The 2-methyl-2,3-benzothiazole derivative [C$_8$H$_7$NS)]$_2^+$[{BuSn(μ-OH)Cl$_3$}$_2$]$^{2-}$·2C$_8$H$_7$NS is also a hydroxo-bridged dimer (Sn–O 2.043 Å, Sn ← O 2.107 Å); two molecules of the organic base are protonated, and all four are hydrogen-bonded to the organotin dimeric anion, **139** [412]. 2[C$_8$H$_7$NS]$^+$[{Bu^{n}Sn(μ-OH)Cl$_2$·H$_2$O]$_2$ has a similar structure [413].

139

a = 2.043 Å
b = 2.107

The original paper describes this compound as an "outer sphere complex in which the ligands are not coordinated directly to the tin atom but are held in position by hydrogen-bonds". The compound is a good example of self-assembly simultaneously involving different types of intermolecular force.

Alkoxo-bridged triorganotin derivatives form zigzag chains, **140**, by Sn ← O dative bonding. Trimethyltin methoxide, Me$_3$SnOMe, is a polymer with alternating Me$_3$Sn moieties and MeO bridges and basically equalized tin–oxygen bonds (Sn–O 2.20 Å, Sn ← O 2.26 Å) [414]. Monoorganotin trialkoxides, RSn(OR')$_3$, are also associated, as suggested by their low volatility; crystal-structure determination of Pr^{i}Sn(OPr^i)$_3$ revealed a dimeric structure with an asymmetric Sn$_2$O$_2$ nucleus formed by dative bonding (Sn–O 2.026 Å, O → Sn 2.273 Å) [415].

Mössbauer and ^{119}Sn NMR spectroscopic studies of several diorganotin alkoxides, R$_2$Sn(OR')$_2$, and triorganotin alkoxides, R$_3$SnOR', indicate self-assembly

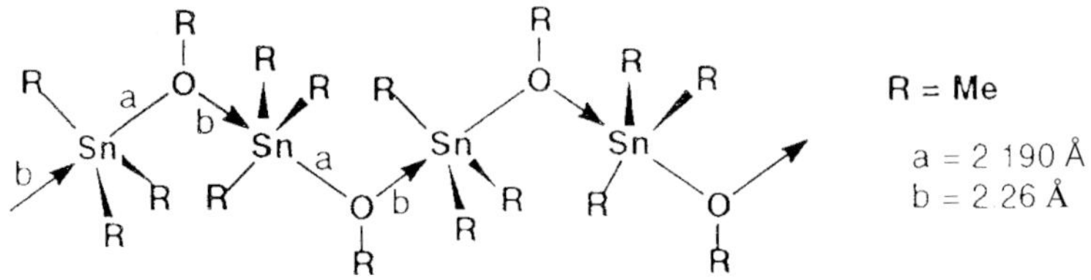

156   3 Supramolecular Self-Assembly by Dative Bonding

140

in the solid state and in solution. The extent of association depends upon the size and number of OR groups and is sensitive to dilution and temperature [416].

Cyclohexanoneoximato trimethyltin, $Me_3SnON=C_6H_{10}$, forms zigzag chains with oxygen bridges, **141**; in this compound the donor–acceptor Sn ← O bonds seem to be weaker (only 2.48 Å compared with Sn–O 2.190 Å) and it might be considered as approaching a secondary Sn···O bond [417].

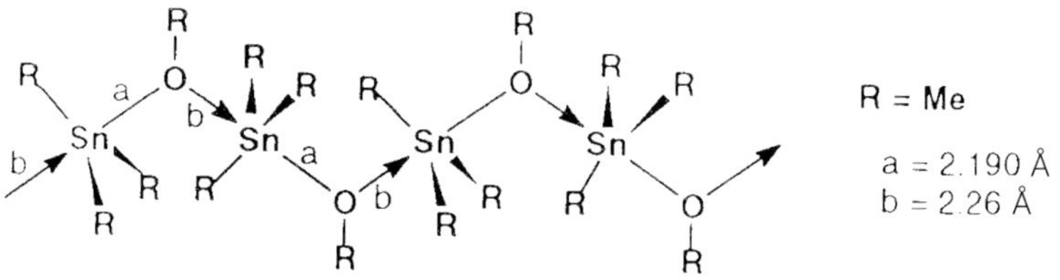

141

Alkoxo bridges are also present in the dimeric xanthate $[(Pr^iOCS_2)Me_2Sn-(\mu\text{-}OEt)OSnMe_2]_2$ (Sn–O 2.129 and 2.152 Å in the bridge) [418].

3.3.1.4 Organotin carboxylates

Organotin carboxylates are seldom monomeric compounds. They tend to self-assemble into supramolecular structures by tin–oxygen donor–acceptor bonding. In addition, weaker interactions might also be present and as a result three sets of tin–oxygen interatomic distances can be observed in these compounds – primary Sn–O covalent bonds (ca 2.0 Å), slightly longer dative Sn ← O bonds (ca 2.2–2.3 Å) and Sn···O secondary interactions (>2.5 Å). In carboxylates it is usually possible to distinguish between single C–O bonds (in the C–O–Sn moiety), and double C=O bonds (involved in donation to tin C=O → Sn).

The structural chemistry of organotin carboxylates has been thoroughly reviewed [419, 420] and only a brief survey of self-assembled supramolecular systems will be presented here.

Triorganotin carboxylates tend to be monomers (**142** and **143**) only with aromatic carboxylic acids (substituted benzoates) or with bulky substituents at tin (e.g. Ph or Cy) [420]. Most other organotin carboxylates are supramolecular aggregates, usually chain polymers with carboxylato bridging.

3.3 Group 14 Metals – Tin and Lead   157

142   **143**

In the polymers tin becomes five-coordinated, trigonal bipyramidal, with the oxygen atoms in axial positions. Intra-unit Sn···O distances are ca. 3.0 Å, which suggests very weak interaction, and in many crystal structure determinations they were simply ignored. They usually distort the trigonal bipyramidal geometry, however.

It has been shown by ^{119}Sn Mössbauer spectral studies [416] that the supramolecular self-assembly is maintained in solution.

A selection of structurally characterized triorganotin carboxylates is listed in Table 3.13, in order of complexity of the organic groups at tin. The Sn–O and Sn ← O bond lengths and O–Sn ← O axial bond angles are given.

It has been shown on the basis of crystal-structure investigation of seventeen triorganotin carboxylates that the repeating tin–tin (non-bonding) distance along a crystal axis in supramolecular arrays of $[R_3SnOOCR']_x$, **144**, is virtually independent of both R and R' and is close to 5.185 Å. Increasing the size of R and R' only changes the distances between chains in the unit cell of the crystal [450].

144

Some triorganotin carboxylate structures, in particular those containing two R_3Sn moieties, deserve additional comment because their structures are more varied. For example, rare types of organotin carboxylates include derivatives of carboranecarboxylic acids $[\{(1,7-C_2B_{10}H_{11}-1-COO)Bu_2Sn\}_2O]_2$ (structure of type A) [451]. The bis(triphenyltin) derivative of phenylmaleic acid, **145**, contains only four-coordinate tetrahedral tin (Sn–O 2.077 and 2.090 Å) and is not associated, but in the bis(triphenyltin) citraconate, **146**, one tin atom participates in the supramolecular association, **147**, and polymer-chain formation and becomes five-coordinate (trigonal pyramidal; Sn–O 2.193 Å, Sn ← O 2.397 Å) whereas the second remains four-coordinate (Sn–O 2.089 Å), as a part of a dangling side chain [441].

Table 3.13. Triorganotin carboxylates, $R_3SnO(O)CR'$.

R	R'	Sn–O bond length (Å)	Sn←O bond length (Å)	O–Sn←O bond angle (°)	Ref.
Me	Me	2.205	2.391	171.6	[421]
Me	CF_3	2.177	2.458	174.8	[421]
Me	C_6H_4Cl-2	2.200	2.414	174.2	[422]
Me	C_6H_4OMe-2	2.208	2.381	170.8	[422]
Me	$C_6H_4NMe_2$-2	2.201	2.426	170.0	[422]
Me	$C_6H_4NH_2$-4·C_6H_6	2.169	2.477	173.5	[422]
		2.168	2.416	174.1	
Me	C_4H_3O (furyl)	2.191	2.430	172.4	[423]
Me	C_4H_3S (thienyl)	2.149	2.482	173.4	[424]
Me	CH(OH)Ph (mandelate)				[425]
Me	C_6H_4-N=N-C_6H_3-(OH-2)(Me-5)	2.139	2.497	174.3	[426]
Me	$CH(NH_3^+)CH_2CH_2COO^-$	2.221	2.301	174.8	[427]
Vinyl	Me	2.20	2.33	172.7	[428]
Vinyl	CH_2Cl	2.210	2.338	174.2	[428]
Vinyl	CCl_3	2.17	2.49	173.4	[429]
Vinyl	$C_5H_4FeC_5H_5$ (ferrocenyl)	2.12	2.42	172.5	[430]
Me_2Ph	Me	2.201	2.370	174.0	[431]
Bu^n	$CHPh_2$	2.222	2.390	175.0	[432]
Bu^n	$CH_2CH_2SC(S)$-NC_5H_{10}	2.210	2.399	n.a.	[433]
Bu^n	C_5H_5N-2-COO^- $[Cy_2NH_2]^+$ (pyridylcarboxylate)	2.26	2.31	173.2	[434]
Bu^n	$CH_2(3$-$C_9H_6N)$ (indolyl)	2.199	2.524	173.5	[435]
Bu^n	C_6H_4-2-$(SO_3^-)[Cy_2NH_2]^+$	2.169	2.392	170.7	[436]
Ph	H (formate)	2.219	2.317	173.6	[437]
		2.219	2.318	173.1	
Ph	Me	2.185	2.349	173.6	[438]
Ph	CH_2Cl	2.20	2.372	174.7	[439]
Ph	C_6H_4Cl-2	2.201	2.384	173.8	[440]
Ph	CMe=CH-$COOSnBu_3$	2.193	2.397	170.4	[441]
		2.089 (lateral)			
Ph	CMe=CH-CO-N	2.139	2.418	176.9	[442]
Ph	CH=N-OMe	2.185	2.367	173.2	[443]
Ph	$(CH_2)_5$-N=CH-C_6H_4OH-2	2.148	2.328	176.2	[444]
Ph	$(CH_2)_2$-$C(O)Ph$	2.266	2.246	174.6	[445]
Ph	CH_2-$SC(S)NMe_2$	2.199	2.307	172.5	[446]
$PhCH_2$	Me	2.14	2.65	168.6	[447]
		2.131	2.556	169.5	[448]
$ClMe_2Sn$-OOC-C_6H_4N-2 (pyridyl)		2.125	2.285	175.3	[449]
		2.076	2.375	175.9	

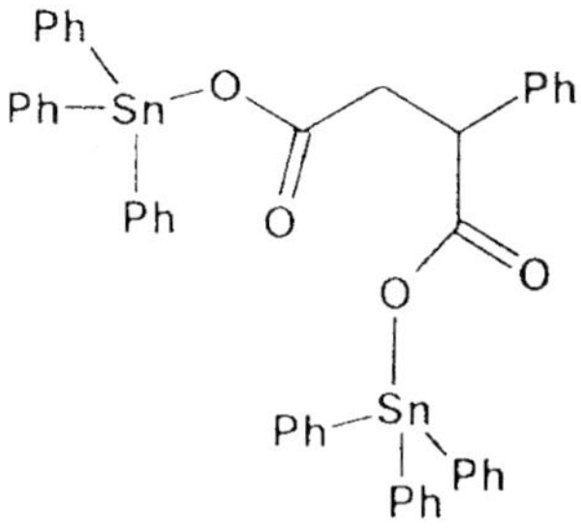

145

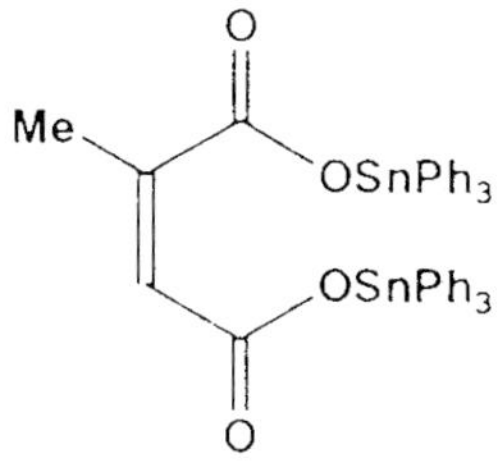

146

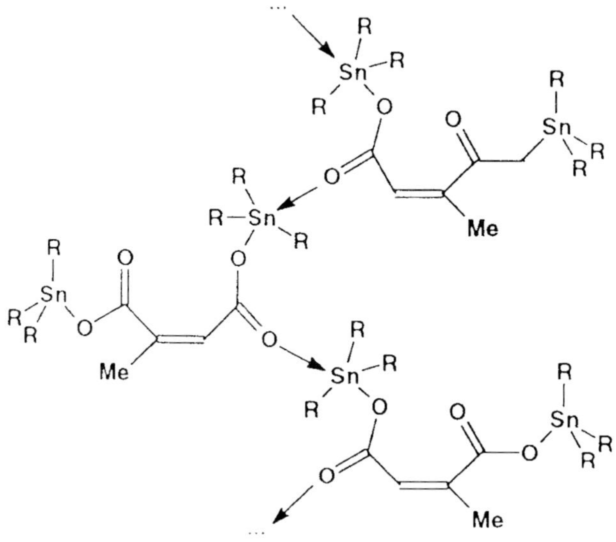

147 R = Ph

Trinuclear self-assembled supermolecules, **148**, have been identified in diorganotin salicylaldoximates [452–454, 454a].

In the supramolecular assembly of the tetranuclear complex, **149**, bis-(dicyclohexylammonium) tris(malonato)tetrakis(tributylstannate), 2[Cy$_2$NH$_2$]$^+$·[{Bu$_3$SnOOCCH$_2$COOSnBu$_3$}{Bu$_3$SnOOCCH$_2$COO$^-$}$_2$]·2EtOH all four tin atoms are five-coordinate (trigonal bipyramidal) and the bis(tributyltin) unit bridges two mono(tributyltin)units. The structure also contains hydrogen-bonds to the cations and Sn ← O bonds from the terminal ethanol molecules [455].

A hexanuclear supramolecular aggregate, **150**, is present in the structure of [bis-(triphenyltin)succinate·2quinoline-*N*-oxide]·2[bis(triphenyltin)succinate], [Ph$_3$SnOOCCH$_2$CH$_2$COOSnPh$_3$·2(quinoline-*N*-oxide)][Ph$_3$SnOOCCH$_2$CH$_2$COOSnPh$_3$], all six tin atoms being five-coordinate [456].

Bis(trimethyltin)-2,2'-bipyridyl-4,4'-dicarboxylate, Me$_3$SnOOC-[2,2-bipy]-

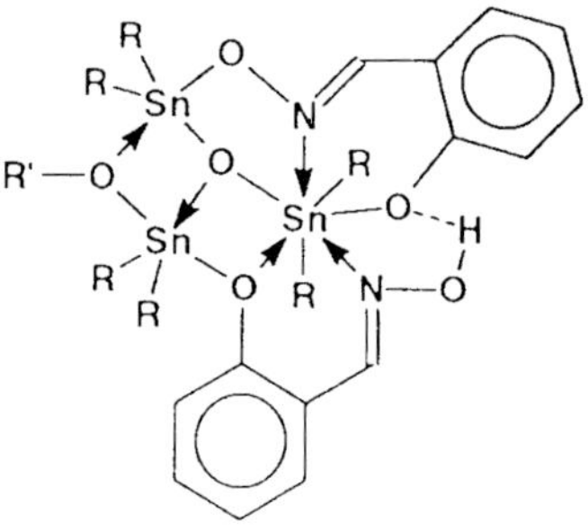

148

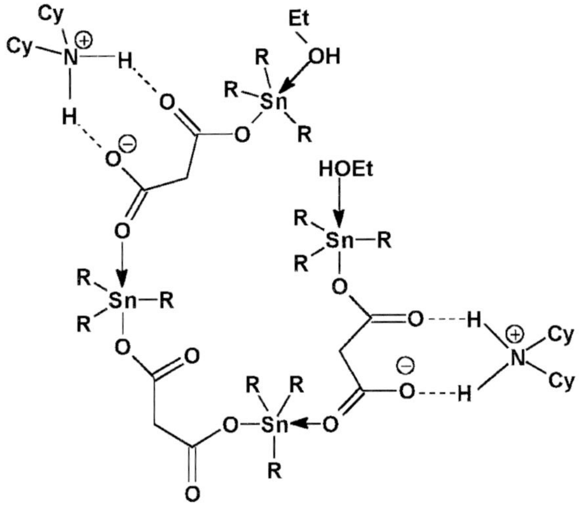

149

R = Bu^n

150

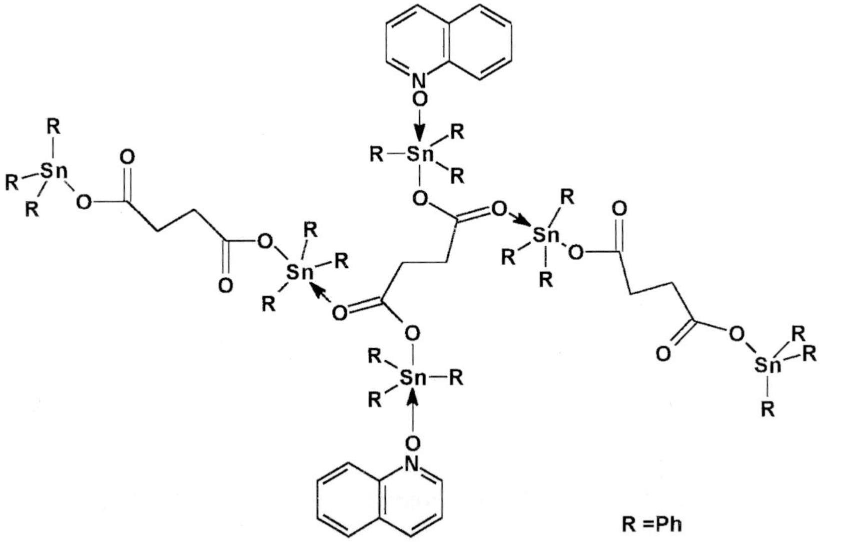

R = Ph

151

COOSnMe$_3$, **151**, has a two-dimensional, supramolecular, layered structure comprising sheets interconnected by trigonal bipyramidal, five-coordinate trimethyltin moieties (Sn–O 2.145, Sn ← O 2.519 Å, O–Sn–O 175.1°) [457].

Bis(trimethyltin)malonate, Me$_3$SnOOCCH$_2$COOSnMe$_3$, **152**, forms a three-dimensional supramolecular network, **153**, by interaction of pentacoordinated, trigonal-bipyramidal Me$_3$Sn units (Sn–O 2.15 and 2.17 Å, Sn ← O 2.44 and 2.46 Å, O–Sn–O 172.5 and 178.5°). Each SnMe$_3$ group is connected to two malonate units and each malonate is connected through four Me$_3$Sn groups to four other neighboring dicarboxylate fragments, to form a 3D structure, finally consisting of four helices connected such as to form a 24-membered ring [458].

Another three-dimensional supramolecular structure, **154**, is found in bis-(triphenyltin) succinate-[bis(triphenyltin)succinate·2DMF], Ph$_3$SnOOCCH$_2$CH$_2$-COOSnPh$_3$·[DMF·Ph$_3$SnOOCCH$_2$CH$_2$COOSnPh$_3$·DMF], all four tin atoms being five-coordinate trigonal bipyramidal [459].

The participation of a second functional group is illustrated by triphenyltin salicylidene-6-aminohexanoate, **155**, in which the five-coordinate tin is bonded to an oxygen from the carboxylic group (Sn–O 2.148 Å) and to the phenolic oxygen of another molecule [460].

The self-assembly of triorganotin carboxylates containing (OR′)$_2$P=O donor binding sites, e.g. Ph$_3$SnOOC(CH$_2$)$_n$P(OEt)$_2$O with $n = 1$ [461] and $n = 2$ [462] is also observed. The tin–oxygen bonds in these cases seem to be slightly weaker, as

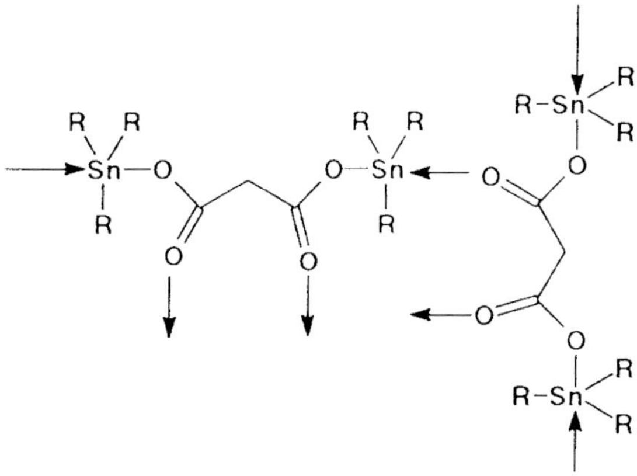

152

R = Me

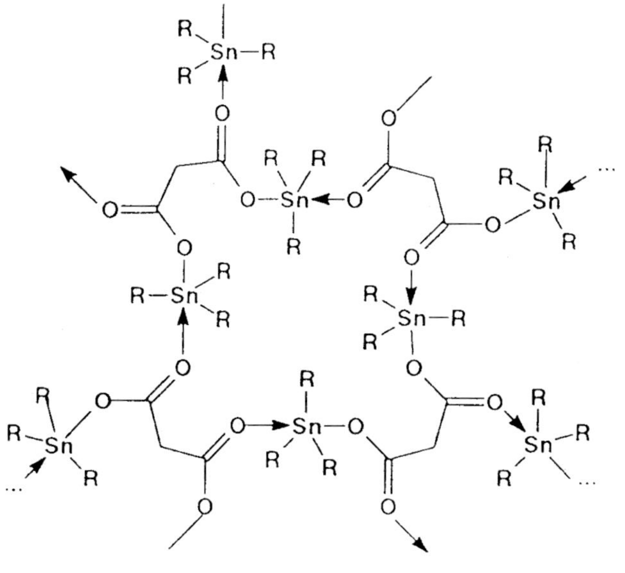

153

illustrated by the O → Sn interatomic distances, 2.420 and 2.397 Å, respectively, longer than the bonds to carboxylate oxygens (2.129 and 2.116 Å). In dimeric [ClPh$_2$SnCH$_2$P(=O)Ph$_2$]$_2$, **156**, the intermolecular tin–oxygen bonds leading to self-assembly are shorter (2.170 and 2.291 Å) and fall in the normal range [462a].

There are numerous other examples of self-assembly induced by an additional donor contained in the R′ group of [R$_3$SnOOCR′]$_x$ supramolecular triorganotin carboxylates.

Diorganotin dicarboxylates, R$_2$Sn(OOCR′)$_2$, are extremely moisture-sensitive and are difficult to obtain in pure state. Some have, however, been isolated, e.g. mono-

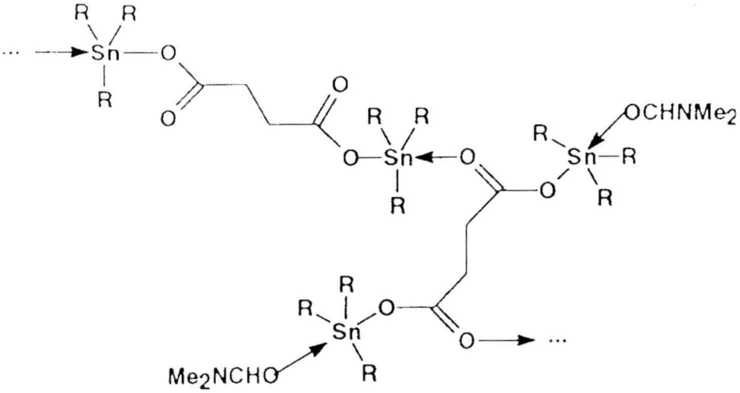

154

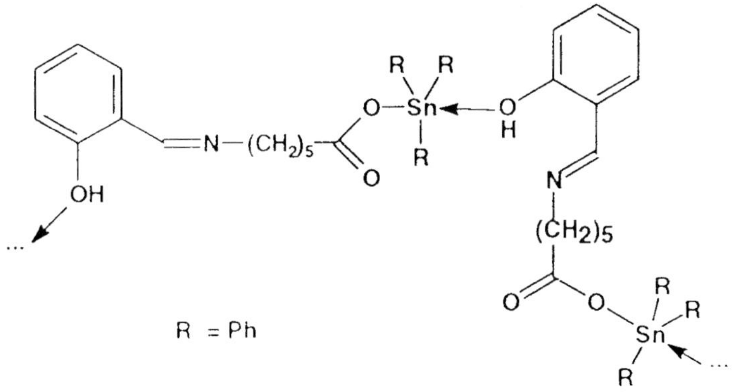

155

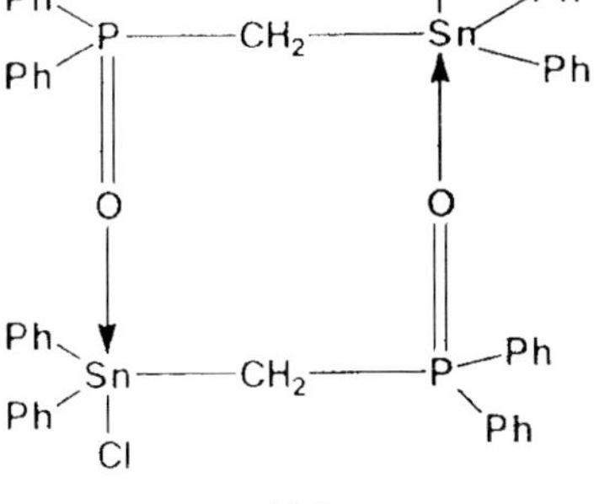

156

meric Me$_2$Sn(OOCCH$_3$)$_2$, 3D-polymeric [Me$_2$Sn(OOCCF$_3$)$_2$]$_x$ [463], monomeric bis-chelated (2-amino-5-chlorobenzoato)dibutyltin [464], polymeric dimethyltin formate, [Me$_2$Sn(OOCH)$_2$]$_x$ [465], and others [466–468]. When, occasionally, the organic groups contain an additional donor function this can form intramolecular dative bonds and prevent supramolecular association, as in dibutyltin bis(picolinate), $Bu^n_2Sn(OOCC_5NH_4)_2$, **157** [469].

157

Diorganotin dicarboxylates tend to hydrolyze and to form distannoxane carboxylates (diorganotin oxocarboxylates) $[\{R_2Sn(OOCR')\}_2O]_2$, usually based on planar Sn_2O_2 cores. Two additional tin atoms connected to this core lead to the formation of an $R_8Sn_4O_2$ central skeleton, common to all diorganotin oxocarboxylates. The carboxylate groups can function as bridges between two tin atoms, thus helping to hold together the supramolecular structure, and/or act as mono- or bidentate chelating groups attached only to one (terminal) tin atom. Additional weak interactions between the second oxygen of the carboxyl group (C=O) might contribute to slight variations in the structure and coordination geometry of tin, or might lead to intermolecular association to form more complex supramolecular structures, in which the tetratin aggregates are further associated. The supramolecular self-assembly is maintained in solution, as has been demonstrated by a comparison of solution and solid-state ^{119}Sn NMR spectra [470]. Many of these compounds have promising antitumor properties, with high in vitro activity [471, 472] (few in vivo investigations have been reported).

There are four main types of tetranuclear diorganotin oxocarboxylate aggregates (Scheme 3.2), depending upon the number of carboxylato bridges. The most common is type A, with two bridging carboxylato groups and two monodentate or chelating to a single tin atom each. In these compounds it is more difficult to distinguish the arrangement of primary covalent Sn–O bonds and dative Sn ← O bonds, but interatomic distances between 2.0 and 2.2 Å are observed.

A list of bis(diorganotin)oxide carboxylates of type A is given in Table 3.14.

A second type of structure (B), containing three bridging carboxylato groups and one basically monodentate group, is found in the dimethyltin acetato complex, $[\{Me_2Sn(OAc)\}_2O]_2$ (Sn–O 2.14, 2.24, 2.25 Å, Sn···O 2.56 Å) [487], and in bis-(5-methoxysalicylato-di-n-butyltin)oxide (Sn–O 2.040 and 2.147 Å) [488].

Type C, with all four carboxylato groups bridging, is the structure of dimethyltin trichloroacetato and *tert*-butylcarboxylato derivatives $[\{Me_2Sn(OOCR')\}_2O]_2$, R' = CCl_3 [489] and Bu^t [490]. A variation is found in $[\{Me_2Sn(OOCPr)\}_2O]_2$.

In type D there is no bridging carboxylato group, two are anisobidentate chelating and two are basically monodentate with some weaker Sn···O interaction to a second tin atom. This structure is found in $[\{R_2Sn(OOCR')\}_2O]_2$ with R = Me, R' = CH_2Cl [491]; R = Me, R' = $C_6H_4NH_2$-p [492]; R = Bu, R' = CH_2N =

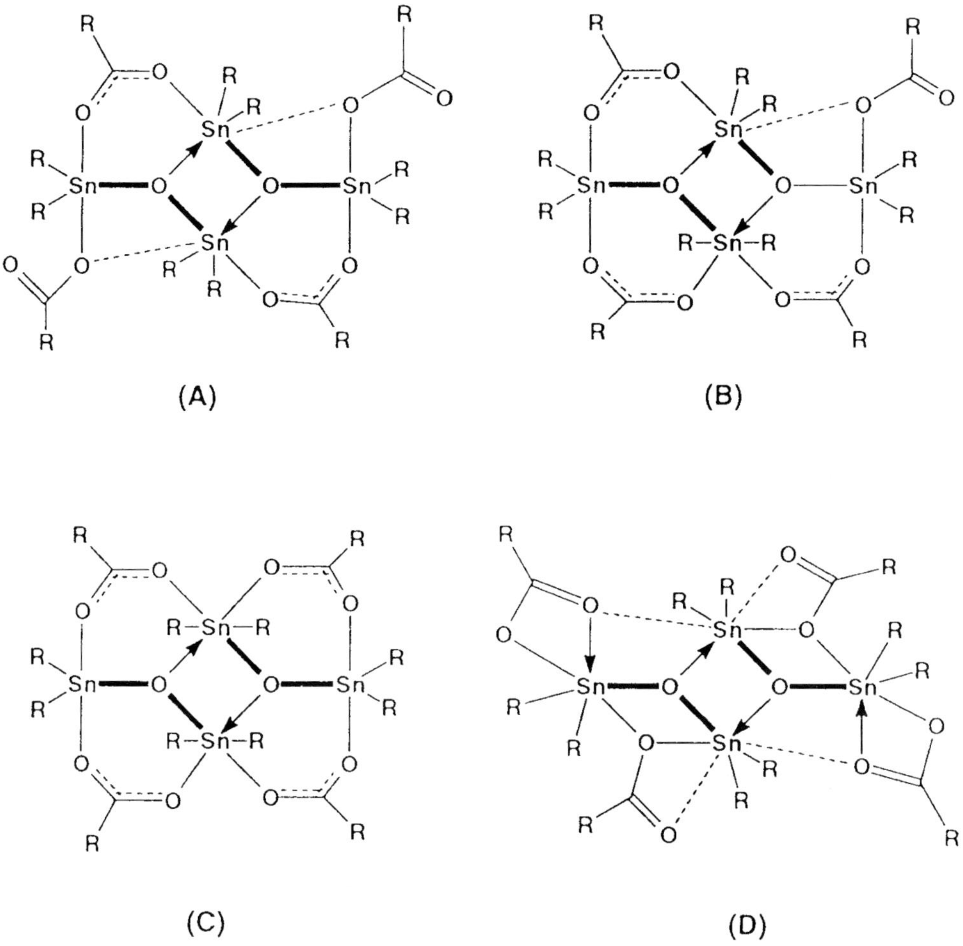

Scheme 3.2. Structural types of diorganotin oxocarboxylates.

CHC$_6$H$_4$OH [493]; R = Bu^n, R′ = 2-pyridyl (picolinate) [494]; R = Bu^n, R′ = CH$_2$CH$_2$C(O)Ph [495]; and R = Bu, R′ = η^5-C$_5$H$_5$FeC$_5$H$_4$ (ferrocenyl) [496]. In the picolinato derivative cited the nitrogen atoms are coordinated to tin, increasing its coordination number to six.

Some organotin carboxylates have structures totally different from those just discussed. For example, Bu$_2$Sn(OOC$_6$H$_4$Br-2)$_2$ is a weakly bridged dimer [497], whereas di-*n*-butyltin cyclobutane-dicarboxylate, Bu^n$_2$Sn(OOC)$_2$C$_4$H$_6$, forms a zigzag polymer [498]. In both compounds tin is six-coordinate octahedral.

A novel type of diorganotin carboxylate, **158**, has been obtained with trichloroacetic acid; the compound [Bu$_4$Sn$_2$(O$_2$CCCl$_3$)$_3$(μ-OH)] contains one Sn–OH–Sn bridge (Sn–O 2.031 and 2.113 Å), one bridging carboxylato group (Sn–O 2.255 and 2.278 Å) and two monodentate carboxylato ligands [499]. The phenyl derivative

Table 3.14. Bis(diorganotin)oxide carboxylate supermolecules of type A.

R	R'	Sn–O bond a	lengths (Å) b	Ref.
Me	CF$_3$	2.039	2.137	[473]
Me	CCl$_3$	2.115	2.069	[474]
Me	C$_6$H$_5$	2.043	2.154 (2 independent molecules)	[475]
		2.140	2.037	
Bu	C$_6$H$_2$(OMe)$_3$	2.166	2.151	[476]
Me	CH-N	2.204	2.034	[477]
Bu^n	C$_6$F$_5$	2.035	2.034	[478]
Vinyl	CF$_3$	2.11	2.08	[479]
Pr^n	CH$_2$SPh	2.062	2.166	[480]
Bu^n	CH$_2$SPh	2.049	2.163	[480]
Bu^n	CCl$_3$	2.12	2.03	[481]
Bu^n	C$_4$H$_4$S (thienyl)	2.177	2.034	[482]
Bu^n	C$_6$H$_4$OMe-2	2.161	2.041 (2 independent molecules)	[483]
		2.034	2.163	
Bu^n	C$_6$H$_4$NH$_2$-2	2.17	2.09 (2 independent molecules)	[484]
		2.01	2.17	
Bu^n	C$_6$H$_4$NO$_2$	2.149	2.053	[485]
Ph	CCl$_3$ (isomer A)	2.154	2.062	[486]

[Ph$_8$Sn$_4$(OOCCCl$_3$)$_6$(μ-OH)$_2$] also has a structure which differs from those of other diorganotin carboxylates [499a]. These structural changes are the result of introducing OH groups instead of oxo groups. The process cannot be controlled and the isolation of the last two compounds is serendipitous.

158

Other structures

Several other supramolecular structures, in which oxygen–tin donation is the driving force in the supramolecular association of organotin compounds, are mentioned here.

Di-*n*-butyltin thiosalicylate molecules, **159**, are self-organized into hexameric aggregates, **160**, involving both primary Sn–O bonds and donation from carboxylic oxygens (Sn–O 2.21–2.24 Å), internal chelating Sn···O bonds (3.07–3.16 Å) and tin–sulfur bonds (2.40–2.41 Å) [500].

Di-*n*-butyltin 3-thiolopropionate, **161**, also forms hexameric aggregates, **162**, in the solid state as a result of tin–oxygen donation (Sn ← O 2.18–2.20 Å), primary Sn–O and Sn–S bonds being involved only endocyclic in the building unit [501]. The compound is monomeric in solution.

168   3 Supramolecular Self-Assembly by Dative Bonding

Supramolecular self-assembly is observed in diethyltin 3-(2-pyridyl)-2-thiolopropenoate, which forms linear, zigzag polymeric arrays with O → Sn bonds connecting the chelate C$_2$OSSn rings (intrachelate Sn–O 2.226 Å and Sn–S 2.424 Å, intermolecular Sn–O 2.219 Å). The pyridine units do not participate in any coordination to tin [501a].

Dinuclear, **163**, and tetranuclear, **164**, supramolecular aggregates have been suggested on the basis of molecular-weight determinations and ^{119}Sn NMR spectra (indicating six-coordinate tin) for some diorganotin derivatives of [4-(1-phenyl-3-methyl-5-pyrazolone)]dioxoalkanes [502].

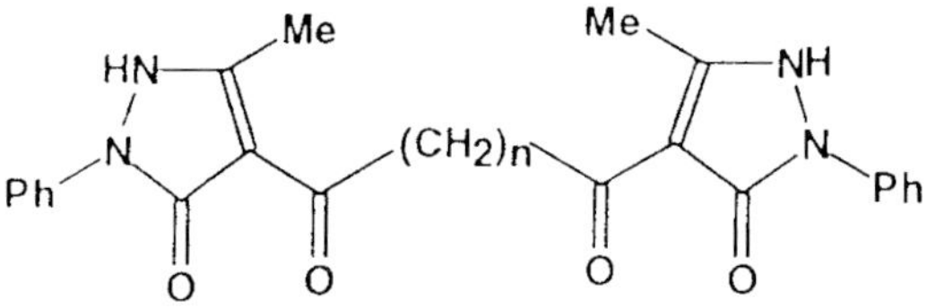

163a

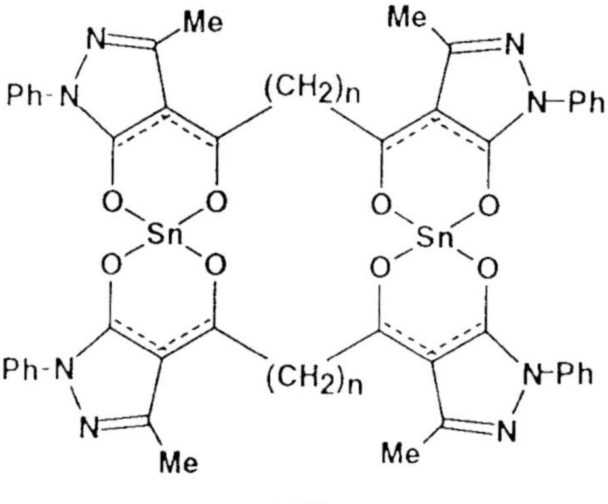

163b

The diversity of the structures cited suggest that the subject of supramolecular association in organotin carboxylato derivatives is far from being exhausted, and novel interesting structures can be expected from the use of new carboxylic acids.

3.3.1.5 Monoorganotin oxo-carboxylates

As the number of organic groups connected to tin decreases the structures become more complex because more sites are available for primary covalent and dative bonding. Monoorganotin tris(carboxylates), RSn(OOCR′)$_3$, have been little investigated (although reported quite early [503, 504]); much better known are the monoorganotin oxo carboxylates of general composition RSnO(OOCR′). In very early studies of these compounds molecular weight determination revealed their association in solution, and various trimeric and hexameric structures were suggested [505–

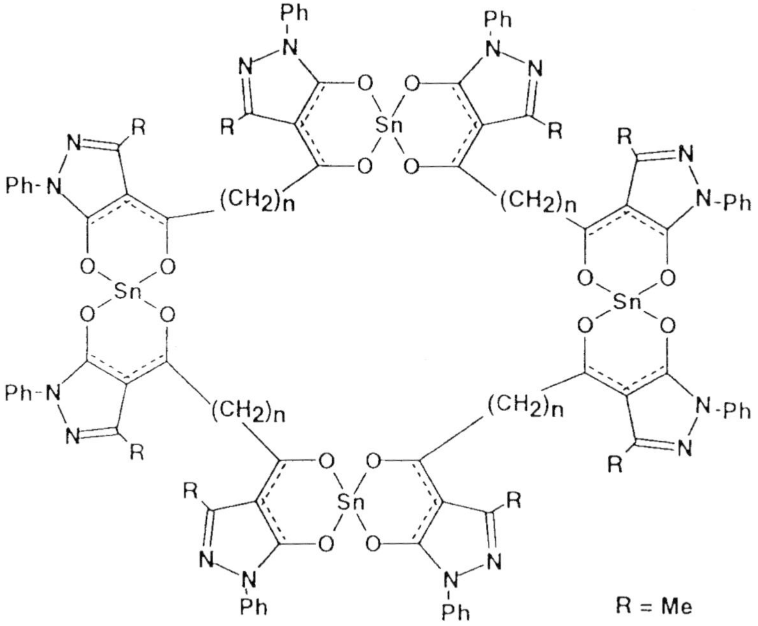

3.3 *Group 14 Metals – Tin and Lead*   169

164

509]. Not until X-ray diffraction studies were performed was their structure finally clarified. Now, two types of hexanuclear organotin oxo-carboxylates are known – hexamers of composition [RSnO(OOCR′)]$_6$, which have a hexagonal prismatic cage structure, **165**, (or 'drum' structure) and 'ladder' type compounds, **166**, of composition [{RSnO(OOCR′)}$_2${RSn(OOCR′)$_2$X}] (X = R′COO or Cl); both contain Sn$_2$O$_2$ aggregates. The compounds have been reviewed [510].

In the 'drum' hexamers [RSnO(OOCR′)]$_6$ structurally characterized so far (R = R′ = Me [511]; R = Bu^n, R′ = Pr^i, Bu^t [512]; *cyclo*-C$_5$H$_9$ and *cyclo*-C$_6$H$_{11}$ [513]; C$_6$H$_4$NO$_2$-2 [514]; and R = Ph, R′ = *cyclo*-C$_6$H$_{11}$ [515]) one can distinguish

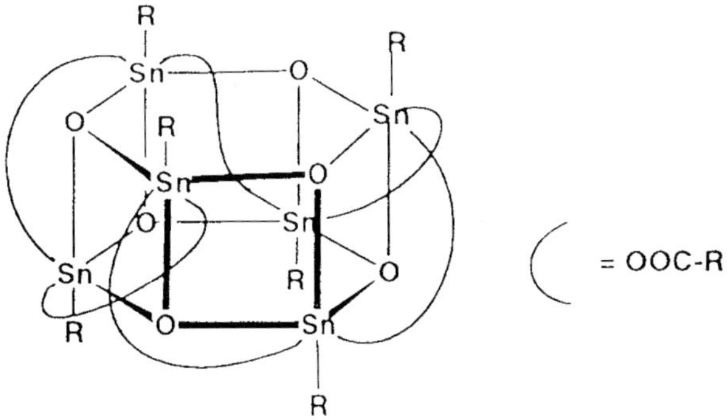

165

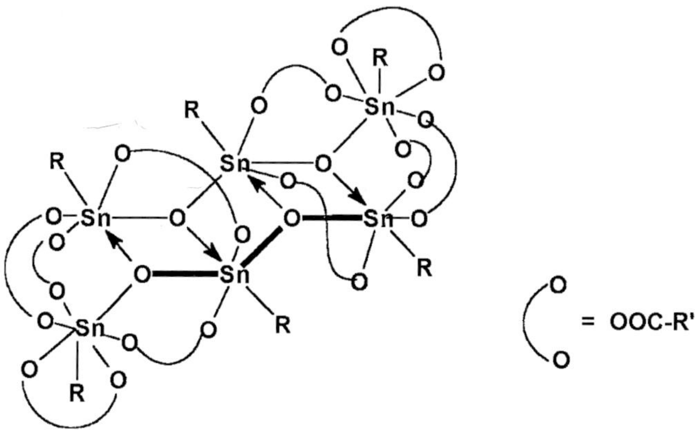

170   3 *Supramolecular Self-Assembly by Dative Bonding*

166

$\overset{O}{\underset{O}{\diagdown}}$ = OOC-R'

six Sn_2O_2 moieties connected in a circle; alternatively the structure can be described as consisting of two six-membered Sn_3O_3 rings connected by dative $Sn \leftarrow O$ bonds. Each Sn_2O_2 face of the resulting hexagonal prism is bridged by a carboxylato group connecting two tin atoms. All tin atoms are six-coordinate, with distorted octahedral geometry.

One variety of the ladder hexanuclear tin oxo-carboxylates, **167**, contains three Sn_2O_2 moieties fused into an open structure, with carboxylate bridging and halogen atoms at tin.

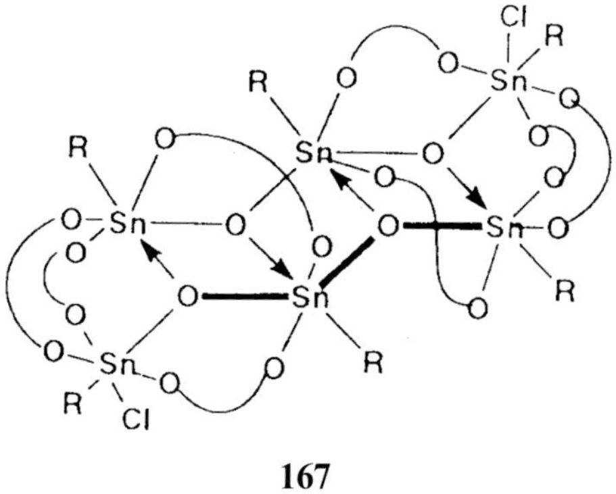

167

The compounds $[\{RSnO(OOCR')\}_2\{RSn(OOCR')_3\}]$ with R = Me, R' = *cyclo*-C_6H_{11} [514]; R = Bu^n, R' = Me, Ph [514]; and *cyclo*-C_6H_{11} [513] and the chloro derivative $[\{Bu^nSnO(OOCPh)\}_2\{Bu^nSn(OOCPh)_2Cl\}]$ [513] have been structurally characterized. In the latter, the chlorine substituent replaces a chelating (non-bridging) carboxylato group.

The 'drum' and 'ladder' types are interconvertible, as demonstrated by ^{119}Sn NMR spectroscopy in solution. The most stable form is the hexagonal prism, with the incompletely hydrolyzed 'ladder' probably being an intermediate.

3.3.1.6 Organotin derivatives of inorganic acids

Like carboxylates, numerous organotin derivatives ('esters') of inorganic and element-organic acids (carbonates, nitrates, phosphates, phosphonates, phosphinates and their arsenic analogs, sulfonates, sulfinates, etc.), are characterized by supramolecular self-assembly in solution and/or supramolecular self-organization in solid state.

Carbonates

A spectroscopic investigation (^{119}Sn NMR and Mössbauer) of bis(triorganotin) carbonates, $(R_3Sn)_2CO_3$ (R = Me, Bu^n) indicated the presence of both four and five coordinate tin [516]. Single-crystal X-ray diffraction analysis of the two compounds [517, 518] confirmed this finding and established a supramolecular structure, **168**. This consists of a helical chain containing trigonal bipyramidal five-coordinate tin (alternating with carbonato groups) and dangling –OSnR$_3$ side-groups, containing four-coordinate tin. Solid state and solution ^{119}Sn NMR and ^{13}C NMR spectroscopy revealed a dynamic process by which the tin atoms interchange their positions.

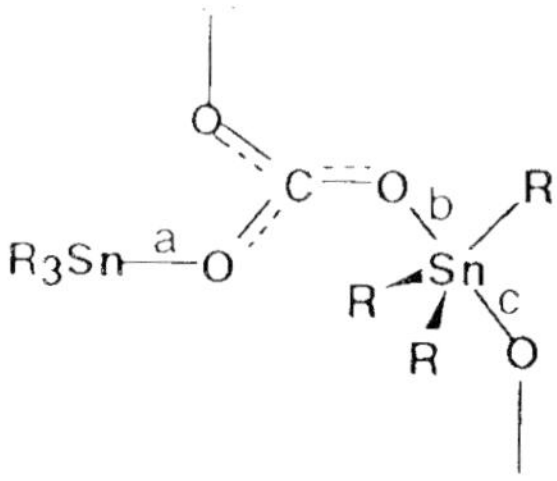

R = Me
a = 2.031 Å
b = 2.248 Å
c = 2.258 Å

R = Bu^n
a = 2.063 2.014 Å
b = 2.253 2.268 Å
c = 2.272 2.258 Å

168

Nitrates

The first probable structures of organotin nitrates were suggested on the basis of ^{119}Sn Mössbauer and vibrational spectroscopic data. A polymeric structure with bridging NO$_3$ groups has been suggested for trimethyltin nitrate, Me$_3$SnNO$_3$, **169**.

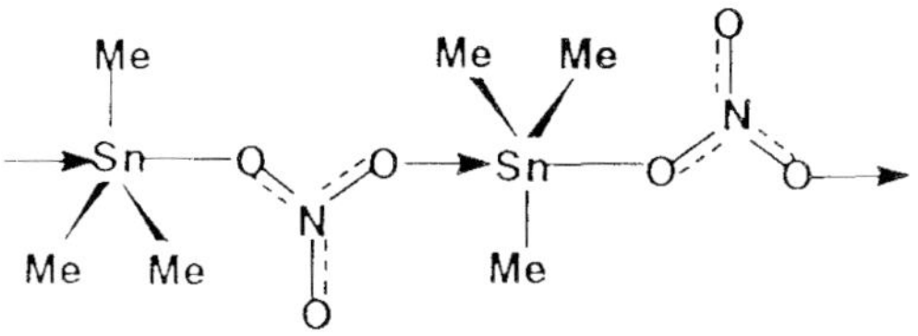

169

3 Supramolecular Self-Assembly by Dative Bonding

Single-crystal X-ray diffraction analysis of MeSn(NO$_3$)$_3$ indicated a tendency of nitrato groups to act as bidentate chelating rather than bridging ligands [519]; this tendency has also been observed in the structure of Ph$_2$Sn(NO$_3$)$_2$(OPPh$_3$); both compounds are unassociated [520]. Unlike other diorganotin derivatives of monobasic acids, R$_2$SnX$_2$ (X = F, Cl, SO$_3$F, CN, SCN) which contain six-coordinate tin because of intermolecular supramolecular self-assembly, at least in the solid state, the diorganotin nitrates, e.g. Me$_2$Sn(NO$_3$)$_2$ [521], seem to prefer intramolecular chelation of the ligand, and avoid self-assembly.

Supramolecular self-organization in the crystal, with bridging nitrato groups [522], **170**, is observed for the unusual compound [Sn(NO$_3$)SnPh$_3$]; this is indicative of the tendency of some organotin nitrates to associate. The long tin–oxygen distances observed might suggest ionic character rather than dative or secondary bonding.

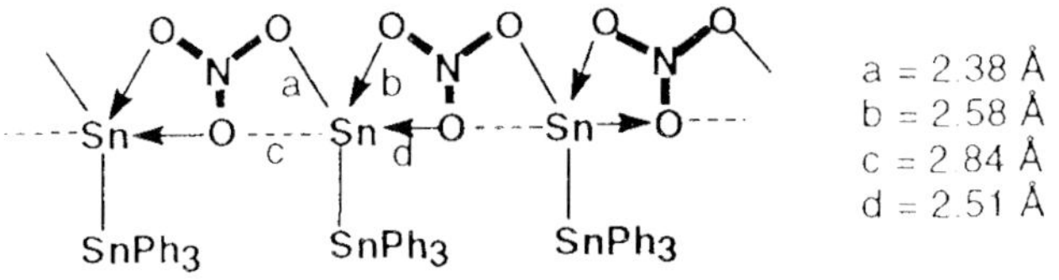

a = 2.38 Å
b = 2.58 Å
c = 2.84 Å
d = 2.51 Å

170

A completely different structure has been found for Me$_2$Sn(OH)(NO$_3$); this is a hydroxo-bridged dimer containing an Sn$_2$(μ-OH)$_2$ core, **171** [523], a structure also found in the anion, **172**, present in [Ag(As(PPh$_3$)$_4$]$_2$[Ph$_2$Sn(μ-OH)(NO$_3$)$_2$]$_2^{2-}$·2MeCN [524] and in [imidazolinium]$_2$[Me$_2$Sn(μ-OH)(NO$_3$)$_2$] [524a].

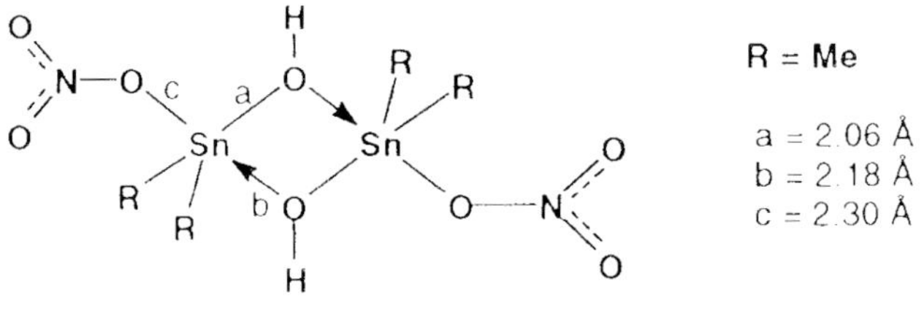

R = Me

a = 2.06 Å
b = 2.18 Å
c = 2.30 Å

171

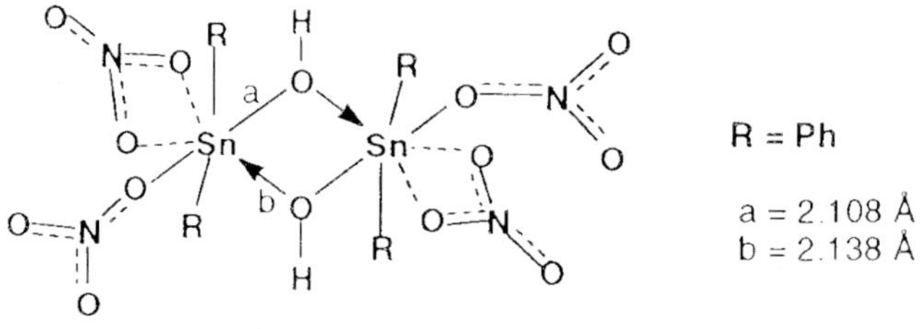

R = Ph

a = 2.108 Å
b = 2.138 Å

172

Phosphates and derivatives of other phosphorus-containing acids

Organotin orthophosphates have been little investigated. The tris(dimethyltin) derivative is a strongly hydrated compound, $(Me_2Sn)_3(PO_4)_2 \cdot 8H_2O$, with a supramolecular, self-organized structure, **173**, containing six-coordinate octahedral tin. The alternation of Me_2Sn and PO_4 groups leads to the formation of infinite ribbons made up of eight-membered $Sn_2O_4P_2$ rings, connected by hydrogen-bonding of the water molecules in a complex three-dimensional architecture [525].

173

Introduction of organic groups on the phosphorus atom reduces the connectivity of the ligand and less complex, yet supramolecular, structures are formed. Thus, triphenyltin diphenylphosphate, $Ph_3SnO_2P(OPh)_2$ [526], and triphenyltin (O-methyl)methyl phosphonate, $Ph_3SnO_2P(OMe)Me$ [527] are cyclic hexamers, **174**, and trimethyltin diphenylphosphinate, $Me_3SnO_2PPh_2$ [528] is a cyclic tetramer, **175**.

Other triorganotin phosphinates self-assemble into chain-like, supramolecular helical arrays, **176**; these include trimethyltin dimethylphosphinate, $Me_3SnO_2PMe_2$, and dichlorophosphate, $Me_3SnO_2PCl_2$ [529], and tri-*n*-butyltin diphenylphosphinate, $Bu_3SnO_2PPh_2$ [530]. The self-assembly of tributyltin derivatives of phosphorus oxyacids (orthophosphates, phosphonates, and phosphinates), $(Bu_3Sn)_nO_nP(O)$-Ph_{3-n}, $n = 0-2$, has been investigated by ^{119}Sn and ^{31}P NMR and ^{119}Sn Mössbauer spectroscopy; results are indicative of intermolecular association both in solution and in the solid state [531]. On the basis of spectroscopic evidence associated structures in the solid state and in solution have also been suggested for tri-

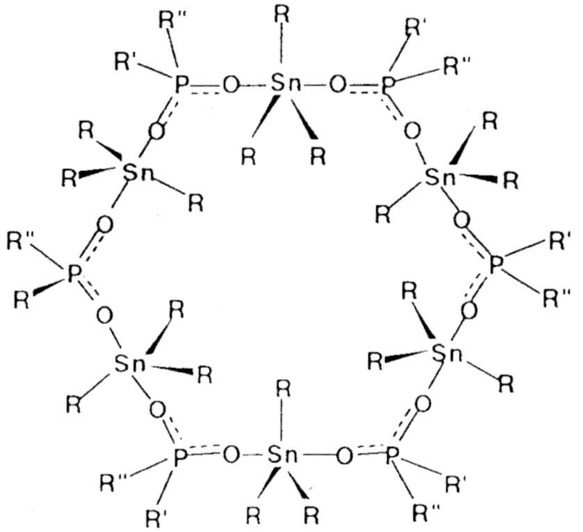

R = Ph, R' = R" = OPh

R = Ph, R' = Me, R" = OMe

174

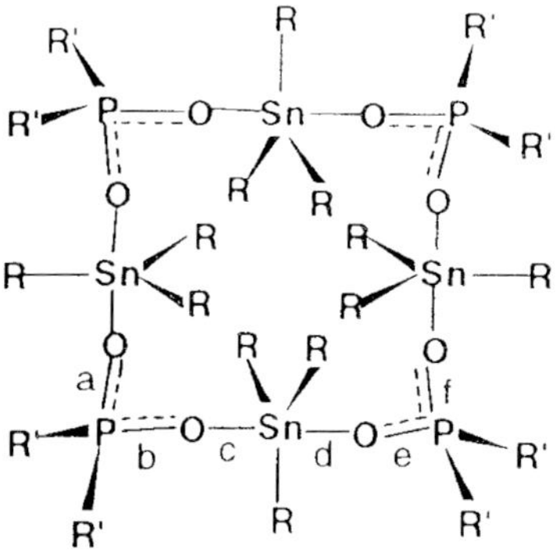

R = Me, R' = Ph

a = 1.483 Å
b = 1.494 Å
c = 2.243 Å
d = 2.245 Å
e = 1.486 Å
f = 1.509 Å

175

organotin *O*-alkyl phosphonates, $R_3SnOP(O)H(OR')$ (R = Me, Bu^n, Ph; R' = Me and Et) [532].

Diorganotin derivatives of phosphorus oxyacids form supramolecular linear arrays containing double bridges or bidimensional layers. Structurally characterized self-assembled compounds include the phosphinates $Et_2Sn(O_2PPh_2)_2$, **177** [533], $Et_2Sn(O_2PMe_2)_2$ and $Ph_2Sn(O_2PMe_2)_2$ [534], and $[ClEt_2Sn(O_2PMe_2)]_x$ [535].

Occasionally supramolecular self-assembly occurs without participation of the organophosphorus ligand and involves only $Sn_2(\mu\text{-}OH)_2$ four-membered cyclic moieties, also known in other hydroxo tin derivatives. This is so for the monothiophosphates $[Ph_2Sn(\mu\text{-}OH)OSP(OEt)_2]_2$ [536], $[Ph_2Sn(\mu\text{-}OH)\{OSP(OPh)_2\}]_2$ [537], and $[Bu^t_2Sn(\mu\text{-}OH)\{OSP(OEt)_2\}]_2$ [538], and for the monothiophosphinate $[Ph_2Sn(\mu\text{-}OH)(OSPPh_2)]_2$ [539].

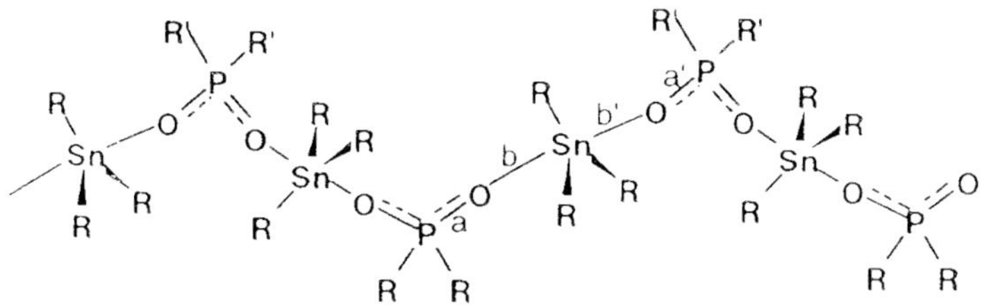

R = Me, R' = Me
 a = 1.45 Å; b = 2.265 Å

R = Me, R' = Cl
 a = 1.45 Å; b = 2.265 Å

R = Bu^n, R' = Ph
 a = 1.480 Å; b = 2.220 Å
 a' = 1.495 Å; b' = 2.220 Å

176

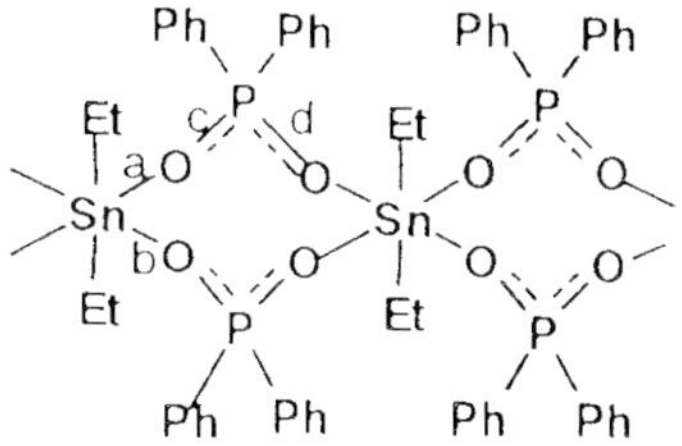

a = 2.229 Å
b = 2.210 Å
c = 1.512 Å
d = 1.496 Å

177

Monoorganotin derivatives of organophosphorus acids form oxotin cluster supermolecules by self-assembly [540, 541]. Examples include the dinuclear ('butterfly') compound [Bu^{n}Sn(OH)(O$_2$PCy$_2$)$_2$)]$_2$, **178** [542], the trinuclear compound [{Bu^{n}Sn(OH)O$_2$PPh$_2$}$_3$O][Ph$_2$PO$_2$], **179** [543], the cubane clusters [Bu^{n}Sn(O)-O$_2$PCy$_2$]$_4$ [544], [Bu^{n}Sn(O)O$_2$PBu$^t_2$]$_4$, and [Bu^{n}Sn(O)O$_2$P(CH$_2$Ph)$_2$]$_4$, **180** [542], the tetranuclear [{Bu^{n}Sn(O)O$_2$PBu$^t_2$}{Bu^{n}Sn(OH)$_2$O$_2$PBu$^t_2$}]$_2$, **181**, [(Bu$^n_2$Sn)$_2$(OH)-(O$_2$PPh$_2$)(OSPPh$_2$)$_2$]$_2$, **182** [545], and [Me$_2$Sn$_2$(OH){O$_2$P(OPh)$_2$}$_3$(O$_3$POPh)]$_2$, **183**, and [R'$_2$Sn$_2$O{O$_2$P(OH)R}$_4$]$_2$ (R = Et, Bu^t, R' = Bu^n), **184** [546], and the hexanuclear prismane ('drum') assemblies of [{MeSn(O)OOCMe}{MeSn(O)O$_2$PBu$^t_2$}]$_3$ and [Bu^{n}Sn(O)O$_2$P(OPh)$_2$]$_6$, **185** [511, 546a] among others [547].

Organotin derivatives of sulfur-containing acids

Although very few organotin sulfates seem to have been structurally characterized, spectroscopic evidence can be accepted as proof of supramolecular self-assembly of

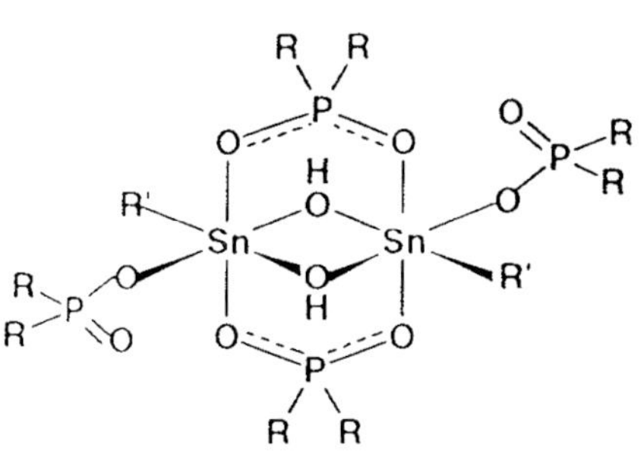

3 Supramolecular Self-Assembly by Dative Bonding

R = Cy, R' = Bu^n
178

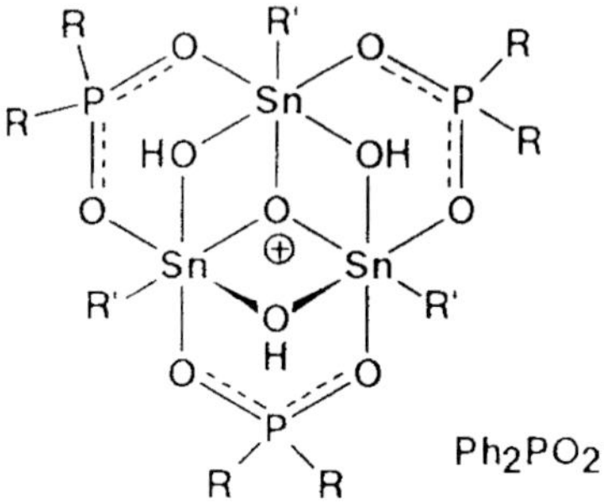

R = Ph, R' = Bu^n
Ph$_2$PO$_2^-$
179

⌒ = O$_2$PR'$_2$

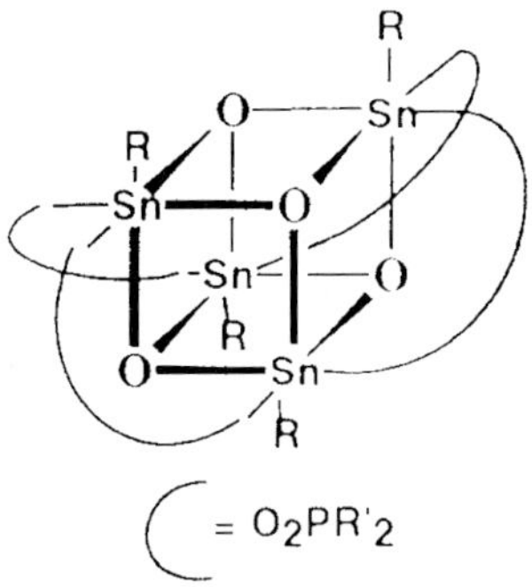

R = Bu^n R' = Cy, Bu^t or CH$_2$Ph
180

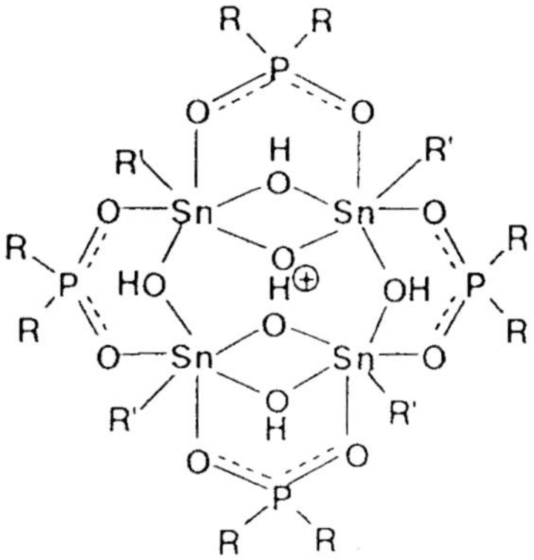

R = Bu^t, R' = Me or Bu^n
181

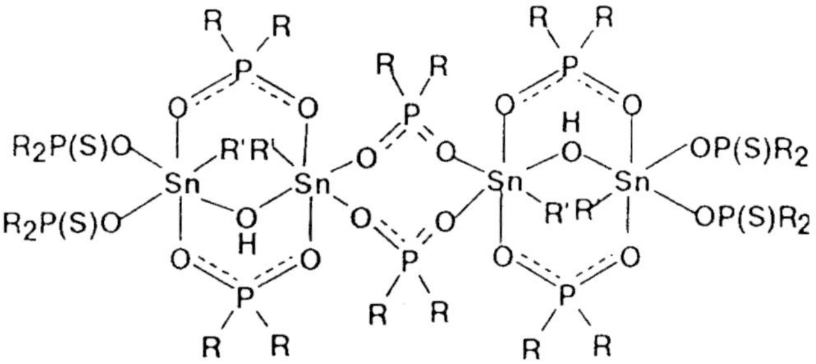

R = Ph, R' = Bu^n
182

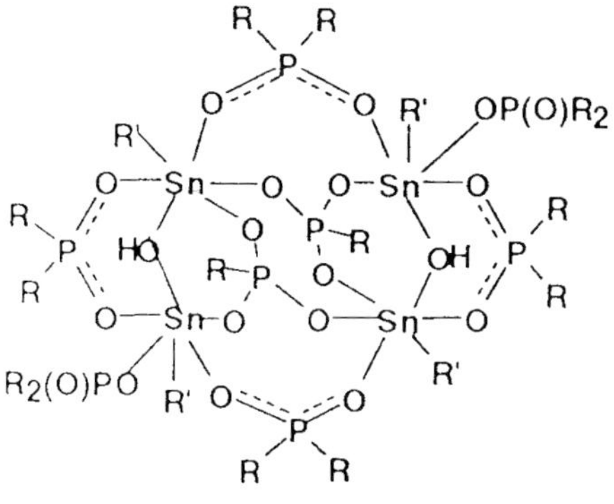

183
R = OPh
R' = Me

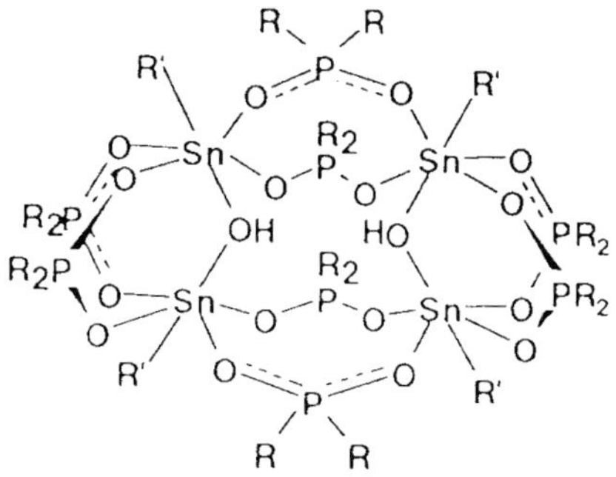

184
R = Bu^t
R' = Bu^n

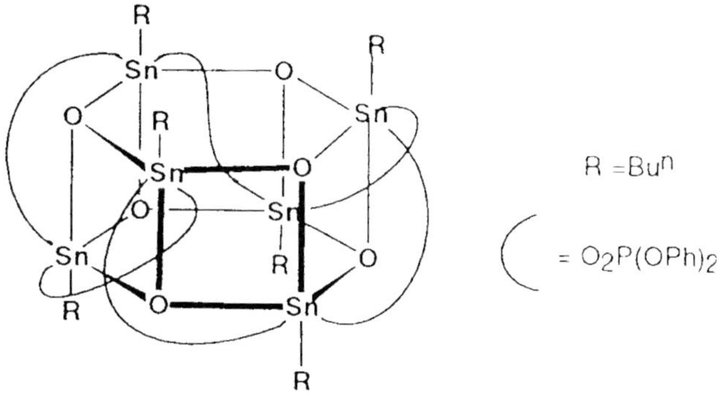

185
R = Bu^n

◠ = O$_2$P(OPh)$_2$

these compounds. Mössbauer and infrared spectral data support a polymeric structure for triorganotin sulfates, (R$_3$Sn)$_2$SO$_4$ (R = Me, Bu^n) [548]. The hydrated sulfate (Me$_3$Sn)$_2$SO$_4$·H$_2$O, is a dinuclear species containing five-coordinate tin [549].

The sulfite (Ph$_3$Sn)$_2$SO$_3$ is a self-assembled chain array in the solid state (Sn–O 2.032, 2.252, and 2.264 Å) [550].

Trimethyltin sulfinates, Me$_3$SnO$_2$SR, **186**, (R = Me [551, 552], CH$_2$CCH [553]) formed by insertion of sulfur dioxide into the Sn–C bond, are self-organized, helical, supramolecular arrays.

Dimethyltin bis(trifluorosulfate), Me$_2$Sn(SO$_3$F)$_2$, is strongly associated and contains six-coordinate tin with *trans* methyl groups. It forms bidimensional sheets in the solid state [554]. Dimethyltin thiosulfate, Me$_2$SnS$_2$O$_3$, also forms a bidimensional supramolecular structure [555].

186

Derivatives of selenium-containing acids

Organotin derivatives of seleninic acids, R_3SnO_2SeR' ($R = R' =$ Me [556], $R = R' =$ Ph [557]) are self-organized into helical chains, **187**.

Hydrated trimethyltin selenite, [$(Me_3Sn)_2SeO_3(H_2O)$], forms a complex supramolecular structure containing Sn–O–Se–O–Sn chains with $(H_2O)Me_3Sn$ side groups, linked by hydrogen-bonds in a tridimensional network [558].

R = Me	R = Ph
a = 2.23 Å	a = 2.223 Å
b = 2.24 Å	b = 2.225 Å
c = 1.76 Å	c = 1.696 Å
d = 1.68 Å	d = 1.682 Å

187

Organotin derivatives of other inorganic oxoacids

Tris(trimethyltin) chromate hydroxide, $(Me_3Sn)_3CrO_4(OH)$, has a complicated three-dimensional structure containing hydroxo and chromato bridges and trigonal-bipyramidal five-coordinate tin atoms. In the crystal, chains containing $SnMe_3$-$OCrO_2OSnMe_3OCrO_2O$ and $SnMe_3OHSnMe_3OCrO_2O$ sequences can be distinguished along the y and z axes, respectively, with Sn–O interatomic distances in the range 2.14–2.17 Å to OH groups and 2.48 and 2.51 Å to CrO_4 groups [559]. The long distances probably correspond to electrostatic interactions (ionic bonds).

Dimethyltin molybdate, Me_2SnMoO_4, has a tridimensional supramolecular structure consisting of tetrahedral MoO_4 groups connecting Me_3Sn moieties, in which the tin atom becomes six-coordinate by formation of four Sn–O bonds; the Sn–O interatomic distances are in the range 2.17–2.34 Å and the methyl groups are in *trans* positions [560]. The diphenyltin derivative is obtained as a hydroxo-bridged dimeric anion [$(Ph_2Sn)_2(\mu\text{-}OH)_2(\mu\text{-}MoO_4)_2$]$^{2-}$, **188** (isolated and structurally characterized as the tetra-*n*-butylammonium salt) and also contains molybdato bridges. The Sn–O bond lengths are in the range 2.115–2.189 Å [561].

188 **189** R = Ph

Trimethyltin perrhenate, Me$_3$SnReO$_4$, forms a supramolecular array based upon Sn–O–Re–O zigzag chains, with trigonal bipyramidal, five-coordinate tin and oxygen atoms in axial positions; the Sn–O interatomic distances are 2.293 and 2.302 Å [562]. The diphenyltin derivative of perrhenic acid is a ladder type tetranuclear supermolecule, [(ReO$_4$)SnPh$_2$OSnPh$_2$OH]$_2$, **189**, as already encountered in other hydroxo tin derivatives. The Sn–O interatomic distances are in the range 2.008–2.238 Å [563].

It seems that most organotin derivatives of oxoacids have a strong tendency to self-organize into supramolecular architectures in which the tin atom achieves a preferred five- or six-coordinate geometry.

References

[1] J. Boersma and J. G. Noltes, *Tetrahedron Letters*, **1966**, 1521.
[2] P. T. Moseley and H. M. M. Shearer, *J. Chem. Soc. Chem. Comm.*, **1966**, 877; *J. Chem. Soc. Dalton Trans.* **1973**, 65.
[3] A. Müller, B. Neumüller and K. Dehnicke, *Angew. Chem.* **1997**, *109*, 2447; Angew. Chem. Int. Ed. Engl. **1997**, 36, 2350.
[4] G. E. Coates and D. Ridley, *J. Chem. Soc. A*, **1966**, 1064.
[5] J. G. Noltes and J. Boersma, *J. Organomet. Chem.* **1968**, *12*, 425.
[6] M. Parvez, G. L. Bergstresser and H. G. Richey, Jr., *Acta Cryst.* **1992**, *C48*, 641.
[7] M. M. Olmstead, P. P. Power and S. C. Shoner, *J. Am. Chem. Soc.* **1991**, *113*, 3379.
[8] S. S. Al–Juaid, N. H. Buttrus, C. Eaborn, P. B. Hitchcock, A. T. L. Robert, J. D. Smith and A. C. Sullivan, *J. Chem. Soc. Chem. Comm.*, **1986**, 908.
[9] R. M. Fabicon, M. Parvez and A. G. Richley, Jr., *J. Am. Chem. Soc.* **1991**, *113*, 1412.
[10] M. R. P. van der Vliet, G. van Koten, P. Buysingh, J. T. B. H. Jastrzebski and A. L. Spek, *Organometallics*, **1987**, *6*, 537.
[11] J. T. B. H. Jastrzebski, J. Boersma, G. van Koten, W. J. J. Smets, and A. L. Spek, *Recl. Trav. Chim. Pays-Bas*, **1988**, *107*, 263.
[12] M. Kitamura, S. Okada, S. Suga and R. Noyori, *J. Am. Chem. Soc.* **1989**, *111*, 4028.
[13] D. Steinborn, M. Rausch, U. Baumeister, I. Potocnak, D. Miklos and M. Dunaj-Jurco, *Z. Anorg. Allg. Chem.* **1996**, *622*, 1941.

[14] A. L. Spek, *Cryst. Struct. Commun.* **1973**, *3*, 535.
[15] J. Boersma, F. Verbeek and J. G. Noltes, *J. Organomet. Chem.* **1971**, *33*, C53.
[16] J. Dekker, A. Schouten, P. H. M. Budzelaar, J. Boersma and G. J. M. Van der Kerk, *J. Organomet. Chem.* **1987**, *320*, 1.
[17] I. B. Gorrell, A. Looney, G. Parkin and A. L. Rheingold, *J. Am. Chem. Soc.* **1990**, *112*, 4068.
[18] H. M. M. Shearrer and C. B. Spencer, *Chem. Commun.* **1966**, 194; *Acta Cryst.* 1980, *B36*, 2046.
[19] W. A. Herrmann, S. Bogdanovich, J. Behm and M. Denk, *J. Organomet. Chem.* 1992, *430*, C33.
[20] F. Schindler, H. Schmidbaur and U. Krüger, *Angew. Chem.* **1965**, *4*, 876.
[21] G. E. Coates and D. Ridley, *J. Chem. Soc.* **1965**, 1870.
[22] M. M. Olmstead, P. P. Power and S. C. Shoner, *J. Am. Chem. Soc.* **1991**, *113*, 3379.
[23] W. H. Eisenhuth and J. R. Van Wazer, *J. Am. Chem. Soc.* **1968**, *90*, 5397.
[24] M. L. Ziegler and J. Weiss, *Angew. Chem.* **1970**, 82, 931; *Angew. Chem. Int. Ed. Engl.* **1970**, *9*, 905.
[25] M. Ishimori, T. Hagiwara, T. Tsuruta, Y. Kai, N. Yasuda and N. Kasai, *Bull. Chem. Soc. Japan*, **1976**, *49*, 1165.
[26] J. Boersma, A. L. Spek and J. G. Noltes, *J. Organomet. Chem.* **1974**, *81*, 7.
[27] F. H. Van der Steen, J. Boersma, A. L. Spek and G. van Koten, *J. Organomet. Chem.* **1990**, *390*, C21; *Organometallics*, **1991**, *10*, 2467.
[28] M. B. Hursthouse, M. A. Malik, M. Motevalli and P. O'Brien, *J. Chem. Soc. Chem. Comm.* **1991**, 1690.
[29] I. Abrahams, M. A. Malik, M. Motevalli and P. O'Brien, *J. Chem. Soc. Dalton Trans.* **1995**, 1043.
[30] D. M. Knotter, M. D. Janssen, D. M. Grove, W. J. J. Smets, E. Horn, A. L. Spek and G. van Koten, *Inorg. Chem.* **1991**, *30*, 4361.
[31] M. B. Hursthouse, M. A. Malik, M. Motevalli and P. O'Brien, *Organometallics*, 1991, *10*, 730.
[32] M. A. Malik and P. O'Brien, *Mater. Chem.* **1991**, *3*, 999.
[33] G. W. Adamson, N. A. Bell and H. M. M. Shearer, *Acta Cryst.* **1982**, *B38*, 462.
[34] G. W. Adamson and H. M. Shearer, *J. Chem. Soc. Chem. Commun.* **1969**, 897.
[35] M. B. Hursthouse, M. A. Malik, M. Motevalli and P. O'Brien, *J. Mater. Chem.* **1992**, *2*, 949.
[36] M. A. Malik, M. Motevalli, J. R. Walsh and P. O'Brien, *Organometallics*, 1992, *11*, 3136.
[37] K. Ruhlandt-Senge and P. P. Power, *Inorg. Chem.* **1993**, *32*, 4505.
[38] N. A. Bell, H. M. M. Shearer and C. B. Spencer, *Acta Cryst.* **1983**, *C39*, 1182.
[39] J. G. Noltes and J. Boersma, *J. Organomet. Chem.* **1969**, *16*, 345.
[40] M. R. P. van Vliet, G. van Koten, J. F. Modder, J. A. M. van Beek, W. J. Klaver, K. Goubitz and C. H. Stam, *J. Organomet. Chem.* **1987**, *319*, 285.
[41] M. G. Davidson, D. Elilio, S. Liless, A. Martin, P. R. Raithby, R. Snaith and D. S. Wright, *Organometallics*, **1993**, *12*, 1.
[42] F. A. J. J. van Santvoort, H. Krabbendam, A. L. Speek and J. Boersma, *Inorg. Chem.* **1978**, *17*, 388.
[43] N. A. Bell, H. M. M. Shearer and C. B. Spencer, *Acta Cryst.* **1984**, *C40*, 613.
[44] M. J. Almond, M. P. Beer, M. G. B. Drew and D. A. Rice, *J. Organomet. Chem.* **1991**, *421*, 129.
[45] S. S. Al–Juaid, N. H. Buttrus, C. Eaborn, P. B. Hitchcock, J. D. Smith and K. Tavakkoli, *J. Chem. Soc. Chem. Comm.* **1988**, 1389.
[46] M. Weidenbruch, M. Herrndorf, A. Schafer, S. Pohl and W. Saak, *J. Organomet. Chem.* **1989**, *361*, 139.
[47] A. K. Duhme and H. Strasdeit, *Eur. J. Inorg. Chem.* **1998**, 657.
[48] C. C. Cummins, R. R. Schrock and W. M. Davis, *Organometallics*, **1991**, *10*, 3781.
[49] B. L. Benac, A. H. Cowley, R. A. Jones, C. M. Nunn and T. C. Wright, *J. Am. Chem. Soc.* **1989**, *111*, 4986.
[49a] M. Green and P. O'Brien, *Advan. Mater.* **1998**, *10*, 527.

[50] N. C. Baenziger, R. M. Flynn, D. C. Swenson and N. L. Holy, *Acta Cryst.* **1978**, *B34*, 2300.
[51] N. C. Baenziger, R. M. Flynn, and N. L. Holy, *Acta Cryst.* **1980**, *B36*, 1642.
[52] B. Neumüller, *Coord. Chem. Rev.* **1997**, *158*, 69.
[53] C. E. Holloway and M. Melnik, *J. Organomet. Chem.* **1997**, *543*, 1.
[54] A. J. Downs (Editor), *Chemistry of Aluminium, Gallium, Indium and Thallium*, Chapmann & Hall, London, New York, **1993**.
[55] G. H. Robinson (Editor), *Coordination Chemistry of Aluminum*, Wiley-VCH, New York, **1993**.
[56] A. V. Grosse and J. Mavity, *J. Org. Chem.* **1940**, *5*, 106.
[57] A. W. Laubengayer and G. F. Lengnik, *Inorg. Chem.* **1966**, *5*, 503.
[58] J. J. Eisch and W. C. Kaska, *J. Am. Chem. Soc.* **1966**, *88*, 2976.
[59] J. Weidlein and V. Krieg, *J. Organomet. Chem.* **1969**, *17*, 41.
[60] V. Krieg and J. Weidlein, *J. Organomet. Chem.* **1970**, *21*, 281.
[61] K. Brendhaugen, A. Haaland and D. P. Novak, *Acta Chem. Scand.* **1974**, *28A*, 45.
[62] G. Allegra, G. Perego and A. Immirzi, *Makromol. Chem.* **1963**, *61*, 69.
[63] C. N. McMahon, J. A. Francis and A. R. Barron, *J. Chem. Cryst.* **1997**, *27*, 191.
[64] M. S. Lalama, J. Kampf, D. G. Dick and J. P. Oliver, *J. Organomet. Chem.* **1995**, *14*, 495.
[64a] M. D. Healy, J. W. Ziller and A. R. Barron, *Organometallics* **1992**, *11*, 3041.
[64b] W. Uhl, A. Vester and W. Hiller, *Z. Anorg. Allg. Chem.* **1990**, *589*, 175.
[65] M. A. Petrie, P. P. Power, H. V. Rasika Dias, K. Ruhlandt-Senge, K. M. Waggoner and R. J. Wehmschulte, *Organometallics*, **1993**, *12*, 1086.
[66] P. R. Schonberg, R. T. Paine, C. F. Campana and E. N. Duesler, *Organometallics* **1982**, *1*, 799.
[67] P. R. Schonberg, R. T. Paine and C. F. Campana, *J. Am. Chem. Soc.* **1979**, *101*, 7726.
[67a] H. J. Koch, S. Schultz, H. W. Roesky, M. Noltemeyer, H.-G. Schmidt, A. Heine, R. Herbst-Irmer, D. Stalke and G. M. Sheldrick, *Chem. Ber.* **1992**, *125*, 1107.
[68] A. Herzog, H. W. Roesky, F. Jaeger, A. Steiner and M. Noltemeyer, *Organometallics*, **1996**, *15*, 909.
[69] S. D. Waezsada, F. Q. Liu, C. A. Barnes, H. W. Roesky, M. L. Montero and I. Uson, *Angew. Chem. Int. Ed. Engl.* **1997**, *36*, 2625.
[70] A. W. Apblett, A. C. Warren and A. R. Barron, *Chem. Mater.* **1992**, *4*, 167.
[71] C. Landry, J. A. Davis, A. W. Apblett and A. R. Barron, *J. Mater. Chem.* **1993**, *3*, 597.
[72] J. J. Eisch, in *Comprehensive Organometallic Chemistry II*, Pergamon Press, London, **1995**, Vol. 1, p. 431.
[73] J. P. Oliver, R. Kumar, *Polyhedron*, **1990**, *9*, 409.
[74] S. Pasynkiewicz, *Polyhedron* **1990**, *9*, 429.
[75] E. G. Hoffmann and W. Tornau, *Angew. Chem.* **1961**, *73*, 578.
[76] I. Haiduc, *The Chemistry of Inorganic Ring Systems*, Wiley-Interscience, London, **1970**, Vol. 2, p. 1019.
[77] B. Cetinkaya, P. B. Hitchcock, H. A. Jasim, M. F. Lappert, and H. D. Williams, *Polyhedron* **1990**, *9*, 239.
[78] R. Kumar, M. L. Sierra, V. Srini, J. de Mel and J. P. Oliver, *Organometallics*, **1990**, *9*, 484.
[79] J. J. Eisch, in *Comprehensive Organometallic Chemistry* (G. Wilkinson, F. G. A. Stone and E. W. Abel), Pergamon Press, **1982**, Vol. 1, 555.
[80] M. Boleslawski, S. Pasynkiewicz, A. Kunicki and J. Serwatowski, *J. Organomet. Chem.* **1976**, *116*, 285.
[81] M. R. Mason, J. M. Smith, S. G. Bott and A. R. Barron, *J. Am. Chem. Soc.* **1993**, *115*, 4971.
[82] P. J. Wheatley, *J. Chem. Soc.* **1963**, 2562.
[83] D. G. Hendershot, R. Kumar, M. Barber and J. P. Oliver, *Organometallics*, **1991**, *10*, 1917.
[84] C. N. McMahon, S. G. Bott and A. R. Barron, *J. Chem. Soc. Dalton Trans.* **1997**, 3129.
[84a] J. L. Atwood, D. C. Hrncir, R. D. Priester and R. D. Rogers, *Organometallics* **1983**, *2*, 985.
[85] J. L. Atwood and M. J. Zaworotko, *J. Chem. Soc. Chem. Commun.* **1983**, 302.
[85a] C. Schnitter, H. W. Roesky, T. Albers, H.-G. Schmidt, L. Rupken, E. Parisini and G. M. Sheldrick, *Chem. Eur. J.* **1997**, *3*, 1783.
[86] R. J. Wehmschulte and P. P. Power, *J. Am. Chem. Soc.* **1997**, *119*, 8387.
[87] D. G. Hendershot, M. Barber, R. Kumar and J. P. Oliver, *Organometallics*, **1991**, *10*, 3302.

[88] A. Ecker, R. Koppe, C. Ufling and H. Schnöckel, *Z. Anorg. Allg. Chem.* **1998**, *624*, 817.
[89] J. Lewinski, J. Zachara and I. Justyniak, *Organometallics* **1997**, *16*, 4597.
[90] W. Liu, A. Hassan and S. Wang, *Organometallics* **1997**, *16*, 4257.
[91] E. Hecht, T. Gelbrich, K. H. Thiele and J. Sieler, *Z. Anorg. Allg. Chem.* **1998**, *624*, 315.
[92] M. Boleslawski and J. Servatowski, *J. Organomet. Chem.* **1983**, *255*, 269.
[93] C. J. Harlan, M. R. Mason and A. R. Barron, *Organometallics*, **1994**, *13*, 2957.
[94] C. J. Harlan, S. G. Bott, B. Wu, R. W. Lenz and A. R. Barron, *Chem. Commun.* **1997**, 2183.
[95] J. Storre, A. Klemp, H. W. Roesky, H.-G. Schmidt, M. Noltemeyer, R. Fleischer and D. Stalke, *J. Am. Chem. Soc.* **1996**, *118*, 1380.
[96] W. Uhl, I. Hahn, M. Koch and M. Layh, *Inorg. Chim. Acta* **1996**, *249*, 33.
[97] C. Schnitter, H. W. Roesky, T. Albers, H. G. Schmidt, C. Röpken, E. Parisini, G. M. Sheldrick, *Chem. Eur. J.* **1997**, *3*, 1783.
[98] C. C. Landry, C. J. Harlan, S. G. Bott, A. R. Barron, *Angew. Chem.* **1995**, *107*, 1315; *Angew. Chem. Int. Ed. Engl.* **1995**, *34*, 1201.
[99] J. A. Tossell, *Inorg. Chem.* **1998**, *37*, 2223.
[100] C. E. Bethley, C. L. Aitken, C. J. Harlan, Y. Koide, S. G. Bott and A. R. Barron, *Organometallics* **1997**, *16*, 329.
[101] J. Lewinski, J. Zachara and I. Justyniak, *Organometallics* **1997**, *16*, 3859.
[101a] J. Lewinski, J. Zachara and K. B. Starowieyski, *J. Chem. Soc. Dalton Trans.* **1997**, 4217.
[102] Y. Koide and A. R. Barron, *Organometallics*, **1995**, *14*, 4026.
[103] J. M. Corker, D. J. Browning and M. Webster, *Acta. Cryst.* **1996**, *C52*, 583.
[104] G. E. Coates and R. N. Mukherjee, *J. Chem. Soc.* **1964**, 1295.
[105] Y. Yang, H. G. Schmidt, M. Noltemeyer, J. Pinkas and H. W. Roesky, *J. Chem. Soc. Dalton Trans.* **1996**, 3609.
[106] M. G. Walawalkar, R. Murugavel, H. W. Roesky and H.-G. Schmidt, *Inorg. Chem.* **1997**, *36*, 4202.
[107] Y. Yang, M. G. Walawalkar, J. Pinkas, H. W. Roesky and H.-G. Schmidt, *Angew. Chem.* **1998**, *100*, 101; *Angew. Chem. Int. Ed. Engl.* **1998**, *37*, 96.
[108] M. R. Mason, R. M. Mathews, M. S. Mashuta and J. F. Richardson, *Inorg. Chem* **1997**, *36*, 6476.
[108a] M. R. Mason, *J. Cluster. Sci.* **1998**, *9*, 1.
[108b] Y. Yang, J. Pinkas, M. Noltemeyer and H. W. Roesky, *Inorg. Chem.* **1998**, *37*, 6404.
[109] D. J. Brauer and G. D. Stucky, *J. Am. Chem. Soc.* **1969**, *91*, 5462.
[110] A. Haaland, O. Stokkeland and J. Weidlein, *J. Organomet. Chem.* **1975**, *94*, 353.
[111] V. S. J. de Mel, R. Kumar and J. P. Oliver, *Organometallics*, **1990**, *9*, 1303.
[112] M. Tagiof, M. J. Heeg, M. Bailey, D. G. Dick, R. Kumar, D. G. Hendershot, H. Rahbanoohi and J. P. Oliver, *Organometallics*, **1995**, *14*, 2903.
[113] J. P. Oliver, *J. Organomet. Chem.* **1995**, *500*, 269.
[114] K. Ruhlandt-Senge and P. P. Power, *Inorg. Chem.* **1991**, *30*, 2633.
[114a] C. J. Harlan, E. G. Gillan, S. G. Bott and A. R. Barron, *Organometallics* **1996**, *15*, 5479.
[115] R. Kumar, D. G. Dick, S. U. Ghazi, M. Taghiof, M. J. Heeg and J. P. Oliver, *Organometallics*, **1995**, *14*, 1601.
[116] A. H. Cowley, R. A. Jones, P. R. Harris, D. A. Atwood, L. Contreras and C. J. Burek, *Angew. Chem.* **1991**, *103*, 1164; *Angew. Chem. Int. Ed.* **1991**, *30*, 1143.
[117] T. D. Getman and G. W. Franklin, *Comments Inorg. Chem.* **1995**, *17*, 79.
[118] I. Haiduc, *The Chemistry of Inorganic Ring Systems*, Wiley-Interscience, London, **1970**, Vol. 2, p. 1025.
[119] M. F. Lappert, P. P. Power, A. R. Sanger and R. C. Srivastava, *Metal and Metalloid Amides*, Ellis Horwood-Wiley, New York, **1980**.
[120] T. Mole and E. A. Jeffery, *Organoaluminum Compounds*, Elsevier, Amsterdam, **1972**.
[121] M. Cesari and S. Cucinella, in *The Chemistry of Inorganic Ring Systems*, Edited by I. Haiduc and D. B. Sowerby, Academic Press, Vol. 1, **1987**, Chapter 6, p. 167.
[122] D. C. Bradley, I. S. Harding, I. A. Maia and M. Motevalli, *J. Chem. Soc. Dalton Trans.* **1997**, 2969.
[123] S. D. Waezsada, F. Q. Liu, E. F. Murphy, H. W. Roesky, M. Treichert, I. Uson, H.-G. Schmidt, T. Albers, E. Parisini and M. Noltemeyer, *Organometallics* **1997**, *16*, 1260.

[124] S. D. Waezsada, C. Rennekamp, H. W. Roesky, C. Ropken and E. Parisini, *Z. Anorg. Allg. Chem.* **1998**, *624*, 987.
[125] S. Kuhner, K. W. Klinkenhammer, W. Schwarz and J. Weidlein, *Z. Anorg. Allg. Chem.* **1998**, *624*, 1051.
[126] L. V. Interrante, G. A. Sigel, M. Garbauskas, C. Hejna and G. A. Slack, *Inorg. Chem.* **1989**, *28*, 252.
[127] G. M. McLaughlin, G. A. Sim and J. D. Smith, *J. Chem. Soc. Dalton Trans.* **1972**, 2197.
[128] D. A. Atwood and D. Rutherford, *Main Group Chem.* **1996**, *1*, 431.
[129] M. A. Petrie, K. Ruhland-Senge and P. P. Power, *Inorg. Chem.* **1993**, *32*, 1135.
[130] W. J. Evans, M. A. Ansari, J. W. Ziller and S. I. Khan, *Inorg. Chem.* **1996**, *35*, 5435.
[131] P. B. Hitchcock, M. F. Lappert and H. D. Williams, *Polyhedron* **1990**, *9*, 245.
[132] Y. Kim, J. H. Kim, J. E. Park, H. Song and J. T. Park, *J. Organomet. Chem.* **1997**, *545/546*, 99.
[133] S. G. Bott, Y. Koida and A. R. Barron, *J. Chem. Cryst.* **1996**, *26*, 563.
[134] J. Müller and K. Dehnicke, *J. Organomet. Chem.* **1968**, *12*, 37.
[135] V. Krieg and J. Weidlein, *Z. Anorg. Allg. Chem.* **1969**, *368*, 44.
[136] J. Müller and K. Dehnicke, *Z. Anorg. Allg. Chem.* **1966**, *348*, 261.
[137] N. Wiberg, W. C. Joo and H. Henke, *Inorg. Nucl. Chem. Lett.* **1967**, *3*, 267.
[138] W. Uhl, R. Gerding, S. Pohl and W. Saak, *Chem. Ber.* **1995**, *128*, 81.
[139] N. Emig, F. P. Gabai, H. Krautscheid, R. Reau and G. Bertrand, *Angew. Chem. Int. Ed. Engl.* **1998**, *37*, 989.
[139a] A. Miehr, O. Ambacher, T. Metzger, E. Born and A. Fischer, *Chem. Vap. Deposition* **1996**, *2*, 51.
[140] K. M. Waggoner, H. Hope and P. P. Power, *Angew. Chem. Int. Ed. Engl.* **1988**, *27*, 1699.
[141] K. M. Waggoner and P. P. Power, *J. Am. Chem. Soc.* **1991**, *113*, 3385.
[142] G. Del Piero, M. Cesari, G. Dozzi and A. Mazzei, *J. Organomet. Chem.* **1977**, *129*, 281.
[143] T. R. R. McDonald and W. S. McDonald, *Acta Cryst.* **1972**, *B28*, 1619.
[143a] T. Belgardt, S. D. Waezada, H. W. Roesky, M. Gornitzka, L. Haming and D. Stalke, *Inorg. Chem.* **1994**, *33*, 6247.
[144] R. D. Davy and H. F. Schaeffer III, *Inorg. Chem.* **1998**, *37*, 2291.
[145] M. Cesari, G. Perego, G. Del Piero, S. Cucinella and E. Cernia, *J. Organomet. Chem.* **1974**, *78*, 203.
[146] G. Del Piero, M. Cesari, G. Perego, S. Cucinella and E. Cernia, *J. Organomet. Chem.* **1977**, *129*, 289.
[147] G. Del Piero, S. Cucinella, and M. Cesari, *J. Organomet. Chem.* **1979**, *173*, 263.
[148] G. Del Piero, G. Perego, S. Cucinella, M. Cesari and A. Mazzei, *J. Organomet. Chem.* **1977**, *136*, 13.
[149] P. B. Hitchcock, J. D. Smithand K. M. Thomas, *J. Chem. Soc. Dalton Trans.* **1976**, 1433.
[150] G. Del Piero, M. Cesari, G. Perego, S. Cucinella and E. Cernia, *J. Organomet. Chem.* **1977**, *129*, 289.
[151] U. Thewalt and I. Kawada, *Chem. Ber.* **1970**, *103*, 2754.
[152] G. Perego, G. Del Piero, M. Cesari, A. Zazzetta and G. Dozzi, *J. Organomet. Chem.* **1975**, *87*, 53.
[153] G. Perego and G. Dozzi, *J. Organomet. Chem.* **1981**, *205*, 21.
[154] S. Amirkhalili, P. B. Hitchcock and J. D. Smith, *J. Chem. Soc. Dalton Trans.* **1979**, 1206.
[155] C. Schnitter, S. D. Waezsada, H. W. Roesky, M. Teichert, I. Uson, and E. Parizini, *Organometallics*, **1997**, *16*, 1197.
[155a] S. Horchler, E. Parizini, H. W. Roesky, H.-G. Schmidt and M. Noltemeyer, *J. Chem. Soc. Dalton Trans.* **1997**, 2761.
[156] U. Wannagat, T. Bluementhal, D. J. Brauer and H. Burger, *J. Organomet. Chem.* **1983**, *249*, 33.
[157] I. Haiduc, *The Chemistry of Inorganic Ring Systems*, Wiley-Interscience, London, **1970**, Vol. 2, p. 1033.
[158] G. E. Coates and J. Graham, *J. Chem. Soc.* **1963**, 233.
[159] E. Hey-Hawkins, M. F. Lappert, J. L. Atwood and S. G. Bott, *J. Chem. Soc. Dalton Trans.* **1991**, 939.

[160] O. T. Beachley and G. E. Coates, *J. Chem. Soc.* **1965**, 3241.
[161] J. F. Janik, E. N. Duesler, W. F. McNamara, M. Westerhausen and R. T. Paine, *Organometallics*, **1989**, *8*, 506.
[162] A. H. Cowley, R. A. Jones, M. A. Mardones, S. G. Bott and J. L. Atwood, *Angew. Chem.* **1990**, *102*, 1504; *Angew. Chem. Int. Ed. Engl.* **1990**, *29*, 1409.
[163] D. E. Heaton, R. A. Jones, K. B. Kidd, A. H. Cowley and C. M. Nunn, *Polyhedron* **1988**, *7*, 1901.
[164] R. L. Wells, A. T. McPhail and T. M. Speer, *Organometallics* **1992**, *11*, 960.
[165] J. A. L. Cooke, A. P. Purdy, R. L. Wells and P. S. White, *Organometallics*, **1996**, *15*, 84.
[166] P. G. Perkins and M. E. Twentiman, *J. Chem. Soc.* **1965**, 1038.
[167] A. P. Purdy, R. L. Wells, A. T. McPhail and C. G. Pitt, *Organometallics* **1987**, *6*, 2099.
[168] A. H. Cowley, R. A. Jones, M. A. Mardones, J. L. Atwood and S. G. Bott, *Heteroatom Chem.* 1991, *2*, 581.
[169] O. T. Beachley Jr., R. B. Hallock, H. M. Zhang and J. L. Atwood, *Organometallics* **1985**, *4*, 1675.
[170] D. C. Bradley, H. Chudzynska, M. M. Factor, D. M. Frigo, M. B. Hursthouse, B. Hussein and L. M. Smith, *Polyhedron* **1988**, *7*, 1289.
[171] D. A. Atwood, A. H. Cowley, R. A. Jones, M. A. Mardones, J. L. Atwood and S. G. Bott, *J. Coord. Chem.* **1992**, *25*, 233.
[172] M. B. Power, W. M. Cleaver, A. W. Apblett, A. R. Barron and J. W. Ziller, *Polyhedron* **1992**, *11*, 477.
[173] B. Werner, T. Krauter and B. Neumüller, *Inorg. Chem.* **1996**, *35*, 2977.
[174] B. Neumüller and F. Gahlmann, *Angew. Chem. Int. Ed. Engl.* **1993**, *32*, 1701.
[175] M. R. Kopp, T. Krauter, B. Werner and B. Neumüller, *Z. Anorg. Allg. Chem.* **1998**, *624*, 881.
[176] G. G. Hoffmann, D. Hellert and M. Warren, *J. Organomet. Chem.* **1993**, *444*, 21.
[177] A. R. Barron, *Chem. Soc. Rev.* **1993**, *22*, 93.
[178] M. D. Healy, M. B. Power and A. R. Barron, *Coord. Chem. Rev.* **1994**, *130*, 63.
[179] A. R. Barron, *Comments Inorg. Chem.* **1993**, *14*, 123.
[180] M. E. Kenney and A. W. Laubengayer, *J. Am. Chem. Soc.* **1954**, *76*, 4839.
[181] G. S. Smith and J. L. Hoard, *J. Am. Chem. Soc.* **1959**, *81*, 3907.
[182] A. H. Cowley, P. R. Harris, R. A. Jones and C. M. Nunn, *Organometallics* **1991**, *10*, 652.
[183] A. A. Naiini, V. Young, H. Han, M. Akinc and J. G. Verkade, *Inorg. Chem.* **1993**, *32*, 3781.
[184] D. A. Atwood, A. H. Cowley, P. R. Harris, R. A. Jones, S. U. Koschmieder, C. M. Nunn, J. L. Atwood and S. G. Bott, *Organometallics* **1993**, *12*, 24.
[185] Q. Zhao, H. Sun, W. Chen, C. Duan,Y. Liu, Y. Pan and X. You, *Organometallics*, **1998**, *17*, 156.
[185a] P. D. Croucher, A. Drljaca, S. Papadopoulos and C. L. Raston, *Chem. Commun.* **1999**, 153.
[186] C. Schnitter, H. W. Roesky, T. Albers, H.-G. Schmidt, C. Ropken, E. Parizini and G. M. Sheldrick, *Chem. Eur. J.* **1997**, *3*, 1783.
[187] J. Storre, T. Belgardt, D. Stalke and H. W. Roesky, *Angew. Chem.* **1994**, *106*, 1365; *Angew. Chem. Int. Ed. Engl.* **1994**, *33*, 1244.
[188] W. M. Cleaver and A. R. Barron, *Organometallics* **1993**, *12*, 1001.
[189] M. A. Petrie, M. M. Olmstead and P. P. Power, *J. Am. Chem. Soc.* **1991**, *113*, 8704.
[190] G. E. Coates and R. G. Hayter, *J. Chem. Soc.* **1953**, 2519.
[191] H. Schmidbaur, *Angew. Chem.* **1965**, *77*, 169.
[192] H. Schmidbaur and F. Schindler, *Chem. Ber.* **1966**, *99*, 2178.
[192a] M. B. Power, W. M. Cleaver, A. W. Apblett, A. R. Barron and J. W. Ziller, *Polyhedron* **1992**, *11*, 477.
[192b] W. M. Cleaver, A. R. Barron, A. R. McGufey and S. G. Bott, *Polyhedron* **1994**, *13*, 2831.
[193] A. H. Cowley, S. K. Mehrotra, J. L. Atwood and W. E. Hunter, *Organometallics*, **1985**, *4*, 1115.
[194] E. Hecht, T. Gelbrich, S. Wernik, R. Weimann, K. H. Thiele, J. Sieler and H. Schumann, *Z. Anorg. Allg. Chem.* **1998**, *624*, 1061.
[195] S. J. Rettig, A. Storr and J. Trotter, *Can. J. Chem.* **1975**, *53*, 58.
[196] J. Storre, A. Klemp, H. W. Roesky, R. Fleischer and D. Stalke, *Organometallics*, **1997**, *16*, 3074.
[197] H. D. Hausen, K. Sille, J. Weidlein and W. Schwarz, *J. Organomet. Chem.* **1978**, *160*, 411.

[198] F. E. Hahn and B. Schneider, *Z. Naturforsch.* **1990**, *B45*, 134.
[199] D. J. Browning, J. M. Corkers and M. Webster, *Acta Crystallogr.* **1996**, *C52*, 882.
[200] C. C. Landry, A. Hynes, A. R. Barron, I. Haiduc and C. Silvestru, *Polyhedron* **1996**, *15*, 391.
[200a] A. Keys, S. Bott and A. R. Barron, *Chem. Commun.* **1996**, 2339.
[201] M. G. Walawalkar, R. Murugavel, A. Voigt, H. W. Roesky and H.-G. Schmidt, *J. Am. Chem. Soc.* **1997**, *119*, 4656.
[202] R. Mason, H. S. Masuta and J. F. Richardson, *Angew. Chem.* **1997**, *109*, 249; *Angew. Chem. Int. Ed. Engl.* **1997**, *36*, 239.
[203] M. G. Walawalkar, R. Murugavel, H. W. Roesky, I. Uson and R. Kraetzner, *Inorg. Chem.* **1998**, *37*, 473.
[203a] M. R. Mason, A. M. Parkins, R. M. Matthews, J. D. Fischer, M. S. Mashuta and A. Vij, *Inorg. Chem.* **1998**, *37*, 3734.
[204] S. U. Ghazi, R. Kumar, M. J. Heeg and J. P. Oliver, cited by J. P. Oliver, *J. Organomet. Chem.* **1995**, *500*, 269.
[205] D. G. Hendershot, R. Kumar, M. Barber and J. P. Oliver, *Organometallics* **1991**, *10*, 1917.
[206] M. B. Power and A. R. Barron, *J. Chem. Soc. Chem. Commun.* **1991**, 1315.
[207] G. Hoffmann and C. Burschka, *J. Organomet. Chem.* **1984**, *267*, 229.
[208] A. Boardman, S. E. Feffs, R. W. H. Small and I. J. Worrall, *Inorg. Chim. Acta* **1985**, L39.
[209] S. U. Ghazi, M. J. Heeg and J. P. Oliver, *Inorg. Chem.* **1994**, *33*, 4517.
[210] A. N. McInnes, M. B. Power and A. R. Barron, *Chem. Mater.* **1992**, *4*, 11.
[211] W. M. Cleaver, M. Späth, D. Hnyk, G. McMurdo, M. B. Power, M. Stuke, D. W. H. Rankin and A. R. Barron, *Organometallics* **1995**, *14*, 690.
[212] M. B. Power, J. W. Ziller and A. R. Barron, *Organometallics* **1992**, *11*, 2783.
[213] M. B. Power, J. W. Ziller, A. N. Tyler and A. R. Barron, *Organometallics* **1992**, *11*, 1055.
[214] R. Kumar, D. G. Dick, S. U. Ghazi, M. Taghiof, M. J. Heeg and J. P. Oliver, *Organometallics* **1995**, *14*, 1601.
[215] H. Rahbarnoohi, R. L. Wells, L. M. Liable-Sands, G. P. Yap and A. L. Rheingold, *Organometallics* **1997**, *16*, 3959.
[216] G. C. Hoffmann and R. Fischer, *Inorg. Chem.* **1989**, *28*, 4165.
[217] N. S. Vyazankin, M. N. Bochkarev and A. I. Charov, *J. Organomet. Chem.* **1971**, *27*, 175.
[218] S. Schulz, E. G. Gillan, J. L. Ross, L. M. Rogers, R. D. Rogers and A. R. Barron, *Organometallics*, **1996**, *15*, 4880.
[219] M. A. Banks, O. T. Beachley Jr., H. J. Gysling and H. R. Luss, *Organometallics* **1990**, *9*, 1979.
[220] W. Uhl, M. Layh, G. Becker, K. W. Klinkhammer and T. Hildebrand, *Chem. Ber.* **1992**, *125*, 1547.
[221] G. E. Coates, *J. Chem. Soc.* **1951**, 2003.
[222] K. M. Waggoner and P. P. Power, *J. Am. Chem. Soc.* **1991**, *113*, 3385.
[223] S. J. Schauer, C. H. Lake, C. L. Watkins, L. K. Krannich and D. H. Powell, *J. Organomet. Chem.* **1997**, *549*, 31.
[224] A. Storr, *J. Chem. Soc. A*, **1968**, 2605.
[225] J. Müller and K. Dehnicke, *J. Organomet. Chem.* **1968**, *12*, 37.
[226] J. T. Park, Y. Kim, J. Kim, K. Kim and Y. Kim, *Organometallics* **1992**, *11*, 3320.
[227] W. R. Nutt, R. E. Stimson, M. F. Leopold and B. H. Rubin, *Inorg. Chem.* **1982**, *21*, 1909.
[228] O. T. Beachley Jr., M. J. Noble, M. R. Churchill and C. H. Lake, *Organometallics* **1992**, *11*, 1051.
[229] D. A. Atwood, R. A. Jones, A. H. Cowley, S. G. Bott and J. L. Atwood, *Polyhedron* **1991**, *10*, 1897.
[230] D. C. Bradley, H. M. Dawes, M. B. Hursthouse, L. M. Smith and M. Thornton-Pett, *Polyhedron* **1990**, *9*, 343.
[231] T. Belgardt, H. W. Roesky, M. Noltemeyer and H.-G. Schmidt, *Angew. Chem.* **1993**, *105*, 1101; *Angew. Chem. Int. Ed. Engl.* **1993**, *32*, 1056.
[231a] F. Cordeddu, H. D. Hausen and J. Weidlein, *Z. Anorg. Allg. Chem.* **1996**, *622*, 573.
[232] D. A. Peters, M. P. Power, E. D. Bourret and J. Arnold, *Chem. Commun.* **1998**, 753.
[233] D. A. Atwood, R. Jones, A. H. Cowley, J. L. Atwood and S. G. Bott, *J. Organomet. Chem.* **1990**, *394*, C6.

[234] D. Boyer, R. Gassend, J. C. Maire and J. Elguero, *J. Organomet. Chem.* **1981**, *215*, 157.
[235] D. F. Rendle, A. Storr and J. Trotter, *Can. J. Chem.* **1975**, *53*, 2930.
[236] D. F. Rendle, A. Storr and J. Trotter, *Can. J. Chem.* **1975**, *53*, 2944.
[237] A. H. Cowley and R. A. Jones, *Angew. Chem.* **1989**, 101, 1235; *Angew. Chem. Int. Ed. Engl.* **1989**, *28*, 1208.
[238] A. H. Cowley, *J. Organomet. Chem.* **1990**, *400*, 71.
[239] R. L. Wells, *Coord. Chem. Rev.* **1992**, 273.
[240] Y. Pan and P. Boudjouk, *Main Group Chem.* **1995**, *1*, 61.
[240a] D. J. Cole-Hamilton, *Chem. Commun.* **1999**, 759.
[241] G. G. Hoffmann, R. Fischer, U. Schubert and B. Hirle, *J. Organomet. Chem.* **1992**, *441*, 7.
[242] A. M. Arif, B. L. Benac, A. H. Cowley, R. Geerts, R. A. Jones, K. B. Kidd, J. M. Power and S. T. Schwab, *J. Chem. Soc. Chem. Comm.* **1986**, 1543.
[243] D. A. Atwood, A. H. Cowley, P. R. Harris, R. A. Jones, S. U. Koschmieder and C. M. Nunn, *J. Organomet. Chem.* **1993**, *449*, 61.
[244] D. E. Heaton, R. A. Jones, K. B. Kidd, A. H. Cowley, *Polyhedron* **1988**, *7*, 1901.
[245] M. A. Banks, O. T. Beachley Jr., L. A. Buttery, M. R. Churchill and J. C. Fettinger, *Organometallics*, **1991**, *10*, 1901.
[246] K. E. Lee, K. T. Higa, R. A. Nissan and R. J. Butcher, *Organometallics*, **1992**, *11*, 2816.
[247] G. H. Robinson, J. A. Burns and W. T. Pennington, *Main Group Chem.* **1995**, *1*, 153.
[248] O. T. Beachley Jr., T. L. Royster Jr., J. R. Arhar and A. L. Rheingold, *Organometallics*, **1993**, *12*, 1976.
[249] A. H. Cowley, R. A. Jones, M. A. Mardones, and C. M. Nunn, *Organometallics*, **1991**, *10*, 1635.
[250] A. H. Cowley, P. R. Harris, R. A. Jones and C. M. Nunn, *Organometallics*, **1991**, *10*, 652.
[251] O. T. Beachley Jr., J. D. Maloney and R. D. Rogers, *J. Organomet. Chem.* **1993**, *449*, 69.
[252] R. L. Wells, S. R. Aubuchon, M. F. Self, J. P. Jasinski, R. C. Woudenberg and R. J. Butcher, *Organometallics*, **1992**, *11*, 3370.
[253] R. L. Wells, A. T. McPhail, J. W. Pasterczyk and A. Alvanipour, *Organometallics* **1992**, *11*, 226.
[254] K. T. Higa and C. George, *Organometallics*, **1990**, *9*, 275.
[255] E. K. Byrne, L. Parkanyi and K. H. Theopold, *Science* **1988**, *241*, 332.
[256] A. Dashti-Mommertz, B. Werner and B. Neumüller, *Polyhedron* **1998**, *17*, 523.
[257] A. M. Arif, B. L. Benac, A. H. Cowley, R. Geerts, R. A. Jones, K. B. Kidd, J. M. Power and S. T. Schwab, *J. Chem. Soc. Chem. Commun.* **1986**, 1543.
[258] J. C. Pazik, C. George and A. Berry, *Inorg. Chim. Acta* **1991**, *187*, 207.
[259] R. L. Wells, J. W. Pasterczyk, A. T. McPhail, J. D. Johansen and A. Alvanipour, *J. Organomet. Chem.* **1991**, *407*, 17.
[260] R. L. Wells, A. P. Purdy, A. T. McPhail and C. G. Pitt, *J. Organomet. Chem.* **1986**, *308*, 281.
[261] A. P. Purdy, R. L. Wells, A. T. McPhail and C. G. Pitt, *Organometallics*, **1987**, *6*, 2099.
[262] O. T. Beachley Jr., J. D. Maloney and R. D. Rogers, *J. Organomet. Chem.* **1993**, *449*, 69.
[263] W. K. Holley, R. L. Wells, S. Shafieezad, A. T. McPhail and C. G. Pitt, *J. Organomet. Chem.* 1990, *381*, 15.
[264] R. L. Wells, W. K. Holley, S. Shafieezad, A. T. McPhail and C. G. Pitt, *Phosphorus, Sulfur, Silicon* **1989**, *41*, 15.
[265] R. L. Wells, A. T. McPhail, and A. Alvanipour, *Polyhedron* **1992**, *11*, 839.
[266] R. L. Wells, E. E. Foss, P. S. White, A. L. Rheingold and L. M. Liable-Sands, *Organometallics*, **1997**, *16*, 4771.
[267] R. A. Baldwin, E. E. Foss, R. L. Wells, P. S. White, A. L. Rheingold and G. P. A. Yap, *Organometallics* **1996**, *15*, 5035.
[268] A. H. Cowley, R. A. Jones, K. B. Kidd, C. M. Nunn and D. L. Westmoreland, *J. Organomet. Chem.* **1988**, *341*, C1.
[269] T. Krauter and B. Neumüller, *Z. Anorg. Allg. Chem.* **1995**, *621*, 597.
[270] J. T. Leman and A. R. Barron, *Organometallics*, **1989**, *8*, 2214.
[271] B. Neumüller, *Z. Naturforsch.* **1991**, *46b*, 1539.
[272] O. T. Beachley Jr., J. D. Maloney, M. R. Churchill and C. H. Lake, *Organometalllics*, **1991**, *10*, 3568.

[273] J. T. Leman, J. W. Ziller and A. R. Barron, *Organometallics* **1991**, *10*, 1766.
[274] S. L. Stoll, S. G. Bott and A. R. Barron, *Polyhedron* **1997**, *16*, 1763.
[275] B. Neumüller and F. Gahlmann, *J. Organomet. Chem.* **1991**, *414*, 271.
[276] I. Haiduc, *The Chemistry of Inorganic Ring Systems*, Wiley-Interscience, London, **1970**, vol. 2, p. 1056.
[277] D. C. Bradley, D. M. Frigo, M. B. Hursthouse and B. Hussain, *Organometallics* **1988**, *7*, 1112.
[278] W. M. Cleaver and A. R. Barron, *J. Am. Chem. Soc.* **1989**, *111*, 8966.
[279] N. W. Alcock, I. A. Degnan, S. M. Roe and M. G. H. Wallbridge, *J. Organomet. Chem.* **1991**, *414*, 285.
[280] A. J. Canty, L. A. Titcombe, B. W. Skelton and A. H. White, *Inorg. Chim. Acta* **1986**, *117*, L35.
[281] A. J. Canty, R. Colton and I. M. Thomas, *J. Organomet. Chem.* **1993**, *455*, 283.
[282] S. S. Al–Juaid, N. H. Buttrus, C. Eaborn, P. B. Hitchcock, A. L. Roberts, J. D. Smith and A. C. Sullivan, *J. Chem. Soc. Chem. Commun.* **1986**, 908.
[283] W. Uhl and M. Pohlmann, *Chem. Commun.* **1998**, 451.
[284] A. Blaschette, A. Michalides and P. G. Jones, *J. Organomet. Chem.* **1991**, *411*, 57.
[285] O. T. Beachley, J. C. Lee, Jr., H. J. Gysling, S. H. L. Chao, M. R. Churchill, and C. H. Lake, *Organometallics*, **1992**, *11*, 3144.
[286] H. Rahbarnoohi, R. L. Wells, L. M. Liable-Sands, G. P. A. Yap and A. L. Rheingold, *Organometallics* **1997**, *16*, 3959.
[287] H. Rahbarnoohi, M. Taghiof, M. J. Heeg, D. G. Dick and J. P. Oliver, *Inorg. Chem.* **1994**, *33*, 6307.
[288] S. U. Ghazi, M. J. Heeg and J. P. Oliver, *Inorg. Chem.* **1994**, *33*, 4517.
[289] O. T. Beachley, J. C. Lee, Jr., H. J. Gysling, S. H. L. Chao, M. R. Churchill and C. H. Lake, *Organometallics* **1992**, *11*, 3144.
[290] H. Rahbarnoohi, R. Kumar, M. J. Heeg and J. P. Oliver, *Organometallics* **1995**, *14*, 502.
[291] D. C. Bradley, H. Dawes, D. M. Frigo, M. B. Hursthouse and B. Hussain, *J. Organomet. Chem.* **1987**, *325*, 55.
[292] K. Mertz, W. Schwarz, B. Eberwein, J. Weidlein, H. Hess and H. D. Hausen, *Z. Anorg. Allg. Chem.* **1977**, *429*, 99.
[293] J. D. Aitchison, J. D. J. Backer-Dirks, D. C. Bradley, M. M. Faktor, D. M. Frigo, M. B. Hursthouse, B. Hussain and R. L. Short, *J. Organomet. Chem.* **1989**, *366*, 11.
[294] G. Rossetto, D. A. N. Brianese, U. Casellato, F. Ossola, M. Porchia, A. Vittadini, P. Zanella and R. Graziani, *Inorg. Chim. Acta* **1990**, *170*, 95.
[295] M. Trapp, H. D. Hausen, G. Weckler and J. Weidlein, *J. Organomet. Chem.* **1993**, *450*, 53.
[296] A. M. Arif, D. C. Bradley, H. Haves, D. M. Frigo, M. B. Hursthouse and B. Hussain, *J. Chem. Soc. Dalton Trans.* **1987**, 2159.
[297] O. T. Beachley Jr., C. Bueno, M. R. Churchill, R. B. Hallock and R. G. Simmons, *Inorg. Chem.* **1981**, *20*, 2423.
[298] B. Neumüller, *Z. Naturforsch.* **1990**, *45b*, 1559.
[299] B. Neumüller, *Chem. Ber.* **1989**, *122*, 2283.
[300] F. Weller and U. Müller, *Chem. Ber.* **1979**, *112*, 2039.
[301] B. Neumüller, *Z. Naturforsch.* **1991**, *46b*, 753.
[302] J. Blank, H. D. Hausen and J. Weidlein, *J. Organomet. Chem.* **1993**, *444*, C4.
[303] G. E. Coates and R. A. Whitcombe, *J. Chem. Soc.* **1956**, 3351.
[304] O. T. Beachley Jr., C. Bueno, M. R. Churchill, R. B. Hallock and R. G. Simmons, *Inorg. Chem.* **1981**, *20*, 2423.
[305] O. T. Beachley, Jr., G. E. Coates and L. Kohnstam, *J. Chem. Soc.* **1965**, 3248.
[306] G. Schmidt, S. Kuhner, H. D. Hausen and J. Weidlein, *Z. Anorg. Allg. Chem.* **1997**, *623*, 1499.
[307] R. A. Fischer, H. Sussek, A. Miehr, H. Pritzkow and E. Herdtweck, *J. Organomet. Chem.* **1997**, *548*, 73.
[308] M. Veith, H. Lange, O. Recktenwald and W. Frank, *J. Organomet. Chem.* **1985**, *294*, 273.
[309] S. W. Haggata, M. A. Malik, M. Motevalli and P. O'Brien, *J. Organomet. Chem.* **1996**, *511*, 199.

[310] A. H. Cowley, R. A. Jones, M. A. Mardones and C. M. Nunn, *Organometallics* **1991**, *10*, 1635.
[311] C. J. Carrano, A. H. Cowley, D. M. Giolando, R. A. Jones, C. M. Nunn and J. M. Power, *Inorg. Chem.* **1988**, *27*, 2709.
[312] N. W. Alcock, I. A. Degnan, M. G. H. Wallbridge, H. R. Powell, M. McPartlin and G. M. Sheldrick, *J. Organomet. Chem.* **1989**, *361*, C33.
[313] T. Douglas and K. H. Theopold, *Inorg. Chem.* **1991**, *30*, 596.
[314] O. T. Beachley Jr., J. P. Kopasz, H. Zhang, W. E. Hunter and J. L. Atwood, *J. Organomet. Chem.* **1987**, *325*, 69.
[315] O. T. Beachley, Jr., J. D. Maloney, M. A. Banks and R. D. Rogers, *Organometallics* **1995**, *14*, 3448.
[316] R. L. Wells, A. T. McPhail and M. F. Self, *Organometallics* **1992**, *11*, 221.
[317] R. L. Wells, A. T. McPhail, L. J. Jones III and M. F. Self, *J. Organomet. Chem.* **1993**, *449*, 85.
[318] R. L. Wells, A. T. McPhail, L. J. Jones and M. F. Self, *Polyhedron* **1993**, *12*, 141.
[319] J. A. Burns, M. D. B. Dillingham, H. J. Byers, K. D. Gripper, W. T. Pennington, and G. H. Robinson, *Organometallics* **1994**, *13*, 1514.
[320] B. Werner and B. Neumüller, *Organometallics*, **1996**, *15*, 4258.
[321] M. A. Banks, O. T. Beachley, L. A. Buttre, M. R. Churchill and J. C. Fettinger, *Organometallics* **1991**, *10*, 1901.
[322] D. A. Atwood, A. H. Cowley, R. A. Jones and M. A. Mardones, *J. Organomet. Chem.* **1993**, *449*, C1.
[323] D. C. Bradley, H. Chudzynska, M. M. Faktor, D. M. Frigo, B. Hursthouse, B. Hussain and L. Smith, *Polyhedron* **1988**, *7*, 1289.
[324] F. Maurs and G. Constant, *Polyhedron*, **1984**, *3*, 581.
[325] T. Douglas and K. H. Theopold, *Inorg. Chem.* **1991**, *30*, 594.
[326] S. M. Stuczynski, R. L. Opila, P. Marsh, J. G. Brennan and M. L. Steigerwald, *Chem. Mater.* **1991**, *3*, 379.
[327] A. H. Cowley, R. A. Jones, K. B. Kidd, C. M. Nunn and D. L. Westmoreland, *J. Organomet. Chem.* **1988**, *341*, C1.
[328] R. L. Wells, L. J. Jones, A. T. McPhail and A. Alvanipour, *Organometallics* **1991**, *10*, 2345.
[329] R. L. Wells, A. T. McPhail, L. J. Jones and M. F. Self, *Polyhedron* **1993**, *12*, 141.
[330] A. Dashti-Mommertz, B. Werner and B. Neumüller, *Polyhedron*, **1998**, *17*, 525.
[331] A. H. Cowley, R. A. Jones, K. B. Kidd, C. M. Nunn and D. L. Westmoreland, *J. Organomet. Chem.* **1988**, *341*, C1.
[332] H. Luth and M. R. Truter, *J. Chem. Soc. A*, **1970**, 1287.
[333] P. J. Burke, L. A. Gray, P. J. C. Hayward, R. W. Matthews, M. McPartlin and D. G. Gillies, *J. Organomet. Chem.* **1977**, *136*, C7.
[334] H. Kurosawa, K. Yasuda and R. Okawara, *Bull. Chem. Soc. Japan*, **1967**, *40*, 861.
[335] G. M. Sheldrick and J. P. Yesinowski, *J. Chem. Soc. Dalton Trans.* **1975**, 870.
[336] H. W. Roesky, M. Scholz, M. Noltemeyer and F. T. Edelmann, *Inorg. Chem.* **1989**, *28*, 3829.
[337] Y. M. Chow and D. Britton, *Acta Cryst.* **1975**, *B31*, 1929.
[338] G. H. W. Millburn and M. R. Truter, *J. Chem. Soc. A*, **1967**, 648.
[338a] R. T. Griffin, K. Henrick, R. T. Matthews and M. McPartlin, *J. Chem. Soc. Dalton Trans.* **1980**, 1551.
[339] K. Henrick, M. McPartlin, G. B. Deacon and R. J. Phillips, *J. Organomet. Chem.* **1981**, *204*, 287.
[340] K. Henrick, R. W. Matthews and P. A. Tasker, *Acta Cryst.* **1978**, *B34*, 935.
[341] N. Playa, A. Macias, J. M. Varela, A. Sanchez, J. S. Casas and J. Sordo, *Polyhedron* **1991**, *10*, 1465.
[342] K. Henrick, R. W. Mathews and P. A. Tasker, *Acta Cryst.* **1978**, *B34*, 1347.
[343] T. L. Blundell and H. M. Powell, *Chem. Commun.* **1967**, 54.
[344] A. Blaschette, P. G. Jones, A. Michalides and M. Näveke, *J. Organomet. Chem.* **1991**, *415*, 25.
[345] R. C. Poller, *J. Organomet. Chem.* **1965**, *3*, 321.

[346] H. Reuter and H. Puff, *J. Organomet. Chem.* **1989**, *379*, 223.
[347] S. Calogero, P. Ganis, V. Peruzzo, G. Tagliavini and G. Valle, *J. Organomet. Chem.* **1981**, *220*, 11.
[348] D. Tudela, R. Fernandez, V. K. Belsky and V. E. Zavodnik, *J. Chem. Soc. Dalton Trans.* **1996**, 2123.
[349] H. C. Clark, R. J. O'Brien and J. Trotter, *J. Chem. Soc. A* **1964**, 2332.
[350] H. Reuter and H. Puff, *J. Organomet. Chem.* **1989**, *379*, 223.
[351] D. Tudela, E. Gutièrrez-Puebla and A. Monge, *J. Chem. Soc. Dalton Trans.* **1992**, 1069.
[352] L. N. Zakharov, Yu. T. Struchkov, E. A. Kuz'mina and B. I. Petrov, *Kristallografiya (Russ.)*, **1983**, *28*, 271.
[353] D. Williams, *Austral. Def. Stand Labs. Ann. Rept.* **1967/1968**, 37.
[354] S. S. Al Juaid, S. M. Dhaher, C. Eaborn, P. B. Hitchcock and J. D. Smith, *J. Organomet. Chem.* **1987**, *325*, 117.
[355a] H. Bai, R. K. Harris and H. Reuter, *J. Organomet. Chem.* **1991**, *408*, 167.
[355b] H. Bai and R. K. Harris, *J. Mag. Res.* **1992**, *96*, 24.
[356] D. Dakternieks and H. Zhu, *Inorg. Chim. Acta* **1992**, *196*, 19.
[357] E. O. Schlemper and W. C. Hamilton, *Inorg. Chem.* **1966**, *5*, 995.
[358] T. H. Lambertsen, P. G. Jones and R. Schmutzler, *Polyhedron*, **1992**, *11*, 331.
[359] A. J. Gibbons, A. K. Sawyer and A. Ross, *J. Org. Chem.* **1961**, *26*, 2304.
[360] C. K. Chu and J. D. Murray, *J. Chem. Soc. A*, **1971**, 360.
[361] P. G. Harrison, M. J. Begley and K. C. Molloy, *J. Organomet. Chem.* **1980**, *186*, 213.
[362] R. Graziani, U. Casellato and G. Plazzogna, *Acta Cryst.* **1983**, *C 39*, 1188.
[363] D. Dakternieks, R. W. Gable and B. F. Hoskins, *Inorg. Chim. Acta*, **1984**, *85*, L 43.
[364] H. Puff, E. Friedrichs and F. Vizel, *Z. Anorg. Allg. Chem.* **1981**, *477*, 50.
[365] R. Hämäläinen and U. Turpinen, *J. Organomet. Chem.* **1987**, *333*, 323.
[366] H. Puff, I. Bung, E. Friedrichs and A. Jansen, *J. Organomet. Chem.* **1983**, *254*, 23.
[367] E. R. T. Tiekink, *Acta Cryst.* **1991**, *C47*, 661.
[368] R. A. Kresinski, R. L. Staples and J. P. Fackler, Jr., *Acta Cryst.* **1994**, *C50*, 40.
[369] P. G. Harrison, M. J. Begley and K. C. Molloy, *J. Organomet. Chem.* **1980**, *186*, 213.
[370] J. Beckmann, M. Biesemans, K. Hassler, K. Jurkschat, J. C. Martins, M. Schürmann and R. Willem, *Inorg. Chem.* **1998**, *37*, 4891.
[371] D. Dakternieks, K. Jurkschat, S. van Dreumel and E. R. T. Tiekink, *Inorg. Chem.* **1997**, *36*, 2023.
[372] J. A. Blair, R. A. Howie, J. L. Wardell and P. J. Cox, *Polyhedron*, **1997**, *16*, 881.
[373] C. Vatsa, V. K. Jain, T. K. Das and E. R. T. Tiekink, *J. Organomet. Chem.* **1991**, *418*, 329.
[374] M. J. Cox, M. I. Mohamed-Ibrahim and E. R. T. Tiekink, *Main Group Met. Chem.* **1997**, *20*, 207.
[375] R. Okawara, *Proc. Chem. Soc.* **1961**, 383.
[376] D. L. Allenston, A. G. Davies, M. Hancock and R. F. M. White, *J. Chem. Soc.* **1963**, 5469.
[377] W. J. Considine, J. J. Ventura, A. J. Gibbons and A. Ross, *Can. J. Chem.* **1963**, *41*, 1239.
[378] A. G. Davies, L. Smith, P. J. Smith and W. McFarlane, *J. Organomet. Chem.* **1971**, *29*, 245.
[379] Y. M. Chow, Acta Cryst. *Inorg. Chem.* **1971**, *10*, 673.
[380] D. Dakternieks, K. Jurkschat, D. Schollmeyer and H. Wu, *Organometallics* **1994**, *13*, 4121.
[381] M. Mehring, M. Schürmann, H. Reuter, D. Dakternieks and K. Jurkschat, *Angew. Chem.* **1997**, *109*, 1150.
[382] M. Hill, M. F. Mahon and K. C. Molloy, *Main Group Chem.* **1996**, *1*, 309.
[383] B. G. Aggarwal, Y. P. Singh, R. Bohra, G. Srivastava and A. K. Rai, *J. Organomet. Chem.* **1993**, *444*, 47.
[384] A. G. Davies and S. Slater, *Silicon, Germanium, Tin and Lead*, **1986**, *9*, 87.
[385] N. Kasai, K. Yasuda and R. Okawara, *J. Organomet. Chem.* **1965**, *3*, 172.
[386] E. G. Perevalova, M. D. Reshetova, P. N. Ostapchuk, Yu. L. Slovokhotov, Yu. T. Struchkov, F. M. Spiridonov, A. V. Kizin and I. G. Yukhno, *Metallorg. Khim. (Russ.)* **1990**, *3*, 100; *Organomet. Chem. USSR*, **1990**, *3*, 53.
[387] G. B. Deacon, E. Lawrenz, K. T. Nelson and E. R. T. Tiekink, *Main Group Met. Chem.* **1993**, *16*, 265.

[388] C. Glidewel and D. C. Liles, *Acta Cryst.* **1978**, *B34*, 129.
[389] U. Wannagat, V. Damrath, W. Huch, M. Veith and U. Harder, *J. Organomet. Chem.* **1993**, *443*, 153.
[390] T. P. Lockhart, *J. Organomet. Chem.* **1985**, *287*, 179.
[391] H. Reuter and H. Puff, *J. Organomet. Chem.* **1989**, *379*, 223.
[392] J. B. Hall and D. Britton, *Acta Cryst.* **1972**, *28B*, 2133.
[393] A. Blaschette, E. Wieland, P. G. Jones and I. Hippel, *J. Organomet. Chem.* **1993**, *445*, 55.
[394] I. Lange, P. G. Jones and A. Blaschette, *J. Organomet. Chem.* **1995**, *485*, 179.
[394a] A. Wirth, D. Henschel, A. Blaschette and P. G. Jones, *Z. Anorg. Allg. Chem.*, **1997**, *623*, 587.
[395] A. Wirth, D. Henschel, P. G. Jones and A. Blaschette, *J. Organomet. Chem.* **1996**, *525*, 167.
[396] H. Puff, H. Hevendehl, K. Höfer, H. Reuter and W. Schuh, *J. Organomet. Chem.* **1985**, *287*, 163.
[396a] K. Jurkschat, C. Krüger and J. Meunier-Piret, *Main Group Met. Chem.* **1992**, *15*, 61.
[397] R. A. Kresinski, R. J. Staples and J. P. Fackler, Jr., *Acta Cryst.* **1994**, *C50*, 40.
[398] C. Vatsa, V. K. Jain, T. K. Das and E. R. T. Tiekink, *J. Organomet. Chem.* **1991**, *418*, 329.
[399] P. Brown, M. F. Mahon and K. C. Molloy, *J. Chem. Soc. Dalton Trans.* **1992**, 3503.
[400] F. Cervantes, H. K. Sharma, I. Haiduc and K. H. Pannell, *J. Chem. Soc. Dalton Trans.* **1998**, 1.
[401] J. Beckmann, K. Jurkschat, B. Mahieu and M. Schurmann, *Main Group Metal Chem.* **1998**, *21*, 113.
[402] G. D. Smith, P. E. Fenwick and I. P. Rothwell, *Acta Crystallogr.* **1995**, *51C*, 2501.
[403] V. B. Mokal, V. K. Jain and E. R. T. Tiekink, *J. Organomet. Chem.* **1992**, *431*, 283.
[403a] J. B. Lambert, B. Kuhlmann and C. L. Stern, *Acta Crystallogr.* **1993**, *C49*, 887.
[404] O. S. Jung, J. H. Jeong and Y. S. Sohn, *J. Organomet. Chem.* **1992**, *439*, 23.
[405] C. Lecomte, J. Protas and M. Devaud, *Acta Cryst.* **1976**, *B32*, 923.
[406] R. R. Holmes, S. Shafieezad, V. V. Chandrasekar, J. M. Holmes and R. O. Day, *J. Am. Chem. Soc.* **1988**, *110*, 1744.
[407] H. Puff and H. Reuter, *J. Organomet. Chem.* **1989**, *368*, 173.
[408] H. Reuter and A. Sebald, *Z. Naturforsch.* **1993**, *48b*, 195.
[408a] D. Dakternieks, H. Zhu, E. R. T. Tiekink and R. Colton, *J. Organomet. Chem.* **1994**, *476*, 33.
[408b] F. Banse, F. Ribot, P. Toledano, J. Maquet and C. Sanchez, *Inorg. Chem.* **1995**, *34*, 6371.
[408c] F. Ribot, F. Banse, F. Diter and C. Sanchez, *New J. Chem.* **1995**, *19*, 1145.
[408d] C. Eychene-Baron, F. Ribot and C. Sanchez, *J. Organomet. Chem.* **1998**, *567*, 137.
[409] P. Jaumier, B. Jousseaume, M. Lahcini, F. Ribot and C. Sanchez, *Chem. Commun.* **1998**, 369.
[410] H. Reuter, *Angew. Chem.* **1991**, *103*, 1487.
[411] F. Ribot, C. Sanchez, R. Willem, J. C. Martins and M. Biesemans, *Inorg. Chem.* **1998**, *37*, 911.
[412] S. B. Teo, S. G. Teoh, R. C. Okechukwu and H. K. Fun, *J. Organomet. Chem.* **1993**, *454*, 67.
[413] W. Chen, *J. Organomet. Chem.* **1994**, *471*, 69.
[414] A. M. Domingos and G. M. Sheldrick, *Acta Cryst.* **1974**, *B 30*, 519.
[415] H. Reuter and D. Schröder, *J. Organomet. Chem.* **1993**, *455*, 83.
[416] P. J. Smith, R. F. M. White and L. Smith, *J. Organomet. Chem.* **1972**, *40*, 341.
[417] P. F. R. Ewings, P. G. Harrison, T. J. King, R. C. Phillips and J. A. Richards, *J. Chem. Soc. Dalton Trans.* **1975**, 1950.
[418] M. J. Cox, M. I. Mohamed-Ibrahim and E. R. T. Tiekink, *Main Group Metal Chem.* **1997**, *20*, 208.
[419] E. R. T. Tiekink, *Appl. Organomet. Chem.* **1991**, *5*, 1.
[420] E. R. T. Tiekink, *Trends Organomet. Chem.* **1994**, *1*, 71.
[421] H. Chih and B. R. Penfold, *J. Cryst. Mol. Str.* **1973**, *3*, 285.
[422] P. J. Smith, R. O. Day, V. Chandrasekar, J. M. Holmes and R. R. Holmes, *Inorg. Chem.* **1986**, *25*, 2495.
[423] R. R. T. Tiekink, G. K. Sandhu and S. P. Verma, *Acta Cryst.* **1989**, *C45*, 1810.
[424] G. K. Sandhu, S. P. Verma and E. R. T. Tiekink, *J. Organomet. Chem.* **1990**, *393*, 195.

[425] T. V. Sizova, N. S. Yashina, V. S. Petrosyan, A. V. Yatsenko, V. V. Chernyshev and L. A. Aslanov, *J. Organomet. Chem.* **1993**, *453*, 171.
[426] T. S. B. Baul and E. R. T. Tiekink, *Acta Cryst.* **1996**, *C52*, 1428.
[427] F. Huber, B. Mundus-Glowacki and H. Preut, *J. Organomet. Chem.* **1989**, *365*, 111.
[428] G. Vale, V. Peruzzo, D. Martin and P. Granis, *Cryst. Struct. Comm.* **1982**, *11*, 595.
[429] S. Calogero, D. A. Clemento, V. Peruzzo and G. Tagliavini, *J. Chem. Soc. Dalton Trans.* **1979**, 1172.
[430] R. Graziani, U. Casellato and G. Plazogna, *J. Organomet. Chem.* **1980**, *187*, 381.
[431] M. M. Amini, S. W. Ng, K. A. Fidelis, M. J. Heeg, C. R. Muchmore, D. van der Helm and J. J. Zuckerman, *J. Organomet. Chem.* **1989**, *365*, 103.
[432] A. S. Heinehen, J. W. Bruno and J. C. Huffman, *J. Organomet. Chem.* **1990**, *382*, 361.
[433] J. Lokaj, V. Vrabel, E. Kellö and V. Ratay, *Coll. Czech Chem. Comm.* **1989**, *54*, 684.
[434] S. W. Ng, V. G. Kumar Das and E. R. T. Tiekink, *J. Organomet. Chem.* **1991**, *403*, 111.
[435] K. C. Molloy, T. G. Purcell, M. F. Mahon, and E. Minshall, *Applied Organomet. Chem.* **1987**, *1*, 507.
[436] D. W. Allen, I. W. Nowell, J. S. Brooks and R. W. Clarkson, *J. Organomet. Chem.* **1981**, *219*, 29.
[437] K. C. Molloy, K. Quill and I. W. Nowell, *J. Chem. Soc. Dalton Trans.* **1987**, 101.
[438] K. C. Molloy, T. G. Purcell, K. Quill and I. W. Nowell, *J. Organomet. Chem.* **1984**, *267*, 237.
[439] S. W. Ng, K. L. Chin, C. Wei, V. G. Kumar Das and R. J. Butcher, *J. Organomet. Chem.* **1989**, *376*, 277.
[440] R. R. Holmes, R. O. Day, V. Chandrasekar, J. F. Vollano and J. M. Holmes, *Inorg. Chem.* **1986**, *25*, 2490.
[441] A. Samuel-Lewis, P. J. Smith, J. H. Aupers, D. Hampson and D. C. Povey, *J. Organomet. Chem.* **1992**, *437*, 131.
[442] A. Samuel-Lewis, P. J. Smith, J. H. Aupers, and D. C. Povey, *J. Organomet. Chem.* **1991**, *402*, 319.
[443] K. M. Lo, S. W. Ng, A. Wei and V. G. Kumar Das, *Acta Cryst.* **1992**, *C48*, 1657.
[444] Y. C. Toong, S. P. Tai, M. C. Pun, R. C. Hynes. L. E. Khoo and F. E. Smith, *Can. J. Chem.* **1992**, *70*, 2683.
[445] S. W. Ng, V. G. Kumar Das and A. Syed, *J. Organomet. Chem.* **1989**, *364*, 353.
[446] S. W. Ng and V. G. Kumar Das, *Acta Cryst.* **1995**, *51C*, 2489.
[447] N. W. Alcock and R. E. Timms, *J. Chem. Soc. A*, **1968**, 1873.
[448] G. Fergusson, T. Spalding, A. T. O'Dowd and K. C. O'Shea, *Acta Cryst.* **1995**, *C51*, 2546.
[449] I. W. Nowell, J. S. Brooks, G. Beech and R. Hill, *J. Organomet. Chem.* **1983**, *244*, 119.
[450] S. W. Ng, C. Wei and V. G. Kumar Das, *J. Organomet. Chem.* **1988**, *345*, 59.
[451] M. Gielen, A. Bouhdid, R. Willem, V. I. Bregadze, L. V. Ermanson and E. R. T. Tiekink, *J. Organomet. Chem.* **1995**, *501*, 277.
[452] F. Kayser, M. Biesemans, M. Boualam, E. R. T. Tiekink, A. El Khloufi, J. Meunier-Piret, A. Bouhdid, K. Jurkschat, M. Gielen and R. Willem, *Organometallics*, **1994**, *13*, 1098, 4126.
[453] R. Willem, A. Bouhdid, F. Kayser, A. Delmotte, M. Gielen, J. C. Martins, M. Biesemans, B. Mahieu, and E. R. T. Tiekink, *Organometallics*, **1996**, *15*, 1920.
[454] R. Willem, A. Bouhdid, A. Meddour, C. Camacho-Camacho, F. Mercier, M. Gielen, M. Biesemans, F. Ribot, C. Sanchez and E. R. T. Tiekink, *Organometallics*, **1997**, *16*, 4377.
[454a] M. Meddour, F. Mercier, J. C. Martins, M. Gielen, M. Biesemans and R. Willem, *Inorg. Chem.* **1997**, *36*, 5712.
[455] S. W. Ng, V. G. Kumar Das, B. W. Skelton and A. H. White, *J. Organomet. Chem.* **1992**, *430*, 139.
[456] S. W. Ng and V. G. Kumar Das, *Z. Kristallogr.* **1995**, *210*, 133.
[457] G. Stocco, G. Guli, M. A. Girasolo, G. Bruno, F. Nicolo and R. Scopelitti, *Acta Cryst.* **1996**, *C52*, 829.
[458] U. Schubert, *J. Organomet. Chem.* **1978**, *155*, 285.
[459] S. W. Ng and V. G. Kumar Das, *Acta Cryst.* **1993**, *C 49*, 754.
[460] Y. C. Tong, S. P. Tai, M. C. Pun, R. C. Hynes, L. E. Khoo and F. E. Smith, *Can. J. Chem.* **1992**, *70*, 2683.

[461] S. W. Ng and V. G. Kumar Das, *J. Chem. Cryst.* **1994**, *24*, 337.
[462] S. W. Ng and W. G. Kumar Das, *Acta Cryst.* **1996**, *C52*, 1373.
[462a] J. P. Fackler, Jr., G. Garzon, R. A. Kresinski, H. G. Murray and R. G. Raptis, *Polyhedron* **1994**, *13*, 1705.
[463] F. Mistry, S. J. Rettig, J. Trotter and F. Aubke, *Z. Anorg. Allg. Chem.* **1995**, *621*, 1875.
[464] E. S. Looi, C. A. Keok, S. B. Teo and J. P. Declerq, *J. Organomet. Chem.* **1996**, *523*, 75.
[465] F. Mistry, S. J. Rettig, J. Trotter and F. Aubke, *Acta Crystallogr.* **1990**, *C46*, 2091.
[466] T. P. Lockhart and F. Davidson, *Organometallics* **1987**, *6*, 2471.
[467] V. Chandrasekar, R. O. Day, J. M. Holmes and R. R. Holmes, *Inorg. Chem.* **1988**, *27*, 958.
[468] M. F. Garbauskas and J. M. Wengrovius, *Acta Crystallogr.* **1971**, *C47*, 1969.
[469] K. Jurkschat and E. R. T. Tiekink, *Main Group Chem.* **1994**, *17*, 659.
[470] F. Ribot, C. Sanchez, A. Meddour, M. Gielen, E. R. T. Tiekink, M. Biesemans and R. Willem, *J. Organomet. Chem.* **1998**, *552*, 177
[471] (a) M. Gielen, A. El Khloufi, D. de Vos, H. J. Kolker, J. M. Schellens and R. Willem, *Bull. Soc. Chim. Belg.* **1993**, *102*, 761; (b) M. Gielen, A. El Khloufi, M. Biesemans and R. Willem, *Appl. Organomet. Chem.* **1993**, *7*, 119 and 201; c) M. Gielen, M. Biesemans, A. El Khloufi, J. Meunier-Piret, F. Kayser and R. Willem, *J. Fluorine Chem.* **1993**, *64*, 279.
[472a] M. Gielen, *Coord. Chem. Rev.* **1996**, *151*, 41.
[472b] M. Gielen, *Main Group Met. Chem.* **1994**, *17*, 1.
[472c] Valle, V. Peruzzo, G. Tagliavini and P. Ganis, *J. Organomet. Chem.* **1984**, *276*, 325.
[473] R. Faggiani, J. P. Johnson, I. D. Brown and T. Birchall, *Acta Cryst.* **1978**, *B34*, 3743.
[474] T. Birchall, C. S. Frampton and J. Johnson, *Acta Cryst.* **1987**, *C43*, 1492.
[475] C. Vatsa, V. K. Jain, T. K. Das and E. R. T. Tiekink, *J. Organomet. Chem.* **1991**, *421*, 21.
[476] M. Gielen, J. Meunier-Piret, M. Biesemans, R. Willem and A. El Khloufi, *Appl. Organomet. Chem.* **1992**, *6*, 59.
[477] G. K. Sandhu, R. Hundal and E. R. T. Tiekink, *J. Organomet. Chem.* **1992**, *430*, 15.
[478] M. Gielen, T. Mancilla, R. Willem, M. Gielen, E. R. T. Tiekink, A. Bouhdid, D. de Vos, M. Biesemans, I. Verbruggen and R. Willem, *Appl. Organomet. Chem.* **1995**, *9*, 639.
[479] C. D. Garner, B. Hughes and T. J. King, *Inorg. Nucl. Chem. Letters* **1976**, *12*, 859.
[480] G. K. Sandhu, N. Sharma and E. R. T. Tiekink, *J. Organomet. Chem.* **1991**, *403*, 119.
[481] R. Graziani, G. Bombieri, E. Forselini, P. Furlan, V. Peruzzo and G. Tagliavini, *J. Organomet. Chem.* **1977**, *125*, 43.
[482] C. Vatsa, V. K. Jain, T. K. Das and E. R. T. Tiekink, *J. Organomet. Chem.* **1990**, *396*, 9.
[483] C. S. Parulekar, V. K. Jain, T. Kesavadas and E. R. T. Tiekink, *J. Organomet. Chem.* **1990**, *387*, 163.
[484] S. P. Narula, S. K. Bharadwaj, H. K. Sharma, G. Mairesse, P. Barbier and G. Nowogrocki, *J. Chem. Soc. Dalton Trans.* **1988**, 1719.
[485] S. P. Narula, S. K. Bharadwaj, Y. Sharda, D. C. Povey and G. W. Smith, *J. Organomet. Chem.* **1992**, *430*, 167.
[486] R. Graziani, G. Bombieri, E. Forselini, P. Furlan, V. Peruzzo and G. Tagliavini, *J. Organomet. Chem.* **1977**, *125*, 43.
[487] T. P. Lockhardt, W. T. Manders and E. M. Holt, *J. Am. Chem. Soc.* **1986**, *108*, 6611.
[488] M. Boualam, R. Willem, M. Biesemans, B. Mathhieu, J. Meunier-Piret and M. Gielen, *Main Group Main Chem.* **1991**, *14*, 41.
[489] N. W. Alcock and S. M. Roe, *J. Chem. Soc. Dalton Trans.* **1989**, 1589.
[490] C. Vatsa, V. K. Jain, T. Kesavadas and E. R. T. Tiekink, *J. Organomet. Chem.* **1991**, *408*, 157.
[491] G. Valle, V. Peruzzo, G. Tagliavini and P. Ganis, *J. Organomet. Chem.* **1984**, *276*, 325.
[492] V. Chandrasekar, R. O. Day, J. M. Holmes and R. R. Holmes, *Inorg. Chem.* **1988**, *27*, 958.
[493] L. E. Khoo, Z. H. Zhou and T. C. W. Mak, *J. Cryst. Spectr. Res.* **1993**, *23*, 153.
[494] C. S. Parulekar, V. K. Jain, T. K. Das, A. R. Gupta, B. F. Hoskins and E. R. T. Tiekink, *J. Organomet. Chem.* **1989**, *372*, 193.
[495] S. W. Ng, C. Wei and V. G. Kumar Das, *J. Organomet. Chem.* **1991**, *412*, 39.
[496] J. Tao, W. Xiao and Q. Yang, *J. Organomet. Chem.* **1997**, *531*, 223.
[497] S. W. Ng, V. G. Kumar Das, W. H. Yip, R. J. Wang and T. C. W. Mak, *J. Organomet. Chem.* **1990**, *393*, 201.

[498] H. Preut, F. Huber and M. Gielen, *Acta Cryst.* **1990**, *C46*, 2071.
[499] S. G. Teoh, S. H. Ang, E. S. Looi, C. A. Keok, S. B. Teo and H. K. Fun, *J. Organomet. Chem.* **1997**, *527*, 15.
[499a] N. W. Alcock and S. M. Roe, *J. Chem. Soc.* **1989**, 1589.
[500] J. Meunier-Piret, M. Boualam, R. Willem and M. Gielen, *Main Group Metal Chem.* **1993**, *16*, 329.
[501] T. P. Lockhart, *Organometallics* **1988**, *7*, 1438.
[501a] J. S. Casas, A. Castineiras, M. D. Couce, N. Playa, U. Russo, J. Sordo and J. M. Varela, *J. Chem. Soc. Dalton Trans.* **1998**, 1513.
[502] C. Pettinari, F. Marchetti, A. Cingolani, D. Leonesi, G. G. Lobbia and A. Lorenzotti, *J. Organomet. Chem.* **1993**, *454*, 59.
[503] H. H. Anderson, *Inorg. Chem.* **1964**, *3*, 912.
[504] J. Otera, T. Yano, E. Kunimoto and T. Nakata, *Organometallics*, **1984**, *3*, 426.
[505] H. Lambourne, *J. Chem. Soc.* **1922**, *121*, 2533.
[506] H. Lambourne, *J. Chem. Soc.* **1924**, *125*, 2013.
[507] B. F. E. Ford, B. V. Liengme and J. R. Sams, *J. Organomet. Chem.* **1969**, *19*, 53.
[508] R. C. Poller, J. N. R. Ruddick, B. Taylor and D. L. B. Toley, *J. Organomet. Chem.* **1970**, *24*, 341.
[509] A. Roy and A. K. Gosh, *Inorg. Chim. Acta Letters* **1977**, *24*, L 89; **1978**, *29*, L 275.
[510] R. R. Holmes, *Acc. Chem. Res.* **1989**, *22*, 190.
[511] R. O. Day, V. Chandrasekar, K. C. Kumara Swamy, J. M. Holmes, S. D. Burton and R. R. Holmes, *Inorg. Chem.* **1988**, *27*, 2887.
[512] V. B. Mokal, V. K. Jain and E. R. T. Tiekink, *J. Organomet. Chem.* **1991**, *407*, 173.
[513] V. Chandrasekar, C. G. Schmid, S. D. Burton, J. M. Holmes, R. O. Day and R. R. Holmes, *Inorg. Chem.* **1987**, *26*, 1050.
[514] R. R. Holmes, C. G. Schmid, V. Chandrasekar, R. O. Day and J. M. Holmes, *J. Am. Chem. Soc.* **1987**, *109*, 1408.
[515] V. Chandrasekar, R. O. Day and R. R. Holmes, *Inorg. Chem.* **1985**, *24*, 1970.
[516] S. J. Blunden, R. Hill and J. N. R. Ruddick, *J. Organomet. Chem.* **1984**, *267*, C5.
[517] E. R. T. Tiekink, *J. Organomet. Chem.* **1986**, *302*, C1.
[518] J. Kummerlen, A. Sebald and H. Reuter, *J. Organomet. Chem.* **1992**, *427*, 309.
[519] D. Potts, H. D. Sharma, A. J. Carty and A. Walker, *Inorg. Chem.* **1974**, *13*, 1205.
[520] M. Nardelli, C. Pelizzi and G. Pelizzi, *J. Chem. Soc. Dalton Trans.* **1978**, 131.
[521] J. Hilton, E. K. Nunn and S. C. Wallwork, *J. Chem. Soc. Dalton Trans.* **1973**, 173.
[522] M. Nardelli, C. Pelizzi, G. Pelizzi and P. Tarasconi, *Z. Anorg. Allg. Chem.* **1977**, *431*, 250.
[523] A. M. Domingos and G. M. Sheldrick, *J. Chem. Soc. Dalton Trans.* **1974**, 475.
[524] A. Bonardi, A. C. Antoni, C. Pelizzi, G. Pelizzi and P. Tarasconi, *J. Organomet. Chem.* **1991**, *402*, 281.
[524a] A. Sanchez-Gonzalez, A. Castineiras, J. S. Casas, J. Sordo and U. Russo, *Inorg. Chim. Acta* **1994**, *216*, 257.
[525] J. P. Ashmore, T. Chivers, K. A. Kerr and J. H. G. Van Roode, *J. Chem. Soc. Chem. Commun.* **1973**, 173; *Inorg. Chem.* **1977**, *16*, 191.
[526] K. C. Molloy, F. A. K. Nasser, C. L. Barnes, D. van der Helm and J. J. Zuckerman, *Inorg. Chem.* **1982**, *21*, 960.
[527] J. G. Masters, F. A. K. Nasser, M. B. Hossain, A. P. Hagen, D. van der Helm and J. J. Zukerman, *J. Organomet. Chem.*, **1990**, *385*, 39.
[528] M. G. Newton, I. Haiduc, R. B. King and C. Silvestru, *J. Chem. Soc. Chem. Commun.* **1993**, 1229.
[529] F. Weller and A. F. Shihada, *J. Organomet. Chem.* **1987**, *322*, 185.
[530] A. F. Shihada and F. Weller, *Z. Naturforsch.* **1995**, *50b*, 1343.
[531] S. J. Blunden, R. Hill and D. G. Gillies, *J. Organomet. Chem.* **1984**, *270*, 39.
[532] J. N. Pandey and G. Srivastava, *J. Organomet. Chem.* **1988**, *354*, 301.
[533] A. F. Shihada and F. Weller, *Z. Naturforsch.* **1996**, *51b*, 1111.
[534] A. F. Shihada and F. Weller, *Z. Naturforsch.* **1997**, *52b*, 587.
[535] A. F. Shihada and F. Weller, *Z. Naturforsch.* **1998**, *53b*, 699.

[536] F. A. K. Nasser, M. B. Hossain, D. van der Helm and J. J. Zuckerman, *Inorg. Chem.* **1983**, *22*, 3107.
[537] V. B. Mokal, V. K. Jain and E. R. T. Tiekink, *J. Organomet. Chem.* **1994**, *471*, 53.
[538] V. B. Mokal, V. K. Jain and E. R. T. Tiekink, *J. Organomet. Chem.* **1993**, *448*, 75.
[539] A. Silvestru, C. Silvestru, I. Haiduc, J. E. Drake, J. Yang and F. Caruso, *Polyhedron*, **1997**, *16*, 949.
[540] R. R. Holmes, *Acc. Chem. Res.* **1989**, *22*, 190.
[541] K. C. Kumara Swamy, R. O. Day and R. R. Holmes, *Phosphorus, Sulfur & Silicon*, **1989**, *41*, 291.
[542] R. R. Holmes, K. C. Kumara Swamy, C. G. Schmidt and R. O. Day, *J. Am. Chem. Soc.* **1988**, *110*, 7060.
[543] R. O. Day, J. M. Holmes, V. Chandrasekar and R. R. Holmes, *J. Am. Chem. Soc.* **1987**, *109*, 940.
[544] K. C. Kumara Swamy, R. O. Day and R. R. Holmes, *J. Am. Chem. Soc.* **1987**, *109*, 5546.
[545] K. C. Kumara Swamy, C. G. Schmidt, R. O. Day and R. R. Holmes, *J. Am. Chem. Soc.* **1988**, *110*, 7067.
[546] K. C. Kumara Swamy, C. G. Schmidt, R. O. Day and R. R. Holmes, *J. Am. Chem. Soc.* **1990**, *112*, 223.
[546a] V. B. Mokal, V. K. Jain and E. R. T. Tiekink, *J. Organomet. Chem.* **1991**, *407*, 173.
[547] K. C. Kumara Swamy, R. O. Day and R. R. Holmes, *Inorg. Chem.* **1992**, *31*, 4184.
[548] R. E. B. Garrod, R. H. Pratt and J. R. Sams, *Inorg. Chem.* **1971**, *10*, 424.
[549] K. C. Molloy, K. Quill, D. Cunningham, P. Mc Ardle and T. Higgins, *J. Chem. Soc. Dalton Trans.* **1989**, 267.
[550] M. Herberhold, S. Gerstmann, W. Milius and B. Wrackmayer, *Z. Naturforsch.* **1997**, *52b*, 1278.
[551] R. Hengel, U. Kunze and J. Straehle, *Z. Anorg. Allg. Chem.* **1976**, *423*, 35.
[552] G. M. Sheldrick and R. Taylor, *Acta Cryst.* **1977**, *B33*, 135.
[553] D. Ginderow and M. Huber, *Acta Cryst.* **1973**, *B29*, 560.
[554] F. H. Allen, J. A. Lerbscher and J. Trotter, *J. Chem. Soc.* **1971**, 2507.
[555] V. I. Scherbakov, N. A. Sarycheva, I. K. Grigoreva, R. I. Bochkova and G. A. Razuvaev, *Metalloorg. Khim. (Russ.)* **1990**, *3*, 352; *Organomet. Chem. USSR (Engl. transl.)* **1990**, *3*, 168.
[556] U. Ansorge, E. Lindner and J. Straehle, *Chem. Ber.* **1978**, *111*, 3048.
[557] V. Chandrasekar, M. G. Muralidhara, K. R. J. Thomas, E. R. T. Tiekink, *Inorg. Chem.* **1992**, *31*, 4707.
[558] A. Diasse-Sarr, L. Diop, M. F. Mahon and K. C. Molloy, *Main Group Met. Chem.* **1997**, *20*, 223.
[559] A. M. Domingos and G. M. Sheldrick, *J. Chem. Soc. Dalton Trans.* **1974**, 477.
[560] Y. Sasaki, H. Imoto and O. Nagano, *Bull. Chem. Soc. Japan* **1984**, *57*, 1417.
[561] B. Krebs, B. Lettmann, H. Pohlmann and R. Froehlich, *Z. Krist.* **1991**, *196*, 231.
[562] E. Herdtweck, P. Kiprof, W. A. Herrmann, J. G. Kuchler and I. Degnan, *Z. Naturforsch.* **1990**, *45b*, 937.
[563] T. Schoop, H. W. Roesky, M. Noltemeyer and H.-G. Schmidt, *Organometallics* **1993**, *12*, 571.

4 Supramolecular Self-Assembly by Formation of Secondary Bonds

4.1 Homoatomic Interactions

Self-assembly as a result of secondary bonding interactions is quite common in simple compounds of the elements of Groups 15 and 16, when bulky substituents do not prevent it. It has been noted for binuclear compounds of the types R_2EER_2 (E = P, As, Sb, Bi) and REER (E = S, Se, Te) on going from third-period elements to their heavier congeners in Groups 15 and 16 that the intermolecular forces holding the molecules together in the solid state gradually change from van der Waals to increasingly strong secondary bond interactions [1, 2].

In Group 15, the phosphorus compounds do not seem to be associated by P···P secondary bonding in any of their compounds, but methylarsenic $(MeAs)_x$ has been described as a ladder (or ribbon) polymer, **1**, with an As–As interatomic distance of 2.40 Å and As···As 2.90 Å. It is not clear how this structure agrees with a trivalent state of arsenic [3, 4]. The sum of the covalent radii for arsenic is 2.390 Å and the sum of van der Waals radii is 3.70 Å.

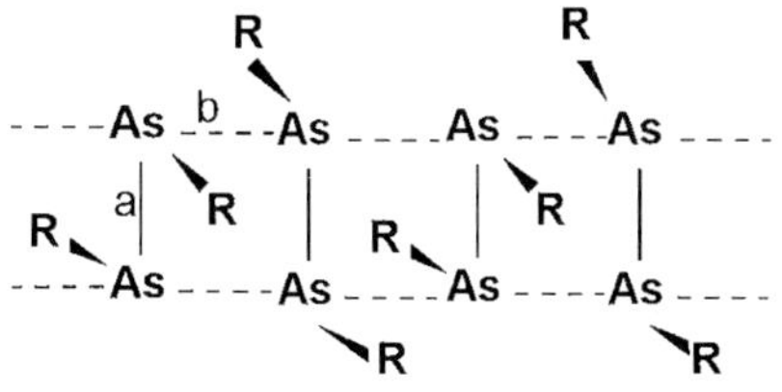

$R = CH_3$
$a = 2.40$ Å
$b = 2.90$ Å

1

Tetramethyldiarsane, $Me_2AsAsMe_2$, is associated in solid state and forms chains, **2**, with alternating primary covalent As–As bonds (2.43 Å) and weak secondary As···As bonds (3.70 Å) whereas tetraphenyldiarsane, $Ph_2AsAsPh_2$ [1] and $(Me_3Si)_2AsAs(SiMe_3)_2$ [5] are monomeric and unassociated. In 2,2'5,5'-diarsole the molecules are self-assembled (primary As–As 2.436 Å, secondary As···As 3.353 Å)

[6]. Pentamethylcyclopentadienyl arsenic diiodide, σ-$C_5Me_5AsI_2$, self-assembles into dimeric pairs via As···As secondary bonds (3.37 Å) rather than arsenic–iodine interactions as one might expect [7].

```
       R  R            R  R
        \ /             \ /
    a    As      b       As
----As------As--------As------As------
    / \             / \
   R   R           R   R

                2
```

R = CH$_3$
a = 2.43 Å
b = 3.70 Å

The intermolecular contacts and the reciprocal arrangement of molecules in crystals of η^5-$C_5H_5(CO)_2CrAs_3$ are also indicative of supramolecular self-organization. Redrawing of the structure with published atomic coordinates reveals that the molecular tectons form layers with shortest intermolecular As···As contacts at 3.515 Å within the layer, and additional As···As secondary interactions connecting the layers into double sheets. A lateral view of a layer illustrates the structure (Figure 4.1) [8].

Similar self-assembly via As···As secondary interactions occurs in η^5-C_5Me_5·$(CO)_2CrAs_3$ [9] and in η^5-$C_5H_5(CO)_2MoAs_3$ (Figure 4.2.) [10]. In the original paper only the molecular structure is given, but representation of the crystal packing using published atomic coordinates reveals a secondary supramolecular structure [11].

The sum of the covalent radii for antimony, i.e. the expected Sb–Sb primary bond length, is 2.74 Å, and the sum of van der Waals radii is Sb···Sb 4.10 Å. If experimental data are compared with these values several supramolecular self-assembled distibanes can be identified.

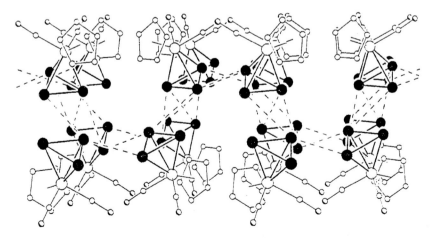

Fig. 4.1. The supramolecular structure of η^5-$C_5H_5(CO)_2CrAs_3$.

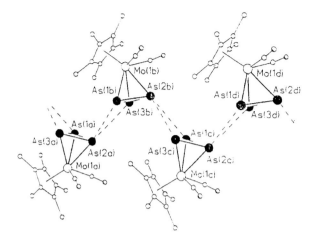

Fig. 4.2. The supramolecular structure of η^5-$C_5H_5(CO)_2MoAs_3$.

In the solid state molecules of the simplest diantimony organo-derivative, $Me_2SbSbMe_2$, are self-organized by Sb···Sb secondary bonding interactions (3.645 Å) which alternate with primary bonds (Sb–Sb 2.862 Å) in a linear chain, **3** (Sb–Sb···Sb 179.2°) [12, 13]. Several other distibanes are similarly associated. These compounds, which have been the subject of a review [14], include the trimethylsilyl derivative, $(Me_3Si)_2SbSb(SiMe_3)_2$ (Sb–Sb 2.867 Å, Sb···Sb 3.99 Å) [15], the trimethylgermyl derivative, $(Me_3Ge)_2SbSb(GeMe_3)_2$ [16], and the trimethylstannyl derivative, $(Me_3Sn)_2SbSb(SnMe_3)_2$ (Sb···Sb 3.879 and 3.81.1 Å at −120 °C; 3.890 Å at room temperature) [17, 18]; all are self-assembled in chain-like structures. Tetraphenyldistibane, $Ph_2SbSbPh_2$, (with Sb···Sb contacts at 4.29 Å) is not associated [19]. The intermolecular assembly of distibanes can be detected by Raman spectroscopy [20].

```
         R   R              R   R
          \ /                \ /
    ----Sb —a— Sb ---b--- Sb —— Sb ------       R = CH3
          / \                / \
         R   R              R   R                a = 2.84 Å
                                                 b = 3.68 Å
                       3
```

A similar type of self-assembly has been observed in 2,2′,5,5′-tetramethyldistibole [21, 22]. The molecular tectons are aligned in chains with a colinear Sb–Sb···Sb–Sb sequence, **4**. Note that the arsenic analog contains discrete molecules in the crystal with no secondary supramolecular structure.

All these compounds are intensely colored and thermochromic. It is believed that these properties arise because of the intermolecular association through secondary bonds [23], and this idea is supported by theoretical studies [24].

Surprisingly, cyclotetrastibanes are also self-organized in the solid state. Thus, the mesityl derivative (benzene solvate), [MesSb]$_4 \cdot$C$_6$H$_6$, in the solid state has a supramolecular structure, **5**, in which the rings are connected by Sb$\cdots$Sb secondary bonds (3.88 Å) to form infinite chain-like arrays [25]. Benzene molecules in the crystal are η^6-connected to antimony. The *tert*-butyl derivative [SbBu^t]$_4$ has not been described as associated [26]. In the hexamer [SbPh]$_6$ Sb$\cdots$Sb distances are longer (4.21 Å) and the compound is probably best described as unassociated [27].

The cage compound MeC(CH$_2$)$_3$Sb$_3$, which contains a Sb$_3$ homocycle, **6a**, in the solid state has a beautiful self-organized two-dimensional supramolecular structure, based upon multiple secondary-bond intermolecular interactions, as illustrated in **6b** (only the Sb$_3$ rings are shown for clarity) [28].

Supramolecular self-assembly as a result of Sb···Sb interactions has been discovered in [C$_5$H$_5$(CO)$_2$MoSb$_3$] (Sb···Sb 3.838, 4.040, 4.083, and 4.214 Å and in [(C$_5$Me$_5$)(CO)$_2$MoSb$_3$] (Sb···Sb 3.745 and 3.834 Å) [29].

Dibismuthanes are associated by secondary Bi···Bi bonds. Thus, tetramethyldibismuthane, Me$_2$BiBiMe$_2$, and tetrakis(trimethylsilyl)dibismuthane, (Me$_3$Si)$_2$BiBi(SiMe$_3$)$_2$ [30, 31], have Bi···Bi 3.58 Å (in the methyl derivative) and 3.804 Å (in the trimethylsilyl derivative) intermolecular contacts. Tetraphenyldibismuthane, Ph$_2$BiBiPh$_2$ (intramolecular Bi–Bi 2.990 Å), is, however, not associated [32].

In Group 16 the tendency of self-assembly increases from sulfur to the heavier elements. Thus molecules of dimethyldisulfane, MeSSMe (α-form) are arranged in chains, but the intermolecular S···S distances (3.78 Å) are larger than the sum of van der Waals radii (3.60 Å). There is a crystalline β-form of the same compound, in which the molecules are grouped in pairs, loosely connected at S···S 3.74 Å. The molecules of dimethyldiselane, MeSeSeMe, and dimethylditellane, MeTeTeMe, are associated in chains (packed in turn as sheets) with stronger intermolecular secondary bonds (Se–Se 2.31 Å, Se···Se 3.55 Å and Te–Te 2.71 Å, Te···Te 3.74 Å, respectively) [33]. By contrast, diphenyldisulfane, PhSSPh (S–S 2.03 Å) [34] and diphenyldiselane, PhSeSePh (2.29 Å) [35] are discrete isolated molecules in the crystal, with no secondary association. Diphenylditellane, PhTeTePh (Te–Te 2.712 Å), is better described as a molecular, unassociated compound, the shortest intermolecular Te···Te distances being 4.255 Å [36]. The expected interatomic distances from the sum of covalent radii are Se–Se 2.406 Å and Te–Te 2.782 Å, whereas the sum of van der Waals radii is Se···Se 3.80 Å and Te···Te 4.12 Å.

Other self-organized structures based upon secondary bonds in organotellurium chemistry include tetratellurotetracene (forming a helical Te–Te···Te–Te chain with Te–Te 2.67 Å, Te···Te 4.055 Å and stacking of the molecules with inter-stack Te···Te distances of 3.701 Å) [37, 38], and bis(4-methoxyphenyl)ditellane, 4-MeOC$_6$H$_4$TeTeC$_6$H$_4$OMe [39], and 2,3,6,7-tetramethylnaphto[1,8-cd:4,5-c'd']-bis[1,2]tellurole, **7a**, containing eight-membered Te$_8$ quasi-rings in a three-dimensional structure, **7b** (Te···Te 4.019 Å) [40].

1-Telluracyclohexane-3,5-dione forms tetrameric supermolecules, **8**, by Te···Te secondary bonding (3.95–4.18 Å) to form a unique cyclic arrangement [41, 42].

A metastable compound triphenyltelluronium phenyltelluride (formally 'diphenyl tellurium') Ph$_3$Te$^+$PhTe$^-$, is a dimeric assembly (described as a 'pair of ion pairs') in which the cations and anions are associated by Te···Te secondary bonds to form a square-planar quasicyclic supermolecule, **9**, with Te···Te distances in the range 3.323–3.596 Å [43].

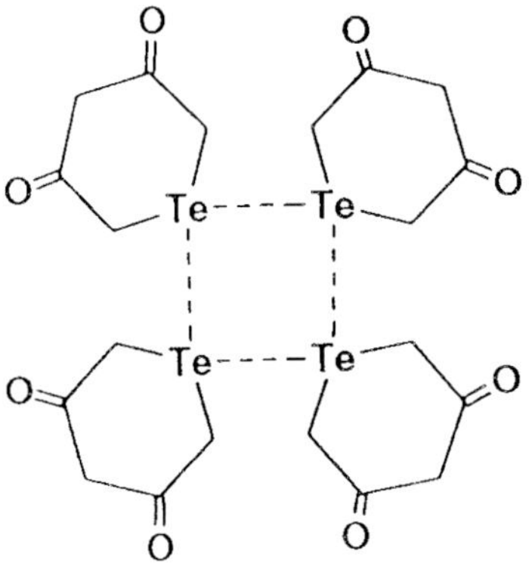

$$a = 3.526 \text{ Å}$$
$$b = 3.379 \text{ Å}$$
$$c = 3.323 \text{ Å}$$
$$d = 3.596 \text{ Å}$$

8 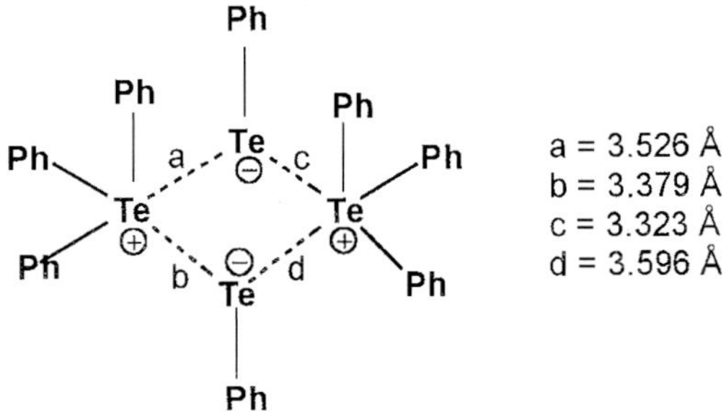   **9**

Secondary-bond self-organization has also been found in bis(diphenyldithiophosphinato)ditellane, $Te_2(S_2PPh_2)_2$, in which the quasi-bicyclic molecules (with Te–Te 2.723 Å) are associated by secondary bonding (Te···Te 3.514 and 3.668 Å) into linear arrays, **10** [44].

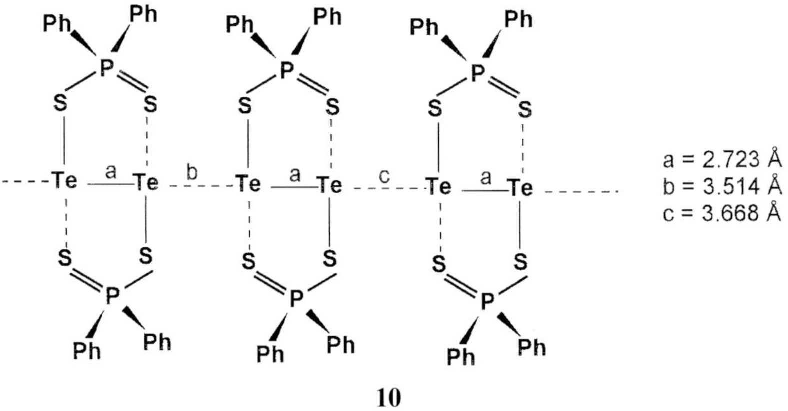

$$a = 2.723 \text{ Å}$$
$$b = 3.514 \text{ Å}$$
$$c = 3.668 \text{ Å}$$

10

It is quite probable that careful analysis of molecule packings and intermolecular distances in crystals will result in the discovery of many more self-organized structures based upon Te···Te secondary interactions.

Unexpected homoatomic secondary bonds, referred to as d^{10}–d^{10} interactions, occur in some organometallic derivatives of Group 13 metals. Thus, pentamethylcyclopentadienyl gallium is a hexameric supermolecule $Ga_6(C_5H_5)_6$ with an octahedral arrangement of gallium atoms held together at quite long Ga···Ga distances of 4.073–4.172 Å; not surprisingly, in solution such a hexamer is readily disassembled into monomeric species [45], thus justifying its description as a self-assembled supermolecule formed by secondary bonding. Much shorter gallium–gallium bonds are observed in the tetramers $Ga_4[C(SiMe_3)_3]_4$ (Ga···Ga 2.688 Å) [46], $Ga_4[C(SiMe_2Et)_3]_4$ (Ga···Ga 2.710 Å) [47], and $Ga_4[Si(SiMe_3)_3]_4$ (Ga···Ga 2.582 Å) [48].

Similar self-assembly has also been observed in organoindium derivatives. Thus, the sulfido compound $In_4S[C(SiMe_3)_3]_3$ contains an In_3S tetrahedron with In···In distances of 3.387 Å (In–In covalent bonds are 2.838 Å), as in the previously reported $In_6(C_5Me_5)_6$, which contains a self-assembled In_6 octahedron formed by weak In···In bonding (whereas the Al–Al bonds in $Al_4(C_5Me_5)_4$ are strong [49]).

This type of self-assembly is more frequent in organothallium derivatives. For example, $C_5(CH_2Ph)_5Tl$ is a dimer formed by Tl···Tl secondary bonding (or d^{10}–d^{10} interactions) [50] and a supramolecular polymeric form [51] is also known. Recently, more examples of self-assembly by secondary Tl···Tl bonding have been identified. Thus, $TlC(SiMe_3)_3$ is monomeric in solution, but in the solid state it self-assembles into tetrameric supermolecules, based upon Tl_4 tetrahedra, with Tl···Tl 3.335–3.638 Å [52]. In thallium(I) tris(pyrazolyl)borate, $Tl[HB(pyz)_3]$ (pyz = pyrazolyl) the molecules are arranged in a chain-like array, but the interatomic distances between thallium atoms (Tl···Tl 4.80 and 5.12 Å) are too long to suggest metal–metal interactions [53]. Surprisingly in this context, Tl···Tl interatomic distances of 3.647 Å in a Tl_4 tetrahedron, indicating secondary bonds, have been found in a tetrathallium tris(cyclopropylpyrazolyl)borate [54]. The nature of the Tl···Tl interactions has been the subject of considerable debate and several theoretical studies on these weak interactions have been published; the whole story has been reviewed [55–59].

An unprecedented supramolecular structure made up of infinite double-strand ribbons formed by Tl···Tl secondary interactions (Tl···Tl 2.490 Å) observed recently in a thallium–nitrogen compound [60] suggests that this type of interaction can be expected to play an important role in thallium(I) structural chemistry and might lead to novel, unexpected architectures.

Gold–gold interactions, known as aurophilicity [61–63], lead to the formation of supramolecular chains, **11**, in isonitrile gold(I) chlorides $[Au(CNBu^t)Cl]$ (Au···Au 3.695 Å) [64], nitrates $[Au(CNBu^t)(NO_3)]$ (Au···Au 3.295 and 3.324 Å) [65], cya-

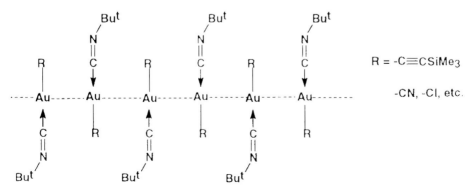

11

nides [Au(CNBu^t)(CN)] (Au···Au 3.568 Å) [66], and in σ-alkynyl gold derivatives [Au(CCSiMe$_3$)(CNBu^t)] [67], or to the formation of a gold catenane in [{Au$_6$(CCBu^t)$_6$}$_2$] [68].

4.2 Group 12 Metals – Zinc, Cadmium, Mercury

In this triad only mercury seems to form definite supramolecular assemblies by secondary bonding interactions. The supramolecular chemistry of zinc and cadmium (based upon dative bond self-assembly) has been discussed in Sections 3.1.1 and 3.1.2.

4.2.1 Self-assembly of organomercury compounds

Mercury is a 'soft' metal and as such it is expected to form secondary bonds, most readily with sulfur, selenium, and other heavy non-metals. The situation is, however, more complex and secondary interactions with other electronegative atoms have also been observed in the solid state. Interatomic distances longer than the expected van der Waals distances are, moreover, sometimes observed between molecules orientated such that weak interactions lead to particular arrangements in the crystal [69]. The are numerous examples of secondary bonds in organomercury chemistry; although most are intramolecular there are several examples of intermolecular secondary bonds leading to supramolecular self assembly. A review has been published on this subject [70] and many new examples have subsequently been reported.

The structural chemistry of mercury(II) is dominated by its tendency to maintain the linear coordination which results from sp hybridization, and to use the non-hybridized vacant p orbitals in further bonding. Secondary bonds to mercury will distort this linearity, but not to a great extent. Only stronger bonds (of the dative type) will change the hybridization from sp to sp^3 (in tetrahedral geometries). Distinction between dative and secondary bonds is very difficult for mercury, because the whole range of interatomic distances, between the sum of covalent radii and the expected van der Waals distances, is covered by the known structures.

4.2.1.1 Organomercury halides

The sum of the van der Waals radii for the mercury–chlorine pair is ca. 3.90 Å (the value of the van der Waals radius of mercury is still controversial but a value of 1.73 Å has been suggested [71]) and distances shorter than this will be considered in discussing supramolecular self assembly. A good illustration of supramolecular dimer formation is that of the anions [(CF$_3$)$_2$Hg(μ-F)$_2$Hg(CF$_3$)$_2$]$^{2-}$, **12**, (Hg–F 2.395 Å, Hg···F 2.418 Å, probably a dative bond) [72] and [(CF$_3$)$_2$Hg(μ-Cl)$_2$-

Hg(CF$_3$)$_2$]$^{2-}$, **13** [73]. In the latter dinuclear anion two basically linear F$_3$CHgCF$_3$ moieties (C–Hg–C 160.5°) are bridged by two chloride ions at longer Hg···Cl distances of 2.823 and 2.837 Å (the estimated length of the covalent Hg–Cl bond is ca. 2.35 Å).

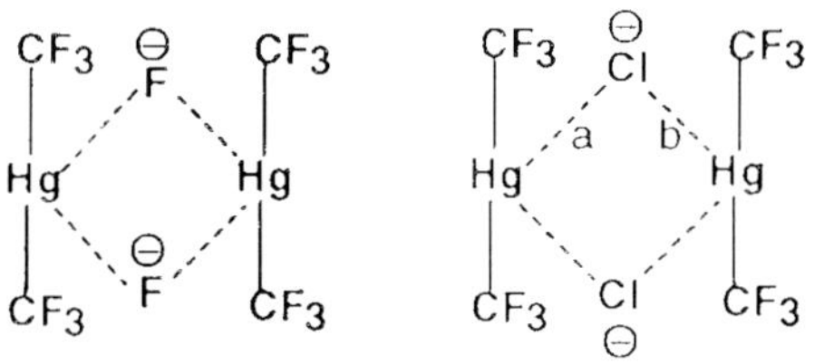

12

13

a = 2.823 Å
b = 2.837 Å

Two non-equivalent mercury–chlorine bonds indicating complete separation of covalent and secondary bonding, were found in dimeric (phenylazophenyl-C,N')mercury chloride, **14**. In addition to the bond leading to dimerization, another secondary interaction with a chlorine of a different molecule occurs at 3.632 Å [74].

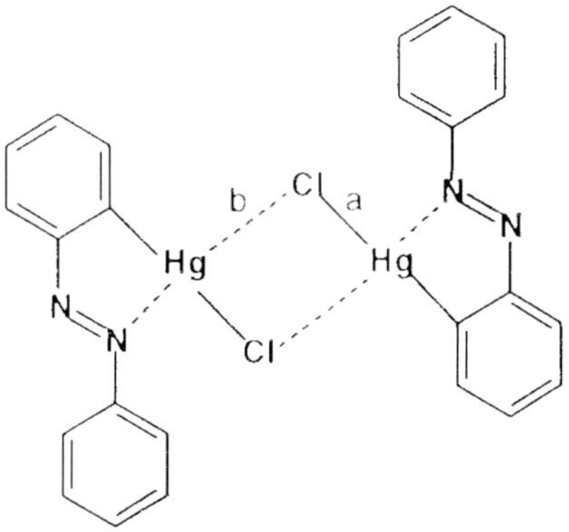

14

a = 2.309 Å
b = 3.362 Å

The four-membered ring unit Hg$_2$Cl$_2$ can occur as a part of a ladder structure. Thus, (2-pyridylphenyl)mercury(II) chloride, **15a**, is a tetramer, with the skeleton shown in **15b** (only atoms directly bonded to mercury are shown for clarity). The mercury–chlorine interatomic distances are in the range 3.184–3.442 Å [75].

Substituted cyclopentadienylmercury(II) chloride, (σ-C$_5$Me$_4$SiMe$_2$Bu^t)HgCl, forms tetrameric supermolecules, **16**, containing an eight-membered Hg$_4$Cl$_4$ quasi-ring (Hg–Cl 2.333 and 2.336 Å, Hg···Cl 3.105 and 3.117 Å) [76], whereas the mesitylamine bis(chloromercuryethyl) derivative [2,4,6-Me$_3$C$_6$H$_2$N{(CH$_2$)$_2$HgCl}$_2$] forms a centrosymmetric dimer by secondary Hg···Cl bonding (primary Hg–Cl

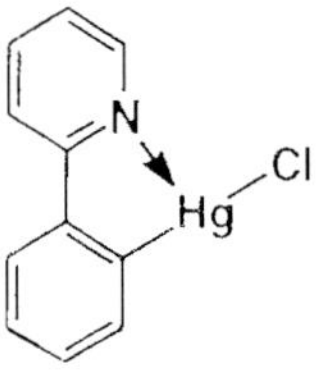

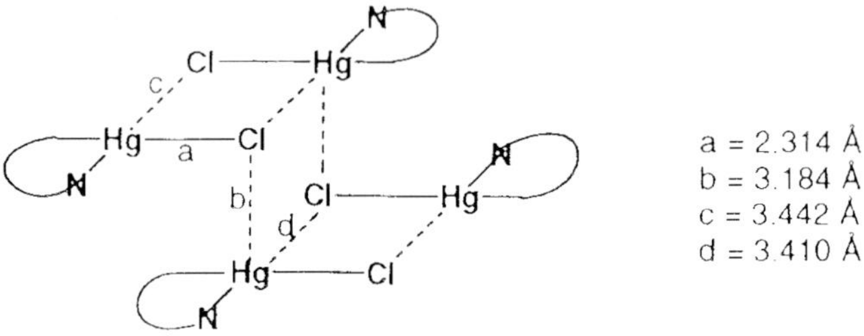

15a

15b

a = 2.314 Å
b = 3.184 Å
c = 3.442 Å
d = 3.410 Å

2.34 Å and secondary Hg···Cl 3.20–3.38 Å) [77]. Benzylmercury chloride, PhCH$_2$HgCl, forms a ladder-type supramolecular array [78] and pentamethylcyclopentadienyl mercury(II)chloride, σ-C$_5$Me$_5$HgCl, is a self-organized ribbon (Hg–Cl 2.350–2.361 Å, Hg···Cl 3.098–3.238 Å) running as a zigzag ladder in the crystal [79].

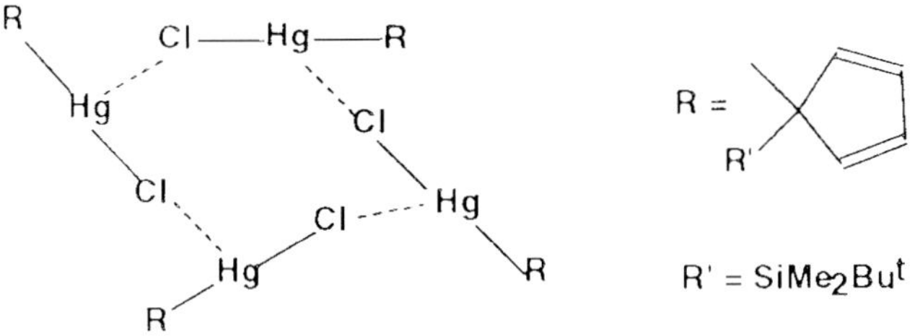

16

R' = SiMe$_2$Bu^t

Another ladder containing Hg–Cl primary bonds (2.329 and 2.345 Å) and secondary Hg···Cl bonds (3.303–3.566 Å) has been reported in meso-α,α-bis-(chloromercurio)-α,α'-bis(trimethylsilyl)-m-xylene [80], **17**, and an intricate compact

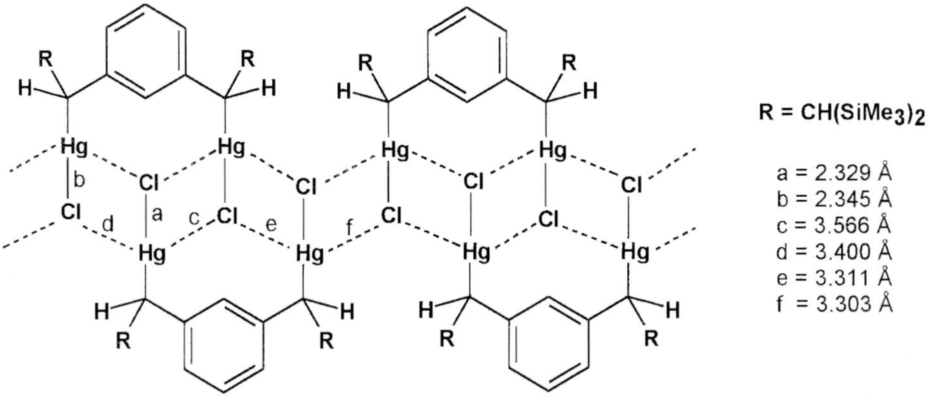

17

R = CH(SiMe$_3$)$_2$

a = 2.329 Å
b = 2.345 Å
c = 3.566 Å
d = 3.400 Å
e = 3.311 Å
f = 3.303 Å

three-dimensional network formed via secondary Hg···Cl interactions (primary Hg–Cl 2.345 and 2.353 Å, secondary Hg···Cl 3.211–3.350 Å) has been reported for α,α'-bis(chloromercurio)-*m*-xylene [81].

In the dinuclear compound ClHgCH$_2$HgCl the molecules are held together by interatomic Hg···Cl contacts in the range close to the van der Waals distance (3.318–3.450 Å) and even longer, to form zigzag chains along the *c* axis of the crystal [82].

In (chloromercurio)acetaldehyde, ClHgCH$_2$CHO, both intermolecular mercury–chlorine (Hg···Cl 3.370 Å – compare with the covalent Hg–Cl bond of 2.326 Å in the same molecule) and mercury–oxygen (Hg···O 2.84 and 2.87 Å) secondary bonds are present [83]. Similarly, in (methoxycarbonyl)mercury(II) chloride, ClHgC(=O)OMe, intermolecular distances Hg···Cl 3.21 Å and Hg···O 3.01 Å hold the molecular tectons together in a supramolecular structure [84].

Supramolecular self-assembly by mercury–chlorine bridging leading to a two-dimensional network occurs in [EtNH$_3$][Hg(CN)$_2$Cl$_2$], containing six-coordinate octahedral mercury; the CN ligands are monodentate and non-bridging and the Hg–Cl distances are in the range 2.901–3.432 Å [85].

With the Cl$^-$ anion dichloro-1,2-phenylenedimercury, 1,2-C$_6$H$_4$(HgCl)$_2$, forms dimeric species [{C$_6$H$_4$(HgCl)$_2$}$_2$Cl]$^-$ (Hg–Cl 2.310–2.344 Å; Hg···Cl 3.35 Å) which are further associated in the crystal by Hg···Cl secondary bonding (3.361 and 3.998 Å). The 1:1 complex of Cl$^-$ with 1,2-C$_6$H$_4$(HgCl)$_2$ is also a supramolecular structure in the solid state and forms infinite ribbons by secondary bonding (Hg–Cl 2.334 and 2.346 Å; Hg···Cl 3.454, 3.296, and 3.586 Å) [87].

4.2.1.2 Organomercury–oxygen compounds

The methylmercury(II) trifluoroacetate 1-methylpyridine (α-picoline) adduct self-assembles into dimers via mercury–oxygen bonds, to form an unsymmetric four-membered ring, **18** (Hg···O 2.668 and 2.805 Å). These bonds are shorter than the van der Waals distance (3.0 Å). The methylmercury–picoline cation is nearly linear (C–Hg–N 169.5°) [88].

a = 2.668 Å
b = 2.805 Å

18

Although phenylmercury(II) acetate, PhHgOOCCH$_3$, has been described as consisting of discrete molecules in the solid state, intermolecular mercury–oxygen contacts of 2.819 Å and 2.907 Å (shorter than the van der Waals distance) suggest a certain degree of self-organization via intermolecular interaction. In phenylmercury(II) trifluoroacetate, PhHgOOCCF$_3$, mercury–oxygen intermolecular contacts leading to self-assembly are observed at 2.828 Å and 2.952 Å. Intramolecular short bonds from mercury to carboxylic oxygens are observed at 2.084 Å and 2.121 Å [89,90]. (2-Pyridin-2-yl)phenylmercury acetate is a dimer formed by Hg···O secondary bonding (primary Hg–O 2.005 Å, secondary Hg···O 2.870 Å) and additional Hg···C(π) intermolecular interactions [91].

Phenylmercury quinaldinate forms dimers, **19**, with loose Hg···O bridges (Hg–O 2.16 Å; Hg···O 2.79 Å in the four-membered ring) and phenylmercury oxinate forms infinite helical arrays (Hg···O 3.33 and 3.37 Å) [92].

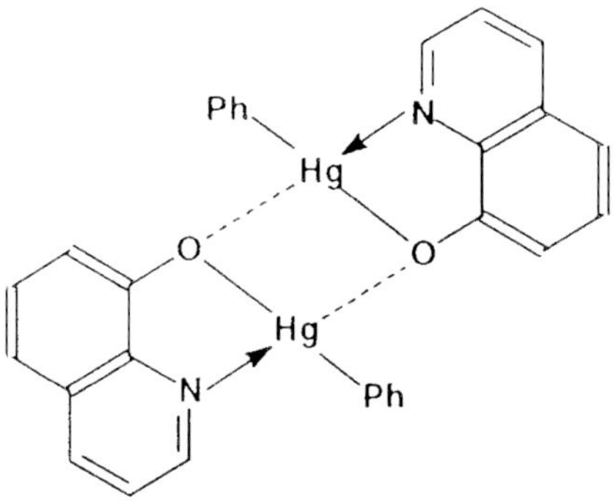

19

A different mode of self-assembly occurs in the π-complex of hexamethylbenzene with mercury(II) trifluoroacetate, **20**. In this compound four trifluoroacetato

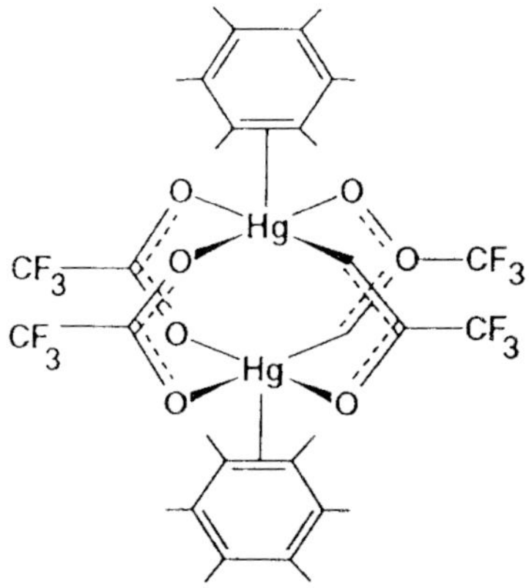

20

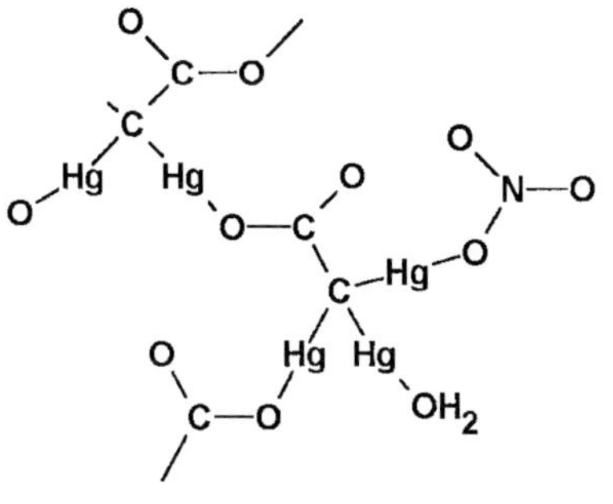

$[Hg(H_2OHg)(HgNO_3)CCOO]_n^{n+}$

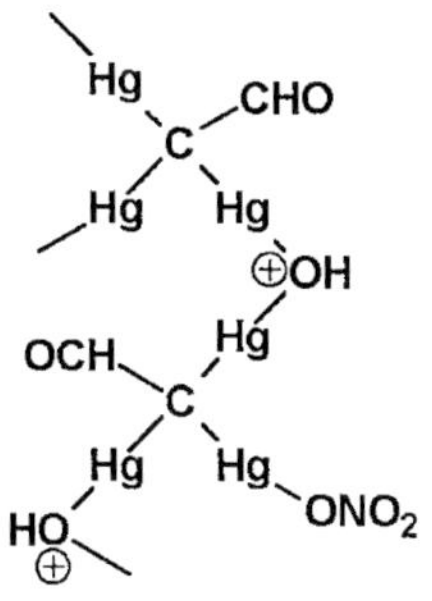

$[HOHg_2(NO_3Hg)CCHO]_n^{n+}$

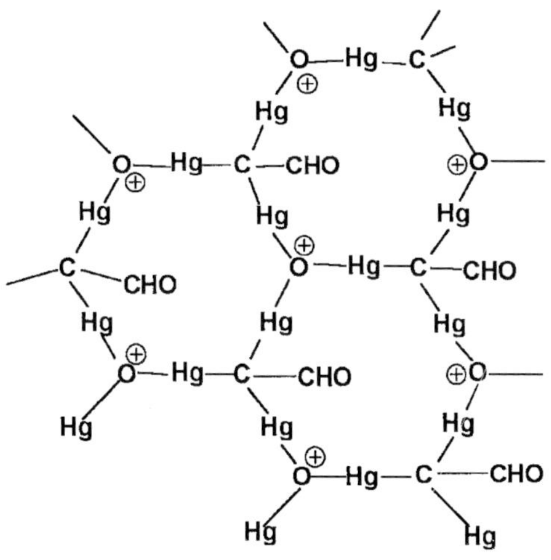

$[OHg_3CCHO]_n^{n+}$

Scheme 4.1

groups form bridges between the two mercury ions, each bonded as η^2-C_6Me_6 to the aromatic molecule [93].

The so-called trimercurate acetic acid derivatives $[Hg(H_2OHg)(NO_3Hg)CCOO]$-NO_3 and $2(NO_3Hg)_3CCOOH \cdot HNO_3$ [94], and the trimercurated acetaldehyde derivatives $2[OHg_3CCHO]NO_3 \cdot HNO_3$, $[HOHg_2(NO_3Hg)CCHO]NO_3$ [95], and $[OHg_3CCHO]NO_3 \cdot H_2O$ [96], are a series of unusual mercury–oxygen supramolecular compounds; their polymeric structures are shown in Scheme 4.1.

Phenylene-bis(trifluoroacetato)dimercury, 1,2-$C_6H_4(HgOOCCF_3)_2$, forms chains of stacked molecular tectons associated by Hg···O bonds (2.844 and 2.878 Å) further interconnected by Hg···O 3.057 Å [97].

The methylmercury(II) moiety, CH$_3$Hg, is one of the most toxic forms of mercury; it causes irreversible damage to the central nervous system and is a species of interest in the environmental chemistry of this metal. It is now well known that it forms strong bonds with sulfhydryl groups of biological molecules [98, 99]. Not surprisingly, many methylmercury derivatives of different functional compounds have been structurally investigated in relation to studies on chelation therapy for methylmercury(II) poisoning [100]; some are characterized by supramolecular self-organization in the solid state. Although the metal in the MeHg group has a strong tendency to linear coordination, it maintains some residual Lewis acidity [101] leading to complexation in solution and supramolecular self-organization in the solid state, usually via secondary bonding. Hg···O secondary bond self-assembly has been reported for methylmercury DL-alaninate (Hg···O 2.92 and 3.05 Å) [102], methylmercury L-serinate (Hg···O 2.890 and 3.00 Å) [103], methylmercury adeninates (MeHg)$_5$Ad$_2$ (Hg···O 2.67–3.20 Å) [104] and (MeHg)$_3$-(Ad-2H)]·0.5H$_2$O [105], μ-(1-methylcytosinato-N3,N4) bis(methylmercury) nitrate (Hg···O 2.83–2.98 Å) [106], and (4-nitroimidazolato)methylmercury (Hg···O 3.07 and 3.19 Å, also Hg···N 2.82 Å) [107]. The methylmercury derivative of 1-methylimidazoline-2-thione, [MeHg(C$_4$H$_6$N$_2$S)]NO$_3$, is associated via nitrato bridges (Hg···O 2.833 and 3.153 Å) [108] and the complex of methylmercury trifluoroacetate with α-picoline forms dimers by OCOCF$_3$ bridging (Hg–O 2.668 and Hg···O 2.805 Å) [109].

Methylmercury 2-diphenylphosphinylbenzoate, MeHgOOCC$_6$H$_4$PPh$_2$-2, dimerizes via Hg···O interactions (primary Hg–O 2.086 Å, secondary Hg···O 2.874 Å) [110].

A beautiful self-organized supramolecular structure, **21**, was found for methylmercury(II) diphenylmonothiophosphinate, MeHgSP(=O)Ph$_2$; it is associated by Hg···O secondary bonding into quasi-cyclic dimers which are connected, again via

R = Me
a = 2.897 Å
b = 2.831 Å

21

secondary Hg···O bonding, into double chains with an inorganic (polar) core wrapped with lipophilic phenyl groups [111].

4.2.1.3 Organomercury–sulfur compounds

Because sulfur is a soft donor it is expected to participate in Hg···S secondary interactions and this is observed quite frequently. The expected van der Waals interatomic distance between sulfur and mercury is 3.5 Å, and intermolecular distances below this value are often observed.

Methylmercury(II) diethyldithiocarbamate, $MeHgS_2CNEt_2$, **22**, (primary Hg–S 2.418 Å, secondary Hg···S 3.147 Å) [112], and phenylmercury(II) diethyldithiocarbamate, $PhHgS_2CNEt_2$ (primary Hg–S 2.387 Å, secondary Hg···S 3.133 and 3.398 Å) [113], are both assembled into dimeric species by secondary Hg···S bonding. The xanthates $PhHgS_2COR$ (R = Me and Pr^i) also contain Hg···S intermolecular contacts, but these are significantly longer (Hg–S 2.396 Å and Hg···S 3.567 Å in $MeHgS_2COMe$ [114]; Hg···S 3.445 and 3.537 Å in $PhHgS_2COMe$).

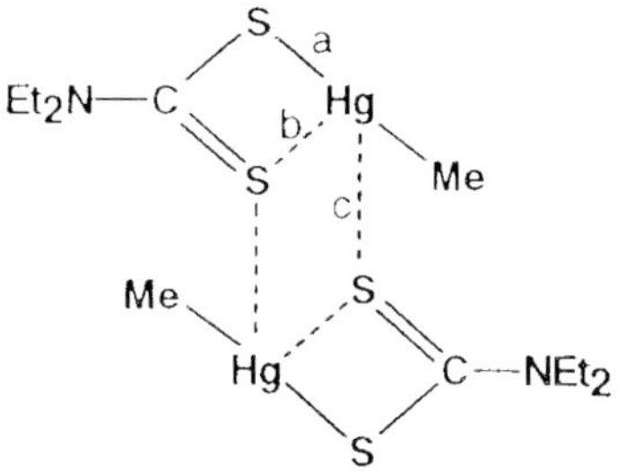

a = 2.418 Å
b = 2.964 Å
c = 3.147 Å

22

Other self-organized mercury-sulfur compounds include $MeHg(SC_6H_2Pr^i_3\text{-}2,4,6)$ (primary Hg–S 2.344 Å, secondary Hg···S 3.269 Å) [115], dimeric MeHg(2,5-dimercapto-1,3,4-thiazolyl) (Hg···S 3.35 Å) [116], polymeric (chain-like) trans-1,2-dimercaptocyclohexane-bis[methylmercury(II)] (Hg···S 3.295, 3.499, and 3.689 Å) [117], and polymeric $HgCH_2P(=S)Ph_2$ (Hg···S 3.17 Å), **23** [118].

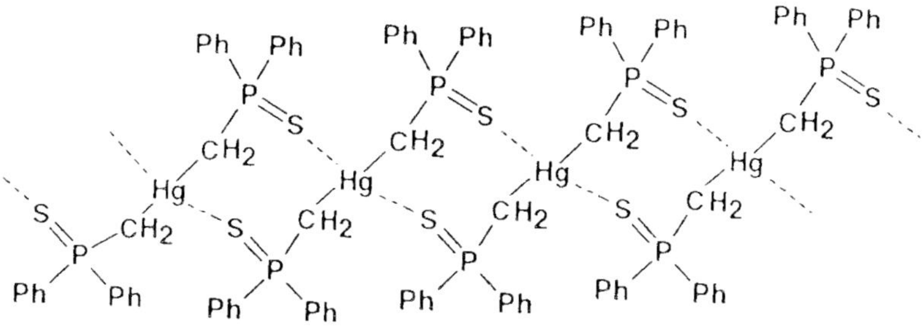

23

In methylmercury *o*-nitrophenylthiolate, MeHg(SC$_6$H$_4$NO$_2$-*o*), the supramolecular architecture, **24**, contains Hg$_2$S$_2$ units (Hg–S 2.379 and 2.366 Å; Hg$\cdots$S 3.322 and 3.539 Å) and additional Hg$\cdots$ONO interactions (3.48 Å) [119].

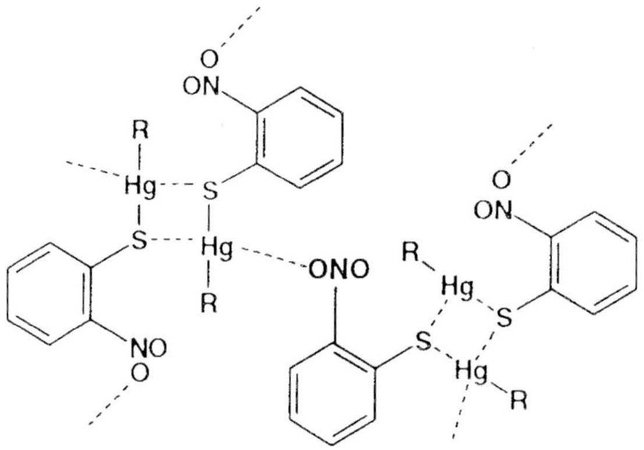

24

Methylmercury- and trifluoromethylmercury L-cysteinato complexes, RHgSCMe$_2$CH(NH$_3^+$)COO$^-$·$\frac{1}{2}$H$_2$O (R = Me, CF$_3$) form supramolecular arrays by stacking the RHgS segments (Hg–S 2.35–2.38 Å, Hg$\cdots$S 3.35–3.46 Å) [120].

The organomercury dithiophosphates and dithiophosphinates form a series of self-organized compounds. Thus, phenylmercury diethyldithiophosphate, PhHgS$_2$P(OEt)$_2$ (Hg–S 2.383 Å, Hg$\cdots$S 3.323 Å) [121], and methylmercury(II) diphenyldithiophosphinate, MeHgS$_2$PPh$_2$ (Hg–S 2.379 Å, Hg$\cdots$S 3.152 Å) [122] are centrosymmetric quasi-cyclic dimers, **25**, whereas phenylmercury(II) diethyldithiophosphinate, PhHgS$_2$PEt$_2$, is a polymer, **26** (Hg–S 2.375 Å, Hg$\cdots$S 3.183 Å) [123].

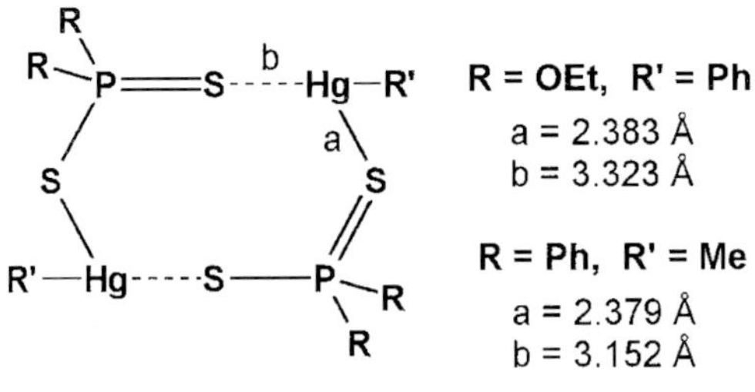

R = OEt, R' = Ph
a = 2.383 Å
b = 3.323 Å

R = Ph, R' = Me
a = 2.379 Å
b = 3.152 Å

25

The compound (η^5-C$_5$H$_5$)$_2$Mo(HgSEt)$_2$ has a unique supramolecular structure, **27**. It occurs in an insoluble form containing dimeric units (Hg–S 2.43 Å, Hg$\cdots$S

3.17 Å) and a soluble orange form consisting of infinite supramolecular arrays (Hg···S 3.32 Å) [124].

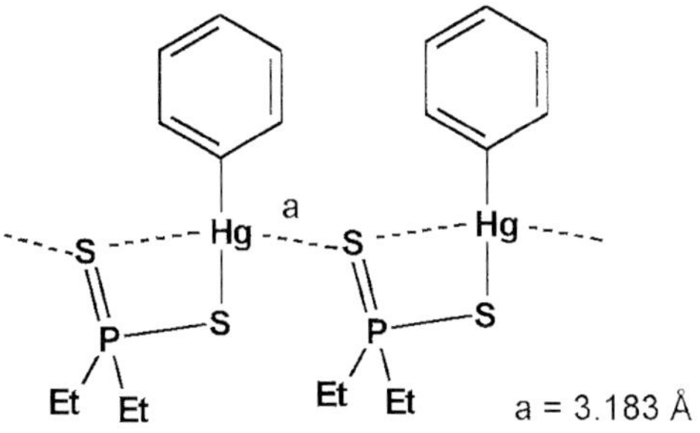

26

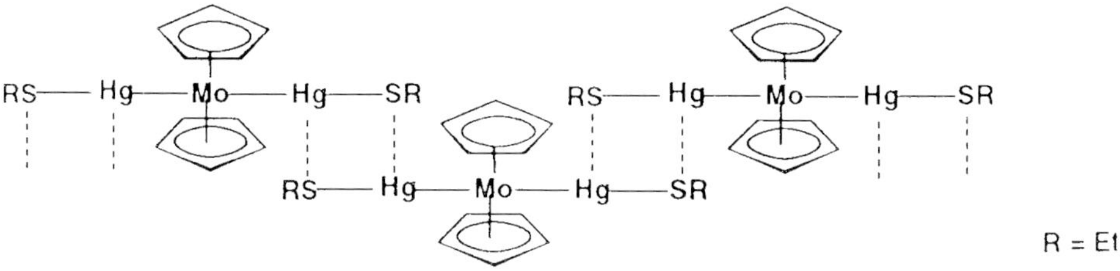

a = 3.183 Å

27

R = Et

Bis(methylmercury)-1,2-cyclohexanedithiolate has a 3D supramolecular architecture with intramolecular Hg–S 2.363–2.367 Å and Hg···S 2.857 Å, and intermolecular Hg···S secondary bonds (3.295–3.859 Å) [125].

In some self-organized organomercury compounds the intermolecular distances are long, close to or even larger than van der Waals distances. The molecular packing, however, suggests there is some interaction. For example, between the molecules of (2-mercaptobenzothiazolato)methylmercury [126] Hg···S distances of 3.36, 3.50, 3.75, and 3.79 Å are observed but the reciprocal orientation of the molecules can be described as self-organized. Long intermolecular Hg···S distances have also been measured in methylmercury(II) and phenylmercury(II) dithizonates (3.69 and 3.58 Å, respectively) [127], and (4-amino-5-methyl-2-pyrimidinethiolato)-methylmercury(II) (Hg···S 3.67 Å) [128].

Supramolecular self-organization can also be expected in organomercury–selenium compounds, both soft elements, which have a strong tendency to participate in closed-shell bonding interactions. Thus, methylmercury 2,4,6-tri-*tert*-butylphenylselenolate, MeHgSe($C_6H_2Bu^t_3$-2,4,6) forms loosely associated dimers by mercury–selenium secondary bonding (Hg–Se 2.460 Å, Hg···Se 3.380 Å) [129]. The formation of some inorganic cyclic tetramers [{HgCl(py)(SeEt)}$_4$] and [{HgCl(py)$_{0.5}$(SeBu^t)}$_4$] (Hg–Se 2.543, 2.551 Å and Hg···Se 2.579 and 2.589 Å) and of polymeric Hg(SeMe)$_2$ (Hg···Se 2.764 Å) occurs through similar interactions [130].

4.2.1.4 Organomercury–nitrogen compounds

Methylmercury(II) azide, MeHgN$_3$ [131], and trifluoromethylmercury(II) azide, **28**, CF$_3$HgN$_3$, and trifluoromercury(II) cyanate, CF$_3$HgNCO, **29** [132], are all self-organized supramolecular compounds. Methylmercury azide forms layers (Hg–N 2.22 Å, Hg$\cdots$N 2.87, 2.95, 3.21, and 3.17 Å), trifluoromethylmercury azide forms centrosymmetric dimers linked into layers (Hg–N 2.02 Å, Hg$\cdots$N 2.74 Å) and the cyanate forms dimers (Hg–N 2.03 Å, Hg$\cdots$N 2.88 Å) connected by Hg$\cdots$O interactions.

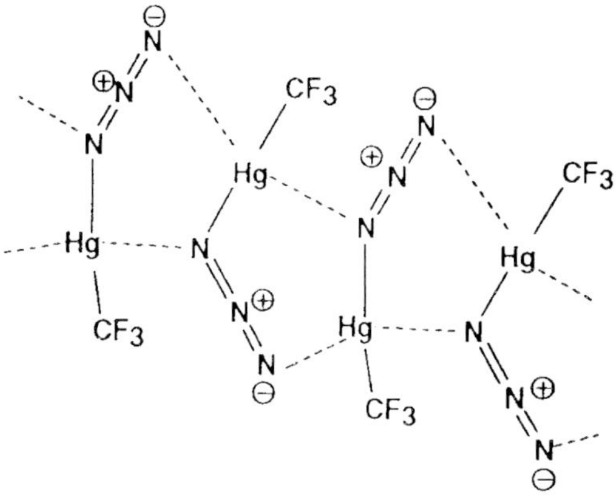

28

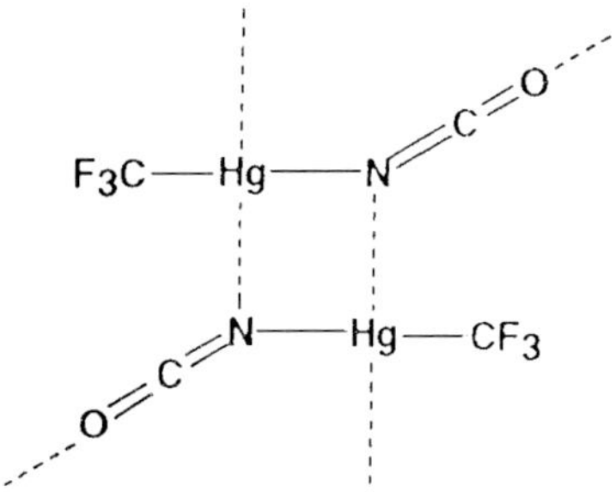

29

In the cyanide, MeHgCN, the Hg$\cdots$N intermolecular distances (3.15 and 3.26 Å), are comparable with the value expected for the Hg$\cdots$N van der Waals distance (3.2 Å), but the reciprocal orientation of the molecules suggests interaction and self-organization [133].

Secondary mercury–nitrogen bonds leading to supramolecular self-assembly have been found in methylmercury-2-S-methylthiouracylate (Hg···N 3.03 Å) [134] and in μ-adeninato-μ-N^6,N^6-bis(methylmercury) (Hg–N 2.057–2.063 Å, Hg···N 2.80–2.87 Å) [135].

4.3 Group 13 Metals – Gallium, Indium, Thallium

4.3.1 Self-assembly of organogallium compounds

In group 13 self-assembly by secondary bonding is predominantly observed for the heavier elements in association with soft donors, such as halogens (except fluorine), sulfur, and selenium; it is rarely observed for oxygen compounds. Occasionally, however, the large differences between interatomic distances indicate exceptions and suggest that secondary bonding should be considered. Although organogallium–oxygen compounds are associated mostly by dative bonding, in some organogallium chelates the self-assembly is perhaps better described as occurring as a result of secondary bonding. Thus, dimethylgallium tropolonate forms a dimer, **30**, [Me$_2$Ga(C$_7$H$_5$O$_2$)]$_2$, with short (1.963 Å) and long (2.545 Å) gallium–oxygen bonds in the four-membered Ga$_2$O$_2$ ring, compared with the dative bond (2.020 Å) in the chelate ring [136].

a = 1.963 Å
b = 2.020 Å
c = 2.545 Å

30

Short (1.957 and 2.046 Å) and long (2.521 Å) gallium–oxygen interatomic distances are observed in [Me$_2$Ga(μ-OC$_6$H$_4$Me-2)]$_2$ also [137]. Such examples are, however, atypical.

Gallium–halogen secondary bond self-assembly occurs in the so-called 'Menschutkin-type complexes', e.g. [{(η^6-C$_6$H$_6$)$_2$}$_2$Ga}$^+$GaBr$_4^-$]$_2$·3C$_6$H$_6$ [137a], [(η^6-1,2,4,5-C$_6$H$_2$Me$_4$)(η^6-C$_6$H$_5$Me)Ga$^+$GaCl$_4^-$]$_2$ [137b], [(η^6-C$_6$H$_3$Me$_3$)$_2$Ga$^+$GaCl$_4^-$]$_2$ [137c], and polymeric ([2.2]paracyclophane)gallium tetrabromogallate [137d].

4.3.2 Self-assembly of organoindium compounds

4.3.2.1 Organoindium halides

Supramolecular self-organization, with different amounts of complexity based upon dative bonds (see Section 3.2.3.1) and secondary bonds, is observed for organoindium halides. Long indium–halogen interatomic distances (although shorter than the expected van der Waals distance (ca. 3.8 Å for In···Cl)) are observed for most structurally characterized organoindium halides.

Dimethylindium chloride, Me_2InCl, forms double-chain ribbons which interact via additional secondary bonds, resulting in a supramolecular two-dimensional array, **31** [138]. The In–Cl bonds within the ribbon are somewhat elongated (primary In–Cl 2.673 Å; secondary In···Cl 2.945 and 2.954 Å) and the interatomic distances between ribbons are rather long (3.450 Å), indicating secondary bonding character.

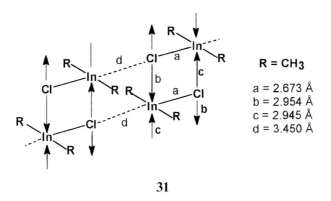

31

In the neopentyl-trimethylsilylmethyl derivative $[(Bu^tCH_2)(Me_3SiCH_2)InCl]_n$ dimeric tectons formed by dative bonding (In–Cl 2.572 and 2.659 Å) are self-organized in ladder supramolecular arrays, **32**, via secondary bonding (In···Cl 3.528 Å) [139]. The indium atom is five-coordinate (distorted trigonal bipyramidal).

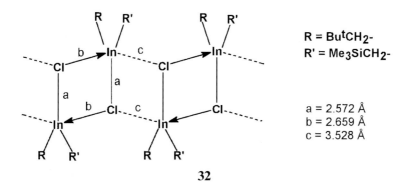

32

A slightly different structural motif is observed in the solid state structure of [MeInCl$_2$]$_x$. In this compound the supramolecular self-assembly is maintained only by secondary In···Cl bonds which connect the tectons in a ladder-type ribbon, **33** [140].

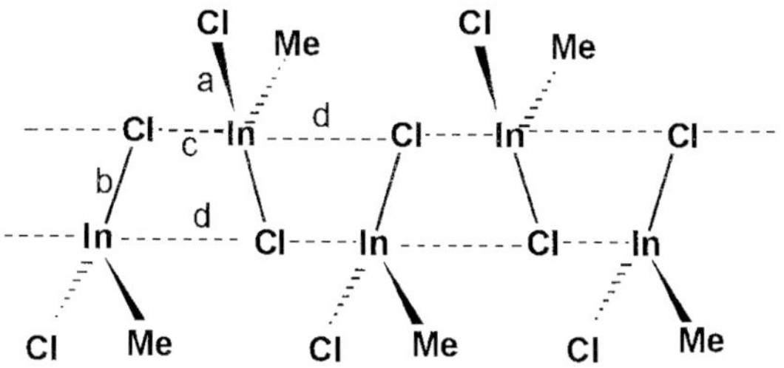

a = 2.384 Å
b = 2.400 Å
c = 3.203 Å
d = 3.799 Å

33

In mesitylindium diiodide, [MesInI$_2$]$_2$, both halogens are involved in the intermolecular association; this results in a new supramolecular motif, **34** [141].

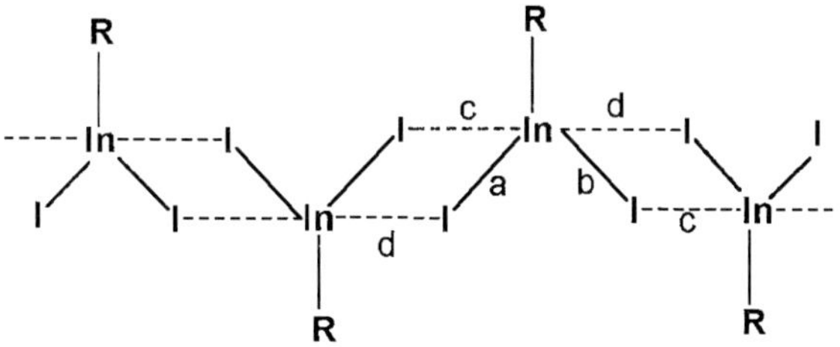

R = Mes
a = 2.760 Å
b = 2.772 Å
c = 3.183 Å
d = 3.192 Å

34

The structure of neopentylindium dichloride [Bu^{t}CH$_2$InCl$_2$]$_x$ is similarly built but the orientation of the molecular tectons is different, as shown in, **35** [142].

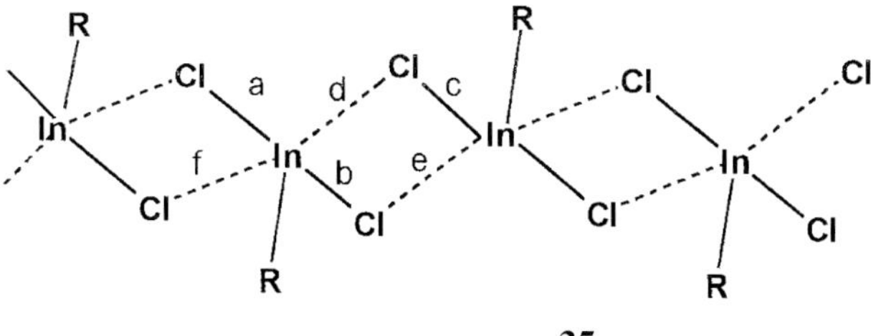

R = Bu^{t}CH$_2$-
a = 2.410 Å
b = 2.436 Å
c = 2.452 Å
d = 2.700 Å
e = 2.701 Å
f = 2.821 Å

35

With very large organic or bulky ligands on indium the complexity is low;

examples of dimers formed by dative bonding are cited in Section 3.2.3.1. Only secondary bonds participate in the formation of dimeric supermolecules, **36**, in the compound [{H$_2$B(pz)$_2$}(Me)InCl]$_2$, where the bulky bis(pyrazolylborato) ligand prevents a close approach of the binding sites.

Molecule A
a = 2.410 Å
b = 3.206 Å

Molecule B
a = 2.439 Å
b = 3.066 Å

36

Two independent molecules in the crystal have two sets of differing interatomic distances – primary In–Cl 2.410 Å and secondary In···Cl 3.203 Å in molecule A and primary In–Cl 2.439 Å and secondary In···Cl 3.066 Å in molecule B [143]. It is interesting to note that an increase of the In–Cl bond length in molecule B results in a decrease of the secondary bond distance.

Dimeric and polymeric (dimethylaminomethyl)phenylindium iodides associated by asymmetric iodine bridges (In–I 2.838 Å, In···I 3.405 and In–I 2.837 and In···I 4.262 Å) have been reported [143a].

4.3.2.2 Organoindium–oxygen derivatives

Organoindium carboxylates are expected to undergo self-assembly. Indeed, dimethylindium acetate forms a supramolecular array in which Me$_2$InOAc chelate tectons (In–O 2.371 Å) are connected in chain-like arrays, **37**, via longer intermolecular In–O contacts (2.600 Å) [144]. A similar structure has been found for diethylindium acetate, [Et$_2$InOAc]$_x$, with intra-tecton In–O 2.44 Å and inter-tecton In···O 2.67 Å [145]. Perhaps these longer interatomic distances should be regarded as strong secondary bonds or as weak dative bonds.

In diethylindium thioacetate, [Et$_2$InOSCMe]$_x$, the chelate molecular tectons are associated by In···O bonding (2.450 Å) into linear arrays, **38** [146]. Although the In···S distances are not reported, it seems that the structure also contains secondary In···S bonds.

Surprisingly few organoindium–sulfur compounds have been structurally characterized, but secondary bond self-assembly in this family of compounds can be expected.

4.3.2.3 Organoindium–nitrogen derivatives

Organoindium amides are associated by dative bonding in various supramolecular structures (see Section 3.2.3.4); secondary In···N bonds seem to be rare. One example is the adduct $Me_3In·N(CH_2CH_2)_3N$, for which an unusually long In···N interatomic distance of 2.62 Å has been measured [147]. The compound has an infinite chain structure, built from planar (!) $InMe_3$ and diazabicyclooctane molecules, alternating in a supramolecular array, **39**.

218   4 Supramolecular Self-Assembly by Formation of Secondary Bonds

In the compound [Cs{(PhCH$_2$)$_3$InN$_3$}]$_x$ anionic tectons are connected in chain-like arrays by weak In···N secondary bonds (2.905 Å); additional Cs···N and Cs–π-aryl interactions hold the supramolecular structure, **40**, together [148].

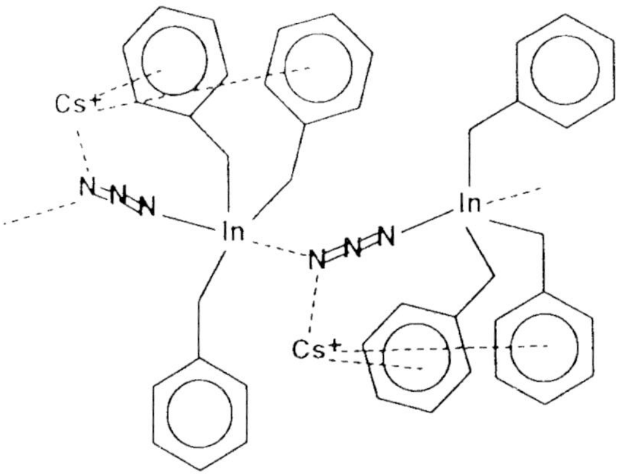

40

4.3.3 Self-assembly of organothallium compounds

4.3.3.1 Organothallium halides

Molecular weight determinations of diorganothalium(III) halides in solution indicate dimeric association, but only a relatively small number of such compounds have been subjected to X-ray crystal structure determinations.

Bis(trimethylsilylmethyl)thallium(III) chloride, (Me$_3$SiCH$_2$)$_2$TlCl, is dimeric, **41**, with non-equivalent thallium–chlorine bridges and four-coordinate thallium [149].

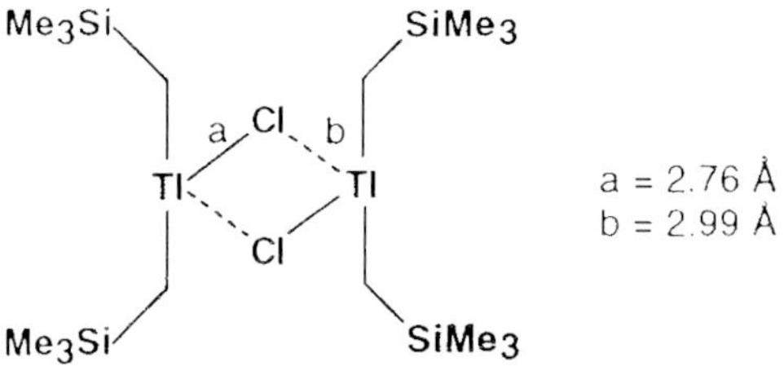

a = 2.76 Å
b = 2.99 Å

41

Bis(p-tetrafluorophenyl)thallium(III) bromide, (4-HC$_6$F$_4$)$_2$TlBr, forms supramolecular arrays, **42**, in the solid state, with unsymmetrical bromide bridging [150]; it is believed that similar structures are likely for R$_2$TlBr with R = C$_6$F$_5$ and

2-HC$_6$F$_4$, on the basis of similar vibrational spectra, and also for (C$_6$F$_5$)$_2$TlCl, on the basis of isomorphism with the corresponding bromide. In benzene solution the polymer breaks down to dimeric species.

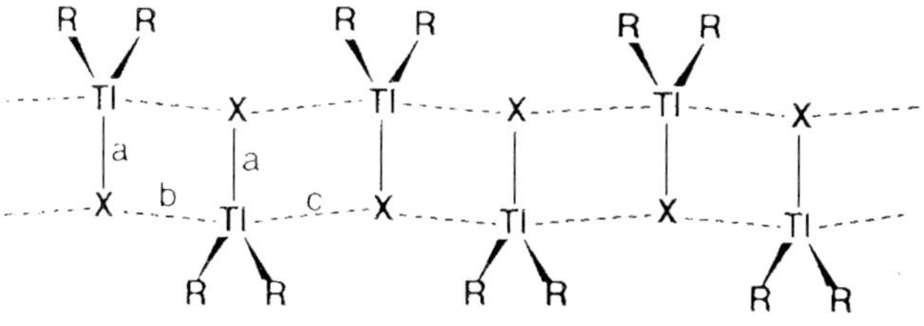

R = C$_6$F$_5$, X = Br

a = 2.734 Å
b = 3.016 Å
c = 3.214 Å

42

The same effect is produced by donors forming diorganothallium(III) halide adducts, R$_2$TlX·L, with L = oxygen or nitrogen donors (OPPh$_3$, OAsPh$_3$, pyridine, etc.), which are dimeric, e.g. **43**, in solution [151].

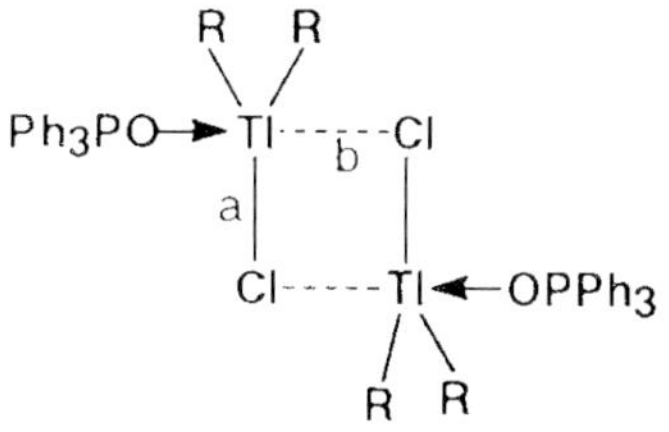

R = 2,3,5,6-C$_6$F$_4$H

a = 2.541 Å
b = 2.936 Å

43

A structurally characterized halogen-bridged dimer, **44**, 2-Me$_2$NC$_6$H$_4$TlCl$_2$,

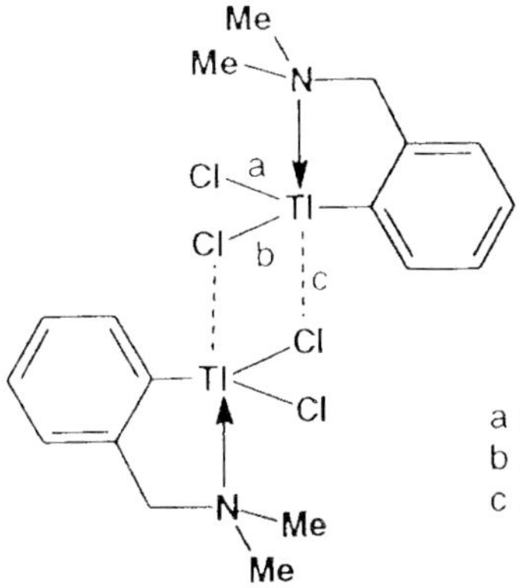

a = 2.392 Å
b = 2.530 Å
c = 2.937 Å

44

which contains an internal N–Tl dative bond, still retains Lewis acid properties to form a dimer with an asymmetric double chlorine bridge [152]. The terminal Tl–Cl bond is shorter than the bond involved in bridging.

4.3.3.2 Organothallium–sulfur compounds

Numerous examples of supramolecular self-organization can be found among thallium–sulfur compounds, both 'soft' elements with a high tendency to form Tl···S secondary bonds. The association of organothallium–sulfur compounds has been first detected on the basis of molecular weight determinations, which showed that Me_2TlSMe and its selenium analog are dimeric in boiling benzene [153, 154]. The structure of Me_2TlSPh contains asymmetric thallium–sulfur bridges (Tl–S 2.748 and 2.991 Å) [155]. Other similar compounds investigated include diphenylthallium diethyldithiocarbamate and dimethylthallium dipropyldithiocarbamate (Tl···S 3.41 Å) [157].

Dimethyl(diphenylphosphinodithioformato-S,S')thallium(III), $Me_2Tl(S_2CPPh_2)$ ·THF, forms supramolecular arrays, **45**, via Tl···S secondary bonds (Tl···S 3.21 Å) slightly longer than the chelating Tl–S primary bonds (2.98 and 3.04 Å) [158].

a = 3.04 Å
b = 2.98 Å
c = 3.21 Å

45

Organothallium(III) thiosemicarbazone derivatives are usually self-organized in solid state supramolecular architectures. An example is (*p*-anisaldehyde thiosemicarbazonato) dimethylthallium(III), which contains four-membered chelate rings (with Tl–N 2.56 Å and Tl–S 2.991 Å) interconnected in chains by Tl···S secondary bonds (Tl···S 3.304 Å) [159].

Several organothallium(III) dithiophosphates and dithiophosphinates are also assembled in the solid state as supramolecular architectures. These include dimethylthallium(III) dialkyl dithiophosphates, $Me_2TlS_2P(OR)_2$, with R = Me, Et, Cy [160], dimethylthallium diethyldithiophosphinate, $Me_2TlS_2PEt_2$ (Tl–S 2.981 and

2.991 Å, Tl···3.334 Å), diphenylthallium diethyldithiophosphinate, $Ph_2TlS_2PEt_2$ (Tl–S 2.854 and 2.933 Å, Tl···S 3.321 and 3.729 Å) [161], dimethylthallium diphenyldithiophosphinate, $Me_2TlS_2PPh_2$ [162], diphenylthallium(III) dicyclohexyldithiophosphinate, $(Ph_2TlS_2PCy_2)_2$ [163], and anionic $[Ph_2Tl(S_2PPh_2)_2]^-$ (as the NEt_4^+ salt) [164]. In all these compounds the chelate tectons are connected by Tl···S secondary bonds. An illustration is provided for $(Ph_2TlS_2PCy_2)_2$, which forms double bridged ladders, **46**, but other modes of self-organization are also possible.

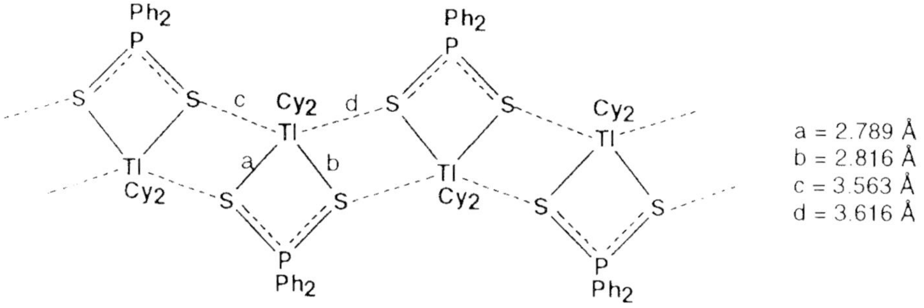

46

a = 2.789 Å
b = 2.816 Å
c = 3.563 Å
d = 3.616 Å

Occasionally the self-assembly modes are more complex and can simultaneously involve Tl···S and Tl···O secondary bonds. Thus, dimethylthallium(III) methylxanthate, $Me_2Tl(S_2COMe)$, is associated into a complex architecture, **47**, by Tl···S bonding (3.19 and 3.35 Å, longer than the chelate bonds Tl–S 2.96 Å) and Tl···O bonding from the methoxy group (Tl···O 3.13 Å) [165].

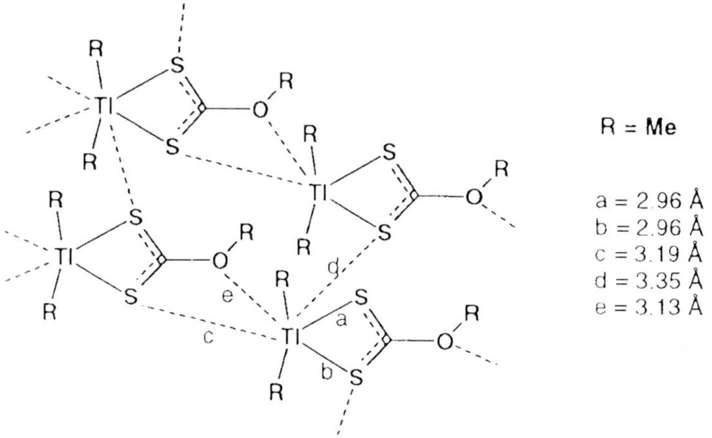

R = Me

a = 2.96 Å
b = 2.96 Å
c = 3.19 Å
d = 3.35 Å
e = 3.13 Å

47

Similarly, dimethylthallium(III) pyridine-1-oxide 2-thiolate, contains both intermolecular Tl···S bonds (3.284 Å) and Tl$_2$O$_2$ four-membered quasi-cyclic moieties formed by Tl···O bonding (2.849 Å). These connect the chelate tectons into infinite chains, **48**, running along the *c* axis of the crystal [166]. Dimethylthallium(III) pyridine-2-thiolate is also a chain-like supramolecular array in which the tectons are connected by Tl···S (2.870 and 3.160 Å) and Tl···N (2.494 Å) bonds [166a].

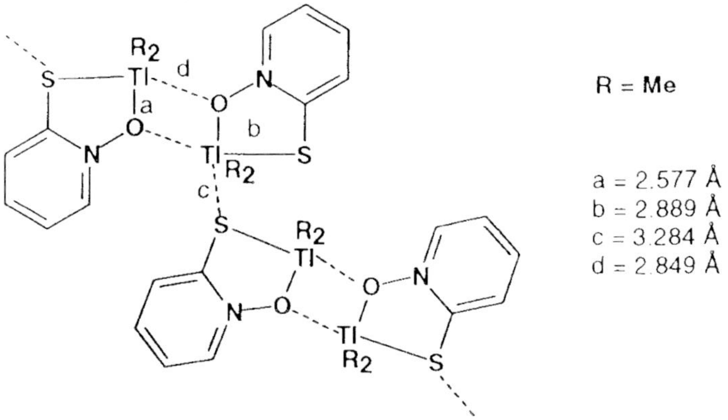

R = Me

a = 2.577 Å
b = 2.889 Å
c = 3.284 Å
d = 2.849 Å

48

Other examples include dimethylthallium(III) *p*-dimethylaminobenzylidene rhodaninate [167] and dimethylthallium(III) 2-thioorotate [168].

Unusual supramolecular self-assembly by Tl···S bonding has been observed in Tl{(η^5-C$_5$Me$_5$)Ti(1,2-S$_2$C$_6$H$_4$)$_2$} [169].

Polymeric ribbons of self-assembled chelate molecules, with participation of dative thallium–oxygen and secondary thallium–sulfur intermolecular bonds are present in dimethylthallium(III) diethylmonothiocarbamate, Me$_2$Tl[S(O)CNEt$_2$], **49** [170], and in diphenylthallium(III) diphenylmonothiophosphinate, Ph$_2$Tl[S(O)-PPh$_2$], **50** [171].

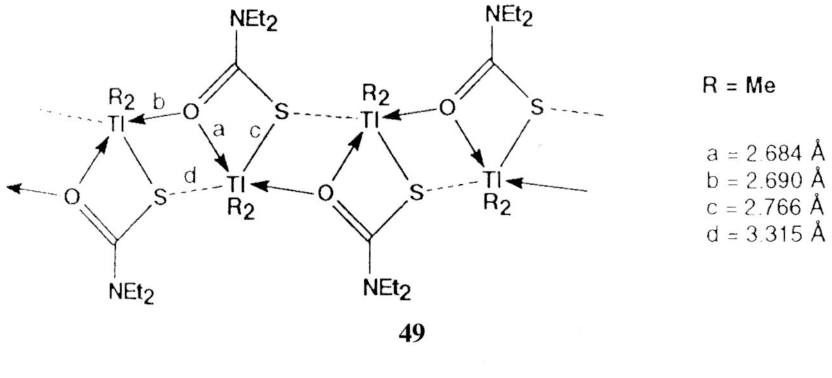

R = Me

a = 2.684 Å
b = 2.690 Å
c = 2.766 Å
d = 3.315 Å

49

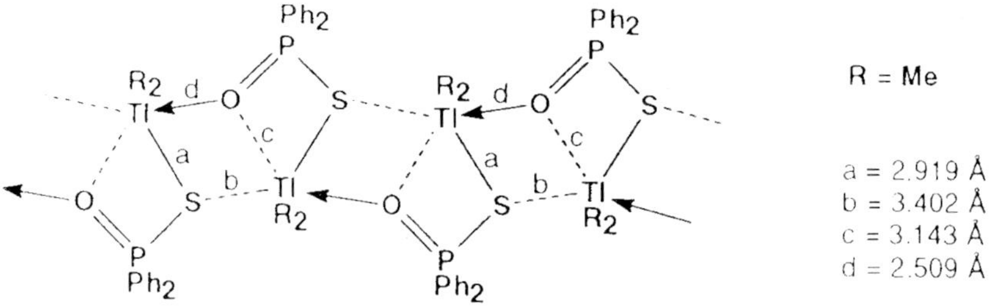

4.3 Group 13 Metals – Gallium, Indium, Thallium

R = Me

a = 2.919 Å
b = 3.402 Å
c = 3.143 Å
d = 2.509 Å

50

Other chelate complexes containing TlR$_2$ coordination centers also tend to self-organize in the solid state in a complex manner involving different types of intermolecular bond. Thus, the dimethylthallium(III) complex of 2-furanthiocarbohydrazide is self-assembled in a more complex way, including Tl···S bonds (in a four-membered quasi-cyclic moiety), Tl···O bonds (from the furan ring of one molecule to the thallium atom of a neighboring molecule), and also Tl···N intermolecular connections [172].

Dimethylthallium 4-amino-5-mercapto-3-trifluoromethyl-1,2,4-triazolate has a mixed type self-organized structure in which strips of molecules dimerized by Tl···S (3.18 Å) and Tl···N (2.61 Å) bonding are interconnected along the c axis by N–H···N bonds between the NH$_2$ groups and N heteroatoms of neighboring molecules [173].

An intricate system of Tl···S and Tl···N intermolecular bonds connects the molecular tectons of the dimethylthallium(III) bismuthiol complex (bismuthiol = 2,5-dimercapto-1,3,4-thiadiazole) into a complex supramolecular architecture, **51** [174].

a = 3.08 Å
b = 3.45 Å
c = 3.39 Å
d = 2.70 Å
e = 2.70 Å
f = 3.05 Å

51

Current experience suggests that secondary bond self-assembly can be expected in almost all diorganothallium–sulfur derivatives.

4.3.3.3 Organothallium–nitrogen compounds

Self-assembly of organothallium derivatives via thallium–nitrogen bonds is rarely observed. Dimethylthallium(III) pyrazolides are associated into dimers even in benzene solution [175], but no crystal structure determination is available for the solid-state compound. The bis(pentafluorophenyl)thallium(III) 2,2′-dipyridylamido complex is also dimerized; strong (probably dative) Tl–N bonds (2.27 Å) connect the amido nitrogen directly to thallium (primary Tl–N bond) and the dimerization is achieved by weaker and longer secondary Tl···N bonding (2.46 and 2.57 Å) [176]. Still longer Tl···N distances (3.18 and 3.32 Å) are observed between the amido nitrogen and thallium atoms of neighboring molecular tectons, resulting in the formation of an internal four-membered Tl_2N_2 quasi-ring, **52**.

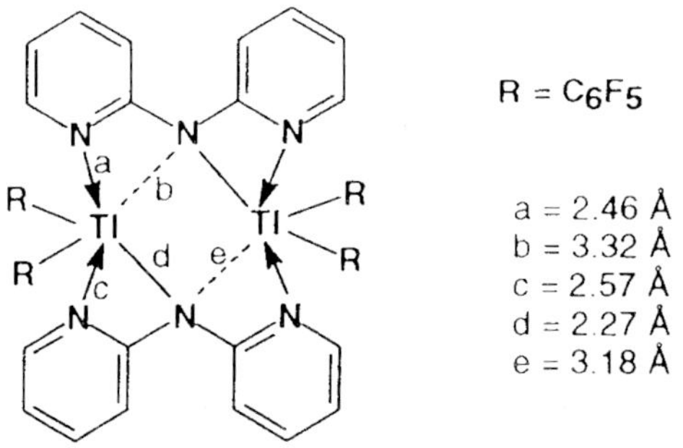
N
N
N
a
b
R
Tl
Tl
R
R
R
d
e
c
N
N
N
R = C6F5
a = 2.46 Å
b = 3.32 Å
c = 2.57 Å
d = 2.27 Å
e = 3.18 Å

52

Complex supramolecular architectures have been established for dimethylthallium(III) derivatives of polyfunctional nitrogen ligands, e.g. the cyanide, the azide, the cyanate, the thiocyanate [177], the dicyanamide, and the tricyanomethide [178]. The cyanide, Me_2TlCN, **53**, contains layers of nearly coplanar CN groups

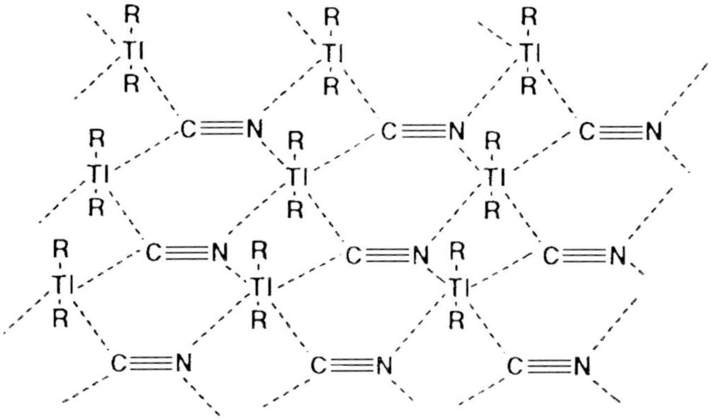
R
Tl
R
C≡N

$[Me_2TlCN]_x$

53

and Tl atoms, with the Tl–Me bonds perpendicular to the plane. In the azide, **54**, the planarity is disturbed by the failure of the azide to be coplanar with all four neighboring thallium atoms. In the thiocyanato and cyanato complexes, **55** and **56**, the structures are more complex because of the coordination of both nitrogen and sulfur or oxygen ends of the ligands, resulting in an architecture based upon Tl$\cdots$N and Tl$\cdots$S or Tl$\cdots$O bonds. In addition, polymorphism observed in the thiocyanate and cyanate leads to a more diverse structural picture. Although in the original paper the interactions between the linear Me$_2$Tl coordination centers and the ligands are described as ionic, these can probably be considered as secondary bonding interactions, in line with the description of other supramolecular structures of organothallium compounds.

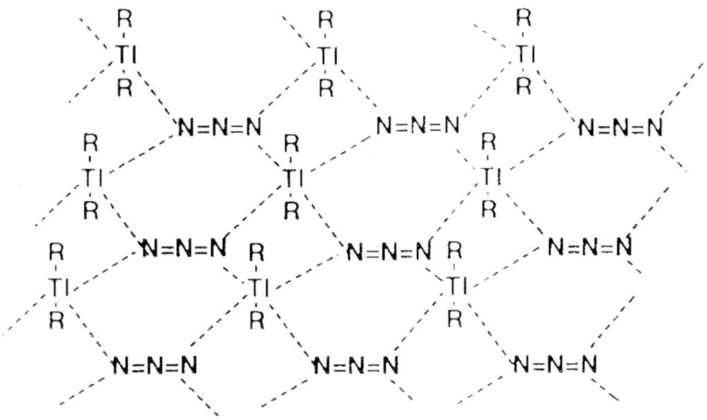

[Me$_2$TlN$_3$]$_x$

54

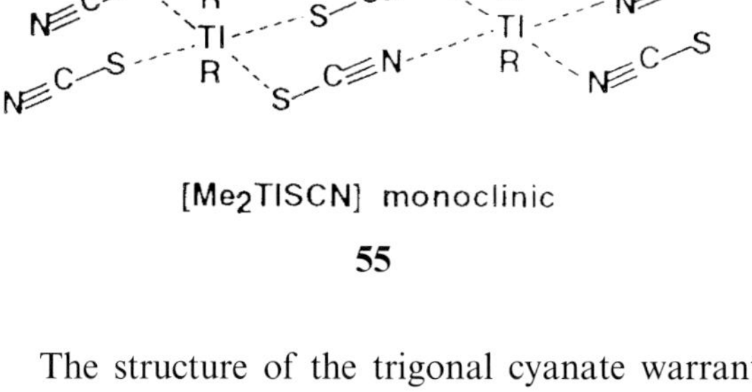

[Me$_2$TlSCN] monoclinic

55

The structure of the trigonal cyanate warrants special comment. It is different

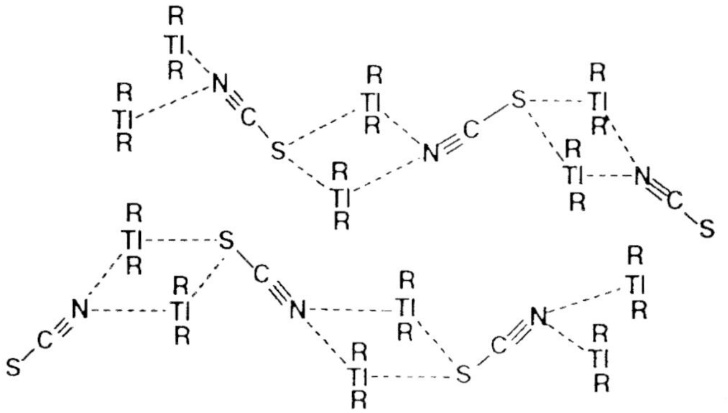

[Me₂TlSCN]ₓ orthorhombic

56

from the rhombic cyanate. The latter forms infinite polymeric Tl···N···Tl···N chains connected into layers by Tl···O interactions whereas the trigonal cyanate involves four-membered Tl$_2$N$_2$ quasi-cyclic tectons connected into a three-dimensional network by Tl···O interactions.

In the dicyanamide, **57**, and the tricyanomethide, **58**, dimethylthallium(III) derivatives the structure of the ligands determines the formation of large rings, as part of complex supramolecular architectures.

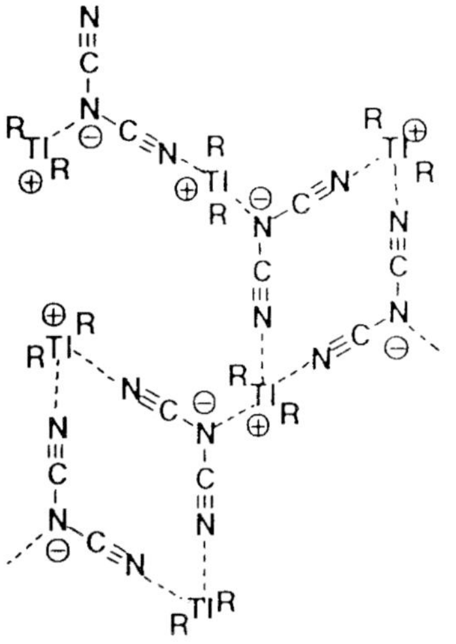

[Me₂TlN(CN)₂]

57

[Me₂TlC(CN)₃]
58

The crystal structures of organothallium(III) complexes illustrate the great diversity possible in the building of supramolecular architectures from relatively simple polyfunctional building blocks and using metal coordination centers of low valence but high coordination numbers, e.g. thallium in diorgano-substituted moieties, TlR$_2$.

Finally, we mention the supramolecular structure of an inorganic (naked) thallium(I) derivative of the ferrocenyltris(pyrazolyl)borate ligand [179]. In this complex two pyrazole groups close a chelate ring with thallium(I) (Tl–N 2.638 and 2.676 Å) and the third forms an intermolecular Tl···N bond (2.780 Å) leading to self-assembly into infinite chains.

4.4 Group 14 – Tin and Lead

4.4.1 Self-assembly of organotin compounds

4.4.1.1 Organotin halides

With the exception of organotin fluorides, all other organotin halides tend to self-associate in the solid state via secondary bonding, unless bulky substituents prevent the process. There is, however, one exception. A long tin–fluorine interatomic distance (3.451 Å) suggesting secondary interactions, has been measured in the trimethylstannyl derivative of 1,4-bis(perfluoroisopropoxy)tetrafluorobenzene, $Me_3SnOC(CF_3)_2C_6F_4C(CF_3)_2OSnMe_3$. These weak interactions lead to the formation of polymeric sheets [180].

Like the fluorides, other *triorganotin halides* are monomeric, **59a**, when the organic groups are bulky, but with smaller substituents linear polymeric chains, **59b** or **59c**, are formed. In the supramolecular polymers the tin atom is five-coordinate and has trigonal bipyramidal geometry, with the halogen atoms in axial positions.

59a **59b** **59c**

One short (primary, covalent) and one long (secondary) tin–halogen bond are observed in each trigonal bipyramidal unit. The short bonds are within the normal limits expected from the sum of the covalent radii and are only slightly longer than the standard values measured in tetrahedral monomeric compounds. The interatomic distances for the secondary bonds vary within broader limits, and sometimes are close to the sum of the van der Waals radii.

Trimethyltin chloride has a self-organized chain structure of type **59c**, with a bending angle Sn–Cl···Sn of 150.5°, and basically linear Cl–Sn···Cl (176.85°) coordination. The tin–chlorine interatomic distances are distinctly different (Sn–Cl 2.430 Å and Sn···Cl 3.269 Å) [181, 182], the longer being significantly shorter than the sum of van der Waals radii (estimated as 3.85 Å). It is interesting to compare these values with the axial Sn–Cl bond length in the anionic complex $[Sn(CH_3)_3Cl_2]^{2-}$, which is longer (2.572 and 2.696 Å) [183] than the primary bond in trimethyltin chloride.

A recent structure determination of $[Me_3SnCl \cdot Me_3SnF \cdot TaF_5]_n$ reveals a new chain supramolecular array with long bonds (Sn···Cl 2.555 and 2.562 Å, Sn···F 2.514 Å) [185a].

Tribenzyltin chloride, $(PhCH_2)_3SnCl$ (Sn–Cl 2.387 Å, Sn···Cl 3.531 Å and Cl–Sn···Cl 180°) [184] and Cy_3SnCl (Sn–Cl 2.407 Å, Sn···Cl 3.306 Å, Cl–Sn···Cl 180°) [185] are also self-organized supramolecular chains in the solid state.

Some other triorganotin monochlorides are not associated in the solid state. Triphenyltin monochloride (Sn–Cl 2.353 Å) [186] was described as discrete unassociated in the solid state and even tricyclohexyltin chloride was first described as monomeric [187], like the triorganotin bromides and iodide investigated so far (see Table 4.1). Tris(*m*-tolyl)tin chloride (Sn–Cl 2.379 Å) and tris(3,5-dimethylphenyl)tin chloride (Sn–Cl 2.357 Å) are also discrete monomers [187a].

An unusual pair of compounds $ClM^+[Ph_2P(CH_2)_2SnMe_2Cl_2^-][Ph_2P(CH_2)_2$-$SnMe_2Cl]$ with M = Pd or Pt, are self-organized into supramolecular chain-like assemblies, **60**, via secondary Sn···Cl bonds (M = Pd, Sn···Cl 3.232 Å; M = Pt 3.248 Å). The structure also contains intramolecular secondary Sn···Cl bonds [192].

4.4 Group 14 – Tin and Lead   229

Table 4.1 Triorganotin halides.

Compound and ref.	Form of association	Sn–X (Å)	Sn···X (Å)	Sn–X–Sn (°)
SnMe$_3$Cl [181, 182]	Polymer	2.430	3.269	150.5
SnMe$_2$(Mes)Cl [188]	Monomer	2.408	–	–
SnPh$_3$Cl [186, 189]	Monomer	2.3620	–	–
		2.3692		
SnCy$_3$Cl [185, 189]	Monomer	2.20	–	–
	Polymer	2.407	3.306	180
Sn(CH$_2$Ph)$_3$Cl [184]	Polymer	2.387	3.531	180
Sn(C$_6$H$_4$Me-3)$_3$Cl [190]	Monomer	2.379	–	–
SnPh$_3$Br [191]	Monomer	2.495	–	–
SnCy$_3$Br [187]	Monomer	2.52	–	–
SnCy$_3$I [187]	Monomer	2.54	–	–

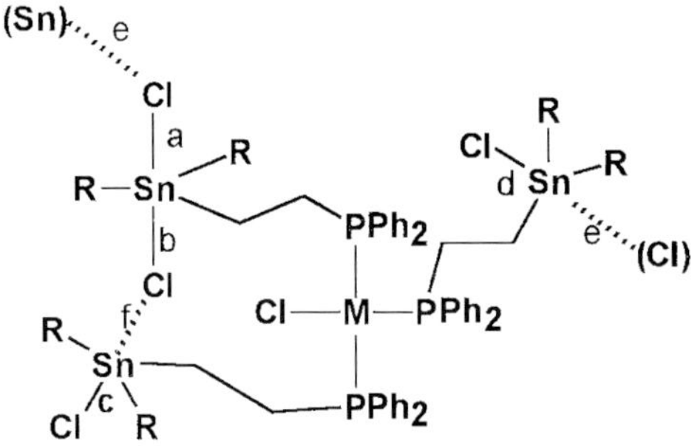

R = Me

	M = Pd	M = Pt
a =	2.637 Å	2.630 Å
b =	2.679 Å	2.683 Å
c =	2.441 Å	2.445 Å
d =	2.440 Å	2.436 Å
e =	3.232 Å	3.248 Å
f =	3.234 Å	3.223 Å

60

Diorganotin dihalides have a greater tendency to self-organize in polymeric structures. The several possibilities are illustrated in **61**.

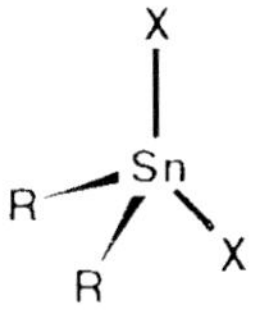

61a 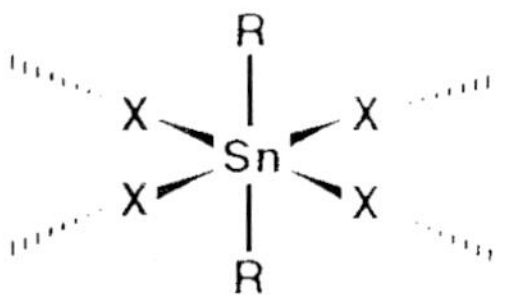   **61b**

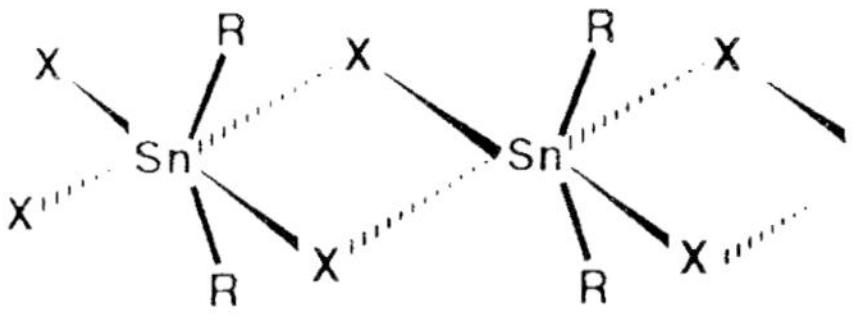

61c 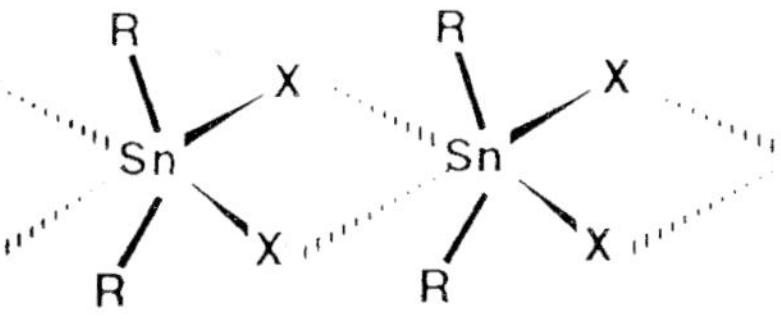   **61d**

Several diorganotin dichlorides have been structurally characterized and all are self-organized chain polymers in the solid state, with a strongly distorted octahedral coordination geometry at tin. The tin atoms are connected by double bridges of chlorine atoms. In structure **61c** the double bridge consists of a primary Sn–Cl bond and a secondary Sn···Cl bond. This type was found in dimethyltin dichloride (Sn–Cl 2.40 Å; Sn···Cl 3.54 Å) [193]; the tin and chlorine atoms are coplanar, and the methyl groups are located above and below the plane. Gas-phase electron-diffraction data show that the compound is a tetrahedral monomeric molecule [194]. Similar structures of type **61c** have been reported for R_2SnX_2 with R = Et, X = Cl and Br [195]; R = Bu^n, X = Cl [196], R = Cy, X = Cl, and Br [197], and MePhSnCl$_2$ [198a].

The second structural type, **61d**, in which both primary Sn–Cl bonds in the bridge are attached to the same chlorine atom, was found in $(ClCH_2)_2SnCl_2$ (Sn–Cl 2.37 Å, Sn···Cl 3.71 Å) [199] and in Et_2SnI_2 [195].

A controversial report on the structure of diphenyltin dichloride, Ph_2SnCl_2, describes it as a monomeric compound of type **61a** [200], but according to other authors the compound is a tetramer, **62**, formed from two octahedral and two trigonal bipyramidal units containing, respectively, six- and five-coordinated tin, with secondary bond distances Sn···Cl 3.77–3.78 Å [199]. The dilemma remains unsolved.

a = 2.336 Å
b = 2.353 Å
c = 2.335 Å
d = 3.780 Å
e = 3.770 Å

62

Some molecular structure parameters of diorganotin dihalides are collected in Table 4.2.

The trinuclear complex $[Sn_3Me_6Cl_8]^{2-}$, **63**, prepared electrolytically, can be re-

a = 2.42 Å
b = 2.96 Å
c = 3.02 Å
d = 2.68 Å
e = 2.68 Å
f = 2.63 Å
g = 2.65 Å
h = 3.03 Å
i = 3.10 Å
j = 2.42 Å

R = Me, X = Cl

63

Table 4.2 Diorganotin dihalides.

Compound and ref.	Form of association	Sn–X (Å)	Sn···X (Å)
SnMe$_2$Cl$_2$ [193]	Polymer	2.40	3.54
SnEt$_2$Cl$_2$ [195]	Polymer	2.385	3.483
		2.384	3.440
Sn(CH$_2$Cl)$_2$Cl$_2$ [199]	Polymer	2.37	3.71
SnMePhCl$_2$ [198a]	Polymer	2.335	3.422
		2.36	3.81
SnBu$^n{}_2$Cl$_2$ [196]	Polymer	2.372	3.544
		2.388	3.514
SnPh$_2$Cl$_2$ [199]	Tetramer	2.357	3.780
		2.336	3.770
SnPh$_2$Cl$_2$ [200]	Monomer	2.353	3.77
		2.336	
SnCy$_2$Cl$_2$ [197]	Polymer	2.400	3.54
		2.393	
SnCy$_2$Cl$_2$ [198]	Polymer	2.407	3.332
		2.371	3.976

garded as a fragment of the polymeric structure of dimethyltin dichloride [201]. It is the product of assembling two R$_2$SnCl$_2$ molecules to a central six-coordinate anion, [R$_2$SnCl$_4$]$^{2-}$.

Tin–chlorine interatomic distances cover a broad range, but can be clearly grouped in two sets – normal covalent single Sn–Cl bonds (2.42–2.68 Å) and secondary Sn···Cl bonds (2.96–3.10 Å). The latter are significantly shorter than the sum of the van der Waals radii.

It should be mentioned that the dimeric dichlorodistannoxanes (ClR$_2$SnOSnR$_2$Cl)$_2$ (associated by dative bonding) are self-organized in the solid state into supramolecular arrays by tin–chlorine secondary bonds (R = Me, Sn···Cl 2.78 Å; R = Et, Sn···Cl 2.73 Å) [201a].

Few *monoorganotin trihalides* have been structurally characterized. Methyltin trichloride, MeSnCl$_3$ is a tetrahedral molecule in the gas phase (electron diffraction) [202], but in the crystal both the trichloride [203] and the tribromide, MeSnBr$_3$ [204], are weakly associated by double bridges, **64**. The bromide forms centrosymmetric dimers [MeSnBr$_3$]$_2$ (Sn–Br 2.454 Å and Sn···Br 3.785 Å). The iodide,

64

MeSnI$_3$, is not associated even in solid state [205]. This confirms the decreasing tendency of self-assembly in organotin halides in the order Sn–F > Sn–Cl > Sn–Br > Sn–I.

Organotin(II) halides with bulky organic groups can be prepared. These are associated into dimeric structures with bridging Sn–Cl bonds, which are longer than in organotin(IV) halides. Thus, MesSnCl (Sn–Cl 2.600 and 2.685 Å) [206] and (PhMe$_2$Si)$_3$CSnCl (2.596 and 2.779 Å) are such dimers [207].

A particularly interesting type of self-organized supramolecular structure is observed for some half-sandwich organotin(II) halides.

Cyclopentadienyltin(II) chloride consists of a double-chain ribbon, **65**, with unsymmetrical tin–chloride bridges connecting the C$_5$H$_5$SnCl units (Sn–Cl 2.679 Å, Sn$\cdots$Cl 3.242 and 3.262 Å) [208]. The η^5-C$_5$Me$_5$ derivative is a dimer, **66**, [η^5-C$_5$Me$_5$SnCl]$_2$ with bridging chlorines which form primary and secondary tin–halogen bonds (Sn–Cl 2.693 and Sn$\cdots$Cl 3.444 Å; Sn–Cl 2.657 and Sn$\cdots$Cl 3.413 Å in two independent molecules) [209].

65

66

A polynuclear supermolecule [Sn$_9$(η^5-C$_5$Me$_4$SiMe$_2$Bu^t)$_6$Cl$_{12}$] formed by self-assembly of (η^5-C$_5$Me$_4$SiMe$_2$Bu^t)SnCl with SnCl$_2$ by secondary Sn$\cdots$Cl interactions has a disparate set of tin–chlorine interatomic distances (Sn–Cl 2.474, 2.822, 2.963, 3.092, 3.486, and 3.538 Å) [210].

An unusual compound is a Sn(II) cationic complex derived from dimethylbis(σ-tetramethylcyclopentadienyl)silane, in which the tetrafluoroborate counter ion bridges two stannocene units to form a self-organized dimer, **67** [211]. The tin–fluorine bonds (2.82 and 2.74 Å) are longer than normal covalent bonds and this was assigned to the weak nucleophilic character of the tetrafluoroborate anion. Perhaps this is another example of tin–fluorine secondary bonds.

Arene complexes, illustrated by (η^6-C$_6$H$_6$)SnCl(AlCl$_4$), have unusual self-organized chain structures [212, 213]. This compound contains dimeric, planar Sn$_2$Cl$_2$ units bridged by tetrachloroaluminate AlCl$_4$ groups, and forms infinite chains, **68**, by secondary Sn$\cdots$Cl bonding. In the chain four-membered Sn$_2$Cl$_2$ rings alternate with eight-membered quasi-rings, and two sets of tin–chlorine interatomic distances illustrate single covalent and secondary bonds. The *p*-xylene analog has a similar structure.

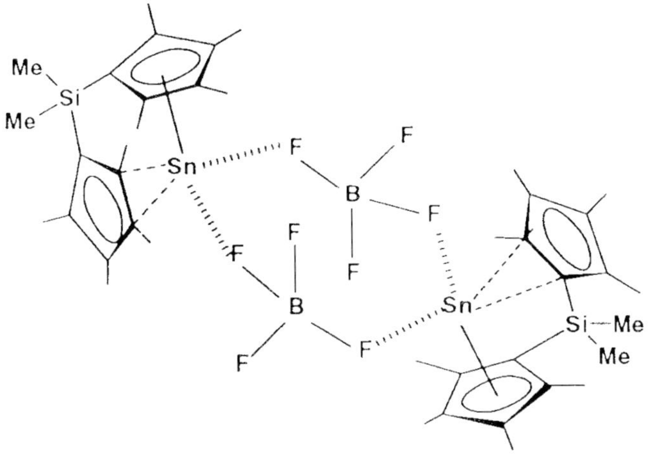

67

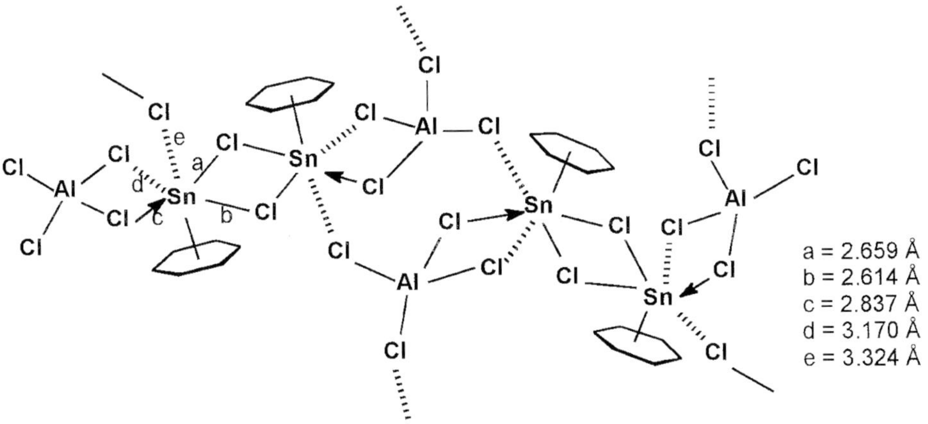

a = 2.659 Å
b = 2.614 Å
c = 2.837 Å
d = 3.170 Å
e = 3.324 Å

68

Two toluene, [(CH$_3$C$_6$H$_4$SnCl)(AlCl$_4$)] (α and β isomers), and mesitylene, [{(CH$_3$)$_3$C$_6$H$_3$SnCl}(AlCl$_4$)], derivatives have also been prepared and have similar self-organized chain structures [214]. Careful analysis of the structures of [(AreneSnCl)(AlCl$_4$)] complexes showed that all have a similar dimeric building unit containing a central Sn$_2$Cl$_2$ ring, with the η^6-arene attached to tin and with an AlCl$_4$ group chelated to each tin through two chloride ions; these chelating groups are on opposite sides of the Sn$_2$Cl$_2$ ring, as illustrated in **69**.

[Structure diagram labeled **69** showing a Sn–Cl bridged complex with two AlCl$_4$ groups and two arene (η^6) ligands, with bonds labeled a, b, c, d.]

69

In all these compounds the central core contains the short bonds (a and b); the chelating Sn–Cl bond (c) and (d) are somewhat longer, and secondary Sn···Cl bonds (e) are the longest. The Sn$_2$Cl$_2$ is approximately in a plane perpendicular to the paper. Secondary bonds from tin to chlorine atoms of the AlCl$_4$ groups of adjacent units connect them into a chain supramolecular structure. All these self-organized structures are externally wrapped with organic molecules of arenes and in the crystals they are packed like rods, with only van der Waals interactions between them.

Hexamethylbenzene forms a tetrameric compound [(η^6-C$_6$Me$_6$)SnCl(AlCl$_4$)]$_4$ in which secondary Sn···Cl bonds hold the self-organized structure together; it has been prepared from Sn(AlCl$_4$)$_2$ and hexamethylbenzene. The structure is centrosymmetric and consists of two dimeric tectons [(η^6-C$_6$Me$_6$)SnCl(AlCl$_4$)]$_2$ interconnected through Sn···Cl–Al bridges; a central eight-membered quasi-ring Al$_2$Sn$_2$Cl$_4$ is fused to a complex polycyclic system, **70**, made of four-membered Sn$_2$Cl$_2$ and SnCl$_2$Al rings [215].

A simpler compound, the dimer [(η^6-C$_6$H$_6$)Sn(AlCl$_4$)$_2$]$_2$, **71**, was obtained from anhydrous SnCl$_2$, AlCl$_3$, and hexamethylbenzene in 1:2:1 molar ratio, with benzene as solvent. In this compound two aryltin moieties are bridged by two AlCl$_4$ groups, and other two AlCl$_4$ groups each chelate a tin atom. An eight-membered Al$_2$Sn$_2$Cl$_4$ quasi-ring is also present as a central core [216]. A related compound [(η^6-C$_6$H$_6$)$_2$SnCl(AlCl$_4$)]$_2$ has been formed from molten Sn[AlCl$_4$]$_2$ and benzene under reflux. In this compound two [(η^6-C$_6$H$_6$)$_2$Sn units are bridged by two chlorine atoms and by two AlCl$_4$ groups [217].

In the hexamethylbenzene derivative the bulky arene probably prevents the extension of the structure into a polymeric array.

A fascinating graphite-like supramolecular structure based upon Sn···Cl bonds (3.086 Å) has been recently reported for [217a].

4.4.1.2 Organotin–oxygen compounds

Compounds containing at least one tin–oxygen bond in a saturated five-membered ring have a tendency to self-assemble by intermolecular tin–oxygen bonding. Such compounds include dioxastannolannes (a class of antitumor-active compounds)

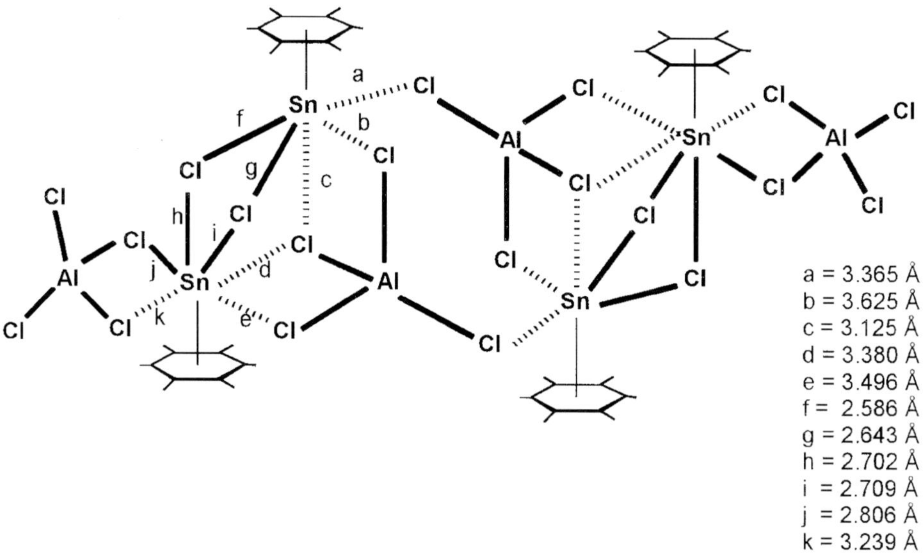

4.4 Group 14 – Tin and Lead

70

a = 3.365 Å
b = 3.625 Å
c = 3.125 Å
d = 3.380 Å
e = 3.496 Å
f = 2.586 Å
g = 2.643 Å
h = 2.702 Å
i = 2.709 Å
j = 2.806 Å
k = 3.239 Å

71

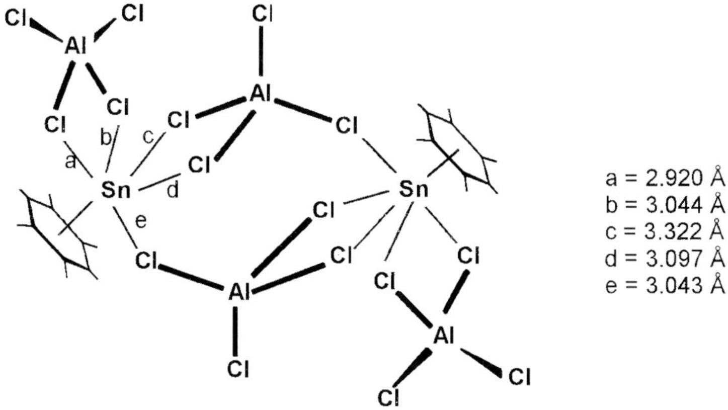

a = 2.920 Å
b = 3.044 Å
c = 3.322 Å
d = 3.097 Å
e = 3.043 Å

[218] and organotin derivatives of sugars. The tendency to self-assemble is influenced by steric factors.

2,2-Di-*n*-butyl-1,3,2-dioxastannolane exists as a dimer in solution, but the monomer and higher oligomers have also been detected [219–222]. Early ^{119}Sn NMR studies supported the conclusion that the dimers contain five-coordinate tin [223]. An X-ray diffraction crystal structure determination established that in the solid state the compound is associated via tin–oxygen bonds to form an infinite ribbon, **72**, in which each ring is doubly connected to two neighbors by tin–oxygen bonds. The resulting polymer contains six-coordinate tin in a very distorted octahedral environment. Within the five-membered ring the Sn–O bonds are shorter (1.975 and

2.097 Å) as expected for covalent single bonds, and the intermolecular contacts are significantly longer (2.495 and 2.520 Å) [224, 225], and can probably be reasonably assumed to be strong secondary bonds. Clear-cut distinction between dative and secondary bonds is difficult and the intermolecular tin–oxygen distances in associated dioxastannolanes cover a broad range, with small values close to dative bonds in some compounds and large values which can be assigned to secondary bonds in other compounds. For convenience all self-organized dioxastannolanes will be described together in this section.

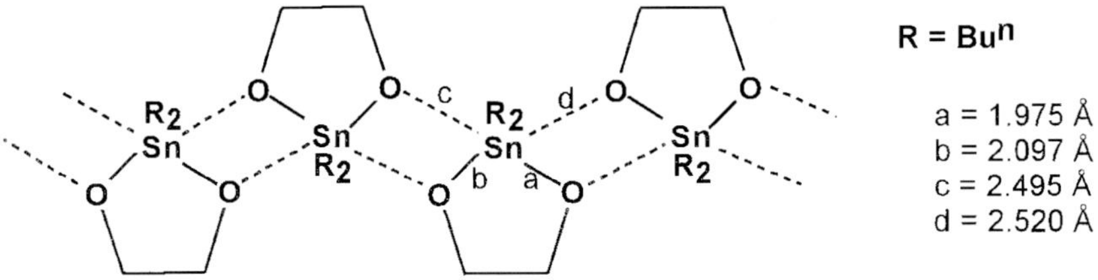

R = Bu^n

a = 1.975 Å
b = 2.097 Å
c = 2.495 Å
d = 2.520 Å

72

Because acyclic dialkyltin dialkoxides have no similar tendency to self-organization, it has been suggested [225] that the intermolecular association which changes the hybridization of tin from sp^3 towards sp^{3}d, relieves the strain in the five-membered ring by changing the O–Sn–O bond angle from 109 to ca. 90° between apical and equatorial positions in a trigonal bipyramid.

Solution and solid-state studies with ^{119}Sn and ^{13}C NMR spectroscopy [226–228] and Mössbauer spectroscopy [229] were performed to gain some insight into the structure and dynamics of these compounds. It was concluded that in solution dimeric, **73**, trimeric, **74**, tetrameric, **75**, and even pentameric species are present.

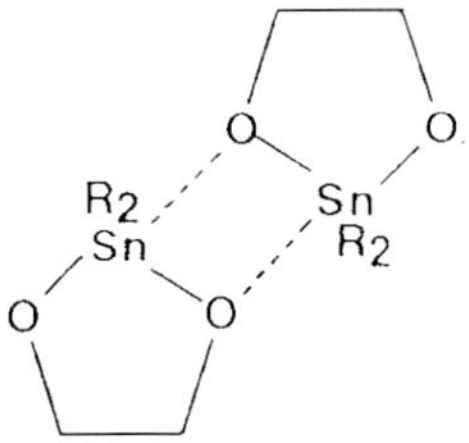

dimer

73

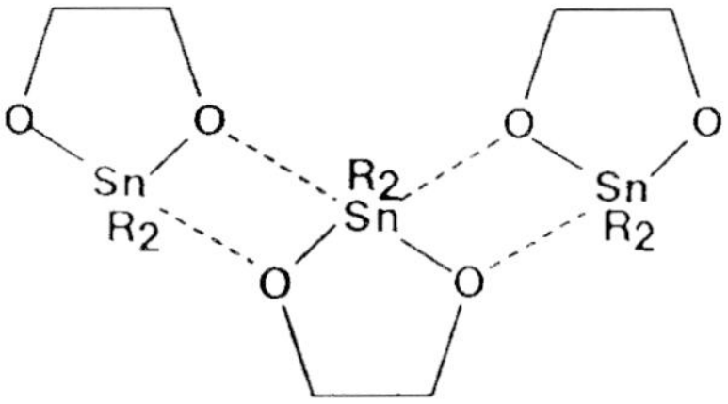

trimer

74

The role of steric factors is illustrated by the association of the di-*tert*-butyl derivative of dioxastannolane as dimer only, **76**, both in solution and in the solid state [230]. In the dimer the intermolecular tin–oxygen distances suggest strong (perhaps dative) bonding.

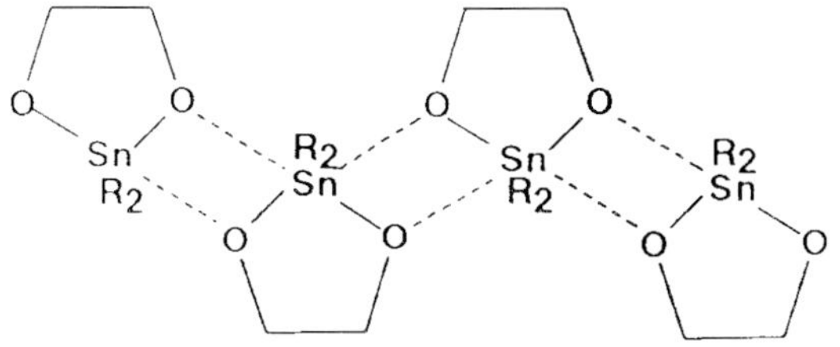

tetramer
75

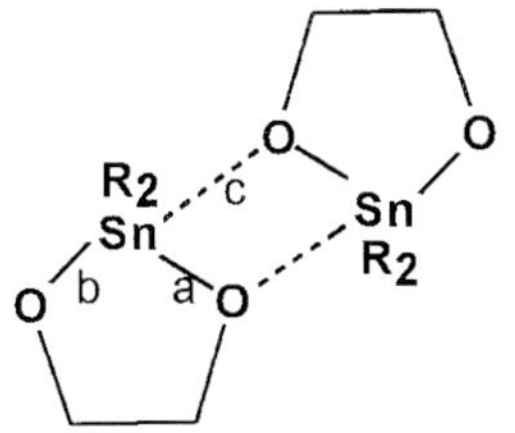

R = Bu^t

a = 2.086 Å
b = 2.049 Å
c = 2.253 Å

76

In the context of dimeric association of dioxastannolanes, it is interesting to mention that a ten-membered ring, **77**, containing two tin and two oxygen atoms contains *intra*molecular, transannular dative (or strong secondary) bonds, Sn–O 2.268 Å, i.e. weaker than the ring Sn–O bonds (Sn–O 2.060 Å), as suggested by the interatomic distances [231]. Compare this structure with that of the *tert*-butyl derivative, **76**.

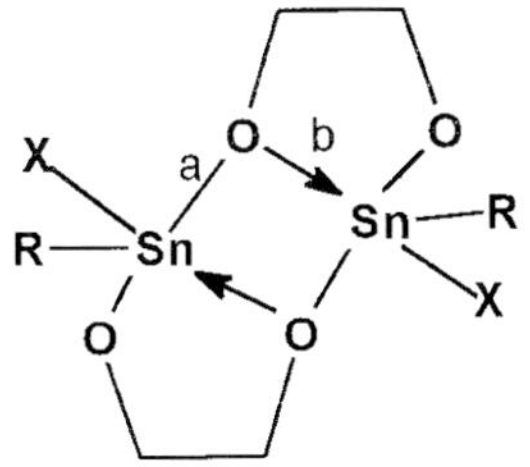

R = Ph, X = I

a = 2.060 Å
b = 2.268 Å

77

Sugar derivatives investigated by X-ray crystallography include methyl 4,6-di-*O*-benzylidene-2,3-*O*-dibutylstannylene-α-D-glucopyranoside (a dimeric derivative of glucose, with five-coordinate tin [232, 233]) and the corresponding mannose derivative (a linear pentamer, **78**, containing two five-coordinate tin atoms in the terminal tectons and three six-coordinate tin atoms in the three middle tectons) [234]. The different extents of association are because of steric peculiarities of the two molecules.

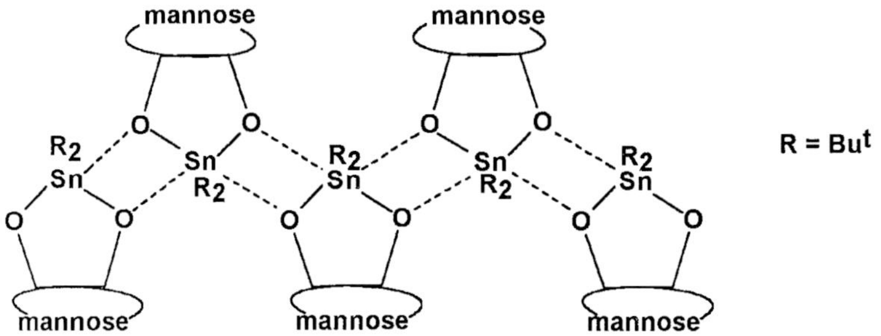

78

R = Bu^t

Dialkyl-1,2,6-stannadioxacyclohexanes also self-assemble, at least in the solid state. An X-ray structure determination of the di-*n*-butyl derivative showed that the molecules are associated in infinite ribbons, **79**, via pairs of secondary Sn···O bonds (2.57 Å) significantly longer than the ring Sn–O bonds (2.05 Å) [235].

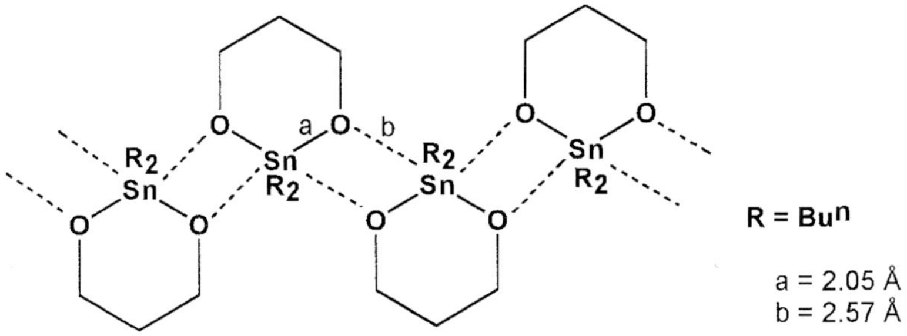

79

R = Bu^n

a = 2.05 Å
b = 2.57 Å

2,2-Dialkyl-1,3,2-oxathiastannolanes are associated in the solid state to form infinite chain-like arrays, **80**, as suggested by Mössbauer and ^{13}C and ^{119}Sn NMR spectra. X-Ray diffraction analysis of the 2,2-di-*n*-butyl derivative shows that only Sn···O bonds (2.353 Å) participate in the intermolecular association and the tin atom becomes five-coordinate [236].

In solution the compounds with R = Me, Pr, Bu, and PhCH$_2$ are dimers. With larger R groups on tin the extended association is sterically hindered and the 2,2-di-*tert*-butyl derivative was found to be a dimer, **81**, in the solid state, associated via rather strong (possibly dative) tin–oxygen bonds (Sn–O 2.086 and 2.049 Å; Sn···O 2.253 Å) [237]. 2,2-Diphenyl-1,3-2-oxathiastannolane is also a dimer, **81**, (Sn–O 2.072 and Sn···O 2.248 Å) [238].

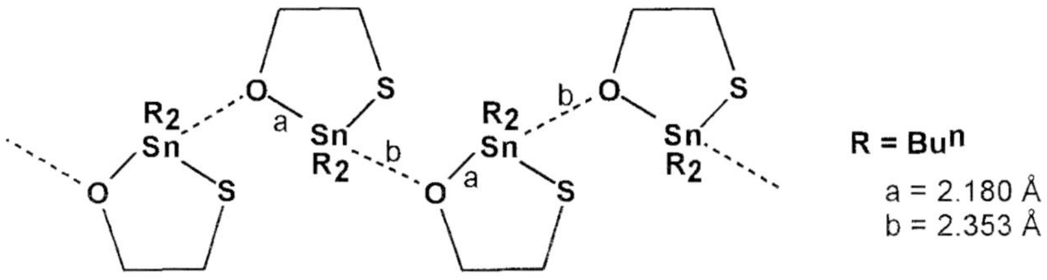

80

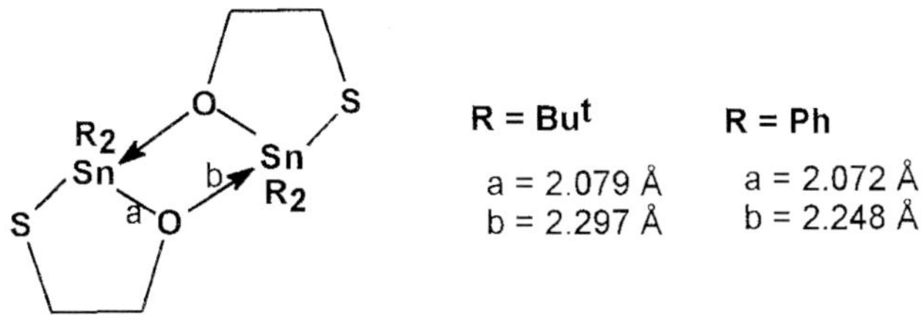

R = Bu^n

a = 2.180 Å
b = 2.353 Å

81

R = Bu^t

a = 2.079 Å
b = 2.297 Å

R = Ph

a = 2.072 Å
b = 2.248 Å

A dimeric structure, **82**, has been inferred from ^{13}C and ^{119}Sn NMR spectra (indicating five-coordinate tin) for the metal carbonyl complexes of 5-aza-2,8-dioxa-1-stannocane [239]; an unusual and unexpected linear trimeric structure, **83**, has been established by X-ray diffraction for a methylstannatrane [240].

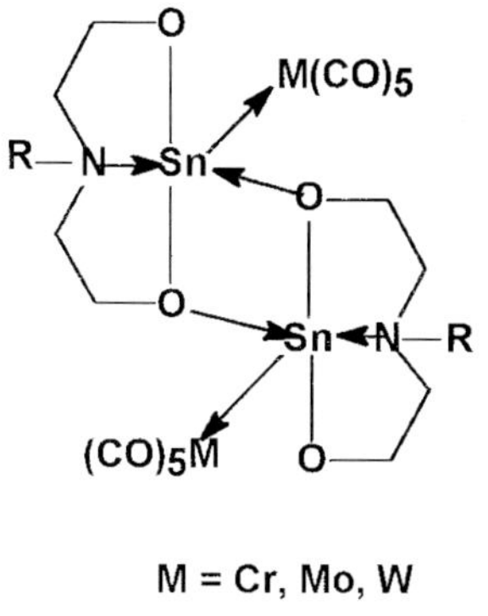

M = Cr, Mo, W

82

It seems that when oxygen atoms doubly bonded to carbon (C=O) and coordinated to tin also participate in intermolecular self-assembly by connecting a second tin atom, the bond leading to dimerization is a secondary interaction, characterized by a long tin–oxygen interatomic distance. An example is the diorganotin complex of maltol (3-hydroxy-2-methyl-4H-pyran-4-one), **84a**, which forms a dimer, **84b**, by Sn···O secondary bonding (3.435 Å) [241].

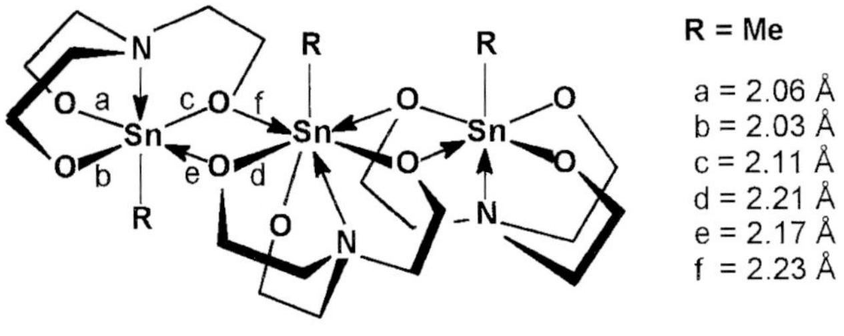

83

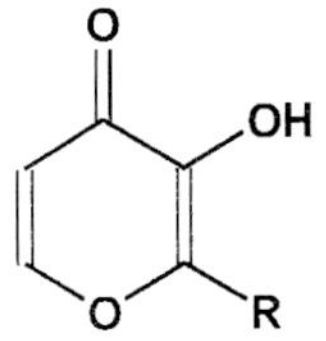

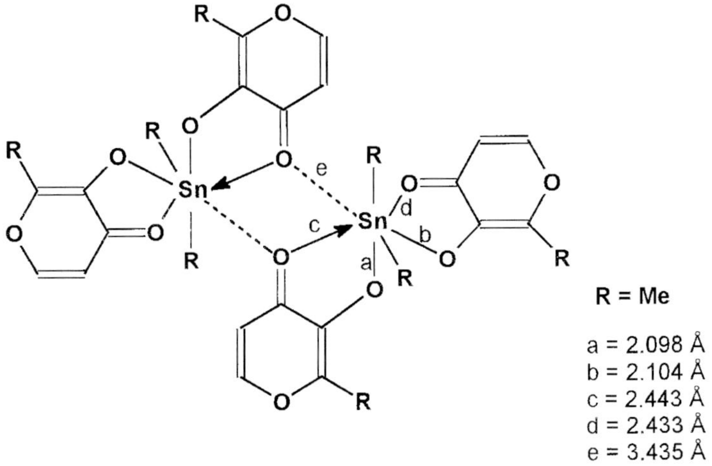

R = Me
a = 2.06 Å
b = 2.03 Å
c = 2.11 Å
d = 2.21 Å
e = 2.17 Å
f = 2.23 Å

84a **84b**

R = Me
a = 2.098 Å
b = 2.104 Å
c = 2.443 Å
d = 2.433 Å
e = 3.435 Å

Similarly, the fungicidal triphenyltin derivative of salicylaldehyde, **85a**, is a polymer, **85b**, associated via secondary Sn···O bonding (2.427 and 2.459 Å) [242].

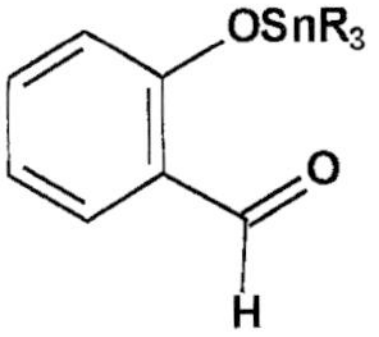

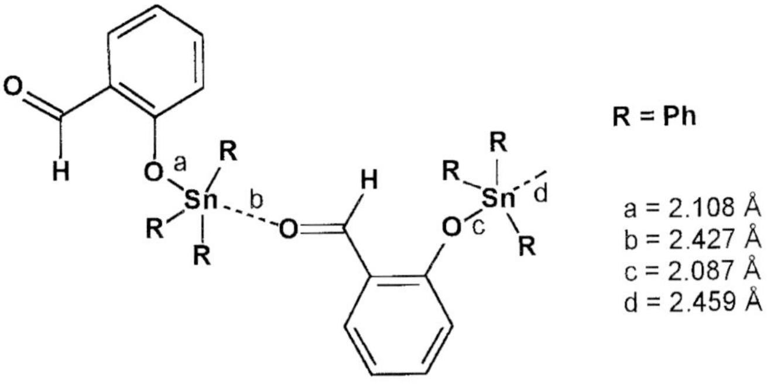

85a **85b**

R = Ph
a = 2.108 Å
b = 2.427 Å
c = 2.087 Å
d = 2.459 Å

Supramolecular self-assembly is observed for some diorganotin sulfanyl prope-

4.4 Group 14 – Tin and Lead

noates. The pyridyl derivative, **86a** (R = Et, R′ = 2-C$_5$H$_5$N), forms zigzag chain-like arrays, **86b**, similar to those of triorganotin carboxylates [R$_3$SnOOCR′]$_x$, with one R replaced by sulfur in the equatorial plane [243].

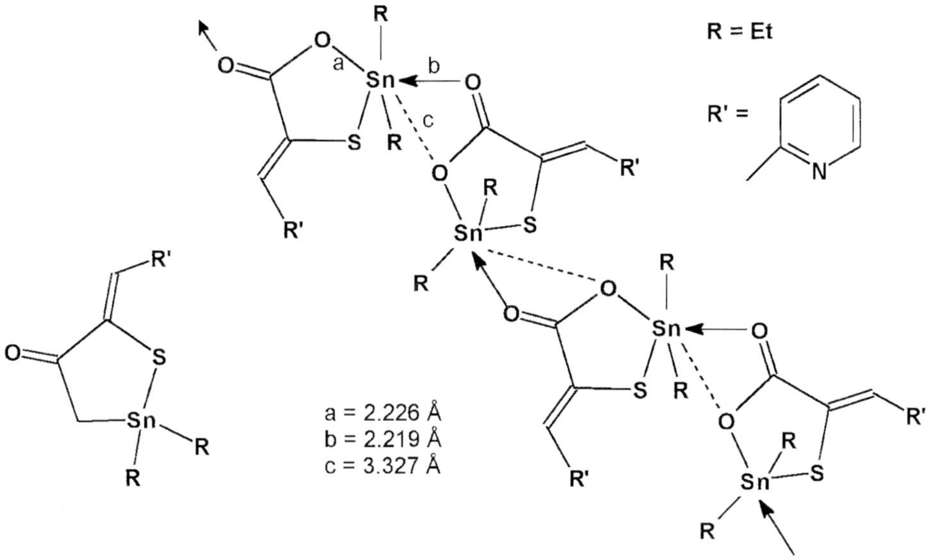

R = Et

R′ =

a = 2.226 Å
b = 2.219 Å
c = 3.327 Å

86a **86b**

Another example is provided by some diorganotin derivatives of 2-(2-methoxy-carbonyl)hydrazonopropionic acid, with Sn···O 2.571 Å (R = Bu^n) and 2.721 Å (R = CH$_2$CMe$_2$Ph), **87** [244].

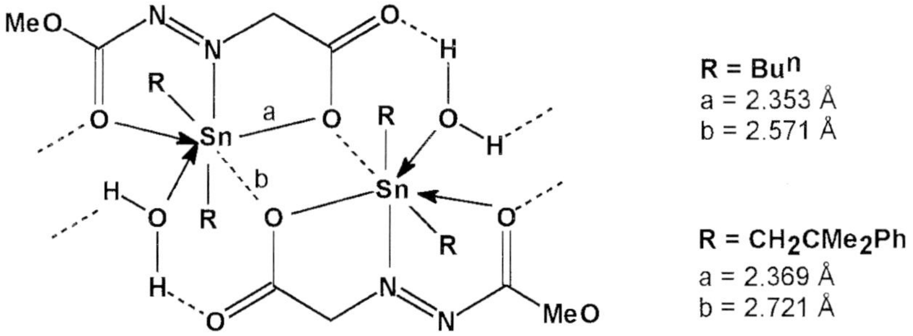

R = Bu^n
a = 2.353 Å
b = 2.571 Å

R = CH$_2$CMe$_2$Ph
a = 2.369 Å
b = 2.721 Å

87

N-Substituted triorganotin derivatives of acetamide, MeC(=O)NMeSnMe$_3$ and MeC(=O)N(SnMe$_3$)$_2$, are also associated by Sn···O secondary bonds to form chain-like arrays **88** and **89** [245].

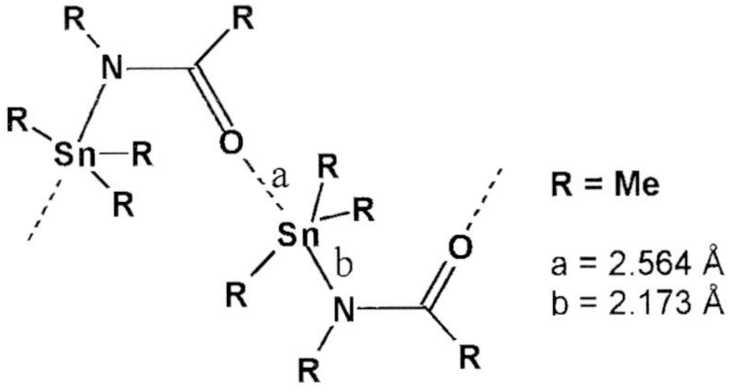

88
R = Me
a = 2.564 Å
b = 2.173 Å

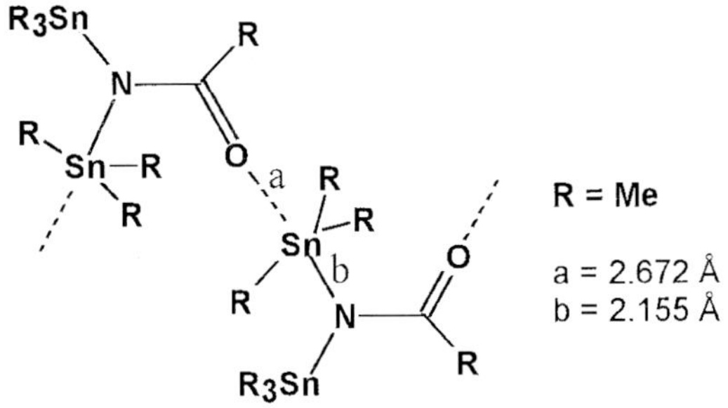

89
R = Me
a = 2.672 Å
b = 2.155 Å

Tributyl- and triphenyltin esters of 2-benzylbenzoic acid, **90** and **91**, are self-assembled supramolecular chains formed by Sn···O secondary bonding (2.675 and 2.880 Å, respectively) [246].

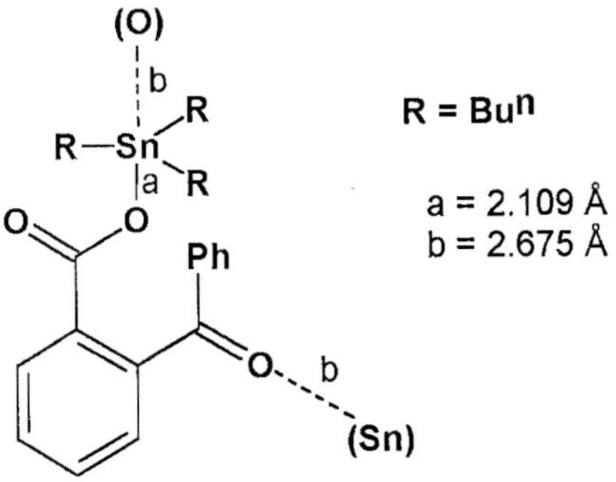

90
R = Bu^n
a = 2.109 Å
b = 2.675 Å

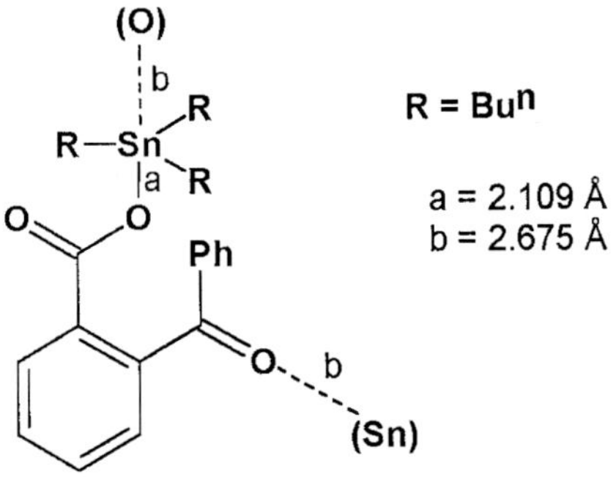

91
R = Bu^n
a = 2.109 Å
b = 2.675 Å

Bis(trimethylstannyl)-2,2′-bipyridyl-4,4′-dicarboxylate forms a two-dimensional layered supramolecular structure with normal short (2.145 Å) and (secondary) long (2.519 Å) bonds [247].

Diorganotin derivatives of pyridinedicarboxylic acid are also associated via Sn···O secondary bonds. Thus, aquadimethyl(2,6-pyridinedicarboxylato)tin and the dibutyltin derivative, **92**, and dimethyltin *N*-methyliminodiacetate, **93**, are supramolecular dimers formed via Sn···O secondary bonding, whereas bis(dimethyltin)-ethylenediamine tetraacetate is a polymer [248, 249].

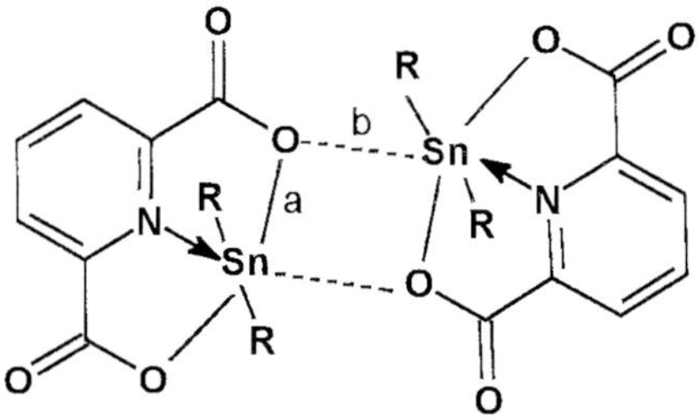

R = Me   R = Bu^n

a = 2.471 Å   a = 2.422 Å
b = 2.593 Å   b = 2.783 Å

92

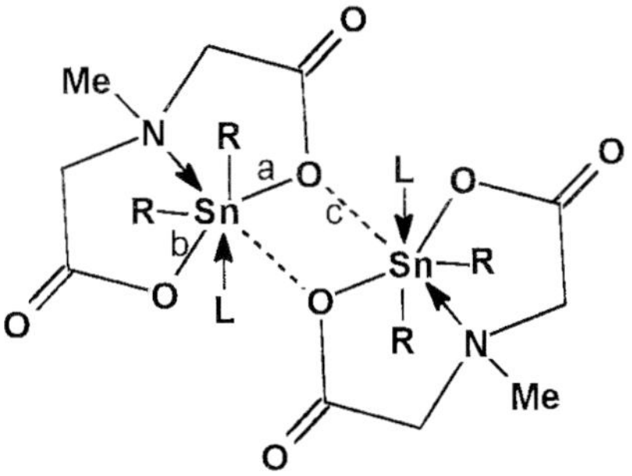

R = Me, L = H$_2$O

a = 2.353 Å
b = 2.188 Å
c = 2.735 Å

93

[*trans*-Dimethyl-*cis*-dichloro-bis{*meso*-1,2-diphenyl- and dipropylsulfinyl)ethane]-tin forms chain-like arrays, **94**, via (S=)O···Sn secondary bonds (R = Ph, 2.397 and 2.404 Å; R = Pr 2.316 and 2.320 Å) [250, 250a].

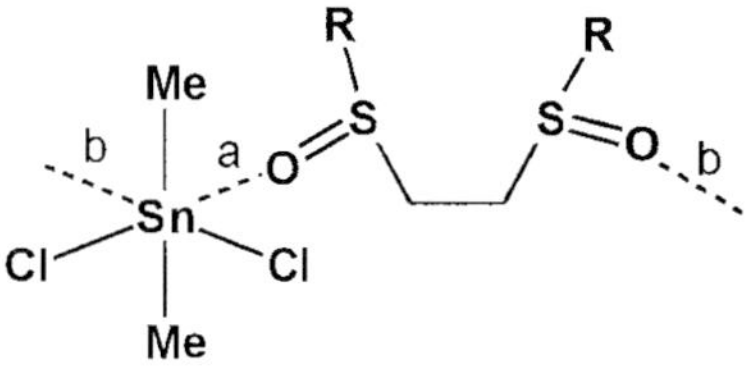

R = Ph

a = 2.397 Å
b = 2.404 Å

94

2-*N*-Triphenylstannyl-5-phenylisothiazol-3(2*H*)-one 1,1-dioxide, **95**, and 2-triphenylstannyl-4,5-tetramethyleneisothiazol-3(2*H*)-one 1,1-dioxide, **96**, are self-assembled via (S=)O···Sn secondary bonds [251].

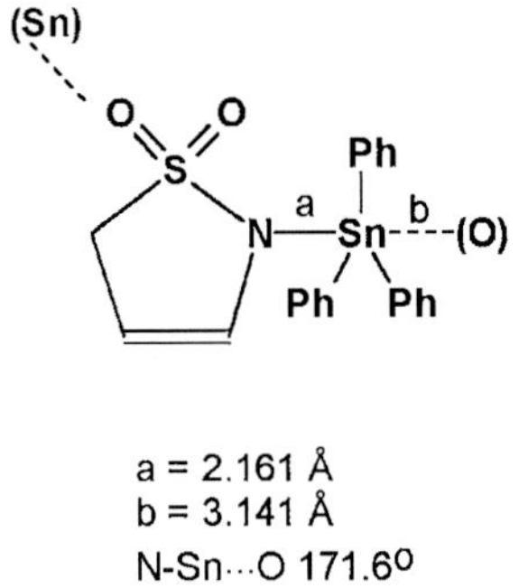

95
a = 2.161 Å
b = 3.141 Å
N-Sn···O 171.6°

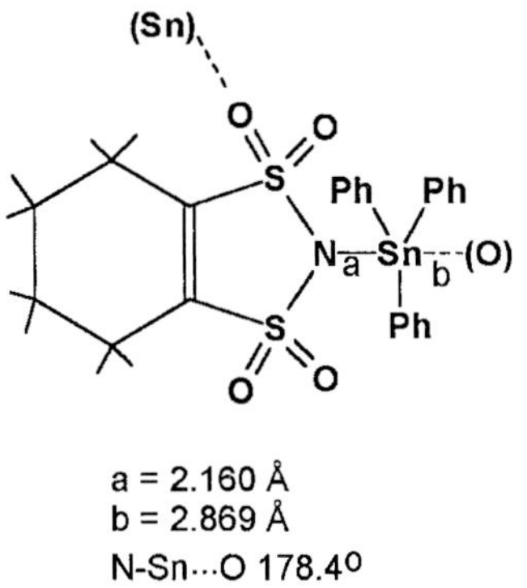

96
a = 2.160 Å
b = 2.869 Å
N-Sn···O 178.4°

Tin-bonded phosphoryl oxygens (P=)O···Sn also tend to form longer bonds and the self-assembly of the tetraphenyldiphosphinoethane dioxide adduct of dibutyltin dichloride, **97**, is an example of secondary bond association (Sn···O 2.386 and 2.640 Å) of this type [252]. Similarly, in O,O-diethylphosphonoacetato triphenyltin there is an Sn···O(=P) bond (2.420 Å) much longer than the carboxylic Sn–O(C) bond (2.129 Å) [253].

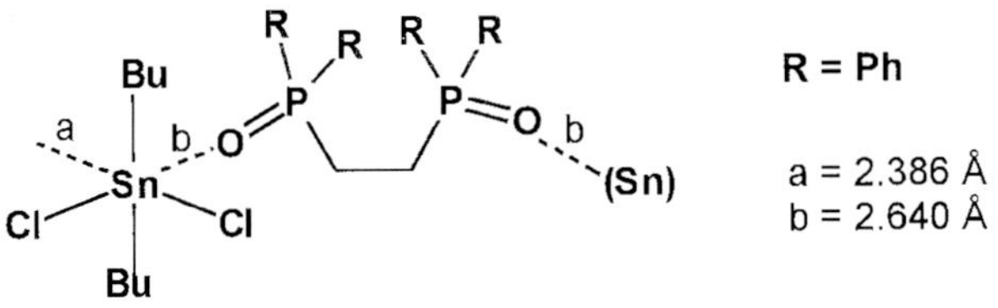

97
R = Ph
a = 2.386 Å
b = 2.640 Å

Trichloro(4-acetoxybutyl)tin, $CH_3COO(CH_2)_4SnCl_3$, forms a macrocyclic dimer by O···Sn donation (2.463 Å) [254] and ω-hydroxyalkyltrichlorostannanes, $HO(CH_2)_nSnCl_3$ ($n = 3$–5) self-assemble into chains by O–Sn donation (Sn–O 2.447 Å) in the solid state, whereas in solution they degrade to monomers [255].

A long O···Sn bond (2.822 Å) indicating secondary interaction, has been observed in a self-assembled trimethyltin derivative, **98**, of a sulfur–nitrogen ring [256].

4.4.1.3 Organotin–sulfur compounds

Self-assembly via secondary Sn···S bonding occurs in several sulfur-containing organotin compounds, but it is not a general phenomenon.

Diorganodithiastannolanes, $R_2Sn(SCH_2CH_2)_2$, illustrate the process and the

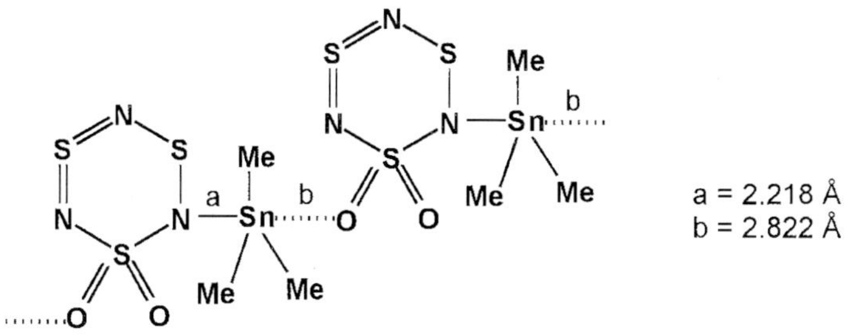

98

a = 2.218 Å
b = 2.822 Å

steric influences of the organic groups attached to tin. For example, 2,2-dimethyl-1,3,2-dithiastannolane molecules are associated in infinite arrays, **99**, by secondary Sn···S bonds (3.181 Å). This seems to relieve the strain in the five-membered ring, because the tin atom becomes five-coordinate with more favorable bond angles. Only one sulfur atom of the ring participates in the intermolecular bonding; the second is at a distance of 4.70 Å from the tin atom of the adjacent molecule and is therefore not involved [257, 258].

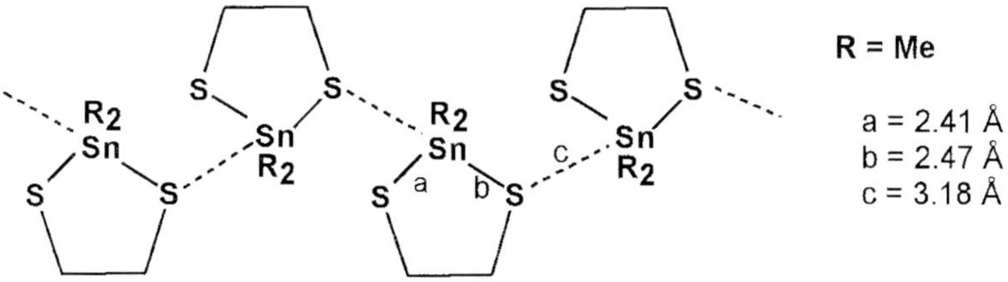

99

R = Me
a = 2.41 Å
b = 2.47 Å
c = 3.18 Å

The analogous di-n-butyl derivative, **100**, is also associated, but the Sn···S secondary bonds are weaker (3.688 Å) and the intermolecular connections are different – the five membered rings are doubly connected via secondary bonds and tin becomes six-coordinate. In the ring the Sn–S bonds have the normal lengths (2.414 Å) expected for single bonds. In this example both sulfur atoms participate in the

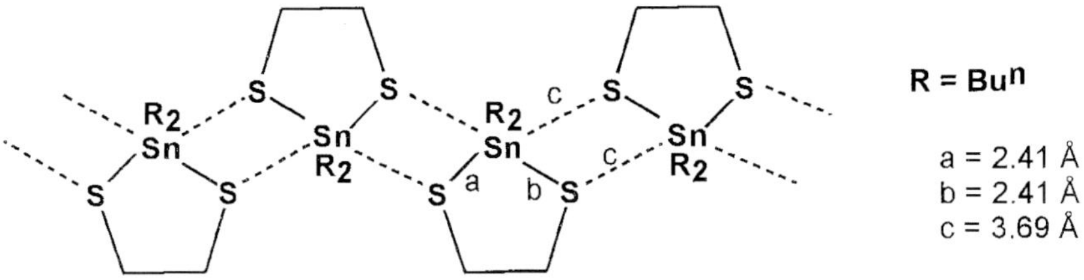

100

R = Bu^n
a = 2.41 Å
b = 2.41 Å
c = 3.69 Å

intermolecular association. Comparison of solution and solid-state ^{13}C and ^{119}Sn NMR spectra show that the ethyl and isopropyl analogs are also similarly associated in the solid state. All are monomeric in solution [259]. The *tert*-butyl derivative is a monomeric molecule (Sn–S 2.411 Å) both in solution and in the crystalline state [260], showing that steric effects play an important role in this sort of intermolecular self-organization.

Spiro-*bis*(ethanedithiolate), $(CH_2CH_2S)_2Sn$, is also associated, **101**, via pairs of Sn$\cdots$S bonds (Sn$\cdots$S 3.764 and 3.811 Å compared with Sn–S 2.405 and 2.388 Å) [261] and the Mössbauer spectrum of the compound also indicates sixfold coordination at tin [262]. 2,2-Dialkyl-1,3,2-oxathiastannolanes are associated only through S$\cdots$O bonds (see Section 3.3.1) illustrating the stronger Lewis base character of oxygen towards tin [263].

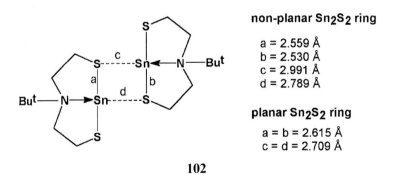

a = 2.405 Å
b = 2.388 Å
c = 3.764 Å
d = 3.811 Å

101

An interesting example is provided by 5-*tert*-butyl-5-aza-2,8-dithia-1-stann(II)-ocane, which is associated into dimeric supermolecules, **102**, in the solid state. The unit cell contains four asymmetric dimers, with non-planar Sn_2S_2 four-membered quasi-rings, and two centrosymmetric dimers, with planar four-membered Sn_2S_2 quasi-rings. The interatomic distances are shown in the structure diagrams [264].

non-planar Sn_2S_2 ring

a = 2.559 Å
b = 2.530 Å
c = 2.991 Å
d = 2.789 Å

planar Sn_2S_2 ring

a = b = 2.615 Å
c = d = 2.709 Å

102

Self-assembly by Sn$\cdots$S secondary bonding is observed in trithiastannocanes, which are loosely associated to form chain-like arrays, **103** (Sn···S 3.68 and 4.10 Å) [264a].

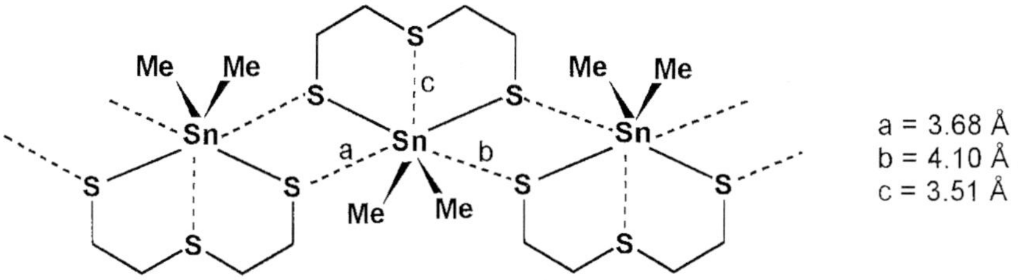

a = 3.68 Å
b = 4.10 Å
c = 3.51 Å

103

Similar self-assembly is observed in (not containing Sn–C bonds) tin(II) derivative of a phosphine thiol, **104** [264b].

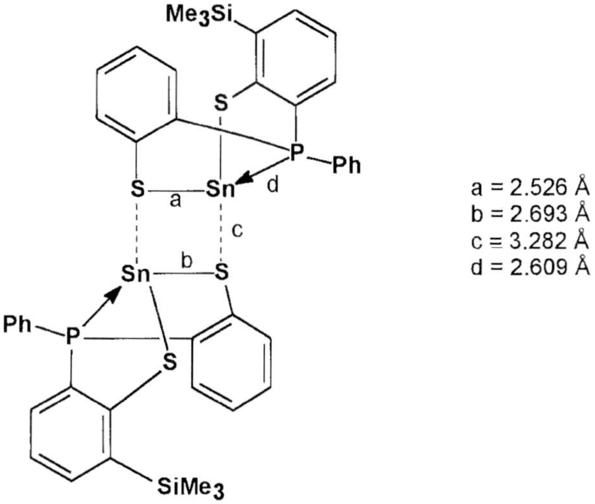

a = 2.526 Å
b = 2.693 Å
c = 3.282 Å
d = 2.609 Å

104

Unexpected supramolecular self-organization was observed in trimethyltin dithiotetraphenylimidodiphosphinate, Me$_3$Sn(SPh$_2$PNPPh$_2$S), **105** [265]. The ligand, which normally closes six-membered chelate rings, **106** (e.g. with dimethyltin coordination centers [266]), in this compound forms bridges connecting the trimethyltin groups in a helical polymer. The coordination geometry around the metal is trigonal bipyramidal and the tin–sulfur interatomic distances clearly show primary (Sn–S 2.517 Å) and secondary (Sn···S 3.627 Å) bond character.

Trimethyltin(IV) isothiocyanate, Me$_3$SnNCS, also forms infinite zigzag chains,

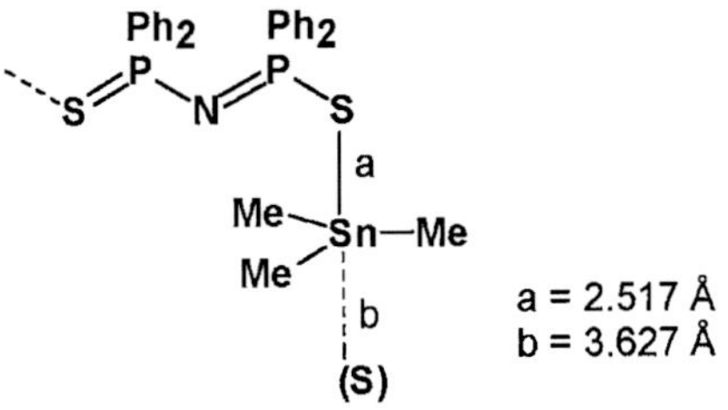
Ph2 Ph2
P N P S
a
Me Sn Me
Me
b
(S)
a = 2.517 Å
b = 3.627 Å

105

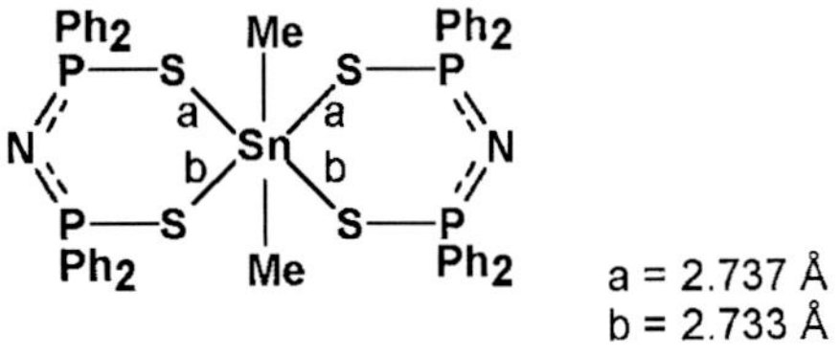
Ph2 Me Ph2
Sn
Ph2 Me Ph2
a = 2.737 Å
b = 2.733 Å

106

107a, with primary bonds apparently to nitrogen (Sn–N 2.15 Å) and secondary bonds to sulfur (Sn···S 3.13 Å). Tin is again five-coordinate and the zigzag chain is linear along the S–Sn–S–N=C=S sequence, with the bending occurring at sulfur, with a C–S···Sn angle of 97° [267]. Triphenyltin(IV) isothiocyanate, Ph_3SnNCS, exists in monoclinic [268] and orthorhombic [269] forms. The supramolecular chain-like structure, **107b**, is very similar to that of the trimethyltin derivative (Sn–N 2.22 Å, Sn···S 2.92 Å, C–S···Sn 97° in the monoclinic form, and Sn–N 2.26 Å, Sn···S 2.904 Å, C–S···Sn 107.6° in the orthorhombic form).

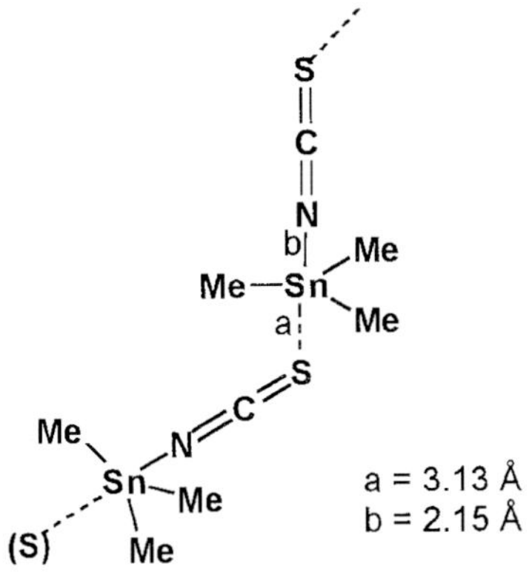
a = 3.13 Å
b = 2.15 Å

107a

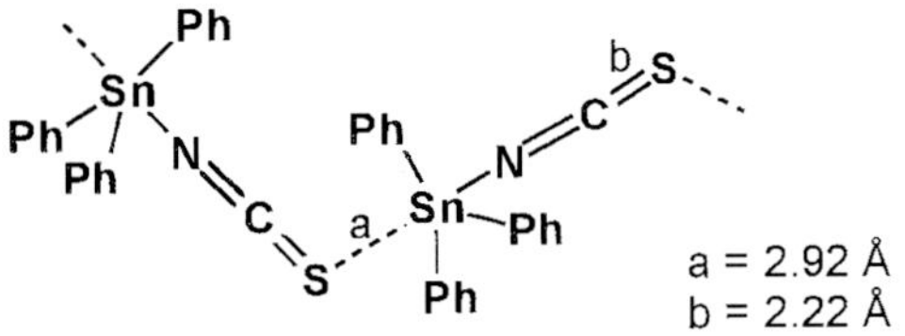
a = 2.92 Å
b = 2.22 Å

107b

In dimethyltin(IV) diisocyanate, $Me_2Sn(NCS)_2$, the primary bonds are between tin and nitrogen and self-assembly occurs via Sn···S interactions (Sn–N 2.139 Å, Sn···S 3.202 Å) with formation of infinite chains connected in two-dimensional layers, **108** [270, 271].

In view of the considerations described above it is surprising that triphenyltin methylthiolate, Ph_3SnSMe, is non-associated and contains discrete monomeric molecules in the crystal [272].

Rather strong intermolecular Sn···S secondary interactions occur in the helical supramolecular chains found in $[Ph_3SnO(S)PPh_2]_x$ (2.785 Å) [273] and in

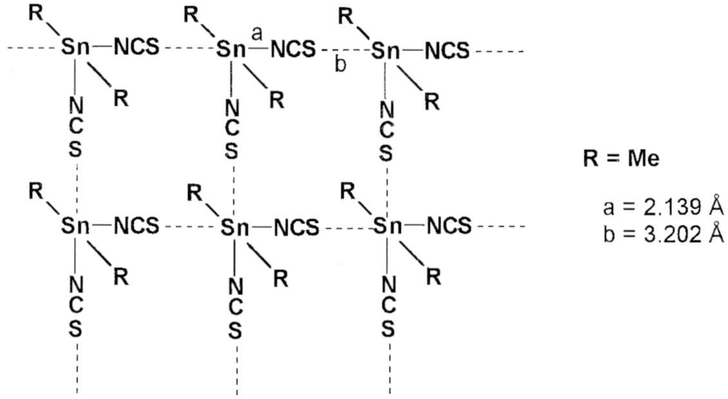

[diagram of structure **108** showing a 2D grid of Sn—NCS units with R groups, with a = 2.139 Å, b = 3.202 Å, R = Me]

$[Me_3SnO(S)PMe_2]_x$ (2.737 Å) [274] whereas in $[Me_2Sn(OSPPh_2)_2]_x$ the intermolecular secondary bonds Sn···S are weaker (3.901 Å) [275].

4.4.1.4 Organotin–nitrogen compounds

Tin–nitrogen bond lengths cover a large range of values. The shortest are close to the sum of the covalent radii (ca. 2.09 Å) and are observed in cyclodi- and -tristannazanes. Primary Sn–N bonds (as found in a cyclostannazane wrapped in *tert*-butyl groups which prevent any intermolecular interactions and make self-assembly impossible), are much shorter. Thus, the Sn–N bonds are 2.030 Å in $(Bu^t_2SnNH)_3$, **109**, and 2.05 Å in $(Bu^t_2SnNBu^t)_2$, **110**; they are 2.085 and 2.092 Å in $(Bu^t_2SnNSO_2Me)_2$ [276].

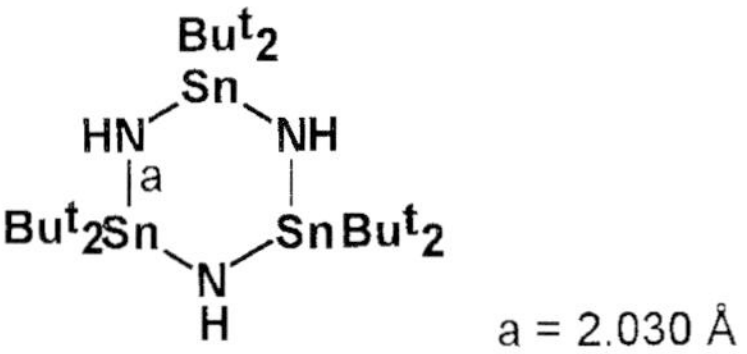

[structures **109** (six-membered Sn₃N₃ ring with Bu^t₂ on Sn and H on N, a = 2.030 Å) and **110** (four-membered Sn₂N₂ ring with Bu^t₂ on Sn and Bu^t on N, a = 2.05 Å)]

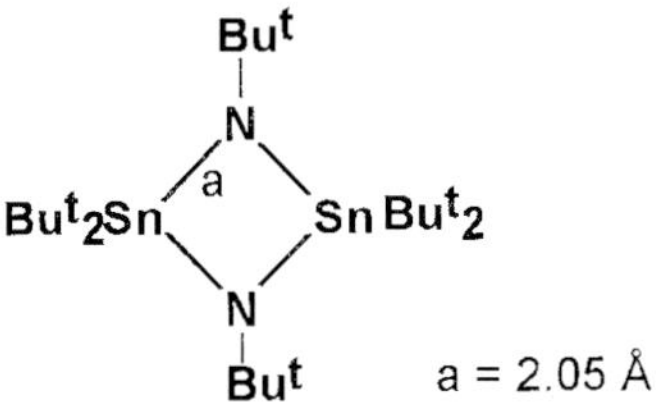

Much longer tin–nitrogen *intra*molecular distances have been observed in some internally chelated organotin derivatives of nitrogen-containing donors. It is perhaps useful at this point to cite some structures of organotin compounds containing *intra*molecular tin–nitrogen bonds, for comparison with the interatomic distances mentioned in the previous normal covalent compounds and with the forthcoming supramolecular self-assembled arrays. Thus, in pyridyl- and oxazolinyl-substituted stannylated thiophenes **111a** and **111b** the *intra*molecular tin–nitrogen secondary bonds are 2.720 Å in the pyridyl derivative and 2.580 and 2.525 Å (two independent molecules) in the oxazolinyl derivative [277]. In [2-(4,4-dimethyl-2-oxazolinyl)-phenyl]dimethyltin bromide (Sn···N 2.414 Å) and bis[3-(2-pyridyl)-2-thienyl]-diphenyltin (Sn···N 2.580 Å) the secondary Sn···N bonds are shorter [278].

Tin–nitrogen dative bonds in the range 2.511–2.56 Å were measured in some internally chelated organotin derivatives containing *ortho*-dimethylaminomethylene donor groups, **112a** and **112b** [279, 280].

Experimentally measured tin–nitrogen intermolecular distances cover a range from values rather close to those of primary bonds (ca. 2.2 Å, probably dative bonds) to very large values, close to 2.8 Å (secondary bonds). The sum of the covalent bond radii for the Sn–N pair is 2.092 Å and the sum of the van der Waals radii is 3.65 Å. Distinguishing between dative and secondary tin–nitrogen bonds is difficult and for convenience all supramolecular assemblies based upon tin–nitrogen interactions will be discussed in this section, with the understanding that shorter Sn–N bonds should perhaps be regarded as dative.

Triorganotin derivatives

Trimethyltin(IV) azide, Me$_3$SnN$_3$ [281], is a linear polymer, **113**, consisting of planar SnMe$_3$ groups linked in zigzag chains by nitrogen atoms of the azido moiety, with only one nitrogen participating, giving trigonal bipyramidal coordination geometry at tin. The two tin–nitrogen bonds are equal (2.386 Å) and no distinction can be made between primary and secondary interactions, but the bond lengths measured are longer than those of covalent Sn–N bonds. Thus, one effect of bond symmetrization (equalization) occurs here, as in some other bridged compounds. The sterically crowded tri-*tert*-butyl azide, Bu$^t_3$SnN$_3$, is a monomer with Sn–N 2.101 Å; the diorganotin compound Bu$^t_2$Sn(N$_3$)$_2$ (Sn–N 2.141 and 2.156 Å) is also monomeric in the crystal [281a].

113

Trimethyltin(IV) cyanide, Me$_3$SnCN, has a similar chain structure, **114**, with trigonal bipyramidal tin and equalized Sn–C and Sn–N bond distances; this leads to the interpretation that "this structure is best described as an arrangement of Me$_3$Sn$^+$ and CN$^-$ ions with the CN ordered in direction along an axis, but disordered in orientation along this axis, with perhaps a small amount of covalent character to account for the orientation of the CN group" [282]. Triethyltin(IV) cyanide, Et$_3$SnCN, has a similar structure [283].

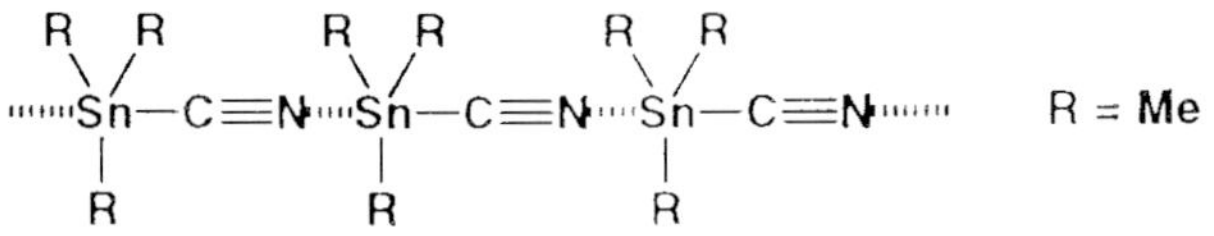

$$\cdots\text{Sn}\overset{R\ R}{\underset{R}{\diagup}}\text{—C}\equiv\text{N}\cdots\text{Sn}\overset{R\ R}{\underset{R}{\diagup}}\text{—C}\equiv\text{N}\cdots\text{Sn}\overset{R\ R}{\underset{R}{\diagup}}\text{—C}\equiv\text{N}\cdots \qquad R = Me$$

114

Trimethylgermanium(IV) cyanide, Me$_3$GeCN, mentioned here for comparison, is monomeric and unassociated, because the shortest intermolecular contact between germanium and nitrogen (3.57 Å) is the value expected for the van der Waals distance [284].

Bis(trimethyltin(IV)) carbodiimide, Me$_3$SnNCNSnMe$_3$, forms an infinite helical network, **115**, of planar SnMe$_3$ groups linked by linear N=C=N units (Sn···N 2.47 Å) [285]. The observed tin–nitrogen distance suggests that this results from secondary-bond association between SnMe$_3^+$ groups and NCN$^-$ anions.

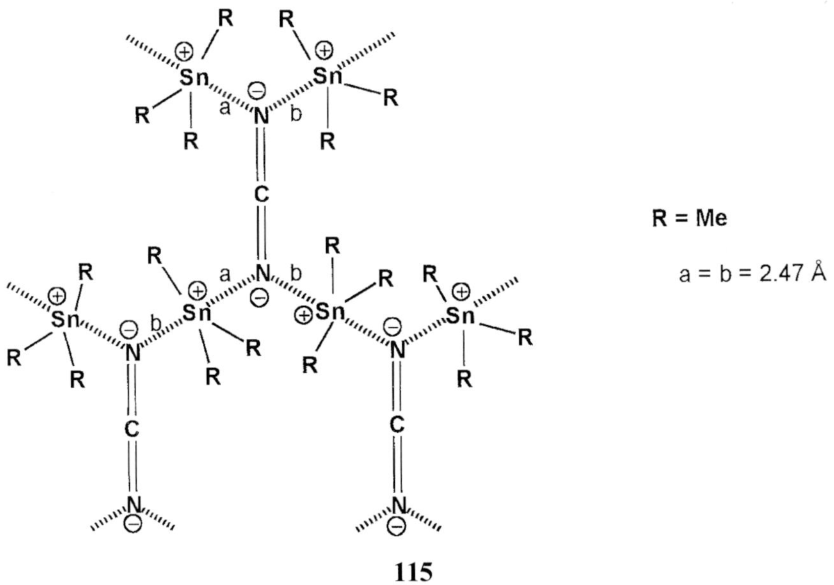

R = Me

a = b = 2.47 Å

115

Trimethyltin(IV) dicyanamide, Me$_3$Sn[N(CN)$_2$], and dimethyltin(IV) bis-(dicyanamide), Me$_2$Sn[N(CN)$_2$]$_2$, are also self-organized into supramolecular arrays [286]. The trimethyltin derivative contains Sn···NCNCN··· chains, **116**, with five-coordinate, trigonal bipyramidal tin), whereas the dimethyltin compound forms an infinite two-dimensional layer, **117**, of six-coordinate octahedral tin atoms bridged by NCNCN groups. The tin–nitrogen bonds are longer than calculated single-covalent bond distances, which led to the interpretation that this is an ionic bond. It is probably better to consider them as secondary bonds between the tin atoms and the nitrogen sites of the anionic ligands.

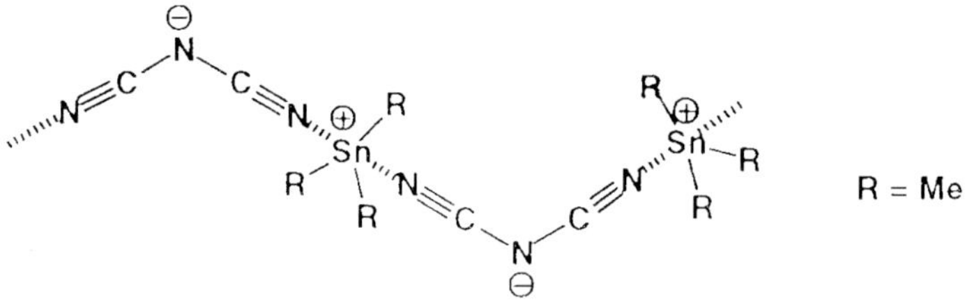

R = Me

116

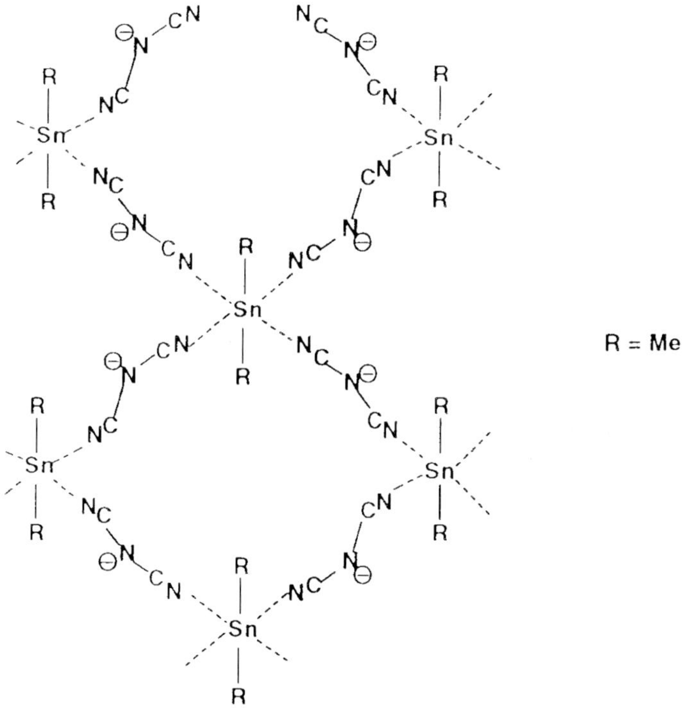

R = Me

117

N-Trimethylstannyl)-*N*-nitromethylamine, Me$_3$SnN(Me)NO$_2$, contains both Sn···N (2.33 Å) and Sn···O (2.39 Å) bonds in a zigzag chain, **118**, containing five-coordinate, trigonal bipyramidal tin [287].

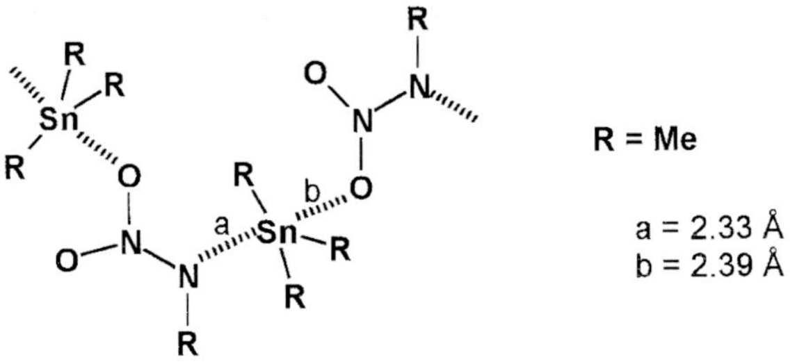

118

R = Me

a = 2.33 Å
b = 2.39 Å

Poly[(μ-4,4′-bipyridine)trimethyltin]dimesylamide (acetonitrile solvate), Me₃SnN-(SO₂Me)₂·4,4′-bipy, forms infinite chains, **119**, as a result of 4,4′-bipyridine bridging, leading to the formation of a polymeric cation with the anions non-coordinated. The Sn···N bonds (2.411 and 2.420 Å) are axial in the trigonal bipyramidal coordination at tin (N···Sn···N 176.8°) [288].

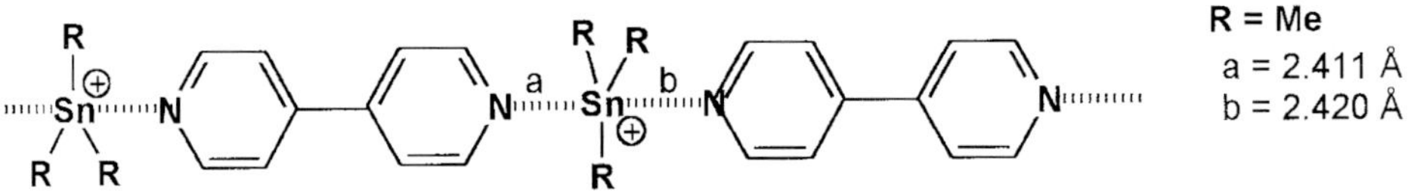

119

R = Me
a = 2.411 Å
b = 2.420 Å

Supramolecular association of triorganotin imidazoles and triazoles is inferred from vibrational spectra [288a] and supported by Mössbauer spectra [288b].

Triorganotin triazole derivatives, e.g. tricyclohexyltin-1,2,4-triazole (a commercial fungicide known under the trade name of Peropal) [289] and tribenzyltin-1,2,4-triazole [290] are supramolecular chain-like arrays, **120**, formed as a result of tin–nitrogen intermolecular interactions and containing trigonal bipyramidal five-coordinate tin with the nitrogen in axial positions. Other organotin triazoles, e.g. 1-(halodiphenylstannylmethyl) derivatives [290a], were found to be dimeric or tetrameric.

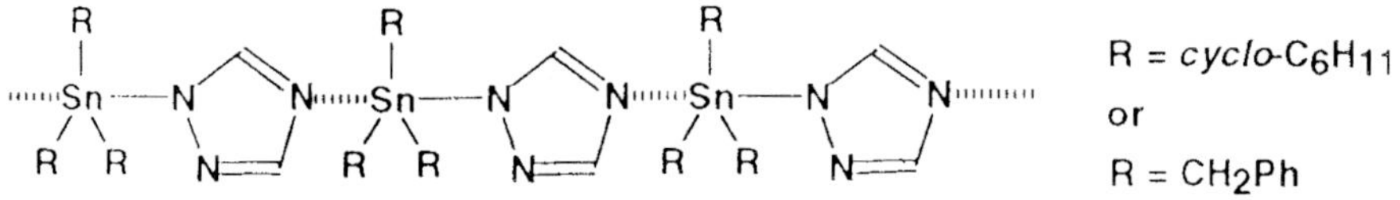

120

R = cyclo-C₆H₁₁
or
R = CH₂Ph

Trimethyltin(IV) 1-phenyl-1H-tetrazole-5-thiolate, Me₃SnSCN₄Ph, is self-assembled into trimeric supermolecules, **121**. The primary bond is between tin and sulfur (Sn–S 2.565 Å), and there is an intermolecular shorter Sn···N secondary bond (2.747 Å) and a longer chelating Sn···N secondary bond (3.285 Å) between the nitrogen heteroatom and tin [291, 292]. The tributyltin analog, Bu^{n}₃SnSCN₄Ph, forms a linear supramolecular array, **122** (Sn···N 2.559 and 2.676 Å) [292a].

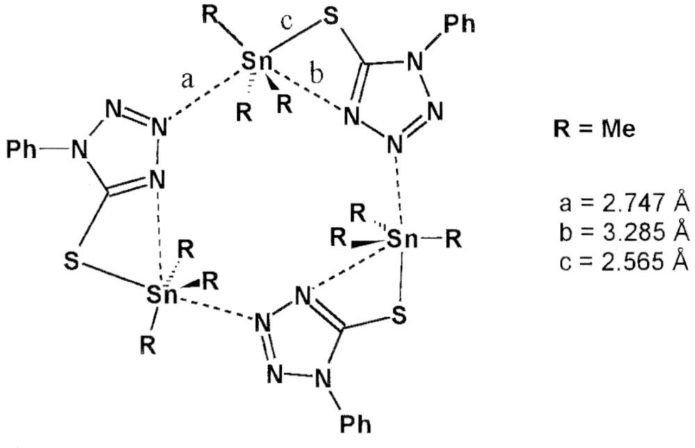

121

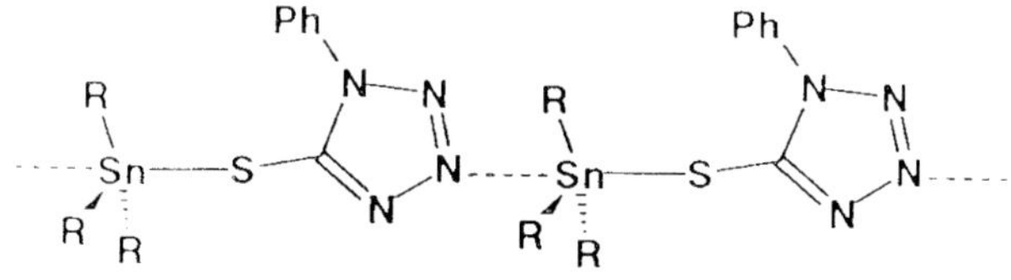

122

The triorganotin(IV) 1-phenyl-1H-tetrazole-5-thiolates, R₃SnSCN₄R, (R = Me, Bu, Ph, R' = Me, Ph) were found to be monomeric in solution, but spectroscopic data suggest that all are self-assembled in the solid state [293].

Fascinating supramolecular structures have been identified in the solid state of triorganotin tetrazoles [294–296]. Nitrogen–tin interactions between adjacent molecules lead to supramolecular self-assembly of triorganotin derivatives into helical chains, **123**, and of functionalized polytetrazoles, e.g. **124**, into two-dimensional layers and three-dimensional networks with large channels.

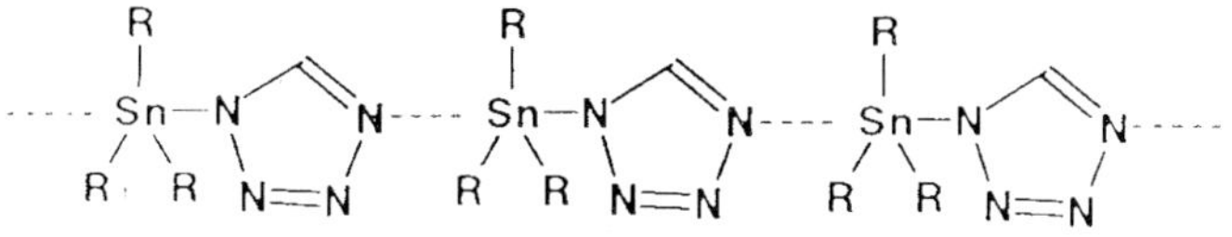

123

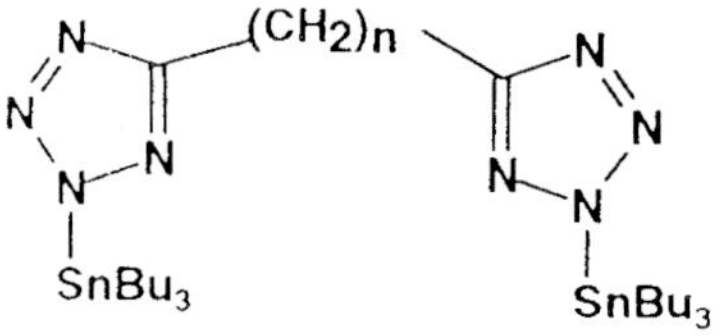

124

Diorganotin derivatives

Some diorganootin derivatives have already been mentioned above, in relation to their triorganotin counterparts.

Dimethyltin(IV) dicyanide, $Me_2Sn(CN)_2$, is self-organized in the solid state, occurring as two-dimensional layers, because of Sn···N interactions (2.68 Å); it is interesting to note that weak Si···N and Ge···N interactions were inferred from the structures of analogous dimethylsilicon and dimethylgermanium dicyanides (and in the dimethyllead analog with stronger Pb···N interactions). The Me_2Sn moieties are bent (148.7°) and the NC–Sn–CN angle is 85.3°, illustrating strong distortions from tetrahedral coordination as a result of the polymeric self-assembly [297].

Diorganotin(IV) dichloride pyrazine adducts are associated, as shown by spectroscopic data and X-ray analysis of the Ph_2SnCl_2–pyrazine adduct, **125** [298, 299]. The tin–nitrogen bonds, described in the original paper as "abnormally long" (2.961 and 2.783 Å) are, obviously, secondary Sn···N bonds. It is interesting to underscore that in the structure of the diphenyltin compound layers of zigzag polymeric chains $(Ph_2SnCl_2·pyz)_x$ (containing six-coordinate, octahedral tin) alternate with layers of molecules containing only five-coordinate tin, i.e. Ph_2SnCl_2·pyz·Ph_2SnCl_2. This leads to the overall composition Ph_2SnCl_2·0.75pyz. In the polymer the chlorine atoms are in *cis* positions, and this results in the bending of the chains at tin.

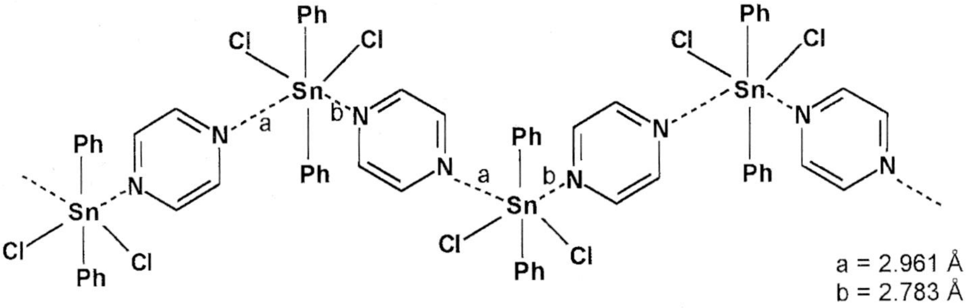

a = 2.961 Å
b = 2.783 Å

125

The dimeric association of dimethylcyclostannadithiadiazene, $Me_2SnS_2N_2$, **126**, involves shorter tin–nitrogen bonds than those in previous examples, but a difference can, however, be observed between intra- and intermolecular bonds (Sn–N 2.135 Å in the ring and Sn···N 2.316 Å between rings) [300]. Perhaps in this compound the bonds leading to dimerization could be described as electron pair dative bonds rather than secondary Sn···N bonds.

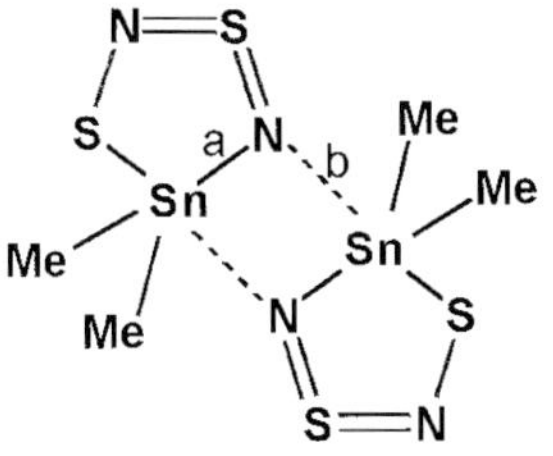

a = 2.135 Å
b = 2.316 Å

126

Some four- and six-membered tin–nitrogen–phosphorus inorganic heterocycles dimerize via rather strong tin–nitrogen bonds (2.333 Å in **127** and 2.241 Å in **128**) [301]. These bonds are probably best described as dative.

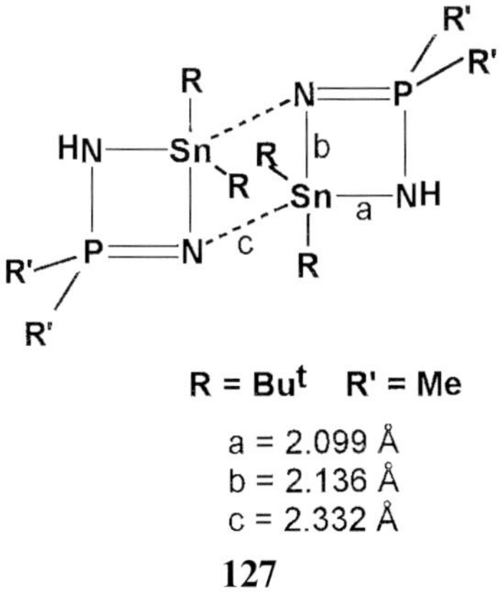

R = Bu^t R' = Me

a = 2.099 Å
b = 2.136 Å
c = 2.332 Å

127

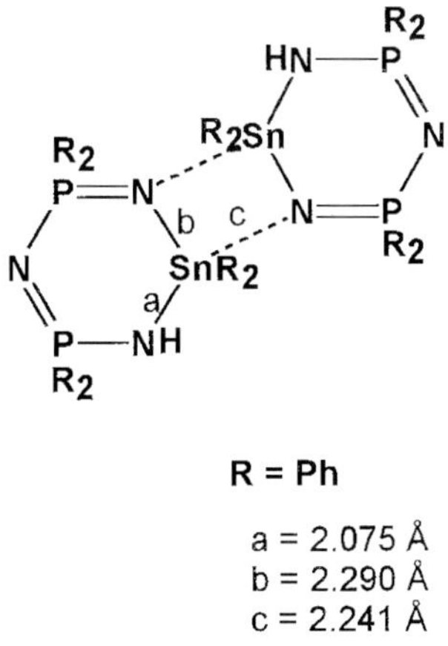

R = Ph

a = 2.075 Å
b = 2.290 Å
c = 2.241 Å

128

Triorganotin derivatives of amino acids are sometimes associated by both tin–oxygen bonding (like all carboxylates) and tin–nitrogen bonding. For example, trimethyltin glycinate, $Me_3SnOOCCH_2NH_2$, forms a complex supramolecular array, **129**, based upon Sn–O (2.21 Å) and Sn···N (2.46 Å) interactions, supplemented by additional C=O···HN hydrogen-bonds [302].

Similarly, triphenyltin nicotinate (3-pyridylcarboxylate), $Ph_3SnOOCC_6H_4N$, forms supramolecular chain-like arrays, **130**, with Sn–O 2.137 Å and Sn···N 2.568 Å [303].

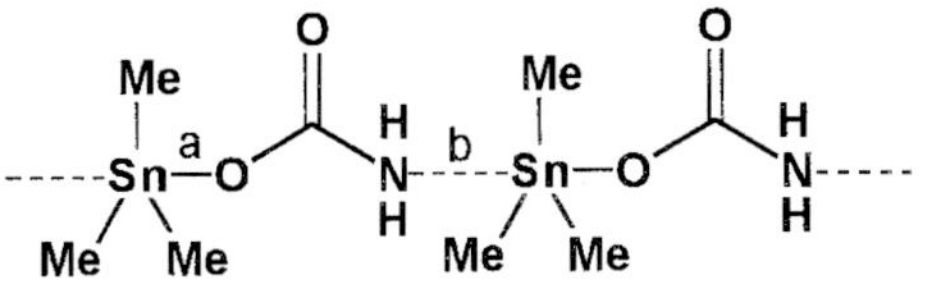

129

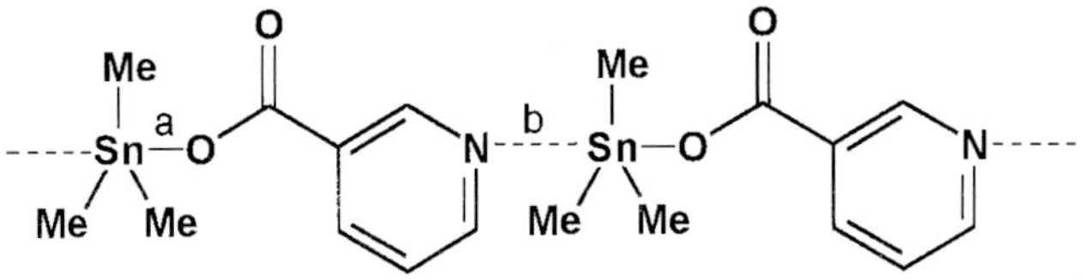

a = 2.21 Å
b = 2.46 Å

130

a = 2.137 Å
b = 2.568 Å

More similar examples can be found among the structures of other nitrogen-functionalized organotin carboxylates.

Organotin cyanometallates

A peculiar class of supramolecular materials involving Sn···N interactions include triorganotin and related cyanometallates, based upon a linear sequence, **131**, of metal centers alternating with cyanide bridges connected through organometallic moieties.

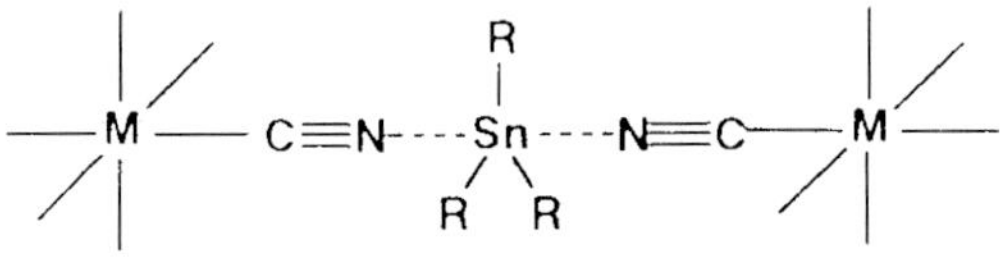

131

M is usually a six-coordinate metal and the resulting architecture is tridimensional with large channels, but four-coordinate metal centers have also been used. Thus, the planar [Ni(CN)$_4$] module leads to the formation of a macrocyclic system, **132** (R = Me, Sn–O 2.149 Å, Sn···N 2.529 Å) [304].

Extensive preparative, spectral, and structural studies have been conducted on compounds of the type [(R$_3$Sn)$_n$M(CN)$_6$]·xL (n = 3 or 4, M = Fe [305–322], Ru, Os [323–327], Co [328, 328a] (R = Me, Pr^n, Bu^n, Ph, L = H$_2$O, DMSO, etc.), [(R$_3$Sn)$_2$M(CN)$_4$] (M = Cu [329, 329a], [(R$_3$Sn)Au(CN)$_2$] [328a], Ni, Pd, Pt [328b, 330], R = Me, Et), [(Me$_3$Sn)Cu(CN)$_2$](0.5bipy) [330a], and [(R$_3$Sn)$_4$M(CN)$_8$] (M = Mo, W) [331]. The class can be extended by using thiocyanato ligands, as

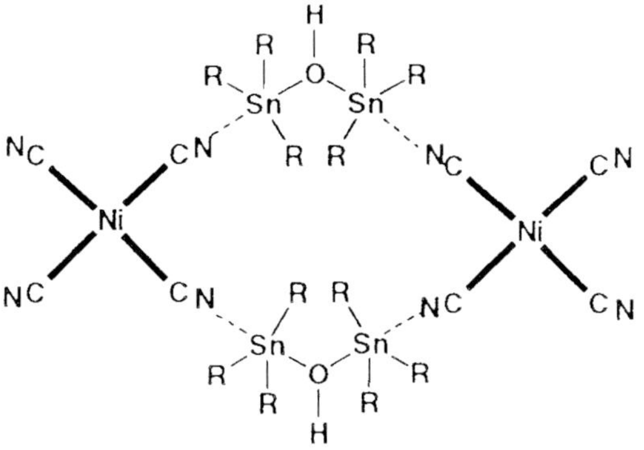

132

in [(R₃Sn)₃Rh(SCN)₆] [332]. These are interesting and promising materials for a variety of reasons. They may act as hosts for various organic and organometallic guests (e.g. ferrocene [305], and cobaltocenium cations [306, 318]) to form intercalation compounds, and some have semiconducting and ferromagnetic properties. Some of these materials have been structurally characterized by X-ray diffraction, and solid state ^{119}Sn and ^{13}C NMR and Mössbauer spectroscopy have also been widely applied for their study.

4.4.2 Self-assembly in organolead compounds

Compared with organotin compounds, organolead organometallics have been much less investigated by X-ray diffraction. There are, however, some well documented examples of supramolecular self-organization in organolead chemistry.

Several organolead halides have been structurally characterized. Trimethyllead chloride, Me₃PbCl forms zigzag chains with Pb–Cl 2.764 Å and Pb···Cl 2.814 Å. The coordination geometry of lead is distorted trigonal bipyramidal [333]. The small differences between lead–chlorine bond distances perhaps suggest dative-bond character. Trimethyllead iodide molecules, Me₃PbI, are self-organized in zigzag chains, **133**, with linear I–Pb–I moieties (176.25°) as parts of a trigonal bipyramid, and bent Pb–I···Pb fragments. The lead–iodine bonds are distinctly non-identical and primary (Pb–I 3.038 Å) and secondary (Pb···I 3.360 Å) bonds can be distinguished [334].

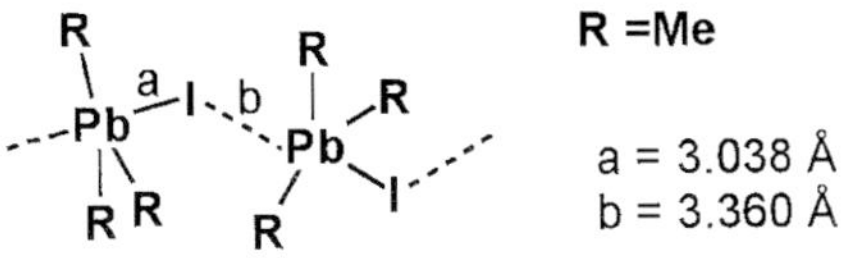

R = Me
a = 3.038 Å
b = 3.360 Å

133

Similar chain association has been also reported for triphenyllead chloride, Ph$_3$PbCl (Pb–Cl 2.706 Å, Pb···Cl 2.947 Å, Cl–Pb···Cl 179.5°), bromide, Ph$_3$PbBr (Pb–Br 2.852 Å, Pb···Br 3.106 Å, Br–Pb···Br 173.8°) [335], and benzyldiphenyllead bromide, BzPh$_2$PbBr (Pb–Br 2.885 Å, Pb···Br 2.985 Å, Br–Pb···Br 173.61°, Pb–Br···Pb 122.73°) [336]. In all these compounds the coordination geometry of lead is trigonal bipyramidal, with the halogen atoms in axial positions, and the chains are bent at the halogen. Five-coordination in triorganolead halides was also inferred from vibrational spectroscopy studies of trimethyllead [337] and triphenyllead halides [338, 339]. The Pb–Br bond lengths in the halogen-bridged polymer are somewhat longer than in monomeric organolead bromides, e.g. Pb–Br 2.754 Å in Ph$_3$PbBr·OPPh$_3$, in which trigonal bipyramidal geometry is achieved by coordination of triphenylphosphine oxide, also in the axial position [340].

Diphenyllead dichloride, Ph$_2$PbCl$_2$, is associated in doubly bridged chains, **134** (Pb–Cl 2.795 Å) and the metal is six-coordinate (distorted octahedral geometry). The phenyl groups are perpendicular to the plane containing the Pb and Cl atoms [341].

Few organolead(II) halides are known. Some with bulky organic groups have been isolated, and intermolecular association can occur. Thus, trimeric tris-(trimethylsilyl)methyllead(II) chloride, [(Me$_3$Si)$_3$CPbCl]$_3$ forms a six-membered ring supermolecule, **135**, in a distorted boat conformation with Pb–Cl in the range 2.71–2.74 Å [342].

Dimeric association through Pb···Cl secondary bonds has been found in ClPb[C(SiMe$_3$)$_2$CPhNSiMe$_3$] (primary Pb–Cl 2.609 Å, secondary Pb···Cl 3.276 Å) [343].

Cryoscopic measurements suggest that (1,3-di-*tert*-butylcyclopentadienyl)lead(II) tetrafluoroborate, (η^5-Bu$^t_2$C$_5$H$_3$)$_2$Pb[BF$_4$], is associated as a cyclic trimer, **136** [344], whereas an X-ray crystal structure determination shows that the (pentamethylcyclopentadienyl)lead(II) tetrafluoroborate, (η^5-C$_5$Me$_5$)Pb[BF$_4$], is associated into dimeric supermolecules, **137** [345]; association of both compounds occurs as a result of tetrafluoroborate bridging.

In the arene complex, (η^6-C$_6$H$_6$)Pb[AlCl$_4$]$_2$·C$_6$H$_6$, one tetrachloroaluminato group is chelating, the second is bridging, resulting in a self-organized chain structure, **138**. The Pb···Cl interatomic distances are longer in the chain (3.021–3.218 Å) than in the chelate ring (2.854 and 2.952 Å), and all are longer than expected from

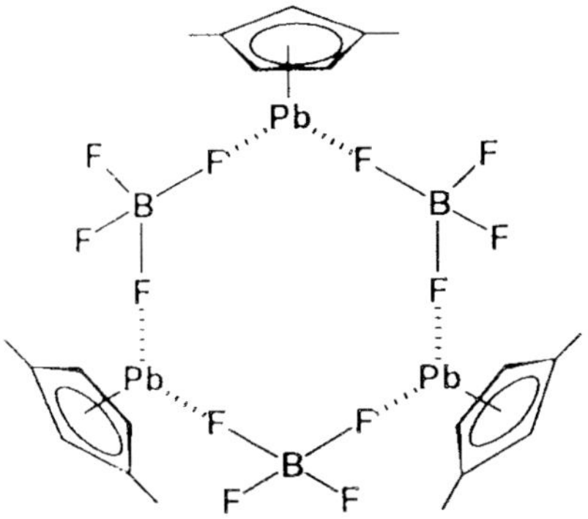

136 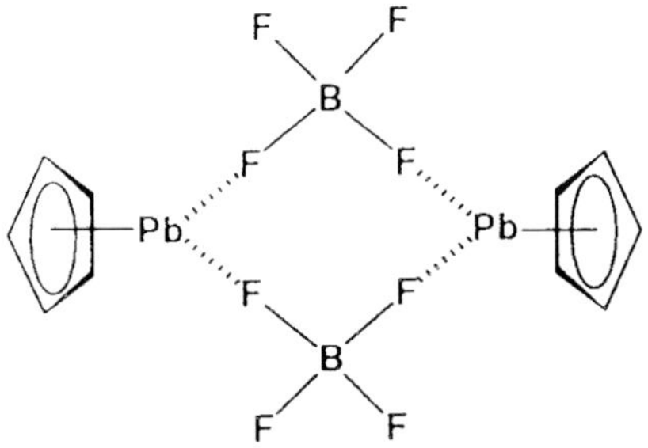 **137**

the sum of the covalent radii (2.50 Å) but significantly shorter than the sum of the van der Waals radii (4.0 Å) [346].

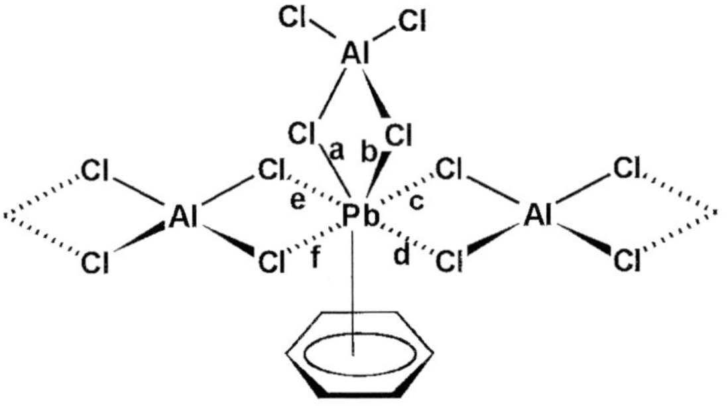

a = 2.854 Å
b = 2.952 Å
c = 3.021 Å
d = 3.084 Å
e = 3.218 Å
f = 3.111 Å

138

Triptycene forms [(C$_{20}$H$_{14}$)Pb(AlCl$_4$)$_2$·0.5C$_6$H$_6$] which also has a supramolecular structure [347].

Self-organization can be expected in organolead hydroxides, alkoxides, carboxylates, and other oxygen–ligand derivatives. Few have been structurally characterized. Triphenyllead(IV) hydroxide, Ph$_3$PbOH [348], forms a chain supramolecular array. Similarly, trimethyllead(IV) acetate, Me$_3$Pb(OAc), forms infinite zigzag chains, **139**, of planar Me$_3$Pb groups bridged by acetato groups, resulting in a trigonal bipyramidal coordination (Pb–O 2.327 and Pb···O 2.555 Å; O–Pb–O 169.7°). The compound is isostructural with trimethyltin acetate [349].

Trimethyllead(IV) furoate, Me$_3$Pb(OOCC$_4$H$_3$), is a similarly self-assembled zigzag polymer, with distorted trigonal bipyramidal geometry (Pb–O 2.353 and 2.534 Å, O–Pb–O 169.4°) and weak chelation [350].

The lead(II) derivative of a cobalt cluster carboxylate, [Pb{(CO)$_9$Co$_3$-(μ_3-CCOO)}$_2$], has an unusual supramolecular structure. It is made up of chains,

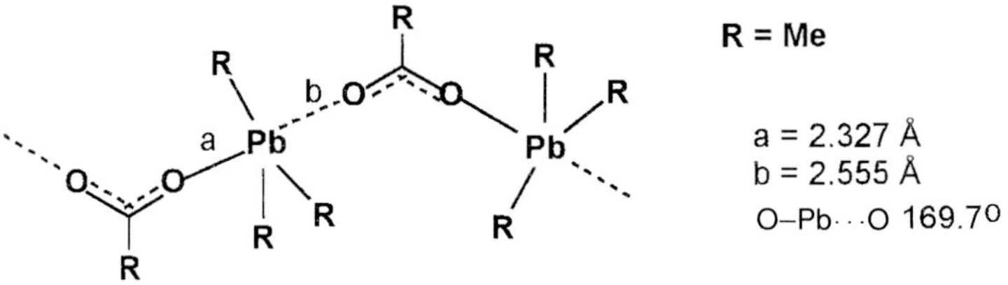

R = Me

a = 2.327 Å
b = 2.555 Å
O–Pb···O 169.7°

139

140, in which the lead atoms are doubly bridged by oxygens of the chelating carboxylate, with the cluster units arranged on both sides of the chain [351].

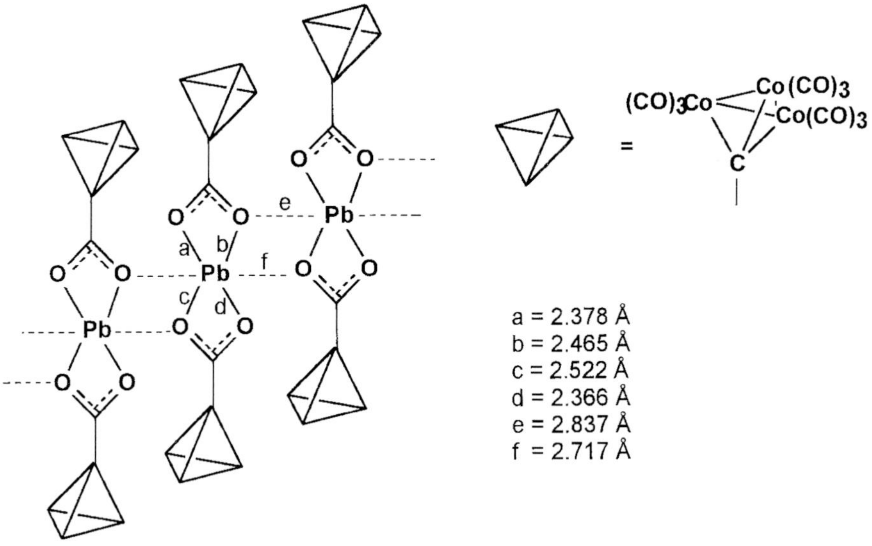

$(CO)_3Co$—$Co(CO)_3$
         $Co(CO)_3$
= C

a = 2.378 Å
b = 2.465 Å
c = 2.522 Å
d = 2.366 Å
e = 2.837 Å
f = 2.717 Å

140

The lead–oxygen chain is embedded into a 'glove' of cluster units exposing only carbonyl groups to the exterior. It is worth mentioning that one of the pyrolysis products of this material has catalytic activity in the hydrogenation of 2-butenal.

Diphenyllead(IV) diacetate, $Ph_2Pb(OAc)_2$, forms a polymeric chain with one acetato group bridging and another chelating and Pb–O bond lengths of 2.348 and 2.547 Å in the chain and 2.354 and 2.364 Å in the chelate rings [352]. A different type of self-assembly is observed for the monohydrate, $Ph_2Pb(OAc)_2 \cdot H_2O$. It contains two molecular tectons connected through a weak Pb···O bond (2.72 Å) and a water molecule, which participates by hydrogen-bonding, as shown in **141** [353]. Each lead atom is seven-coordinate, pentagonal bipyramidal, with five oxygens in the equatorial plane and the phenyl groups in axial positions.

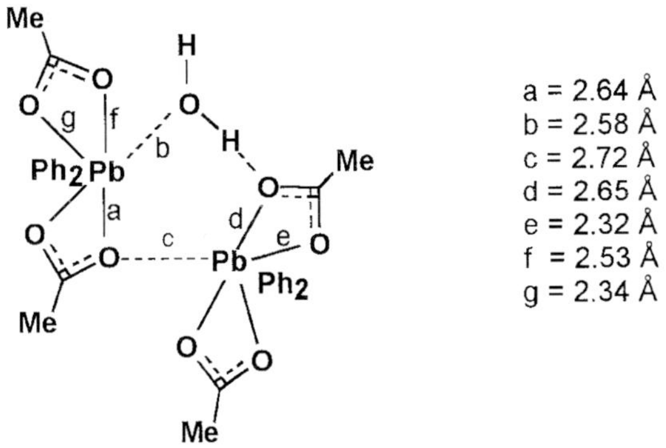

a = 2.64 Å
b = 2.58 Å
c = 2.72 Å
d = 2.65 Å
e = 2.32 Å
f = 2.53 Å
g = 2.34 Å

141

Triphenyllead 2-fluoro-4-nitrosophenolate is associated in chain-like supramolecular arrays, **142**, with secondary oxygen–lead bonds from the *p*-nitroso substituent [354].

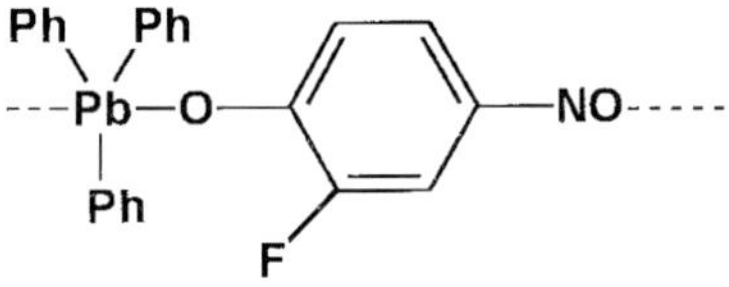

142

In the trimethyllead(IV) dimesylamide derivative, $Me_3PbN(SO_2Me)_2$, infinite chains are formed by Pb···O intermolecular contacts (Pb···O 2.653 Å) from an oxygen atom of the ligand. Weak chelating occurs from another oxygen. The compound contains a lead–nitrogen bond (Pb–N 2.484 Å) and the metal has distorted trigonal bipyramidal coordination geometry, with oxygen and nitrogen in axial positions (O–Pb–N 169.3°) [355]. A similar type of self-organization occurs in $[Me_3PbN(SO_2F)_2]_x$ (Pb···O 2.615 Å, Pb···N 2.603 Å) [355a].

Supramolecular self-assembly can be expected in organolead–sulfur compounds. X-Ray diffraction studies have revealed that triphenyllead methylthiolate, Ph_3PbSMe (Pb–S 2.489 Å) [356], and triphenyllead(IV) benzenethiolate, Ph_3PbSPh, (Pb–S 2.515 Å) are molecular monomers. Triphenyllead(IV) diethyldithiophosphate, $Ph_3PbS_2P(OEt)_2$, contains monodentate ligands (Pb–S 2.554 Å) and is also unassociated in solid state. Molecules of diphenyllead(IV) bis(dibenzyldithiophosphate), $Ph_2Pb[S_2P(OCH_2Ph)_2]_2$, on the other hand, self-assemble into dimers, by secondary lead–sulfur bonding (Pb···S 3.69 Å) making the metal seven-coordinate, with two phenyl groups in axial positions and five sulfur atoms in equatorial positions because of the anisobidentate character of the ligands [357]. Triphenyllead(IV) dimethyldithiophosphinate, $Ph_3PbS_2PMe_2$, forms a linear array, with bridging dithiophosphinato groups, **143**, linking the molecules via secondary Pb···S bonds [358].

$$\text{Structure 143:}$$

Ph Ph ... Pb ... S=P(Me)(Me)-S ... Pb(Ph)(Ph) ... S-P(Me)(Me)=S ... (a = 2.708 Å, b = 3.028 Å)

143

This is in agreement with the various modes of supramolecular self-organization in the solid state observed for all structurally characterized lead(II) dithiophosphates and dithiophosphinates (reviewed elsewhere [359,360]). The self-organization of inorganic lead(II) thio compounds can be illustrated by the three polymeric structures formed as a result of secondary Pb···S interactions – lead(II) bis(dimethyldithiophosphinate), $Pb(S_2PMe_2)_2$ [361], lead(II) bis(diphenyldithiophosphinate), $Pb(S_2PPh_2)_2$ [362], and lead(II) methylxanthate, $Pb(S_2COMe)_2$ [363]. In the dimethyldithiophosphinate, the molecules form eight-membered quasi-cyclic dimers, further assembled in infinite chains, whereas in the diphenyldithiophosphinate the pairs of molecules are also assembled in chairs, but in a different mode. In the xanthate yet another new mode of bridging is observed in the self-organized chain polymer.

Five-membered dithiaplumbolane, $(SCH_2CH_2S)PbPh_2$, is self-assembled into chains, **144**, via Pb···S secondary bonding (3.550 Å) [364, 365].

144 (a = 3.550 Å)

Dimeric self-assembly, **145**, is observed in the triclinic form of 5,5-diphenyl-1,4,6,5-trithiaplumbocane, the molecules of which are associated via secondary Pb···S interactions (Pb···S 3.75 Å) [366].

A number of organolead(IV) compounds are self-organized into chain structures via lead–nitrogen bonds. Such compounds include chain-like trimethyllead(IV) azide, Me_3PbN_3 [367, 368], triphenyllead(IV) azide, Ph_3PbN_3 (Pb···N 2.54 Å) [369], trimethyllead(IV) cyanide, Me_3PbCN [370], triphenyllead(IV)cyanate, Ph_3PbNCO (Pb–N 2.38 Å, Pb–O 2.65 Å) [371], and dimethyllead(IV) dicyanide,

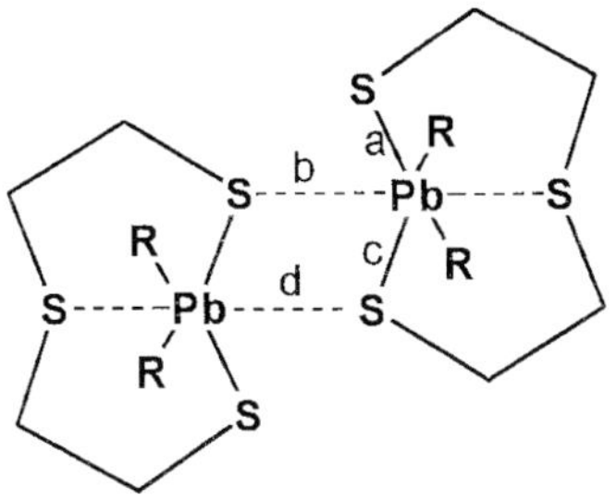

	R = Ph (two independent molecules)
	Molecule A Molecule B
	a = 2.514 Å a = 2.518 Å
	b = 3.750 Å b = 4.110 Å
	c = 2.519 Å c = 2.491 Å
	d = 3.750 Å d = 4.110 Å

145

Me$_2$Pb(CN)$_2$ [372]. Triphenyllead pyridine-4-thiolate, Ph$_3$PbSC$_5$H$_4$N, forms chains, **146**, via Pb···N secondary bonding [373].

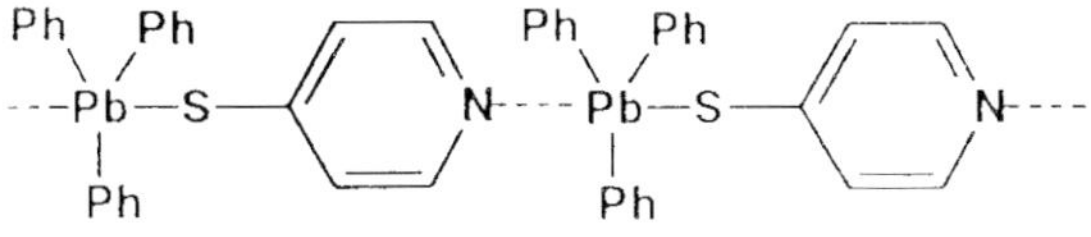

146

Tridimensional polymeric trimethyllead(IV) tetracyanometallates, [(Me$_3$Pb)$_2$·M(CN)$_4$] (M = Fe, Co), similar to the organotin analogs, have also been investigated [314, 374, 375]. In their structures they have extended channels which can serve as hosts for various molecules, e.g. pyridine.

More lead-containing supramolecular self-assembled structures with soft donors can be expected, but the interest in toxic organolead compounds is currently rather limited.

4.5 Group 15 – Arsenic, Antimony, Bismuth

4.5.1 Self-assembly in organoarsenic compounds

Supramolecular self-assembly frequently occurs in organoarsenic–sulfur compounds. Thus, chlorodithiaarsolidine, ClAsSCH$_2$CH$_2$S, forms dimers, **147a**, by secondary As···S bonding (3.430 Å); additional As···Cl secondary interactions (As···Cl 3.432 and 3.650 Å) connect the molecules in the crystal. Chloride-ion abstraction with AlCl$_3$ or GaCl$_3$ produces dithiaarsolidinium cations, which also self-assemble into dimers, **147b** and **147c**.

The As···S secondary bonds become shorter with halogen removal, i.e. the interactions between the cations are stronger. The As···S distance is reduced to

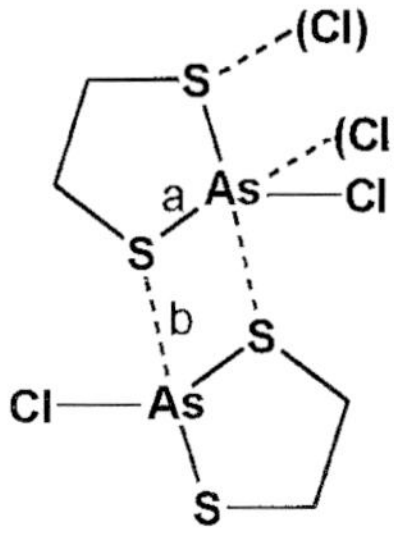

147a
a = 2.252 Å
b = 3.430 Å

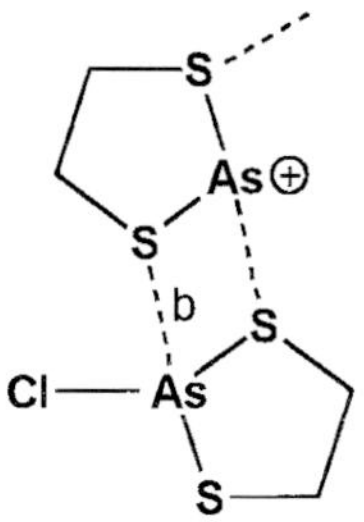

147b
b = 2.523 Å

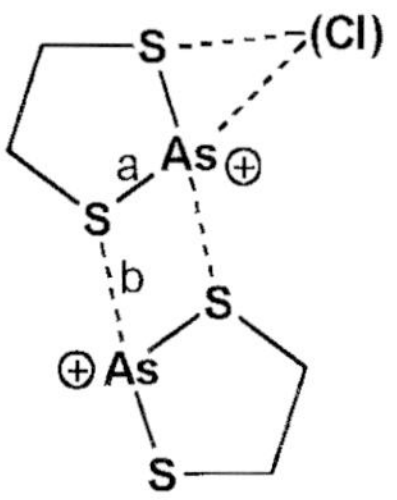

147c
a = 2.181 Å
b = 2.442 Å

2.523 Å in the monochloro dimer and to 2.442 Å in the unsubstituted dithiaarsolidinium dimer. Secondary bonds to chlorine atoms of the anions (As···Cl 3.180–3.772 Å and between sulfur and chlorine, S···Cl 3.437–3.637 Å) are also present in the crystal of dithiaarsolidinium tetrachlorogallate, **147**. Similar self-assembly occurs in the nitrogen analog, **148**, with As···N 2.103 Å [376].

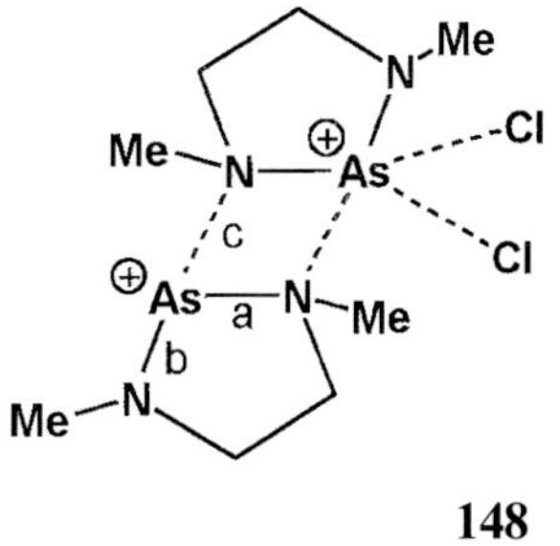

148
a = 1.955 Å
b = 1.763 Å
c = 2.103 Å

An unexpected self-assembly, **149**, with dimer formation, was observed in phenoxarsine dithiophosphinates, $O(C_6H_4)_2AsS_2PR_2$. With R = Ph [377] and R = Et [378] dimers, **149**, are formed by secondary As···S bonding (primary As–S 2.315 Å, intermolecular secondary As···S 3.402 Å, intramolecular secondary As···S 3.402 Å when R = Ph; primary As–S 2.318 Å, intermolecular secondary As···S 3.440 Å and intramolecular secondary As···S 3.505 Å when R = Et). The methyl derivative (R = Me) is, however, an unassociated monomer. The secondary As···S bonds observed in these compounds are longer (weaker) than the intramolecular secondary bonds measured in a phenylarsenic dithiophosphate, $PhAs[S_2P(OPr^i)_2]_2$, investigated earlier (primary As–S 2.310 and 2.317 Å; intramolecular secondary As···S 3.135 and 3.187 Å) [379].

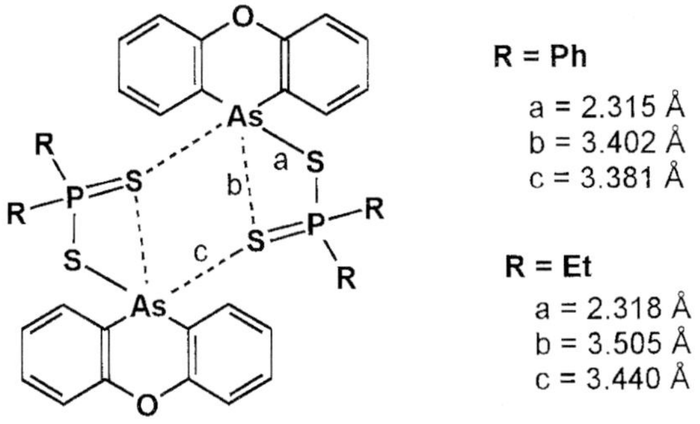

149

R = Ph
a = 2.315 Å
b = 3.402 Å
c = 3.381 Å

R = Et
a = 2.318 Å
b = 3.505 Å
c = 3.440 Å

In the solid state, the dinuclear anions [Me$_2$As$_2$Cl$_5$]$^-$ are self-assembled dimeric supermolecules, **150**, with As···Cl secondary bonds (3.126 and 3.365 Å) [380].

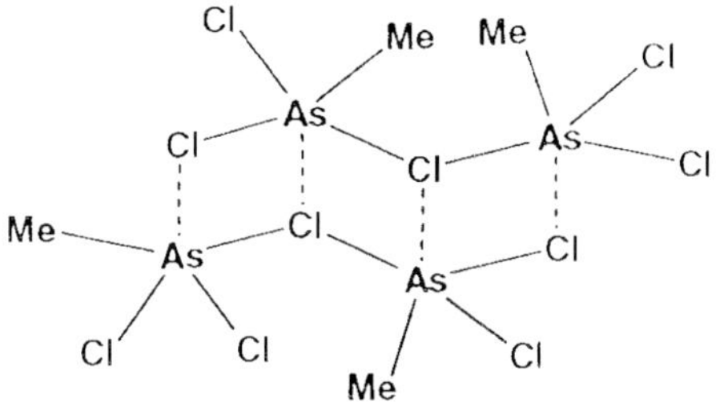

150

A supramolecular self-assembly, **151**, has been reported for dimethylarsenic cyanide, Me$_2$AsCN, which forms chains via As···N secondary bonds (As···N 3.18 Å) [381]. Methylarsenic dicyanide, MeAs(CN)$_2$, is also associated into supramolecular arrays [382].

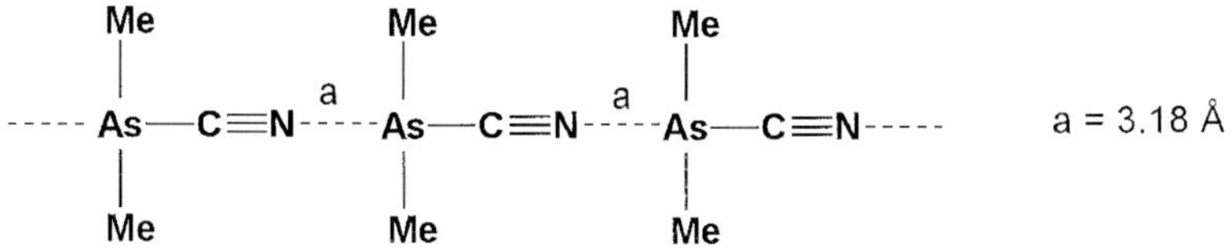

a = 3.18 Å

151

Methylarsenic dimesylamide chloride, MeAs{N(SO$_2$Me)$_2$}Cl, is associated in the solid state, forming double chains connected by two weak As···Cl secondary bonds (3.520 and 3.682 Å) [383].

Menschutkin-type supramolecular compounds [(AsBr$_3$)$_2$(μ-η^6-C$_6$Et$_6$)]$_n$ and [(AsCl$_3$)$_2$(μ-η^6-C$_6$Et$_6$)]$_n$ have been structurally characterized [384, 385].

4.5.2 Self-assembly in organoantimony compounds

4.5.2.1 Organoantimony halides

Among the tetraorganoantimony(V) halides, only the fluoride has been reported to self-assemble, with formation of polymeric chains by fluorine bridging. Tetramethylantimony fluoride, SbMe$_4$F, forms helical chains, **152**, in which the antimony atom is six-coordinated with distorted octahedral geometry. As in other polymeric self-assembled organometallic fluorides, the tendency of bond equalization is also obvious in this compound, and the two cis-oriented Sb–F interatomic distances are basically equal, Sb–F 2.369 and 2.382 Å, perhaps suggesting that these should be regarded as dative bonds. They are, however, longer than the Sb–F (primary) normal covalent Sb–F bonds measured in monomeric (five-coordinate, trigonal bipyramidal) trimethylantimony difluoride, Me$_3$SbF$_2$, with axial Sb–F 1.993 and 2.004 Å [386].

152

Triorganoantimony(V) dihalides seem to be all monomeric, because of the preferred five-coordinate, trigonal bipyramidal geometry with the halogen atoms in axial positions. This is observed for Me$_3$SbF$_2$ cited above and for Ph$_3$SbX$_2$ (X = Cl, Br), all of which have been structurally characterized [387]. The antimony–halogen interatomic distances in these compounds (Sb–Cl 2.458 and 2.468 Å; Sb–Br 2.632 Å) should be regarded as normal covalent single bonds and can serve as references for comparisons with bond lengths measured in other (self-assembled) compounds to be discussed below.

In the solid state the 1:1 adducts of triorganoantimony(V) dihalides with antimony(III) halides, Me$_3$SbX$_2$·SbX$_3$ (X = Cl, Br) have a complex self-assembled structure, **153**, in which pairs of the two molecules associated by Sb···X secondary bonding (e.g. Sb···Cl 3.203 Å) are interlinked by additional Sb···X secondary bonds (Sb···Cl 3.308 Å) to form double chains [388, 389]. Only the antimony atoms of the SbX$_3$ molecules participate in these secondary bonds, each becoming six-coordinate by forming three primary Sb–X bonds and three secondary Sb···X bonds, as shown for the chloride. The five-coordinate, trigonal bipyramidal SbR$_3$X$_2$ tectons participate in the supramolecular architecture through their chlorine atoms, connected (as Sb···Cl) to the antimony atoms of the SbX$_3$ moieties. The phenyl derivative, Ph$_3$SbCl)$_2$·SbCl$_3$, is also supramolecular [390].

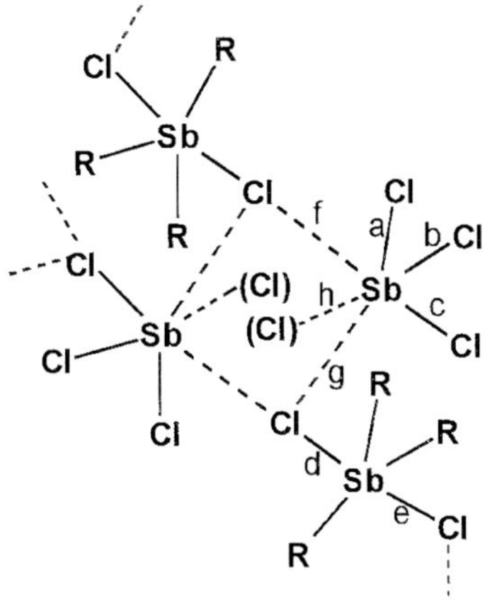

4.5 Group 15 – Arsenic, Antimony, Bismuth

R = Me

a = 2.362 Å
b = 2.380 Å
c = 2.382 Å
d = 2.572 Å
e = 2.473 Å
f = 3.203 Å
g = 3.308 Å
h = 3.154 Å

153

Diorganoantimony(V) trihalides seem to prefer six-coordinate, octahedral geometry, and as a consequence will associate into dimers. Thus, dimethylantimony(V) trichloride, Me_2SbCl_3, was found to be a dimer $[Me_2SbCl_2(\mu\text{-}Cl)]_2$, **154**, with bridging chlorines, Sb–Cl 2.35 Å and Sb···Cl 2.80 Å. A second isomeric form of this compound exists as an ion pair, $[SbMe_4]^+[SbCl_6]^-$; this illustrates the coordination preferences of antimony [391].

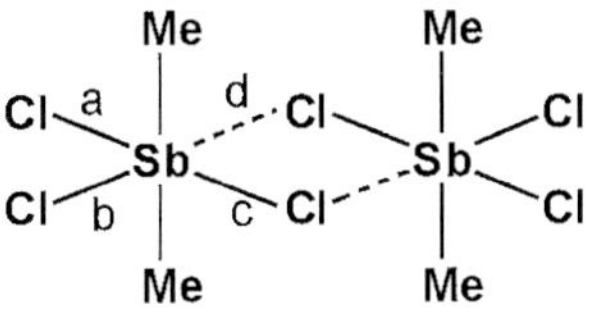

a = 2.353 Å
b = 2.536 Å
c = 2.798 Å
d = 2.801 Å

154

Diphenylantimony(V) trichloride is a chlorine-bridged dimeric supermolecule, $[Ph_2SbCl_2(\mu\text{-}Cl)]_2$ (terminal Sb–Cl 2.388 Å, bridging Sb–Cl 2.620 Å, Sb···Cl 2.839 Å) [392]. The adduct $Ph_3SbCl_2 \cdot SbCl_3$ is a tetranuclear self-assembled supermolecule, **155**, also formed via secondary Sb···Cl interactions (Sb–Cl 2.306–2.354 Å, Sb···Cl 3.236 and 3.287 Å) [393].

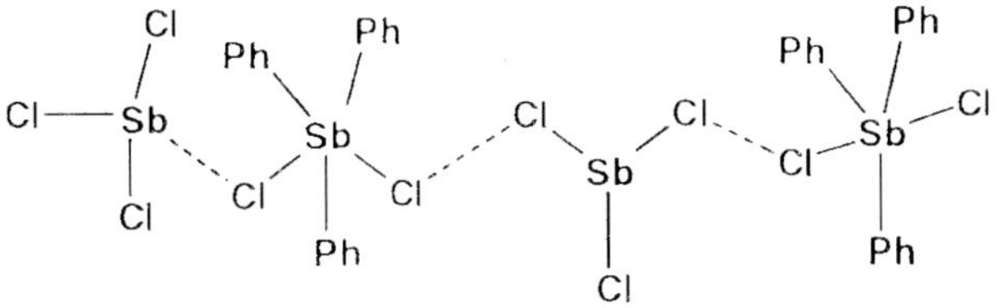

155

Methyldichlorostibacyclopentane forms cyclic tetramers, **156**, in the crystal [394]. Another supermolecular structure is illustrated by (2-pyridyl)(Me$_3$Si)–CSbCl, in which internally chlorine-bridged Sb$_2$C$_2$ rings are interconnected into a linear array, **157** (Sb···Cl 3.054 Å) [395].

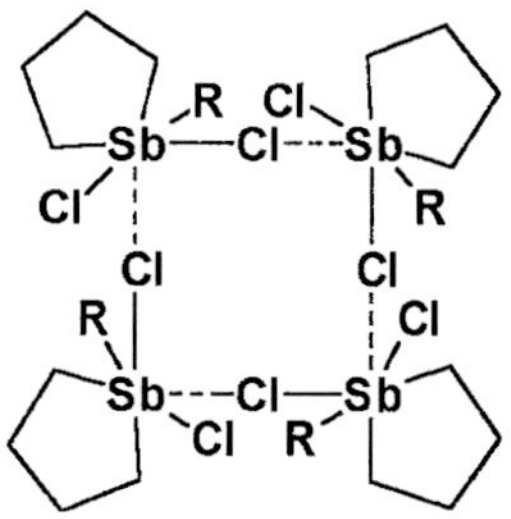

R = Me

Sb-Cl 2.485 Å
       2.519 Å
Sb···Cl 3.613 Å

156

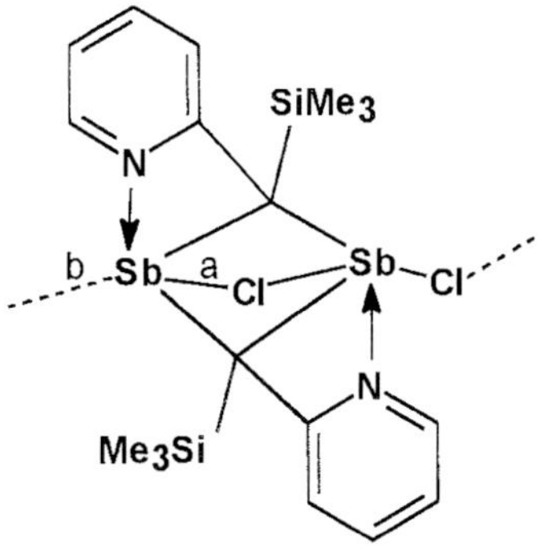

a = 2.385 Å
b = 3.054 Å

157

Diorganoantimony(III) halides can also be associated. The fluoride Ph$_2$SbF molecules are self-organized in zigzag chains, **158**, with strong, nearly equal (probably best described as dative) Sb–F bonds (2.166 and 2.221 Å) [396]; these are, however, longer than in SbF$_3$ (1.92 Å) or Na[SbF$_4$] (1.94–2.08 Å), which again shows the tendency of equalization of bridging metal–fluorine bonds. In other organoantimony halides the association occurs via secondary bonding.

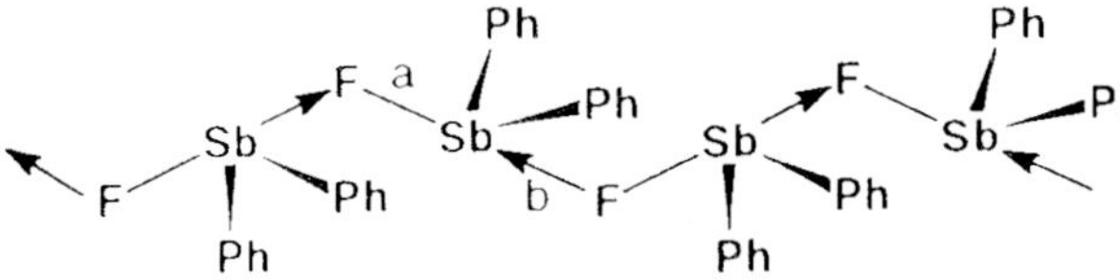

a = 2.166 Å
b = 2.221 Å

158

The anion [Ph$_2$SbI$_2$]$^-$ is a dimer (Sb–I 2.925 and 3.109 Å) consisting of two square pyramids joined by a double Sb–I–Sb bridge [397].

4.5 Group 15 – Arsenic, Antimony, Bismuth

Monoorganoantimony(III) halides have the greatest tendency to self-assemble. Methylantimony diiodide, $MeSbI_2$, forms double bridged chains, **159**, with all the methyl groups on the same side of the chain and alternating short (ave 2.78 Å) and long (ave 3.42 Å) antimony–iodine bond lengths [398]. The packing of the chains in the crystal is interesting; the chains are arranged 'face-to-face'. In contrast to the above cited organoantimony(III) halides, $(Me_3Si)_2CHSbCl_2$ is monomeric because the bulkiness of substituent prevents self-assembly [399].

a = 2.799 Å
b = 2.761 Å
c = 3.467 Å
d = 3.398 Å

159

Phenylantimony(III) dihalides, $PhSbX_2$ (X = Cl, Br, I), have isotypical structures and are all self-assembled via secondary bonding to form two-dimensional sheets, with all the phenyl groups on the same side of the plane [400]. In these compounds η^3-$C_6H_5\cdots$Sb interactions between adjacent molecules are also present. The interatomic distances are: intramolecular (primary, covalent bonds) Sb–Cl 2.376 and 2.411 Å; Sb–Br 2.526 and 2.563 Å; Sb–I 2.738 and 2.753 Å; secondary bonds Sb$\cdots$Cl 3.443 and 3.865 Å; Sb$\cdots$Br 3.620 and 4.058 Å; Sb$\cdots$I 3.81 and 4.07 Å.

Phenylantimony(III) diiodide forms an unusual complex, **160**, in which four $PhSbI_2$ molecules are self-assembled around a central I^- ion via primary (Sb–I 2.8 Å) and secondary (Sb$\cdots$I 3.6 Å) bonds. The complex was analyzed theoretically, by the extended Hückel method, as a $[I_2Sb(\mu\text{-}I)]_4$ cyclic matrix, incorporating a central I^- as host [401].

160

The monophenylantimony anions $[PhSbX_3]^-$ (X = Cl [402, 403], Br [404], I [405]) and the mixed halide anion $[PhSbBr_2Cl]^-$ [406], are all dimers self-assembled by secondary bonding. The structures are not isotypical. In the chloride dimer (bipyridinium salt), **161a**, the dimer contains two square pyramidal units joined by a double bridge, with the phenyl groups on alternate sides of the Sb_2Cl_2 plane; additional secondary Sb$\cdots$Cl bonds (3.756 Å), *trans* to the phenyl groups assemble the

272   4 Supramolecular Self-Assembly by Formation of Secondary Bonds

dimers into a more complex structure, making the antimony atom six-coordinate (distorted octahedron). The tetramethylammonium salt, **161b**, contains discrete dimeric supermolecules with five-coordinate square pyramidal antimony.

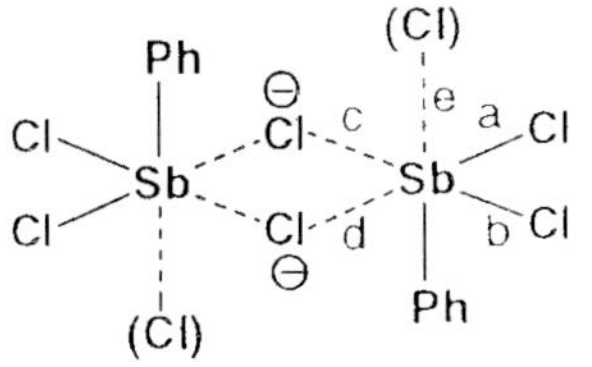

$a = 2.432$ Å
$b = 2.444$ Å
$c = 3.007$ Å
$d = 3.103$ Å
$e = 3.756$ Å

[bipyH]+ salt

161a

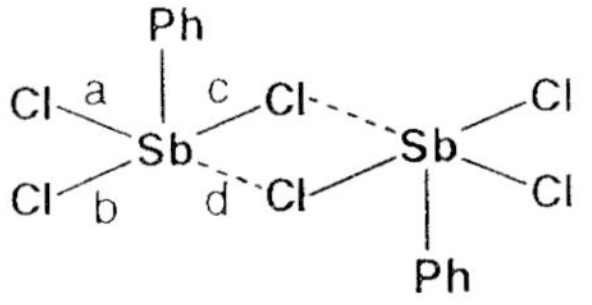

$a = 2.423$ Å
$b = 2.540$ Å
$c = 2.657$ Å
$d = 3.121$ Å

[NMe4]+ salt

161b

In the iodo derivative (tetraethylammonium salt), **162**, the central Sb_2I_2 moiety is non-planar, the phenyl groups are on the same side and the antimony atoms can be regarded as five-coordinate (square pyramidal). The mixed halide anion, $[PhSbBr_2Cl]^-$, **163**, is bromine-bridged, and the phenyl groups are on alternate sides of the Sb_2Br_2 plane. The coordination geometry is square-pyramidal and there are no further intermolecular interactions between the anions.

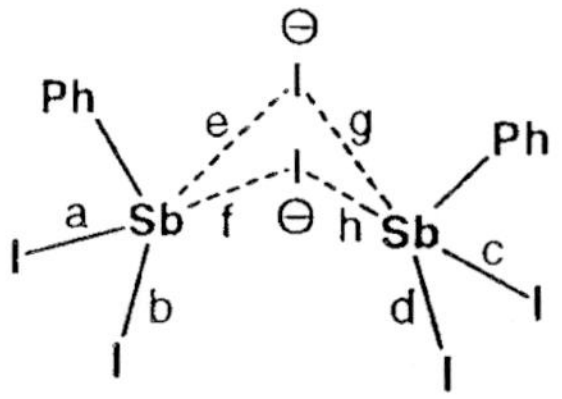

$a = 2.826$ Å
$b = 2.868$ Å
$c = 2.841$ Å
$d = 2.890$ Å
$e = 3.219$ Å
$f = 3.305$ Å
$g = 3.187$ Å
$h = 3.240$ Å

[NEt4]+ salt

162

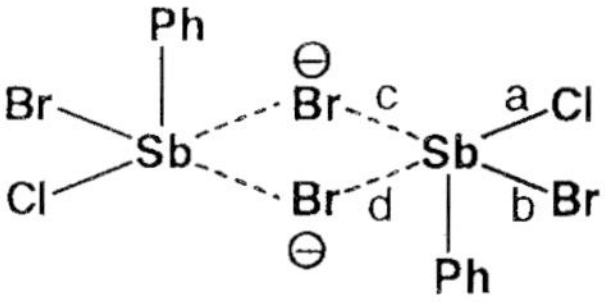

4.5 Group 15 – Arsenic, Antimony, Bismuth

[Structure 163 with labels:]

a = 2.545 Å
b = 2.650 Å
c = 2.925 Å
d = 3.168 Å

163

Interesting examples of self-assembly are provided by the arene complexes of antimony halides (so-called Menschutkin-type complexes). The addition compound between benzene and antimony trichloride, $C_6H_6 \cdot 2SbCl_3$, has been known for a long time [407, 408], but its crystal structure has been determined by X-ray diffraction only recently [409]. It consists of $SbCl_3$ molecules π-bonded to both faces of the benzene ring, to form an 'inverse' sandwich moiety, **164**. Further secondary Sb···Cl bonds connect these tectons into a self-organized layer structure.

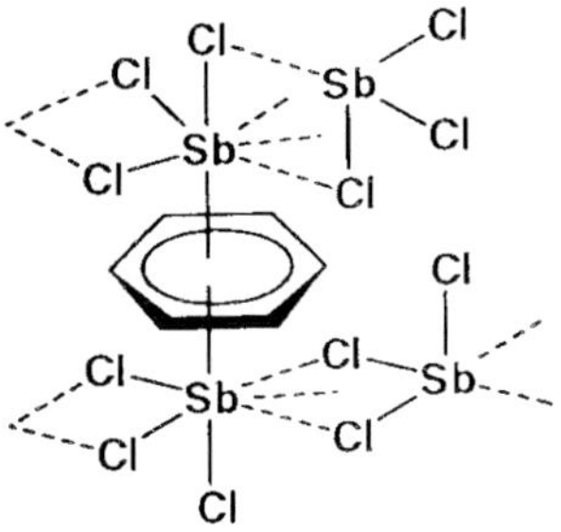

164

The coordination geometry around antimony is pentagonal bipyramidal, with a benzene molecule occupying one axial site; another axial site is occupied by a chlorine atom bonded to antimony by a primary Sb–Cl bond. In the equatorial plane there are two primary (normal covalent) Sb–Cl bonds (in the range 2.340–2.378 Å) whereas three other secondary Sb···Cl interactions (range 3.401–4.04 Å) distort the geometry and contribute to the formation of the supramolecular architecture. The supramolecular assembly exists only in the solid state; in solution or melt disassembly occurs.

Similar 1:2 complexes of hexamethylbenzene with SbX_3 (X = Cl, Br) have also been investigated [410, 411]. The structures of $C_6Me_6 \cdot 2SbX_3$ are built up of tetrameric cyclic Sb_4X_{12} units, with the arenes alternating on the two sides of the ring and η^6-connected to antimony. Each metal atom is bonded to three halogen atoms by primary Sb–X bonds, and a network of secondary Sb···X bonds gives rise to a three-dimensional supramolecular architecture. The short (primary) Sb–X bonds are in the normal range (Sb–Cl 2.319–2.373 Å; Sb–Br 2.481–2.528 Å) and the secondary Sb···X bonds cover a broader range (Sb···Cl 2.895–3.675 Å; Sb···Br 2.990–3.833 Å).

With mesitylene, 1:1 complexes of the type $C_6H_3Me_3 \cdot SbX_3$ are formed. Their structure is made up of hexameric tectons, connected into two-dimensional layers by secondary Sb$\cdots$X bonds. Again, arene molecules alternate on both sides of the layer. The layers are arranged in the crystal such that only mesitylene molecules are adjacent, the inorganic components being sandwiched between the organic coordinated molecules [412]. Each antimony atom is coordinated in a pentagonal bipyramidal arrangement, with one arene molecule in an apical position and three primary bonds to halogens (Sb–Br 2.502–2.554 Å); further secondary bonds lead to the supramolecular self-organization (Sb$\cdots$Br 3.510–3.994 Å).

A similar 1:1 complex, $C_6Et_6 \cdot SbCl_3$, is formed by hexaethylbenzene with antimony trichloride; surprisingly this was found to be monomeric, with no supramolecular self-organization [413]. There is no obvious reason for this behavior.

Some other arene–antimony halide complexes have been structurally characterized; all have intermolecular secondary Sb$\cdots$X interactions and supramolecular self-organization. Examples are naphthalene$\cdot 2SbCl_3$ [414], phenanthrene$\cdot 2SbCl_3$ [415], pyrene$\cdot 2SbBr_3$ [416], and $[(SbCl_3)_2(\mu\text{-}\eta^6\text{-}[2^3](1,4)\text{cyclophane}]x \cdot 0,5C_6H_6$ [417]. In the naphthalene complex [414] pairs of $SbCl_3$ molecules are interconnected in planar stacks, with shorter primary bonds (axial Sb–Cl 2.367 Å, equatorial Sb–Cl 2.347 Å) and intermolecular distances Sb$\cdots$Cl 3.581–3.832 Å.

More recently the complexes phenanthrene$\cdot 2SbBr_3$ and pyrene$\cdot 2SbCl_3$ were added to the list of structurally characterized compounds [418]. The phenanthrene$\cdot 2SbBr_3$ complex develops a monodimensional self-organized structure by Sb$\cdots$Br secondary bonding (3.428–3.824 Å).

The phenanthrene$\cdot SbCl_3$ complex contains trinuclear aggregates with intermolecular distances Sb$\cdots$Cl 3.410 and 3.260 Å [415]. In the pyrene complex the $SbCl_3$ molecules (Sb–Cl 2.346–2.38.8 Å) are coordinated on both sides of the arene and are further connected by secondary Sb$\cdots$Cl (3.515, 3.652, and 4.000 Å) bonds to form inorganic layers with the organic molecules sandwiched between them. Structurally characterized complexes 2,2'-dithienyl$\cdot 2SbCl_3$ (Sb–Cl 2.356–2.374 Å, Sb$\cdots$Cl 3.552–3.761 Å) and benzo[b]thiophene$\cdot 2SbCl_3$ (Sb–Cl 2.352–2.367 Å, Sb$\cdots$Cl 3.546–3.847 Å) show that aromatic heterocycles can behave similarly [419].

4.5.2.2 Organoantimony–oxygen compounds

Several organoantimony compounds have a self-organized supramolecular structure assembled with the aid of secondary Sb$\cdots$O bonds. By comparison with other main group metals of comparable Lewis acid properties one can expect more examples of self-assembly through secondary bond interactions in organoantimony compounds containing oxygen-functional groups.

Tetraorganoantimony(V) alkoxides and triorganoantimony(V) dialkoxides are monomeric, five-coordinate compounds, e.g. Ph_4SbOMe and $Ph_3Sb(OMe)_2$ [420]. This is not surprising, because the trigonal bipyramidal geometry found in these compounds is a preferred coordination mode of antimony(V). Despite this, triphenyl(4-nitrocatecholato)antimony(V) was found to dimerize via Sb$\cdots$O secondary bonds, **165** (Sb–O 2.078 and 2.039 Å, Sb···O 3.341 Å) [421]. The sum of the van der Waals radii for the antimony–oxygen pair is estimated to be 3.57 Å.

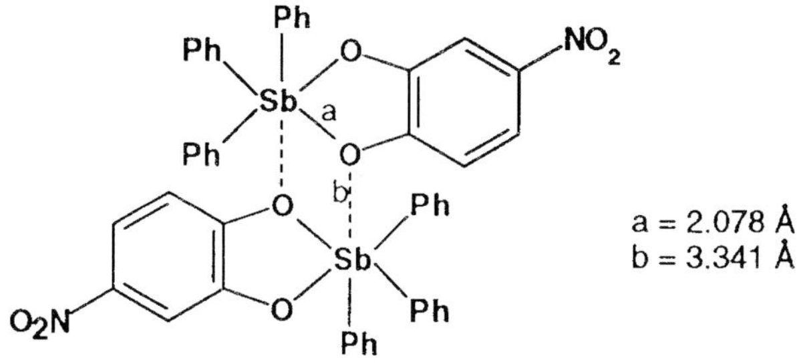

a = 2.078 Å
b = 3.341 Å

165

One structurally characterized monoorganoantimony alkoxide is the dimer [MeSbBr$_2$(μ-OMe)]$_2$, with identical bridging Sb–O(Me) interatomic distances (Sb–O 2.135 Å) which are, however, longer than the terminal Sb–O(Me) distances (1.962 Å) [422].

Organoantimony carboxylates can be expected to self-assemble. Diphenylantimony(III) acetate consists of self-assembled chains, **166**, with bridging acetato groups and pseudo-trigonal bipyramidal coordination [423].

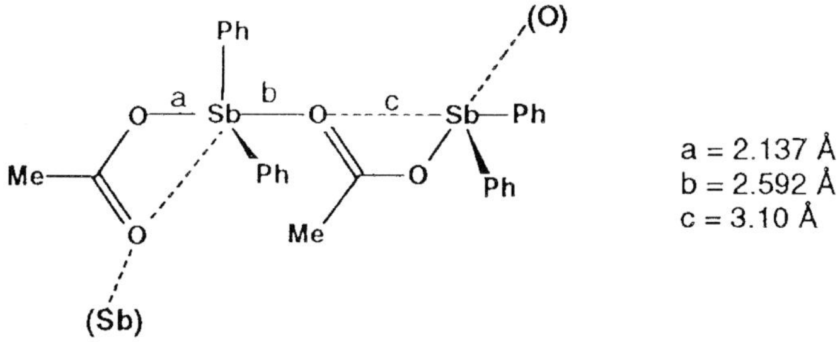

a = 2.137 Å
b = 2.592 Å
c = 3.10 Å

166

Surprisingly few organoantimony carboxylates have been structurally characterized. One might expect that further examples of self-assembly in this class of compounds will be identified if systematic structure investigation (as performed for organotin carboxylates) were to be conducted.

Organophosphorus acids can play a similar role – bridging antimony atoms with the formation of self-assembled structures. The monothiophosphinate, Me$_4$SbOP(S)Me$_2$, is one example. On the basis of Mössbauer spectra a dimeric structure has been postulated for this compound [424]; an X-ray diffraction study revealed Sb···O interatomic distances of 2.532 and 2.749 Å, compared with 2.07 Å expected from the sum of the covalent radii of a normal Sb–O bond; some distortion of the coordination geometry around antimony from tetrahedral towards a pseudo-trigonal bipyramid are also in agreement with the self-assembly interpretation [425].

In agreement with these expectations, diphenylantimony(III) diphenylphosphinate, Ph$_2$SbO$_2$PPh$_2$, was found to be a self-organized polymer, **167**, with phosphinate bridging [426]. The coordination at antimony is pseudo-trigonal bipyramidal. It can be argued whether the observed Sb···O distances of 2.23 and 2.29 Å should be regarded as secondary bonds or as dative bonds, but both are longer than single covalent Sb–O bonds (expected to be 2.0 Å).

167

a = 2.23 Å
b = 2.29 Å

On the basis of the data presented it would be highly desirable to investigate more organoantimony derivatives of carboxylic, organophosphoric, and other acids, with very good chances of discovering new supramolecular, self-organized structures.

4.5.2.3 Organoantimony–sulfur compounds

This type of self-assembly is rather common, because antimony and sulfur form a soft acid–base pair. The Sb–S covalent bond is expected to be ca. 2.4 Å (from the sum of the covalent radii) and the van der Waals interatomic distance has been estimated as ca. 4.0 Å; several self-assembled compounds with interatomic Sb···S distances from 3.0 to slightly less than 4.0 Å will be discussed in this section.

A first example is 2-phenyl-1,3-dithiastibolane, in which the five-membered rings are assembled into helical chains, **168**, by secondary Sb···S bonding (3.34 Å) [427]. The structure is similar to that of 2-chloro-1,3-dithiastibolane, a compound without an Sb–C bond [428].

168

a = 2.458 Å
b = 2.426 Å
c = 3.340 Å

Double bridged self-assembly occurs in 1-chloro-4,5-dimethyl-4,5-didehydro-2-thiastibolane, **169**, and 1-bromo-4,5-diphenyl-4,5-didehydro-2-thiastibolane, **170**, to form dimeric molecules in the solid state. In benzene solution the dimers dissociate into monomers [429].

Self-assembly also occurs in trithiastibocanes. Thus, 2-phenyl-1,3,6,2-trithiastibocane molecules are associated via Sb···S secondary bonds (3.594 and 3.868 Å) into linear tetramers, **171**, whereas the oxo analog, 5-phenyl-1,4,6,5-oxadithiastibocane is similarly self-assembled into double chains, **172** (Sb···S 3.558 and 3.862 Å) [430]. The compounds also contain intramolecular, transannular Sb···S and Sb···O secondary bonds, respectively.

The extent of association can vary with the nature of the organic groups attached to the metal, probably because of electronic influences which modify the Lewis acid–base properties of the metal. Thus, *p*-nitrophenyl trithiastibocane molecules assemble into dimers only, **173** [431].

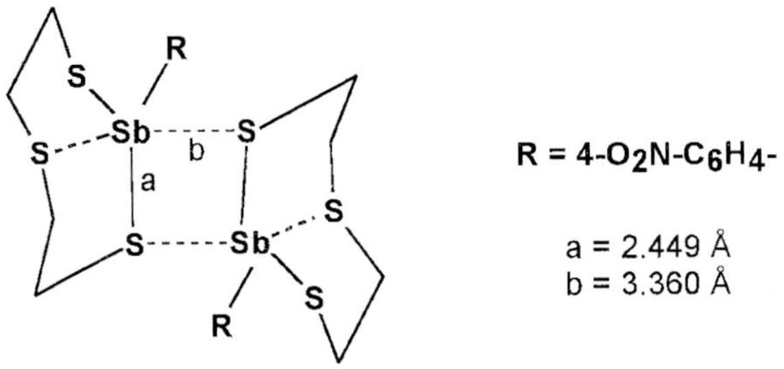

R = 4-O$_2$N-C$_6$H$_4$-

a = 2.449 Å
b = 3.360 Å

173

Surprisingly, the competition for secondary bonds between the ring heteroatoms and exocyclic substituents such as dithiophosphinato ligands, does not prevent self-assembly by secondary bonding. Thus, the diphenyldithiophosphinato derivative of the oxadithiastibocane self-assembles into dimers, **174**, and the related trithiastilbocane derivative forms infinite chains, **175**. In both compounds the coordination number of antimony is increased to six by participation of transannular and intermolecular secondary bonds [432].

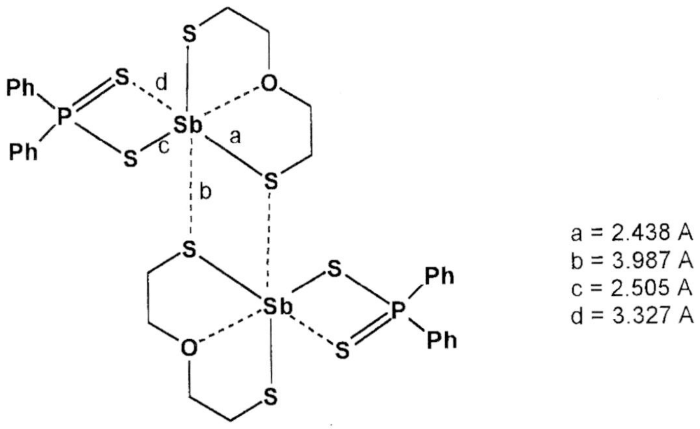

a = 2.438 A
b = 3.987 A
c = 2.505 A
d = 3.327 A

174

Dimeric self-assembly was also found in the dimethyldithiophosphinato derivatives of the same oxadithia and trithiastibocanes [433].

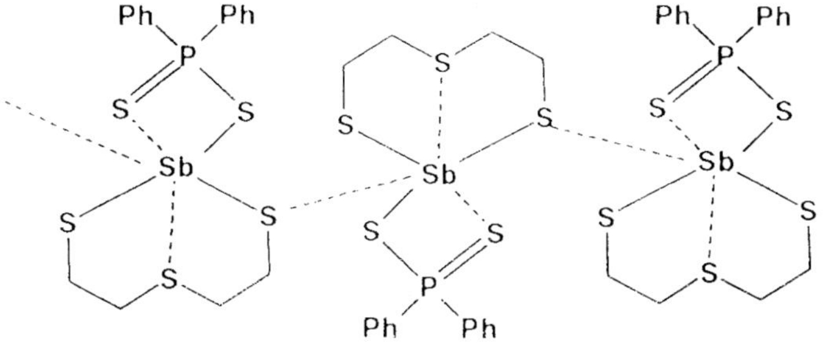

4.5 Group 15 – Arsenic, Antimony, Bismuth   279

175

Other sulfur–ligand organoantimony(III) derivatives also have self-organized structures. Thus, methylantimony(III) bis(ethylxanthate), MeSb(S$_2$COEt)$_2$, **176**, (Sb–S 2.581–2.904 Å; Sb···S 3.353 Å) [434], phenylantimony(III) bis(monothioacetate), PhSb[S(O)CMe]$_2$ [435], and methylantimony(III) bis(diethyldithiocarbamate), MeSb(S$_2$CNEt$_2$)$_2$, **177**, (Sb–S 2.538 and 2.554 Å; Sb···S 3.847 Å) [436] all associate as dimeric supermolecules.

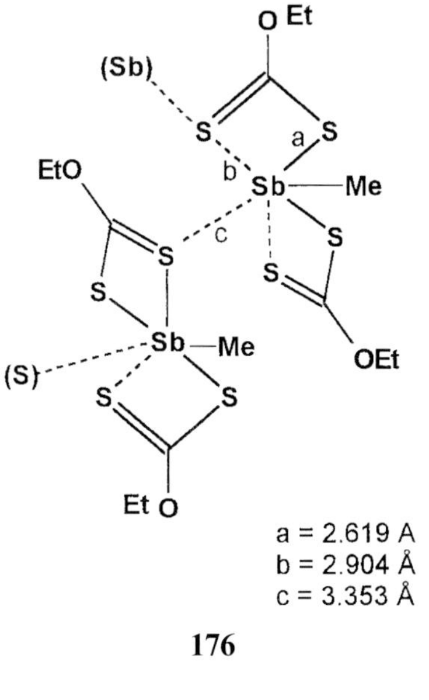

176
a = 2.619 Å
b = 2.904 Å
c = 3.353 Å

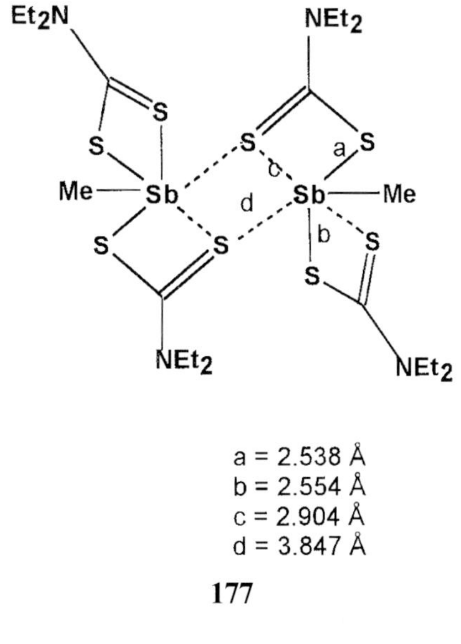

177
a = 2.538 Å
b = 2.554 Å
c = 2.904 Å
d = 3.847 Å

A spectacular example of self-organization is observed for diphenylantimony(III) thiocyanate, Ph$_2$SbSCN [437], which forms helical chains, **178**. There are three different modes of coordination around antimony – in one antimony is connected to two sulfurs, in another to two nitrogens, and in the third to a sulfur and a nitrogen atom.

The geometric factors here, i.e. the bond angles at antimony (nearly linear) and at

280 4 Supramolecular Self-Assembly by Formation of Secondary Bonds

```
                    SCN----(Sb)
             Ph\   /
               Sb
            a /   \
             S     Ph
             ‖
             C
             ‖
             N
              \d
               Sb—Ph
              / \e
             Ph  N
                 ‖ a = 2.700 Å
                 C    b = 2.831 Å
            Ph  S    c = 2.842 Å
             \  b    d = 2.364 Å
      Ph      \/     e = 2.304 Å
       |    f Sb     f = 2.273 Å
       |     / \Ph   g = 2.700 Å
(Sb)—NCS--g-Sb-c--SCN
           |
           Ph
```

178

sulfur and nitrogen (bent) favor the development of a 'triangular spiral' structure of the polymeric chain formed through self-assembly.

A significant number of examples is known in which organothiophosphorus acid derivatives are self-organized by Sb···S secondary bonding [438]. Thus, diphenylantimony(III) diphenylmonothiophosphinate, $Ph_2SbOP(S)Ph_2$, consists of helical chains, **179**, formed via Sb···S secondary bonds, leading to pseudo-trigonal bipyramidal coordination at antimony [439].

```
    Ph  Ph      Ph
     \ /         |       a
      P  b       Sb    O       S
     ‖ \    a   / \   /       ‖
     O  S------    O-P
Ph\                    /  \
  Sb—Ph    Ph         Ph  Ph
```
a = 2.299 Å
b = 2.753 Å

179

Diorganoantimony(III) dithiophosphates and dithiophosphinates assume a variety of self-organized structures by changing the organic groups either at the metal or at phosphorus. Thus, dimeric diphenylantimony(III) diphenyldithiophosphinate, $[Ph_2SbS_2PPh_2]_2$, **180** (Sb–S 2.490 Å, Sb···S 3.474 Å) and the dithioarsinate analog, $[Ph_2SbS_2AsPh_2]_2$, **181** (Sb–S 2.486 Å, Sb···S 3.369 Å), self-organize into quasi-tricyclic systems containing both intramolecular Sb···S secondary bonds (leading to chelation) and intermolecular Sb···S secondary bonds (leading to dimerization) [440]. Surprisingly, the dimeric bis(*p*-tolyl)antimony(III) diethyldithiophosphinate,

[(4-MeC$_6$H$_4$)$_2$SbS$_2$PEt$_2$]$_2$, has a larger transannular antimony–sulfur distance (in the range of the van der Waals distance) and negligible, if any, interaction; it should be regarded as a quasi-monocyclic eight-membered ring, **182** [441]. A similar structure was found in [Ph$_2$SbS$_2$PMe$_2$]$_2$ (Sb–S 2.439 Å, Sb···S 3.450 Å) [442].

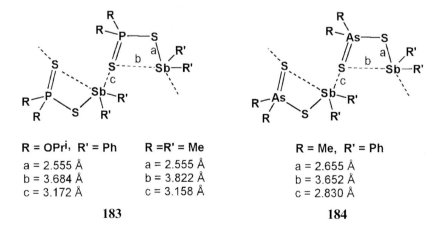

a = 2.490 Å
b = 3.474 Å
c = 3.440 Å

180

a = 2.486 Å
b = 3.369 Å
c = 3.500 Å

181

Ar = p-tolyl
a = 2.485 Å
b = 3.556 Å
c = 4.013 Å

182

Analogous compounds self-organize into helical chain-like arrays. Thus, diphenylantimony(III) isopropyldithiophosphate, Ph$_2$SbS$_2$P(OPr^i)$_2$ [443], dimethylantimony(III) dimethyldithiophosphinate, Me$_2$SbS$_2$PMe$_2$ [444], and the related dimethyldithioarsinate, Ph$_2$SbS$_2$AsMe$_2$ [445], are all helical polymers (**183** and **184**) in which the molecular tectons are connected by secondary Sb···S bonds.

R = OPr^i, R' = Ph
a = 2.555 Å
b = 3.684 Å
c = 3.172 Å

R = R' = Me
a = 2.555 Å
b = 3.822 Å
c = 3.158 Å

183

R = Me, R' = Ph
a = 2.655 Å
b = 3.652 Å
c = 2.830 Å

184

4.5.2.4 Organoantimony–nitrogen compounds

It might be expected that antimony(III)–nitrogen compounds could self-assemble into supramolecular structures, but rather few compounds have been structurally characterized. Dimethylantimony(III) azide, Me_2SbN_3, is associated in the solid state to form zigzag chains (Sb–N 2.322 and 2.434 Å), **185**, bent at nitrogen (Sb--N–Sb 126.0°) and almost linear at antimony (N–Sb–N 169.9°) [446]. The small difference between the Sb–N bond lengths suggests that dative bonding might also be considered in the description of the self-assembly.

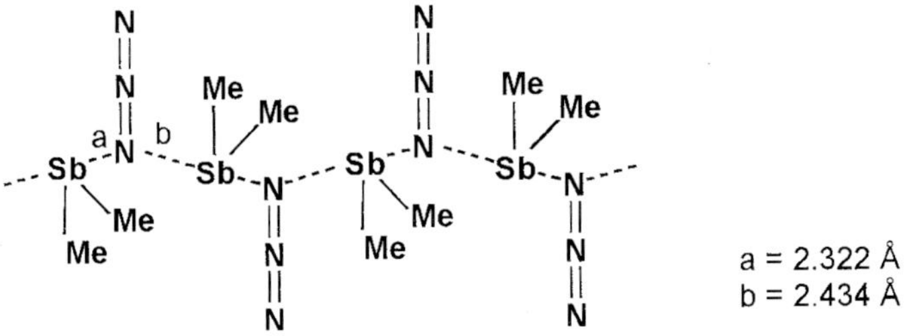

185

$a = 2.322$ Å
$b = 2.434$ Å

4.5.3 Self-assembly in organobismuth compounds

4.5.3.1 Organobismuth–halogen compounds

It has been illustrated by structural studies that the organobismuth halides tend to self-organize in the solid state. Their number is, however, limited to simpler representatives. Sterically demanding groups probably prevent association.

Diphenylbismuth chloride, Ph_2BiCl, is one of the very few structurally characterized diorganobismuth halides [447]. The compound forms zigzag chains, **186**, with almost linear Cl–Bi···Cl units (Cl–Bi···Cl 175.6°) as parts of a pseudo-trigonal bipyramidal coordination geometry; the chain is bent at the halogen (Bi–Cl···Bi 100.6°). The Bi–Cl bonds are practically equal in length (2.746 and 2.763 Å) but longer than normal covalent Bi–Cl bonds (ca. 2.40 Å). Surprisingly, even sterically crowded diarylbismuth halides, such as 2,4,6-$Ph_3C_6H_2BiCl_2$, are associated. 2,4,6-$Ph_3C_6H_2BiCl_2$ is an unsymmetrical dimer with Bi–Cl 2.530 Å and Bi···Cl 3.074 Å [447a].

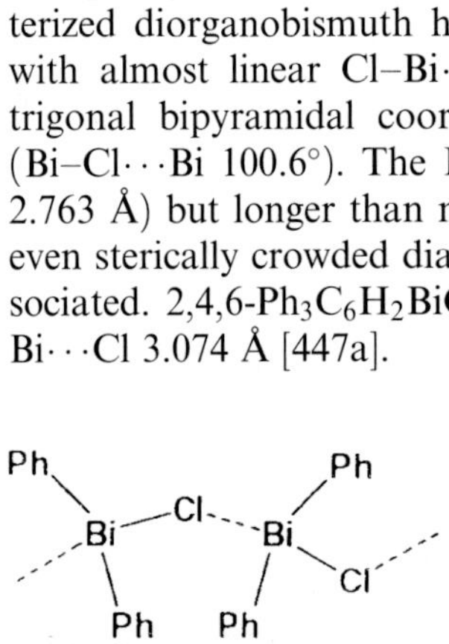

186

4.5 Group 15 – Arsenic, Antimony, Bismuth   283

Monoorganobismuth dihalides, RBiX$_2$, have enjoyed more attention from crystallographers. The best studied is methylbismuth diiodide, which has a one-dimensional chain-like structure, **187**, with the bismuth atoms doubly bridged by iodine atoms, and all methyl groups on the same side of the chain. The bismuth atom is five-coordinate, in square pyramidal geometry, with the methyl groups in apical positions. The Bi–I interatomic distances do not differ much, although shorter (3.086 and 3.087 Å) and longer (3.122 and 3.128 Å) bonds can be distinguished as alternating in the chains. The interaction between chains is weak. The chains are packed in pairs, such that inorganic sides are in contact, as are the sides carrying the methyl groups. A detailed theoretical analysis of this compound has been performed by use of the extended Hückel method; this showed that the compound is electronically one-dimensional [448]. Other alkylbismuth diiodides [448a] and PhBiI$_2$ [448b] have similar supramolecular, one-dimensional structures.

$$a = 3.087 \text{ Å}$$
$$b = 3.128 \text{ Å}$$

187

If the organic groups on bismuth are replaced by organometallic moieties, self-assembly is limited to cyclic structures. Thus, the cyclopentadienylmanganese dicarbonyl derivative, [(η^5-C$_5$H$_5$)Mn(CO)$_2$]$_2$BiCl is a chloride bridged dimer, **188** [449], and the methylcyclopentadienyliron dicarbonyl derivative, [(η^5-C$_5$H$_4$Me)-Fe(CO)$_2$]$_2$BiCl, is a cyclic trimer, **189**, based upon a Bi$_3$Cl$_3$ six-membered ring [450, 451]. The Bi–Cl bond lengths in the six-membered ring are very close, but seem to indicate a tendency to alternate. The bond angles have strange values for a six-membered ring – very large at bismuth, Cl–Bi–Cl 154.9° (suggesting some tendency towards pseudo-trigonal bipyramidal coordination), and acute at chlorine; Bi–Cl–Bi is only 86.0°.

$$a = 2.914 \text{ Å}$$
$$b = 2.852 \text{ Å}$$
$$c = 2.955 \text{ Å}$$

ML = Mn(CO)$_2$(η^5-C$_5$H$_5$)

188

ML = Fe(CO)$_2$(η^5-C$_5$H$_4$Me)

189

Other structurally characterized monoorganobismuth halides include PhBiCl$_2$-

·THF, **190**, PhBiBr$_2$·THF, and PhBiI$_2$·THF, **191** [452–454]. All are polymeric zigzag chains, bent at the halogen sites, with bismuth atoms in the center of a trigonal bipyramid.

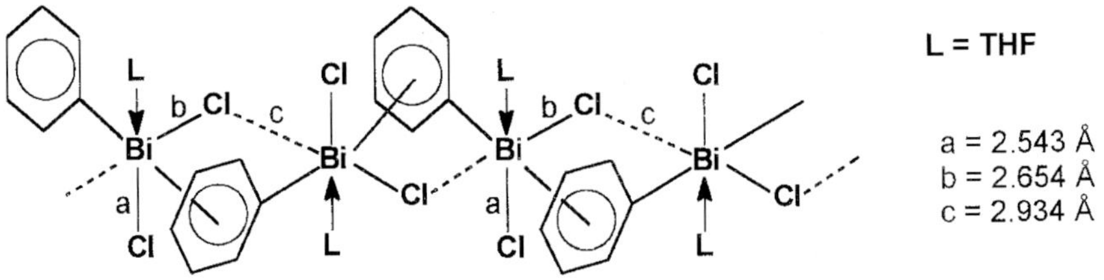

L = THF

a = 2.543 Å
b = 2.654 Å
c = 2.934 Å

190

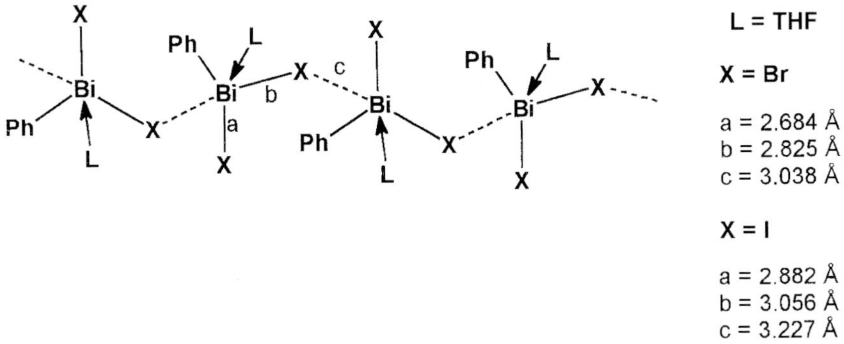

L = THF

X = Br

a = 2.684 Å
b = 2.825 Å
c = 3.038 Å

X = I

a = 2.882 Å
b = 3.056 Å
c = 3.227 Å

191

The same workers [452-454] described the structure of anionic PhBiX$_3^-$ (X = Br, I) which are halogen bridged dimers, **192** and **193**. Not unexpectedly, the diphenyldihalo anions [Ph$_2$BiX$_2$]$^-$ are pseudo-trigonal bipyramidal monomers (unassociated).

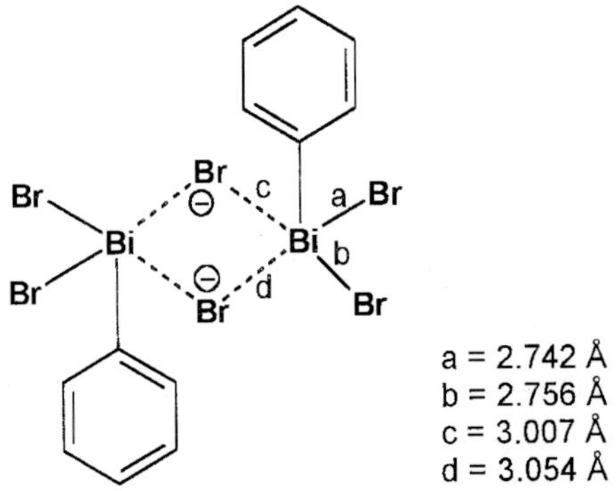

a = 2.742 Å
b = 2.756 Å
c = 3.007 Å
d = 3.054 Å

192

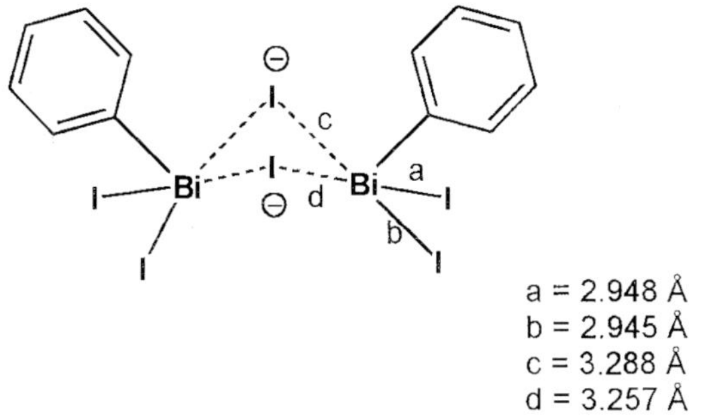

a = 2.948 Å
b = 2.945 Å
c = 3.288 Å
d = 3.257 Å

193

A chain-like polymeric self-organization, **194**, is observed in π-cyclopentadienyl bismuth dichloride, $C_5H_5BiCl_2$. Pseudorotational disorder results in a mixture of η^2 and η^3 bonding of the C_5H_5 ring to the metal atom. The chains are formed by double chloride bridging and the bismuth–chlorine interatomic distances differ [455].

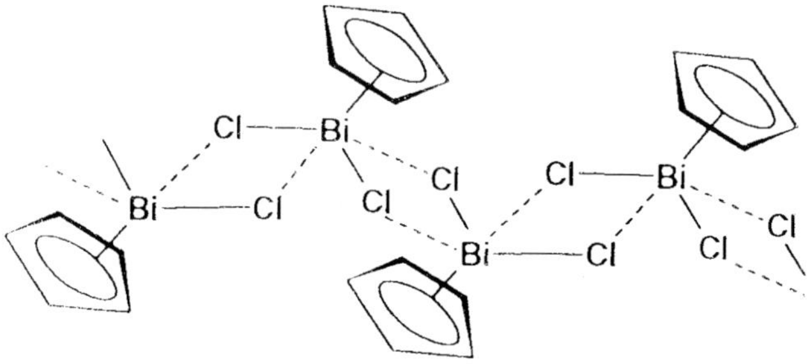

194

Several arene–bismuth halide complexes, reminiscent of the antimony Menshutkin complexes, have been structurally characterized and found to be self-assembled supermolecules. Thus, the benzene complex $C_6H_6 \cdot BiCl_3$ is made up of halogen-bridged dimeric units, interconnected by Bi$\cdots$Cl secondary bonds into a two-dimensional layer [456]. Each bismuth atom forms three short (normal, covalent) Bi–Cl bonds (in the range 2.444–2.486 Å), two longer bridging bonds (Bi$\cdots$Cl 3.246 and 3.219 Å) within the dimeric moiety and is interconnected by two still longer secondary bonds (i.e. 3.415 and 3.621 Å for one Bi atom and 3.578 and 3.483 Å for a second Bi atom) into the two-dimensional layer. The hexamethylbenzene complex $[\eta^6\text{-}C_6Me_6BiCl_2]^+[AlCl_4]^-$ is a dimer self-assembled by Bi$\cdots$Cl secondary bonding [456a].

The three xylene derivatives, $C_6H_4(CH_3)_2 \cdot BiCl_3$, are self-organized in chain structures, based upon dinuclear tectons [457]. In all compounds the arene is bonded in η^6-fashion. In the derivatives of *meta-* and *para-*xylene the chains are formed by double-halogen bridging, with each bismuth atom forming three normal covalent Bi–Cl bonds (one terminal, ca. 2.40 Å, and two bridging Bi–Cl bonds within the dimeric Bi$_2$Cl$_2$ unit, ca. 2.48 Å). In the Bi$_2$Cl$_2$ unit the bridges are asymmetric, and Bi···Cl secondary bonds (3.5–3.20 Å) are formed between adjacent units, **195**.

195

In the *ortho-*xylene derivative the bridging is different, making the bismuth atom trigonal bipyramidal. The chain-like arrays in this compound are formed via triple bismuth–halogen bridges.

In the 1:1 complexes of mesitylene with BiX$_3$ (X = Cl and Br) self-assembly leads to formation of two-dimensional sheets containing six-membered Bi$_6$X$_6$ moieties and a complex array of primary Bi–X bonds (three at each bismuth atom, ca. 2.5 Å) and secondary Bi···X bonds (in the range 3.1–4.0 Å) [458].

The 1:2 complexes of hexamethylbenzene with BiX$_3$ (X = Cl and Br) are 'inverse-sandwich' compounds, with two bismuth atoms on each side of the arene plane. The supramolecular architecture is formed from tetranuclear tectons based upon eight-membered Bi$_4$X$_4$ quasi-rings, interconnected into a three-dimensional architecture [459].

An interesting [2.2]paracyclophane–bismuth bromide 1:2 adduct has been investigated. It forms self-organized extended (BiBr$_3$)$_x$ chains, in which each bismuth atom participates in two Bi–Br bridges (with two short bonds Bi–Br 2.676 and 2.663 Å, and two secondary Bi···Br bonds 3.325 and 3.402 Å) and is connected to a terminal bromine (Bi–Br 2.595 Å). The chains are interconnected by the paracyclophane molecules, each one connected in a η^6 fashion to two adjacent chains [460].

The broad diversity of bismuth–halogen associations promises further interesting results in this area.

4.5.3.2 Organobismuth–oxygen compounds

Although few crystal structures of organobismuth–oxygen compounds have been determined, they suggest that self-assembly through Bi···O bonds might be a widespread phenomenon.

Two organobismuth(III) alkoxides, Et_2BiOAr, with Ar = Ph and C_6F_5, self-assemble as chiral helical chains in the solid state, **196**. The Bi–O distances in the chain are identical, making formal classification of these compounds difficult within our framework. In Et_2BiOPh the distance Bi–O is 2.4105 Å whereas in $Et_2BiOC_6F_5$ it is 2.382 Å. The coordination geometry of bismuth is pseudo-trigonal bipyramidal, with the oxygens in axial positions and linear disposition – O–Bi–O 179.54° when R = Ph and 179.0° when R = C_6F_5 [461].

196

It can be seen that, as in many other supramolecular structures of organometallic compounds, the lipophilic organic groups wrap the inorganic chain, which is embedded in the interior of the helix.

Bridging by OC_6F_5 has also been found in dimeric $[Bi(OC_6F_5)_3]_2$ [462], a compound which does not contain a direct Bi–C bond.

By analogy with other organometallic derivatives, organobismuth carboxylates would be expected to self-assemble into oligomeric or polymeric supramolecular structures. Crystal structure determinations are, however, scarce in this area. Polymeric association of diphenylbismuth carboxylates has been inferred from spectroscopic and solubility properties [463]. The likelihood of carboxylate bridging has been confirmed by an X-ray diffraction analysis of diphenylbismuth(III) N-benzoylglycinate [464]. In this compound bridging carboxylate groups link diphenylbismuth units into a monodimensional array (Bi–O 2.396 Å). The structure, **197**, is, however, more complex, because additional secondary Bi···O bonds (3.267 Å) are formed to the benzoyl oxygen of an adjacent molecule. A chelate ring from the carboxylato bridge (Bi···O 3.297 Å) is closed within the tecton.

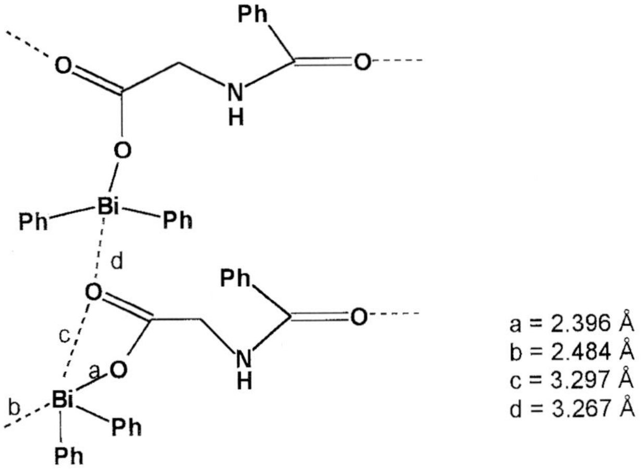

197

a = 2.396 Å
b = 2.484 Å
c = 3.297 Å
d = 3.267 Å

Secondary Bi···O bonds (3.37 Å) induce association of the molecules of bis(1-oxopyridine-2-thiolato)phenylbismuth into dimers [465]. The mesylamide derivative, [Ph$_2$BiN(SO$_2$Me)$_2$]$_2$, is also a cyclic dimer, associated via Bi···O bonds [466].

4.5.3.3 Organobismuth–sulfur compounds

This type of association is better demonstrated in organobismuth chemistry. Thus, the molecules of diphenylbismuth(III) isopropylxanthate, Ph$_2$BiS$_2$COPr^i, are self-organized in helical chains, **198**, with distinct primary (Bi–S 2.66 Å) and secondary (intermolecular Bi···S 3.23 Å) bismuth–sulfur bonds [467].

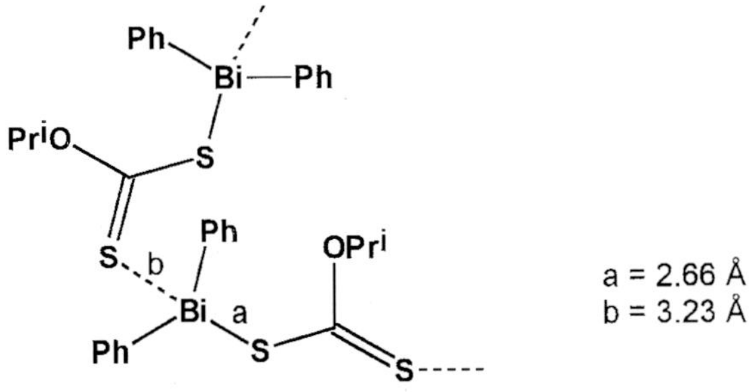

198

a = 2.66 Å
b = 3.23 Å

In a similar manner, methylbismuth(III) diethyldithiocarbamate, MeBi(S$_2$CNEt$_2$)$_2$, contains intramolecular primary Bi–S bonds (ave 2.97 Å) and intermolecular secondary Bi···S bonds (Bi···S 3.27 and 3.36 Å) leading to self assembly into dimeric structures, **199** [468]. Phenylbismuth(III) bis(methylxanthate), PhBi(S₂COMe)₂, is, however, unassociated and its crystal consists of discrete molecules [469].

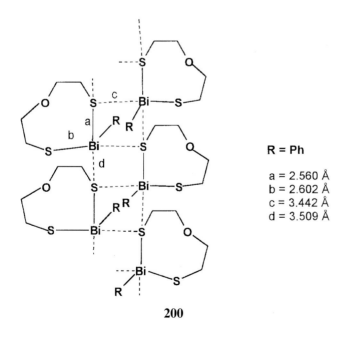

a = 2.98 Å
b = 3.36 Å
c = 3.27 Å
d = 2.96 Å

199

Oxadithiabismocane molecules are self-assembled into double chains by secondary bonding, to form a polymeric architecture, **200**, isostructural with that of the antimony analog. There is a clear distinction between normal covalent Bi–S bonds (intramolecular, 2.560 and 2.602 Å) and secondary Bi···S bonds (intermolecular, 3.440 and 3.590 Å) [470].

R = Ph

a = 2.560 Å
b = 2.602 Å
c = 3.442 Å
d = 3.509 Å

200

The tendency of bismuth–sulfur compounds to self-assemble via Bi···S secondary bonds is also illustrated by the dimerization of some compounds without direct Bi–C bonds, such as bismuth(III) thiolates, e.g. Bi(SC$_6$F$_5$)$_3$ (Bi–S 2.584 Å, Bi···S 3.323 Å) [471], bismuth(III) dithiophosphinates, e.g. Bi(S$_2$PMe$_2$)$_3$ [472] and Bi(S$_2$PPh$_2$)$_3$·C$_6$H$_6$ [473], and dithiophosphates, e.g. Bi[S$_2$P(OMe)$_2$]$_3$ [474], although monomeric, unassociated representatives of these classes are also known. The dithioarsinate Bi(S$_2$AsMe$_2$)$_3$ is also dimeric in the solid state [475]. This suggests that related organobismuth(III) derivatives, R$_2$Bi(S$_2$PR'$_2$), should also be self-assembled, at least in the solid state.

4.5.3.4 Organobismuth–nitrogen compounds

Few organobismuth–nitrogen compounds have been structurally characterized. The azide Me$_2$BiN$_3$ forms supramolecular chain-like arrays, similar to those described for arsenic and antimony. Again, the bridging bonds are equalized (Bi–N 2.49 and 2.50 Å). The chains are bent at nitrogen (Bi–N–Bi 123°) and almost linear at bismuth (N–Bi–N 169°) [476].

4.6 Group 16 – Selenium and Tellurium

4.6.1 Self-assembly in organoselenium compounds

Several inorganic compounds of selenium are known to contain secondary bonds, but only organoselenium derivatives will be covered here. Among these, the organoselenium halides are of interest, because selenium iodides were – for many years – considered not to exist; the compounds have recently become available [477].

The first structurally characterized arylselenium compound, 2,4,6-Bu$^t_3$C$_6$H$_2$SeI, was found to be monomeric in the crystal, without significant intermolecular Se···I, I···I, or Se···Se contacts [478]. Attempts to grow crystals of the related 2,4,6-Pr$^i_3$C$_6$H$_2$SeI produced an adduct of the corresponding diselenide with molecular iodine, **201**. The adduct is the product of self-assembly by secondary bonding between iodine and two selenium atoms from adjacent molecules.

The compound has been described (not quite appropriately) as an "intercalation compound" of solid diselenide with half equivalents of iodine.

More interesting is the structure of the diphenyldiselane–diiodine adduct, which is an eight-membered quasi-cyclic compound, **202**, formed by secondary bonding (Se···I 2.992 and 3.588 Å) [479]. The sum of the van der Waals radii for selenium and iodine is 3.88 Å.

Mesitylselenium iodide, 2,4,6-Me$_3$C$_6$H$_2$SeI, has a supramolecular self-organized architecture, **203**, based on secondary bonds between iodine and iodine, and between iodine and selenium pairs. Similar self-organization by secondary bonding

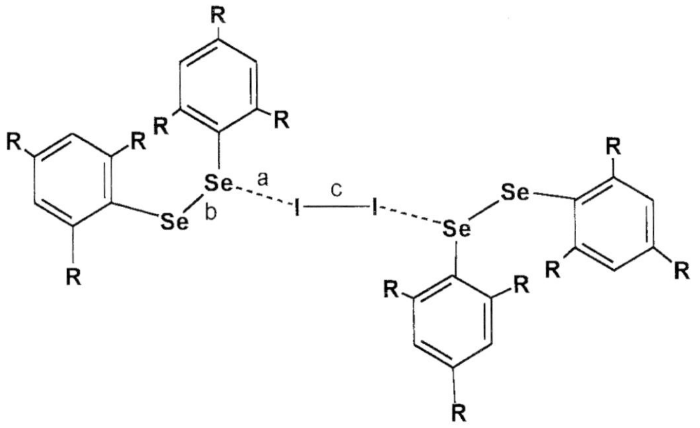

4.6 Group 16 – Selenium and Tellurium

201

R = Pr^i

a = 3.483 Å
b = 2.353 Å
c = 2.722 Å

202

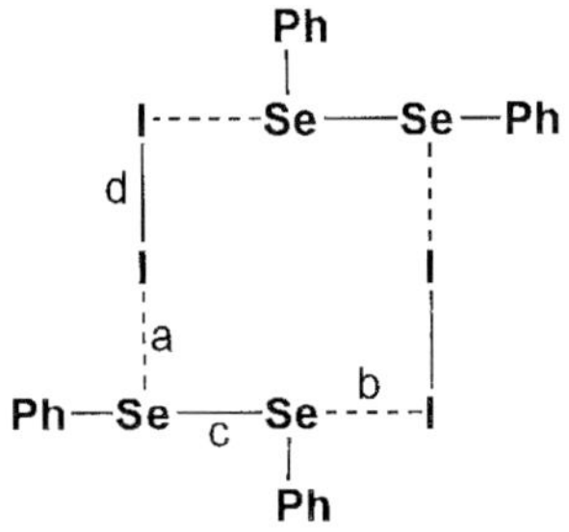

a = 2.992 Å
b = 3.588 Å
c = 2.347 Å
d = 2.775 Å

was also found in 1,2,4,5-tetramethylphenylselenium iodide, in which secondary bonds between selenium and iodine lead to formation of inorganic channels in the crystal, surrounded by the aromatic groups.

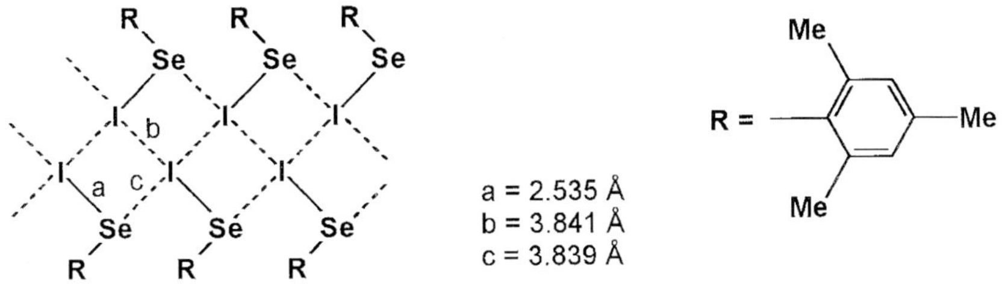

203

a = 2.535 Å
b = 3.841 Å
c = 3.839 Å

The structure of 4,4-dibromo-1-thia-4-selenacyclohexane [480], is dimeric, **204**, with primary Se–Br (2.545 and 2.548 Å) and secondary Se···Br (3.567 and 3.588 Å) as bridges.

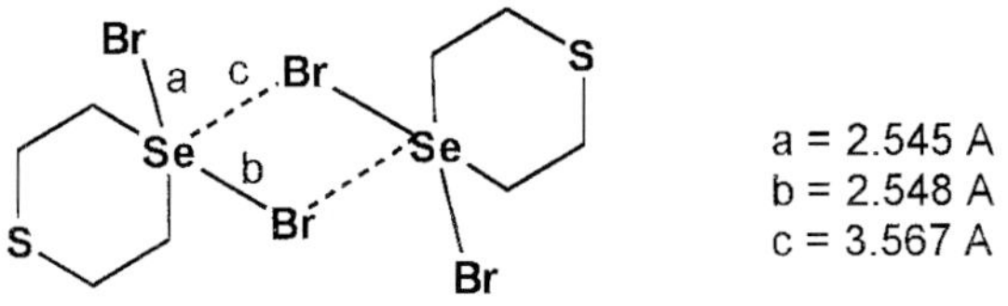

$$\text{Structure 204}$$

a = 2.545 Å
b = 2.548 Å
c = 3.567 Å

204

1,4-Bis(selenocyanato)benzene, p-NCSeC$_6$H$_4$SeCN, forms a supramolecular structure, **205**, in which the selenocyanato groups interact by secondary Se···N bonding (3.06 and 3.32 Å) to form ribbons running in the crystal [481]. The sum of the van der Waals radii for selenium and nitrogen is 3.5 Å.

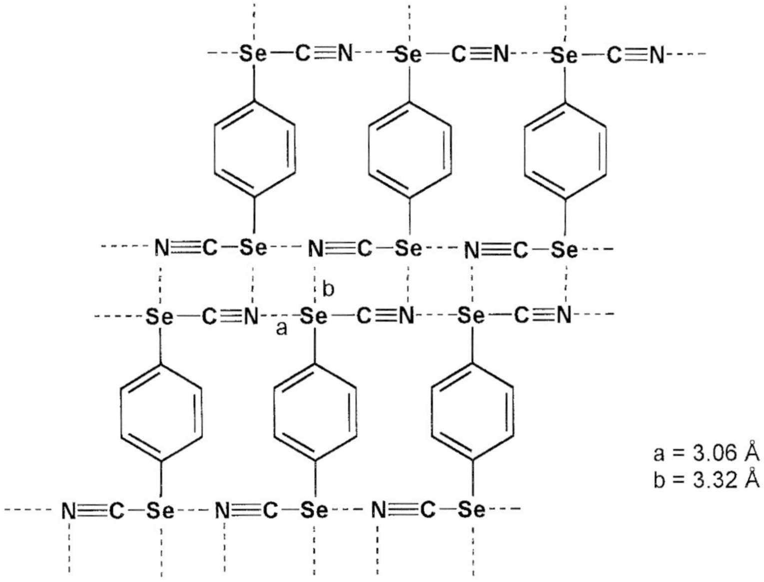

a = 3.06 Å
b = 3.32 Å

205

The compounds Se(SCN)$_2$ and Se(SeCN)$_2$ also form supramolecular arrays, **206**, via Se···N secondary bonds. The interatomic Se···N distances are 2.98, 3.03, and 3.32 Å in Se(SCN)$_2$ and 3.16, 3.08, and 3.27 Å in Se(Se(CN)$_2$ [481].

The few examples cited suggest that an interesting supramolecular chemistry of selenium might emerge from further structural investigation of a broader range of compounds.

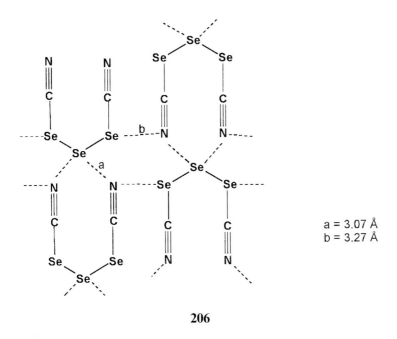

206

4.6.2 Self-assembly in organotellurium compounds

4.6.2.1 Organotellurium–halogen compounds

As a 'soft' element tellurium becomes involved in secondary bonding in many of its compounds and interesting self-assemblies are produced leading to supramolecular architectures. Tellurium is attractive because of its diversity of oxidation states and coordination numbers.

Several structurally characterized organotellurium halides have been found to be associated in the solid state. For example, α-dimethyltellurium dichloride, Me_2TeCl_2, forms a polymeric array (Te–Cl 2.541 and 2.480 Å; Te···Cl 3.46 and 3.52 Å) [482]. α-Me_2TeI_2 is a self-organized polymer (Te–I 2.854–2.994 Å, Te···I 3.659–4.030 Å) [483] whereas β-Me_2TeI_2 is a ionic compound $[Me_3Te]^+[MeTeI_4]^-$ with cation–anion secondary bond interactions (Te···I 3.84–3.400 Å) [484].

Diphenyltellurium dihalides, Ph_2TeX_2, with X = F (Te–F 2.006 Å, Te···F 3.208 Å) [485], X = Cl (Te–Cl 2.482 and 2.529 Å, Te···Cl 3.677 Å) [486], X = Br (Te–Br 2.682 Å, Te···Br 3.93 Å) [487], and X = I (Te–I 2.928 Å, Te···I 3.955 Å) [488], contain pseudo-trigonal bipyramidal tellurium coordination centers, with halogens in axial positions and secondary interactions between tellurium and the halogen atoms of adjacent molecules; this intermolecular association is also apparent from the ^{125}Te Mössbauer spectra [489]. Bis(pentafluorophenyltellurium) dihalides, $(C_6F_5)_2TeX_2$, with X = F (Te–F 1.990 Å, Te···F 2.952 Å) [490], Cl (Te–Cl 2.485

Å, Te···Cl 3.589 Å) and Br (Te–Br 2.650 Å and Te···Br 3.848 Å) [491] have been investigated. All contain distinct primary and secondary tellurium–halogen bonds, as suggested by the interatomic distances. The sum of the atomic radii for Te–Cl is 2.414 Å and the mean value calculated for 22 Te–Cl compounds was found to be 2.52 Å [492]. The sum of the van der Waals radii for Te···Cl is 3.81 Å. For the Te–Br bond the sum of atomic radii is 2.59 Å and van der Waals radii 3.91 Å.

Perfluoroethyltellurium trifluoride, $C_2F_5TeF_3$ is associated into infinite chains (terminal Te–F 1.870, 1.874 Å, bridging Te–F 2.190, 2.196 Å, Te···F 3.168 Å). The self-organized structure of $(C_6F_5)_2TeCl_2$ is shown in **207** and that of the corresponding bromide in **208**.

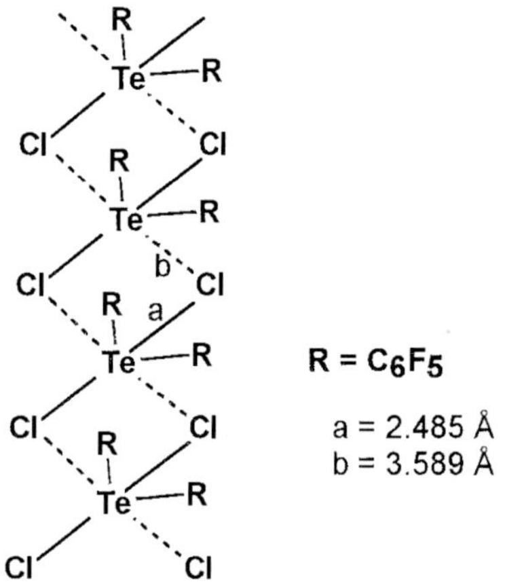

207 $R = C_6F_5$ $a = 2.485$ Å $b = 3.589$ Å

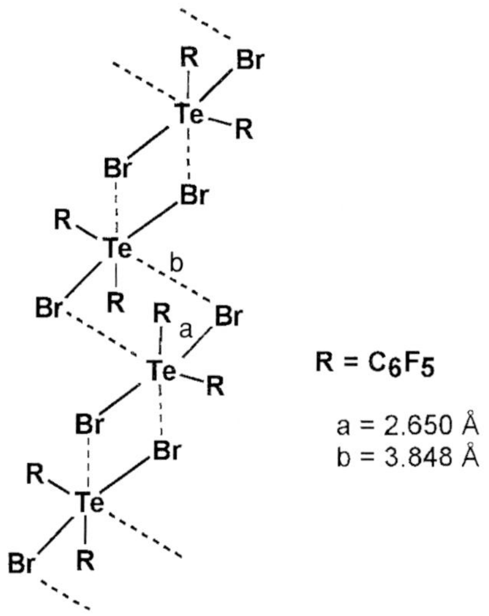

208 $R = C_6F_5$ $a = 2.650$ Å $b = 3.848$ Å

Weak Te···Cl bonds lead to intermolecular association of phenylazophenyltellurium chloride [493]. Phenoxatellurin-10,10-dichloride is a tetramer (Te–Cl 2.478, 2.576 Å, Te···Cl 3.504, 3.368 Å) [494] whereas the 10,10-diiodide is an iodine-bridged polymeric chain-like array (Te–I 2.941, 2.945 Å, Te···I 3.739, 3.788 Å) [495].

Bis(p-trimethylsilylphenyl)tellurium dichloride, $(4-Me_3SiC_6H_4)TeCl_2$, is another example of a self-organized supramolecular structure, **209** [496].

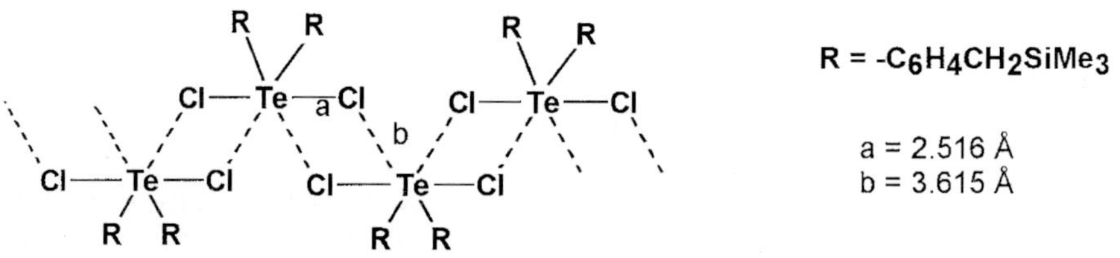

$R = -C_6H_4CH_2SiMe_3$

$a = 2.516$ Å
$b = 3.615$ Å

209

Similar intermolecular bonding is present in halogen derivatives of telluraheterocycles, such as 4,4-dibromo-1-thia-4-telluracyclohexane [497], 4,4-diiodo-1-thia-4-telluracyclohexane [498], 4,4-diiodotellura-1-oxacyclohexane [499], diiododibenzotellurophene [500], and 1,1-diiodo-3,4-benzo-1-telluracylopentane [501]. In 4,4-dibromo-1-thia-4-telluracyclohexane, $S(CH_2CH_2)_2TeBr_2$, the coordination geometry about tellurium is distorted octahedral and consists of two basically perpendicular Te–C bonds, two axial Te–Br bonds (Te–Br 2.657 and 2.689 Å) and two weak secondary bonds with a bromine atom of an adjacent molecule (Te···Br 3.591 Å) and with a sulfur atom of a third molecule (Te···S 3.588 Å). This leads to a supramolecular structure, **210**, made up of interconnected helices.

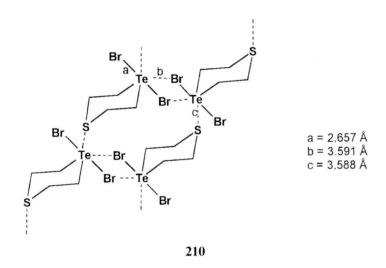

210

A different type of self-organization has been found for the analogous diiodo derivative, $S(CH_2CH_2)_2TeI_2$. In this compound the coordination geometry about tellurium is trigonal bipyramidal with iodine atoms in axial sites and the Te–C annular bonds in equatorial positions. The molecular tectons are associated by secondary bonding, with the formation of two sets of linear polymers and cross-linking between the sets (Te–I 2.851 Å, Te···I 2.985 Å). Both Te···I and I···I secondary interactions are present in the supramolecular architecture, each set containing ···I–Te–I···I–Te–I···moieties. The intermolecular I···I distances are 3.66 Å in one set and 3.90 Å in another set (the sum of the van der Waals radii is 4.30 Å). In each molecule the iodine atoms form a covalent bond with tellurium and a secondary bond with a tellurium atom of a neighboring molecule, in addition to the secondary I···I intermolecular contact, which results in a nearly planar group of two iodine and two tellurium atoms [498].

1,1-Dichloro-1-telluracyclohexane-3,5-dione forms a complex supramolecular structure, **211**, by secondary Te···Cl bonding, in which six-membered Te_3Cl_3 quasi-rings can be distinguished (Te–Cl 2.453–2.521 Å, Te···Cl 3.409, 3.650 Å) [502].

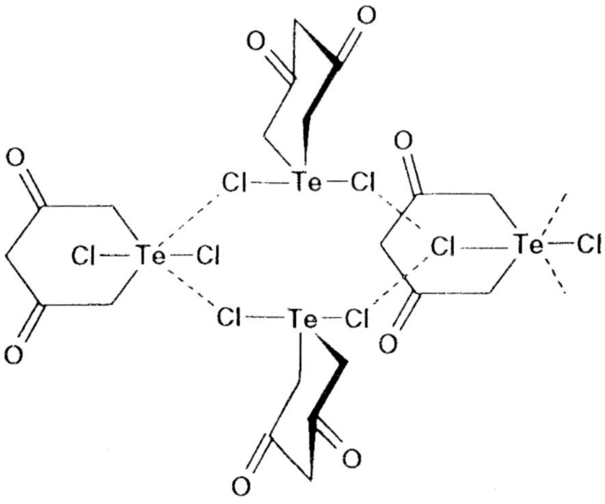

211

Halogen-bridged self-assembly is also observed in monoorganotellurium derivatives. It has been reported for dimeric p-ethoxyphenyltellurium trihalides [4-EtOC$_6$H$_4$TeX$_3$]$_2$ with X = Cl, Te–Cl 2.397, 2.395 Å, Te···Cl 2.740 and 2.757 Å; X = Br, Te–Br 2.527 and 2.509 Å, Te···Br 2.992 and 2.940 Å, and X = I, Te–I 2.776 and 2.797 Å, Te···I 3.097 and 3.192 Å) [503, 504], p-phenoxyphenyltellurium trichloride, [4-PhOC$_6$H$_4$TeCl$_3$]$_2$ [505], phenyltellurium trichloride (Te–Cl 2.369, 2.3762, Te···Cl 3.809, 4.118) [486], phenyltellurium triiodide (Te–I 2.775–2.792 Å, Te···I 3.152–3.285 Å) [488], and biphenylyltellurium trihalides, [2-PhC$_6$H$_4$TeX$_3$]$_n$ (X = Br, dimeric, Te–Br 2.490, 2.647, and 2.675 Å, Te···Br 3.713 Å [506]; and X = I (polymeric α- and β-forms)) [507, 508].

Other halogen-bridged supramolecules are the dimers p-methoxyphenyltellurium(IV) dimethyldithiophosphate dibromide, [4-MeOC$_6$H$_4$Te{S$_2$P(OMe)$_2$}-Br(μ-Br)]$_2$ (terminal Te–Br 2.616, bridging Te–Br 2.677 Å, Te···Br 3.810 Å), **212** [509], phenyltellurium(IV) diethyldithiocarbamate diiodide, [PhTe(S$_2$CNEt$_2$)-I(μ-I)]$_2$, **213**, p-methoxytellurium(IV) diethyldithiocarbamate bromide–iodide, [4-MeOC$_6$H$_4$Te(S$_2$CNEt$_2$)(Br$_{0.4}$I$_{0.6}$)]$_2$ [510], and [TeMe$_2$(S$_2$CNC$_5$H$_{11}$)(μ-Cl)]$_2$ (Te–Cl

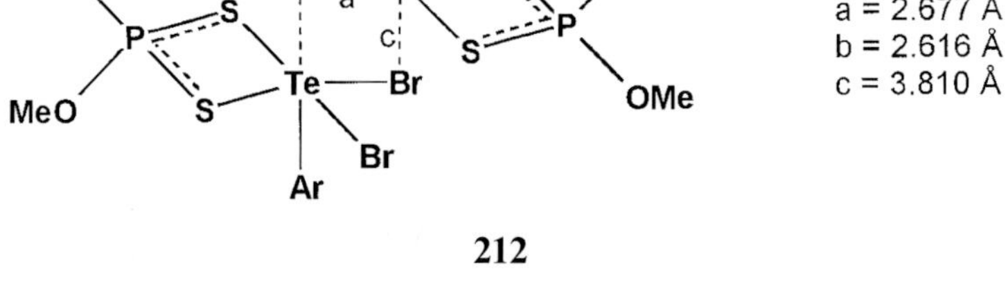

a = 2.677 Å
b = 2.616 Å
c = 3.810 Å

212

2,672 Å, Te···Cl 3.601 Å), whereas other compounds of similar compositions form chain-like arrays, e.g. [TeMe$_2$(S$_2$CNC$_5$H$_{11}$)(μ-I)]$_x$, **214** (Te–I 3.052 Å, Te···I 3.872 Å) [511]. The chain structure of the iodide is reminiscent of the helical supramolecular structure of the inorganic compound Te(S$_2$CNEt$_2$)I, associated through iodine bridging (Te–I 3.108 Å, Te···I 3.279 Å) [512].

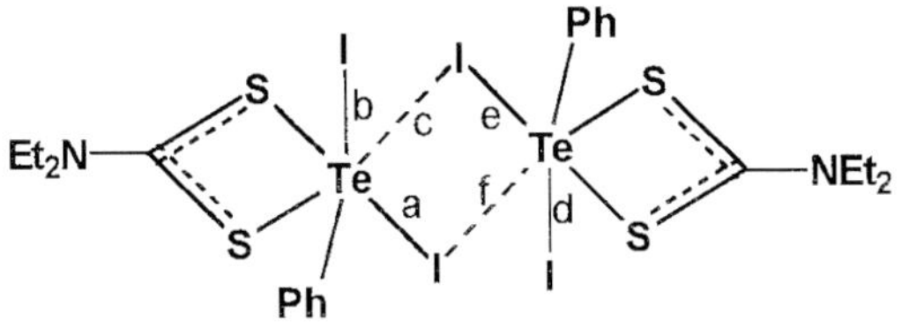

a = 2.952 Å
b = 2.942 Å
c = 4.233 Å
d = 2.948 Å
e = 2.941 Å
f = 4.192 Å

213

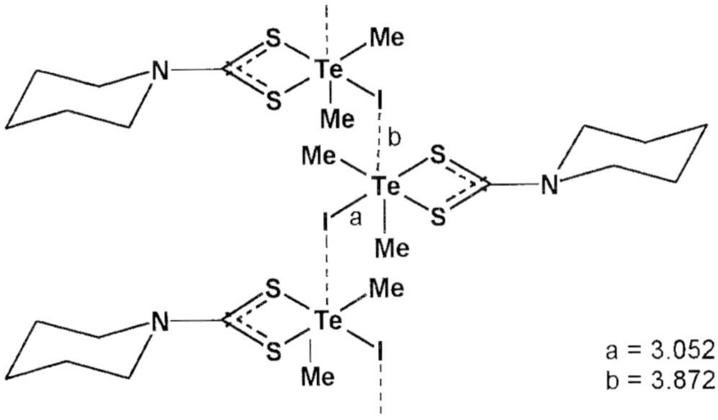

a = 3.052 Å
b = 3.872 Å

214

An interesting supermolecule is the dimer **215**, formed by hexafluoroantimonate bridging of two trimesityl ditellurium cations [513].

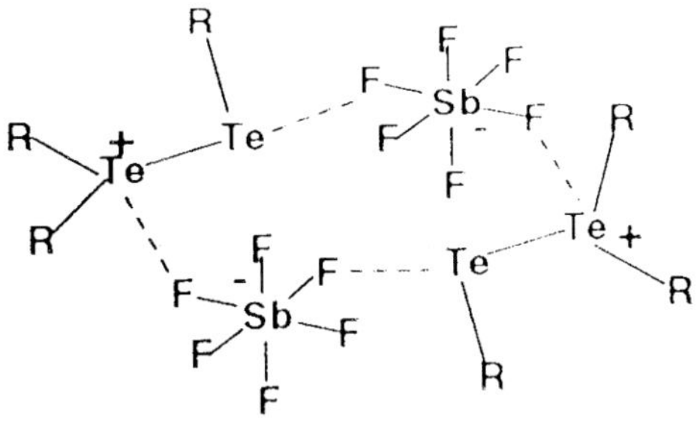

R = mesityl

215

Triethyltelluronium halides, [Et$_3$TeX]$_4$ are tetrameric cubane supermolecules, **216**, held together by secondary bonds (Te···Cl 3.448 Å, Te···I 3.813–3.861 Å) between tellurium and halogen in addition to the electrostatic attraction. Triphenyltelluronium iodide, Ph$_3$TeI forms dimers associated into two-dimensional sheets [514, 515]. The chloride Ph$_3$TeCl is a dimer (Te···Cl 3.142–3.234 Å) [516].

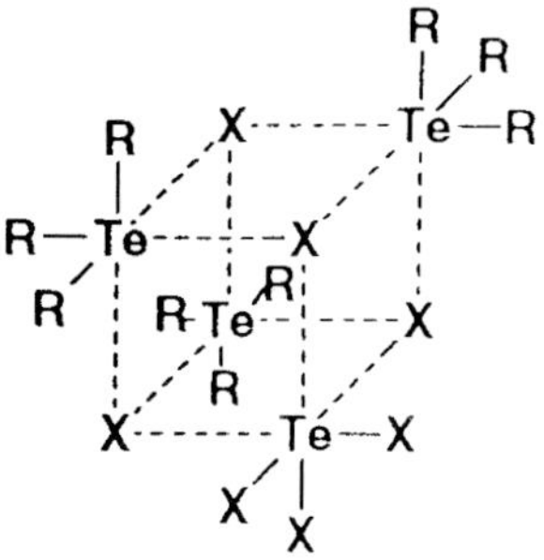

216

4.6.2.2 Organotellurium–oxygen compounds

The trichloro(ethane-1,2-diolato-*O,O'*)tellurate(IV) anion [OCH$_2$CH$_2$OTeCl$_3$]$_2$$^{2-}$, **217**, self-assembles into dimeric pairs, via Te···O secondary bonds (Te–O 1.946 and 1.968 Å, Te···O 2.764 Å) [517]. Oxygen-bridging is also present in bis(*p*-methoxyphenyl)tellurium(IV) diacetate [(4-MeOC$_6$H$_4$)$_2$Te(OOCMe)$_2$]$_2$ [518], and in diphenyltellurium(IV) bis(trichloroacetate), [Ph$_2$Te(OOCCCl$_3$)$_2$]$_2$, **218** [519].

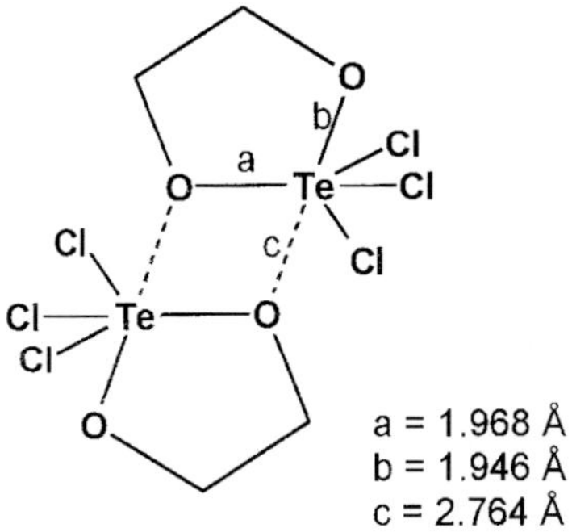

a = 1.968 Å
b = 1.946 Å
c = 2.764 Å

217

Tellurium–oxygen secondary bonds (2.762 and 2.787 Å) were found in dimeric 2,2'-biphenylene-2-biphenyltellurium 2,4,6-trichlorophenoxide [520].

4.6.2.3 Organotellurium–sulfur compounds

Tellurium–sulfur secondary bonds can also be a source of supramolecular self-organization [521]. These are common in organotellurium derivatives of thiophosphorus acids. Thus, in aryltellurium(II) dimethyldithiophosphates, ArTe-[S$_2$P(OMe)$_2$], where Ar = Ph [522], 4-MeOC$_6$H$_4$ [523], and EtOC$_6$H$_4$ [524], the molecules are assembled into chain structures, **219**, via dithiophosphate bridging.

Phenyltellurium(II) diphenyldithiophosphinate, PhTeS$_2$PPh$_2$, is thermochromic (yellow at 173 K and red at 273 K) and both forms contain zigzag chains, **220**, assembled as a result of Te···S secondary bonding (Te···S 3.383 Å in the yellow form and Te···S 3.422 Å in the red form) [525, 526].

A heterocyclic tellurium(IV) derivative, benzotelluracyclopentane dithiophosphate [527], contains both intra- and intermolecular Te···S secondary bonds, the latter leading to chain association of the molecular tectons, **221**.

It should be noted that in the examples cited, the dithiophosphorus ligands display three different modes of bridging (**222–224**).

Phenyltellurium(II) diphenyldithioimidodiphosphinate, PhTe(SPh$_2$PNPPh$_2$S), is a quasi-cyclic dimer, **225**, formed by Te···S secondary bonding (primary Te–S 2.557 Å, secondary Te···S 2.843 Å) [528].

In triphenyltelluronium compounds, the cation is frequently associated with the anion by secondary bonding. This has been observed in quasicyclic compounds

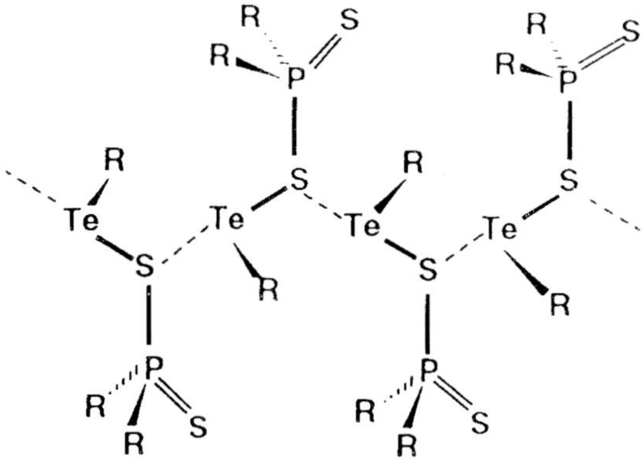

300   4 *Supramolecular Self-Assembly by Formation of Secondary Bonds*

220

221

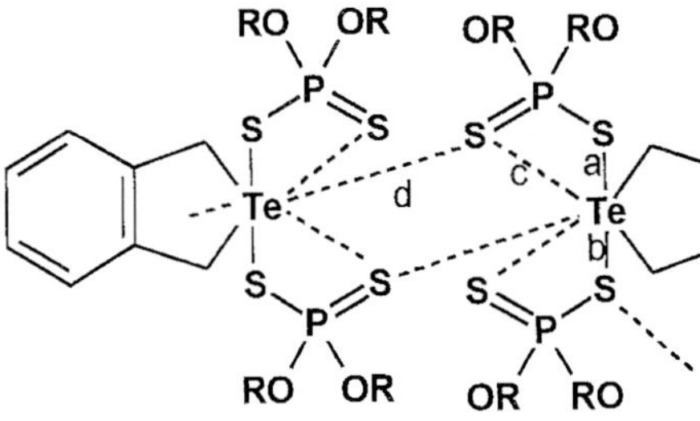

a = 2.621 Å
b = 2.627 Å
c = 3.447 Å
d = 3.393 Å

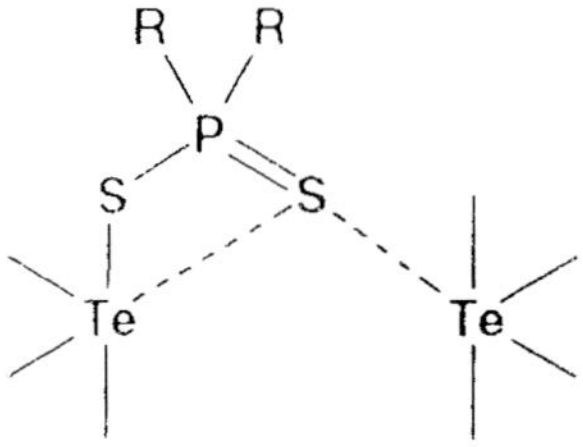

222 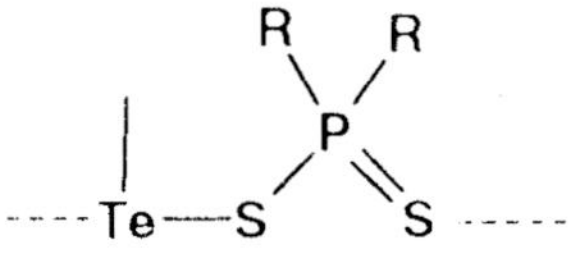   **223** 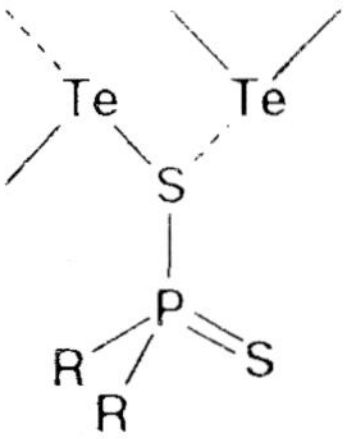   **224**

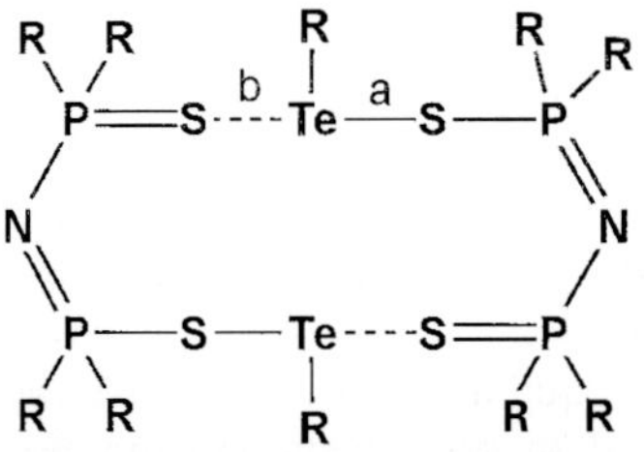

225

R = Ph

a = 2.557 Å
b = 2.843 Å

[Ph$_3$Te]$^+$[Ph$_2$PS$_2$]$^-$, **226**, and [Ph$_3$Te]$^+$[SPh$_2$PNPPh$_2$S]$^-$, **227**. Similarly, dimeric assemblies are formed by secondary bonding in [Ph$_3$Te]$^+$[PhSe]$^-$, **228**, (Te···Se 3.151–3.339 Å) [529].

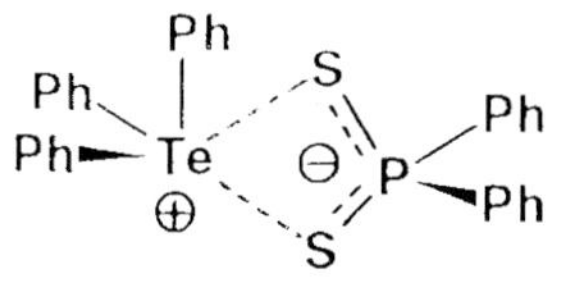

226

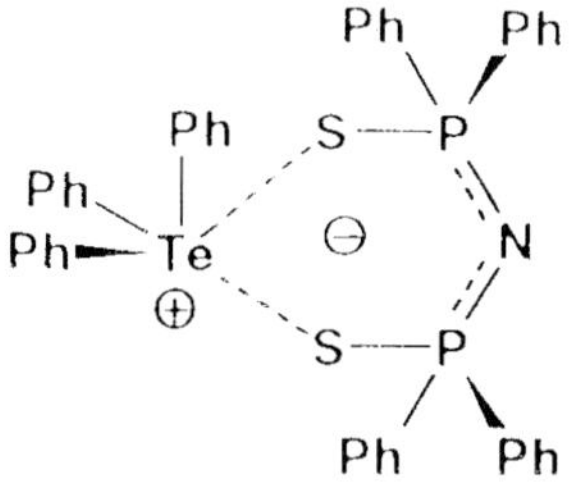

227

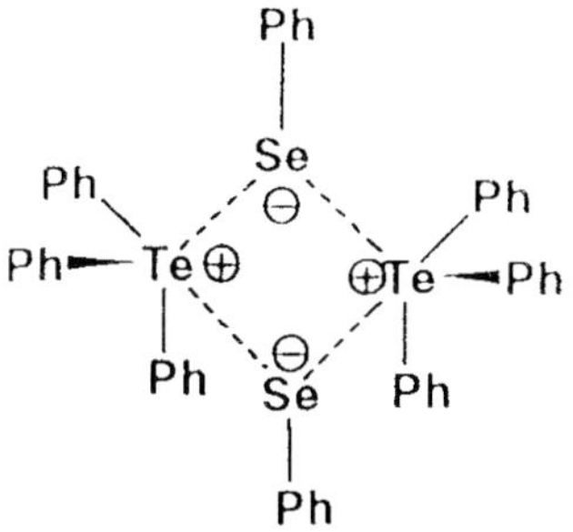

228

Organotellurium thiocarbonates also form supramolecular assemblies. Me$_2$Te(SOCOPr^i)$_2$ is a self-assembled tellurium–sulfur dimer and Me$_2$Tl-(SOCOPr^i)$_2$ is a chain-like array [530]. Organotellurium dithiocarbamates are also associated. Interestingly, some halo organotellurium dithiocarbamates, e.g. [TeMe$_2$I(S$_2$CNC$_4$H$_8$)]$_2$, **229**, and [TeMe$_2$Br(S$_2$CNC$_4$H$_8$)]$_x$, **230**, are supramolecular aggregates formed via tellurium–sulfur secondary bonds rather than tellurium–halogen bridging.

4.6.2.4 Organotellurium–nitrogen compounds

Dimeric self-assembly via tellurium–nitrogen secondary bonding has been reported in (2-phenylazophenyl-C,N')tellurium(II) thiocyanate (Te···N 3.535 Å), **231** [531]. The NCS group is linear (N–C–S 179.3°) and is oriented approximately perpendicular to the tellurium–organic ligand plane (Te–S–C 98.4°). In the bis-(thiocyanato) diphenyltelluriumoxide, (NCSTePh$_2$)$_2$O, however, the chain-like supramolecular self-assembly occurs as a result of tellurium–sulfur secondary bonds (Te···S 3.416 Å) [532].

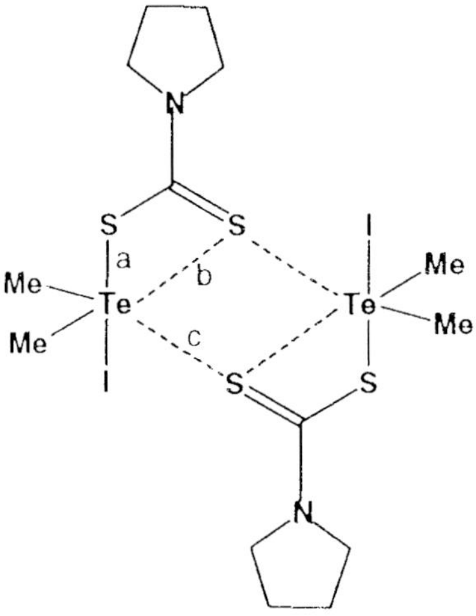

a = 2.501
b = 3.281
c = 3.562

229

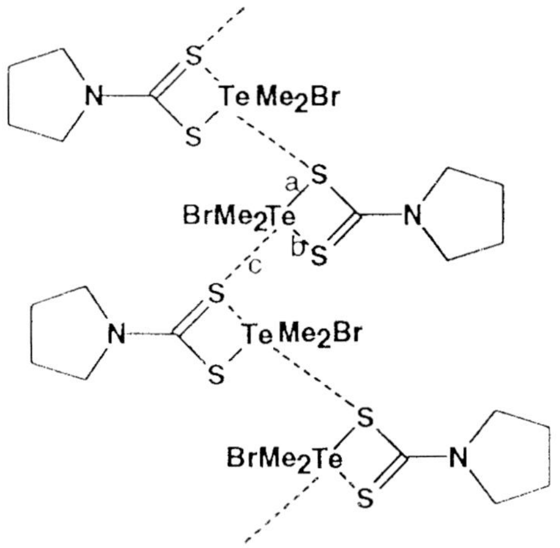

a = 2.519
b = 3.232
c = 3.562

230

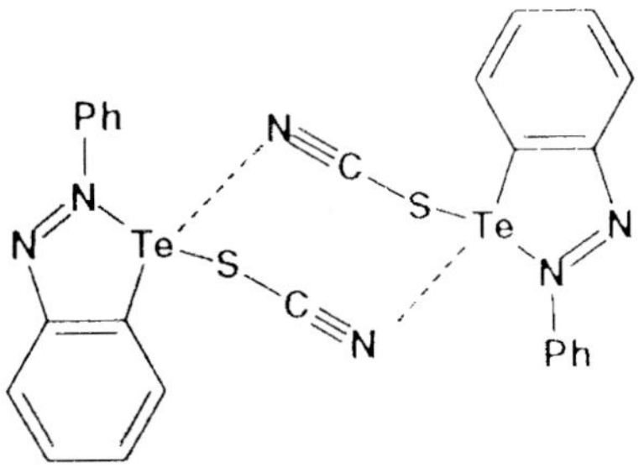

231

Polymeric self-assembly occurs in 1,2,5-telluradiazole, **232** [533] and in phenanthro[9,10c]-1,2,5-telluradiazole (Te···N 2.834 Å) [534], but di-(*tert*-butyl)-benzotelluradiazole associates only as dimers, **233** [535]. The parent telluradiazole is a high-melting solid, poorly soluble in organic solvents, because of strong inter-molecular bonds [536].

The triphenyltelluronium cyanate, $Ph_3Te(NCO)$, and thiocyanate, $Ph_3Te(SCN)$

are associated by Te···N, Te···O, and Te···S secondary bonds. The cyanate forms tetramers [537] and the thiocyanate forms dimers and tetramers [538].

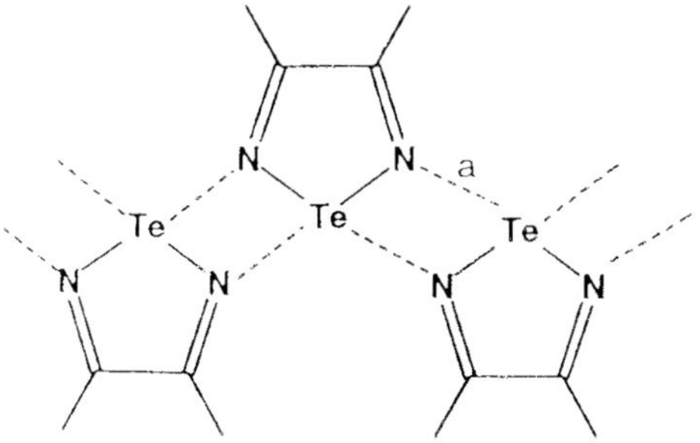

R = H   a = 2.764

232

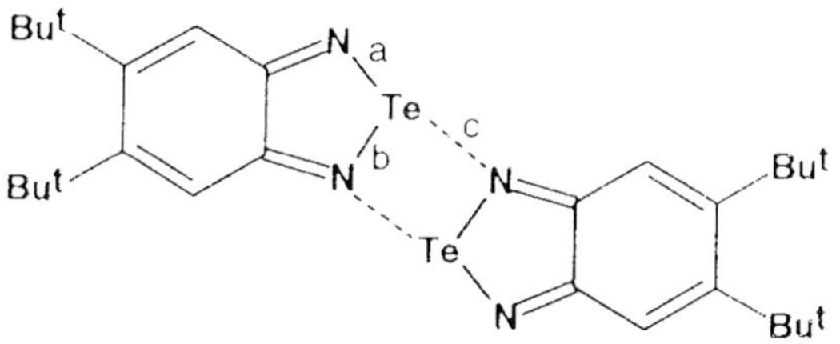

a = 2.006
b = 2.002
c = 2.628

233

References

[1] G. Becker and O. Mundt, in *"Unkonventionelle Wechselwirkungen in der Chemie metallischer Elemente"*, Ed. B. Krebs, VCH Weinheim, **1992**, p. 199.
[2] P. Pyykkö, *Chem. Rev.* **1997**, *97*, 597.
[3] J. J. Daly and F. Sanz, *Helv. Chim. Acta* **1970**, *53*, 1879.
[4] P. Choudhury and A. L. Rheingold, *Inorg. Chim. Acta* **1978**, *28*, L 127.
[5] G. Becker, G. Gutekunst and C. Witthauer, *Z. Anorg. Allg. Chem.* **1982**, *486*, 90.
[6] A. J. Ashe III, W. M. Butler and T. R. Diephouse, *Organometallics*, **1983**, *2*, 105.
[7] E. V. Atomonov, K. Megges, S. Wocadlo and J. Lorbert, *J. Organomet. Chem.* **1996**, *524*, 253.

[8] W. Chen, R. C. S. Wong and L. Y. Goh, *Acta Cryst.* **1994**, *C50*, 998.
[9] O. J. Scherer, W. Wiedemann and G. Wolmershäuser, *Chem. Ber.* **1990**, *123*, 3.
[10] I. Bernal, H. Brunner, W. Meier, H. Pfisterer, J. Wachter and M. L. Ziegler, *Angew. Chem. Int. Ed. Engl.* **1984**, *23*, 438.
[11] R. Rösler, *private communication*.
[12] G. Mundt, H. Riffel, G. Becker and A. Simon, *Z. Naturforsch.* **1984**, *B39*, 317.
[13] A. J. Ashe III, E. G. Ludwig Jr., J. Oleksyszyn and J. C. Huffman, *Organometallics*, **1984**, *3*, 337.
[14] A. J. Ashe III, *Adv. Organomet. Chem.* **1990**, *30*, 77.
[15] G. Becker, H. Freundenblum and C. Witthauer, *Z. Anorg. Allg. Chem.* **1982**, *492*, 37.
[16] S. Röller, M. Dräger, H. J. Breunig, M. Ates and J. Gülec, *J. Organomet. Chem.* **1989**, *378*, 327.
[17] S. Röller, M. Dräger, H. J. Breunig, M. Ates and S. Gülec, *J. Organomet. Chem.* **1987**, *329*, 319.
[18] G. Becker, M. Meiser, O. Mundt and J. Weidlein, *Z. Anorg. Allg. Chem.* **1989**, *569*, 62.
[19] K. von Deuten and D. Rehder, *Cryst. Struct. Commun.* **1980**, *9*, 167.
[20] H. Bürger, R. Eujen, G. Becker, O. Mundt, M. Westerhausen and C. Whittaker, *J. Mol. Struct.* **1983**, *98*, 265.
[21] A. J. Ashe III, W. M. Butler and T. R. Diephouse, *J. Am. Chem. Soc.* **1981**, *103*, 207.
[22] A. J. Ashe III, W. M. Butler and T. R. Diephouse, *Organometallics*, **1983**, *2*, 1005.
[23] H. J. Breunig and S. Gülec, in *"Unkonventionelle Wechselwirkungen in der Chemie metallischer Elemente"*, Ed. B. Krebs, VCH, Weinheim, **1992**, p. 218.
[24] T. Hughbanks, R. Hoffmann, M. H. Whangbo, K. R. Stewart, O. Eisenstein and E. Canadell, *J. Am. Chem. Soc.* **1982**, *104*, 3876.
[25] M. Ates, H. J. Breunig, S. Gülec, W. Offermann, K. Häberle and M. Dräger, *Chem. Ber.* **1989**, *122*, 473.
[26] O. Mundt, G. Becker, H. J. Wessely, H. J. Breunig and H. Kischkel, *Z. Anorg. Allg. Chem.* **1982**, *486*, 70.
[27] H. J. Breunig, A. Soltani-Neshan, K. Häberle and M. Dräger, *Z. Naturforsch.* **1986**, *41b*, 327.
[28] J. Ellermann, E. Köck and H. Burzlaff, *Acta Cryst.* **1985**, *C41*, 1437.
[29] H. J. Breunig, R. Rössler and E. Lork, *Angew. Chem.* **1997**, *109*, 2941; *Angew. Chem. Int. Ed. Engl.* **1997**, *36*, 2819.
[30] O. Mundt, G. Becker, M. Rössler and C. Whittaker, *Z. Anorg. Allg. Chem.* **1983**, *506*, 42.
[31] O. Mundt, G. Becker, M. Rössler and C. Whittaker, *Z. Anorg. Allg. Chem.* **1982**, *486*, 90.
[32] F. Calderazzo, R. Poli and G. Pelizzi, *J. Chem. Soc. Dalton Trans.* **1984**, 2365.
[33] G. Becker, J. Baumgarten, O. Mundt, H. Riffel and A. Simon, *unpublished results quoted in ref. 1*, p. 199.
[34] J. D. Lee and M. W. R. Bryant, *Acta Cryst.* **1969**, *B55*, 2094.
[35] R. E. Marsh, *Acta Cryst.* **1952**, *5*, 458.
[36] G. Llabres, O. Dideberg and L. Dupont, *Acta Cryst.* **1972**, *B28*, 2438.
[37] D. J. Sandman, J. C. Stark and B. M. Foxman, *Organometallics*, **1982**, *1*, 739.
[38] R. P. Shibaeva and V. F. Kaminski, *Cryst. Struct. Comm.* **1981**, *10*, 663.
[39] S. Ludlow and A. E. McCarthy, *J. Organomet. Chem.* **1981**, *219*, 169.
[40] E. Arai, H. Fujiwara, H. Kobayashi, K. Takimiya, O. Otsubo and F. Ogura, *Inorg. Chem.* **1998**, *37*, 2850.
[41] C. L. Raston, R. J. Secomb and A. H. White, *J. Chem. Soc. Dalton Trans.* **1976**, 2307.
[42] J. C. Dewan and J. Silver, *Acta Cryst.* **1977**, *B33*, 1469.
[43] J. Jeske, W. W. du Mont and P. G. Jones, *Angew. Chem.* **1996**, *108*, 2822; *Angew. Chem. Int. Ed. Engl.* **1996**, *35*, 2653.
[44] M. G. Newton, R. B. King, I. Haiduc and A. Silvestru, *Inorg. Chem.* **1993**, *32*, 3795.
[45] D. Loos, E. Baum, A. Ecker, H. Schnöckel and A. J. Downs, *Angew. Chem.* **1997**, *109*, 894; *Angew. Chem. Int. Ed. Engl.* **1997**, *36*, 860.
[46] W. Uhl, W. Hiller, M. Laugh and W. Schwarz, *Angew. Chem.* **1992**, *104*, 1378; *Angew. Chem. Int. Ed. Engl.* **1992**, *31*, 1364.
[47] W. Uhl and A. Jantschak, *J. Organomet. Chem.* **1998**, *555*, 263.

[48] G. Linti, *J. Organomet. Chem.* **1996**, *520*, 107.
[49] W. Uhl, R. Graupner, W. Hiller and M. Neumayer, *Angew. Chem.* **1997**, *109*, 62.
[50] H. Schumann, C. Janiak, J. Pickardt and V. Börner, *Angew. Chem.* **1987**, *99*, 788; *Angew. Chem. Int. Ed. Engl.* **1987**, *27*, 789.
[51] H. Schumann, C. Janiak, M. A. Khan and J. J. Zuckerman, *J. Organomet. Chem.* **1988**, *354*, 7.
[52] W. Uhl, S. U. Keimling, K. W. Klinkhammer and W. Schwarz, *Angew. Chem.* **1997**, *109*, 64; *Angew. Chem. Int. Ed. Engl.* **1977**, *36*, 64.
[53] C. Janiak, S. Temizdemir and T. G. Scharmann, *Z. Anorg. Allg. Chem.* **1998**, *624*, 755.
[54] A. L. Rheingold, L. M. Liable-Sands and S. Trofimenko, *Chem. Commun.* **1997**, 1691.
[55] C. Janiak and R. Hoffmann, *J. Am. Chem. Soc.*, **1990**, *112*, 5924.
[56] C. Janiak, *Coord. Chem. Rev.* **1997**, *163*, 107.
[57] C. Janiak and R. Hoffmann, *Angew. Chem.* **1989**, *101*, 1706; *Angew. Chem. Int. Ed. Engl.* **1989**, *28*, 1688.
[58] P. H. M. Budzelaar and J. Boersma, *Recl. Trav. Chim. Pays-Bas*, **1990**, *109*, 187.
[59] P. Schwerdtfeger, *Inorg. Chem.* **1991**, *30*, 1660.
[60] K. W. Hellmann, L. H. Gade, R. Fleischer and D. Stalke, *Chem. Commun.* **1997**, 527.
[61] H. Schmidbaur, *Chem. Soc. Rev.* **1995**, *24*, 383
[62] D. M. P. Mingos, *J. Chem. Soc. Dalton Trans.* **1996**, 561
[63] S. S. Pathaneni and G. R. Desiraju, *J. Chem. Soc. Dalton Trans.* **1993**, 319
[64] D. S. Eggleton, D. F. Chodosh, R. L. Webb and L. L. Davies, *Acta Cryst.* **1986**, *C42*, 36.
[65] T. J. Mathieson, A. G. Langdon, N. B. Milestone and B. K. Nicholson, *Chem. Commun.* **1998**, 371
[65a] T. J. Mathieson, A. G. Langdon, N. B. Milestone and B. K. Nicholson, *Chem. Commun.* **1999**, 201
[66] C. M. Che, H. K. Yip, W. T. Wong and T. F. Lai, *Inorg. Chim. Acta* **1992**, *197*, 177.
[67] J. Vicente, M. T. Chicote, M. D. Abrisqueta, R. Guerrero and P. G. Jones, *Angew. Chem.* **1997**, 109, 1252; *Angew. Chem. Int. Ed. Engl.* **1997**, *36*, 1203.
[68] D. M. P. Mingos, S. Menzer, D. J. Williams and J. Yau, *Angew. Chem.* **1995**, *107*, 2045; *Angew. Chem. Int. Ed. Engl.* **1995**, *34*, 1894.
[69] C. E. Holloway and M. Melnik, *J. Organomet. Chem.* **1995**, *495*, 1.
[70] L. G. Kuz'mina and Y. T. Struchkov, *Croatica Chemica Acta*, **1984**, *57*, 701.
[71] A. J. Canty and G. B. Deacon, *Inorg. Chim. Acta* **1980**, *45*, L 225.
[72] D. Viets, E. Lork, P. G. Watson and R. Mews, *Angew. Chem.* **1997**, *109*, 655; *Angew. Chem. Int. Ed. Engl.* **1997**, *36*, 623.
[73] B. Korpar-Colig, Z. Popovic, M. Bruvo and I. Vickovic, *Inorg. Chim. Acta*, **1988**, *150*, 113.
[74] M. Ali, W. R. McWhinnie and T. A. Hamor, *J. Organomet. Chem.* **1989**, *371*, C37.
[75] E. C. Constable, T. A. Leese and D. A Tocher, *J. Chem. Soc. Chem. Comm.* **1989**, 570.
[76] P. B. Hitchcock, J. M. Keates and G. A. Lawless, *J. Am. Chem. Soc.* **1998**, *120*, 599.
[77] G. B. Deacon, B. M. Gatehouse, L. W. Guddat and S.C. Ney, *J. Organomet. Chem.* **1989**, *375*, C1.
[78] R. G. Gerr, M. Yu. Antipin, N. G. Furmanova and Yu. T. Struchkov, *Sov. Phys. Crystallogr.* **1979**, *24*, 543.
[79] J. Lorberth, T. F. Berlitz and W. Massa, *Angew. Chem.* **1989**, *104*, 623; *Angew. Chem. Int. Ed. Engl.* **1989**, *28*, 611.
[80] M. Tschinkl, A. Schier, J. Reide, G. Mehltretter and F. P. Gabbai, *Organometallics* **1998**, *17*, 2924.
[81] M. Tschinkl, A. Schier, J. Riede and F. P. Gabbai, *Inorg. Chem.* **1997**, *36*, 5706.
[82] K. P. Jensen, D. K. Breitinger and W. Kress, *Z. Naturforsch.* **1981**, *36b*, 188.
[83] J. Halfpenny and R. W. H. Small, *Acta Cryst.* **1979**, *B35*, 1239.
[84] T. C. W. Mak and J. Trotter, *J. Chem. Soc. A*, **1962**, 3243.
[85] R. Zouari, A. Ben Salah, A. Daoud, W. Rothammel and H. Burzlaff, *Acta Cryst.* **1993**, *C49*, 1596.
[86] A. L. Beauchamp, M. J. Olivier, J. D. Wuest and B. Zacharie, *J. Am. Chem. Soc.* **1986**, *108*, 73.
[87] A. L. Beauchamp, M. J. Olivier, J. D. Wuest and B. Zacharie, *Organometallics*, **1987**, *6*, 153.

[88] R. D. Bach, H. B. Vardhan, A. F. M. Maqsudur Rahman and J. P. Oliver, *Organometallics* **1985**, *4*, 846.
[89] B. Kamenar and M. Penavic, *Inorg. Chim. Acta*, **1972**, *6*, 191.
[90] B. Kamenar, M. Penavic and A. Hergold-Brundic, *Croatica Chemica Acta*, **1984**, *57*, 145.
[91] J. S. Casas, E. E. Castellano, M. S. Garcia-Tasende, A. Sanchez, J. Sordo, E. M. Vazquez-Lopez and J. Zukerman-Schpector, *J. Chem. Crystallogr.* **1996**, *26*, 123.
[92] C. Raston, B. W. Skelton and A. H. White, *Aust. J. Chem.* **1978**, *31*, 537.
[93] W. Lau and J. K. Kochi, *J. Org. Chem.* **1986**, *51*, 1801.
[94] D. Grdenic, M. Sikirica and D. Matkovic-Calogovic, *J. Organomet. Chem.* **1986**, *306*, 1.
[95] D. Grdenic, D. Matkovic and M. Sikirica, *J. Organomet. Chem.* **1987**, *319*, 1.
[96] D. Grdenic, M. Sikirica, D. Matkovic-Calogovic and A. Nagl, *J. Organomet. Chem.* **1983**, 283.
[97] S. Deguire and A. L. Beauchamp, *Acta Cryst.* **1990**, *C46*, 27.
[98] D. L. Rabenstein, *J. Chem. Educ.* **1978**, *55*, 292.
[99] D .L. Rabenstein and C. A. Evans, *Bioinorg. Chem.* **1978**, *8*, 107.
[99a] A. Sigel and H. Sigel (Editors), *Metal Ions in Biological Systems. Vol. 34. Mercury and its Effects on Environmental Biology*, M. Dekker, Inc., New York, **1994**.
[100] A. P. Arnold, A. J. Canty, P. W. Moors and G. B. Deacon, *J. Inorg. Biochem.* **1983**, *19*, 319.
[101] D. L. Rabenstein, *Acc. Chem. Res.* **1978**, *11*, 100.
[102] M. C. Corbell and A. L. Beauchamp, *Can. J. Chem.* **1986**, *64*, 1876.
[103] M. C. Corbell and A. L. Beauchamp, *Can. J. Chem.* **1988**, *66*, 1379.
[104] J. P. Charland, J. F. Britten and A. L. Beauchamp, *Inorg. Chim. Acta*, **1986**, *124*, 161.
[105] J. P. Charland and A. L. Beauchamp, *Inorg. Chem.* **1986**, *25*, 4870.
[106] L. Prizant, R. Rivest and A. L. Beauchamp, *Can. J. Chem.* **1981**, *59*, 2290.
[107] A. R. Norris, R. Kumar and A. L. Beauchamp, *Inorg. Chim. Acta* **1989**, *162*, 139.
[108] A. R. Norris, S. E. Taylor, E. Buncel, F. Belanger-Gariepy and A. L. Beauchamp, *Can. J. Chem.* **1983**, *61*, 1536.
[109] R. H. Bach, H. B. Vardham, A. F. M. Maqsudur Rahman and J. P. Oliver, *Organometallics* **1985**, *4*, 846.
[110] P. Barbaro, C. A. Ghilardi, S. Midollini and G. Scapacci, *J. Organomet. Chem.* **1998**, *555*, 255.
[111] J. S. Casas, A. Castineiras, I. Haiduc, A. Sanchez, J. Sordo and E. M. Vazquez-Lopez, *Polyhedron*, **1997**, *16*, 781.
[112] C. Chieh and L. P. C. Leung, *Can. J. Chem.* **1976**, *54*, 3077.
[113] E. R. T. Tiekink, *J. Organomet. Chem.* **1987**, *322*, 1.
[114] E. R. T. Tiekink, *Inorg. Chim. Acta* **1986**, *112*, L 1.
[115] E. Block, M. Brito, M. Gernon, D., McGowty, H. Kang and J. Zubieta, *Inorg. Chem.* **1990**, *29*, 3172.
[116] M. V. Castano, M. M. Plasencia, A. Macias, J. S. Casas, J. Sordo and E. E. Castellano, *J. Chem. Soc. Dalton Trans.* **1989**, 1409.
[117] N. W. Alcock, P. A. Lampe and P. Moore, *J. Chem. Soc. Dalton Trans.* **1980**, 1471.
[118] S. Wang and J. P. Fackler, Jr., *Inorg. Chem.* **1989**, *28*, 2615.
[119] J. H. Aupers, R. A. Howie and J. L. Wardell, *Polyhedron* **1997**, *16*, 2283.
[120] A. J. Canty, *ACS Symp. Ser.* **1978**, *82*, 339.
[121] E. M. Vazquez-Lopez, A. Castineiras, A. Sanchez, J. Casas and J. Sordo, *J. Cryst. Spectr. Res.* **1992**, *22*, 403.
[122] J. Zukerman-Schpector, E. M. Vazquez-Lopez, A. Sanchez, J. S. Casas, and J. Sordo, *J. Organomet. Chem.* **1991**, *405*, 67.
[123] J. S. Casas, A. Castineiras, A. Sanchez, J. Sordo and E. M. Vazquez-Lopez, *J. Organomet. Chem.* **1994**, *468*, 1.
[124] M. M. Kubicki, R. Kergoat, J. S. Guerchais, I. Bkouche-Waksman, C. Bois and P. L. Hardon, *J. Organomet. Chem.* **1981**, *219*, 329.
[125] N. W. Alcock, P. A. Lampe and P. Moore, *J. Chem. Soc. Dalton Trans.* **1980**, *1471*.
[126] J. Bravo, J. S. Casas, M. V. Castano, M. Gayoso, Y. P. Mascarenhas, A. Sanchez, C. de O. P. Santos and J. Sordo, *Inorg. Chem.* **1985**, *24*, 3435.

[127] A. T. Hutton, H. M. N. H. Irving, L. R. Nassimbeni and G. Gafner, *Acta Cryst.* **1980**, *B36*, 2064.
[128] D. A. Stuart, L. R. Nassimbeni, A. T. Hutton and K. R. Koch, *Acta Cryst.* **1980**, *B36*, 2227.
[129] M. Bochmann, A. P. Coleman and A. K. Powell, *Polyhedron*, **1992**, *11*, 507.
[130] A. P. Arnold, A. J. Cany, B. W. Skelton and A. H. White, *J. Chem. Soc. Dalton Trans.* **1982**, 607.
[131] U. Müller, *Z. Naturforsch.* **1973**, *28b*, 426.
[132] D. J. Brauer, H. Burger, G. Pawelke, K. H. Flegler and A. Haas, *J. Organomet. Chem.* **1978**, *160*, 389.
[133] J. C. Mills, H. S. Preston and C. H. L Kennard, *J. Organomet. Chem.* **1968**, *14*, 33.
[134] M. S. Garcia-Tasende, M. I. Suarez-Gimeno, A. Sanchez, J. S. Casas and E. E. Castellano, *J. Organomet. Chem.* **1990**, *384*, 19.
[135] J. P. Charland, *Inorg. Chim. Acta* **1987**, *135*, 191.
[136] S. J. Rettig, A. Storr and J. Trotter, *Can. J. Chem.* **1983**, *62*, 1705.
[137] D. G. Hendershot, M. Barber, R. Kumar and J. P. Oliver, *Organometallics* **1991**, 10.
[137a] M. Uson-Finkelzeller, W. Bublak, B. Huber, G. Müller and H. Schmidbaur, *Z. Naturforsch.* **1986**, *41b*, 346.
[137b] H. Schmidbaur, R. Nowak, B. Huber and G. Müller, *Polyhedron* **1990**, *9*, 283.
[137c] H. Schmidbaur, U. Thewalt and T. Zafiropolos, *Chem. Ber.* **1984**, *117*, 338.
[137d] H. Schmidbaur, W Bublak, B. Huber and G. Müller, *Organometallics* **1986**, *5*, 1647.
[138] H. D. Hausen, K. Mertz, E. Veigel and J. Weidlein, *Z. Anorg. Allg. Chem.* **1974**, *410*, 156.
[139] O. T. Beachley Jr., J. D. Maloney, M. R. Churchill and C. H. Lake, *Organometallics* **1991**, *10*, 3568.
[140] K. Mertz, W. Schwarz, F. Zettler and H. D. Hausen, *Z. Naturforsch.* **1975**, *30b*, 159.
[141] J. T. Leman, J. W. Ziller and A. R. Barron, *Organometallics* **1991**, *10*, 1766.
[142] O. T. Beachley Jr., E. F. Spiegel, J. P. Kopasz and R. D. Rogers, *Organometallics*, **1989**, *8*, 1915.
[143] D. L. Reger, S. J. Knox, A. L. Rheingold and B. S. Haggerty, *Organometallics* **1990**, *9*, 2581.
[143a] A. H. Cowley, F. P. Gabai, H. S. Isom, A. Decken, and R. D. Culp, *Main Group Met. Chem.* **1995**, *1*, 9.
[144] F. W. B. Einstein, M. M. Gilbert and D. G. Tuck, *J. Chem. Soc. Dalton Trans.* **1973**, 248.
[145] H. D. Hausen and H. U. Schwering, *Z. Anorg. Allg. Chem.* **1973**, *398*, 119.
[146] H. D. Hausen, *Z. Naturforsch.* **1972**, *27b*, 82; H. D. Hausen and H. J. Guder, *J. Organomet. Chem.* **1973**, *57*, 243.
[147] D. C. Bradley, H. Dawes, D. M. Frigo, M. B. Hursthouse and B. Hussain, *J. Organomet. Chem.* **1987**, *325*, 55.
[148] M. R. Kopp and B. Neumüller, *Organometallics* **1997**, *16*, 5623.
[149] F. Brady, K. Henrick, R. W. Matthews and D. G. Gillies, *J. Organomet. Chem.* **1980**, *193*, 21.
[150] G. B. Deacon, R. J. Phillips, K. Henrick and M. McPartlin, *Inorg. Chim. Acta* **1979**, *35*, L 335.
[151] G. B. Deacon and R. J. Phillips, *J. Organomet. Chem.* **1980**, *199*, 15.
[152] J. Vicente, J. A. Abad, J. F. Gutierrez-Jugo and P. G. Jones, *J. Chem. Soc. Dalton Trans.* **1989**, 2241.
[153] G. E. Coates and R. A. Whitcombe, *J. Chem. Soc.* **1956**, 3351.
[154] G. D. Shier and R. S. Drago, *J. Organomet. Chem.* **1966**, *5*, 339.
[155] P. J. Burke, L. A. Gray, P. J. C. Hayward, R. W. Matthews, M. McPartlin and D. G. Gillies, *J. Organomet. Chem.* **1977**, *136*, C7.
[156] R. T. Griffin, K. Henrich, R. W. Matthews and M. McPartlin, *J. Chem. Soc. Dalton Trans.* **1980**, 1551.
[157] M. V. Castano, C. Freire, A. Sanchez, J. S. Casas, J. Sordo, E. E. Castellano and J. Zukerman-Schpector, *unpubl. work cited in* ref. [158].
[158] E. M. Vazquez-Lopez, A. Sanchez, J. S. Casas, J. Sordo and E. E. Castellano, *J. Organomet. Chem.* **1992**, *438*, 29.

[159] J. S. Casas, E. E. Castellano, A. Macias, M. C. Rodriguez-Argüeles, A. Sanchez and J. Sordo, *J. Chem. Soc. Dalton Trans.* **1993**, 353.
[160] E. M. Vazquez-Lopez, *Ph.D. Thesis, Universiy of Santiago de Compostela, Spain,* **1993** (Thesis supervisor Prof. J. S. Casas), *private communication.*
[161] R. Carballo, J. S. Casas, E .E. Castellano, A. Sanchez, J. Sordo, E. M. Vazquez-Lopez and J. Zukerman-Schpector, *Polyhedron* **1997**, *16*, 3609.
[162] J. S. Casas, A. Sanchez, J. Sordo, E. M. Vazquez-Lopez, E. E. Castellano and J. Zukerman-Schpector, *Polyhedron* **1992**, *11*, 2889
[163] J. S. Casas, E. E. Castellano, A. Castineiras, A. Sanchez, J. Sordo, E. M. Vazquez-Lopez and J. Zukerman-Schpector, *J. Chem. Soc. Dalton Trans.* **1995**, 1403.
[164] J. S. Casas, A. Sanchez, R. Carballo, and C. Maichle-Mössmer, *Polyhedron,* **1996**, *15*, 861.
[165] W. Schwarz, G. Mann and J. Weidlein, *J. Organomet. Chem.* **1976**, *122*, 303.
[166] M. V. Castano, A. Sanchez, J. S. Casas, J. Sordo, J. L. Briasno, J. F. Piniella, X. Solans, G. Germain, T. Debaerdemaeker and J. Glaser, *Organometallics*, **1988**, *7*, 1897.
[166a] M. V. Castano, A. Macias, A. Castineiras, A. Sanchez-Gonzales, E. Garcia-Martinez, J. S. Casas, J. Sordo, W. Hiller and E. E. Castellano, *J. Chem. Soc. Dalton Trans.* **1990**, 1001.
[167] J. S. Casas, A. Macias, N. Playa, A. Sanchez, J. Sordo and J. M. Varela, *Polyhedron* **1992**, *11*, 2231.
[168] M. S. Garcia-Tasende, B. E. Rivero, A. Castineiras, A. Sanchez, J. S. Casas, J. Sordo, W. Hiller and J. Strähle, *Inorg. Chim. Acta* **1991**, *181*, 43.
[169] M. A. Spence, G. M. Rosair and W. E. Lindsell, *J. Chem. Soc. Dalton Trans.* **1998**, 1581.
[170] J. S. Casas, A. Castineiras, A. Sanchez, J. Sordo and E. M. Vazquez-Lopez, *Main Group Met. Chem.* **1996**, *19*, 231.
[171] J. S. Casas, A. Castineiras, I. Haiduc, A. Sanchez, J. Sordo and E. M. Vazquez-Lopez, *Polyhedron* **1994**, *13*, 1805.
[172] J. S. Casas, A. Castineiras, A. Macias, M. C. Rodriguez-Argüelles, A. Sanchez and J. Sordo, *Inorg. Chim. Acta* **1992**, *197*, 163.
[173] Y. P. Mascarenhas. I. Vencato, M. C. Carrascal, J. M. Varela, J. S. Casas and J. Sordo, *J. Organomet. Chem.* **1988**, *344*, 137.
[174] M. V. Castano, A. Sanchez, J. S. Casas, J. Sordo and E. E. Castellano, *Inorg. Chim. Acta* **1992**, *201*, 83.
[175] P. K. Byers, A. J. Canty, K. Mills and L. Titcombe, *J. Organomet. Chem.* **1985**, *295*, 401.
[176] G. B. Deacon, S. J. Faulks, B. M. Gatehouse and A. J. Jozsa, *Inorg. Chim. Acta* **1977**, *21*, L1.
[177] Y. M. Chow and D. Britton, *Acta Cryst.* **1975**, *B31*, 1922.
[178] Y. M. Chow and D. Britton, *Acta Cryst.* **1975**, *B31*, 1934.
[179] F. Jäkle, K. Polborn and M. Wagner, *Chem. Ber.* **1996**, *129*, 603.
[180] A. Vij, R. L. Kirchmeier, R. D. Willett and J. M. Shreeve, *Inorg. Chem.* **1994**, *33*, 5456.
[181] J. L. Lefferts, K. C. Molloy, M. B. Hossain, D. van der Helm and J. J. Zuckerman, *J. Organomet. Chem.* **1982**, *240*, 349.
[182] M. B. Hossain, J. L. Lefferts, K. C. Molloy, D. van der Helm and J. J. Zuckerman, *Inorg. Chim. Acta,* **1979**, *36*, L409.
[183] P. J. Vergamini, H. Varenkamp and L. F. Dahl, *J. Am. Chem. Soc.* **1971**, *93*, 6327.
[184] S. W. Ng, *Acta Cryst.* **1997**, *C53*, 56.
[185] S. Calogero, P. Ganis, V. Peruzzo and G. Tagliavini, *J. Organomet. Chem.* **1979**, *179*, 145.
[185a] O. I. Guzyr, M. Schormann, J. Schimkowiak, H. W. Roesky, C. Lehmann, M. G. Walawalkar, R. Murugavel, H.-G. Schmidt and M. Noltemeyer, *Organometallics* **1999**, *18*, 832.
[186] J. S. Tse, F. L. Lee and E. J. Gabe, *Acta Cryst.* **1986**, *C42*, 1876.
[187] S. Calogero, P. Ganis, V. Peruzzo, G. Tagliavini and G. Valle, *J. Organomet. Chem.* **1981**, *220*, 11.
[187a] I. Wharf and M. G. Simard, *J. Organomet. Chem.* **1997**, *532*, 1.
[188] T. Krauter and B. Neumüller, *Z. Naturforsch.* **1998**, *53b*, 503.
[189] N. G. Bokii, G. N. Zakharova and Yu. T. Struchkov, *J. Struct. Chem.USSR (Engl. transl.)* **1971**, *11*, 828.
[190] I. Wharf and M. G. Simard, *J. Organomet. Chem.* **1997**, *532*, 1.
[191] H. Preut and F. Huber, *Acta Cryst.* **1979**, *B35*, 744.

[192] T. Schulz, *Dissertation, Martin-Luther Universität Halle-Wittenberg*, **1996** (Thesis supervisor Prof. H. Weichmann)
[193] A. G. Davies, H. J. Milledge, D. C. Puxley and P. J. Smith, *J. Chem. Soc. A*, **1970**, 2862.
[194] H. Fujii and M. Kimura, *Bull. Chem. Soc. Japan*, **1971**, *44*, 2643.
[195] N. W. Alcock and J. F. Sawyer, *J. Chem. Soc. Dalton Trans.* **1977**, 1090.
[196] J. F. Sawyer, *Acta Cryst.* **1988**, *C44*, 633.
[197] P. Ganis, G. Valle, D. Furlani and G. Tagliavini, *J. Organomet. Chem.* **1986**, *302*, 165.
[198] K. C. Molloy, K. Quill and I. W. Nowell, *J. Organomet. Chem.* **1985**, *289*, 271.
[198a] M. M. Amini, E. M. Holt and J. J. Zuckerman, *J. Organomet. Chem.* **1987**, *327*, 147.
[199] N. G. Bokii and Yu. T. Struchkov, *J. Struct. Chem. USSR (Engl. transl.)* **1972**, *13*, 619; *Zh. Strukt. Khim.* **1972**, *13*, 665.
[200] P. T. Greene and R. F. Bryan, *J. Chem. Soc. A*, **1971**, 2549.
[201] R. Shimizu, G. E. Matsubayashi and T. Tanaka, *Inorg. Chim. Acta*, **1986**, *122*, 37.
[201a] P. G. Harrison, M. J. Begley and K. C. Molloy, *J. Organomet. Chem.* **1980**, *186*, 213.
[202] B. Beagley, K. McAloon and J. M. Freeman, *Acta Cryst.* **1974**, *B30*, 444.
[203] W. Frank, G. J. Reiss and D. Kuhn, *Acta Cryst.* **1994**, *C50*, 1904.
[204] D. Zhang, S. Q. Dou and A. Weiss, *Z. Naturforsch.* **1991**, *46A*, 337.
[205] J. S. Tse, M. J. Collins, F. L. Lee and E. J. Gabe, *J. Organomet. Chem.* **1986**, *310*, 169.
[206] R. S. Simons, L. Pu, M. H. Olmstead and P. P. Power, *Organometallics* **1997**, *16*, 1920.
[207] C. Eaborn, P. B. Hitchcock, J. D. Smith and S. E. Sözerli, *Organometallics* **1997**, *16*, 5653.
[208] K. D. Boss, E. J. Bulten, J. G. Noltes and A. L. Spek, *J. Organomet. Chem.* **1975**, *99*, 1975.
[209] S. P. Constantine, G. M. De Lima, P. B. Hitchcock, J. M. Keats, G. A. Lawless and I. Marziano, *Organometallics*, **1997**, *16*, 793.
[210] S. P. Constantine, G. M. de Lima, P. B. Hitchcock, J. M. Keites and G. A. Lawless, *Chem. Commun.* **1996**, 2337.
[211] F. X. Kohl, R. Dickbreder, P. Jutzi, G. Müller and B. Huber, *Chem. Ber.* **1989**, *122*, 871.
[212] M. S. Weininger, P. F. Rodesiler, A. G. Gash and E. L. Amma, *J. Am. Chem. Soc.* **1972**, *94*, 2135.
[213] M. S. Weininger, P. F. Rodesiler and E. L. Amma, *Inorg. Chem.* **1979**, *18*, 751.
[214] W. Frank, *Z. Anorg. Allg. Chem.* **1990**, *585*, 121.
[215] H. Schmidbaur, T. Probst, B. Huber, G. Müller and C. Krüger, *J. Organomet. Chem.* **1989**, *365*, 53.
[216] H. Schmidbaur, T. Probst, O. Steigelmann and G. Müller, *Z. Naturforsch.* **1989**, *44b*, 1175.
[217] H. Schmidbaur, T. Probst, B. Huber, O. Steigelmann and G. Müller, *Organometallics* **1989**, *8*, 1567.
[217a] B. Rähe, P. Müller, H. W. Roesky and I. Usón, *Angew. Chem..* **1999**, *111*, 2069; *Angew. Chem. Int. Ed.* **1999**, *38*, 2050.
[218] S. C. Ng, P. G. Parsons, K. Y. Sim, C. J. Tranter, R. H. White and D. J. Young, *Appl. Organomet. Chem.* **1997**, *11*, 577.
[219] R. C. Mehrotra and V. D. Gupta, *J. Organomet. Chem.* **1965**, *4*, 145.
[220] W. J. Considine, *J. Organomet. Chem.* **1966**, *5*, 263.
[221] J. C. Pommier and J. Valade, *J. Organomet. Chem.* **1968**, *12*, 433.
[222] G. Domazetis, R. J. Magee and B. D. James, *J. Inorg. Nucl. Chem.* **1979**, *41*, 1546.
[223] P. J. Smith, R. F. M. White and L. Smith, *J. Organomet. Chem.* **1972**, *40*, 341.
[224] A. G. Davies, A. L. Price, H. M. Dawes and M. B. Hursthouse, *J. Organomet. Chem.* **1984**, *270*, C1.
[225] A. G. Davies, A. L. Price, H. M. Dawes and M. B. Hursthouse, *J. Chem. Soc. Dalton Trans.* **1986**, 297.
[226] T. B. Grindley, R. E. Wasylishen, R. Thangarasa, W. P. Power and R. D. Curtis, *Can. J. Chem.* **1992**, *70*, 205.
[227] T. B. Grindley and R. Thangarasa, *Can. J. Chem.* **1990**, *68*, 1007.
[228] T. B. Grindley and R. Thangarasa, *J. Am. Chem. Soc.* **1990**, *112*, 1364.
[229] R. H. Herber, A. Shanzer and J. Libman, *Organometallics* **1984**, *3*, 586.
[230] P. A. Bates, M. B. Hursthouse, A. G. Davies and S. Slater, *J. Organomet. Chem.* **1989**, *363*, 45.

[231] A. R. Forrester, S. J. Garden, R. A. Howie and J. L. Wardell, *J. Chem. Soc. Dalton Trans.* **1992**, 2615.
[232] S. David, C. Pascard and M. Cesario, *Nouv. J. Chim.* **1979**, *3*, 63.
[233] T. S. Cameron, P. K. Bakshi, T. Thangarasa and T. B. Grindley, *Can. J. Chem.* **1992**, *70*, 1623.
[234] C. W. Holzapfel, J. M. Koekemoer, C. F. Morris, G. J. Kruger and J. A. Pretorius, *S. Afr. J. Chem.* **1982**, *35*, 80.
[235] J. C. Pommier, E. Mendes and J. Valade, *J. Organomet. Chem.* **1973**, *55*, C 19.
[236] P. A. Bates, M. B. Hursthouse, A. G. Davies and S. D. Slater, *J. Organomet. Chem.* **1987**, *325*, 129.
[237] P. A. Bates, M. B. Hursthouse, A. G. Davies and S. D. Slater, *J. Organomet. Chem.* **1989**, *363*, 45.
[238] G. Engel, K. Fütterer and G. Mattern, *Z. Kristallogr.* **1992**, *199*, 113.
[239] A. Zschunke, M. Scheer, M. Völtzke, K. Jurkschat and A. Tzschach, *J. Organomet. Chem.* **1986**, *308*, 325.
[240] R. G. Swisher, R. O. Day and R. R. Holmes, *Inorg. Chem.* **1983**, *22*, 3692.
[241] S. Battacharyia, N. Seth, V. D. Gupta, H. Nöth, K. Philborn, M. Thomas and H. Schwenk, *Chem. Ber.* **1994**, *127*, 1895.
[242] B. D. James, L. M. Kivlington, B. W. Skelton and A. H. White, *Appl. Organomet. Chem.* **1998**, *12*, 13.
[243] J. S. Casas, A. Castineiras, M. D. Couce, N. Playa, U. Russo, A. Sanchez, J. Sordo and J. M. Varela, *J. Chem. Soc. Dalton Trans.* **1998**, *1513*.
[244] S. Knoll, F. Tschwatschal, T. Gelbrich, T. Ristau and R. Borsdorf, *Z. Anorg. Allg. Chem.* **1998**, *624*, 1015.
[245] S. Geetha, M. Ye and J. G. Verkade, *Inorg. Chem.* **1995**, *34*, 6158.
[246] L. E. Khoo, N. K. Goh, L. L. Koh, Y. Xu, S. L. Bao and T. C. W. Mak, *Polyhedron* **1995**, *14*, 2281.
[247] G. Stocco, G. Guli, M. A. Girasolo, G. Bruno, F. Nicolo and R. Scopellitti, *Acta Cryst.* **1996**, *C52*, 829.
[248] F. Huber, H. Preut, E. Hoffmann and M. Gielen, *Acta Cryst.* **1989**, *C45*, 51.
[249] S. I. Aizawa, T. Natsume, K. Hatano and S. Funahashi, *Inorg. Chim. Acta* **1996**, *248*, 215.
[250] C. C. Carvalho, R. H. P. Francisco, M. T. P. Gambardella, G. F. de Sousa and C. A. L. Filgueiras, *Acta Cryst.* **1996**, *C52*, 1627.
[250a] C. C. Carvalho, R. H. P. Francisco, M. T. P. Gambardella, G. F. de Sousa and C. A. L. Filgueiras, *Acta Cryst.* **1996**, *C52*, 1629.
[251] S. W. Ng, V. G. Kumar Das and B. Schulze, *Malaysian J. Sci.* **1995**, *B16*, 89.
[252] P. G. Harrison, N. W. Sharpe, C. Pelizzi, G. Pelizzi and P. Tarasconi, *J. Chem. Soc. Dalton Trans.* **1983**, 921.
[253] S. W. Ng and V. G. Kumar Das, *J. Chem. Crystallogr.* **1994**, *24*, 337.
[254] P. Jaumier, B. Jousseaume, E. R. T. Tiekink, M. Biesemans, R. Willen, *Organometallics* **1997**, *16*, 5124.
[255] M. Biesemans, R. Willem, S. Damoun, P. Geerlings, E. R. T. Tiekink, P. Jaumier, M. Labcini and B. Jausseaume, *Organometallics* **1998**, *17*, 90.
[256] H. W. Roesky, M. Witt, M. Diehl, J. W. Bats and H. Fuess, *Chem. Ber.* **1979**, *112*, 1372.
[257] M. Dräger, *Z. Anorg. allg. Chem.* **1981**, *477*, 154.
[258] A. S. Secco and J. Trotter, *Acta Cryst.* **1983**, *C39*, 451.
[259] A. G. Davies, S. D. Slater, D. C. Povey and G. W. Smith, *J. Organomet. Chem.* **1988**, *352*, 283.
[260] P. A. Bates, M. B. Hursthouse, A. G. Davies and S. D. Slater, *J. Organomet. Chem.* **1989**, *363*, 45.
[261] C. A. Mackay, *Thesis, London*, **1973** (cited by A. G. Davies et al. ref. [259]) **1988**, *352*, 282).
[262] R. H. Herber and M. F. Leahy, *Adv. Chem. Ser.* **1976**, *157*, 155.
[263] P. A. Bates, M. B. Hursthouse, A. G. Davies and S. D. Slater, *J. Organomet. Chem.* **1987**, *325*, 129.
[264] K. Jurkschat, M. Scheer, A. Tzschach, J. Meunier-Piret and M. van Meerssche, *J. Organomet. Chem.* **1985**, *281*, 173.

[264a] U. Kolb, M. Beuter, M. Gerner and M. Dräger, *Organometallics* **1994**, *13*, 4413.
[264b] N. Froehlich, P. B. Hitchcock, J. Hu, M. F. Lappert and J. R. Dilworth, *J. Chem. Soc. Dalton Trans.* **1996**, 1941.
[265] K. C. Molloy, M. F. Mahon, I. Haiduc and C. Silvestru, *Polyhedron* **1995**, *14*, 1169.
[266] I. Haiduc, C. Silvestru, H. W. Roesky, H. G. Schmidt and M. Noltemeyer, *Polyhedron* **1993**, *12*, 69.
[267] R. A. Forder and G. M. Sheldrick, *Chem. Commun.* **1969**, 1125; *J. Organomet. Chem.* **1970**, *21*, 115.
[268] A. M. Domingos and G. M. Sheldrick, *J. Organomet. Chem.* **1974**, *67*, 257.
[269] L. E. Khoo, X. M. Chen and T. C. W. Mak, *Acta Cryst.* **1991**, *C47*, 2647.
[270] R. A. Forder and G. Sheldrick, *J. Organomet. Chem.* **1970**, *22*, 611.
[271] Y. M. Chow, *Inorg. Chem.* **1970**, *9*, 794.
[272] G. D. Andretti, G. Bocelli, G. Calestani and P. Sgarabotto, *J. Organomet. Chem.* **1984**, *273*, 31.
[273] A. Silvestru, J. E. Drake and J. Yang, *Polyhedron* **1997**, *16*, 4113.
[274] A. F. Shihada, I. A. A. Jassim and F. Weller, *J. Organomet. Chem.* **1984**, *268*, 125.
[275] C. Silvestru, I. Haiduc, F. Caruso, M. Rossi, B. Mathieu and M. Gielen, *J. Organomet. Chem.* **1993**, *448*, 75.
[276] H. Puff, D. Hänssgen, N. Beckermann, A. Roloff and W. Schuh, *J. Organomet. Chem.* **1989**, *373*, 37.
[277] K. M. Lo, S. Selvaratnam, S. W. Ng, C. Wei and V. G. Kumar Das, *J. Organomet. Chem.* **1992**, *430*, 149.
[278] V. G. Kumar Das, K. M. Lo, C. Wei and T. C. W. Mak, *Organometallics*, **1987**, *6*, 10.
[279] G. van Koten, J. G. Noltes and A. L. Spek, *J. Organomet. Chem.* **1976**, *118*, 183.
[280] G. van Koten, J. T. B. H. Jastrzebski, J. G. Noltes, A. L. Spek and J. C. Schoone, *J. Organomet. Chem.* **1978**, *148*, 233.
[281] R. Allmann, R. Hohlfeld, A. Waskowska and J. Lorberth, *J. Organomet. Chem.* **1980**, *192*, 353.
[281a] D. Hänssgen, M. Jansen, C. Leben and T. Oster, *J. Organomet. Chem.* **1995**, *494*, 223.
[282] E. O. Schlemper and D. Britton, *Inorg. Chem.* **1966**, *5*, 507.
[283] Y. M. Chow and D. Britton, *Acta Cryst.* **1971**, *B27*, 856.
[284] E. O. Schlemper and D. Britton, *Inorg. Chem.* **1966**, *5*, 511.
[285] R. A. Forder and G. M. Sheldrick, *J. Chem. Soc. A*, **1971**, 1107.
[286] Y. M. Chow, *Inorg. Chem.* **1971**, *10*, 1938.
[287] A. M. Domingos and G. M. Sheldrick, *J. Organomet. Chem.* **1974**, *69*, 207.
[288] I. Lange, E. Wieland, P. G. Jones and A. Blaschette, *J. Organomet. Chem.* **1993**, *458*, 57.
[288a] J. G. A. Luijten, M. J. Jansen and G. J. M. van der Kerk, *Rec. Trav. Chim. Pays-Bas* **1962**, *81*, 203; *J. Organomet. Chem.* **1964**, *1*, 286.
[288b] R. Gassend, M. Delmas, J. C. Maire, Y. Richard and C. More, *J. Organomet. Chem.* **1972**, *42*, C 29; R. Gassend, J. C. Maire and J. C. Pommiere, *J. Organomet. Chem.* **1977**, *132*, 69.
[289] I. Hammann, *Nachrichten Bayer* **1978**, *31*, 61.
[290] X. Xie, C. Chen, Q. Xie and X. Xu, *Yingyong Huaxue*, **1992**, *9*, 52; *Chem. Abstr.* **1993**, *118*, 169232t.
[290a] P. J. Cox, S. M. S. V. Dodge-Harrison, R. A. Owie and J. L. Wardell, *J. Chem. Res.* **1994**, S 162.
[291] R. Cea-Olivares, O. Jimenez-Sandoval, G. Espinosa-Perez and C. Silvestru, *J. Organomet. Chem.* **1994**, *484*, 33.
[292] O. Jimenez-Sandoval, R. Cea-Olivares, I. Haiduc, C. Silvestru and G. Espinosa-Perez, *Phosphorus, Sulfur and Silicon*, **1994**, *93/94*, 387.
[292a] R. Cea-Olivares, O. Jimenez-Sandoval, G. Espinosa-Perez and C. Silvestru, *Polyhedron* **1994**, *13*, 2818.
[293] R. J. Deeth, K. C. Molloy, M. F. Mahon and S. Whittaker, *J. Organomet. Chem.* **1992**, *430*, 25.
[294] S. J. Blunden, M. F. Mahon, K. C. Molloy and P. C. Wakefield, *J. Chem. Soc. Dalton Trans.* **1994**, 2135.

[295] A. Goodger, M. Hill, M. F. Mahon, J. McGinley and K. C. Molloy, *J. Chem. Soc. Dalton Trans.* **1996**, 847.
[296] M. Hill, M. F. Mahon, J. McGinley and K. C. Molloy, *J. Chem. Soc. Dalton Trans.* **1996**, 835.
[297] J. Konnert, D. Britton and Y. M. Chow, *Acta Cryst.* **1972**, *B28*, 180.
[298] D. Cunningham, P. McArdle, J. McManus, T. Higgins and K. Molloy, *J. Chem. Soc. Dalton Trans.* **1988**, 2621.
[299] D. Cunningham, J. McManus and M. J. Hynes, *J. Organomet. Chem.* **1990**, *393*, 69.
[300] H. W. Roesky, *Z. Naturforsch.* **1976**, *31b*, 680.
[301] U. Doehring, D. Hansgen M. Jansen, M. Nieger and M. Tellenbach, *Z. Anorg. Allg. Chem.* **1998**, *624*, 963.
[302] B. Y. K. Ho, K. C. Molloy, J. J. Zuckerman, F. Reidinger and J. A. Zubieta, *J. Organomet. Chem.* **1980**, *187*, 213.
[303] S. W. Ng, V. G. Kumar Das, F. van Meurs, J. D. Schagen and L. H. Straver, *Acta Cryst.* **1989**, *C45*, 570.
[304] T. M. Soliman, S. E. H. Etaiw, G. Fendesak and R. D. Fischer, *J. Organomet. Chem.* **1991**, *415*, C1.
[304a] E. Siebel, R. D. Fischer, J. Kopf, N. A. Davies, D. C. Apperley and R. K. Harris, *Inorg. Chem. Commun.* **1998**, *1*, 346.
[305] P. Brandt, A. K. Brimah and R. D. Fischer, *Angew. Chem.* **1988**, *100*, 1578; *Angew. Chem. Int.Ed. Engl.* **1988**, *27*, 1521.
[306] S. Eller, P. Brandt, A. K. Brimah, P. Schwarz and R. D. Fischer, *Angew. Chem.* **1989**, *101*, 1274; *Angew. Chem. Int. Ed. Engl.* **1989**, *28*, 1263.
[307] M. Adam, A. K. Brimah, R. D. Fischer and X. F. Li, *Inorg. Chem.* **1990**, *29*, 1595.
[308] S. Eller, M. Adam and R. D. Fischer, *Angew. Chem.* **1990**, *102*, 1157; *Angew. Chem. Int. Ed. Engl.* **1990**, *29*, 1126.
[309] A. Bonardi, C. Carini, C. Pelizzi, G. Pelizzi, G. Predieri, P. Tarasconi, M. A. Zorroddu and K. C. Molloy *J. Organomet. Chem.* **1991**, *401*, 283.
[310] C. Carini, C. Pelizzi, G. Pelizzi, G. Predieri, P. Tarasconi and F. Vitali, *J. Chem. Soc. Chem. Commun.* **1990**, 613.
[311] M. Adam, A. K. Brimah, R. D. Fischer and L. Xing-Fu, *Inorg. Chem.* **1990**, *29*, 1595.
[312] U. Behrens, A. K. Brimah and R. D. Fischer, *J. Organomet. Chem.* **1991**, *411*, 325.
[313] R. D. Fischer, G. R. Sienel, D. Lambright, D. H. Oh, S. Balasubramanian, B. Hedman and K. O. Hodgson, *Inorg. Chem.* **1991**, *30*, 1441.
[314] U. Behrens, A. K. Brimah, T. M. Soliman, R. D. Fischer, D. C. Apperley, N. A. Davies and R. K. Harris, *Organometallics*, **1992**, *11*, 1718.
[315] P. Brandt, U. Illgen, R. D. Fischer, E. Sanchez-Martines and R. Diaz-Calleja, *Z. Naturforsch.* **1993**, *48b*, 1565.
[316] S. Eller, P. Schwarz, A. K. Brimah, R. D. Fischer, D. C. Apperley, N. A. Davies and R. K. Harris, *Organometallics*, **1993**, *12*, 3232.
[317] S. E. H. Etaiw and A. M. A. Ibrahim, *J. Organomet. Chem.* **1993**, *456*, 229.
[318] S. E. H. Etaiw and A. M. A. Ibrahim, *J. Organomet. Chem.* **1996**, *522*, 77.
[319] P. Schwarz, E. Siebel, R. D. Fischer, D. C. Apperley, N. A. Davies and R. K. Harris, *Angew. Chem.* **1995**, *107*, 1311; *Angew. Chem. Int. Ed. Engl.* **1995**, *34*, 1197.
[320] J. Lu, W. T. A. Harrison and A. J. Jacobson, *Angew. Chem.* **1995**, *34*, 2557.
[321] J. Lu, W. T. A. Harrison and A. J. Jacobson, *Inorg. Chem.* **1996**, *35*, 4271.
[322] A. M. A. Ibrahim and S .E. H. Etaiw, *Polyhedron*, **1997**, *16*, 1585.
[323] S. Eller, M. Adam and R. D. Fischer, *Angew. Chem.* **1990**, *99*, 1157; *Angew. Chem. Int. Ed. Engl.* **1990**, *29*, 1126.
[324] D. C. Apperley, N. A. Davies, R. K. Harris, S. Eller, P. Schwarz and R. D. Fischer, *J. Chem. Soc. Chem. Commun.* **1992**, 740.
[325] R. K. Harris, D. C. Apperley, N. A. Davies and R. D. Fischer, *Bull. Magn. Res.* **1993**, *15*, 22.
[326] R. K. Harris, M. Maral-Sunnetcioglu and R. D. Fischer, *Spectrochim. Acta* **1994**, *50A*, 2069.
[327] P. Schwarz, S. Eller, E. Siebel, T. M. Soliman, R. D. Fischer, D. C. Apperley, N. A. Davies and R. K. Harris, *Angew. Chem.* **1996**, *108*, 1611; *Angew. Chem. Int. Ed. Engl.* **1996**, *35*, 1525.

[328] K. Yünlü, N. Hoch and R. D. Fischer, *Angew. Chem.* **1985**, *97*, 863; *Angew. Chem. Int. Ed. Engl.* **1985**, *24*, 879.
[328a] D. C. Apperley, N. A. Davies, R. K. Harris, A. K. Brimach, S. Eller and R. D. Fischer, *Organometalics* **1990**, *9*, 2672.
[328b] R. Uson, J. Fornies, M. A. Uson and E. Lalinde, *J. Organomet. Chem.* **1980**, *185*, 359.
[329] A. K. Brimach, E. Siebel, R. D. Fischer, N. A. Davies, D. A. Apperley and R. K. Harris, *J. Organomet. Chem.* **1994**, *475*, 85.
[329a] A. M. A. Ibrahim, *J. Organomet. Chem.* **1998**, *556*, 1.
[330] R. Uson, J. Fornies, M. A. Uson and E. Lalinde, *J. Organomet. Chem.* **1980**, *185*, 359.
[330a] A. M. A. Ibrahim, E. Siebel and R. D. Fischer, *Inorg. Chem.* **1998**, *37*, 3521.
[331] J. U. Schütze, R. Eckhardt, R. D. Fischer, D. C. Apperley, N. A. Davies and R. K. Harris, *J. Organomet. Chem.* **1997**, *534*, 187.
[332] E. Siebel and R. D. Fischer, *Chem. Eur. J.* **1997**, *3*, 1987.
[333] D. Zhang, S. Q. Du and A. Weiss, *Z. Naturforsch.* **1991**, *46a*, 337.
[334] R. Hillwig, F. Kunkel, K. Harms, B. Neumüller and K. Dehnicke, *Z. Naturforsch.* **1997**, *52b*, 149.
[335] H. Preut and F. Huber, *Z. Anorg. Allg. Chem.* **1977**, *435*, 234.
[336] U. Fahrenkampf, M. Schürmann and F. Huber, *Acta Cryst.* **1994**, *C50*, 1252.
[337] R. J. H. Clark, A. G. Davies and R. J. Puddephatt, *J. Am. Chem. Soc.* **1968**, *90*, 6923.
[338] R. J. H. Clark, A. G. Davies and R. J. Puddephatt, *Inorg. Chem.* **1969**, *8*, 457.
[339] I. Wharf, R. Cuenca, E. Besso and M. Onyszchuk, *J. Organomet. Chem.* **1984**, *277*, 245.
[340] H. J. Eppley, J. L. Ealy, C. H. Yoder, J. N. Spencer and A. L. Rheingold, *J. Organomet. Chem.* **1992**, *431*, 133.
[341] M. Mammi, V. Busetti and A. Del Pra, *Inorg. Chim. Acta* **1967**, *1*, 419.
[342] C. Eaborn, P. B. Hitchcock, J. D. Smith and S. E. Sözerli, *Organometallics* **1997**, *16*, 5653.
[343] P. B. Hitchcock, M. F. Lappert and M. Layh, *Inorg. Chim. Acta* **1998**, *269*, 181.
[344] P. Jutzi and R. Dickbreder, *J. Organomet. Chem.* **1989**, *373*, 301.
[345] P. Jutzi, R. Dickbreder and H. Nöth, *Chem. Ber.* **1989**, *122*, 865.
[346] A. G. Gash, P. F. Rodesiler and E. L. Amma, *Inorg. Chem.* **1974**, *13*, 2429.
[347] H. Schmidbaur, T. Probst and O. Steigelmann, *Organometallics* **1991**, *10*, 3176.
[348] C. Glidewell and D. C. Liles, *Acta Cryst.* **1978**, *B34*, 129.
[349] G. M. Sheldrick and R. Taylor, *Acta Cryst.* **1975**, *B31*, 2740.
[350] H. Preut, P. Röhm and F. Huber, *Acta Cryst.* **1986**, *C42*, 657.
[351] X. J. Lei, M. Y. Shang, A. Patil, E. E. Wolf and T. P. Fehlner, *Inorg. Chem.* **1996**, *35*, 3217.
[352] M. Schurmann and F. Huber, *J. Organomet. Chem.* **1997**, *530*, 121.
[353] C. Gaffney, P. G. Harrison and T. J. King, *J. Chem. Soc. Dalton Trans.* **1982**, *1061*.
[354] N. G. Bokii, A. I. Udelnov, Yu. T. Struchkov, D. N. Kravtsov and V. M. Pacherskaya, *Zh. Strukt. Khim.* **1977**, *18*, 1025.
[355] A. Blaschette, T. Hammann, A. Michalides and P. G. Jones, *J. Organomet. Chem.* **1993**, *456*, 49.
[355a] O. Hiemisch, D. Henschel, A. Blaschette and P. G. Jones, *Z. Anorg. Allg. Chem.* **1997**, *623*, 324.
[356] G. D. Andretti, G. Bocelli, G. Calestani and P. Sgarabotto, *J. Organomet. Chem.* **1984**, *273*, 31.
[357] M. G. Begley, C. Gaffney, P. G. Harrison and A. Steel, *J. Organomet. Chem.* **1985**, *289*, 281.
[358] F. T. Edelmann, I. Haiduc, C. Silvestru, H.-G. Schmidt and M. Noltemeyer, *Polyhedron* **1998**, *17*, 2043.
[359] I. Haiduc, D. B. Sowerby and S. F. Lu, *Polyhedron* **1995**, *14*, 3389.
[360] I. Haiduc and D. B. Sowerby, *Polyhedron* **1996**, *15*, 2469.
[361] C. Silvestru, I. Haiduc, R. Cea-Olivares and S. Hernandez-Ortega, *Inorg. Chim. Acta* **1995**, *233*, 151.
[362] K. H. Ebert, H. J. Breunig, C. Silvestru, I. Stefan and I. Haiduc, *Inorg. Chem.* **1994**, *33*, 1695.
[363] E. R. T. Tiekink, *Acta Cryst.* **1988**, *C44*, 250.

[364] M. Dräger and N. Kleiner, *Z. Kristallogr.* **1979**, *149*, 112.
[365] M. Dräger and N. Kleiner, *Angew. Chem.* **1980**, *92*, 950; *Angew. Chem. Int. Ed. Engl.* **1980**, *19*, 923.
[366] M. Dräger and N. Kleiner, *Z. Anorg. Allg. Chem.* **1985**, *522*, 48.
[367] J. Müller, *Z. Naturforsch.* **1985**, *40b*, 1320.
[368] R. Allmann, A. Waskowska, R. Hohlfeld and J. Lorberth, *J. Organomet. Chem.* **1980**, *198*, 155.
[369] J. Müller, U. Müller, A. Loss, J. Lorberth, H. Donath and W. Massa, *Z. Naturforsch.* **1985**, *40b*, 1320.
[370] Y. M. Chow and D. Britton, *Acta Cryst.* **1971**, *B27*, 856.
[371] T. N. Tarkhova, E. V. Chuprunov, L. E. Nikolaeva, M. A. Simonov and N. V. Belov, *Kristallografiya (USSR)* **1978**, *23*, 506.
[372] J. Konnert, D. Britton and Y. M. Chow, *Acta Cryst.* **1972**, *B28*, 180.
[373] N. G. Furmanova, Yu. T. Struchkov, D. N. Kravtsov and E. M. Rokhlina, *Zh. Strukt. Khim.* **1979**, *20*, 1047.
[374] A. M. A. Ibrahim, S. E. H. Etaiw and T. M. Soliman, *J. Organomet. Chem.* **1992**, *430*, 87.
[375] A. M. A. Ibrahim, T. M. Soliman, S. E. H. Etaiw and R. D. Fischer, *J. Organomet. Chem.* **1994**, *468*, 93.
[375a] A. K. Brimah, P. Schwarz, R. D. Fischer, N. A. Davies and R. K. Harris, *J. Organomet. Chem.* **1998**, *568*, 1.
[376] N. Burford, T. M. Parks, B. W. Royan, B. Borecka, T. S. Cameron, J. F. Richardson, E. J. Gabe and R. Hynes, *J. Am. Chem Soc.* **1992**, *114*, 8147.
[377] R. Cea-Olivares, J. G. Alvarado, G. Espinosa-Pérez, C. Silvestru and I. Haiduc, *J. Chem. Soc. Dalton Trans.* **1994**, 2191.
[378] R. Cea-Olivares, J. G. Alvarado, G. Espinosa-Pérez and S. Hernandez-Ortega, *Inorg. Chim. Acta,* **1997**, *255*, 319.
[379] R. K. Gupta, A. K. Rai, R. C. Mehrotra, V. K. Jain, B. F. Hoskins and E. R. T. Tiekink, *Inorg. Chem.* **1985**, *24*, 3280.
[380] S. Grewe, T. Hausler, M. Mannel, B. Rossenbeck and W. S. Sheldrick, *Z. Anorg. Allg. Chem.* **1998**, *624*, 613.
[381] N. Camerman and J. Trotter, *Can. J. Chem.* **1963**, *41*, 460.
[382] E. O. Schlemper and D. Britton, *Acta Cryst.* **1966**, *20*, 777.
[383] A. Weitze, I. Lange, D. Henschel, A. Blaschette and P. G. Jones, *Phosphorus, Sulfur, Silicon,* **1997**, *122*, 107.
[384] H. Schmidbaur, W. Bublak, B. Huber and G. Müller, *Angew. Chem.* **1987**, *99*, 248; *Angew. Chem. Int. Ed. Engl.* **1987**, *26*, 234.
[385] H. Schmidbaur, R. Novak, O. Steigelmann and G. Müller, *Chem. Ber.* **1990**, *123*, 1221.
[386] W. Schwarz and H. J. Guder, *Z. Anorg. Allg. Chem.* **1978**, *444*, 105.
[387] M. J. Begley and D. B. Sowerby, *Acta Cryst.* **1993**, *C49*, 1044.
[388] J. Werner, W. Schwarz and A. Schmidt, *Z. Naturforsch.* **1981**, *36b*, 556.
[389] A. Almenningen, G. Gunderssen, F. Marchetti and P. F. Zannazzi, *J. Chem. Soc. Chem. Commun.* **1981**, 181.
[390] M. Hall and D. B. Sowerby, *J. Chem. Soc. Dalton Trans.* **1983**, 1095.
[391] W. Schwarz and H. J. Guder, *Z. Naturforsch.* **1978**, *33b*, 485.
[392] J. Bordner, G. O. Doak and J. R. Peters, Jr. *J. Am. Chem. Soc.* **1974**, *96*, 6763.
[393] M. Hall and D. B. Sowerby, *J. Chem. Soc. Dalton Trans.* **1983**, 1095.
[394] H. A. Meinema, J. G. Noltes, A. L. Spek and A. J. M. Duisenberg, *Recl. Trav. Chim. Pays-Bas,* **1988**, *107*, 226.
[395] P. C. Andrews, C. L. Raston, B. W. Skelton, V. A. Tolhurst and A. H. White, *Chem. Commun.* **1998**, 575.
[396] S. P. Bone and D. B. Sowerby, *J. Chem. Soc. Dalton Trans.* **1979**, 1430.
[397] W. S. Sheldrick and C. Martin, *Z. Naturforsch.* **1991**, *46b*, 639.
[398] H. J. Breunig, K. H. Ebert, S. Gülec, M. Dräger, D. B. Sowerby, M. J. Begley and U. Behrens, *J. Organomet. Chem.* **1992**, *427*, 39.
[399] A. H. Cowley, N. C. Norman, M. Pakulski, D. L. Bricker and D. H. Russell, *J. Am. Chem Soc.* **1985**, *107*, 8211.

[400] O. Mundt, G. Becker, H. Stadelmann and H. Thurn, *Z. Anorg. Allg. Chem.* **1992**, *617*, 59.
[401] J. von Seyerl, O. Scheidsteger, H. Berke and G. Huttner, *J. Organomet. Chem.* **1986**, *311*, 85.
[402] M. Hall and D. B. Sowerby, *J. Organomet. Chem.* **1988**, *347*, 59.
[403] H. Preut, F. Huber and G. Alonzo, *Acta Cryst.* **1987**, *C43*, 46.
[404] H. Preut, F. Huber, G. Alonzo and N. Bertazi, *Acta Cryst.* **1986**.
[405] W. S. Sheldrick and C. Martin, *Z. Naturforsch.* **1991**, *46b*, 639.
[406] P. Sharma, N. Rosas, A. Toscano, S. Hernandez, R. Shankar and A. Cabrera, *Main Group Met. Chem.* **1996**, *19*, 21.
[407] B. N. Menshutkin, *Zh. Russ. Fiz. Khim. Obshcestva*, **1911**, *43*, 1298, 1786.
[408] W. Smith and G. W. Davis, *J. Chem. Soc. Trans.* **1882**, *41*, 411.
[409] D. Mootz and V. Händler, *Z. Anorg. Allg. Chem.* **1986**, *533*, 23.
[410] H. Schmidbaur, R. Nowak, A. Schier, J. M. Wallis, B. Huber and G. Müller, *Chem. Ber.* **1987**, *120*, 1829.
[411] H. Schmidbaur, R. Nowak, O. Steigelmann and G. Müller, *Chem. Ber.* **1990**, *123*, 1221.
[412] H. Schmidbaur, J. M. Wallis, R. Nowak, B. Huber and G. Müller, *Chem. Ber.* **1987**, *120*, 1837.
[413] H. Schmidbaur, R. Nowak, B. Huber and G. Müller, *Organometallics*, **1987**, *6*, 2266.
[414] R. Hulme and I. T. Szymanski, *Acta Cryst.* **1969**, *B25*, 753.
[415] A. Demalde, A. Mangia, N. Nardelli, G. Pelizzi and M. E. Vidoni, *Acta Cryst.* **1972**, *B28*, 147.
[416] G. Bombieri, G. Peyronel and I. M. Vezzosi, *Inorg. Chim. Acta* **1972**, *6*, 349.
[417] T. Probst, O. Steigelmann, J. Riedle and H. Schmidbaur, *Chem. Ber.* **1991**, *124*, 1089.
[418] D. Mootz and V. Händler, *Z. Anorg. Allg. Chem.* **1985**, *521*, 122.
[419] L. Korte, A. Lipka and D. Mootz, *Z. Anorg. Allg. Chem.* **1985**, *524*, 157.
[420] K. W. Shen, W. E. McEwen, S. J. La Placa, W. C. Hamilton and A. P. Wolf, *J. Am. Chem. Soc.* **1968**, *90*, 1718.
[421] R. R. Holmes, R. O. Day, V. Chandrasekar and J. M. Holmes, *Inorg. Chem.* **1987**, *26*, 163.
[422] M. Wieber, J. Walz and C. Burschka, *Z. Anorg. Allg. Chem.* **1990**, *585*, 65.
[423] S. P. Bone and D. B. Sowerby, *J. Organomet. Chem.* **1980**, *184*, 181.
[424] J. Pebler, K. Schmidt, K. Dehnicke and J. Weidlein, *Z. Anorg. Allg. Chem.* **1978**, *440*, 269.
[425] W. Schwarz and, H. D. Hausen, *Z. Anorg. Allg. Chem.* **1978**, *441*, 175.
[426] M. J. Begley, D. B. Sowerby, D. M. Wesolek, C. Silvestru and I. Haiduc, *J. Organomet. Chem.* **1986**, *316*, 281.
[427] H. M. Hoffmann and M. Dräger, *J. Organomet. Chem.* **1987**, *329*, 51.
[428] M. A. Bush, P. F. Lindley and P. Woodward, *J. Chem. Soc. A*, **1967**, 1826.
[429] S. L. Buchwald, R. A. Fisher and W. M. Davis, *Organometallics*, **1989**, *8*, 2082.
[430] H. M. Hoffmann and M. Dräger, *J. Organomet. Chem.* **1985**, *295*, 33.
[431] H. M. Hoffmann and M. Dräger, *J. Organomet. Chem.* **1987**, *320*, 273.
[432] M. A. Munoz-Hernández, R. Cea-Olivares and S. Hernandez-Ortega, *Z. Anorg. Allg. Chem.* **1996**, *622*, 1392.
[433] M. A. Munoz-Hernández, R. Cea-Olivares, R. A. Toscano and S. Hernandez-Ortega, *Z. Anorg. Allg. Chem.* **1997**, *623*, 642.
[434] M. Wieber, D. Wirth and C. Burschka, *Z. Anorg. Allg. Chem.* **1983**, *505*, 141.
[435] M. Hall, D. B. Sowerby and C. P. Falshaw, *J. Organomet. Chem.* **1986**, *315*, 321.
[436] M. Wieber, D. Wirth, J. Metter and C. Burschka, *Z. Anorg. Allg. Chem.* **1985**, *520*, 65.
[437] G. E. Forster, I. G. Southerington, M. J. Begley and D. B. Sowerby, *J. Chem. Soc. Chem. Comm.* **1991**, 54.
[438] C. Silvestru and I. Haiduc, *Coord. Chem. Revs.* **1996**, *147*, 117.
[439] M. J. Begley, D. B. Sowerby, D. M. Wesolek, C. Silvestru and I. Haiduc, *J. Organomet. Chem.* **1986**, *316*, 1981.
[440] C. Silvestru, L. Silaghi-Dumitrescu, I. Haiduc, M. J. Begley, M. Nunn and D. B. Sowerby, *J. Chem. Soc. Dalton Trans.* **1986**, 1031.
[441] C. Silvestru, I. Haiduc, R. Kaller, K. H. Ebert and H. J. Breunig, *Polyhedron* **1993**, *12*, 2611.
[442] M. N. Gibbons, D. B. Sowerby, C. Silvestru and I. Haiduc, *Polyhedron* **1996**, *15*, 4573.

[443] C. Silvestru, M. Curtui, I. Haiduc, M. J. Begley and D. B. Sowerby, *J. Organomet. Chem.* **1992**, *426*, 49.
[444] K. H. Ebert, H. J. Breunig, C. Silvestru and I. Haiduc, *Polyhedron* **1994**, *13*, 2531.
[445] D. B. Sowerby, M. J. Begley, L. Silaghi-Dumitrescu, I. Silaghi-Dumitrescu and I. Haiduc, *J. Organomet. Chem.* **1994**, *469*, 45.
[446] J. Müller, U. Müller, A. Loss, J. Loberth, H. Donath and W. Massa, *Z. Naturforsch.* **1985**, *40b*, 1320.
[447] R. Hillwig, F. Kunkel, K. Harms, B. Neumüller and K. Dehnicke, *Z. Naturforsch.* **1996**, *52b*, 149.
[447a] E. V. Avtomonov, X. W. Li and J. Lorberth, *J. Organomet. Chem.* **1997**, *530*, 71.
[448] S. Wang, D. B. Mitzi, G. A. Landrum, H. Genin and R. Hoffmann, *J. Am. Chem. Soc.* **1997**, *119*, 724.
[448a] D. B. Mitzi, *Inorg. Chem.* **1996**, *35*, 7614.
[448b] W. Clegg, *J. Mater. Chem.* **1994**, *4*, 891.
[449] J. von Seyerl and G. Huttner, *J. Organomet. Chem.* **1980**, *195*, 207.
[450] W. Clegg, N. A. Compton, R. J. Errington and N. C. Norman, *Polyhedron* **1987**, *6*, 2031.
[451] J. M. Wallis, G. Müller and H. Schmidbaur, *J. Organomet. Chem.* **1987**, *327*, 159.
[452] W. Clegg, R. J. Errington, G. A. Fisher, R. J. Flynn and N. C. Norman, *J. Chem. Soc. Dalton Trans.* **1993**, 637.
[453] W. Clegg, R. J. Errington, G. A. Fisher, D. C. R. Hockless, N. C. Norman, A. G. Orpen and S. E. Stratford, *J. Chem. Soc. Dalton Trans.* **1993**, 1967.
[454] N. C. Norman, *Phosphorus, Sulfur, Silicon* **1994**, *87*, 167.
[455] W. Frank, *J. Organomet. Chem.* **1990**, *386*, 177.
[456] W. Frank, J. Schneider and S. Müller-Becker, *J. Chem. Soc. Chem. Comm.* **1993**, 799.
[456a] W. Frank, J. Weber, E. Fuchs, *Angew. Chem.* **1987**, *99*, 68; *Angew. Chem. Int. Ed. Engl.* **1987**, *26*, 74.
[457] S. Müller-Becker, W. Frank and J. Schneider, *Z. Anorg. Allg. Chem.* **1993**, *619*, 1073.
[458] H. Schmidbaur, J. M. Wallis, R. Nowak, B. Huber and G. Müller, *Chem. Ber.* **1987**, *120*, 1837.
[459] H. Schmidbaur, R. Nowak, A. Schier, J. M. Wallis, B. Huber and G. Müller, *Chem. Ber.* **1987**, *120*, 1829.
[460] I. M. Vezzosi, L. P. Battaglia and A. B. Corradi, *J. Chem. Soc. Dalton Trans.* **1992**, 375.
[461] K. H. Whitmire, J. C. Hutchison, A. L. McKnight and C. M. Jones, *J. Chem. Soc.* **1992**, 1021.
[462] C. M. Jones, M. D. Burkartz and K. H. Whitmire, *Angew. Chem.* **1992**, *104*, 466; *Angew. Chem. Int. Ed. Engl.* **1992**, *31*, 451.
[463] F. Huber and S. Bock, *Z. Naturforsch.* **1982**, *37b*, 815.
[464] F. Huber, M. Domagala and H. Preut, *Acta Cryst.* **1988**, *C44*, 828.
[465] J. D. Curry and R. J. Jandacek, *J. Chem. Soc. Dalton Trans.* **1972**, 1120.
[466] J. Hillmann, H. D. Hausen, W. Schwarz and J. Weidlein, *Z. Anorg. Allg. Chem.* **1995**, *621*, 1785.
[467] M. Wieber, H. G. Rüdling and C. Burschka, *Z. Anorg. Allg. Chem.* **1980**, *470*, 171.
[468] C. Burschka and M. Wieber, *Z. Naturforsch.* **1979**, *34b*, 1037.
[469] C. Burschka, *Z. Anorg. Allg. Chem.* **1982**, *485*, 217.
[470] M. Dräger and B. M. Schmidt, *J. Organomet. Chem.* **1985**, *290*, 133.
[471] L. J. Farrugia, F. J. Lawlor and N. C. Norman, *Polyhedron* **1995**, *14*, 311.
[472] F. T. Edelmann, M. Noltemeyer, I. Haiduc, C. Silvestru and R. Cea-Olivares, *Polyhedron* **1994**, *13*, 547.
[473] M. J. Begley, D. B. Sowerby and I. Haiduc, *J. Chem. Soc. Dalton Trans.* **1987**, 145.
[474] M. Wieber and M. Schröpf, *Phosphorus, Sulfur & Silicon* **1995**, *102*, 265.
[475] R. Cea-Olivares, K. H. Ebert, L. Silaghi-Dumitrescu and I. Haiduc, *Heteroatom Chem.* **1997**, *8*, 317.
[476] J. Müller, U. Müller, A. Loss, J. Lorberth, H. Donath and W. Massa, *Z. Naturforsch.* **1985**, *40b*, 1320.
[477] W. W. du Mont, *Main Group Chemistry News*, **1995**, *2*, 18.

[478] W. W. du Mont, S. Kubiniok, K. Peters and H. G. von Schnering, *Angew. Chem.* **1987**, *99*, 820; *Angew. Chem. Int. Ed.* **1987**, *26*, 780.
[479] S. Kubiniok, W. W. du Mont, S. Pohl and W. Saak, *Angew. Chem.* **1988**, *100*, 434; *Angew. Chem. Int. Ed. Engl.* **1988**, *27*, 431.
[480] L. Battelle, C. Knobler and J. D. McCullough, *Inorg. Chem.* **1967**, *6*, 958.
[481] W. S. McDonald and L. D. Pettit, *J. Chem. Soc. A*, **1970**, 2044.
[482] G. D. Christofferson, R. A. Sparks, J. D. McCullough, *Acta Cryst.* **1958**, *11*, 782.
[483] L. Y. Y. Chan and F. W. B. Einstein, *J. Chem. Soc. Dalton Trans.*, **1972**, 316.
[484] F. Einstein, J. Trotter and C. Williston, *J. Chem. Soc. A*, **1967**, 2018.
[485] F. Berry and A. J. Edwards, *J. Chem. Soc. Dalton Trans.* **1980**, 2306.
[486] N. W. Alcock and W. D. Harrison, *J. Chem. Soc. Dalton Trans.* **1982**, 251.
[487] G. D. Christofferson and J. D. McCullough, *Acta Cryst.* **1958**, *11*, 249.
[488] N. W. Alcock and W. D. Harrison, *J. Chem. Soc. Dalton Trans.* **1984**, 869.
[489] C. H. W. Jones, R. D. Sharma and D. Naumann, *Can. J. Chem.* **1986**, *64*, 987.
[490] J. Aramini, R. J. Batchelor, C. H. W. Jones, F. W. B. Einstein and R. D. Sharma, *Can. J. Chem.* **1987**, *65*, 2643
[491] D. Naumann, L. Ehmanns, K. F. Tebbe and W. Crump, *Z. Anorg. Allg. Chem.* **1993**, *619*, 1269
[492] F. H. Allen, *J. Chem. Soc. Perkin II*, **1987**, S 1.
[492a] C. Lau, J. Passmore, E. K. Richardson, T. K. Whidden and P. S. White, *Can. J. Chem.* **1985**, *63*, 2273.
[493] R. E. Cobledick, F. W. B. Einstein W. R. Mc Whinnie and F. H. Musa, *J. Chem. Res.* **1979**, *S145*, M 1901.
[494] P. H. Bird, I. Bernal, J. C. Turley and G. E. Martin, *Inorg. Chem.* **1980**, *19*, 2556.
[495] J. D. McCullough, *Inorg. Chem.* **1973**, *12*, 2669.
[496] R. K. Chadha and J. E. Drake, *J. Organomet. Chem.* **1984**, *268*, 141.
[497] C. Knobler and J. D. McCullough, *Inorg. Chem.* **1972**, *11*, 3026.
[498] C. Knobler and J. D. McCullough, *Inorg. Chem.* **1970**, *9*, 797.
[499] H. Hope, C. Knobler and J. D. McCullough, *Inorg. Chem.* **1973**, *12*, 2665.
[500] J. D. McCullough, *Inorg. Chem.* **1975**, *14*, 1142.
[501] C. Knobler and R. F. Ziolo, *J. Organomet. Chem.* **1979**, *178*, 423.
[502] C. L. Raston, R. J. Secomb and A. H. White, *J. Chem. Soc. Dalton Trans.* **1976**, 2307.
[503] L. S. Refaat, K. Maartman-Moe and S. Husebye, *Acta Chem. Scand.* **1984**, *A38*, 147.
[504] P. H. Bird, V. Kumar and B. C. Pant, *Inorg. Chem.* **1980**, *19*, 2487.
[505] R. K. Chadha and J. E. Drake, *J. Organomet. Chem.* **1985**, *293*, 37.
[506] C. Knobler and J. D. McCullough, *Inorg. Chem.* **1977**, *16*, 612.
[507] J. D. McCullough and C. Knobler, *Inorg. Chem.* **1976**, *15*, 2728.
[508] J. D. McCullough, *Inorg. Chem.* **1977**, *16*, 2318.
[509] R. K. Chadha, J. E. Drake, N. T. McManus, B. A. Quinlan and A. B. Sarkar, *Organometallics* **1987**, *6*, 813.
[510] S. Husebye, S. Kudis, S. V. Lindeman and P. Strauch, *Acta Cryst.* **1995**, *C51*, 1870.
[511] J. E. Drake and J. Yang, *Inorg. Chem.* **1997**, *36*, 1890.
[512] M. R. Udupa, M. Seshasayee and T. A. Hamor, *Polyhedron*, **1993**, *12*, 2201.
[513] J. Jeske, W. W. du Mont and P. G. Jones, *Angew. Chem.* **1997**, *109*, 2305.
[514] R. K. Chadha and J. E. Drake, *J. Organomet. Chem.* **1986**, *299*, 331.
[515] R. K. Chadha, J. E. Drake, M. A. Khan and G. Singh, *J. Organomet. Chem.* **1984**, *260*, 73.
[516] R. F. Ziolo and M. Extine, *Inorg. Chem.* **1980**, *19*, 2964.
[517] M. R. Sundberg, R. Uggla, T. Laitalainen and J. Bergman, *J. Chem. Soc. Dalton Trans.* **1994**, 3279
[518] N. W. Alcock, W. D. Harrison and C. Howes, *J. Chem. Soc. Dalton Trans.* **1984**, 1709.
[519] N. W. Alcock, J. Culver and S. M. Roe, *J. Chem. Soc. Dalton Trans.* **1992**, 1477.
[520] S. Sato, N. Kondo and N. Furukawa, *Organometallics* **1995**, *14*, 5393.
[521] I. Haiduc, R. B. King and M. G. Newton, *Chem. Rev.* **1994**, *94*, 301.
[522] D. Lang, *Ph.D. Thesis, Univ. Würzburg*, **1993** (Thesis supervisor Prof. M. Wieber).
[523] S. Husebye, K. Maartmann-Moe and O. Mikalsen, *Acta Chem. Scand.* **1989**, *43*, 868.

[524] S. Husebye, K. Maartmann-Moe and O. Mikalsen, *Acta Chem. Scand.* **1990**, *44*, 464.
[525] A. Silvestru, I. Haiduc, K. H. Ebert and H. Breunig, *Inorg. Chem.* **1994**, *33*, 1253.
[526] A. Silvestru, I. Haiduc, K. H. Ebert and H. Breunig, *J. Organomet. Chem.* **1994**, *482*, 253.
[527] D. Dakternieks, R. Di Giacomo, R. W. Gable and B. F. Hoskins, *J. Am Chem. Soc.* **1988**, *110*, 6753.
[528] S. Husebye, K. Maartmann-Moe and O. Mikalsen, *Acta Chem. Scand.* **1990**, *44*, 802.
[529] J. Jeske, W. W. du Mont and P. G. Jones, *Angew. Chem.* **1996**, *108*, 2822; *Angew. Chem. Int. Ed.* **1996**, *35*, 2653.
[530] J. E. Drake, R. Ratnani and J. Yang, *Can. J. Chem.* **1996**, *74*, 1968.
[531] M. A. K. Ahmed, W. R. McWhinnie and T. A. Hamor, *J. Organomet. Chem.* **1985**, *293*, 219.
[532] C. S. Mancinelli, D. D. Titus and R. F. Ziolo, *J. Organomet. Chem.* **1976**, *10*, 278.
[533] V. Bertini, P. Dapporto, F. Lucchesini, A. Sega, and A. De Munno, *Acta Cryst.* **1984**, *C40*, 653
[534] R. Neidlein and D. Knecht, *Z. Naturforsch.* **1987**, *42B*, 84.
[535] T. Chivers, X. L. Gao and M. Parvez, *Inorg. Chem.* **1996**, *35*, 9.
[536] V. Bertini, F. Luchesini and A. De Munno, *Synthesis* **1982**, 681.
[537] D. D. Titus, J. S. Lee and R. F. Ziolo, *J. Organomet. Chem.* **1976**, *120*, 381.
[538] J. S. Lee, D. D. Titus and R. F. Ziolo, *J. Chem. Soc.* **1976**, 501.

5 Supramolecular Self-Assembly by Hydrogen-Bond Interactions

5.1 Introduction

Hydrogen-bonding is a major driving force behind supramolecular self-assembly in organic chemistry, as mentioned in Chapter 1. The presence of protonated functional groups in many organometallic compounds can also lead to the production of a wide variety of self-organized supramolecular structures, in both main group and transition metal organometallics. It is almost impossible to present here a comprehensive coverage of such supramolecular architectures and the examples selected are intended merely to illustrate their potential diversity. Chemists and crystallographers should bear in mind the possibility of hydrogen-bond self-assembly whenever they analyze the crystal structures of organometallics bearing functional groups which have a propensity for hydrogen-bond formation.

Significant progress towards crystal engineering based upon hydrogen-bonding has been achieved in organic supramolecular chemistry. We are approaching a point of similar understanding which will lead soon to the possibility of building organometallic crystal architectures with pre-established structures.

5.2 Group 14 Elements – Si, Ge, Sn, Pb

5.2.1 Organosilanols and organosiloxanols

The silanols are organosilicon compounds containing Si–OH groups. Because of their tendency to undergo condensation, with formation of siloxanes and water, until relatively recently this class has been considered difficult to handle and silanols were little investigated. The chemistry of organosilanols has been the focus of more attention in recent years and sufficient representatives have been isolated in the solid state, and their structures determined by X-ray diffraction [1, 2], to provide an interesting picture of hydrogen-bond self-assembly and self-organization. One recent development is the isolation of several organosilanetriols and the determination of their molecular and supramolecular structures [3, 4].

The pattern of hydrogen-bond self-assembly in silanols is less predictable than in organic compounds, and depends greatly on the number of Si–OH functions in the molecule and the number and size of organic groups attached to silicon. Supramolecular arrays (linear polymeric chains, two-dimensional sheets or layers, three-dimensional networks) are frequently observed, but bulky organic groups hinder the formation of extended structures, and these are limited to dimeric, tetrameric, or hexameric supermolecules. With few exceptions, there is no general pattern for a given family of silanols, and almost every new compound investigated has a new structure. This makes the synthesis and structural investigation of new organosilanols very attractive. The basicity and propensity of siloxanols toward hydrogen-bond formation has been theoretically investigated by use of an *ab initio* method [5].

Triorganomonosilanols, R_3SiOH, associate as dimers, **1**, tetramers, **2**, or polymers, **3** and **4**. A simple dimer formed as shown in **1** is diphenyl(fluorenyl) silanol, in which the large fluorenyl substituent prevents the approach of more than two molecules at hydrogen-bonding distance. The O$\cdots$O distance is 2.919 Å and the position of hydrogen has not been located [6].

Triphenylsilanol, Ph_3SiOH [7], and di(*tert*-butyl)fluorosilanol, $Bu^t_2Si(F)OH$ [8], form tetramers, **2**. In triphenylsilanol the four oxygen atoms form a flattened tetrahedron, with O$\cdots$O distances in the range 2.47–2.68 Å (the hydrogens were not located). Surprisingly, in the fluorosilanol $Bu^t_2Si(F)OH$ the fluorine atom does not

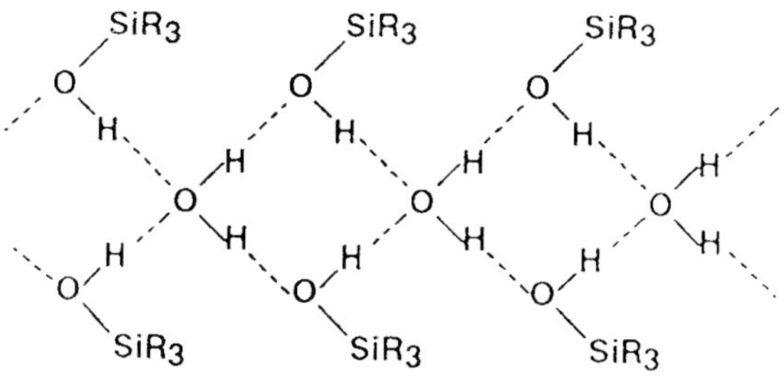

[Structure 4 diagram: chain of hydrogen-bonded SiR3-O-H units]

4

participate in hydrogen-bonding, and only the four Si–OH bonds are associated in a tetrameric unit. The hydrogen atoms were located and the O–H···O linkage is not linear (160°). The four oxygen atoms are not coplanar. The structures of these two silanol tetramers are similar to that of triphenylmethanol, Ph_3COH, which is also a tetramer [9].

The same hydrogen-bonded four-oxygen structural motif is present in the dimer of an organozinc derivative in which two organosilanol end groups participate in both intramolecular hydrogen-bonding (to close an eight-membered C_2Si_2ZnOH quasi-ring) and intermolecular hydrogen-bonding, to form the dimer, **5** [10].

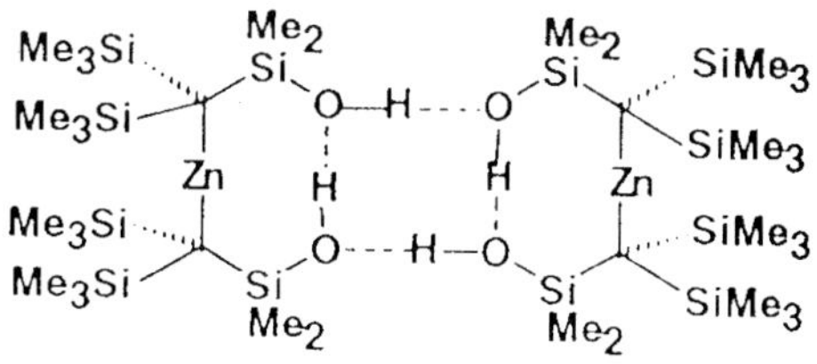

[Structure 5 diagram]

5

The intramolecular O···O distance (2.910 Å) is slightly longer than the intermolecular O···O distance (2.831 Å). The hydrogen atoms were located and both O–H···O bond are bent, the angles being 156° (intramolecular) and 162° (intermolecular).

A linear chain of the type illustrated in **3** was found in *endo*-3-methyl-*exo*-3-hydroxy-3-silabicyclo[3.2.1]octane [11].

A hemihydrate of *tert*-butyldimethylsilanol, $Bu^tMe_2SiOH \cdot H_2O$, an important compound which is formed by hydrolytic cleavage of organosilyl protecting groups in organic synthesis, has a unique structure of the type shown in **4**, with the participation of water molecules. The assembly is a chain polymer of self-assembled rhombohedra, in which two corners are oxygen atoms from the silanol [12].

In monosilanols containing heterocyclic nitrogen, such as biologically active (antimuscarinic) diphenyl(piperidinoethoxymethyl)silanol [13] and diphenyl(piperidinoethyl)silanol [14] strong intermolecular O–H···N bonds connect the molecules into chains.

Diorganosilanediols, $R_2Si(OH)_2$, participate in hydrogen-bonding self-assembly with two Si–OH groups, and give rise to quasi-cyclic dimeric, **6**, or tetrameric, **7**, supermolecules, or single and double chain-like supramolecular arrays, **8–10**.

One dimer of type **6**, with double hydrogen-bond bridges, is the compound

[(Me$_3$Si)$_3$C]PhSi(OH)$_2$, which contains very bulky groups preventing a greater association. As in other associated silanols, the O–H···O bond is non-linear (158.5°) and the O···O distance is 2.86 Å. The hydrogens were located (O–H 0.73 Å). In the related compound (Me$_3$Si)(Ph)Si(OMe)(OH), which is a monosilanol, there is a single hydrogen-bond [15], as in **5**, cited above.

Diorganosilanols can also form tetramers, but the structure is a quasi-tricyclic system, **7**. One representative of this type is bis(σ-pentamethylcyclopentadienyl)-silanediol, (C$_5$Me$_5$)$_2$Si(OH)$_2$. The unit cell contains six different independent molecules, grouped in tetramers. The pentamethylcyclopentadienyl substituents create steric hindrance which prevents the formation of a polymeric structure [16], as happens with silanediols containing smaller organic groups.

The same type of tetranuclear aggregate was found in another sterically hindered silanediol, [(2,4,6-Me$_3$C$_6$H$_2$)(SiMe$_3$)N](Me$_3$SiO)Si(OH)$_2$ [17].

Single chains, **8**, are present in the structure of diethylsilanediol, Et$_2$Si(OH)$_2$, but they are cross-linked by additional hydrogen-bonds, forming a layered structure [18].

The double-chain (or ribbon) structure of type **9** seems to be preferred by several diorganosilanediols, if the organic groups are not too bulky. This type has been identified in di-*iso*-propylsilanediol, Pr$^i_2$Si(OH)$_2$ [19], di-*tert*-butylsilanediol, Bu$^t_2$Si(OH)$_2$ [20], and dicyclohexylsilanediol, Cyh$_2$Si(OH)$_2$ [21]. The structure consists of eight-membered quasi-rings (if the hydrogen is counted as a ring member) further linked into ladder arrays. Some of these diols are practically attractive because of their liquid crystal behavior (e.g. di-*iso*-butylsilanediol forms a thermotropic liquid crystal mesophase).

Diphenylsilanediol, Ph$_2$Si(OH)$_2$, adopts a more intricate form of self-organization by forming six-membered quasi-rings O$_3$H$_3$ with chair conformation, which are further associated in three-dimensional columns by hydrogen-bonding [22, 23].

1,3-Disiloxanediols, HOR$_2$SiOSiR$_2$OH, undergo hydrogen-bond supramolecular self-assembly to form double chains with different amounts of reciprocal 'slippage'. Thus, in **11** a double chain is formed by connecting two chains, and an eight-membered quasi-ring O$_4$H$_4$ is formed. Another mode, in which the two chains are displaced by one step, resulting in a continuous sequence of O–H···

11

O–H···O–H··· bonds is shown in **12**. Finally, a third possibility, in which a further step of chain slippage occurs, results in a different type, **13**, of O–H···O–H···O–H··· succession.

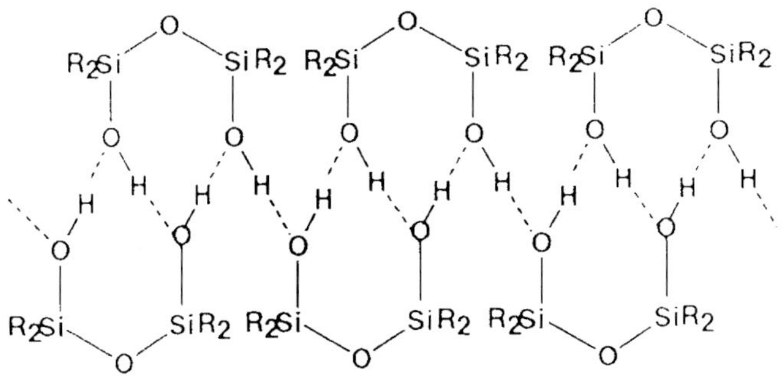

12

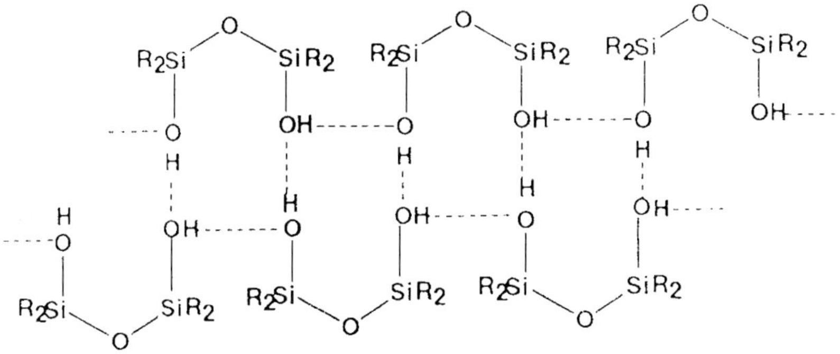

13

The first type, **11**, has been found in a 1,3-dimethyl-2-oxa-1,3-disilacyclohexane-1,3-diol [24]. Type **12** has been found in 1,1,3,3-tetra-*iso*-propyl-1,3-disiloxanediol, HOPr$^i_2$SiOSiPr$^i_2$OH [25]. The third type, **13**, is illustrated by the methyl [26], ethyl [27], *n*-propyl [28], and phenyl derivatives.

A hydrogen-bonded, two-dimensional layer structure was reported for thienyl disiloxane diol [29]. Related diols, the crystal structures of which have been established, include the disilanediol HOBu$^t_2$SiSiBu$^t_2$OH [30] and HOSiMe$_2$C(SiMe$_3$)$_2$-SiMe$_2$OH [17]. Both have a chain-like self-organized structure. In the disilane the six substituents attached to silicon are nearly eclipsed and the OH groups are on opposite sides of the disilane moiety, thus favoring inter- rather than intramolecular hydrogen-bonding.

Bis(hydroxydimethylsilyl)bis(trimethylsilyl)methane, HOSiMe$_2$C(SiMe$_3$)$_2$SiMe$_2$OH, is made up of infinite hydrogen-bonded chains, which contain hydrogen-bonds both within the 'monomeric' building units and intermolecular hydrogen-bonds between these units, in a unique fashion, **14**.

14

Monoorganosilanetriols with three OH groups attached to a silicon atom will generate more intricate hydrogen-bond assemblies. Thus, hexameric cages have been found in $(Me_3Si)_3CSi(OH)_3$ and $(Me_3Si)_3SiSi(OH)_3$. These contain an internal hydrophilic core, as a result of the self-organization of Si–OH groups, wrapped in a lipophilic outer jacket made from the organic groups. Such a structure explains the ready solubility in hydrocarbons, and their volatility (these trisilanols can be sublimed) [31, 32].

Tetrameric and octameric aggregates have been found in some organosilanetriols $RSi(OH)_3$ [33]. *tert*-Butylsilanetriol, $Bu^tSi(OH)_3$ [34], and cyclohexylsilanetriol, $CyhSi(OH)_3$ [35], form two-dimensional layers, as does pentamethylcyclopentadienylsilanetriol, $C_5Me_5Si(OH)_3$ [36]. The highly sterically hindered aminosilanetriol $(2,4,6-Me_3C_6H_2)(Me_3Si)NSi(OH)_3$ forms lipophilically wrapped columns, with the inorganic polar part in the interior [37]. (Trimethylsilylcyclopentadienyl)silanetriol $(C_5H_4SiMe_3)Si(OH)_3$, has been found to have an unprecedented supramolecular solid-state structure consisting of a monodimensional tubular network; it thus represents a new structural type [38].

A series of organocyclosiloxane-polyols has also been structurally characterized. Among those cited in the literature, cyclohexasiloxane-1,7-diol, **15**, forms two-dimensional layers, **16** [39] and 1,1,5,5-tetrakis(hydroxodimethylsiloxy)-3,3,7,7-tetraphenylcyclotetrasiloxane, **17**, forms a complicated three-dimensional network [40].

The branched tetrasilanol, $C(SiMe_2OH)_4$, has a very interesting structure consisting of an infinite three-dimensional network. Each molecule contains two intramolecular hydrogen-bonds; another set of hydrogen-bonds then connects the molecules into chains, crosslinked into sheets via hydrogen-bonds; the sheets are held

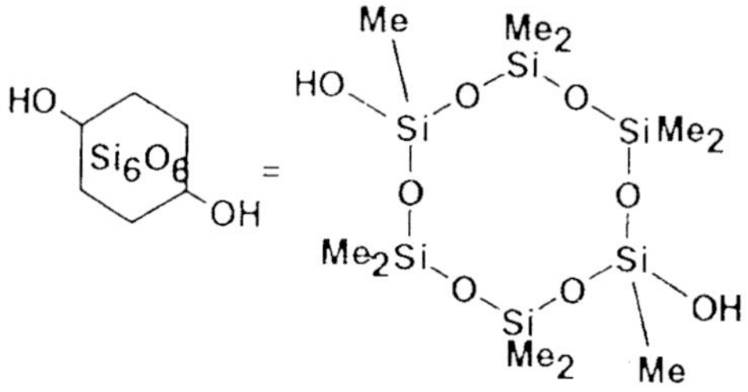

15

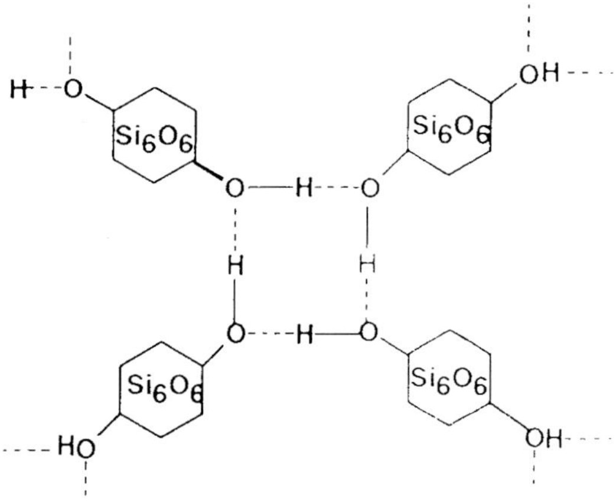

16

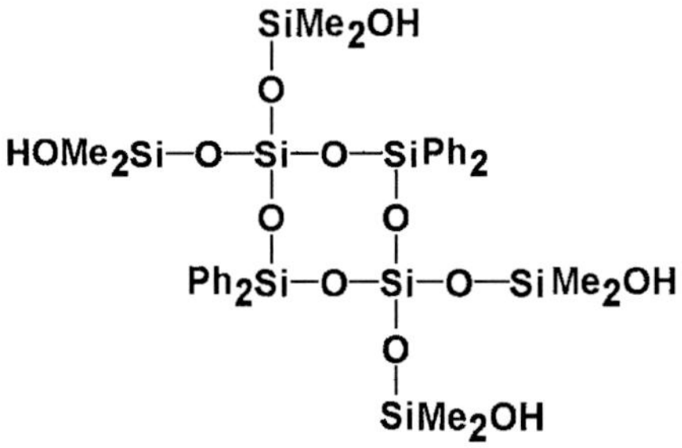

17

together, in a three-dimensional network, in the same way. The asymmetric unit of the crystal contains four different molecules [41].

Heterocyclic organosilicon derivatives, such as dihydroxosiloles, are also associated as tetramers or polymers through hydrogen-bonds [42].

Recently, new interesting results were reported (by electronic publishing) about the structures of hydrogen-bond complexes formed between silanols and amines or nitrogen heterocycles. These compounds can be regarded as models for the interaction between the silicate mineral world and organic (living) matter in nature [43].

Organosilicon compounds containing NH and keto groups in the organic substituents can also form supramolecular structures based upon hydrogen-bonds. A spectacular example is the hollow solid $Si[C_5NH_4(=O)]_4$ (a pyridone derivative) [44].

A new development is the discovery of hydrogen-bonds in phenylsilanols, based upon Si–OH···aryl (described as $\pi \cdots$ HO) interactions. Thus, in $(Me_3Si)CSiPh(X)$-OH, where X = I or OMe, the OH function of the silanol interacts with the aryl group of a neighboring molecule, leading to dimer formation [45]. In the compound with X = OH this $\pi \cdots$ HO interaction is additional to the normal hydrogen-bond detected earlier.

5.2.2 Organogermanium hydroxides

Hydroxy organogermanium compounds (usually referred to in the literature as hydroxides rather than 'germanols') have been less investigated than their organosilicon counterparts. From the examples known to date it seems that there is much similarity between the patterns of hydrogen-bond self-organization in E–OH derivatives of silicon and germanium. Thus, triphenylgermanium hydroxide, Ph_3GeOH, is a hydrogen-bonded tetramer, **18** [46], as its isomorphous organosilicon analog.

Di-*tert*-butylgermanium dihydroxide, $Bu^t{}_2Ge(OH)_2$, forms dimeric quasi-rings, associated into double chains, **19** [47], again like the silicon analog cited above.

An organogermanium compound with no silicon counterpart, namely polymeric carboxyethylgermanium sesquioxide, $(HOOCCH_2CH_2GeO_{1.5})_x$ (known as a potent antitumor agent), consists of germanium–oxygen layers of fused Ge_6O_6 rings crosslinked in a three-dimensional network by carboxyl groups [48]. There are no Ge–OH groups in this compound, and the supramolecular self-assembly occurs through traditional hydrogen-bond association of carboxylic groups.

5.2.3 Organotin compounds

Organotin hydroxides tend to self-organize by formation of dative O → Sn bonds, and so their supramolecular architectures are completely different. Unlike silicon or germanium, tin has a strong tendency to increase its coordination number, which will also influence the structures of Sn–OH organo derivatives.

The stepwise hydrolysis of monoorganotin trihalides can lead to a great variety of compounds. The dative bond hydrated dimers $[RSn(OH)X_2 \cdot H_2O]_2$, **20**, formed in the first step, containing six-coordinate tin, are associated via intermolecular hydrogen-bonds established between the OH bridges and the halogen atoms.

20

The hydrogen-bonds connect the centrosymmetric dimers into chains, which in turn are associated into two-dimensional layers. Several such compounds have been structurally investigated, with R = Et [49], Bu^n [50], Pr^i, and Bu^i [51].

Diorganotin halides also form dimers (five-coordinate tin) in the first step of their hydrolysis, and the mixed halide–hydroxide derivatives undergo self-organization in the solid state through X···H–O hydrogen-bonds. Thus, the halides $Bu^t_2Sn(OH)X$ (X = F, Cl, Br) have been structurally investigated and have similar structures [52]. Each centrosymmetric dimer is associated with two neighbors, forming a double chain, **21**.

Triorganotin functional derivatives have a remarkable tendency to self-organize into long chains, containing five-coordinate tin. If hydroxo groups are present in a chain they will tend to establish N···H–O hydrogen-bond bridges, leading to a supramolecular architecture consisting of paired chains. One example is trime-

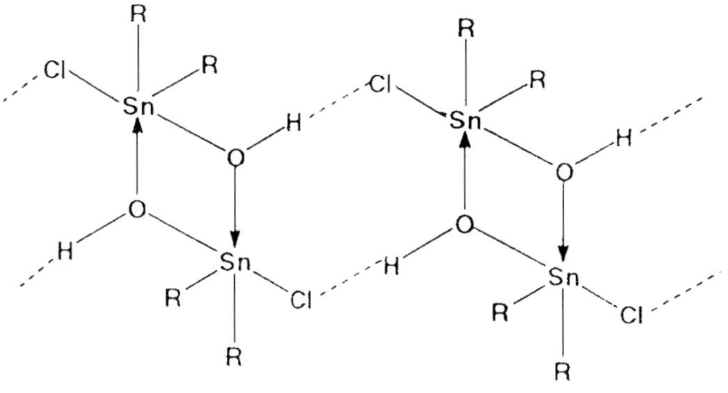

21

thyltin azide–hydroxide, Me₃SnN₃·Me₃SnOH, which has zigzag chains connected in a regular fashion [53]. The compound is isostructural with the isocyanate–hydroxide Me₃SnNCO·Me₃SnOH [54].

Another example is bis(trimethyltin)dimesylamide hydroxide, **22**, which contains antiparallel chains connected by O–H···N bonds to form a ladder structure [55].

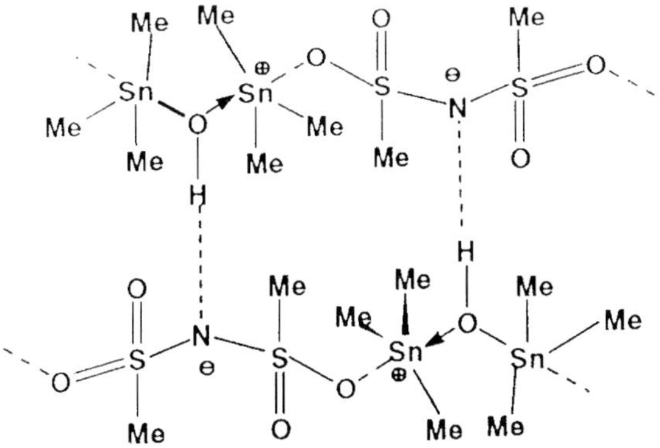

22

Triphenyltin isobutoxide–isobutanol solvate, Ph₃SnOBu^i·Bu^{i}OH, forms polymeric chains, **23**, by hydrogen-bonding between coordinated isobutanol and the oxygen atom of the isobutoxy group [56], and pairwise self-assembly of polymeric chains by hydrogen-bonding occurs in trimethyltin phenyl-α-oxoacetate, **24** [57].

Hydrogen-bond bridges between coordinated water molecules and oxygen atoms from the sulfonic group lead to self-assembly of trimethyltin benzenesulfonate molecules into a supramolecular chain-like array, **25** [58].

Rather similar hydrogen-bond self-assembly is observed in another polymeric five-coordinate trimethyltin cyanamidosulfonate hydrate, 4-MeC₆H₄SO₂NCNSnMe₃, **26**, investigated by X-ray diffraction [59].

An extended net of hydrogen-bonds between cations and anions in the crystal structure of [Me$_3$Sn(OH$_2$)$_2$][N(SO$_2$Me)$_2$] produces a complex supramolecular construction [60]. Bis(aquatrimethylstannyl) oxalate, [Me$_3$Sn(H$_2$O)]$_2$C$_2$O$_4$, forms a three-dimensional supramolecular architecture, **27**, dominated by hydrogen-bonds rather than carboxylate bridges, consisting of tectons made of a SnMe$_3$ group, a monodentate oxalate and a coordinated water molecule. The tectons are held together by hydrogen-bonds between oxalato ligands and coordinated water. It is remarkable that the oxalate anions form hydrogen-bonds in preference to coordination to a second metal atom. The structure contains channels of pseudo fourfold symmetry [61].

The hydrated 1,10-phenanthroline adduct of triphenyltin chlorodifluoroacetate, Ph$_3$Sn(OOCCClF$_2$)·phen·H$_2$O, is a dimeric supermolecule formed by hydrogen-bonding between coordinated water and phenanthroline [62].

There are several known examples of N–H···O hydrogen-bonds in self-organized structures of polymeric organotin compounds. One is dicyclohexylammonium tri-n-butyltin 2-sulfobenzoate, [Cy$_2$NH$_2$]$^+$[Bu$^n_3$Sn(O(O)CC$_6$H$_4$-2-SO$_3$)]$^-$,

27

in which helical chains containing five-coordinate tin connected axially to both carboxylato and sulfonato groups, are bridged by the substituted ammonium cation via N–H···O bridges [63]. Similarly, in bis(dicyclohexylammonium)-tris(oxalato)-tetrakis(tri-n-butyltin) (ethanol solvate) [64], $2[Cy_2NH_2]^+[EtOH·Bu^n_3SnOC(CO_2)(=O)Bu^n_3SnOC(=O)]_2^{2-}$, the substituted ammonium cation forms hydrogen-bond bridges between the tetrametallic chains of the organotin oxalate, to form a three-dimensional network, as shown schematically in **28** (butyl groups omitted for clarity).

A three-dimensional hydrogen-bond lattice is also formed by hydrogen-bond assembly of discrete tetrachlorodimethylstannate anions with bis[2-(aminocarbonyl)-anilinium] cations [65].

From organic chemistry we know that carbamides and other compounds containing the –C(=O)–NH– function, have a great tendency to form hydrogen-bonds. Similarly, in organotin compounds, such derivatives will self-assemble by hydrogen-bonding, to form various types of supramolecular arrays. Thus, triphenyltin 3-ureidopropionate, $Ph_3SnO(O)C(CH_2)_2NHCONH_2$, is a polymer with chains interconnected by hydrogen-bonds to form the spatial arrangement **29** [66].

Adjacent molecules of bis(N-acetylhydroxylamino)dimethyltin, $Me_2Sn(ONHCOMe)_2$, are held together in the crystal by two NH···O=C hydrogen-bonds in a supramolecular structure consisting of infinite linear arrays, **30**.

In the monohydrate, $Me_2Sn(ONHCOMe)_2·H_2O$, layers composed of hydrogen-bonded organotin molecules alternate with layers of water molecules, in a complex three-dimensional structure [67].

A series of bis(pyrazole) and bis(imidazole) complexes of diorganotin dihalides are self-organized chain polymers **31** and **32**, formed by N–H···X bonding (X = Cl, Br). These include $R_2SnX_2·2pyrazole$ (R = Me, X = Cl [68], Br [69]; R = Et, X = Br [70]; R = vinyl, X = Cl [71]; R = Bu, X = Cl [72]), $R_2SnX_2·2imidazole$ (R = Me, X = Cl) [73], and $Me_2SnCl_2·2(2$-chloroimidazole) [74].

In a similar manner NH···Cl hydrogen-bonding results in the self-organization of imidazolethione complexes of organotin dihalides into double chains, **33**. The

28

heterocyclic ligand is in the thione form; one or both NH functions can be involved in hydrogen-bonding (intra and intermolecular) [75–77].

Double-chain supramolecular arrays based upon N–H···Cl hydrogen-bonds are present in (8-aminoquinoline-N,N')dichlorodimethyltin [78].

Hydrogen-bonds also participate in the self-assembly of organotin derivatives of dipeptides, e.g. (L-methionyl-L-methioninato)dimethyltin and (L-alanyl-L-histidinato)dimethyltin [79], in the hydantoic acid derivative of triphenyltin, $Ph_3Sn(OOCCH_2NHCONH_2)$ [80], and in (4-benzoylpyrazol-5-onato)tributyltin [81].

A unique, helical chain structure, **34**, was described for α-(phenylphosphonato)-trimethyltin, obtained from trimethyltin acetate and phenylphosphonic acid, formulated as $[Me_3Sn]^+[Ph(OH)P(O)SnMe_3OP(O)(OH)Ph]^-$ [82]. It contains chains incorporating five-coordinate Me_3Sn units. The chains are linear at tin, but bent at

5.3 Group 15 Elements – As, Sb, Bi

5.3.1 Organoarsenic compounds

Diorganoarsinic acids are hydrogen-bonded dimers, **35** (R = Me [83], *tert*-Bu [84]), or infinite helical chains (R = vinyl or phenyl), associated in the same manner as the carboxylic acids, but with R = CF_3 dimerization occurs at the As=O bond [85]. An unusual tetranuclear aggregate, **36**, consisting of two molecules of diphenylarsinic acid and two molecules of diphenylmonothioarsinic acid, held together by hydrogen-bonds, has also been identified by X-ray crystallography [86].

31

32

33

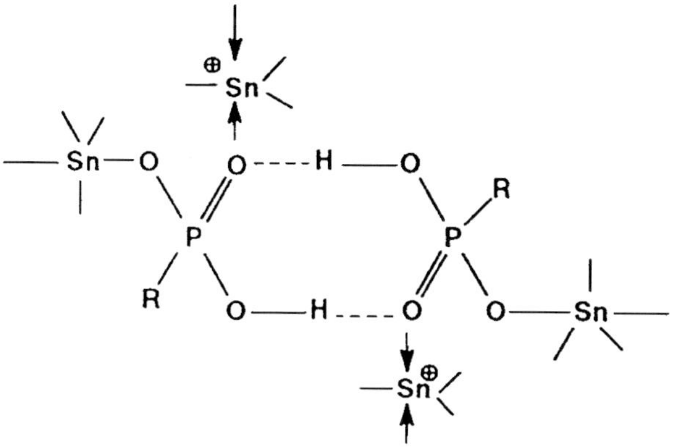

34

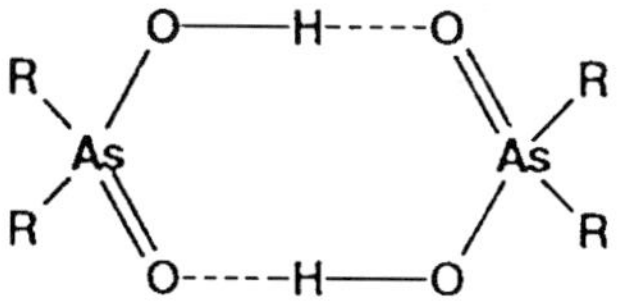

35

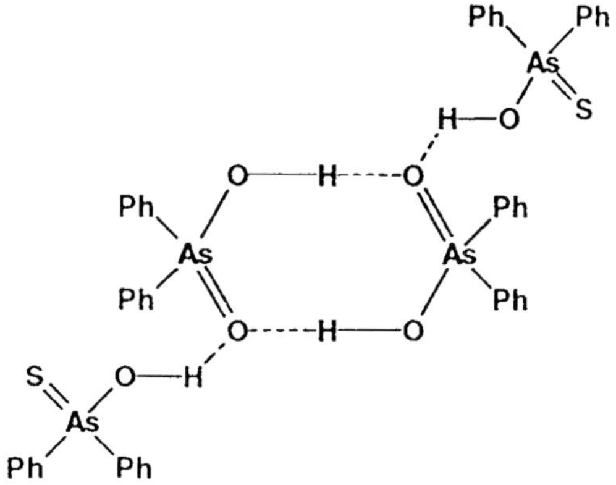

36

Triphenylarsine oxide forms strong symmetrical hydrogen-bonds with very strong acids (e.g. HBF$_4$) [87] to give [Ph$_3$As=O···H$^+$···O=AsPh$_3$] cations, whereas with weaker acids (HCl, HBr) the compound forms strong but asymmetric hydrogen-bond assemblies Ph$_3$As=O···H–X (X = Cl, Br). With very weak protic acids (H$_2$O or PhSO$_2$NH$_2$) centrosymmetric dimeric supermolecules, e.g. **37**, are formed [88, 89].

5.3.2 Organoantimony compounds

Fewer organoantimony compounds containing hydrogen-bonds leading to supramolecular self-organization have been described. It can be expected that many analogies with organotin compounds are possible and perhaps this will stimulate a search for such structures.

Water present as solvate in organometallic compounds might be expected to produce hydrogen-bond aggregates, as seen before. Thus, in apparently simple compounds, such as diphenylantimony trichloride monohydrate, $Ph_2SbCl_3 \cdot H_2O$, water is coordinated to antimony and via $Cl \cdots H-O$ bridges promotes hydrogen-bond self-organization, **38**, in the crystal [90].

A complex supramolecular array involving O_4H_4 hydrogen-bond quasi-rings was found in $Me_3Sb(OH)_2 \cdot H_2O$ (trigonal bipyramidal coordination of antimony with axial OH groups, Sb–O 2.033 and 2.063 Å, and equatorial methyl groups). The carbonate, $[Me_3SbOH]_2CO_3 \cdot 2H_2O$, contains a complex supramolecular array involving an $O_{10}H_{10}$ hydrogen-bond quasi-ring (axial Sb–O 1.988, 2.153 and 2.159, 2.036 Å) [91]. By contrast, crystals of organoantimony(V) dihydroxides with bulky organic groups, such as $Ph_3Sb(OH)_2$ [92] and $(Mes)_3Sb(OH)_2$ (Mes = mesityl) [93] contain discrete, unassociated molecules.

The hydroxophosphinate $[Me_3Sb(OH)(O_2PPh_2)]$ is a chain-like supramolecular array formed by $P=O \cdots H-O$ hydrogen-bonding, **39** [94].

In μ-oxo-bis(trimethylantimony) benzenesulfonate hydrate, $\{[Me_3SbOH_2]O\}(O_3SPh)_2$, **40**, hydrogen-bonds are present between the coordinated water and discrete anions [95].

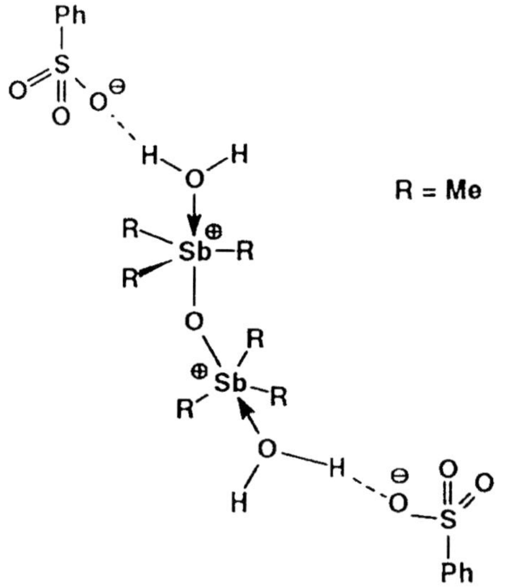

40

In tetraphenylstilbonium benzenesulfonate hydrate, Ph$_4$SbO$_3$SPh·H$_2$O, two hydrogen-bonded water molecules bridge the sulfonic group to form a dimer, **41** [96].

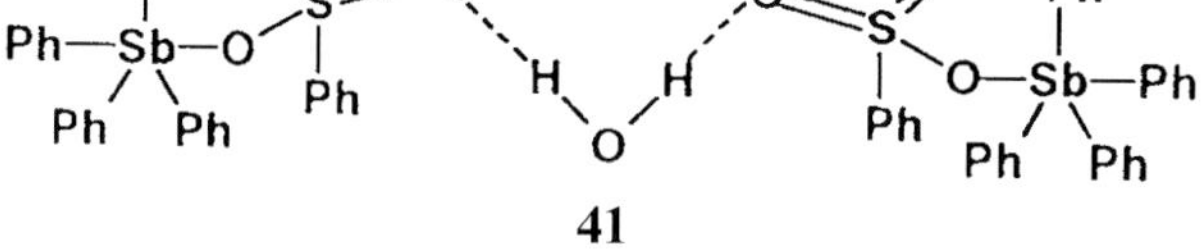

41

The presence of water in organometallic crystal hydrates is almost certain to produce hydrogen-bond self-assembly with formation of supramolecular structures.

5.4 Transition Metals (Organometallic Crystal Engineering)

Logically connected to supramolecular chemistry is crystal engineering, i.e. the rational planning and execution of a crystal structure from its building blocks, molecules, or ions [97, 98]. Factors such as the size, shape, or the propensity of the constituents for peripheral bonding should be taken into account. The most prominent interaction utilized in crystal engineering is hydrogen-bonding; ionic and van der Waals interactions are less important. Traditionally, most work in the field of crystal engineering originated from organic chemistry and even today remains a domain of organic chemistry [99]. Hydrogen-bonding in organic and biological systems has long been the subject of intensive research and is well documented in review articles [100]. Comparable coverage of hydrogen-bonding in organometallic chemistry has been lacking [101]. Organometallic crystal engineering and investigation of hydrogen-bonded transition metal complexes are, however, emerging as promising new fields of research. Only very recently crystal engineering and organometallic architecture have been highlighted in a comprehensive review article [102]. It should be kept in mind that crystals of organometallic complexes or cluster compounds are molecular in nature, just like molecular organic crystals. Thus it is quite clear that interactions holding both types of crystal together are at least similar [103]. It seems that hydrogen-bonding in organotransition metal complexes is a fairly common phenomenon which might in the past have been overlooked. With today's easy access to crystallographic databases [104] it has become possible to analyze large numbers of known crystal structures and draw chemical conclusions.

Hydrogen-bonding is an important type of interaction in supramolecular transition metal complexes because it combines directionality with strength [105]. Three main varieties of hydrogen-bond in organotransition metal compounds will be discussed here. The first part of this section will deal with peripheral hydrogen-bonding, i.e. hydrogen-bonds involving only the coordinated ligands. There is no fundamental difference between this type of hydrogen-bonding and that in the corresponding purely organic systems. The second part of the section will focus on supramolecular transition metal complexes in which the metal atom itself is involved in hydrogen-bonding. Hydrogen-bonds assisted by ionic interactions will be discussed in Section 5.4.3.

5.4.1 Peripheral hydrogen-bonding

5.4.1.1 O–H···O Hydrogen bonds involving carboxyl groups

The dimerization of carboxylic acids is a textbook example of a strongly hydrogen-bonded system [106]. Two O–H···O interactions result in the formation of an eight-membered ring. The same structural motif can also be found in organometallic complexes containing carboxylic acid functions. Their number is somewhat limited because many sensitive metal–ligand combinations do not tolerate the presence of acidic COOH groups in the molecule and so the corresponding carboxylic acid derivatives cannot be synthesized. With more robust organometallic units such as ferrocene, however, organometallic carboxylic acids are readily accessible by standard synthetic procedures. For example, carboxylation of lithioferrocene with CO_2 yields ferrocene carboxylic acid, Fc-COOH, **42**, after acidic work-up [107]. The interactions between carboxylic acid functions are a powerful tool in organometallic crystal engineering [108]. The most interesting result of studies in this area is the finding that there are strict analogies between the hydrogen-bonding patterns in organometallic crystals and those in the corresponding purely organic crystals. With COOH groups acting as hydrogen-bond donors, the acceptor groups are normally carboxyl groups also. Thus the familiar eight-membered ring system of carboxylic acid dimers is the characteristic structural motif of organotransition metal complexes bearing a COOH functional group. This group can be part of an extended π-coordinated ligand system or a σ-bonded ligand, and the metal complex can be mononuclear, dinuclear or even a cluster. Typical examples of the first type of organometallic compound are $(\eta^6\text{-}p\text{-}Bu^tC_6H_4COOH)Cr(CO)_3$, **43** [109], and the sorbic acid complex $(\eta^4\text{-MeCH=CHCH=CHCOOH})Fe(CO)_3$, **44** [110]. The entire ligand system in dimeric **43** is almost exactly planar, just as in free benzoic acid. The η^6-coordinated chromium tricarbonyl fragments are in a *trans* arrangement relative to the central eight-membered ring.

The same type of eight-membered ring is formed upon dimerization of carboxylic functions which are part of a σ-bonded ligand. This may be exemplified by the 'metalloacetic acid' complexes $(C_5H_5)Fe(CO)_2(\sigma\text{-}CH_2COOH)$, **45**, and $(C_5H_5)\text{-}Mo(CO)_3(\sigma\text{-}CH_2COOH)$, **46** [111]. The crystal structures of both compounds are characterized by the presence of discrete dimers which adopt a chair-like arrangement relative to the central eight-membered ring. The carbonyl ligands do not participate in the hydrogen-bonding.

45

340 5 *Supramolecular Self-Assembly by Hydrogen-Bond Interactions*

Structural characterized examples of metal cluster complexes with the same type of self-assembly are $H_3Os_3(CO)_9(\mu_3\text{-CCOOH})$, **47** [112] and $HRu_3(CO)_{10}$-$(\mu\text{-SCH}_2COOH)$, **48** [113]. All three types of metal complex typify the close structural analogy between organometallic and organic carboxylic acids.

Similar considerations apply to organometallic complexes containing two COOH functional groups. There are two main types of free dicarboxylic acids. In the first group self-assembly occurs only via intermolecular hydrogen-bonds, whereas the second group contains an intramolecular O–H···O bond between neighboring carbonyl groups [106]. In these cases also there are striking analogies between the free acids and some of their transition metal complexes. The two compounds (η^2-fumaric acid)Fe(CO)$_4$, **49**, and (η^2-maleic acid)Fe(CO)$_4$, **50**, are especially instructive examples. Association of the COOH functions in **49** results in the formation of a ribbon-like polymer in the solid state [114]. Self-assembly of the fumaric acid ligands is the same as in the free acid [115]. A close similarity between the complexed form and the free acid is also found for the corresponding maleic acid complex **50**. The structure of crystalline maleic acid is characterized by the presence of both intermolecular association and intramolecular hydrogen-bonding [116]. Both types of hydrogen-bonds are simultaneously present also in the iron tetracarbonyl complex of maleic acid; this results in a polymeric chain structure for **50** in the solid state [114b].

49

50

In the dinuclear acetylenedicarboxylic acid complex $[\mu\text{-}\eta^2\text{-}C_2(COOH)_2]Co_2(CO)_6$, **51** [117], intramolecular hydrogen-bonding as in the maleic acid derivatives is not possible because of the tight bending of the ligand in the $\mu\text{-}\eta^2,\eta^2$-coordination mode.

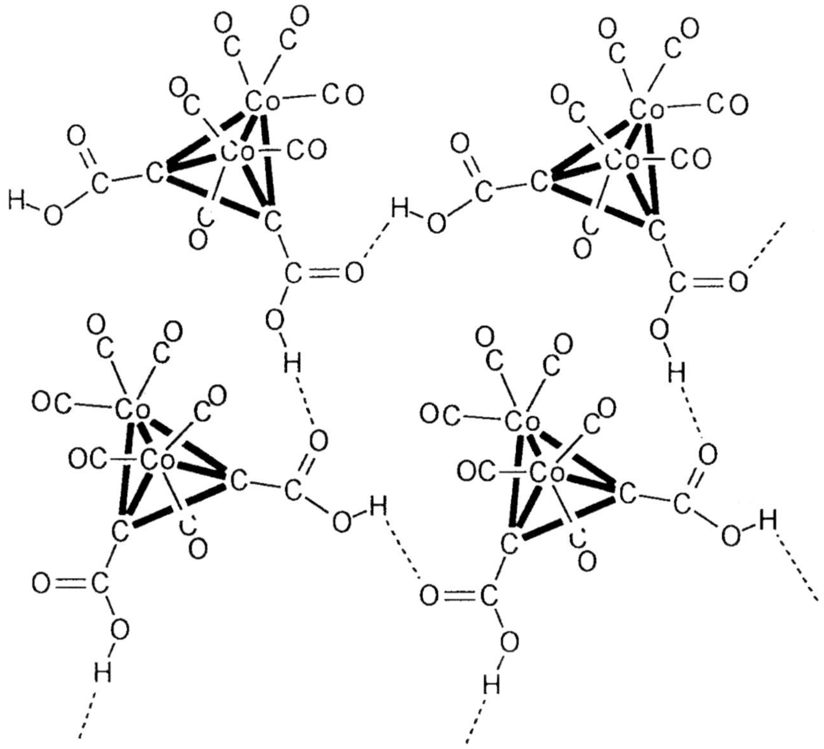

51

This leads to the formation of large rings in the crystal structure and finally results in an extended network of joint ring systems. Each COOH group in **51** interacts with two carboxylic acid functions of neighboring molecules.

Usually the COOH functional groups act both as hydrogen-bond donors and acceptors. The dinuclear tungsten complex $[(C_5H_4COOH)W(CO)_3]_2$, **52** [118], is not a supramolecular compound but it is shown here as an example of hydrogen-bonding between carboxylic acid functions and THF.

Another structurally characterized example of an organometallic dicarboxylic acid is 1,1′-ferrocenedicarboxylic acid, $(C_5H_4COOH)_2Fe$, **53** [119]. Compound **53** crystallizes in two modifications, monoclinic and triclinic, and the triclinic form has been investigated by X-ray diffraction. In this crystal structure the molecules of **53** adopt an eclipsed conformation and associate in the form of discrete hydrogen-bonded dimers.

The crystal structure contains two types of dimers and the corresponding neighboring dimers are connected by weak C–H···O interactions; there are no such interactions between dimers of different types.

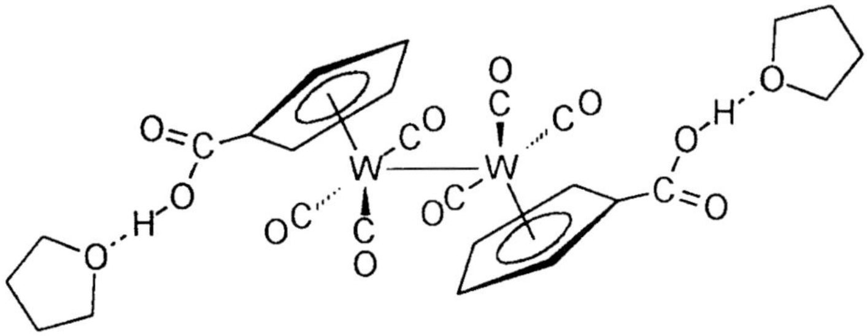

52

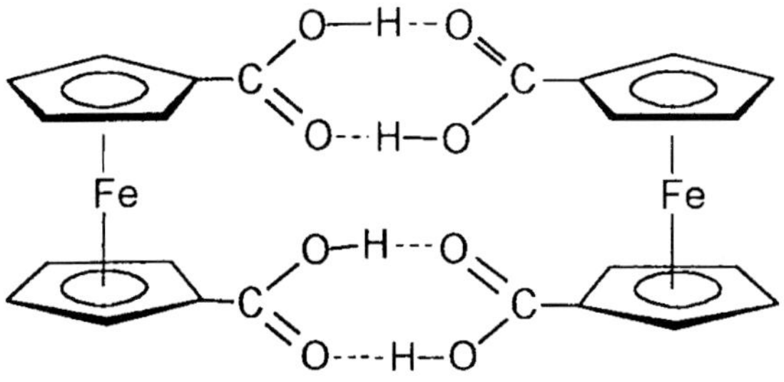

53

5.4.1.2 O–H···O Hydrogen bonds involving hydroxyl groups

Peripheral hydrogen-bonding involving hydroxyl groups has been observed in a number of organometallic alcohols. Hydroxyl groups are unique in the sense that they are approximately equally good hydrogen-bond donors and hydrogen-bond acceptors. Hydrogen-bonds involving OH functions are somewhat weaker than those in which COOH donors participate. There is also more structural flexibility because the OH function is less directed. Often the hydroxyl group acts both as donor and acceptor, but there are also examples in which ester or ketone functional groups are the acceptors. In contrast with the organometallic carboxylic acids discussed above, there are also examples of CO ligands competing with the OH groups as hydrogen-bonding acceptors (vide infra). Three different types of self-assembly have been identified for peripheral hydrogen-bonding involving only hydroxyl groups. In these the structural motifs are often the same as those found in the crystal structures of organic alcohols [120]. In particular, organometallic complexes bearing OH functions have been found to associate as tetramers, **54**, hexamers, **55**, and, occasionally, as polymeric chain structures, **56**.

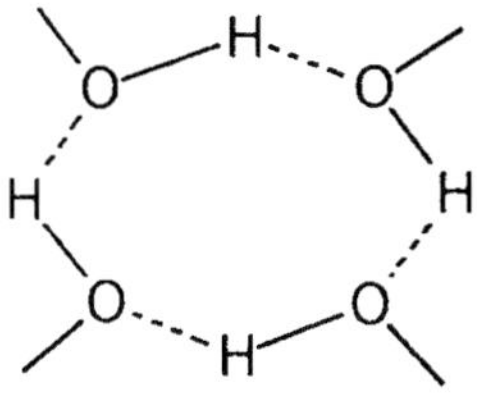

5.4 Transition Metals (Organometallic Crystal Engineering)

54

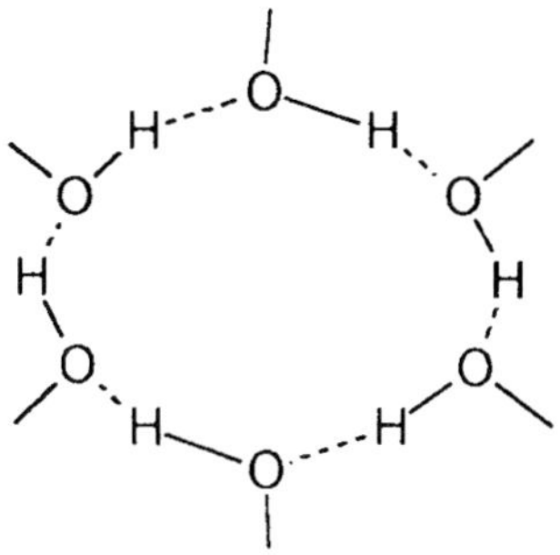

55

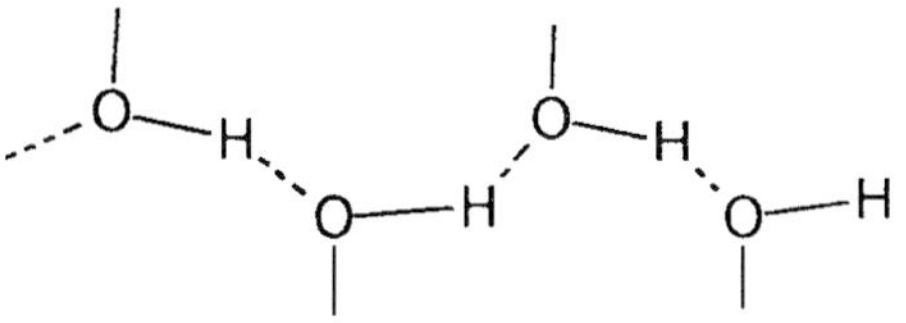

56

The formation of tetramers, **54**, is well established for organic alcohols such as 1-adamantanol, 1-phenyl-1-cyclohexanol, or dicyclohexylmethanol [120]. Structurally characterized examples of organometallic alcohols following this hydrogen-bonding pattern are the glycol derivative $[\eta^6\text{-}C_6H_4(CHMeOH)_2\text{-}1,2]Cr(CO)_3$, **57** [121] and $(C_5H_5)Mo(CO)_2(\eta^3\text{-}C_7H_{10}OH)$, **58** [26]. The latter was prepared by a hydroboration reaction starting from the cycloheptadienyl precursor $(C_5H_5)\text{-}Mo(CO)_2(\eta^3\text{-}C_7H_9)$, **59**. A beautiful supramolecular arrangement has been found in the hexameric (**55**) dinuclear iron complex $[\eta^5\text{-}C_5H_4CH_2CH_2OH)Fe(CO)_2]_2$, **60** [123] (Figure 5.1.). In this large hydrogen-bonded structure six $[\eta^5\text{-}C_5H_4CH_2CH_2OH)Fe(CO)_2]_2$ units are linked by O–H···O hydrogen-bonds to give a central twelve-membered ring.

Surprisingly, the closely related dinuclear molybdenum complex $[\eta^5\text{-}C_5H_4CH_2CH_2OH)Mo(CO)_3]_2$, **61** [124], which contains the same hydroxyethylcyclopentadienyl ligand as in **60**, has been reported to form a chain polymer (**56**) in the solid state. Hydrogen-bonding chains are also present in the compounds $(C_5H_5)Mo(CO)_2(\eta^3\text{-}C_8H_{12}OH)$, **62** [125], and $(\eta^4\text{-}MeC_4H_4CMeOH)Fe(CO)_3$, **63** [126]. As a typical example the crystal structure of **62** is depicted in Figure 5.2. It is especially remarkable that the only chemical difference between tetrameric **58** and polymeric **62** is the slight change in ring size (seven-membered compared with eight-membered ring ligand).

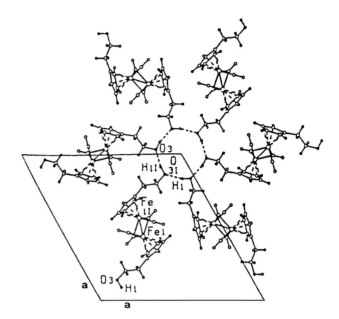

Fig. 5.1. The twelve-membered hydrogen-bonded ring system in hexameric **60** (Reproduced from ref. [123]).

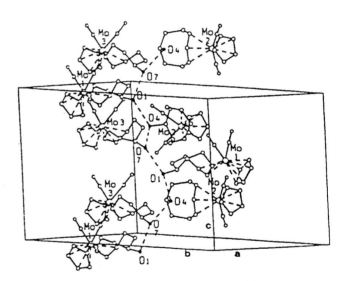

Fig. 5.2. Polymeric chain structure of **62** (Reproduced from ref. [125]).

Nickel and platinum complexes of alkynols and alkynediols are an interesting class of organometallic complex self-associated by O–H···O hydrogen-bond interactions. Their solid-state supramolecular assembly has been the subject of thorough studies [127].

58

Bis(tetramethylbutynediol)nickel(0), **64**, the first homoleptic nickel alkyne complex, was prepared according to Eq.(1) by treatment of bis(cycloocta-1,5-diene)-nickel(0) with two equivalents of the ligand [127a].

$$[Ni(COD)_2] + 2\,HOCMe_2-C\equiv C-CMe_2OH$$

$$\rightarrow [Ni(HOCMe_2-C\equiv C-CMe_2OH)_2] + 2\,COD \tag{1}$$

Compound **64** is a true nickel(0) alkyne complex with no agostic interactions between Ni and the methyl groups of the ligands. In the crystal the molecules are associated by four intermolecular hydrogen-bond interactions between neighboring butynediol ligands. This leads to the formation of chain polymers.

The presence of hydroxyl functions in the molecule is an essential prerequisite for the formation of a stable nickel(0) alkyne complex such as **64**, because it has been shown that similar treatment of Ni(COD)$_2$ with the dimethyl ether of tetramethylbutynediol, MeOCMe$_2$C≡CCMe$_2$OMe, results only in thermal decomposition and formation of metallic nickel. Compound **64** is itself a useful starting material for a variety of heteroleptic nickel alkyne complexes such as (tmeda)Ni(μ-HOCMe$_2$C≡CCMe$_2$OH)Ni(HOCMe$_2$C≡CCMe$_2$OH), **65**, Ni(COD)(HOCMe$_2$C≡CCMe$_2$OH), **66**, and Ni(PPh$_3$)$_2$(HOCMe$_2$C≡CCMe$_2$OH), **67** [127a, c].

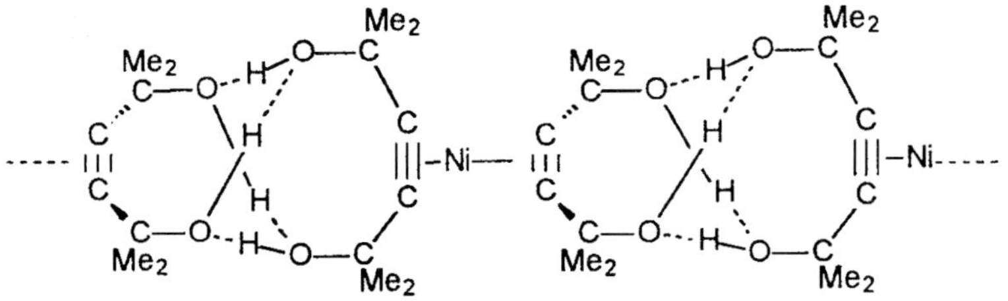

64

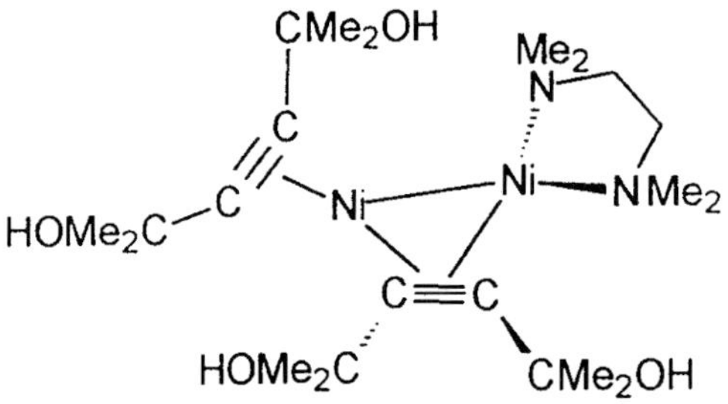

65

In the solid state the binuclear tetramethylethylenediamine derivative **65** forms a supramolecular double-chain structure in which the molecules are linked by four intermolecular O–H···O hydrogen-bonds. There are two additional intramolecular hydrogen-bonds, one within the bridging alkynediol ligand and the other one between the bridging and the terminal alkynediol unit.

In the COD derivative, **65**, self-assembly by hydrogen-bonding leads to a helical chain polymer structure comprising both intra- and intermolecular hydrogen-bond interactions. In a manner similar to that illustrated in Eq.(1) the red trinuclear complex Ni$_3$(Bu^{t}C≡CCMe$_2$OH)$_4$, **68** [127b], was obtained by reaction of Ni(cyclododeca-1,5,9-triene) with two equivalents of the alkynol ligand Bu^{t}C≡CCMe$_2$OH. In the crystal structure of **68** the molecules form a chain polymer. Once again the presence of OH functions and the possibility of forming hydrogen-bonds was found to be essential for the successful synthesis of this unusual metal alkyne complex.

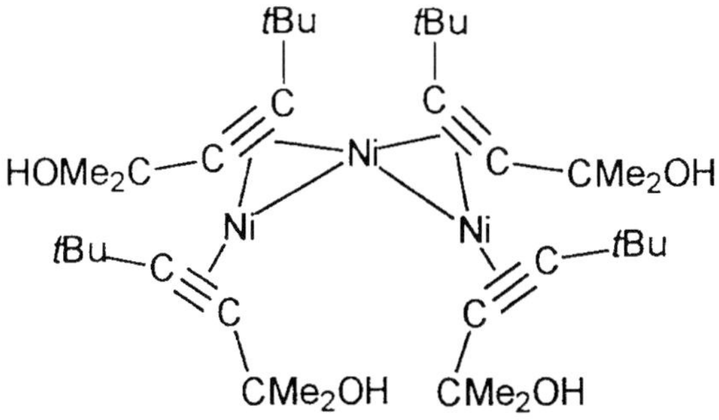

68

Related platinum complexes can be obtained by using Pt(COD)$_2$ as starting material and the range of suitable acetylenic alcohols covers, for example, PhC≡CCMe$_2$OH, Bu^{t}C≡CCEt$_2$OH, and c-C$_6$H$_{11}$(OH)(C≡CBu^t)-1,1 [127e]. It was found that the bis(alkynol)metal(0) complexes of nickel and platinum derived from the latter two alcohols form hydrogen-bridged dimers in the solid state. The circular arrangement of four O–H···O hydrogen-bonds is very similar to that found in the free alkynols [127e, 128]. A typical example is Pt$_2$(Bu^{t}C≡CCEt$_2$OH)$_4$, **69**.

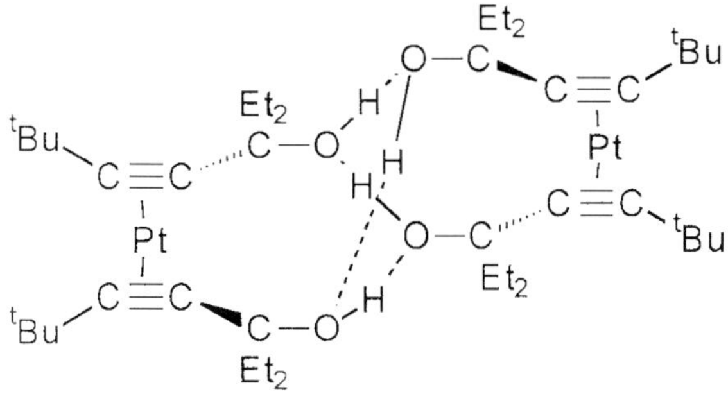

69

Ferrocene-1,1'-diyl-bis(diphenylmethanol), Fe(C$_5$H$_4$CPh$_2$OH)$_2$, **70**, has been reported to form hydrogen-bonded host–guest adducts (diol/guest ratio 1:1 or 1:2) with a variety of hydrogen-bond donors and acceptors including MeOH, EtOH, Me$_2$SO, Me$_2$NCHO, piperazine, and 4,4'-bipyridyl; 1:2 adducts can be isolated with Me$_2$SO, dioxane, pyridine, and piperidine [129]. The 1:1 adduct with methanol, **71**, was found to form a centrosymmetric assembly comprising two molecules of the host **70** and two molecules of the guest MeOH. An array of hydrogen-bonds forms a central twelve-membered (OH)$_6$ ring which adopts a chair conformation. In addition to **70** the crystal structures of various α-ferrocenyl alcohols have been examined for the presence of hydrogen-bonding interactions [130].

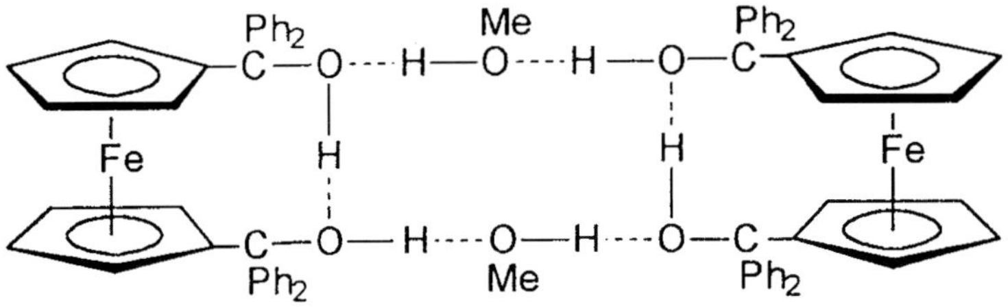

348   5 Supramolecular Self-Assembly by Hydrogen-Bond Interactions

71

Several supramolecular transition metal complexes have been reported which are associated by O–H···O hydrogen-bonding, but in which the acceptor group is a ketonic carbonyl group or an ester group. An early example of such a system is the 1:2 hydroquinone adduct of tricarbonyl(cyclopentadienone)iron(0), {μ-1,4-$C_6H_4(OH)_2$}[$(C_5H_4O)Fe(CO)_3$]$_2$, **72**. This compound is known from early work of Reppe et al. and has been obtained as a by-product from the reaction of acetylene with $Fe(CO)_5$ in aqueous ethanol [131, 132]. The adduct was first assumed to be a charge-transfer complex with five-and six-membered rings in a parallel arrangement. An X-ray study revealed, however, that **72** is a hydrogen-bonded species in which the ketonic groups of the $(C_5H_4O)Fe(CO)_3$ units act as acceptors [133].

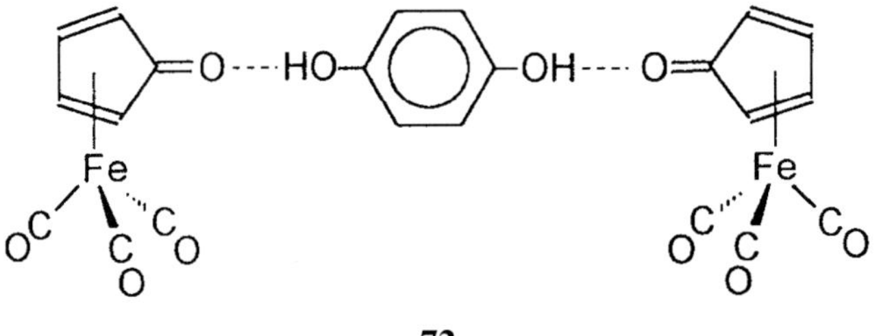

72

A related hydrogen-bonded ruthenium complex of cyclopentadienone has been prepared by an entirely different synthetic route [134]. Hydrolysis of the cationic precursor (η^5-C_5Me_5)Ru(η^5-C_5H_4OH)Br]CF_3SO_3, **73**, with H_2O afforded the neutral cyclopentadienone complex (η^5-C_5Me_5)Ru(η^4-C_5H_4O)Br, **74**. Recrystallization of the product from aqueous acetone afforded the dihydrate (η^5-C_5Me_5)Ru(η^4-C_5H_4O)Br·$2H_2O$, **75**. In the crystal structure of **75** the ketonic oxygen atom of the cyclopentadienone ligand is connected to a pair of H_2O molecules via hydrogen-bonds. The overall structure can be described as a hydrogen-bonded network involving water molecules and the carbonyl group leading to a crystal structure which consists of alternating hydrophilic and hydrophobic layers.

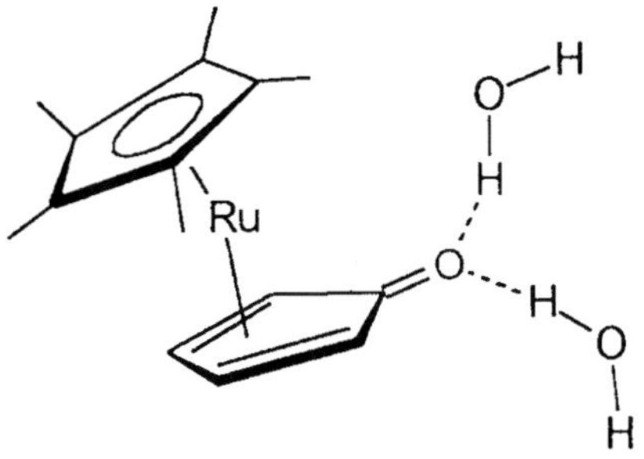

75

An interesting example of an ester group participating in self-assembly by hydrogen-bonding has been reported for two isomeric *anti*-1-hydroxytetrahydronaphthalene Cr(CO)$_3$ complexes **76a** and **76b** both of which have the OH group *anti* to the Cr(CO)$_3$ group [135].

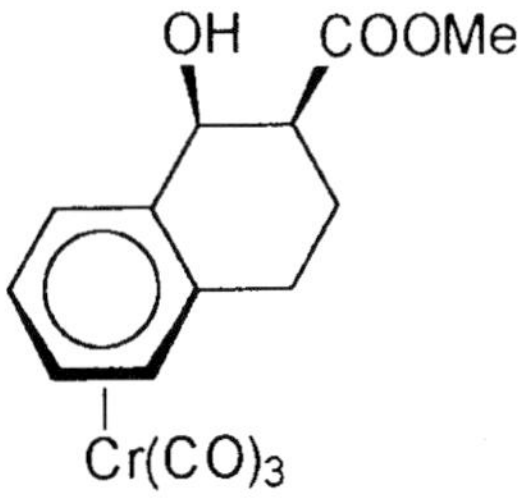

76a 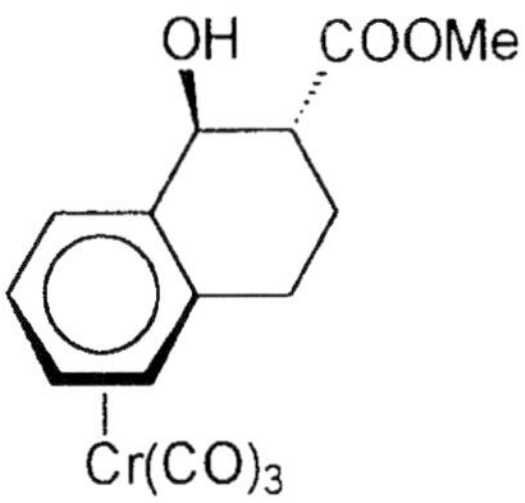   **76b**

In both crystal structures the carbonyl group of the ester function acts as hydrogen-bond acceptor but because of the different stereochemistry (i.e. orientation of the ester group) the hydrogen-bonding network in the crystal structures is quite different. Dimerization of **76a** gives **77a** whereas **76b** forms polymeric chains, **77b** [108].

The last variety of hydrogen-bonding involving hydroxyl groups to be discussed here is the participation of organotransition metal hydroxo complexes. Popular building blocks in this area of crystal engineering are the heterocubane type molecules [Mn(CO)$_3$(μ_3-OH)]$_4$, **78** (M = Mn) and [Re(CO)$_3$(μ_3-OH)]$_4$, **78** (M = Re) which both have T_d symmetry and four rigid hydrogen-bond donor moieties [136].

These compounds have been known for many years and can be prepared in a straightforward manner and in high yields directly from Mn$_2$(CO)$_{10}$ [137] or Re$_2$(CO)$_{10}$ [138], respectively. The bridging hydroxo ligands in these heterocubanes are excellent hydrogen-bond donors, and it has been found that **78** (M = Mn) forms stable adducts even with poor hydrogen-bond acceptors such as toluene or

THF [137, 139]. In the presence of monofunctional hydrogen-bond acceptors the compound cocrystallizes in the form of discrete 1:4 adducts [140]. With multifunctional hydrogen-bond acceptors interesting hydrogen-bonded network solids are obtained. Structural varieties range from one-dimensional strands to two-dimensional grids and three-dimensional diamandoid networks [141]. Typical examples of suitable 'spacer' between the manganese clusters are 1,3-diaminopropane, 4,4'-bipyridyl, and 2,3,5,6-tetramethylpyrazine. As a representative crystal structure the two-dimensional grid formed by **78** (M = Mn) and 1,3-diaminopropane,

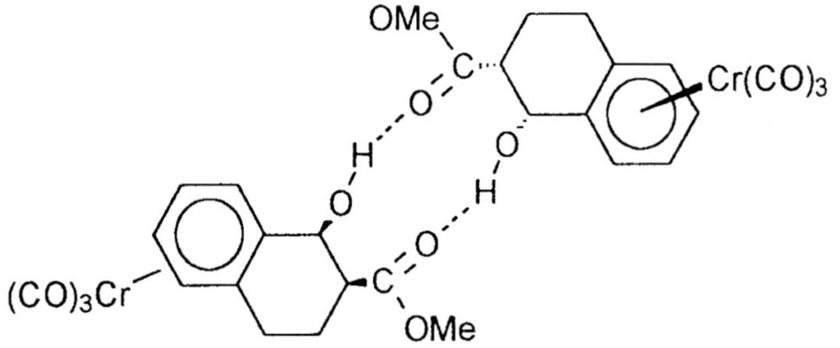

77a

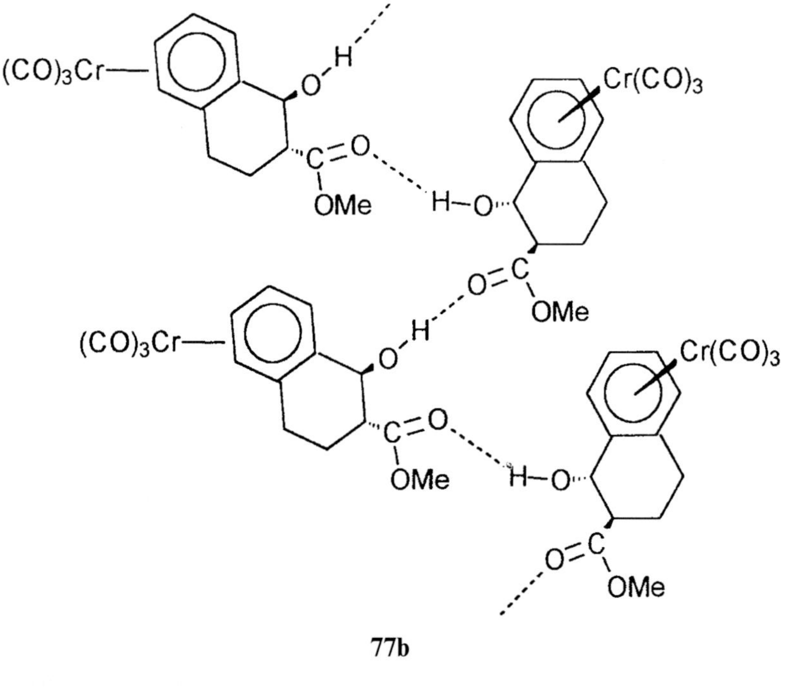

77b

[{Mn(CO)$_3$(μ_3-OH)}$_4$·2H$_2$NCH$_2$CH$_2$CH$_2$NH$_2$]$_n$, **79**, is depicted in Figure 5.3. The hydroxo complex **78** (M = Mn) can be considered a prototypal building block in the design of organometallic solid-state architecture. It is interesting to note that no short intermolecular contacts have been found in the crystal structure of the unsolvated compound [44].

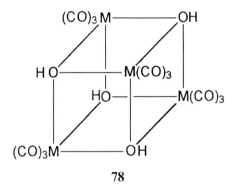

78

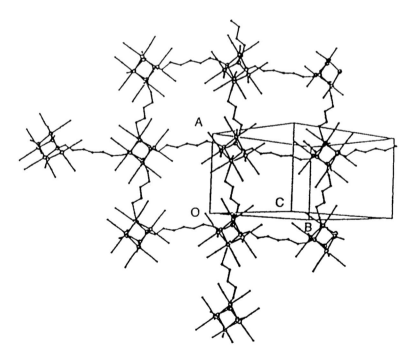

Fig. 5.3. The two-dimensional grid structure of **79** (Reproduced from ref. [141]).

5.4.1.3 N–H···O hydrogen bonds involving amido groups

Various transition metal complexes containing primary or secondary amido functional groups are known. Intermolecular hydrogen-bond distribution and geometry in crystals of such transition metal complexes have been carefully investigated and compared with the hydrogen-bonding patterns in related organic amides [142]. Most of these compounds are not organometallic complexes in the strict sense, however, and will thus not be discussed here. An interesting organometallic amido species with peripheral hydrogen-bonding is the manganese complex $(C_5H_5)Mn(CO)(NO)(\sigma\text{-}CONH_2)$, **80**. This compound is a rare example of an N-unsubstituted aminocarbonyl complex. It was prepared by treatment of the cationic precursor $[(C_5H_5)Mn(CO)_2(NO)]PF_6$, **81**, with liquid ammonia at $-40\,°C$ [143]. In the crystal structure one amide hydrogen forms a cyclic dimer by interaction with the amide oxygen of a second molecule. In addition, the amide oxygen interacts with a hydrogen atom of the cyclopentadienyl ring.

5.4.1.4 C–H···O hydrogen bonds involving carbonyl ligands

Careful analyses of numerous crystal structures of organotransition metal carbonyl complexes have revealed that weak organometallic hydrogen-bonds of the C–H···O type are abundant in these structures [102, 144]. This is not surprising because there are virtually countless metal carbonyl complexes known which con-

80

tain additional hydrocarbon ligands. The C–H···O hydrogen-bond involving terminal or bridging carbonyl ligands is considered a soft intermolecular interaction compared with the hard O–H···O and N–H···O interactions discussed above. There are several special features which determine C–H···O hydrogen-bonding patterns. First of all, the CO ligand is comparable in softness to a C–H group. As a consequence, hydrogen-bonds between these two groups often determine the supramolecular structures of heteroleptic transition metal carbonyl complexes. The C–H···O hydrogen-bonds formed with carbonyl ligands in organometallic complexes are quite directional. Coordination of the CO ligands can be either terminal or doubly or triply bridging (Figure 5.4). In all instances there is a tendency for the C–H···O angle to be approximately 140°. Subtle structural differences can be traced back to minor changes in the basicity of the CO ligands which increases in the order terminal bonding < doubly bridging < triply bridging. For example, because of their higher basicity bridging carbonyl ligands form shorter and more linear hydrogen-bonds than terminal carbonyl ligands. Thus intermolecular interactions with bridging CO ligands have occasionally been observed to be preferred to interactions with terminal ligands.

As mentioned above, potential weak hydrogen-bond donors capable of interacting with CO ligands are abundant in a large variety of organometallic complexes.

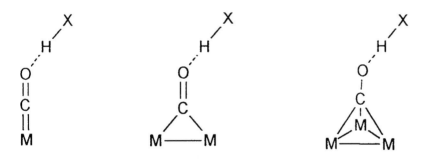

Fig. 5.4. C–H···O hydrogen-bonding involving terminal, doubly, and triply bridging CO ligands.

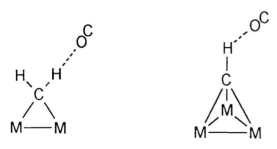

Fig. 5.5. μ-CH$_2$···OC and μ_3-CH···OC hydrogen-bonding interactions in methylene and methylidene metal clusters.

They include, for example, arenes and cyclopentadienyl ligands and, especially in binuclear or cluster complexes, methylene (μ-CH$_2$) and methylidene (μ_3-CH) groups. Figure 5.5 depicts the μ-CH$_2\cdots$OC and μ_3-CH$\cdots$OC bonding interactions in methylene and methylidene metal clusters.

Bifurcation at the acceptor site is another structural feature which is common in certain transition metal carbonyl complexes. In this instance two C–H groups point towards a single oxygen atom of a carbonyl ligand. Bifurcation is not restricted to terminal CO ligands but has also been observed with bridging carbonyls (Figure 5.6). There are also examples of trifurcation.

A typical example of a supramolecular assembly in which terminal CO ligands and hydrogen atoms of cyclopentadienyl ligands participate in C–H$\cdots$O hydrogen-bonds is the iridium cluster compound [(C$_5$H$_5$)Ir(CO)]$_3$, **82** [145]. Other metal carbonyl complexes with similar C–H$\cdots$O hydrogen-bond networks involving CO ligands (either terminal or bridging) and cyclopentadienyl hydrogens are *cis-* and

Fig. 5.6. Bifurcated C–H$\cdots$O hydrogen bonding involving terminal, doubly, and triply bridging CO ligands.

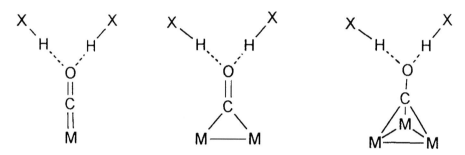

82

trans-[(C$_5$H$_5$)Fe(CO)$_2$]$_2$, **83a, b** [146], and [(C$_5$H$_5$)Cr(CO)$_3$]$_2$, **84** [147]. Hydrogen-bonding networks of the C–H···O type involving carbonyl ligands have also been reported for the cluster molecules (μ_3-η^2:η^2:η^2-C$_6$H$_6$)Ru$_3$(CO)$_9$, **85** [148], (μ_3-S$_3$C$_3$H$_6$)Ru$_3$(CO)$_9$, **86** (S$_3$C$_3$H$_6$ = 1,3,5-trithiacyclohexane) [148], (μ_3-S$_3$C$_3$H$_6$)-Ir$_4$(CO)$_9$, **87** [149], and (μ_3-S$_3$C$_3$H$_6$)Ir$_4$(CO)$_6$(μ-CO)$_3$, **88** [149]. In these cases C–H bonds of coordinated arene or heterocyclic ring ligands act as hydrogen-bond donors.

The formation of a supramolecular assembly by C–H···O hydrogen-bonding between CO ligands and bridging methylene ligands might be exemplified by the dinuclear manganese complex (μ-CH$_2$)[(C$_5$H$_5$)Mn(CO)$_2$]$_2$, **89** [150]. In the crystal structure of **89** the molecules form stacks along the c axis which are held together by two HCH···O and two C–H(Cp)···O interactions per molecule.

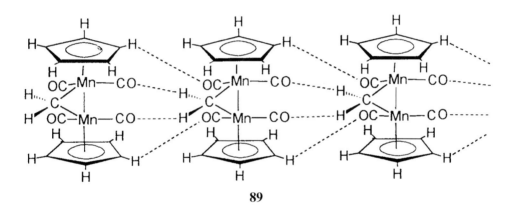

89

The tricobalt derivative (μ_3-CH)Co$_3$(CO)$_9$, **90**, is an instructive example of hydrogen-bonding in a μ_3-methylidene cluster complex [151]. The triply bridging methylidene ligands interact in a trifurcated fashion with three carbonyl oxygens from two neighboring molecules.

Various examples of bi- and trifurcated-acceptor interactions have been found in supramolecular structures of metal carbonyl complexes. In (μ-CO)(μ-CHCH$_3$)-[(C$_5$H$_5$)Fe(CO)]$_2$, **91**, the bridging carbonyl ligand participates in two C–H···O hydrogen-bonding interactions with C–H units of the cyclopentadienyl ligands. As a result the dinuclear molecules are self-assembled into polymeric chains [152].

Other structurally characterized metal carbonyl complexes with bi- and trifurcation at a carbonyl oxygen atom include, for example, the cycloheptatrienyl and cyclooctatetraene cobalt clusters Co$_4$(CO)$_6$(η^5-C$_7$H$_9$)(η_3-C$_7$H$_7$), **92**, Co$_4$(CO)$_6$-(η^4-C$_8$H$_8$)(μ_3-C$_8$H$_8$), **93**, and Co$_4$(CO)$_8$(μ_3-C$_8$H$_8$), **94** [153].

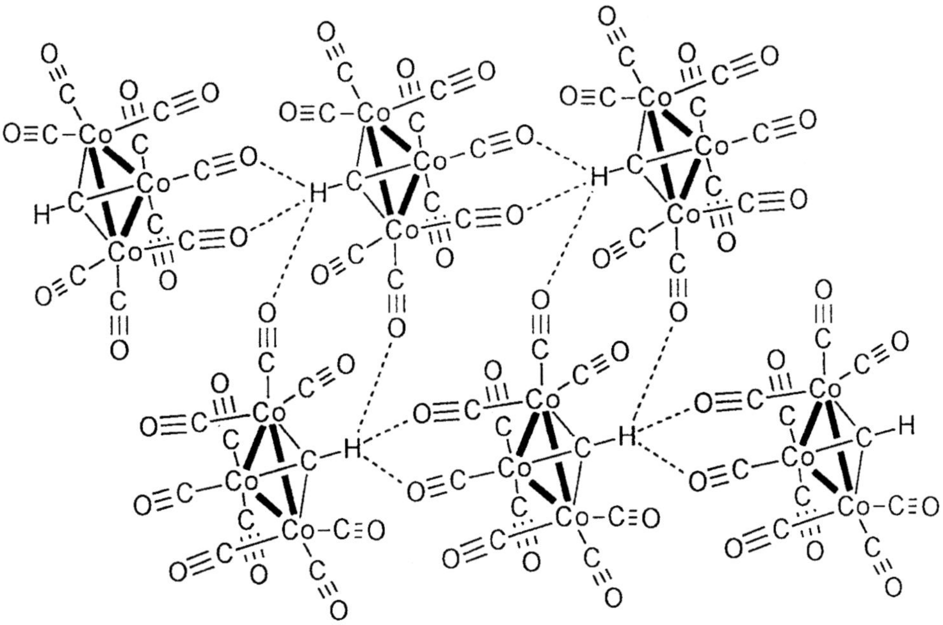

90

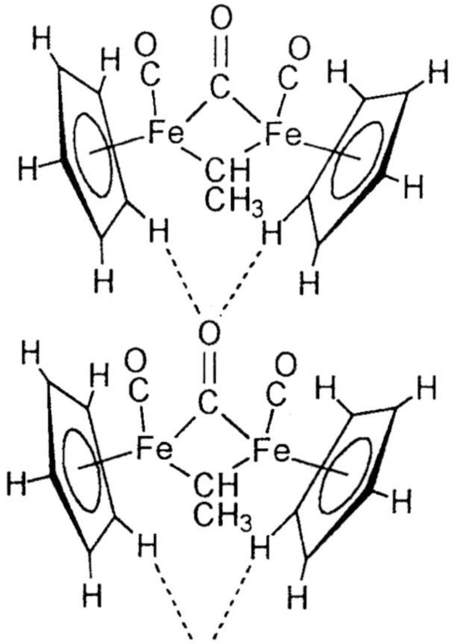

91

Weak hydrogen-bonding involving a terminal alkyne function has been found in the substituted (cycloheptatriene)chromiumtricarbonyl derivative (η^6-7-*exo*-HC≡

CC$_7$H$_7$)Cr(CO)$_3$, **95**, which can be prepared by reacting the cationic precursor [(C$_7$H$_7$)Cr(CO)$_3$]BF$_4$, **96**, with LiC≡CSiMe$_3$ and then desilylating with methanolic KOH [154]. In the solid state the molecules are self-assembled as cyclic dimers with the shortest intermolecular contacts being interactions between the ethynyl group and a carbonyl ligand of the second molecule. The long H···O distance of 2.92 Å has been discussed in terms of weak C–H···O hydrogen-bonding (C–H···O angle 130°), although it is much longer than the sum of the van der Waals radii.

95

5.4.2 Hydrogen bonds involving transition metal atoms

Various hydrogen-bonding interactions are known in which a transition metal atom is directly involved. Most are, however, intramolecular in nature and thus do not play a major role in the self-assembly of supramolecular species. Among these are the well-known M···(H–C) three-center–two-electron agostic interactions between an electron-deficient metal atom and a σ-C–H bond of an organic ligand [155]. Another type of hydrogen-bonding which in most cases is either an intramolecular interaction or a cation–anion interaction is the M–H···H–X type [156]. Thus in this section only intermolecular M–H···O=C interactions between hydride ligand and carbonyl ligands will be discussed.

5.4.2.1 M–H···O≡C intermolecular hydrogen bonds

Hydride ligands directly bound to transition metal atoms can also act as hydrogen-bond donors in supramolecular organotransition metal complexes. In fact, M–H···O≡C intermolecular interactions with carbonyl ligands are quite similar to those involving C–H bonds of organic ligands, i.e. both types of hydrogen bond are comparable in length. It is interesting to note that in self-assembled cluster complexes containing hydride and carbonyl ligands the two intermolecular interactions often coexist and somehow have a cooperative effect in constituting the hydrogen-bonding network. Structurally investigated hydride clusters with both M–H···O≡C and C–H···O contacts in the solid state include $(\mu\text{-H})(\mu\text{-NCHCF}_3)\text{Os}_3(\text{CO})_{10}$, **97** [157], $(\mu\text{-H})_3\text{Os}_3\text{Ni}(\text{CO})_9(\text{C}_5\text{H}_5)_2$, **98** [158], and $(\mu\text{-H})(\mu\text{-}\eta^2\text{-CH})\text{Fe}_4(\text{CO})_{11}(\text{PPh}_3)$, **99** [159]. The supramolecular hydrogen-bonded assembly of **97** is depicted here as a typical example. Two molecules are linked by direct interaction of the bridging hydride with a terminal carbonyl ligand. A ring assembly of four cluster molecules is formed by four additional C–H···O type hydrogen-bonds and the overall supramolecular structure is the result of the cooperative effect of (Os)H···O and C–H···O interactions.

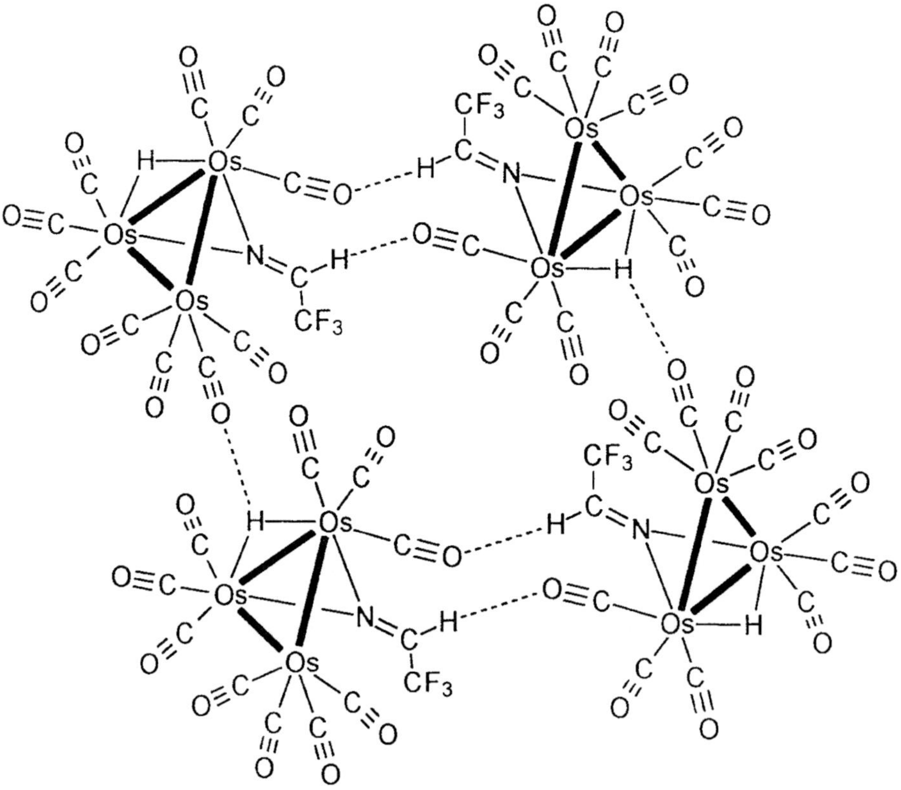

97

5.4.3 Hydrogen-bonds assisted by ionic interactions

A recent development in this field has been appropriately termed "organic–organometallic crystal engineering" [102]. This is an area where the main strategies of organometallic architecture and crystal engineering become most obvious. Current investigations are mainly concerned with supramolecular structures formed by combining metallocene and related sandwich complex cations with suitable organic anions. Three prerequisites are required for the successful design of supramolecular crystal structures:

i) Donor and acceptor sites capable of forming strong hydrogen-bonds should be present in the organic moiety. This will enable self-organization of the organic parts.
ii) The organometallic part should contain a large number of C–H groups capable of forming weak C–H$\cdots$O hydrogen-bonds with the organic moieties. This can be ensured, e.g., by the use of cyclopentadienyl or arene ligands.
iii) The two fragments should have opposite charges to reinforce the weak C–H$\cdots$O hydrogen-bonds.

The overall supramolecular structure is a result of interplay between strong and directional hydrogen-bonds within the organic framework and weak C–H$\cdots$O hydrogen-bonds reinforced by cation–anion interactions. Another important consideration to be made for the successful design of organic–organometallic supramolecular structures is the shape analogy between organometallic molecules and the corresponding free ligands. A highly instructive example is the interaction between 1,3-cyclohexanedione (CHD) with either benzene or the bis(η^6-benzene)chromium cation. In pioneering work by Etter et al. it was shown that crystallization of 1,3-cyclohexanedione from benzene leads to a host–guest complex in which a benzene guest molecule is encapsulated by a ring (cyclamer) of six hydrogen-bonded CHD molecules [160]. The structure of [CHD]$_6 \cdot$C$_6$H$_6$, **100**, is depicted in Figure 5.7.

The interaction of 1,3-cyclohexanedione with the paramagnetic bis(η^6-benzene)chromium cation was investigated following the guiding principle that 'molecules with similar shape and size are organized in the solid state in similar manner, irrespective of the chemical composition'. Oxidation of Cr(η^6-C$_6$H$_6$)$_2$ in THF solution in the presence of CHD afforded the deep yellow salt-like compound [Cr(η^6-C$_6$H$_6$)$_2$][(CHD)$_4$], **101** [161]. According to X-ray structural analysis **101** consists of a bis(benzene)chromium cation encapsulated in an anionic framework derived from 1,3-cyclohexanedione. In the latter the CHD molecules form a 'horse-shoe'-shaped tetrameric aggregate in which the four constituents are linked via C=O$\cdots$HO hydrogen-bonds.

The inner two CHD units form a tautomeric anion to which the outer two CHD molecules are connected in their enol form. The resulting [(CHD)$_4$]$^-$ anion interacts with one bis(η^6-benzene)chromium cation via short C–H$\cdots$OC hydrogen-bonds. Finally two such 'horse-shoe'-shaped systems wrapped around the organometallic cation are related by a center of inversion. This results in an overall supramolecular

Fig. 5.7. Structure of Etter's cyclamer host–guest compound **100** (Reproduced from ref. [160]).

101

structure comprising a nearly planar system formed by two $[Cr(\eta^6\text{-}C_6H_6)_2][(CHD)_4]$ units (Figure 5.8).

Thus the main result of this experiment was the rational design of a self-assembled crystal structure based on the shape analogy between benzene and bis(η^6-benzene)chromium. Under similar reaction conditions bis(η^6-toluene)chromium reacts with CHD to afford $[Cr(\eta^6\text{-}C_6H_5Me)_2][(CHD)_2]$, **102** [161b]. In the solid state **102** forms a C–H···O hydrogen-bonded network in which the bis(η^6-toluene)-chromium cations alternate with 'clamps' formed by the dimeric $[(CHD)_2]^-$ anions.

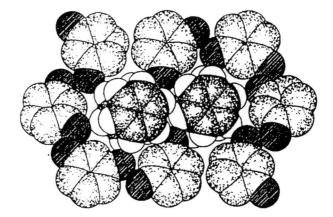

Fig. 5.8. Overall supramolecular structure formed by two units of [Cr(η^6-C$_6$H$_6$)$_2$][(CHD)$_4$] (**101**) (Reproduced from ref. [161]).

102

The design of a related honeycomb-like organic framework wrapped around a metallocene cation was achieved by using the related cobaltocenium cation. Several supramolecular materials were synthesized by reacting the organometallic hydroxide [Co(C$_5$H$_5$)$_2$]OH (made by air-oxidation of cobaltocene in water) with D,L-tartaric acid (H$_2$TA) [162]. In one of the products, [Co(C$_5$H$_5$)$_2$][(HTA)(H$_2$TA)], **103**, an organic 'honeycomb' formed by aggregation of the dimeric [(HTA)(H$_2$TA)]$^-$ anions is molded around the organometallic cation [Co(C$_5$H$_5$)$_2$]$^+$. The organic superanions are assembled via strong O–H$\cdots$O hydrogen-bonds while the entire supramolecular structure is held together by charge-assisted C–H$\cdots$O interactions between the cyclopentadienyl ligands and carbonyl or hydroxyl functions of the organic moieties. A space-filling representation of the tartaric acid template encapsulating the cobaltocenium ion is shown in Figure 5.9.

Several related experiments have been shown to afford layered supramolecular structures. For example, air-oxidation of bis(η^6-benzene)chromium in a two-layer

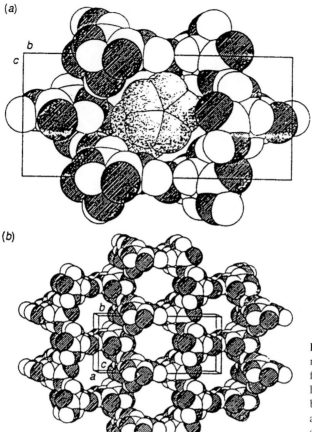

Fig. 5.9. (a) Space-filling representation of **103**; (b) space-filling representation of the honeycomb framework formed by the [(HSQA)(H$_2$SQA)]$^-$ anions in the supramolecular structure of **103** (Reproduced from ref. [162]).

benzene–water system gives the organometallic hydroxide [163] [Cr(η^6-C$_6$H$_6$)$_2$]-[OH]·3H$_2$O, **104**, which can be isolated as a yellow crystalline solid [161b, 164]. The crystal structure of **104** consists of layers of [Cr(η^6-C$_6$H$_6$)$_2$]$^+$ cations intercalated with layers of {[OH]·3H$_2$O}$_n$ units. The overall supramolecular structure is held together by O–H$\cdots$O hydrogen-bonds and C–H$\cdots$O interactions involving the water molecules. Crystalline [Cr(η^6-C$_6$H$_6$)$_2$][CHD]·3H$_2$O, **105**, is prepared by air-oxidation of bis(benzene)chromium in the presence of 1,3-cyclohexanedione. The crystal structure of **105** bears a close resemblance to that of **104**, because in this structure layers of [Cr(η^6-C$_6$H$_6$)$_2$]$^+$ cations are intercalated with strings of [CHD]$^-$·3H$_2$O. Layered supramolecular structures have also been found for the cobaltocenium tartrate [Co(C$_5$H$_5$)$_2$][(HTA)]·H$_2$O, **106** [162], and the squarate derivatives [Co(C$_5$H$_5$)$_2$][(HSQA)], **107**, and [Co(C$_5$H$_5$)$_2$][(HSQA)(H$_2$SQA)], **108** [165]. The

5.4 Transition Metals (Organometallic Crystal Engineering)

last two compounds have been synthesized by reacting the hydroxide [Co(C$_5$H$_5$)$_2$][OH], **109**, with squaric acid (H$_2$SQA) in molar ratios of 1:1 or 1:2, respectively. In **107** the squarate monoanions form ribbons (Figure 5.10, **A**) which alternate with ribbons of cobaltocenium cations. The squarate anions are connected via O–H···O hydrogen-bonds and interactions between the two types of ribbon occurs via charge-assisted C–H···O hydrogen-bonds. Supramolecular anions of composition [(HSQA)(H$_2$SQA)]$^-$ are present in the 1:2 product, **108**. The loss of one proton from every two squaric acid units results in the formation of ribbon-like superanions (Figure 5.10, **B**) held together by negatively charged O–H···O hydrogen-bonds. The resulting ten-membered rings are reminiscent of the cyclic dimers of carboxylic acids. Once again charge-assisted C–H···O hydrogen-bonds lead to a stacking of the two types of ribbon to give the overall supramolecular layer structure as shown in Figure 5.11.

Recent progress in this field has made it clear that "crystal engineering is not an esoteric type of crystallography but rather addresses the problem of intermolecular interactions" [166]. It is a suitable means of purposefully combining physical and chemical features of organometallic and organic molecules or ions in the solid state [162a]. It is not beyond imagination that organic–organometallic crystal engineering will eventually play a major role in the design of intelligent new materials including those with useful magnetic, conducting or superconducting, and non-linear optical properties. In fact, supramolecular bis(benzene)chromium and cobaltocenium salts are quite similar to charge-transfer salts such as [Fe(C$_5$Me$_5$)$_2$][TCNQ], **110** (TCNQ = 7,7,8,8-tetracyano-*p*-quinodimethane), or [Fe(C$_5$Me$_5$)$_2$][TCNE], **111**, (TCNE = tetracyanoethylene) [167]. These are known for their bulk ferromagnetic behavior. Recently the related nickel complex salt [Ni(C$_5$Me$_5$)$_2$][HTCNQF$_4$], **112** (TCNQF$_4$ = perfluorinated 7,7,8,8-tetracyano-*p*-quinodimethane), has been prepared and structurally characterized [168]. The crystal structure contains polymeric anions linked by C–H···N hydrogen-bonds. This example demonstrates that hydrogen-bonds can also play an important role in such charge-transfer salts. It also shows that successful crystal engineering can go hand in hand with the design of promising new materials.

364 5 *Supramolecular Self-Assembly by Hydrogen-Bond Interactions*

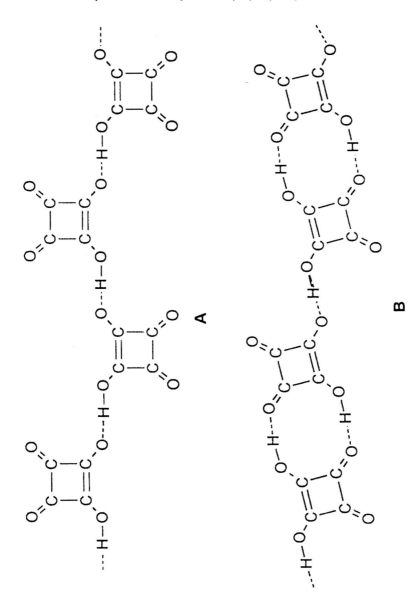

Fig. 5.10. **A**, the [(HSQA)]$_n^-$ ribbon in **107**; **B**, the [(HSQA)(H$_2$SQA)]$_n^-$ ribbon in **108**.

5.4 Transition Metals (Organometallic Crystal Engineering)

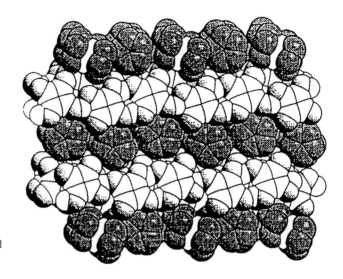

Fig. 5.11. Space-filling representation of the layer structure of **108** (Reproduced from ref. [165]).

References

[1] P. D. Lickiss, *Advan. Inorg. Chem.* **1995**, *42*, 147.
[2] P. D. Lickiss, in vol. *Tailor-made Silicon-Oxygen Compounds. From Molecules to Materials*, R. Corriu and P. Jutzi (Editors), F. Vieweg Verlag, Braunschweig, **1996**, p. 47.
[3] R. Murugavel, V. Chandrasekar and H. W. Roesky, *Acc. Chem. Res.* **1996**, *29*, 183.
[4] R. Murugavel, A. Voigt, M. G. Walawalkar and H. W. Roesky, *Chem. Revs.* **1996**, *96*, 2205.
[5] M. Cypryk, *J. Organomet. Chem.* **1997**, *545/546*, 483.
[6] A. Rengst and U. Schubert, *Chem. Ber.* **1980**, *113*, 278.
[7] H. Puff, K. Braun and H. Reuter, *J. Organomet. Chem.* **1991**, *409*, 119.
[8] N. Buttrus, C. Eaborn, P. B. Hitchcock and A. K. Saxena, *J. Organomet. Chem.* **1985**, *287*, 157
[9] G. Ferguson, J. F. Gallagher, C. Glidewell, J. N. Low and S. N. Scrimgeour, *Acta. Cryst.* **1992**, *C 48*, 1272.
[10] F. I. Aigbirhio, S. S. Al-Juaid, C. Eaborn, A. Habtemarian, P. B. Hitchcock and J. D. Smith, *J. Organomet. Chem.* **1991**, *405*, 149.
[11] M. Haque, W. Horne, S. E. Cremer and C. S. Blakenship, *J. Chem. Soc. Perkin Trans. II*, **1983**, 395.
[12] P. D. Lickiss and K. M. Stubbs, *J. Organomet. Chem.* **1991**, *421*, 171.
[13] R. Tacke, H. Lange, W. S. Sheldrick, G. Lambrecht, U. Moser and E. Mutschler, *Z. Naturforsch.* **1983**, *38b*, 738.
[14] R. Tacke, M. Strecker, W. S. Sheldrick, L. Ernst, E. Heeg, B. Berndt, C. M. Knapstein and R. Niedner, *Chem. Ber.* **1980**, *113*, 1962.
[15] Z. H. Aiube, N. B. Buttrus, C. Eaborn, P. B. Hitchcock and J. A. Zora, *J. Organomet. Chem.* **1985**, *292*, 177.
[16] S. S. Al-Juaid, C. Eaborn, P. B. Hitchcock, P. D. Lickiss, A. Möhrke and P. Jutzi, *J. Organomet. Chem.* **1990**, *384*, 33.

[17] R. Murugavel, A. Voigt, V. Chandrasekhar, H. W. Roesky, H. G. Schmidt and M. Noltemeyer, *Chem. Ber.* **1996**, *129*, 391.
[18] P. E. Tomlins, J. E. Lydon, D. Akrigg and B. Sheldrick, *Acta Cryst.* **1985**, *C41*, 941.
[19] N. H. Buttrus, C. Eaborn, P. B. Hitchcock and P. D. Lickiss, *J. Organomet. Chem.* **1986**, *302*, 159.
[20] N. H. Buttrus, C. Eaborn, P. B. Hitchcock and A. K. Saxena, *J. Organomet. Chem.* **1985**, *284*, 291.
[21] N. H. Buttrus, C. Eaborn, P. B. Hitchcock, P. D. Lickiss, and A. D. Taylor, *J. Organomet. Chem.* **1986**, *309*, 25.
[22] J. K. Fawcett, N. Camerman and A. Camerman, *Canad. J. Chem.* **1977**, *55*, 3631.
[23] L. Párkányi and B. Bocelli, *Cryst. Struct. Commun.* **1978**, *7*, 335.
[24] V. E. Shklover, Yu. T. Struchkov, I. V. Karpova, V. A. Odinets, and A. A. Zhdanov, *Zh. Strukt. Khim.* **1985**, *26*, 125.
[25] W. Clegg, *Acta Cryst.* **1983**, *C39*, 901.
[26] P. D. Lickiss, A. D. Redhouse, R. J. Thompson, W. A. Stanczyk and K. Rozga, *J. Organomet. Chem.* **1993**, *453*, 13.
[27] A. P. Polischuk, M. Yu. Antipin, T. V. Timofeeva, N. N. Makarova, N. A. Golovina and Yu. T. Struchkov, *Sov. Phys.- Cryst.* (Engl. transl) **1991**, *36*, 50.
[28] A. P. Polischuk, N. N. Makarova, M. Yu. Antipin, T. V. Timofeeva, M. A. Kravers and Yu. T. Struchkov, *Sov. Phys.- Cryst.* (Engl. transl) **1990**, *35*, 258.
[29] L. M. Khananashvili, Ts. N. Vardosanidze, G. V. Gridunova, V. E. Shklover, Yu. T. Struchkov, E. G. Markarashvili, M. Sh. Tsutsunova, *Izv. Akad. Nauk Gruz. SSR*, **1984**, *10*, 262.
[30] R. West and E. K. Pham, *J. Organomet. Chem.* **1991**, *403*, 43.
[31] N. H. Buttrus, R. I. Damja, C. Eaborn, P. B. Hitchcock and P. D. Lickiss, *J. Chem. Soc. Chem. Comm.* **1985**, 1385.
[32] S. S. Al-Juaid, N. H. Buttrus, R. I. Damja, Y. Derouiche, C. Eaborn, P. B. Hitchcock and P. D. Lickiss, *J. Organomet. Chem.* **1989**, *371*, 287.
[33] H. W. Roesky, *Chem. Ber.* **1996**, *129*, 391.
[34] N. Wirkhofer, H. W. Roesky, M. Noltemeyer and W. T. Robinson, *Angew. Chem.* **1992**, *104*, 670; *Angew. Chem. Int. Ed. Engl.* **1992**, *31*, 599.
[35] H. Ishida, J. L. Koenig and K. C. Gardner, *J. Chem. Phys.* **1982**, *77*, 5748.
[36] P. Jutzi, G. Strassburger, M. Schneider, H. G. Stammler and B. Neumannn, *Organometallics*, **1996**, *15*, 2842.
[37] R. Murugavel, V. Chandrasekar, A. Voigt, H. W. Roesky, H. G. Schmidt and M. Noltemeyer, *Organometallics*, **1995**, *14*, 5298.
[38] P. Jutzi, M. Schneider, H. G. Stammler and B. Neumann, *Organometallics* **1997**, *16*, 5377.
[39] N. G. Furmanova, V. I. Andrianov, and N. N. Makarova, *J. Struct. Chem. USSR* (Engl. transl.) **1987**, *28*, 256.
[40] I. L. Dubchak, V. E. Shklover, Yu. T. Struchkov, E. S. Khynku and A. A. Zhdanov, *J. Struct. Chem. USSR* (Engl.transl.) **1981**, *22*, 770.
[41] S. A. Al-Juaid, C. Eaborn, P. B. Hitchcock and P. D. Lickiss, *J. Organomet. Chem.* **1988**, *353*, 297.
[42] S. Yamaguchi, R. Z. Jin and K. Tamao, *Organometallics*, **1997**, *16*, 2230.
[43] I. Baxter, L. Cother, C. Dupuy, P. D. Lickiss, A. J. P. White and D. J. Williams, ECTOC-3 [*Electronic Conference on Organometallic Chemistry*], http://www.ch.ic.ac.uk/ectoc/ectoc-3.
[44] J. S. Moore, *Nature*, **1995**, *374*, 495.
[45] S. A. Al-Juaid, A. K. A. Al-Nasr, C. Eaborn and P. B. Hitchcock, *J. Organomet. Chem.* **1992**, *429*, C9.
[46] G. Ferguson, J. F. Gallagher, D. Murphy, T. R. Spalding, C. Glidewell and H. D. Holden, *Acta Cryst.* **1992**, *C48*, 1228.
[47] H. Puff, S. Franken, W. Schuh and W. Schwab, *J. Organomet. Chem.* **1983**, *254*, 33.
[48] Z. X. Xie, S. Z. Hu, Z. H. Chen, S. A. Li and W. P. Shi, *Acta Cryst.* **1993**, *C49*, 1154.
[49] C. Lecomte, J. Protas and M. Devaud, *Acta Cryst.* **1976**, *B32*, 923.

[50] R. R. Holmes, S. Shafieezad, V. Chandrasekar, J. M. Holmes and R. O. Day, *J. Am. Chem. Soc.* **1988**, *110*, 1174.
[51] H. Puff and H. Reuter, *J. Organomet. Chem.* **1989**, *364*, 57.
[52] H. Puff, H. Hevendehl, K. Höfer, H. Reuter and W. Schuh, *J. Organomet. Chem.* **1985**, *287*, 163.
[53] R. Altmann, R. Hohlfeld, S. Olejnik and J. Lorberth, *J. Organomet. Chem.* **1981**, *210*, 51.
[54] J. B. Hall and D. Britton, *Acta Cryst.* **1972**, *B28*, 2133.
[55] A. Blaschette, E. Wieland, P. G. Jones and I. Hippel, *J. Organomet. Chem.* **1993**, *445*, 55.
[56] H. Reuter and D. Schroder, *Acta Cryst.* **1993**, *C 49*, 954.
[57] T. V. Sizova, N. S. Yashina, V. S. Petrosyan, A. V. Yatsenko, V. V. Chernyshev and L. A. Aslanov, *J. Organomet. Chem.* **1993**, *453*, 171.
[58] P. G. Harrison, R. C. Phillips and J. A. Richards, *J. Organomet. Chem.* **1976**, *114*, 47.
[59] L. Jäger, B. Freude, A. Krug and H. Hartung, *J. Organomet. Chem.* **1994**, *467*, 163.
[60] A. Blaschette, D. Schomburg and E. Wieland, *Z. Anorg. Allg. Chem.* **1989**, *571*, 75.
[61] L. Diop, M. F. Mahon, K. C. Molloy and M. Sidibe, *Main Group Metal Chem.* **1997**, *20*, 649.
[62] S. W. Ng, *Acta Cryst.* **1997**, *C 53*, 1059.
[63] S. W. Ng, V. G. Kumar Das and E. R. T. Tiekink, *J. Organomet. Chem.* **1991**, *411*, 121.
[64] S. W. Ng, V. G. Kumar Das, M. B. Hossain, F. Goerlitz and D. van der Helm, *J. Organomet. Chem.* **1990**, *390*, 19.
[65] F. A. K. Nasser, M. B. Hossain, D. van der Helm and J. J. Zuckerman, *Inorg. Chem.* **1984**, *23*, 606.
[66] K. M. Lo, V. G. Kumar Das, W. H. Yip and T. C. W. Mak, *J. Organomet. Chem.* **1991**, *412*, 21.
[67] P. G. Harrison, T. J. King and R. C. Phillips, *J. Chem. Soc. Dalton Trans.* **1976**, 2317.
[68] G. Valle, E. Ettore, V. Peruzzo and G. Plazzogna, *J. Organomet. Chem.* **1987**, *326*, 169.
[69] B. Alberte, A. Sanchez-Gonzalez, E. R. Garcia, and E. E. Castellano, *J. Organomet. Chem.* **1988**, *338*, 187.
[70] A. Sanchez-Gonzalez, B. Alberte, J. S. Casas, J. Sordo, A. Castineiras, W. Hiller and J. Strähle, *J. Organomet. Chem.* **1988**, *353*, 169.
[71] V. Peruzzo, G. Plazzogna and G. Valle, *J. Organomet. Chem.* **1989**, *375*, 167.
[72] A. Sanchez-Gonzales, J. S. Casas, J. Sordo and G. Valle, *J. Organomet. Chem.* **1992**, *435*, 29.
[73] E. Garcia-Martinez, A. Sanchez-Gonzalez, A. Macias, M. V. Castano, J. S. Casas and J. Sordo, *J. Organomet. Chem.* **1990**, *385*, 329.
[74] U. Casellato, R. Graziani and A. Sanchez-Gonzalez, *Acta Cryst.* **1992**, *C48*, 2125.
[75] R. Graziani, V. Peruzzo, G. Plazzogna and U. Casellato, *J. Organomet. Chem.* **1990**, *396*, 19.
[76] G. Bandoli, A. Dolmella, V. Peruzzo and G. Plazzogna, *J. Organomet. Chem.* **1993**, *452*, 47.
[77] E. Garcia Martinez, A. Sanchez Gonzalez, J. S. Casas, G. Sordo, G. Valle and U. Russo, *J. Organomet. Chem.* **1993**, *453*, 47.
[78] A. Hazell, K. F. Thong, J. Ouyang and L. E. Khoo, *Acta Cryst.* **1997**, *C 53*, 1226.
[79] G. S. Tocco, G. Guliand and G. Valle, *Acta Cryst.* **1992**, *C 48*, 2116.
[80] S. K. Kamrudin, T. K. Chattopadhyaya, A. Roy and E. R. T. Tiekink, *Appl. Organomet. Chem.* **1996**, *10*, 513.
[81] M. F. Mahon, K. C. Molloy, B. A. Omotowa and M. A. Mesubi, *J. Organomet. Chem.* **1996**, *511*, 227.
[82] K. C. Molloy, M. B. Hossain, D. van der Helm, D. Cunningham and J. J. Zuckerman, *Inorg. Chem.* **1981**, *20*, 2402.
[83] J. Trotter and T. Zoebl, *J. Chem. Soc.* **1965**, 4466.
[84] M. R. Smith, R. A. Zingaro and E. A. Meyers, *J. Organomet. Chem.* **1969**, *20*, 105.
[85] R. Bohra, H. W. Roesky, M. Noltemeyer and G. M. Sheldrick, *J. Chem. Soc. Dalton Trans.* **1984**, 2011.
[86] L. Silaghi-Dumitrescu, I. Silaghi-Dumitrescu, J. Zuckerman-Schpector, I. Haiduc and D. B. Sowerby, *J. Organomet. Chem.* **1996**, *517*, 101.
[87] C. Glidewell, *J. Fluorine Chem.* **1981**, *18*, 143.

[88] G. Fergusson and E. W. Macaulay, *J. Chem. Soc.* A **1969**, 1.
[89] G. Fergusson, A. J. Lough and C. Glidewell, *J. Chem. Soc.* Perkin II, **1989**, 2065.
[90] T. T. Bamgboye, M. J. Begley and D. B. Sowerby, *J. Organomet. Chem.* **1989**, *362*, 77.
[91] G. Lang, K. W. Klinkenhammer, C. Recker and A. Schmidt, *Z. Anorg. Allg. Chem.* **1998**, *624*, 689.
[92] S. Pankaj, N. Rosas, G. Espinosa-Perez and A. Cabrerra, *Acta Cryst.* **1996**, *C 52*, 889.
[93] T. Westhoff, F. Huber, R. Ruther and H. Preut, *J. Organomet. Chem.* **1988**, *352*, 107.
[94] C. Silvestru, A. Silvestru, I. Haiduc, D. B. Sowerby, K. H. Ebert and H. J. Breunig, *Polyhedron* **1997**, *16*, 2643.
[95] R. Rüther, F. Huber and H. Preut, *J. Organomet. Chem.* **1988**, *342*, 185.
[96] R. Rüther, F. Huber and H. Preut, *J. Organomet. Chem.* **1985**, *295*, 21.
[97] C. V. K. Sharma and G. R. Desiraju, In *Perspectives in Supramolecular Chemistry. The Crystal as a Supramolecular Entity* (Ed.: G. R. Desiraju), Wiley, Chichester, **1996**.
[98] G. R. Desiraju, *Crystal Engineering: The Design of Organic Solids*, Elsevier, Amsterdam, **1989**.
[98a] D. Braga and F. Grepioni, *Chem. Commun.* **1999**, 1.
[98b] D. Braga, F. Grepioni and G. R. Desiraju, *Chem. Rev.* **1998**, *98*, 1375.
[99] G. R. Desiraju, *Angew. Chem.* **1995**, *107*, 2541; *Angew. Chem. Int. Ed. Engl.* **1995**, *34*, 2328.
[100] (a) G. A. Jeffrey and W. Saenger, *Hydrogen Bonding in Biological Structures*, Springer, Berlin, **1991**; (b) G. A. Jeffrey, *An Introduction to Hydrogen Bonding*, Oxford University Press, New York, **1997**.
[101] (a) L. Brammer, D. Zhao, F. T. Lapido and J. Braddock-Wilking, *Acta Crystallogr., Sect B* **1995**, *B51*, 632; (b) D. Braga, F. Grepioni and G. R. Desiraju, *J. Organomet. Chem.* **1997**, *548*, 33.
[102] D. Graga, F. Grepioni and G. R. Desiraju, *Chem. Rev.* **1998**, *98*, 1375.
[103] (a) D. Braga and F. Grepioni, *Acc. Chem. Res.* **1994**, *27*, 51; (b) D. Braga and F. Grepioni, *Chem. Commun.* **1996**, 571.
[104] *Crystallographic Databases* (Eds.: F. H. Allen, G. Bergerhoff and R. Sievers), International Union of Crystallography, Chester, **1987**.
[105] (a) M. C. Etter, *Acc. Chem. Res.* **1990**, *23*, 120; (b) M. C. Etter, *J. Phys. Chem..* **1991**, *95*, 4601.
[106] (a) L. Leiserowitz, *Acta. Crystallogr., Sect. B* **1976**, *B32*, 775; (b) A. Gavezzotti and G. Filippini, *J. Phys. Chem.* **1994**, *98*, 4831.
[107] R. A. Benkeser, D. Goggin and G. Schroll, *J. Am. Chem. Soc.* **1954**, *76*, 4025.
[108] D. Braga, F. Grepioni, P. Sabatino and G. R. Desiraju, *Organometallics* **1994**, *13*, 3532.
[109] F. van Meurs and H. van Koningsveld, *J. Organomet. Chem.* **1974**, *78*, 229.
[110] R. Eiss, *Inorg. Chem.* **1970**, *9*, 1650.
[111] J. K. P. Ariyaratne, A. M. Bierrum, M. L. H. Green, C. K. Prout and M. G. Swanwick, *J. Chem. Soc. A* **1969**, 1309.
[112] J. Krause, D.-Y. Jan and S. G. Shore, *J. Am. Chem. Soc.* **1987**, *109*, 4416.
[113] S. Jeannin, Y. Jeannin and G. Lavigne, *Inorg. Chem.* **1978**, *17*, 2103.
[114] (a) C. Pedone and A. Sirigu, *Inorg. Chem.* **1968**, *7*, 2614; (b) Y. Hsiou, Y. Wang and L.-K. Liu, *Acta Crystallogr., Sect. C* **1989**, *C45*, 721.
[115] (a) C. J. Brown, *Acta Crystallogr.* **1966**, *21*, 1; (b) A. L. Bednowitz and B. Post, *Acta Crystallogr.* **1966**, *21*, 566.
[116] M. N. G. James and G. J. Williams, *Acta. Crystallogr., Sect. B* **1974**, *B30*, 1249.
[117] F. Baert, A. Guelzim and P. Coppens, *Acta. Crystallogr., Sect. B* **1984**, *B40*, 590.
[118] A. Avey, S. C. Tenhaeff, T. J. R. Weakley and D. R. Tyler, *Organometallics* **1991**, *10*, 3607.
[119] F. Takusagawa and T. F. Koetzle, *Acta. Crystallogr., Sect. B* **1979**, *B35*, 2888.
[120] C. P. Brock and L. L. Duncan, *Chem. Mater.* **1994**, *6*, 1307.
[121] Y. Dusausoy, J. Protas, J. Besancon and S. Top, *J. Organomet. Chem.* **1975**, *94*, 47.
[122] A. J. Pearson, S. Mallik, R. Mortezaei, M. W. D. Perry, R. J. Shively and W. J. Youngs, *J. Am. Chem. Soc.* **1990**, *112*, 8034.

5.4 Transition Metals (Organometallic Crystal Engineering) 369

[123] S. C. Tenhaeff, D. R. Tyler and T. J. R. Weakley, *Acta Crystallogr., Sect. C* **1992**, *C48*, 162.
[124] S. C. Tenhaeff, D. R. Tyler and T. J. R. Weakley, *Acta Crystallogr., Sect. C* **1991**, *C47*, 303.
[125] A. J. Pearson, S. Mallik, A. A. Pinkerton, J. P. Adams and S. Zheng, *J. Org. Chem.* **1992**, *57*, 2910.
[126] P. E. Riley and R. E. Davis, *Acta. Crystallogr., Sect. B* **1976**, *B32*, 381.
[127] (a) D. Walther, A. Schmidt, T. Klettke, W. Imhof and H. Görls, *Angew. Chem.* **1994**, *106*, 1421; *Angew. Chem. Int. Ed. Engl.* **1994**, *33*, 1373; (b) D. Walther, T. Klettke and H. Görls, *Angew. Chem.* **1995**, *107*, 2022; *Angew. Chem. Int. Ed. Engl.* **1995**, *34*, 1860; (c) D. Walther, T. Klettke, W. Imhof and H. Görls, *Z. Anorg. Allg. Chem.* **1996**, *622*, 1134; (d) T. Klettke, D. Walther, A. Schmidt, H. Görls, W. Imhof and W. Günther, *Chem. Ber.* **1996**, *129*, 1457; (e) D. Braga, F. Grepioni, D. Walther, K. Heubach, A. Schmidt, W. Imhof, H. Görls and T. Klettke, *Organometallics* **1997**, *16*, 4910.
[128] F. A. J. Singelenberg and B. P. van Eijck, *Acta Crystallogr., Sect. C* **1987**, *C43*, 693.
[129] G. Ferguson, J. F. Gallagher, C. Glidewell and C. M. Zakaria, *J. Chem. Soc., Dalton Trans.* **1993**, 3499.
[130] C. Glidewell, R. B. Klar, P. Lightfoot, C. M. Zakaria and G. Ferguson, *Acta. Crystallogr., Sect. B* **1996**, *B52*, 110.
[131] W. Reppe und H. Vetter, *Liebigs Ann. Chem.* **1953**, *582*, 133.
[132] (a) E. Weiss, R. G. Merenyi and W. Hübel, *Chem. Ber.* **1962**, *95*, 1170; (b) S. C. Wallwork, *J. Chem. Soc.* **1961**, 494.
[133] K.-J. Jens and E. Weiss, *J. Organomet. Chem.* **1981**, *210*, C27.
[134] K. Kirchner, K. Mereiter, K. Mauthner and R. Schmid, *Organometallics* **1994**, *13*, 3405.
[135] E. P. Kündig, G. Bernardinelli and J. Leresche, *J. Chem. Soc., Chem. Commun.* **1991**, 1713.
[136] (a) E. W. Abel, W. Harrison, R. A. N. McLean, W. C. Marsh and J. Trotter, *J. Chem. Soc., Chem. Commun.* **1970**, 1531; (b) M. Herberhold, G. Süß, J. Ellermann and H. Gabelein, *Chem. Ber.* **1978**, *111*, 2931; (c) E. W. Abel, G. Farrow and I. D. H. Towdle, *J. Chem. Soc., Dalton Trans.* **1979**, 71.
[137] M. D. Clerk and M. J. Zaworotko, *J. Chem. Soc., Chem. Commun.* **1991**, 1607.
[138] (a) M. Herberhold and G. Süß, *Angew. Chem.* **1975**, *87*, 710; *Angew. Chem. Int. Ed. Engl.* **1975**, *14*, 700.
[139] K. T. Holman and M. J. Zaworotko, *J. Chem. Crystallogr.* **1995**, *25*, 93.
[140] M. D. Clerk, S. B. Copp, S. Subramanian and M. J. Zaworotko, *Supramolecular Chem..* **1992**, *1*, 7.
[141] (a) S. B. Copp, S. Subramanian and M. J. Zaworotko, *J. Am. Chem. Soc.* **1992**, *114*, 8719; (b) S. B. Copp, S. Subramanian and M. J. Zaworotko, *Angew. Chem.* **1993**, *105*, 755; *Angew. Chem. Int. Ed. Engl.* **1993**, *32*, 706; (c) S. B. Copp, S. Subramanian and M. J. Zaworotko, *J. Chem. Soc., Chem. Commun.* **1993**, 1078; (d) H. Abourahma, S. B. Copp, M.-A. MacDonals, R. E. Meléndez, S. D. Batchilder and M. J. Zaworotko, *J. Chem. Crystallogr.* **1995**, *25*, 731.
[142] K. Biradha, G. R. Desiraju, D. Braga and F. Grepioni, *Organometallics* **1996**, *15*, 1284.
[143] D. Messer, G. Landgraf and H. Behrens, *J. Organomet. Chem.* **1979**, *172*, 349.
[144] D. Braga, F. Grepioni, K. Biradha, V. R. Pedireddi and G. R. Desiraju, *J. Am. Chem. Soc.* **1995**, *117*, 3156.
[145] D. Braga, F. Grepioni, H. Wadepohl, S. Gebert, M. J. Calhorda and L. F. Veiros, *Organometallics* **1995**, *14*, 5350.
[146] R. F. Bryan and P. T. Greene, *J. Chem. Soc. A* **1970**, 3064; (b) R. F. Bryan, P. T. Greene, M. J. Newlands and D. S. Field, *J. Chem. Soc. A* **1970**, 3068; (c) A. Mitschler, B. Rees and M. S. Lehmann, *J. Am. Chem. Soc.* **1978**, *100*, 3390.
[147] R. D. Adams, D. M. Collins and F. A. Cotton, *J. Am. Chem. Soc.* **1974**, *96*, 749.
[148] D. Braga, F. Grepioni, M. J. Calhorda and L. F. Veiros, *Organometallics* **1995**, *14*, **1992**.
[149] D. Braga and F. Grepioni, *J. Chem. Soc., Dalton Trans.* **1993**, 1223.
[150] (a) D. A. Clemente, B. Rees, G. Bandoli, M. Cingi Biagini, B. Reiter and W. A. Herrmann, *Angew. Chem.* **1981**, *93*, 920; *Angew. Chem. Int. Ed. Engl.* **1981**, *20*, 887, (b) D. A. Clemente, M. Cingi Biagini, B. Rees and W. A. Herrmann, *Inorg. Chem.* **1982**, *21*, 3741.

[151] P. Leung, P. Coppens, R. K. McMullan and T. F. Koetzle, *Acta. Crystallogr., Sect. B* **1981**, *B37*, 1347.
[152] A. G. Orpen, *J. Chem. Soc., Dalton Trans.* **1983**, 1427.
[153] H. Wadepohl, S. Gebert, H. Pritzkow, F. Grepioni and D. Braga, *Chem. Eur. J.* **1998**, *4*, 279.
[154] T. Steiner, B. Lutz, J. van der Maas, A. M. M. Schreurs, J. Kroon and M. Tamm, *Chem. Commun.* **1998**, 171.
[155] (a) R. H. Crabtree, E. M. Holt, M. Lavin and S. M. Morehouse, *Inorg. Chem.* **1985**, *24*, **1986**; (b) M. Brookhart, M. L. H. Green and L.-L. Wong, *Progr. Inorg. Chem.* **1988**, *36*, 1; (c) R. Crabtree, *Angew. Chem.* **1993**, *105*, 828; *Angew. Chem. Int. Ed. Engl.* **1993**, *32*, 789, (d) R. H. Crabtree, *Chem. Rev.* **1985**, *85*, 245; (e) R. H. Crabtree, O. Eisenstein, G. Sini and E. Peris, *J. Organomet. Chem.* **1998**, *567*, 7.
[156] (a) D. Milstein, J. C. Calabrese and J. D. Williams, *J. Am. Chem. Soc.* **1986**, *108*, 6387; (b) R. C. Stevens, R. Bau, D. Milstein, O. Blum and T. F. Koetzle, *J. Chem. Soc., Dalton Trans.* **1990**, 1429; (c) Q. Liu and R. Hoffmann, *J. Am. Chem. Soc.* **1995**, *117*, 10108; (d) J. Wessel, J. C. Lee, Jr., E. Peris, G. P. A. Yap, J. B. Fortin, J. S. Ricci, G. Sini, A. Albinati, T. F. Koetzle, O. Eisenstein, A. L. Rheingold and R. H. Crabtree, *Angew. Chem.* **1995**, *107*, 2711; *Angew. Chem. Int. Ed. Engl.* **1995**, *34*, 2507; (e) S. A. Fairhurst, R. A. Henderson, D. L. Hughes, S. K. Ibrahim and C. J. Pickett, *J. Chem. Soc., Chem. Commun.* **1995**, 1569; (f) E. Peris, J. C. Lee, Jr., J. R. Rambo, O. Eisenstein and R. H. Crabtree, *J. Am. Chem. Soc.* **1995**, *117*, 3485; (g) E. S. Shubina, N. V. Belkova, A. N. Krylov, E. V. Vorontsov, L. M. Epstein, D. G. Gusev, M. Niedermann and H. Berke, *J. Am. Chem. Soc.* **1996**, *118*, 1105.
[157] Z. Dawoodi and M. J. Martin, *J. Organomet. Chem.* **1981**, *219*, 251.
[158] M. R. Churchill and C. Bueno, *Inorg. Chem.* **1983**, *22*, 1510.
[159] H. Wadepohl, D. Braga and F. Grepioni, *Organometallics* **1995**, *14*, 24.
[160] M. C. Etter, Z. Urbonczyck-Lipkowska, D. A. Jahn and J. S. Frye, *J. Am. Chem. Soc.* **1986**, *108*, 1084.
[161] (a) D. Braga, F. Grepioni, J. J. Byrne and A. Wolf, *J. Chem. Soc., Chem. Commun.* **1995**, 1023; (b) D. Braga, A. L. Costa, F. Grepioni, L. Scaccianoce and E. Tagliavini, *Organometallics* **1997**, *16*, 2070.
[162] (a) D. Braga, A. Angeloni, F. Grepioni and E. Tagliavini, *Chem. Commun.* **1997**, 1447; (b) D. Braga, A. Angeloni, F. Grepioni and E. Tagliavini, *Organometallics* **1997**, *16*, 5478.
[163] J. W. Gilje and H. W. Roesky, *Chem. Rev.* **1994**, *94*, 895.
[164] D. Braga, A. L. Costa, F. Grepioni, L. Scaccianone and E. Tagliavini, *Organometallics* **1996**, *15*, 1084.
[165] D. Braga and F. Grepioni, *Chem. Commun.* **1998**, 911.
[166] J. A. R. P. Sarma and G. R. Desiraju, *Acc. Chem. Res.* **1986**, *19*, 222.
[167] (a) L. R. Melby, R. S. Harder, W. R. Hertler, R. E. Benson and W. E. Mochel, *J. Am. Chem. Soc.* **1962**, *84*, 3374; (b) G. A. Candela, L. J. Swartzendruber, J. S. Miller and M. J. Rice, *J. Am. Chem. Soc.* **1979**, *101*, 2755; (c) J. S. Miller, J. H. Zhang, W. M. Reif, L. D. Preston, D. A. Dixon, A. H. Reis, Jr., E. Gebert, M. Extine, J. Troup, A. J. Epstein and M. D. Ward, *J. Phys. Chem..* **1987**, *91*, 4344; (d) J. S. Miller, J. C. Calabrese, H. Rommelmann, S. R. Chittapeddi, J. H. Zhang, W. M. Reiff and A. J. Epstein, *J. Am. Chem. Soc.* **1987**, *109*, 769.
[168] X. Wang, L. M. Liable-Sands, J. L. Manson, A. L. Rheingold and J. S. Miller, *Chem. Commun.* **1996**, 1979.

6 Supramolecular Self-Assembly Caused by Ionic Interactions

6.1 Alkali Metals

Supramolecular self-assembly caused by ionic interactions is, as one would expect, found mainly in organometallic compounds of the most electropositive metals, i.e. the alkali and alkaline earth metals. Self-assembly is, in fact, very common in such polar organometallics and the solid-state structures cover the whole range of self-aggregation from dimers to polymers, which can be unsolvated or (more commonly) solvated. The complexity is highly dependent on the metal and on the nature of the solvent and the anionic moiety. Among the structurally characterized examples are various real organometallics, i.e. hydrocarbyls of the alkali and alkaline earth metals. Because of their often comparable nature and reactivity, supramolecular amides, alkoxides, and their higher homologs will also be discussed in this chapter.

6.1.1 Hydrocarbyls

Organoalkali reagents are of fundamental importance both in modern organic synthesis and in organometallic chemistry [1–3]. Such polar organometallics are self-assembling systems par excellence and supramolecular association into higher aggregates is very common. Despite their high reactivity and extreme air- and moisture-sensitivity, the structural chemistry of these species is now well developed and various types of oligomer and polymer have been structurally elucidated. The structures of alkali metal hydrocarbyls and related compounds (amides, alkoxides) have been compiled in an excellent overview by Weiss [4a]. The structural chemistry of organolithium compounds is particularly highly diverse, but investigation of the hydrocarbyls of the heavier alkali metals has also made significant progress in recent years [4b]. Table 6.1 shows the types of self-assembly of various base-free and solvated alkali metal organometallics.

Table 6.1. Self-assembly of organoalkali metal compounds.

Compound number	Name and reference

Dimers

1	[LiC(SiMe$_3$)$_3$]$_2$ [5]
2	[LiBu^t(Et$_2$O)]$_2$ [6]
3	[{LiC$_3$Me$_2$(SiMe$_3$)}(tmeda)]$_2$ [7]
4	[{LiC$_3$Me$_2$(CBu$^t_2$OLi)}(thf)$_2$]$_2$ [8, 9]
5	[{LiC$_3$Me$_2$(CBu$^t_2$OLi)}(tmeda)]$_2$ [9]
6	[{LiC$_3$Me$_2${CBu$^t_2$(NBu^t)Li}}(thf)$_2$]$_2$ [9]
7	[LiPh(tmeda)]$_2$ [10]
8	[LiC≡CPh(tmpda)]$_2$ [11]
9	[Li(C$_6$H$_4$Ph)(tmeda)]$_2$ [12]
10	[LiC$_6$H$_3$Mes$_2$-2,6]$_2$ [13]
11	[LiC$_6$H$_3${CH(Me)NMe$_2$}-1,3]$_2$ [14]
12	[{LiC$_6$H$_3${CH(Et)NMe$_2$}-1,3}(LiBu^n)]$_2$ [14]
13	[LiCH$_2$SMe(tmeda)]$_2$ [15]
14	[LiCH$_2$SPh(thf)$_2$]$_2$ [15]
15	[LiCH$_2$SPh(tmeda)]$_2$ [15]
16	[LiCH$_2$PMe$_2$(tmeda)]$_2$ [16]
17	[LiCH$_2$PPhMe(tmeda)]$_2$ [16]
18	[LiCH$_2$PPhMe(sparteine)]$_2$ [16]
19	[LiCH$_2$PPh$_2$(tmeda)]$_2$ [17, 18]
20	[Li$_2$(CH$_2$NPh$_2$)$_2$(thf)$_3$] [18]
21	[Li{C$_6$H$_2$(CF$_3$)$_3$-2,4,6}(Et$_2$O)]$_2$ [19]
22	[NaC$_6$H$_3$Mes$_2$-2,6]$_2$ [20]
23	[NaPh(pmdta)]$_2$ [21]
24	[NaC$_6$H$_4$Ph(tmeda)]$_2$ [22]
25	[Rb{μ-CH(SiMe$_3$)$_2$}(pmdeta)]$_2$ [23]

Trimer

26	[Li{C$_6$H$_3$(NMe$_2$)$_2$-2,6}]$_3$ [24]

Tetramers

27	(LiMe)$_4$ [25–27]
28	(LiEt)$_4$ [28, 29]
29	(LiBu^t)$_4$ [6]
30	[Li(1-norbornyl)]$_4$ [30]
31	(LiCH$_2$CH$_2$CH$_2$OMe)$_4$ [31]
32	(LiCH$_2$CH$_2$CH$_2$NMe$_2$)$_4$ [32]
33	[LiPh(Et$_2$O)]$_4$ [33]
34	[LiPh(Me$_2$S)]$_4$ [34]
35	[(LiC≡CBu^t)(thf)]$_4$ [35]
36	[Li$_4$(CH$_2$NC$_5$H$_{10}$)$_4$(thf)$_2$] [18]
37	[Ph$_2$P(O)CHLiC(H)MeEt]$_4$ [36]
38	(NaMe)$_4$ [37–39]
39	(NaMe)$_4$·[(LiMe)$_4$]$_x$ ($x \leq 0.333$) [38]
40	[NaCH$_2$Ph(tmeda)]$_4$ [40]
41	[Na(o-xylyl)(tmeda)]$_4$ [41]
42	[NaCHPh$_2$(tmeda)]$_4$ [41]

Table 6.1. (cont.)

Compound number	Name and reference
Hexamers	
43	(LiPr^i)$_6$ [42]
44	(LiBu^n)$_6$ [6]
45	[Li(c-C$_5$H$_9$)]$_6$ [43]
46	[Li(c-C$_6$H$_{11}$)]$_6$ [43]
47	[Li(tetramethylcyclopropylmethyl)]$_6$ [43]
48	[LiC$_6$H$_3$Bu$^t_2$-3,5]$_6$ [44]
49	[Li(CH$_2$SiMe$_3$)]$_6$ [45]
Dodecamer	
50	[(LiC≡CBu^t)$_{12}$(thf)$_4$] [35]
Polymers	
51	[(LiMe)$_4$(tmeda)$_2$]$_n$ [46]
52	[LiC(OEt)=CH$_2$]$_n$ [47]
53	[LiC≡CH(en)]$_n$ [48]
54	[(LiC≡CPh)$_4$(tmhda)$_2$]$_n$ [49]
55	[Li(allyl)(tmeda)]$_n$ [50]
56	[LiCH$_2$Ph(dabco)]$_n$ [51]
57	[LiCH$_2$Ph(Et$_2$O)]$_n$ [52]
58	[LiCH$_2$SMe(thf)]$_n$ [15]
59	[LiCH$_2$PPh$_2$(thf)]$_n$ [53]
60	[Na{μ-CH(SiMe$_3$)$_2$}]$_n$ [23]
61	[NaCH$_2$Ph(pmdta)]$_n$ [42]
62	[KCH$_2$Ph(pmdta)]$_n$ [54]
63	[MCHPh$_2$(pmdta)]$_n$ (M = K, Rb, Cs) [55]
64	[NaCPh$_3$(tmeda)]$_n$ [56]
65	[KCPh$_3$(thf)]$_n$ [57]
66	[KCPh$_3$(diglyme)]$_n$ [57]
67	[K{μ-CH(SiMe$_3$)$_2$}(pmdeta)]$_n$ [23]

6.1.1.1 Base-free alkali metal hydrocarbyls

The structures of base-free hydrocarbyls of the alkali metals have been investigated much less than those of the solvated species [4]. The first example of an unsolvated dimer was [LiC(SiMe$_3$)$_3$]$_2$, **1** [5]. This pyrophoric compound was prepared by reacting the corresponding HgBr derivative with *n*-butyllithium.

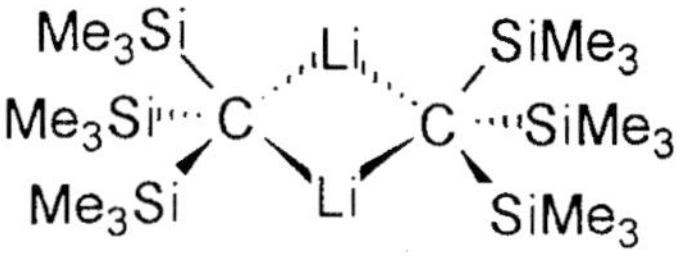

1

More recently other unsolvated dimers became available with the use of sterically highly demanding, lipophilic terphenyl substituents. These enabled the structural characterization of the solvent-free, σ-bonded dimeric lithium aryl [LiC$_6$H$_3$Mes$_2$-2,6]$_2$, **10**, for example [13]. The bulky terphenyl substituent $-$C$_6$H$_3$Mes$_2$-2,6 also played a major role in the successful preparation of the first base-free sodium aryl, [NaC$_6$H$_3$Mes$_2$-2,6]$_2$, **22** [20]. This compound was made by an alkoxide-exchange reaction between **10** and sodium t-butoxide. The crystal structure of **22** consists of metal-bridged dimers in which the sodium ions are primarily coordinated to *ipso*-phenyl and *ipso*-mesityl carbon atoms. Both interactions are almost equally strong.

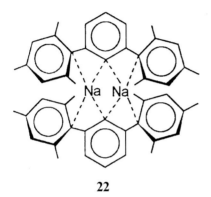

22

The aryllithium compound [Li{C$_6$H$_3$(NMe$_2$)$_2$-2,6}]$_3$, **26**, has been reported to self-assemble in the form of a cyclic trimer [24]. This compound crystallizes without additional donor ligands, because the dimethylamino substituents of the aryl ligands can saturate the coordination sphere of lithium. One of the most thoroughly studied organolithium compounds is tetrameric methyllithium [25–27]. Both (LiMe)$_4$, **27**, and (LiEt)$_4$, **28** [28, 29], form tetrameric aggregates containing an Li$_4$C$_4$ heterocubane core. In methyllithium the tetramers have ideal T_d symmetry and form a cubic lattice. Tetramers have also been found in the crystal structure of methylsodium, **38**. To solve this complicated structure by powder diffraction methods it was necessary to use a novel combination of neutron and synchroton X-ray diffraction techniques and a sample of specially prepared NaCD$_3$ powder [38]. The Na and C positions were determined from synchrotron diffraction data and the positions of the deuterium atoms were obtained from the neutron diffraction study. In the crystal structure one half of the ions form (NaMe)$_4$ tetrahedrons with slightly distorted T_d symmetry. The remaining eight sodium and methyl ions connect the tetramers via Na–C interactions. The overall structure is very similar to that of methyllithium. An even more complicated crystal structure results when methylsodium crystallizes with different amounts of methyllithium. Microcrystalline samples of the composition (NaMe)$_4 \cdot$[(LiMe)$_4$]$_x$ ($x \leq 0.333$), **39**, results from metal-exchange reactions between methyllithium and sodium t-butoxide according to Eq.(1) [38].

$$\text{LiMe} + \text{NaOBu}^t \rightarrow \text{NaMe(LiMe)}_x + \text{LiOBu}^t \tag{1}$$

6.1 Alkali Metals 375

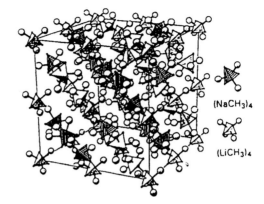

Fig. 6.1. The crystal structure of (NaMe)$_4$·[(LiMe)$_4$]$_x$ ($x \leq 0.333$), **39** (Reproduced from ref. [38]).

Depending on the composition of the solvent mixture the molar ratio NaMe/LiMe can be between 3:1 and 36:1. The unusual structure of this material was determined by X-ray powder diffraction (Figure 6.1).

In a cubic 'giant unit cell' 24 tetrameric (NaMe)$_4$ units form a zeolitic host lattice. This structure is related to that of methyllithium, but the arrangement of the tetramers is more complicated. It results in the presence of large cavities, in which (LiMe)$_4$ units can be intercalated up to the structurally predetermined molar ratio of 3:1. In this sense the host–guest compound (NaMe)$_4$·[(LiMe)$_4$]$_x$ ($x \leq 0.333$), **39**, was termed the 'first organometallic supramolecular compound'.

The higher lithium alkyls (LiBu^t)$_4$, **29** [6], and [Li(1-norbornyl)]$_4$, **30** [30], have also been reported to be associated as tetramers in the solid state. In the crystal, t-butyllithium is a tetramer with approximate T_d symmetry. Each face of the Li$_4$ tetrahedron is capped by a t-butyl unit with the methyl groups being in an ecliptic position relative to the lithium triangle.

29

In (LiBu^n)$_6$, **44** [6], [Li(c-C$_5$H$_{11}$)]$_6$, **45** [43], [Li(c-C$_6$H$_{11}$)]$_6$, **46** [43], [Li(CH$_2$SiMe$_3$)]$_6$, **49** [45], and [Li(tetramethylcyclopropylmethyl)]$_6$, **47** [43], supramolecular self-assembly results in the formation of octahedral cluster molecules.

6 Supramolecular Self-Assembly Caused by Ionic Interactions

The crystal structure determination of *n*-butyllithium can be considered a particularly major achievement because pure *n*-butyllithium is a highly pyrophoric liquid under normal conditions (m.p. $-34(2)\,°C$). Suitable single crystals were obtained by slow crystallization from pentane at $-90\,°C$. In the hexamer six triangles of lithium atoms are capped by *n*-butyl units, and the remaining two triangles are not bridged. The α carbon atoms form short bonds (ave 2.159 Å) to two Li atoms of a triangle and a longer one to the third (ave 2.270 Å). This enables additional electrostatic interaction between one lithium atom in a triangle with the β carbon of the *n*-butyl substituent. Closely related structures have also been reported for the secondary alkyllithium reagent $(LiPr^i)_6$, **43** [42], and the unsolvated lithium aryl $[LiC_6H_3Bu^t_2\text{-}3,5]_6$, **48** [44].

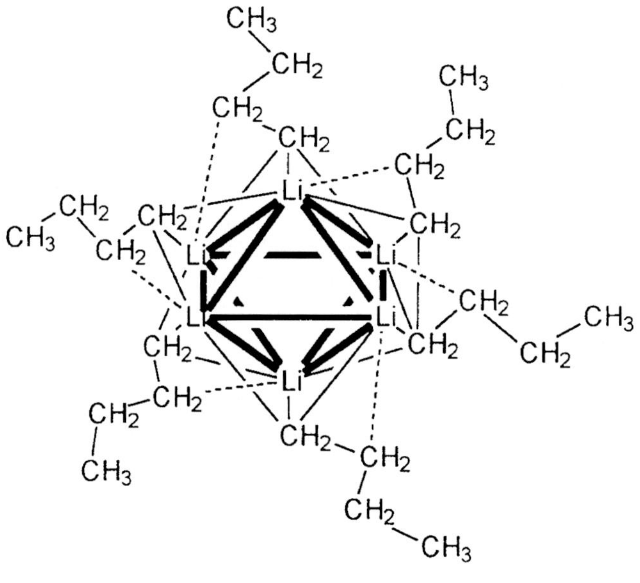

44

Self-assembly to form polymeric aggregates has also been found in the chemistry of base-free organoalkali metal compounds. An interesting example is the crystalline, hexane-soluble sodium alkyl $[Na\{\mu\text{-}CH(SiMe_3)_2\}]_n$, **60**, prepared by reaction of

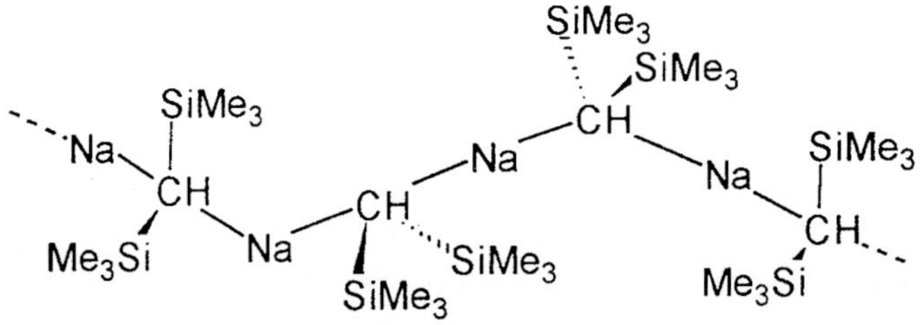

60

the lipophilic and volatile lithium alkyl LiCH(SiMe$_3$)$_2$, **68**, with sodium *t*-butoxide in hexane as solvent [23]. The key to the success of this synthesis is the somewhat greater hexane-solubility of the lithium *t*-butoxide co-product, which enables the less soluble **60** to be readily separated. Insoluble [K{μ-CH(SiMe$_3$)$_2$}]$_n$, **69**, was made analogously. The polymeric structure of **60** is quite remarkable. It consists of chains of alternating cations and anions. The unusually low coordination number of two for the sodium atoms has only a single precedent – in the isoelectronic and isostructural amide [Na{μ-N(SiMe$_3$)$_2$}]$_n$, **70** [58]. Despite its polymeric nature, **60** is hexane-soluble and volatile below 150 °C.

6.1.1.2 Solvated alkali metal hydrocarbyls

A rich body of structural investigations on Lewis base adducts of organoalkali metal compounds has been reported. Although monomeric species such as LiPh(pmdta), **71**, are rare, self-association is quite common in this chemistry [4]. Adduct formation with a variety of solvents is of great importance to the physical and chemical properties of these polar organometallics. Solvation often leads to an increased solubility and enhanced reactivity, and the most common preparative routes to organoalkali metal compounds involve the use of ethereal solvents such as diethyl ether or THF. Normally the extent of self-association is reduced on adduct formation, although many adducts with Lewis bases are still associated as oligomers. Quite often adducts with ethers are thermolabile and release part of the coordinated solvent upon storage of the solid materials. For structural investigations the more robust adducts with chelating amines such as tetramethylethylenediamine (tmeda) or pentamethyldiethylenetriamine (pmdta) have been found to be more useful. Generally the polymeric aggregates of the base-free alkali metal hydrocarbyls are broken up upon treatment with Lewis bases to give well-defined adducts with a less self-assembly. The latter is often dependent on the nature of the coordinating solvent. A nice example for the structure-directing influence of the different Lewis bases is phenyllithium. Polymeric phenyllithium dissolves readily in coordinating solvents or in the presence of lithium halides. With diethyl ether or dimethyl sulfide tetramers of the heterocubane type, [LiPh(Et$_2$O)]$_4$, **33**, and [LiPh(Me$_2$S)]$_4$, **34**, are formed [33, 34], whereas treatment with tmeda yields a dimer, [LiPh(tmeda)]$_2$,

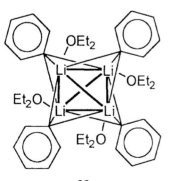

33

7 [10]. Finally, a monomeric molecule, LiPh(pmdta), **71**, is obtained when the tridentate amine ligand pmdta is employed [4].

Dimers are a common structural type in organoalkali metal chemistry. Typical examples, in addition to the tmeda adduct of phenyllithium, are [LiBu^t(Et$_2$O)]$_2$, **2** [6], [LiC≡CPh(tmpda)]$_2$, **8** [11], [Li(C$_6$H$_4$Ph)(tmeda)]$_2$, **9** [12], and some derivatives of the heavier alkali metals, including [NaPh(pmdta)]$_2$, **23** [21], [NaC$_6$H$_4$Ph(tmeda)]$_2$, **24** [22], and [Rb{μ-CH(SiMe$_3$)$_2$}(pmdta)]$_2$, **25** [23]. Alkyl and aryl derivatives are both dimerized in the same way with bridging hydrocarbyl ligands. The central structural feature of all these compounds is a four-membered M$_2$C$_2$ ring (M = alkali metal). In the case of rubidium a tridentate amine ligand is necessary for the formation of a dimeric molecule, whereas for **2** only one diethyl ether ligand is sufficient because of the steric hindrance imposed by the bulky t-butyl groups [6, 23]. In the bridging biphenyl ligands in **24** the two phenyl rings are not coplanar, as was observed also for the lithium compound, **9** [12, 22].

An interesting series of lithiated cyclopropenes was structurally investigated by von Schleyer et al [8]. Such lithiocyclopropenes are useful intermediates in organic syntheses [59], but structural information was lacking until recently. Mono- and dilithiated species were prepared according to Eqs (2–4).

Crystallization of the crude products from hexane–THF or hexane–tmeda afforded the well-defined crystalline solvates **3**–**6**. Although NMR data revealed the presence of monomers in THF solution, all four compounds are dimeric in the solid state. There are, however, interesting structural differences. The monolithiated trimethylsilyl derivative, **3**, forms a dimer like [LiPh(tmeda)]$_2$, **7**. The two cyclopropenyl rings are nearly coplanar and are bridged quite symmetrically by the two Li ions.

The additional chelating ligands in **4**–**6** lead to different types of self-assembly.

These three compounds have been investigated to verify computational predictions of planar tetracoordinate carbon $R^1C^2C_\alpha Li_2$ arrangements in lithiocyclopropene dimers [60]. The C_α and O dilithiated complexes **4** and **5** both have ladder-type dimeric assemblies. Although the puckered ladder backbone in **4** is centrosymmetric, **5** adopts an asymmetric dimeric structure in which all the Li ions have different coordination environments because of different THF solvation. The tetracoordinate carbon $R^1C^2C_\alpha Li_2$ units in both crystal structures are highly planar (twist angles 35–39°) partly because of O-chelation of one of the lithium atoms. The molecular structure of **6** is closely related to that of **5**, although in the former each Li is coordinated by one THF ligand. With twist angles of 17° and 30° the $R^1C^2C_\alpha Li_2$ units come even closer to planarity. An additional unexpected feature of the structure of **6** is the nearly planar coordination of C_β, i.e. the cyclopropene carbon next to the lithiated site.

The structures of functionalized methyllithium derivatives of the type $LiCH_2YR_n$ have been found to be strongly dependent on:

i) the nature of the heteroatom Y;
ii) the type of the substituent R (alkyl or aryl); and
iii) co-ligands which might be present.

Two different types of dimers have been structurally characterized (Figure 6.2) – dimers containing six-membered $Li_2C_2Y_2$ rings (type **A**) have been found in the compounds [LiCH$_2$SPh(tmeda)]$_2$, **15**, [LiCH$_2$PMe$_2$(tmeda)]$_2$, **16**, [LiCH$_2$PPhMe(tmeda)]$_2$, **17**, [LiCH$_2$PPhMe(sparteine)]$_2$, **18**, and [LiCH$_2$PPh$_2$(tmeda)]$_2$, **19**. The derivatives [LiCH$_2$SMe(tmeda)]$_2$, **13**, [LiCH$_2$SPh(THF)$_2$]$_2$, **14**, and [Li$_2$(CH$_2$NPh)$_2$(THF)$_3$], **20**, form dimers with four-membered Li$_2$C$_2$ rings, in which the heteroatom is not coordinated to lithium (type **B**) [15–18].

An unusual dimeric structure was found for the diethyl ether adduct of 2,4,6-tris(trifluoromethyl)phenyllithium, [Li{C$_6$H$_2$(CF$_3$)$_3$-2,4,6}(Et$_2$O)]$_2$, **21** [19]. A pla-

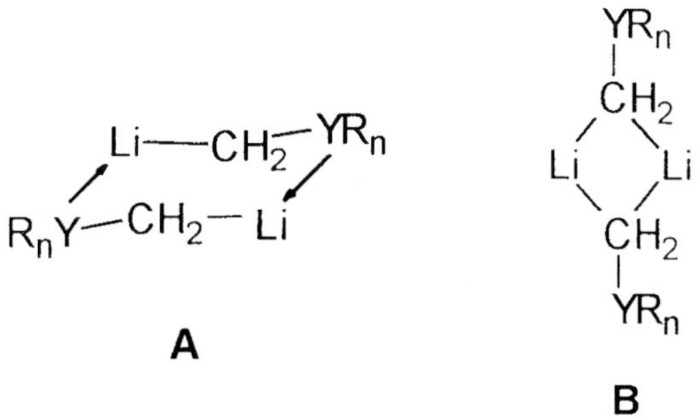

Fig. 6.2. The structures of the six-membered (**A**) and four-membered (**B**) $Li_2C_2Y_2$ rings in dimers of functionalized methyllithium derivatives $LiCH_2YR_n$.

nar Li_2C_2 unit forms the central part of the molecule. The coordination geometry around Li can be described as a distorted trigonal pyramid in which the equatorial positions are occupied by the *ipso* carbon atoms of the aryl rings and an ether oxygen. Two fluorine atoms from *ortho* CF_3 groups, one from each aryl substituent, are coordinated in the axial positions. Although the lithium–fluorine distances are fairly long (ave 2.252 Å), these interactions must be considered the main stabilizing factor in the dimeric molecule.

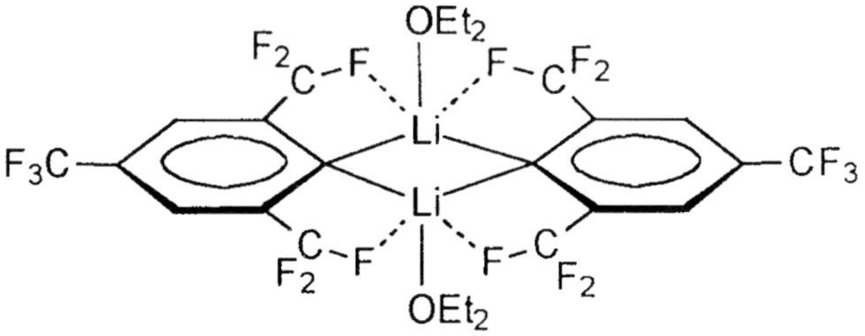

21

Subtle structural differences have been observed in some dimers of lithiated 1,3-bis[(dimethylamino)alkyl]benzenes [14]. Direct lithiation of 1,3-bis[(dimethylamino)*ethyl*]benzene with *n*-butyllithium produced a symmetric dimer, in which the coordination sphere of lithium was saturated by coordination of the chelating amino substituents. An analogous lithiation reaction performed between 1,3-bis[(dimethylamino)*propyl*]benzene and *n*-butyllithium yielded a mixed dimeric aggregate comprising the expected phenyllithium derivative and *n*BuLi in a 1:1 molar ratio. In the molecular structure four lithium ions and four bridging carbon atoms form a ladder-type framework.

Although trimers are quite rare, a more preferred type of supramolecular association in solvated alkali metal hydrocarbyls is the formation of tetramers in the solid state. Typical examples of cyclic tetramers are the sodium *o*-xylyl derivative [Na(*o*-xylyl)(tmeda)]$_4$, **41**, and the diphenylmethyl complex [NaCHPh$_2$(tmeda)]$_4$, **42** [41]. The tetrameric THF adduct of *t*-butylethynyllithium, **35**, forms a heterocubane

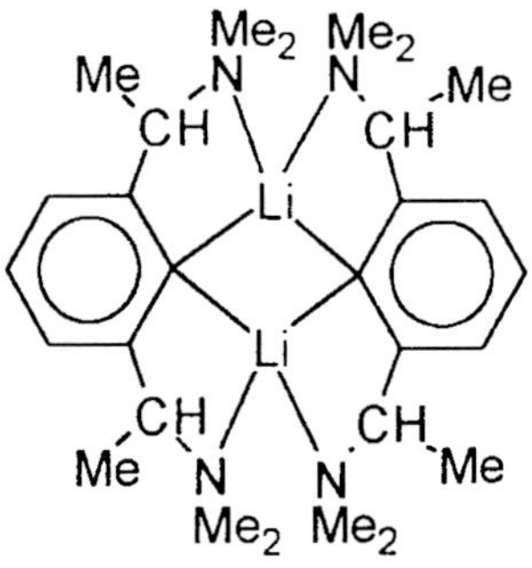

11 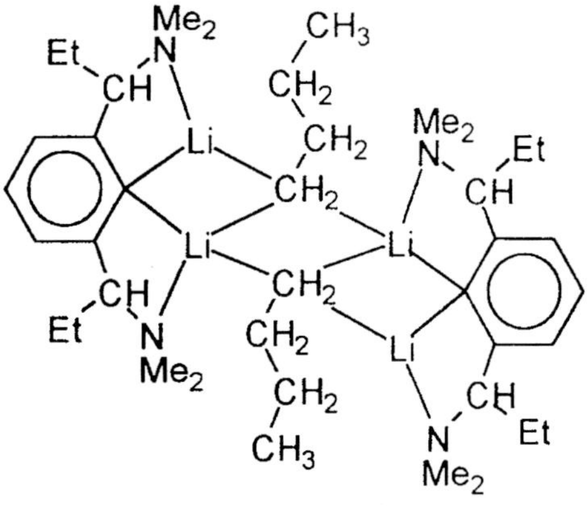   **12**

41   **35**

structure with triply bridging alkynyl ligands [35]. This molecule can be regarded as the result of the stacking of two coordinatively unsaturated dimers.

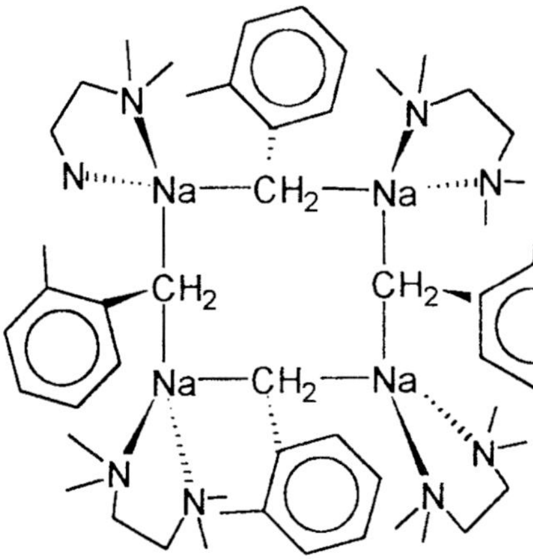

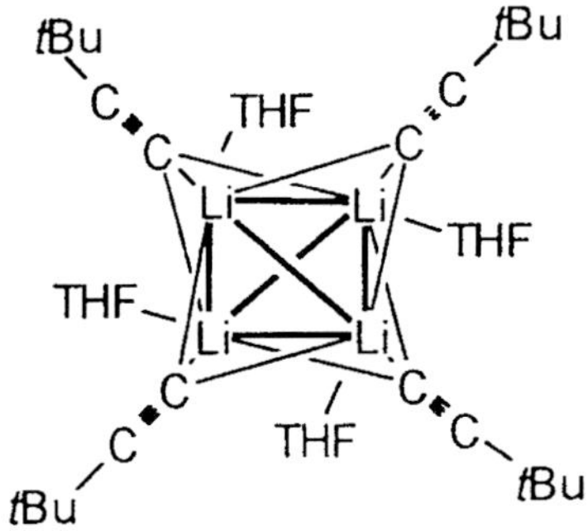

It is interesting to note that, depending on the crystallization conditions, the same compound can also crystallize as a dodecamer, formally formed by stacking of six dimers [35].

Lithiated phosphine oxides, the so-called Horner–Wittig reagents, are important intermediates in organic syntheses, for example in stereoselective syntheses of alkenes and asymmetric syntheses of chiral compounds [61, 62]. Despite the importance of these reagents very little is known about their structures. The first lithiated phosphine oxide containing Li–C bonds has been reported recently. The self-assembled compound [Ph$_2$P(O)CHLiC(H)MeEt]$_4$, **37**, forms a heterocubane-type Li$_4$O$_4$ tetramer without additional donor ligands, despite its preparation in the presence of HMPA [36]. Internal coordination of the carbanions leads to the formation of four-membered OPCLi rings.

Occasionally supramolecular self-assembly of solvated organoalkali metal compounds leads to the formation of complicated polymers, including chain polymers, ladder structures, and three-dimensional networks. Typical examples of chain polymers include [Li(allyl)(tmeda)]$_n$, **55** [48], [LiCH$_2$Ph(dabco)]$_n$, **56** [51], [LiCH$_2$Ph(Et$_2$O)]$_n$, **57** [52], [NaCH$_2$Ph(pmdta)]$_n$, **61** [42], [MCHPh$_2$(pmdta)]$_n$ (M = K, Rb, Cs), **63a–c** [56], [NaCPh$_3$(tmeda)]$_n$, **64** [51], [KCPh$_3$(thf)]$_n$, **65** [57], [KCPh$_3$(diglyme)]$_n$, **66** [57], and [K{μ-CH(SiMe$_3$)$_2$}(pmdta)]$_n$, **67** [23]. In **55** the solvated lithium ions are connected to the terminal CH$_2$ groups of the allyl ions, which are not coordinated via the π-system [50]. Chain polymers involving different hapticities of the bridging organic ligands are common for benzyl, diphenylmethyl, and triphenylmethyl derivatives of the alkali metals. Tmeda-stabilized benzylsodium derivatives such as [NaCH$_2$Ph(tmeda)]$_4$, **40**, and [Na(o-xylyl)(tmeda)]$_4$, **41**, are associated as tetramers [41]. In contrast, a polymeric chain structure is obtained when the tridentate ligand pmdta is used. In the resulting zigzag chains the planar benzyl anions are coordinated to the sodium atoms via the *ipso* carbon atoms [41]. η^2-Benzyl bridges are found in the corresponding lithium compounds [LiCH$_2$Ph(dabco)]$_n$, **56** [51], and [LiCH$_2$Ph(Et$_2$O)]$_n$, **57** [52].

61

Yet another type of hydrocarbyl bridging has been established for some solvated triphenylmethyl derivatives of the heavier alkali metals. [NaCPh$_3$(tmeda)]$_n$, **64**, and [KCPh$_3$(diglyme)]$_n$, **66** [57], are supramolecularly associated into zigzag chains by weak intermolecular sodium–carbon contacts.

In addition to the simple chain structures other types of polymeric associate have also been structurally characterized. Ladder structures have been reported for [LiC≡CH(en)]$_n$, **53** [48], [LiCH$_2$SMe(thf)]$_n$, **58** [15], and [LiCH$_2$PPh$_2$(thf)]$_n$, **59**

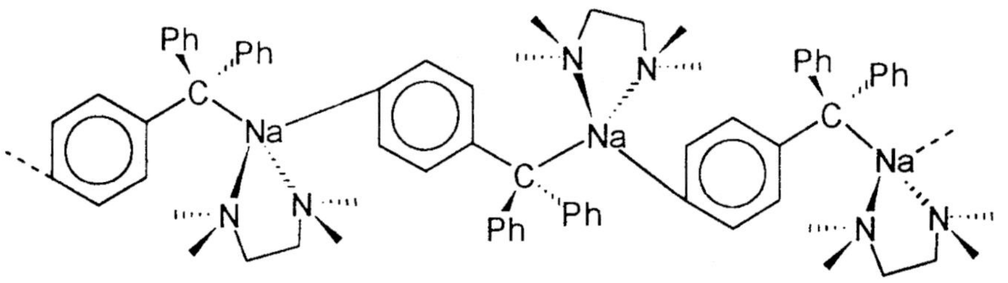

64

[53]. For example, in the ladder-like structure of **59** six-membered $Li_2C_2P_2$ rings with a chair conformation and planar four-membered Li_2C_2 rings are arranged alternately. The interplanar angle between the Li_2C_2 rings and the planar Li_2C_2 units of the neighboring six-membered rings is 145°.

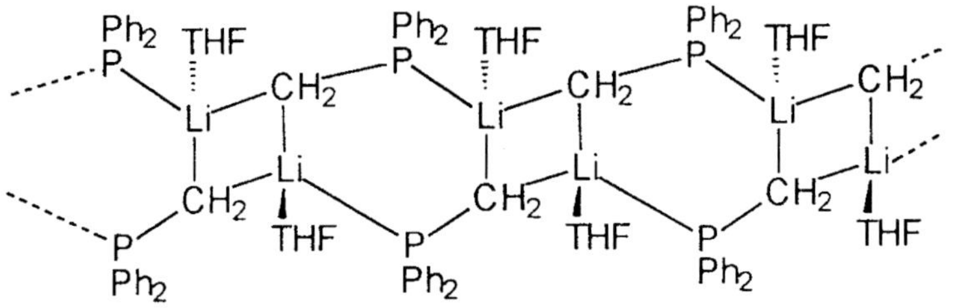

59

The compounds $[(LiMe)_4(tmeda)_2]_n$, **51** [46], and $[(LiC\equiv CPh)_4(tmhda)_2]_n$, **54** [49], are interesting examples of more complicated supramolecular associates. The tetrameric units in base-free methyllithium is quite stable and will not be disrupted upon treatment with ethers or amine ligands. Solvates of methyllithium still contain $(LiMe)_4$ units as part of supramolecular networks. In the tmeda adduct **51** the bidentate amine ligands are not chelating but act as bridges between the tetramers, thus resulting in a three-dimensional network. In **54** the tetrameric subunits are bridged by pairs of tmhda ligands to give a double helix structure. A two-dimensional network was also found in polymeric $[KCPh_3(THF)]_n$, **65** [57].

A complicated polymeric arrangement was also found for base-free α-ethoxyvinyllithium, **52** [47].

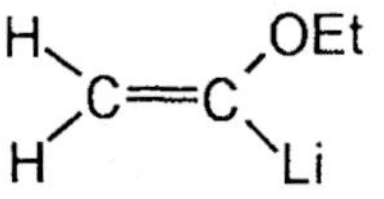

52

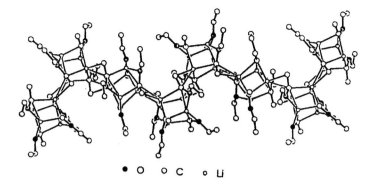

Fig. 6.3. The polymeric chain structure of [LiC(OEt)=CH$_2$]$_n$, **52** (Reproduced from ref. [47]).

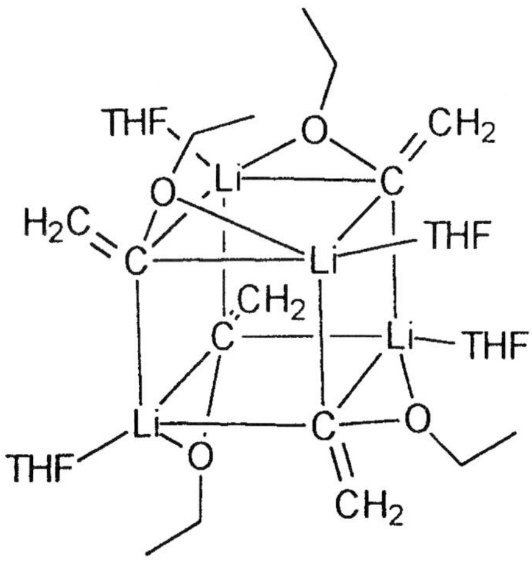

Fig. 6.4. The solution structure of [LiC(OEt)=CH$_2$]$_n$, **52**.

α-Ethoxyvinyllithium, **52**, is of considerable importance as a synthetic equivalent of an acyl anion [63]. Base-free **52** can be obtained by recrystallization from pentane. The crystal structure consists of polymeric chains in which the asymmetric unit contains six H$_2$C=C(Li)OEt molecules. Four form a distorted cubic Li$_4$C$_4$ tetramer and the remaining two participate in a second type of tetrameric aggregate. Adjacent tetrameric units are then linked through η^2-Li–C$_{vinyl}$ interactions (Figure 6.3).

Variable-temperature ^{13}C NMR studies reveal that the polymeric structure does not persist in THF solution. Instead, the data are consistent with the presence of regular heterocubane-type tetramers (Figure 6.4) as in the solvated phenyllithium derivatives **33** and **34**.

Little information has been published on the molecular structures and self-assembly of the higher homologs of the alkali metal hydrocarbyls, i.e. for example

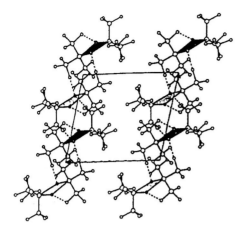

Fig. 6.5. The crystal structure of [NaSi(SiMe$_3$)$_3$]$_n$, **73** (Reproduced from ref. [66]).

silyl and germyl derivatives. Trialkylsilyl lithium compounds can be prepared by reacting elemental lithium with bis(trialkylsilyl)mercury according to Eq.(5) [64].

$$2Li + Hg(SiR_3)_2 \xrightarrow[\text{solvent}]{\text{hydrocarbon}} 2/n(LiSiR_3)_n + Hg \ (R = alkyl) \qquad (5)$$

A hexameric structure has been established for (LiSiMe$_3$)$_6$, **72** [65]. More recently an exciting series of tris(trimethylsilyl)silyl ('hypersilyl') derivatives of the heavier alkali metals has been prepared and their structures elucidated [66]. The base-free compounds [NaSi(SiMe$_3$)$_3$]$_n$, **73**, and [KSi(SiMe$_3$)$_3$]$_n$, **74**, were prepared by reacting bis(hypersilyl)zinc, -cadmium, or -mercury with the appropriate molten alkali metal in boiling heptane. Similar to LiSi(SiMe$_3$)$_3$, **75** [66], both molecules consist of cyclic dimers [MSi(SiMe$_3$)$_3$]$_2$ with almost planar M$_2$Si$_2$ rings. These dimers are further linked to form coordination polymers. They self-assemble to 'super dimers' through short M···CH$_3$ interactions, and these dimers form infinite chain polymers through additional intermolecular agostic interactions (Figure 6.5).

6.1.2 Amides and related compounds

Similar to the hydrocarbyls, alkali metal amides are important as metalating and deprotonating agents. Their structural chemistry is also characterized by extensive self-assembly [2, 67, 68]. Relatively little is known about the crystal structures of unsolvated alkali metal organoamides [69, 70]. The structural chemistry of solvated alkali metal amides is much more developed. Lithium amides and imides, in particular, have been shown to crystallize in a variety of structural types, including ladder structures [71] and stacks of rings [72–74]. Much less information is available on molecular structures of the corresponding alkali metal organophosphides and arsenides, although significant progress in this area has been made in recent years.

Representative examples of structurally characterized self-assembled alkali metal amides and related compounds are listed in Table 6.2.

6.1.2.1 Amides and related nitrogen derivatives

As mentioned above, few studies have been conducted on the molecular structures of unsolvated alkali metal amides. Small degrees of self-assembly can be achieved by use of bulky amide ligands such as –N(SiMe$_3$)$_2$. Cyclic trimers have been found for the silylamides [Li{N(SiMe$_3$)$_2$}]$_3$, **116**, and [Na{N(SiMe$_3$)$_2$}]$_3$, **117** [98, 101]. The trimer **117** contains an almost planar six-membered Na$_3$N$_3$ ring. The sodium ions are not only coordinated by two nitrogens – there are also two relatively short intramolecular Na···H(C) distances (2.384–2.616 Å). Two additional intermolecular Na···H interactions per molecule bring the coordination numbers of Na to 4 and 5, respectively.

One of the first solvent-free alkali metal amides to be investigated was potassium diethylamide, **149** [122]. In the solid state, rings of (KNEt$_2$)$_2$ units form a complicated network. An interesting 'ladder' structure containing amide bridges and π-coordination was found for sodium 2,3,4,5-tetramethylpyrrolide, **150** [123].

The most prominent structural feature in base-adducts of alkali metal amides are dimeric molecules containing a four-membered M$_2$N$_2$ (M = alkali metal) ring. Dimeric structures have been found in various alkali metal diorganoamides containing chelating amines as supporting ligands. Typical examples include [LiNMePh(tmeda)]$_2$, **76**, [LiNPh$_2$(tmeda)]$_2$, **77**, [NaNMe$_2$(pmdta)]$_2$, **92**, [NaNEt$_2$(tmeda)]$_2$, **93**, [NaNPh$_2$(tmeda)]$_2$, **94**, [KNPr^i$_2$(tmeda)]$_2$, **102** [92], and [KNPh$_2$(tmeda)]$_2$, **96** [69, 77], and some adducts of LiN(SiMe$_3$)$_2$ with diethyl ether, THF, and certain esters (e.g. [Li{N(SiMe$_3$)$_2$}{Bu^{t}C(O)OBu^t}]$_2$, **84b**) [81, 82].

A remarkable structural detail found in [KNPr^i$_2$(tmeda)]$_2$, **102**, should be mentioned [92b]. In the crystal the compound forms dimers containing a central four-membered K$_2$N$_2$ ring in which the potassium ions are primarily coordinated by four N atoms. The overall coordination number around potassium is, however, higher than 4, because the structure also includes several secondary intramolecular

Table 6.2. Self-assembly of alkali metal amides and related compounds.

Compound number	Name and reference
Dimers	
76	[LiNMePh(tmeda)]$_2$ [75]
77	[LiNPh$_2$(tmeda)]$_2$ [69]
78	[(LiNPh$_2$)$_3$LiCl(tmeda)$_3$] [76]
79a,b	[LiNHC$_6$R$_5$(thf)$_2$]$_2$ (R = H, F) [77]
80	[LiNH{C$_6$H$_2$Bu$^t_3$-2,4,6}(Et$_2$O)]$_2$ [78]
81	[Li(carbazoly)(thf)$_2$]$_2$ [79]
82	[Li{N(CH$_2$CH$_2$OCH$_2$CH$_2$)$_2$NH}]$_2$ [80]
83	[Li{N(SiMe$_3$)$_2$}(thf)]$_2$ [81]
84a,b	[Li{N(SiMe$_3$)$_2$}{RC(O)OBu^t}]$_2$ (R = Pr^i, Bu^t)[82]
85a,b	[Li{Me$_3$SiNS(Bu^t)NR}]$_2$ (R = SiMe$_3$, Bu^t) [83]
86	[Li{MeC$_6$H$_4$C(NSiMe$_3$)$_2$}(thf)]$_2$ [84]
87	[Li{MeC$_6$H$_4$C(NSiMe$_3$)$_2$}(MeC$_6$H$_4$C≡N)]$_2$ [85]
88	[Li{PhC(NPr^i)$_2$}(thf)]$_2$ [86]
89	[Li$_2${C(NBu^t)$_3$}]$_2$ [87]
90a–c	[RSi{N(Li)SiMe$_3$}$_3$]$_2$ (R = Me, Bu^t, Ph) [88]
91	[Li$_2${Me$_2$SiN(H)Me$_2$SiN}$_2$(thf)$_2$]$_2$ [89]
92	[NaNMe$_2$(pmdta)]$_2$ [69]
93	[NaNEt$_2$(tmeda)]$_2$ [69]
94	[NaNPh$_2$(tmeda)]$_2$ [69]
95	[Na{NPh(2-pyridyl)}(pmdta)]$_2$ [90]
96	[KNPh$_2$(tmeda)]$_2$ [69]
97	[Na(indolyl)(tmeda)]$_2$ [91]
98	[Na(indolyl)(pmdta)]$_2$ [76]
99a,b	[M{Me$_3$SiNS(Ph)NBu^t}(thf)]$_2$ (M = Na, K) [91]
100	[Na$_2${Me$_2$SiN(H)Me$_2$SiN}$_2$(thf)$_3$]$_2$ [89]
101	[K$_2${Me$_2$SiN(H)Me$_2$SiN}$_2$(thf)$_4$]$_2$ [89]
102a,b	[MNPr$^i_2$(tmeda)]$_2$ (M = Na, K) [92]
103	[KNPh$_2$(tmeda)]$_2$ [69]
104	[K(NSiFBu$^t_2$)(thf)$_2$]$_2$ [93]
105	[K{NHC$_6$H$_2$(CF$_3$)$_3$-2,4,6}(thf)$_3$]$_2$ [77]
106	[K{Me$_2$Si(OBu^t)(NSiMe$_3$)}(bipy)]$_2$ [94]
107	[K(carbazolyl)(pmdta)]$_2$ [95]
108a–b	[M{Me$_3$SiNS(Ph)NSiMe$_3$}(thf)]$_2$ (M = Na, K, Rb, Cs) [83]
109	[Cs(carbazolyl)(pmdta)]$_2$ [95]
110	[Li{P(SiMe$_3$)$_2$}(thf)$_2$]$_2$ [96]
111	[LiP{CH(SiMe$_3$)$_2$}$_2$]$_2$ [97]
112a,b	[Bu$^t_2$Si(F)P(Mes)M(thf)$_2$]$_2$ (M = Na, K) [98]
113	[(2,4,6-Pr$^i_3$C$_6$H$_2$)$_2$Si(F)P{SiMe$_2$(CMe$_2$Pr^i)}Na]$_2$ [98]
114	[Li{As(Bu^t)As(Bu^t)$_2$}(thf)]$_2$ [99]
115	[LiAsPh$_2$(Et$_2$O)$_2$]$_2$ [100]
Trimers	
116	[Li{N(SiMe$_3$)$_2$}]$_3$ [101]
117	[Na{N(SiMe$_3$)$_2$}]$_3$ [98, 101]
118	[Li{As(CH$_2$SiMe$_3$)$_2$}$_2$]$_3$ [102]

Table 6.2. (cont.)

Compound number	Name and reference
Tetramers	
119	[{Li(pyrrolidyl)}$_2$(tmeda)]$_2$ [103]
120	[NaN=CBu$^t_2$]$_4$(HN=CBu$^t_2$)$_2$ [104]
121	(CsNHSiMe$_3$)$_4$ [105]
122	[Li$_2$(μ-PBu$^t_2$)(μ_3-PBu$^t_2$)(thf)]$_2$ [106]
123	[RbPH(C$_6$H$_3$Mes$_2$-2,6)]$_4$ [107]
Higher Oligomers	
124	(Ph$_2$NLi).[Ph(C$_6$H$_4$Li)NLi]$_2$·(Bu^{n}Li)$_2$·(Et$_2$O)$_4$ [108]
125	[{Li(pyrrolidyl)}$_3$(pmdta)]$_2$ [103]
126	[LiNHBu^t]$_8$ [109]
127	[(Li$_2$NC$_{10}$H$_7$)$_{10}$(Et$_2$O)$_6$]·(Et$_2$O) [110]
128	[Na$_4$(NMe$_2$)$_4$(HNMe$_2$)(tmeda)]$_2$ [111]
129	[Na$_{10}$(NMe$_2$)$_{10}$(tmeda)$_4$] [111]
130	[Na$_{12}$(NMe$_2$)$_{12}$(tmeda)$_4$] [111]
131	[Na$_{12}$(NMe$_2$)$_{10}$(*p*-xylyl)$_2$(tmeda)$_4$] [111]
132	[Li{P(SiMe$_3$)$_2$}]$_6$ [112]
Polymers	
133	[Li(tetrazolyl)(dmso)]$_n$ [113]
134	[Li(benzotriazolyl)(dmso)]$_n$ [113]
135	[Li(benzimidazolyl)(dmso)$_2$]$_n$ [113]
136	[Li(mercaptopyrimidyl)(hmpa)]$_n$ [114]
137	[Li(NHCH$_2$CH$_2$NH$_2$)]$_n$ [115]
138	[(thf)$_2$Li$_3$(μ_4-N$_3$){(NBu^t)$_3$S}]$_n$ [116]
139	[{LiNH(CH$_2$Ph)}$_2$(PhCH$_2$NH$_2$)]$_n$ [117]
140	[NaN(SiHMe$_2$)$_2$]$_n$ [118]
141	[K$_2$(NPh$_2$)$_2$(tmeda)$_3$]$_n$ [69]
142a,b	[M$_2${Me$_2$SiN(H)Me$_2$SiN}$_2$(thf)$_3$]$_n$ (M = Rb, Cs) [89]
143	[LiPCy$_2$(thf)]$_n$ [119]
144	[LiPPh$_2$(Et$_2$O)]$_n$ [119]
145	[LiPPh$_2$(thf)$_2$]$_n$ [119]
146	[Li{PH(C$_6$H$_2$Me$_3$-2,4,6)(thf)$_2$}]$_n$ [120]
147	[KPH(C$_6$H$_2$Bu$^t_3$-2,4,6)]$_n$ [121]
148	[CsPH(C$_6$H$_3$Mes$_2$-2,6)]$_n$ [107]

K···H(C) interactions involving the methyl groups of the diisopropylamido ligand. This type of interaction has been compared with those found in the alkali metal hypersilanides [MSi(SiMe$_3$)$_3$]$_n$, **73–75** [66].

Systems involving Ph$_2$NLi can be quite complicated, especially when halide ions or excess alkyllithium are present. For example, a large aggregate involving Ph$_2$NLi, Bu^{n}Li, and dilithiated Ph$_2$NH has been reported, **124** [108]. An *in situ* preparation from Ph$_2$NH, [Ph$_2$NH$_2$]Cl, and *n*-butyllithium in the presence of tmeda afforded a 3:1 mixed aggregate of Ph$_2$NLi with lithium chloride. The compound

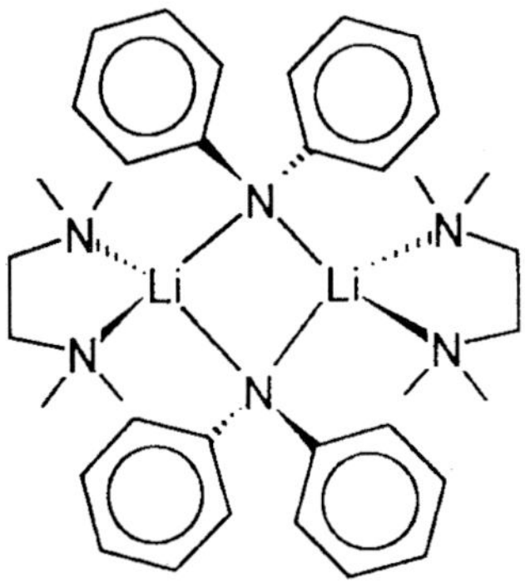

77

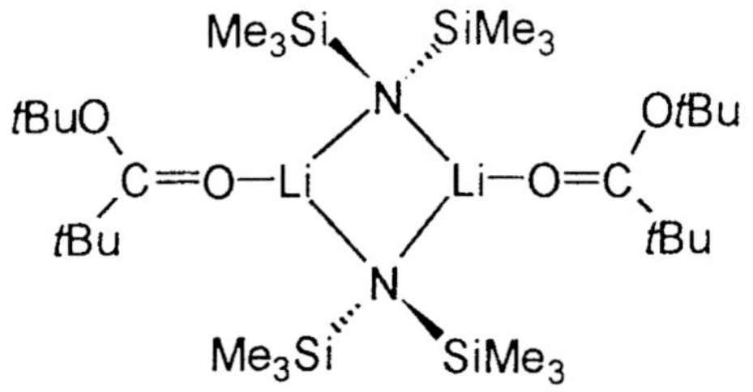

84b

[(Ph$_2$NLi)$_3$LiCl(tmeda)$_3$], **78**, can be viewed as an adduct of a homodimeric lithium diphenylamide and a 1:1 mixed dimer. Notable is the rare Y-shaped, nearly planar (sum of angles 356°) conformation around the chlorine atom. Interestingly, the molecule can be viewed as a crystallographic snapshot of the process of amidolithium ladder structure fragmentation by solvation [76].

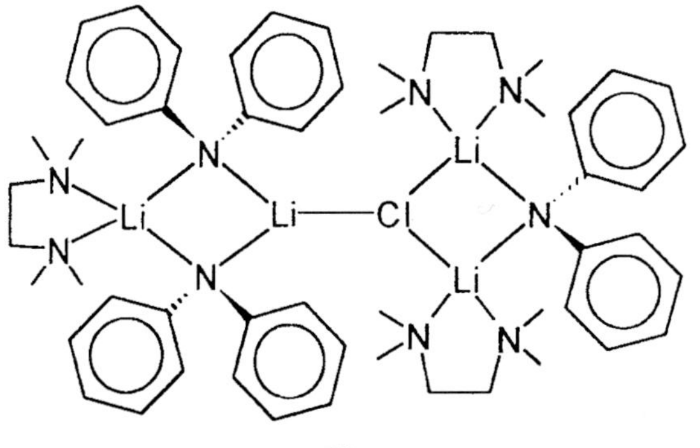

78

Self-organization as a result of the formation of amide-bridged dimers is not limited to *di*organoamides of the alkali metals. Occasionally alkali metal derivatives of primary amines have been shown to be dimeric in the solid state. Structurally characterized examples are [LiNHPh(thf)$_2$]$_2$, **79a** [77a], [LiNHC$_6$F$_5$(thf)$_2$]$_2$, **79b**, and [K{NHC$_6$H$_2$(CF$_3$)$_3$-2,4,6}(thf)$_3$]$_2$, **105** [77b]. These three examples nicely illustrate how the composition of the solvates depends on the size of the alkali metal ion. In **79a** and **79b** two additional THF ligands are sufficient to saturate the coordination sphere of lithium (coordination number 4), whereas the potassium ions in **105** are coordinated by three THF ligands and by two weak K···F interactions.

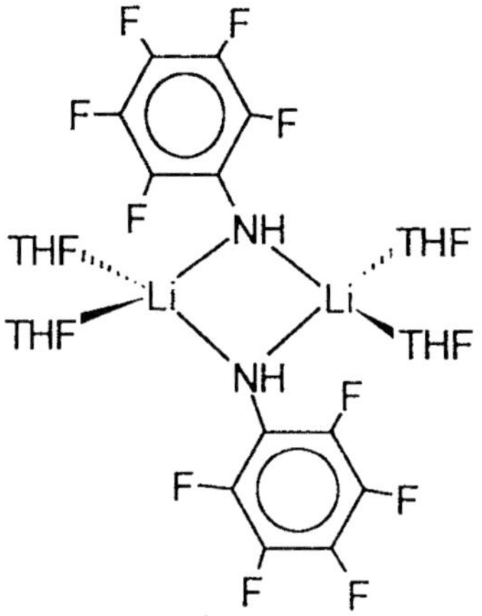

79b

Other types of dimer are formed when the ligands contain additional donor sites and can act as chelating and/or bridging ligands towards the alkali metals. Quite often this results in ladder [71] or step structures in which three four-membered rings are connected via shared edges. Suitable anions include, among others, benzamidinates, diiminosulfinates, and alkoxysilylamides, and typical examples of structurally characterized compounds of this type are [Li{MeC$_6$H$_4$C(NSiMe$_3$)$_2$}(L)]$_2$ (L = THF, **86**, MeC$_6$H$_4$CN, **87**), [Li{Me$_3$SiNS(Bu^t)NR}]$_2$ (R = SiMe$_3$, Bu^t), **85a,b**, [M{Me$_3$SiNS(Ph)NBu^t}(thf)]$_2$ (M = Na, K), **99a,b**, [M{Me$_3$SiNS(Ph)NSiMe$_3$}-(thf)]$_2$ (M = Rb, Cs), **108a,b**, and [K{Me$_2$Si(OBu^t)(NSiMe$_3$)}(bipy)]$_2$, **106** [83–85, 94].

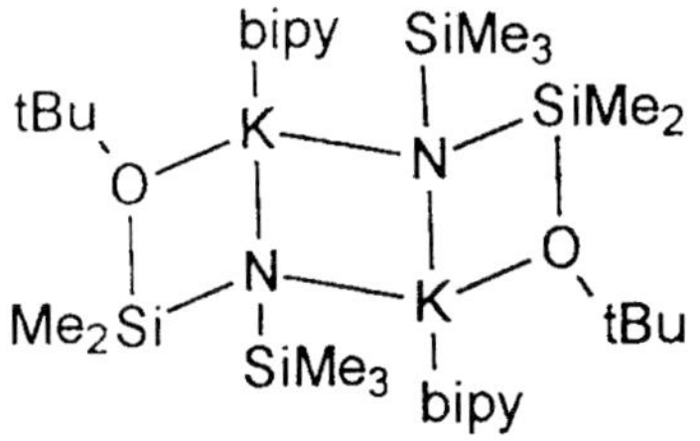

106

Especially noteworthy are the heavier alkali metal diiminosulfinates [M{Me$_3$SiNS(Ph)NSiMe$_3$}(thf)]$_2$, **108a–d**; M = Na, K, Rb, Cs) [83]. All four compounds are isostructural and form ladder-type dimers with a central four-membered M$_2$N$_2$ ring (Figure 6.6). The metal ions are tetracoordinated by three nitrogen atoms and one THF ligand. The most remarkable structural feature of these derivatives is an additional π-interaction of each metal with a phenyl ring of

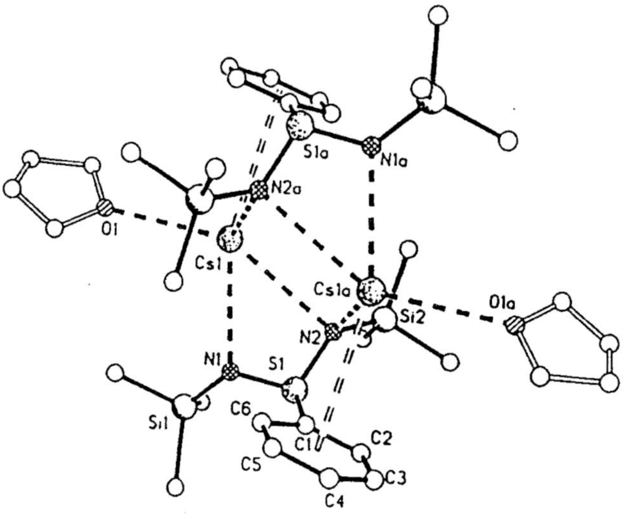

Fig. 6.6. The molecular structure of [Cs{Me$_3$SiNS(Ph)-NSiMe$_3$}(thf)]$_2$, **108d** (Reproduced from ref. [83]).

one of the Me$_3$SiNS(Ph)NSiMe$_3$ ligands. These weak metal-ring interactions have been attributed to a combination of steric and electronic effects.

When the substituents are very bulky unsolvated dimers can be formed. Interesting examples are the lithium salts of diiminosulfinate anions bearing *t*-butyl and trimethylsilyl substituents. In neither compound is additional Lewis base coordinated to lithium, although the compounds are prepared in ethereal solvents. The structures of the resulting dimers depend, moreover, on delicate changes in the electronic environment. Trimethylsilyl substitution at both N atoms leads to the usual step-like structure whereas exchange of one SiMe$_3$ group for *t*-butyl results in the formation of an eight-membered Li$_2$S$_2$N$_4$ ring system [83].

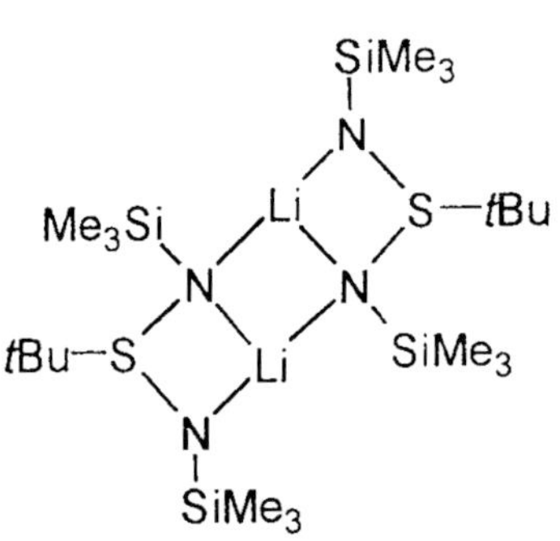

85a 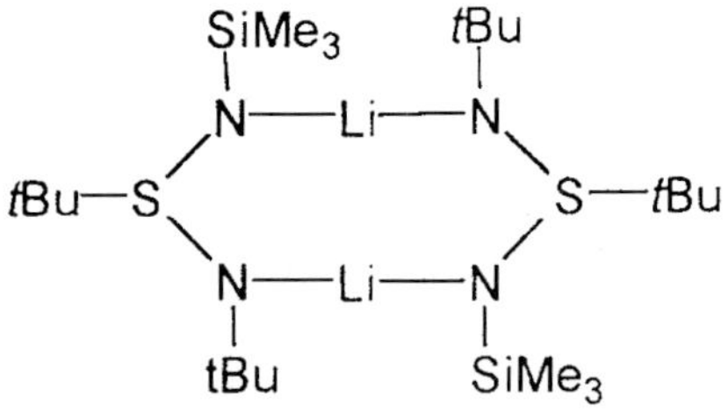   **85b**

Another special example of self-assembly to give a dimer is the lithiation product of a diaza crown ether, **82**. In the product the two lithium ions are bridged by the amide units. Internal coordination by the remaining donor sites results in penta-coordination around Li [80].

82

Interesting cage compounds are obtained by deprotonation of the tris-(organoamino)silanes RSi(NHSiMe$_3$)$_3$ (R = Me, Bu^t, Ph), **151a–c**, with n-butyllithium [88]. The resulting self-assembled products, [RSi{N(Li)SiMe$_3$}$_3$]$_2$, **90a–c**, are dimers of approximate D_{3d} symmetry. The dimers can be described as a trigonal antiprismatic core of six lithium atoms to which two RSi(NHSiMe$_3$)$_3$ fragments are attached. Each lithium and nitrogen atom forms three Li–N bonds, so that the bonding in the Li$_6$N$_6$ cluster is electron-deficient.

An unsolvated dimeric cage structure has been reported for the lithium salt of the novel triazatrimethylenemethane dianion [C(NBu^t)$_3$]$^{2-}$ [87]. The preparation of [Li$_2$ptf{C(NBu^t)$_3$}]$_2$, **89**, was achieved by a two-step procedure outlined in the Eqs(6) and (7).

$$2Bu^tN=C=NBu^t + 2LiNHBu^t \xrightarrow{THF} [Li\{C(NBu^t)_2(HNBu^t)\}]_2(THF) \quad (6)$$
152

$$[Li\{C(NBu^t)_2(HNBu^t)\}]_2(THF) \xrightarrow[-2Bu^nH]{2Bu^nLi} [Li_2\{C(NBu^t)_3\}]_2 \quad (7)$$
89

The unsolvated dimer **89** can be obtained by recrystallization from pentane. A C$_2$N$_6$Li$_4$ cage forms the core of the molecule. Each cage contains two essentially planar [C(NBu^t)$_3$]$^{2-}$ dianions, linked by four lithium atoms. In contrast, the related phenyl derivative [Li$_2${C(NPh)$_3$}]$_2$(thf)$_6$, **153**, forms an acyclic ladder-shaped structure similar to that of various lithium benzamidinates or diiminosulfinates [124].

Two structural types can be distinguished for tetrameric alkali metal amides. [{Li(pyrrolidyl)}$_2$(tmeda)]$_2$, **119**, associates in the form of the familiar ladder or step-like structure containing three fused four-membered rings. The same central core is present in the related hexanuclear adduct [{Li(pyrrolidyl)}$_3$(pmdta)]$_2$, **125**

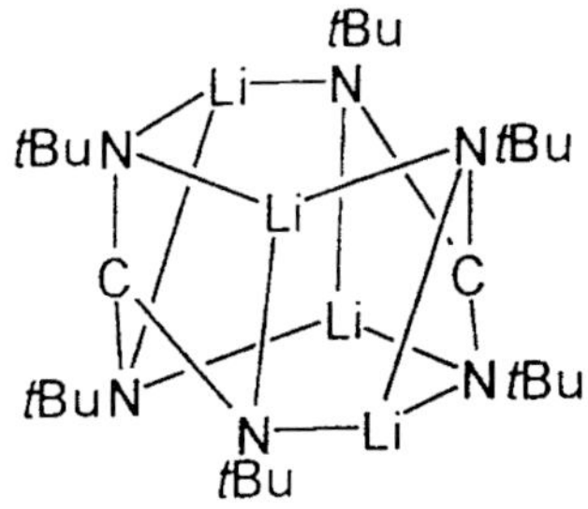

89

[103]. Heterocubane structures have been found in [NaN=CBu$^t_2$]$_4$(HN=CBu$^t_2$)$_2$, **120**, and (CsNHSiMe$_3$)$_4$, **121** [104, 105].

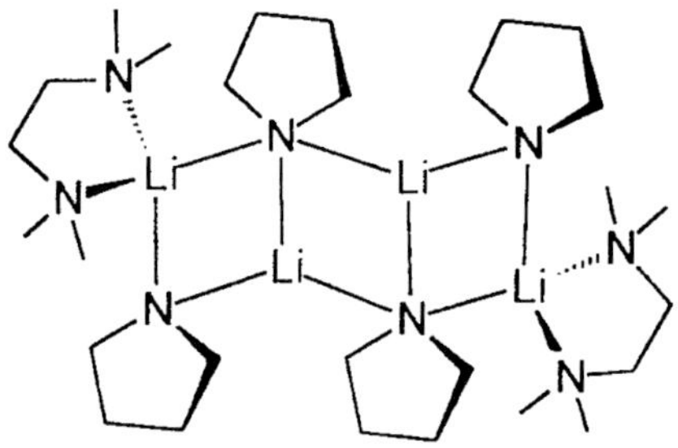

119

Supramolecular self-assembly of lithium amides can occur by a process known as 'ring-laddering' [71] by which ladder structures are generated. An illustrative recent example is octameric [LiNHBu^t]$_8$, **126**, (Figure 6.7) [109].

A fascinating structural variety has been discovered for base adducts of sodium dimethylamide [111]. Several higher oligomers have been isolated from reactions of tmeda with *n*-butylsodium, during which sodium dimethylamide is formed *in situ* (Eqs (8) and (9)):

$$Me_2NCH_2CH_2NMe_2 + Bu^nNa \longrightarrow Me_2NCH_2CH(Na)NMe_2 + C_4H_{10} \quad (8)$$

$$Me_2NCH_2CH(Na)NMe_2 \longrightarrow NaNMe_2 + Me_2NCH=CH_2 \quad (9)$$

The resulting higher oligomers [Na$_4$(NMe$_2$)$_4$(HNMe$_2$)(tmeda)]$_2$, **128**, [Na$_{10}$-(NMe$_2$)$_{10}$(tmeda)$_4$], **129**, and [Na$_{12}$(NMe$_2$)$_{12}$(tmeda)$_4$], **130**, can be isolated as crystalline solids. The decamer, **129**, and dodecamer, **130**, can be regarded as stacks of Na$_2$(NMe$_2$)$_2$ rings. The resulting aggregates are stabilized by terminal tmeda chelating ligands. The mixed-ligand dodecamer [Na$_{12}$(NMe$_2$)$_{10}$(*p*-xylyl)$_2$(tmeda)$_4$],

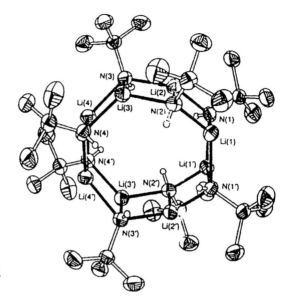

Fig. 6.7. The molecular structure of [LiNHBu^t]$_8$, **126** (Reproduced from ref. [109]).

131, can be isolated when the reaction is performed in the presence of *p*-xylene [111].

Perhaps the most unusual compound among the larger alkali metal amide aggregates is the paramagnetic cluster [(Li$_2$NC$_{10}$H$_7$)$_{10}$(Et$_2$O)$_6$]·(Et$_2$O), **127** [110]. The pale yellow crystalline product is obtained in good yield (68%) by treatment of 1-naphthylamine with two equivalents of *n*-butyllithium in diethyl ether. Its main structural element is a central Li$_4$N$_{10}$ cluster, to which six Li(Et$_2$O)$^+$ units are attached through coordination to nitrogen (Figure 6.8). Each naphthyl N^{2-} unit interacts with four or six Li ions in the central cluster. The origin of the paramagnetism of **127** is unclear, but it has been shown that charge transfer from N to the naphthyl groups is unlikely because no ESR signal attributable to a naphthyl radical anion has been observed.

Polymeric structures are less common in alkali metal amide chemistry, although recently several examples have been reported, including [Li(NHCH$_2$CH$_2$NH$_2$)]$_n$, **137** [115], [{LiNH(CH$_2$Ph)}$_2$(PhCH$_2$NH$_2$)]$_n$, **139** [117], and [NaN(SiHMe$_2$)$_2$]$_n$, **140** [118]. Supramolecular association into polymers seems to become quite important when two nitrogen atoms are involved, e.g. in various heterocycles. Thus polymeric structures have been found in lithiobenzimidazole, lithiomercaptopyrimidine, lithiobenzotriazole, and lithiotetrazole [113, 114]. Generally, the structures of heterocyclic lithioamines are the result of competition between electrostatic and steric effects. Because derivatives with more than one nitrogen atom in the ring form Li–N bonds in the heterocyclic ring plane, additional ligands such as DMSO or HMPA do not interfere with the amide moiety, thus resulting in the electrostatically preferred polymer arrangement – [Li(benzotriazolyl)(dmso)]$_n$, **134**, for example, has a ribbon structure (Figure 6.9), whereas [Li(tetrazolyl)(dmso)]$_n$, **133**, prefers a two-dimensional network (Figure 6.10) [113].

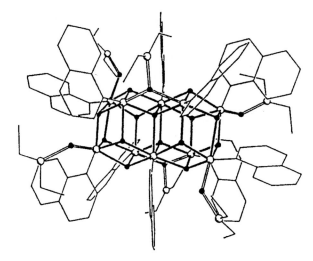

Fig. 6.8. The molecular structure of [(Li$_2$NC$_{10}$H$_7$)$_{10}$(Et$_2$O)$_6$]-(Et$_2$O), **127** (Reproduced from ref. [110]).

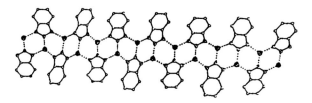

Fig. 6.9. The polymeric structure of [Li(benzotriazolyl)(dmso)]$_n$, **134** (Reproduced from ref. [113]).

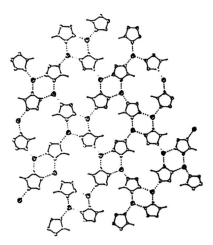

Fig. 6.10. The polymeric structure of [Li(tetrazolyl)(dmso)]$_n$, **133** (Reproduced from ref. [113]).

The structural variety of the homologous series of di-alkali metal octamethylcyclotetrasilazanides is extraordinary [89]. These compounds have been prepared by direct metalation of the parent cyclotetrasilazane with Bu^{n}Li, Na, K, Rb, or Cs in the presence of toluene. The latter reaction involves the formation of a highly reactive benzylcesium intermediate.

$$\left[\begin{array}{c} \text{Me}_2\text{Si}(\text{H})\text{N—SiMe}_2\text{—N—M(THF)}_m \\ \text{M—N—SiMe}_2\text{—N(H)—SiMe}_2 \end{array} \right]_n$$

91: M = Li, m = 2, n = 2
100: M = Na, m = 3, n = 2
101: M = K, m = 4, n = 2
142a: M = Rb, m = 3, n = ∞
142b: M = Cs, m = 3, n = ∞

Whereas the Li, Na, and K derivatives form dimers in the solid state, the rubidium and cesium compounds, **142a** and **142b**, are associated as polymers. In the polymeric chain structure of **142a** cations and anions are arranged in a straight line (Figure 6.11). The eight-membered Si_4N_4 ring approximates a crown conformation. Four- and seven-coordinate rubidium ions alternate in the chain. As an interesting contrast the homologous cesium compound **142b** forms polymeric zigzag chains in the crystalline state (Figure 6.12). Also different from **142a**, the monomeric units are connected via Cs–N bonds only. Both crystal structures are rare examples of bridging THF ligands [89].

Internal coordination of a free amine function was found in the recently described polymer $[Li(NHCH_2CH_2NH_2)]_n$, **137**, the first example of an infinite ladder struc-

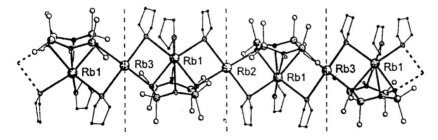

Fig. 6.11. The polymeric structure of $[Rb_2\{Me_2SiN(H)Me_2SiN\}_2(THF)_3]_n$, **142a** (Reproduced from ref. [89]).

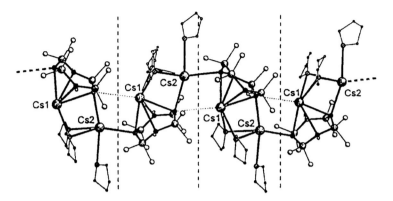

Fig. 6.12. The polymeric structure of $[Cs_2\{Me_2SiN(H)Me_2SiN\}_2(THF)_3]_n$, **142b** (Reproduced from ref. [89]).

ture containing lithium and protic amide moieties. The compound is conveniently prepared by deprotonation of ethylenediamine with lithium nitride (Eq.(10)) [115].

$$Li_3N + 3NH_2CH_2CH_2NH_2 \rightarrow 3Li(NHCH_2CH_2NH_2) + NH_3 \qquad (10)$$

137

The supramolecular structure of **137** consists of infinite one-dimensional ladders of double-edge-sharing LiN_4 tetrahedra with μ-NH_2 and μ_3-NH interactions. No hydrogen-bonding interactions are present between adjacent double chains.

Polymeric $[NaN(SiHMe_2)_2]_n$, **140**, was recently shown to form a related infinite ladder-structure in the solid state [118], whereas $[\{LiNH(CH_2Ph)\}_2(PhCH_2NH_2)]_n$, **139**, exists in the crystal as an infinite, twisted-ladder structure of fused Li_2N_2 rings [117]. The latter compound was made by treatment of benzylamine with Bu^nLi in hexane and addition of a second equivalent of benzylamine to redissolve the initial precipitate. In the Li_2N_2 units which establish the polymer chain only one Li center is solvated by a benzylamine ligand. As a consequence, one ladder edge comprises solvated, four-coordinate, distorted tetrahedral Li ions whereas the other comprises unsolvated, three-coordinate, pyramidal lithium atoms. The overall ladder framework is severely twisted.

6.1.2.2 Phosphides and arsenides

The dimeric lithium diorganophosphides $[LiP(CH_2SiMe_3)_2]_2$, **110** [96], and $[LiP\{CH(SiMe_3)_2\}_2]_2$, **111** [97], have been reported and structurally characterized. The latter compound was the first uncomplexed phosphidolithium to be isolated. It was prepared by the reaction of metallic lithium with $PCl[CH(SiMe_3)_2]_2$ and isolated as a yellow crystalline solid. Its central Li_2N_2 ring is essentially planar.

One of the first structurally characterized lithium organoarsenides, $[Li\{As(Bu^t)As(Bu^t)_2\}(THF)]_2$, **114**, was obtained by reaction of $LiAs(Bu^t)_2$ with $MgBr_2$ in THF solution [99]. $[LiAsPh_2(Et_2O)_2]_2$, **115**, can be prepared by metal-

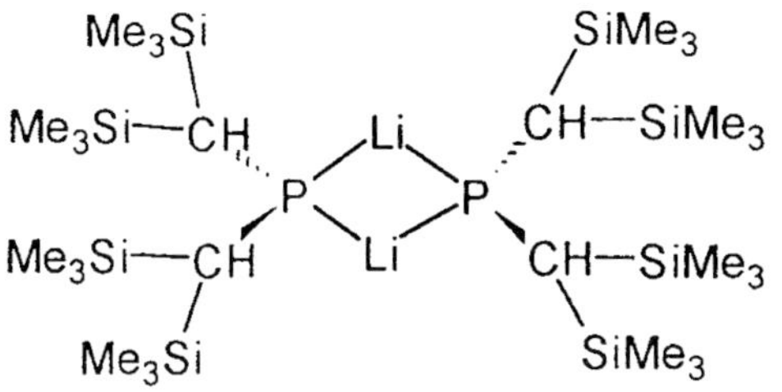

111

ation of the corresponding secondary arsine [100]. Similar to **111** these molecules contain a planar Li_2As_2 core [99, 100].

The formation of dimers has also been established for the alkali metal fluorosilylphosphides $[Bu^t{}_2Si(F)P(Mes)M(THF)_2]_2$ (M = Na, K), **112a,b**, and $[(2,4,6-Pr^i{}_3C_6H_2)_2Si(F)P\{SiMe_2(CMe_2Pr^i)\}Na]_2$, **113** [98]. Such compounds are of interest as precursors for the preparation of phosphasilenes containing an Si=P double bond. Formally the alkali metal fluorosilylphosphides can be regarded as adducts of MF (M = Na, K) and the phosphasilenes. The mesityl-substituted derivatives are dimers with eight-membered $F_2M_2P_2Si_2$ rings whereas a ladder-type tricyclic $F_2Na_2P_2Si_2$ skeleton has been found in the sterically crowded, unsolvated species $[(2,4,6-Pr^i{}_3C_6H_2)_2Si(F)P\{SiMe_2(CMe_2Pr^i)\}Na]_2$, **113**. In this compound the sodium atoms are coordinated to phosphorus and fluorine only.

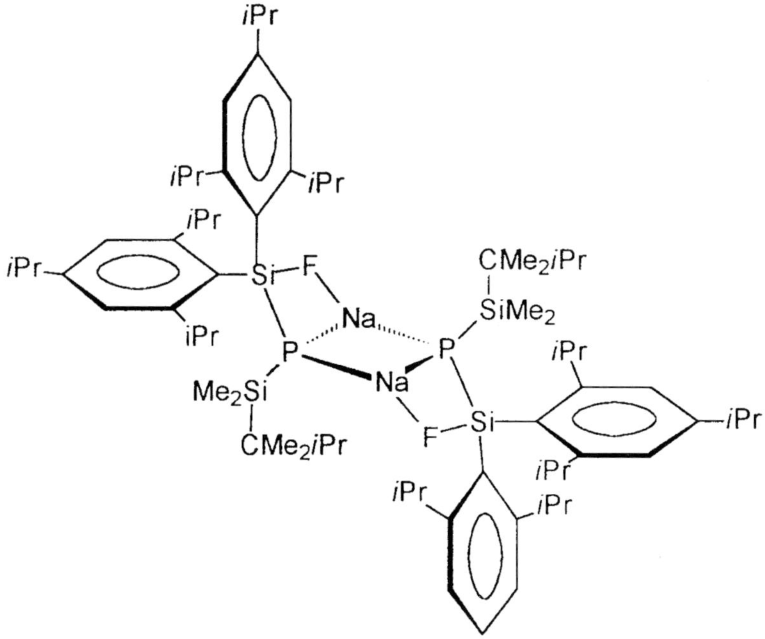

113

The only trimer among the heavier congeners of the lithium amides is [Li{As(CH$_2$SiMe$_3$)$_2$}$_2$]$_3$, **118**, accessible by reaction of metallic Li with ClAs[CH(SiMe$_3$)$_2$]$_2$ [102]. The central Li$_3$As$_3$ ring is not planar but has a boat conformation with average Li–As bond lengths of 2.60(4) Å.

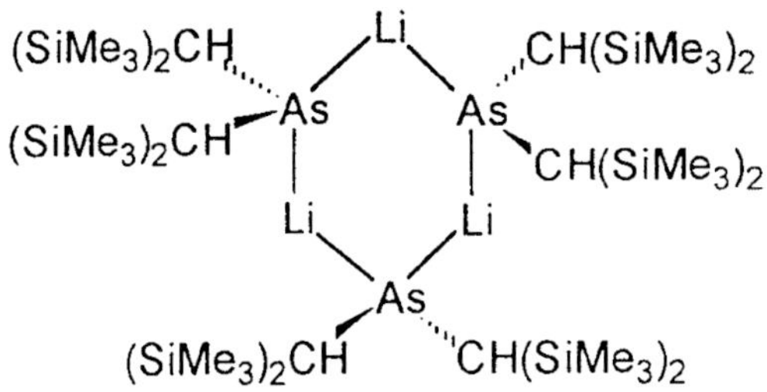

118

The use of secondary phosphines bearing bulky substituents such as *t*-butyl or trimethylsilyl leads to interesting self-assembled oligomers. A tetrameric structure has been reported for [Li$_2$(μ-Bu$^t_2$)(μ$_3$-PBu$^t_2$)(thf)]$_2$, **122**, which contains a 'Z' core of four Li ions [106]. Two of the phosphide units are triply bridging, whereas the other two bridge only two lithium atoms. A THF ligand is coordinated to the end Li ions, and the overall Li$_4$P$_4$ framework is essentially planar. Lithium bis(trimethylsilyl)phosphide, **132**, prepared by metalation of HP(SiMe$_3$)$_2$ with *n*BuLi, can be crystallized solvent-free from toluene [112]. The unusual molecular structure consists of a hexameric arrangement of four five-coordinate and two four-coordinate phosphorus atoms interconnected by four three-coordinate and two dicoordinate lithium ions. The entire Li$_6$P$_6$ framework is coplanar, resulting in a

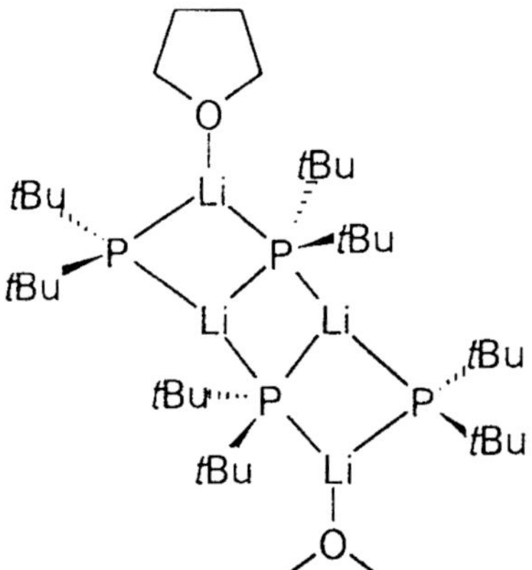

122

'ladder with six Li–P steps'. The compound is a nice example of how the 'laddering principle' in lithium amide chemistry [71] also seems to be applicable to lithium organophosphides.

More recently several alkali metal derivatives of a primary phosphine containing the very bulky 2,6-dimesitylphenyl substituent have been prepared and structurally characterized [107, 121]. The yellow rubidium derivative $[RbPH(C_6H_3Mes_2-2,6)]_4$, **123**, crystallizes solvent-free from toluene [107]. Its crystal structure consists of heterocubane aggregates. In the central Rb_4P_4 core each Rb ion is surrounded by three phosphide ligands. In addition, each rubidium interacts with the π-systems of two mesityl rings from different 2,6-dimesitylphenyl groups.

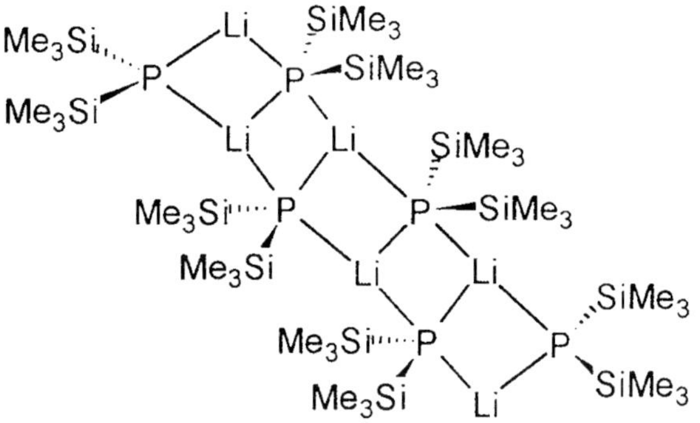

123

The complexes $[LiPCy_2(thf)]_n$, **143**, $[LiPPh_2(Et_2O)]_n$, **144**, and $[LiPPh_2(thf)_2]_n$, **145**, have been synthesized by lithiation of diphenylphosphine and shown to be associated as chain polymers in the solid state [119]. In each case the crystal structure consists of infinite chains of alternating solvated lithium ions and PR_2 units in which the phosphorus has a distorted tetrahedral coordination geometry and Li is either three- (**143**, **144**) or four-coordinate (**145**). A two-dimensional network involving additional π-coordination of mesityl groups to cesium was recently reported for the cesium monoorganophosphide $[CsPH(C_6H_3Mes_2-2,6)]_n$, **148** [107].

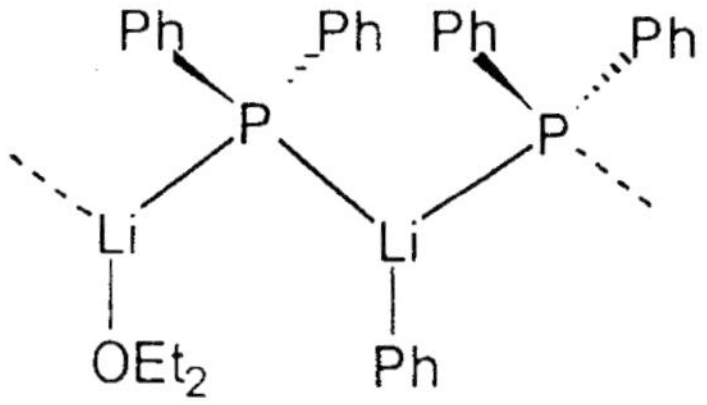

144 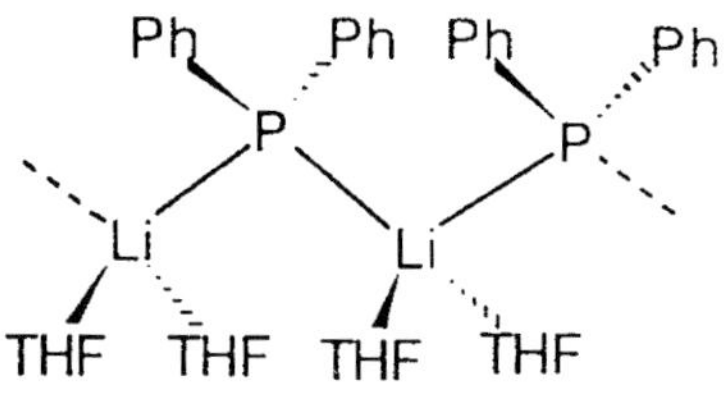   **145**

6.1.3 Alkoxides and their higher homologs

The chemistry and structures of alkali metal alkoxides [125], enolates [126], and phenoxides [127] has been compiled in several comprehensive review articles. Simple alkali metal alkoxides such as methoxides, ethoxides, or isopropoxides, and simple thiolates (LiSMe, NaSMe, KSMe, etc.) crystallize in layered structures. Interesting supramolecular self-assembly is found for other base-free and solvated alkali metal alkoxides, phenoxides, and their sulfur analogs. Typical examples are summarized in Table 6.3.

As in the chemistry of alkali metal amides and related compounds, dimerization is also a common type of self-assembly in alkoxide and thiolate chemistry. The simplest type of dimer contains a central four-membered M_2O_2 ring in which the alkoxide or chalcogenolate ligands act as bridges and the coordination spheres of the metal ions are saturated by an appropriate number of ancillary ligands such as ethers or amines. [Na{$OC_6H_2(CF_3)_3$-2,4,6}(thf)$_2$]$_2$, **162**, and [K{$OC_6H_2(CF_3)_3$-2,4,6}(thf)$_3$]$_2$, **167**, are two interesting examples in which the influence of the size of the metal ion is evident [135]. Whereas sodium derivative **162** contains four supporting THF ligands, the potassium phenoxide **167** has essentially the same structure but with two additional THF ligands in an unusual bridging coordination mode.

162   **167**

The same structural type of dimer involving a central four-membered ring core has been established for bulky thiolate ligands such as the 2,4,6-tris(isopropyl)-thiophenolate anion [138].

6.1 Alkali Metals

Table 6.3. Supramolecular self-assembly of alkali metal alkoxides and related compounds.

Compound number	Name and reference
Dimers	
154	[Li(p-OC$_6$H$_4$Br)(tmeda)]$_2$ [128]
155	[Li(OC$_6$H$_3$Bu$^t_2$-2,6)(Et$_2$O)]$_2$ [129]
156	[Ph$_2$CO{Li(tmeda)}{Li(thf)}]$_2$ [130]
157	[Ph$_4$Si$_2$O(OLiPy)$_2$]$_2$ [131]
158	[Li(μ-OCBu$^t_2$CH$_2$PMe$_2$)]$_2$ [132]
159	[Li{PhCH$_2$C(Ph)SO$_2$CF$_3$}(thf)]$_2$ [133]
160	[Li{SC$_6$H$_4$(CH$_2$N(Me)CH$_2$CH$_2$OMe)-2}]$_2$ [134]
161	[NaOPh(pmdta)]$_2$ [128]
162	[Na{OC$_6$H$_2$(CF$_3$)$_3$-2,4,6}(thf)$_2$]$_2$ [135]
163	[Na{OC$_6$H$_2$(CH$_2$NMe$_2$)$_2$-2,4-Me-4}{HOC$_6$H$_2$(CH$_2$NMe$_2$)$_2$-2,4-Me-4}]$_2$ [136]
164	[NaOC$_6$H$_4$N(C$_6$H$_4$NHCH$_2$CH$_2$OMe)(CH$_2$CH$_2$OMe)]$_2$ [137]
165	[Na(μ-OCBu$^t_2$CH$_2$PMe$_2$)]$_2$ [132]
166	[Na(SC$_6$H$_2$Pr$^i_3$-2,4,6)(pmdta)]$_2$ [138]
167	[K{OC$_6$H$_2$(CF$_3$)$_3$-2,4,6}(thf)$_3$]$_2$ [135]
Trimers	
168	[Li{OC$_6$H$_2$(CH$_2$NMe$_2$)$_2$-2,6-Me-4}]$_3$ [139]
169	[Li(SC$_6$H$_2$Bu$^t_3$-2,4,6)(thf)]$_3$ [140]
170	[Li(SeC$_6$H$_2$Bu$^t_3$-2,4,6)(thf)]$_3$ [140]
Tetramers	
171	[Li(N-methylpseudoephedrate)]$_4$ [141]
172	[Na{OC$_6$H$_2$(CH$_2$NMe$_2$)$_2$-2,6-Me-4}]$_4$ [136]
173a–c	(MOBu^t)$_4$ (M = K, Rb, Cs) [142]
174	[NaOCMe(CF$_3$)$_2$]$_4$ [143]
175a–c	(MOSiMe$_3$)$_4$ (M = K, Rb, Cs) [144]
176	[(C$_6$H$_6$)KOSiMe$_2$Ph]$_4$ [145]
Hexamers	
177	(LiOBu^t)$_6$ [146]
178	(LiOCMe$_2$Ph)$_6$ [147]
179a,b	(MOSiMe$_3$)$_6$ (M = Li, Na) [146]
180	[LiOSi(naphthyl)Me$_2$]$_6$ [148]
181	(LiOSiMe$_2$PEt$_2$)$_6$ [149]
182	[LiOCH(Pr^i)(2-C$_4$H$_3$S)]$_6$ [150]
183	[LiOSiMe$_2$(2-C$_4$H$_3$S)]$_6$ [150]
184	[Li{SC$_6$H$_4$((R)-CH(Me)NMe$_2$)-2}]$_6$ [134]
185	[Li{SC$_6$H$_3$(CH$_2$NMe$_2$)$_2$-2,6}]$_6$ [134]
186	(NaOBu^t)$_6$/(NaOBu^t)$_9$ [151]
187	[NaOC$_6$H$_4$(CH$_2$NMe$_2$)-2]$_6$ [136]
188	[Na$_6$(SC$_6$H$_2$Pr$^i_3$-2,4,6)$_6$(Et$_2$O)$_4$] [138]
189	[K(SC$_6$H$_2$Pr$^i_3$-2,4,6)(thf)]$_6$ [138]
190	[K$_6$(SC$_6$H$_2$Pr$^i_3$-2,4,6)$_6$(tmeda)$_2$(thf)$_2$] [138]
191	[(KSCPh$_3$)$_6$(toluene)$_2$] [152]
192	[(KSCPh$_3$)$_6$(hmpa)$_2$]·2 toluene [152]
193	[(NaSSiPh$_3$)$_6$(toluene)$_2$] [152]

Table 6.3. (cont.)

Compound number	Name and reference
Polymers	
194a,b	(MOPh)$_n$ (M = Li, Na) [153]
195a–c	(MOPh)$_n$ (M = K, Rb, Cs) [153]
196	[Na(SC$_6$H$_2$Pr$^i_3$-2,4,6)(thf)]$_n$ [138]
197	[Na(SC$_6$H$_2$Pr$^i_3$-2,4,6)(tmeda)]$_n$ [138]
198	Na{SC$_6$H$_2$(CF$_3$)$_3$-2,4,6}(thf)$_2$]$_n$ [135]
199	[K(SC$_6$H$_2$Pr$^i_3$-2,4,6)(thf)]$_n$ [138]
200	[K(SC$_6$H$_2$Pr$^i_3$-2,4,6)(pmdta)]$_n$ [138]
201	[K{SC$_6$H$_2$(CF$_3$)$_3$-2,4,6}(thf)]$_n$ [135]

A modified type of dimeric self-assembly is observed when the organic substituents contain 'built-in' donor functions such as pendant methoxyethyl, dimethylaminomethyl, or dimethylphosphinomethyl groups. Suitable ligands have been employed in alkoxide and thiolate chemistry. Structurally characterized examples include [Li(μ-OCBu$^t_2$CH$_2$PMe$_2$)]$_2$, **138** [132], [Li{SC$_6$H$_4$(CH$_2$N(Me)CH$_2$CH$_2$OMe)-2}]$_2$, **160** [134], [Na{OC$_6$H$_2$(CH$_2$NMe$_2$)$_2$-2,4-Me-4}{HOC$_6$H$_2$-(CH$_2$NMe$_2$)$_2$-2,4-Me-4}]$_2$, **163** [136], [NaOC$_6$H$_4$N-(C$_6$H$_4$NHCH$_2$CH$_2$OMe)(CH$_2$CH$_2$OMe)]$_2$, **164** [137], and [Na(μ-OCBu$^t_2$CH$_2$PMe$_2$)]$_2$, **165** [132].

More complicated cases of dimeric self-assembly worth mentioning are [Ph$_4$Si$_2$O(OLiPy)$_2$]$_2$, **157** [131], and [Li{PhCH$_2$C(Ph)SO$_2$CF$_3$}(THF)]$_2$, **159** [133].

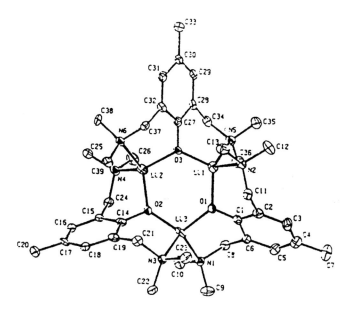

Fig. 6.13. The molecular structure of [Li{OC$_6$H$_2$(CH$_2$NMe$_2$)$_2$-2,6-Me-4}]$_3$, **168** (Reproduced from ref. [139]).

The first compound is a pyridine solvate of dilithiated tetraphenyldisiloxanediol. The dimeric structure is based on a twelve-membered Li$_2$Si$_4$O$_6$ ring system. To some extent the pentacyclic system is related to that in the mineral thortveitite [Sc$_2$Si$_2$O$_7$] [154], which can be viewed as an extended series of interconnecting overlapping twelve-membered Sc$_2$Si$_4$O$_6$ rings with Si$_2$O$_2$ bridges. The most notable structural feature of **157** involves three four-membered Li$_2$O$_2$ rings, which are connected in a folded ladder arrangement. The second dimer, [Li{PhCH$_2$C(Ph)-SO$_2$CF$_3$}(thf)]$_2$, **159** [133], is the THF adduct of a lithiated sulfone, which formally contains a carbanion moiety. Self-association, however, occurs in such a way that the solvated lithium ions are coordinated only via the oxygen atoms of the sulfone unit.

Supramolecular self-assembly of the esthetically appealing trimer [Li{OC$_6$H$_2$(CH$_2$NMe$_2$)$_2$-2,6-Me-4}]$_3$, **168** [139], is assisted by the dangling dimethylaminomethyl functions in the organic ligand. In this molecule adjacent Li atoms are N-coordinated by different CH$_2$NMe$_2$ substituents of the O-bridging phenolate ligands to give a coordination number of 4 around Li (Figure 6.13). The orientation of the phenolate ligands relative to the Li$_3$O$_3$ ring results in a 'propeller-like' molecule with a screw-type chirality. Phenolate anions containing *ortho* amino functions ('pincer phenolates') have been shown to be particularly useful ligands for main group, transition, and lanthanide metals [155].

Rare examples of trimerized thiophenolates and selenophenolates are the 'supermesityl' derivatives [Li(SC$_6$H$_2$Bu^t$_3$-2,4,6)(thf)]$_3$, **169**, and [Li(SeC$_6$H$_2$Bu^t$_3$-2,4,6)-(thf)]$_3$, **170** [140]. The crystal structure of the sulfur derivative, for example, consists of neutral, well-separated trimeric units. The central part of the molecule is a slightly non-planar six-membered Li_3S_3 ring. The coordination of two sulfur atoms and one THF oxygen atom results in an almost undistorted trigonal planar coordination geometry around lithium.

For tetrameric self-assembly of alkali metal alkoxides and related compounds the formation of heterocubane-type aggregates is clearly favored. Heterocubanes containing M_4O_4 units were first found in the alkali metal *t*-butoxides $(MOBu^t)_4$, **173a–c**, and the related trimethylsilanolates $(MOSiMe_3)_4$ (M = K, Rb, Cs), **175a–c** [142, 144]. Because of the approximately spherical structures with the surface having mainly hydrocarbon character these alkali metal compounds are quite volatile and soluble in hydrocarbons. The tetrameric aggregates are retained in solution and in the gas phase. The high solubility as imparted by the supramolecular structure makes the *t*-butoxides especially valuable reagents for the synthesis of organoalkali metal compounds, and useful building blocks and catalysts in organic synthesis [156]. In a corresponding heterocubane-type dimethylphenylsilanolate, $[(C_6H_6)KOSiMe_2Ph]_4$, **176**, the coordination sphere is completed by π-interaction with a benzene ring [145].

175a

For alkali metal *t*-butoxides and trimethylsilanolates the heterocubane structure is found for the heavier alkali metals potassium, rubidium, and cesium. For sodium a related heterocubane derivative can also be obtained with a suitable phenolate ligand providing additional donor sites. One example is tetrameric [Na{OC_6H_2-$(CH_2NMe_2)_2$-2,6-Me-4}]$_4$, **172** [136], which contains tridentate 'pincer phenolate' ligands and pentacoordinate sodium ions.

Intramolecular chelation is also evident in tetrameric [Li(*N*-methylpseudoephedrate)]$_4$, **171** [141]. This was taken as an evidence in support of a proposal that intramolecular lithium chelation occurs in some asymmetric syntheses and also as evidence of transition structures where attack occurs on an edge of a self-assembled species [157–161]. The highly fluorinated sodium alkoxide [NaOCMe(CF_3)$_2$]$_4$, **174**, has interesting structural features [143]. In the solid state this compound forms a slightly distorted Na_4O_4 heterocubane with Na–O distances ranging from 2.250(4) to 2.342(5) Å and O–Na–O angles between 88.11(8) and 91.54(8)° (Figure 6.14). The coordination number around sodium is further

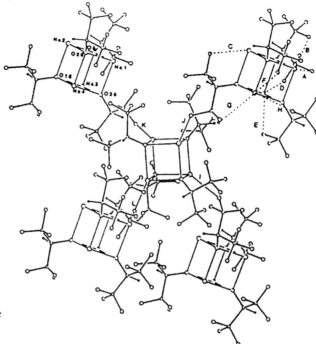

Fig. 6.14. The crystal structure of [NaOCMe(CF$_3$)$_2$]$_4$, **174** (Reproduced from ref. [143]).

increased by the formation of secondary Na···F–C bonds both within cubes and between cubes. The latter connect the heterocubane units into an infinite three-dimensional polymer. The large amount of fluorination of the molecular periphery makes this sodium alkoxide highly volatile (sublimation at 75 °C and 10^{-2} Torr). Chemical vapor deposition (CVD) of **174** on to silica at 285 °C affords NaF via cleavage of C–F bonds.

A prominent type of supramolecular self-association in the chemistry of alkali metal alkoxides and related compounds is the formation of hexameric aggregates. At least four different structural types have been characterized thus far. Several alkali metal derivatives of bulky phenolate and thiophenolate ligands have been found to form box-shaped hexamers which can be described as dimers of trimers or as dimerized finite double ladders. Representative examples are the solvated thiophenolates [Na$_6$(SC$_6$H$_2$Pr$^i_3$-2,4,6)$_6$(Et$_2$O)$_4$], **188** (Figure 6.15), and [K$_6$(SC$_6$H$_2$Pr$^i_3$-2,4,6)$_6$(tmeda)$_2$(thf)$_2$], **190** [138]. In the face-fused cuboidal framework the metal atoms in the central ring are coordinated by four sulfur atoms whereas those in the outer rings are connected with three thiolate ligands. The S–M–S angles in the central M$_2$S$_2$ (M = Na, K) ring are close to 90° or 180°. The coordination sphere of the outer metal atoms is further completed by supporting diethyl ether, THF, or tmeda ligands.

A closely related box-shaped geometry is obtained with phenoxide ligands bearing additional donor functionalities [136]. The sodium phenoxide hexamer

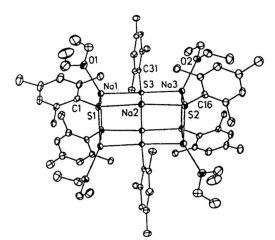

Fig. 6.15. The molecular structure of [Na$_6$(SC$_6$H$_2$Pr$^i_3$-2,4,6)$_6$(Et$_2$O)$_4$], **188** (Reproduced from ref. [138]).

[NaOC$_6$H$_4$(CH$_2$NMe$_2$)-2]$_6$, **187**, contains a unique Na$_6$O$_6$ core forming two face-fused cubes with an inversion point in the center of the fused face, and Na–O–Na and O–Na–O angles of approximately 90° and 180°, respectively. Each of the two phenolate ligands with their oxygen atoms in the inner four-membered Na$_2$O$_2$ ring is O-bridging between four sodium atoms, coordinates with its dimethylamino-methyl substituent to one of these sodium atoms, and has an η^2-arene interaction with another. The remaining amino functions are coordinated to the outer sodium atoms.

The thiolate derivatives [(KSCPh$_3$)$_6$(toluene)$_2$], **191**, [(KSCPh$_3$)$_6$(hmpa)$_2$]· 2toluene, **192**, and [(NaSSiPh$_3$)$_6$(toluene)$_2$], **193**, also contain a box-like M$_6$S$_6$ core [152]. Because of the absence of any coordinating solvent or crown ether molecules, however, these M$_6$S$_6$ aggregates are stabilized by numerous metal-π-interactions with several of the phenyl groups in the periphery of the molecules. Figure 6.16. depicts a typical example.

Formal opening of two opposite metal–chalcogen bonds in the central four-membered ring in the box-shaped hexamers leads to hexagonal prismatic aggregates, which can also be viewed as cyclic ladder structures with three-coordinate metal and four-coordinate chalcogen atoms. Several alkali metal alkoxides and thiolates have been reported to self-assemble in this 'drum-like' form, for example (LiOBu^t)$_6$, **177** [146], (LiOCMe$_2$Ph)$_6$, **178** [147], (MOSiMe$_3$)$_6$ (M = Li, Na), **179a,b** [146], (LiOSiMe$_2$PEt$_2$)$_6$, **181** [149], and [Li{SC$_6$H$_4$((R)CH(Me)NMe$_2$)-2}]$_6$, **184** [134]. The last is an example of intermolecular stabilization as a result of 'built-in' donor functions (Figure 6.17). The ligand used in this complex belongs to a class of tri- or tetradentate arenethiolate anions with potential for intramolecular coordination.

Two thienyl group-containing hexameric lithium alkoxides [LiOCH(Pr^i)-(2-C$_4$H$_3$S)]$_6$, **182**, and [LiOSiMe$_2$(2-C$_4$H$_3$S)]$_6$, **183** [150], have been prepared and structurally characterized to illustrate a phenomenon which the authors call the 'lithio-aversion' of thiophene sulfur atoms. It was found that although lithium–

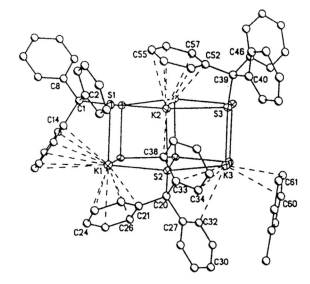

Fig. 6.16. The molecular structure of [(KSCPh$_3$)$_6$(toluene)$_2$], **191** (Reproduced from ref. [152]).

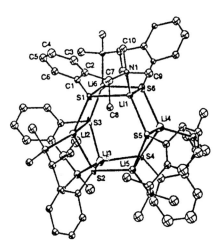

Fig. 6.17. Molecular structure of [Li{SC$_6$H$_4$((R)CH(Me)NMe$_2$)-2}]$_6$, **184** (Reproduced from ref. [134]).

sulfur interactions would result if the thienyl substituents were rotated, no short distances between these atoms are found. Instead, the thienyl conformations are governed by Li–(C=C) π-interactions and the heterocyclic rings apparently try to avoid close proximity between sulfur and lithium.

A unique structural phenomenon was found in the supramolecular crystal structure of sodium *t*-butoxide, **186**. The unit cell of this compound contains both hexameric and nonameric aggregates (Figure 6.18) [151].

Two further structural types of hexameric self-assembly in alkali metal alkoxides are exemplified by the compounds [LiOSi(naphthyl)Me$_2$]$_6$, **180** [148], and [Li{SC$_6$H$_3$(CH$_2$NMe$_2$)$_2$-2,6}]$_6$, **185** [134]. The central part of the former lithium

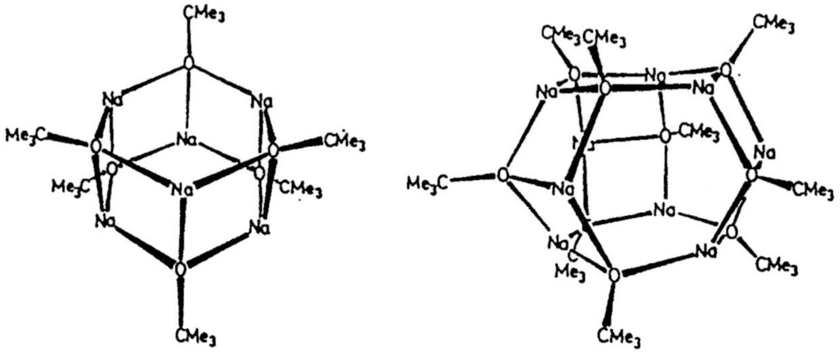

Fig. 6.18. The hexameric and nonameric aggregates of sodium *t*-butoxide, **186** (Reproduced from ref. [151]).

alkoxide consists of a Li_6 ring stabilized by the $OSi(naphthyl)Me_2$ units. Another way of looking at the structure of **180** is that the six lithium atoms form a distorted octahedron in which two opposite facets more broadened than the others are bridged by oxygen. The variety of hexameric structures is further enriched by the beautiful ring structure found in **185** [134]. This unusual supramolecular arrangement has been achieved with the use of the 'pincer arenethiolate' ligand 2,6-bis(dimethylaminomethyl)thiophenolate. This ligand containing two *ortho*-dimethylaminomethyl functions is readily prepared by insertion of elemental sulfur into the Li–C bond of the corresponding aryllithium derivative, according to Eq.(11):

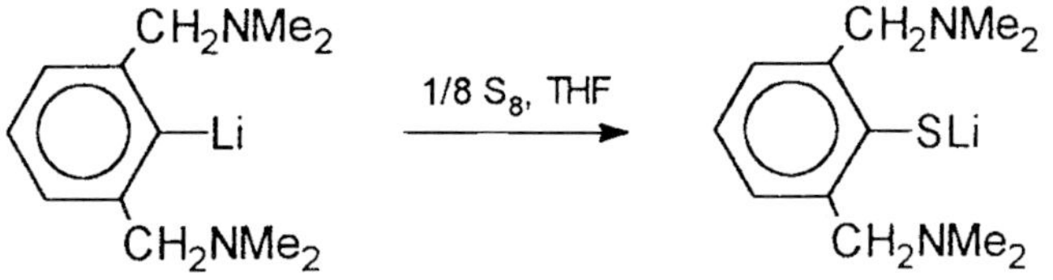

$$\text{2,6-(CH}_2\text{NMe}_2)_2\text{C}_6\text{H}_3\text{-Li} \xrightarrow{1/8 \, S_8, \, THF} \text{2,6-(CH}_2\text{NMe}_2)_2\text{C}_6\text{H}_3\text{-SLi} \quad (11)$$

The molecular structure of **185** consists of an essentially planar twelve-membered Li_6S_6 ring surrounded by the 2,6-bis(dimethylaminomethyl)phenyl substituents (Figure 6.19). Each lithium ion achieves tetracoordination by bonding to two sulfur atoms and two amino functions of neighboring thiophenolate ligands. A noteworthy structural feature is that the hexagonally shaped Li_6S_6 ring contains sulfur atoms that have T-shaped geometry as a result of bonding to two Li atoms and the *ipso* carbon atom of the ligand. The orientation of the aromatic rings relative to the central twelve-membered ring provides an overall 'propeller-like' aggregate with each hexamer having screw-type chirality. Both enantiomers are present in the crystal owing to crystallographic symmetry.

Polymeric self-assembly has frequently been observed in the chemistry of alkali metal alkoxides and related compounds. Not discussed here are simple alkoxides and thiolates such as MOMe (**202**, M = Li–Cs) [162–167], MOPr^i (**203**, M = K–Cs) [168], and MSMe (**204**, M = Li–K) [169]. They form layered structures in the

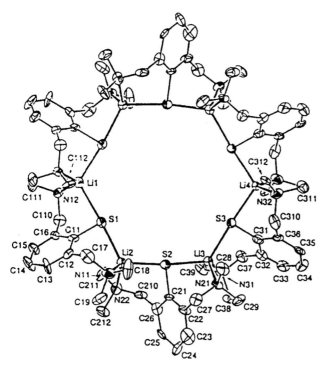

Fig. 6.19. The molecular structure of [Li{SC$_6$H$_3$(CH$_2$NMe$_2$)$_2$-2,6}]$_6$, **185** (Reproduced from ref. [134]).

solid state and should not be described as supramolecular. Interesting examples of self-assembled polymers, however, have been reported for the solid-state structures of base-free alkali metal phenoxides [153, 170]. Unsolvated sodium phenoxide is an important intermediate in the industrial synthesis of sodium salicylate (*o*-hydroxybenzoate) via insertion of CO$_2$ (Kolbe–Schmitt reaction) [171, 172], which in turn is a useful starting material for the production of pharmaceuticals, fertilizers, and pigments. The structure of sodium phenoxide has been determined by both powder and single-crystal X-ray techniques, and high-resolution powder X-ray diffraction has been used to determine the structures of the phenolates of the heavier alkali metals K, Rb, and Cs. All these materials form chain polymers, but the structures of LiOPh, **194a**, and NaOPh, **194b**, differ from those of the higher alkali metals. Sodium phenoxide forms a double chain, in which all the sodium atoms are equally coordinated by three oxygens and the carbon atoms of one phenyl ring (Figure 6.20, **a**). Thus the structure is borderline between ionic bonding and π-coordination. It is important for the mechanism of the Kolbe–Schmitt reaction that the sodium atoms have a low coordination number (3) and are situated at the edges of the chains, making them readily accessible. The Kolbe–Schmitt reaction is a solid-state reaction and from the supramolecular structure of **194b** it is easy to understand why the metal ions are readily attacked by CO$_2$. The higher alkali metal phenolates also have an unusual polymeric chain structure in which there are two

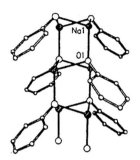
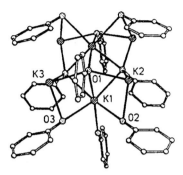

Fig. 6.20. The polymeric structures of $(NaOPh)_n$, **194b**, (left) and $(KOPh)_n$, **195a**, (right) (Reproduced from ref. [153]).

different coordination spheres – a distorted octahedron and tricoordination around the metal atoms (Figure 6.20, **b**). The latter are coordinatively unsaturated and also interact weakly with phenyl rings. The metal–carbon distances correspond approximately to the sum of the van der Waals radii, and so the compounds are not discussed in the section on π-bonded species.

Two types of polymeric self-assembled structures have been reported for alkali metal derivatives of bulky phenoxide and thiophenolate ligands: zigzag chains and ladder polymers. The former have been found in the sodium thiophenolates $[Na(SC_6H_2Pr^i_3\text{-}2,4,6)(thf)]_n$, **196**, (spectroscopic evidence only), $[Na(SC_6H_2Pr^i_3\text{-}2,4,6)(tmeda)]_n$, **197**, and $[Na\{SC_6H_2(CF_3)_3\text{-}2,4,6\}(thf)_2]_n$, **198**, and the potassium derivative $[K(SC_6H_2Pr^i_3\text{-}2,4,6)(pmdta)]_n$, **200** [135, 138]. The most striking structural feature is their one-dimensional polymeric shape, in which the alkali metal thiophenolate units are arranged in infinite zigzag chains, composed of alternating metal and sulfur atoms. The coordination of THF or tmeda as supporting ligands results in a distorted tetrahedral geometry of the alkyl-substituted thiophenolate ligands around the metal. In **198** the sodium atoms are octahedrally coordinated because of two additional Na···F–C secondary bonds to the *ortho*-CF$_3$ groups (Figure 6.21) [135].

The second type of supramolecular self-assembly in alkali metal thiophenolates is the formation of ladder or double-chain polymers. Structurally characterized examples are $[K(SC_6H_2Pr^i_3\text{-}2,4,6)(thf)]_n$, **199**, and $[K\{SC_6H_2(CF_3)_3\text{-}2,4,6\}(thf)]_n$, **201** (Figure 6.22) [135, 138]. Both crystal structures are two-dimensional, puckered, ladder-type polymeric aggregates with alternating potassium and sulfur atoms in nearly square-planar K_2S_2 rings. Each potassium atom is four-coordinate by three sulfur atoms and one oxygen of the THF ligands, resulting in a coordination geometry around potassium which is intermediate between distorted tetrahedral and square planar. The couples $[Na(SC_6H_2Pr^i_3\text{-}2,4,6)(tmeda)]_n$–$[K(SC_6H_2Pr^i_3\text{-}2,4,6)(thf)]_n$ and $[Na\{SC_6H_2(CF_3)_3\text{-}2,4,6\}(thf)_2]_n$–$[K\{SC_6H_2(CF_3)_3\text{-}2,4,6\}\text{-}(thf)]_n$ once again nicely demonstrate how the type of supramolecular self-assembly in these compounds is greatly influenced by the size of the metal ions.

Intermediate between alkali metal alkyls and alkoxides are adducts of these two components. Such mixtures have been shown to be even more reactive than the

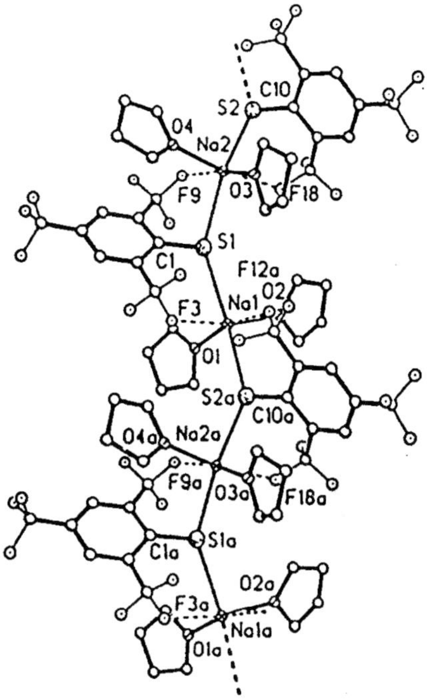

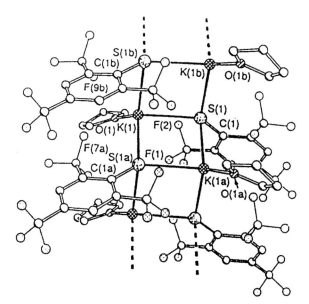

Fig. 6.21. The polymeric chain structure of [Na{SC$_6$H$_2$(CF$_3$)$_3$-2,4,6}(THF)$_2$]$_n$, **198** (Reproduced from ref. [135]).

Fig. 6.22. The polymeric ladder structure of [K{SC$_6$H$_2$(CF$_3$)$_3$-2,4,6}(THF)]$_n$, **201** (Reproduced from ref. [135]).

separate reagents and have found various uses as very strong deprotonating agents (so-called 'superbases', 'complex-bases', or 'LICKOR reagents') [173–176]. For example, benzene does not react with *n*-butyllithium alone, but metalation occurs in the presence of alkali metal *t*-butoxides. Structural information on the reactive species are difficult to obtain. A mixture of trityllithium and cesium 3-ethyl-3-heptoxide has been investigated as a model 'superbase' by use of ^{1}H, ^{6}Li, and ^{133}Cs NMR spectroscopy [176]. Theoretical investigations include model MNDO calculations [176] and *ab initio* studies [177]. Most attempts to isolate crystals from superbasic mixtures suitable for X-ray crystallography have failed. The reason can be found mainly in a preparative problem. When a heavier alkali metal alkoxide is added to a solution of a lithium alkyl in an inert solvent such as hexane, the transmetalation product, i.e. the heavier alkali metal alkyl, is formed as a microcrystalline precipitate and the more soluble lithium alkoxide remains in solution. That the least soluble species finally precipitate from these mixtures, however, does not preclude the formation of mixed aggregates. The first example of a structurally characterized lithium alkoxide–organosodium compound was an intramolecular version of the superbases. The compound lithium (4,6-dimethyl-2-sodiomethylphenoxide) was prepared by deprotonation of sodium 2,4,6-trimethylphenoxide, according to Eq.(12) [178].

$$\text{ArONa} \xrightarrow[\text{hexane} \atop -\text{C}_4\text{H}_{10}]{n\text{-BuLi}} \text{Ar'(Na)(OLi)} \qquad (12)$$

Crystallization in the presence of tmeda affords a tetrameric aggregate of composition [(tmeda)NaCH$_2$C$_6$H$_2$Me$_2$(OLi)]$_4$, **205** [178]. A distorted Li$_4$O$_4$ cube forms the central part of the molecule. Both the lithium and sodium atoms act as bridges between the methylene units and the oxygen atoms. The compound can be regarded as a 'model' intermolecular superbase and overcomes the problem of differing solubilities of lithium alkoxides and the heavier alkali metal hydrocarbyls.

Not a superbase, but closely related, is the adduct of *n*-butyllithium with lithium *t*-butoxide, **206** [179]. Evidence for adduct formation between the two components came initially from NMR studies [180], and it was shown that in benzene solution tetrameric associates are present [181, 182]. Crystallization from hexane gave a well-defined self-assembled tetramer containing two types of differently coordinated Li atom. One set of lithium atoms is bonded to two α-C atoms of *n*-butyl groups and one oxygen, with additional close contact with a β-carbon atom; the other lithium atoms are connected to two oxygens and only one α-carbon [183].

Also related to the above-mentioned superbases and LiR–LiOR adducts is dilithiated benzophenone, **156** [130]. This compound was prepared in the form of deep purple, highly air- and moisture-sensitive crystals by reaction of a THF solution of benzophenone with a large excess of lithium sand followed by crystallization in the presence of tmeda. The resulting tmeda-adduct is associated as a dimer in which a central Li$_2$ unit is bridged by the carbonyl oxygen atoms of the benzophenone dianions. The resulting four-membered Li_2O_2 ring is planar. In addition to the coordination of tmeda the other two Li atoms have an ion-pair interaction with the carbonyl oxygen and three carbon atoms of the benzophenone system.

156

6.2 Alkaline Earth Metals

Although organomagnesium compounds have enormous potential as intermediates in organic syntheses [184], access to unsolvated diorganomagnesium species is limited and structural information is scarce, because such compounds are usually prepared in coordinating solvents such as diethyl ether or THF and desolvation without decomposition is generally difficult. Unsolvated monomers are accessible only by use of very bulky substituents such as tris(trimethylsilyl)methyl [185]. Simple dialkylmagnesium species such as $[Me_2Mg]_n$, **207**, or $[Et_2Mg]_n$, **208**, are supramolecularly associated and form polymeric chains in which the tetracoordinated Mg atoms are connected by pairs of symmetrically bridging alkyl groups. More recently, diphenylmagnesium, **209**, has also been shown to be a polymer. These materials are not volatile and are virtually insoluble in non-coordinating solvents and thus the structures had to be determined by X-ray powder diffraction [186, 187]. In the polymeric chains each magnesium atom is tetrahedrally surrounded by four carbanions. Similar to polymeric dimethylberyllium [188], the carbanions form μ_2-bridges between the Mg atoms.

Yet another phenomenon which complicates full characterization of organomagnesium species, especially Grignard reagents, is the Schlenk equilibrium between organomagnesium halides and mixtures of MgR_2 and magnesium dihalide [189]. Redistribution between alkyl and halide ligands is quite facile. A typical example is the attempted crystallization of dineopentylmagnesium from a toluene–

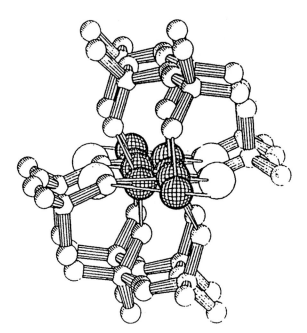

Fig. 6.23. The polymeric structure of [(Np$_2$Mg)$_2$(NpMgBr)$_2$]$_n$, **210** (Reproduced from ref. [190].

heptane solution containing a small amount of neopentylmagnesium bromide, which produced crystals of the 1:1 adduct of the two components [190]. This material forms self-organized polymeric chains of composition [(Np$_2$Mg)$_2$(NpMgBr)$_2$]$_n$, **210** (Np = neopentyl), Figure 6.23. The chain consists of an alternating pattern of two Np$_2$Mg and two NpMgBr units connected by bridging neopentyl and bromide ligands, in which all Mg atoms are tetracoordinate.

Although most structurally characterized Grignard compounds RMgX (X = halide) are monomers in the solid state, many dimers and polymers have also been identified by X-ray analyses. For example, dimerization occurs when the bulky 9-anthracenyl substituent is used. Dimeric anthracenylmagnesiumbromide-dibutyl ether, **211**, contains a planar, centrosymmetric and nearly quadratic four-membered Mg$_2$Br$_2$ ring [191]. In the crystal structure the dimers are stacked along the crystallographic b axis (Figure 6.24).

Dimeric structures have also been found for 'ate'-complexes of magnesium. Bimetallic species such as LiMgPh$_3$ and Li$_2$MgPh$_4$ have been known since the pioneering work of Wittig et al. [192, 193]. Early investigations on these materials include ^{1}H and ^{7}Li NMR measurements and X-ray powder diffraction studies [194]. In the presence of tmeda well defined dimers of composition [Li(tmeda)]$_2$-[Ph$_2$MgPh$_2$MgPh$_2$], **212**, are formed. The central part of the compound consists of a triphenylmagnesate dimer [Ph$_2$MgPh$_2$MgPh$_2$]$^{2-}$ anion, which is isoelectronic and isostructural with Al$_2$Ph$_6$. The two Li(tmeda) moieties are linked to this central unit by two bridging phenyl groups. The compound contains symmetrical phenyl bridges between Mg and Mg, but unsymmetrical phenyl bridges between Mg and Li [195]. The corresponding hexaethyl dimagnesate anion involving distorted tetrahe-

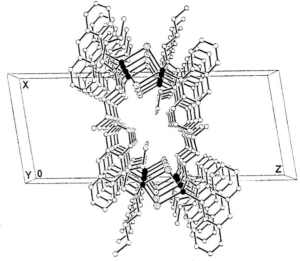

Fig. 6.24. The polymeric structure of anthracenylmagnesiumbromide–dibutyl ether, **211** (Reproduced from ref. [191]).

dral coordination around Mg was found in the "ate-complex" [Mg(neopentyl)-([2.2.1]cryptand)]$_2$[Mg$_2$Et$_6$], **213** [196–198].

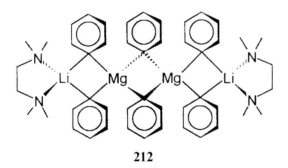

212

Supramolecular association via oxygen bridges has been observed in magnesium chemistry for mono(cyclopentadienyl) magnesium alkoxides. Magnesocene, MgCp$_2$, **214**, reacts with polyfunctional alcohols such as N,N-dimethylethanolamine, 2-ethoxyethanol or 2-(2-ethoxyethoxy)ethanol to afford polymeric cyclopentadienylmagnesium alkoxides, **215**. Various structures may be envisaged for these materials. The simplest involves the presence of linear chains. However, it cannot be ruled out that cross-linking of the chains occurs via 4-coordinate oxygen atoms. Dissolution of the polymers in pyridine yields pyridine adducts of the corresponding monomers [199].

215

A series of other dimerized organomagnesates, including [Li$_2$[MgC≡CPh)$_3$-(tmeda)]$_2$, **216** [200], Na$_2$[Mg(C≡CBu^t)$_3$(tmeda)]$_2$, **217** [201], and Na$_2$[Mg(C≡CBu^t)$_3$(pmdta)]$_2$, **218** [201], contain pentacoordinate magnesium surrounded by three organic substituents and one chelating amine ligand. These compounds will be further discussed in Chapter 7 as they are oligomerized via π-interactions with the acetylide units. Bridging alkali metals induce dimer formation by close contacts with the alkyne units. A more complicated anion was found in the heterometallic benzylmagnesate derivative [Li(tmeda)$_2$][(tmeda)Li(μ_2-CH$_2$Ph)$_2$Mg(CH$_2$Ph)$_2$]$_2$, **219**.

One report in the literature describes the preparation and structural characterization of a compound containing an unusual octaorganotrimagnesate. The compound [Mg$_2$Me$_3$(tacn)$_2$]$_2$[Mg$_3$Me$_8$], **220**, is formed upon treatment of dimethylmagnesium with the cyclic triamine ligand tacn (tacn = N,N′,N″-trimethyl-1,4,7-triaracyclononane) [202].

References

[1] B. J. Wakefield, *Organolithium Methods*, Academic Press, London, **1988**.
[2] A.-M. Sapse and P. von R. Schleyer, *Lithium Chemistry – A Theoretical and Experimental Overview*, Wiley, New York, **1995**.
[3] (a) L. Brandsma and H. Verkruijsse, *Preparative Polar Organometallic Chemistry 1*, Springer, Berlin, **1987**; (b) L. Brandsma and H. Verkruijsse, *Preparative Polar Organometallic Chemistry 2*, Springer, Berlin, **1990**.
[4] (a) E. Weiss, *Angew. Chem.* **1993**, *105*, 1565; *Angew. Chem. Int. Ed. Engl.* **1993**, *32*, 1501; (b) J. D. Smith, *Adv. Organomet. Chem.* **1999**, *43*, 267 and references cited therein.
[5] H. Hiller, M. Layh and W. Uhl, *Angew. Chem.* **1991**, *103*, 339; *Angew. Chem. Int. Ed. Engl.* **1991**, *30*, 324.
[6] T. Kottke and D. Stalke, *Angew. Chem.* **1993**, *105*, 619; *Angew. Chem. Int. Ed. Engl.* **1993**, *32*, 580.
[7] K. Sorger, P. von R. Schleyer and D. Stalke, *J. Am. Chem. Soc.* **1996**, *118*, 1086.
[8] K. Sorger, P. von R. Schleyer and D. Stalke, *J. Chem. Soc., Chem. Commun.* **1995**, 2279.
[9] K. Sorger, P. von R. Schleyer, R. Fleischer and D. Stalke, *J. Am. Chem. Soc.* **1996**, *118*, 6924.
[10] D. Thoennes and E. Weiss, *Chem. Ber.* **1978**, *111*, 3157.
[11] B. Schubert and E. Weiss, *Chem. Ber.* **1983**, *116*, 3212.

[12] W. Bauer, W. Neugebauer, P. von R. Schleyer, H. Dietrich and W. Mahdi, unpublished results, cited in ref. [4].
[13] K. Ruhlandt-Senge, J. Ellison, R. J. Wehmschulte, F. Pauer and P. P. Power, *J. Am. Chem. Soc.* **1993**, *115*, 11353.
[14] J. G. Donkervoort, J. L. Vicario, E. Rijnberg, J. T. B. H. Jastrzebski, H. Kooijman, A. L. Spek and G. van Koten, *J. Organomet. Chem.* **1998**, *463*, 463.
[15] (a) R. Amstutz, T. Laube, W. B. Schweizer, D. Seebach and J. D. Dunitz, *Helv. Chim. Acta.* **1984**, *67*, 224; (b) F. Becke, F. W. Heinemann and D. Steinborn, *Organometallics* **1997**, *16*, 2736.
[16] L. T. Byrne, L. M. Engelhardt, G. E. Jacobsen, W.-P. Leung, R. I. Papasergio, C. L. Raston, B. W. Skelton, P. Twiss and A. H. White, *J. Chem. Soc., Dalton Trans.* **1989**, 105.
[17] S. Blaurock, O. Kühl and E. Hey-Hawkins, *Organometallics* **1997**, *16*, 807.
[18] (a) G. Fraenkel, W. R. Winchester and P. G. Williard, *Organometallics* **1989**, *8*, 2308; (b) F. Becke, D. Steinborn, F. W. Heinemann, T. Rüffer, P. Wiegeleben, R. Boese and D. Bläser, *J. Organomet. Chem.* in print.
[19] D. Stalke and K. H. Whitmire, *J. Chem. Soc., Chem. Commun.* **1990**, 833.
[20] M. Niemeyer and P. P. Power, *Organometallics* **1997**, *16*, 3258.
[21] U. Schümann, U. Behrens and E. Weiss, *Angew. Chem.* **1989**, *101*, 481; *Angew. Chem. Int. Ed. Engl.* **1989**, *28*, 476.
[22] U. Schümann, *Ph.D. Thesis*, Universität Hamburg, **1987**.
[23] P. B. Hitchcock, M. F. Lappert, W.-P. Leung, L. Diansheng and T. Shun, *J. Chem. Soc., Chem. Commun.* **1993**, 1386.
[24] S. Harder, J. Boersma, L. Brandsma, J. A. Kanters, W. Bauer and P. von R. Schleyer, *Organometallics* **1989**, *8*, 1696.
[25] E. Weiss and E. A. C. Lucken, *J. Organomet. Chem.* **1964**, *2*, 197.
[26] E. Weiss and G. Hencken, *J. Organomet. Chem.* **1969**, *21*, 265.
[27] E. Weiss, T. Lambertsen, B. Schubert, J. K. Cockcroft and A. Wiedenmann, *Chem. Ber.* **1990**, *123*, 79.
[28] H. Dietrich, *Acta Crystallogr.* **1963**, *16*, 681.
[29] H. Dietrich, *J. Organomet. Chem.* **1981**, *205*, 291.
[30] V. Jordan, *Diploma Thesis*, Universität Hamburg, **1989**.
[31] G. W. Klumpp, P. J. A. Geurink, N. J. R. van Eikema Hommes, F. J. J. de Kanter, M. Vos and A. L. Spek, *Recl. Trav. Chim. Pays-Bas* **1986**, *10*, 398.
[32] K. S. Lee, P. G. Willard and J. W. Suggs, *J. Organomet. Chem.* **1986**, *299*, 311.
[33] H. Hope and P. P. Power, *J. Am. Chem. Soc.* **1983**, *105*, 5320.
[34] M. M. Olmstead and P. P. Power, *J. Am. Chem. Soc.* **1990**, *112*, 8008.
[35] M. Geissler, J. Kopf, B. Schubert, E. Weiss, W. Neugebauer and P. von R. Schleyer, *Angew. Chem.* **1987**, *99*, 569; *Angew. Chem. Int. Ed. Engl.* **1987**, *26*, 587.
[36] J. E. Davies, R. P. Davies, L. Dunbar, P. R. Raithby, M. G. Russell, R. Snaith, S. Warren and A. E. A. Wheatley, *Angew. Chem.* **1997**, *109*, 2428; *Angew. Chem. Int. Ed. Engl.* **1997**, *36*, 2334.
[37] E. Weiss, G. Sauermann and G. Thirase, *Chem. Ber.* **1983**, *116*, 74.
[38] E. Weiss, S. Corbelin, J. K. Cockcroft and A. N. Fitch, *Angew. Chem.* **1990**, *102*, 728; *Angew. Chem. Int. Ed. Engl.* **1990**, *29*, 650.
[39] E. Weiss, S. Corbelin, J. K. Cockcroft and A. N. Fitch, *Chem. Ber.* **1990**, *123*, 1629.
[40] C. Schade, P. von R. Schleyer, H. Dietrich and W. Mahdi, *J. Am. Chem. Soc.* **1986**, *108*, 2484.
[41] S. Corbelin, N. P. Lorenzen, J. Kopf and E. Weiss, *J. Organomet. Chem.* **1991**, *415*, 293.
[42] U. Siemeling, T. Redecker, B. Neumann and H.-G. Stammler, *J. Am. Chem. Soc.* **1994**, *116*, 5507.
[43] (a) D. Hoffmann, P. von R. Schleyer and D. Stalke, unpublished results; (b) R. Zerger, W. Rhine and G. Stucky, *J. Am. Chem. Soc*, **1974**, *96*, 6048; (c) A. Maercker, M. Bsata, W. Buchmeier and B. Engelen, *Chem. Ber.* **1984**, *117*, 2547.
[44] R. J. Wehmschulte and P. P. Power, *J. Am. Chem. Soc.* **1997**, *119*, 2847.
[45] B. Tecle, A. F. M. M. Rahman and J. P. Oliver, *J. Organomet. Chem.* **1986**, *317*, 267.
[46] H. Köster, D. Thoennes and E. Weiss, *J. Organomet. Chem.* **1978**, *160*, 1.

[47] K. Sorger, W. Bauer, P. von R. Schleyer and D. Stalke, *Angew. Chem.* **1995**, *107*, 1766; *Angew. Chem. Int. Ed. Engl.* **1995**, *34*, 1594.
[48] M. Geissler, *Diploma Thesis*, Universität Hamburg, **1984**.
[49] B. Schubert and E. Weiss, *Angew. Chem.* **1983**, *95*, 499; *Angew. Chem. Int. Ed. Engl.* **1983**, *22*, 496.
[50] H. Köster and E. Weiss, *Chem. Ber.* **1982**, *115*, 3422.
[51] S. P. Patterman, I. L. Karle and G. D. Stucky, *J. Am. Chem. Soc.* **1970**, *92*, 1150.
[52] M. A. Beno, H. Hope, M. M. Olmstead and P. P. Power, *Organometallics* **1985**, *4*, 2117.
[53] D. Steinborn, O. Neumann, H. Weichmann, F. W. Heinemann and J.-P. Wagner, *Polyhedron* **1998**, *17*, 351.
[54] D. Hoffmann, N. J. R. van Eikema Hommes, P. von R. Schleyer, D. Stalke, U. Pieper, D. Wright and R. Snaith, *J. Am. Chem. Soc.* **1995**, *117*, 528.
[55] D. Hoffmann, F. Hampel, W. Bauer and P. von R. Schleyer, unpublished results, cited in ref. [4].
[56] H. Köster and E. Weiss, *J. Organomet. Chem.* **1979**, *168*, 273.
[57] H. Viebrock, *Ph.D. Thesis*, Universität Hamburg, **1992**.
[58] R. Grünong and J. L. Atwood, *J. Organomet. Chem.* **1977**, *137*, 101.
[59] (a) B. Halton and M. G. Banwell, in *The Chemistry of the Cyclopropyl Group* (Ed.: Z. Rappoport), Wiley, New York, **1987**, Part 2; (b) M. S. Baird, *Top. Curr. Chem.* **1988**, *144*, 137.
[60] (a) J. B. Collins, J. D. Dill, E. D. Jemmis, Y. Apeloig, P. von R. Schleyer, R. Seeger and J. A. Pople, *J. Am. Chem. Soc.* **1976**, *98*, 5419; (b) K. Sorger and P. von R. Schleyer, *J. Mol. Struct., Theochem* **1995**, *338*, 317.
[61] (a) L. Horner, H. Hoffmann, H. G. Wippel and G. Klahre, *Chem. Ber.* **1959**, *92*, 2499; (b) B. E. Maryanoff and A. B. Reitz, *Chem. Rev.* **1989**, *89*, 863; (c) J. Clayden and S. Warren, *Angew. Chem.* **1996**, *108*, 261; *Angew. Chem. Int. Ed. Engl.* **1996**, *35*, 24.
[62] (a) S. E. Denmark, N. Chatani and S. V. Pansare, *Tetrahedron* **1992**, *48*, 2191; (b) S. E. Denmark and C.-T. Chen, *J. Am. Chem. Soc.* **1992**, *114*, 10674.
[63] O. W. Lever, *Tetrahedron* **1976**, *32*, 1943.
[64] (a) T. F. Schaaf, W. Butler, M. D. Glick and J. P. Oliver, *J. Am. Chem. Soc.* **1974**, *96*, 7593; (b) W. H. Ilsley, T. F. Schaaf, M. D. Glick and J. P. Oliver, *J. Am. Chem. Soc.* **1980**, *102*, 3769; (c) R. Balasubramanian and J. P. Oliver, *J. Organomet. Chem.* **1980**, *197*, C7.
[65] H. Gilman and F. K. Cartledge, *J. Organomet. Chem.* **1965**, *3*, 255.
[66] (a) K. W. Klinkhammer, *Chem. Eur. J.* **1997**, *3*, 1418; (b) K. W. Klinkhammer and W. Schwarz, *Z. Anorg. Allg. Chem.* **1993**, *619*, 1777.
[67] M. F. Lappert, P. P. Power, A. R. Sanger and R. C. Srivastava, *Metal and Metalloid Amides*, Wiley, New York, **1980**.
[68] (a) R. E. Mulvey, *Chem. Soc. Rev.* **1991**, *20*, 167; (b) K. Gregory, P. von R. Schleyer and R. Snaith, *Adv. Inorg. Chem.* **1991**, *37*, 47.
[69] V. Jordan, *Ph.D. Thesis*, Universität Hamburg, **1992**.
[70] N. Kuhn, G. Henkel and J. Kreuzberg, *Angew. Chem.* **1990**, *102*, 1179; *Angew. Chem. Int. Ed. Engl.* **1990**, *29*, 1143.
[71] D. R. Armstrong, D. Barr, W. Clegg, R. E. Mulvey, D. Reed, R. Snaith and K. Wade, *J. Chem. Soc., Chem. Commun.* **1986**, 869.
[72] K. Gregory, P. von R. Schleyer and R. Snaith, *Adv. Organomet. Chem.* **1992**, *37*, 47.
[73] R. E. Mulvey, *Chem. Soc. Rev.* **1991**, *20*, 167.
[74] M. Veith, *Angew. Chem.* **1987**, *99*, 1; *Angew. Chem. Int. Ed. Engl.* **1987**, *26*, 1.
[75] (a) L. M. Engelhardt, B. S. Jolly, P. C. Punk, C. L. Raston, B. W. Skelton and A. H. White, *Aust. J. Chem.* **1986**, *39*, 1337; (b) D. Barr, W. Clegg, R. E. Mulvey, R. Snaith and J. Wright, *J. Chem. Soc., Chem. Commun.* **1987**, 716.
[76] W. Clegg, A. J. Edwards, F. S. Mair and P. M. Nolan, *Chem. Commun.* **1998**, 23.
[77] (a) T. Kottke, *Ph.D. Thesis*, Universität Göttingen, **1993**; (b) R. von Bülow, H. Gornitzka, T. Kottke and D. Stalke, *Chem. Commun.* **1996**, 1639.
[78] B. Cetinkaya, P. B. Hitchcock, M. F. Lappert, M. C. Misra and A. J. Thorne, *J. Chem. Soc., Chem. Commun.* **1984**, 148.
[79] R. Hacker, E. Kaufmann, P. von R. Schleyer, W. Mahdi and H. Dietrich, *Chem. Ber.* **1987**, *120*, 1533.

References

[80] D. Barr, D. J. Berrisford, L. Méndez, A. M. Z. Slawin, R. Snaith, J. F. Stoddart, D. J. Williams and D. S. Wright, *Angew. Chem.* **1991**, *103*, 97; *Angew. Chem. Int. Engl.* **1991**, *30*, 82.

[81] H. Mack, G. Frenzen, M. Bendikow and M. S. Eisen, *J. Organomet. Chem.* **1997**, *549*, 39.

[82] P. G. Williard, Q.-Y. Liu and L. Lochmann, *J. Am. Chem. Soc.* **1992**, *114*, 348.

[83] (a) F. Pauer and D. Stalke, *J. Organomet. Chem.* **1991**, *418*, 127; (b) S. Freitag, W. Kolodziejski, F. Pauer and D. Stalke, *J. Chem. Soc., Dalton Trans.* **1993**, 3479.

[84] D. Stalke, M. Wedler and F. T. Edelmann, *J. Organomet. Chem.* **1992**, *431*, C1.

[85] M. S. Eisen and M. Kapon, *J. Chem. Soc., Dalton Trans.* **1994**, 3507.

[86] J. Richter, *Ph.D. Thesis*, Universität Göttingen, **1996**.

[87] T. Chivers, M. Parvez and G. Schatte, *J. Organomet. Chem.* **1998**, *550*, 213.

[88] D. J. Brauer, H. Bürger, G. R. Liewald and J. Wilke, *J. Organomet. Chem.* **1985**, *287*, 305.

[89] T. Kottke and D. Stalke, *Organometallics* **1996**, *21*, 4552.

[90] P. C. Andrews, W. Clegg and R. E. Mulvey, *Angew. Chem.* **1990**, *102*, 1480; *Angew. Chem. Int. Ed. Engl.* **1990**, *29*, 1440.

[91] K. Gregory, M. Bremer, W. Bauer, P. von R. Schleyer, N. P. Lorenzen, J. Kopf and E. Weiss, *Organometallics* **1990**, *9*, 1485.

[92] (a) P. C. Andrews, N. D. R. Barnett, R. E. Mulvey, W. Clegg, P. A. O'Neil, D. Barr, L. Cowton, A. J. Dawson and B. J. Wakefield, *J. Organomet. Chem.* **1996**, *518*, 85. (b) W. Clegg, S. Kleditzsch, R. E. Mulvey and P. O'Shaughnessy, *J. Organomet. Chem.* **1998**, *558*, 193.

[93] U. Pieper, D. Stalke, S. Vollbrecht and U. Klingebiel, *Chem. Ber.* **1990**, *123*, 1039.

[94] M. Veith, J. Böhnlein and V. Huch, *Chem. Ber.* **1989**, *122*, 841.

[95] K. Gregory, M. Bremer, P. von R. Schleyer, P. A. A. Klusener and L. Brandsma, *Angew. Chem.* **1989**, *101*, 1261; *Angew. Chem. Int. Ed. Engl.* **1989**, *28*, 1224.

[96] E. Hey, P. B. Hitchcock, M. F. Lappert and A. K. Rai, *J. Organomet. Chem.* **1987**, *325*, 1.

[97] P. B. Hitchcock, M. F. Lappert, P. P. Power and S. J. Smith, *J. Chem. Soc., Chem. Commun.* **1984**, 1669.

[98] (a) M. Andrianarison, D. Stalke and U. Klingebiel, *Chem. Ber.* **1990**, *123*, 71; (b) M. Driess, H. Pritzkow, M. Skipinski and U. Winkler, *Organometallics* **1997**, *16*, 5108; (c) M. Driess, H. Pritzkow, S. Rell and U. Winkler, *Organometallics* **1996**, *15*, 1845; (d) M. Driess, *Adv. Organomet. Chem.* **1996**, *39*, 193.

[99] A. M. Arif, R. A. Jones and K. B. Kidd, *J. Chem. Soc., Chem. Commun.* **1986**, 1440.

[100] R. A. Bartlett, H. V. Rasika Dias, H. Hope, B. D. Murray, M. M. Olmstead and P. P. Power, *J. Am. Chem. Soc.* **1986**, *108*, 6921.

[101] (a) D. Mootz, A. Zinnius and B. Böttcher, *Angew. Chem.* **1969**, *81*, 389; *Angew. Chem. Int. Ed. Engl.* 1969, *8*, 378; (b) R. D. Rogers, J. L. Atwood and R. Grüning, *J. Organomet. Chem.* **1978**, *157*, 229; (c) J. Knizek, I. Krossing, H. Nöth, H. Schwenk and T. Seifert, *Chem. Ber.* **1997**, *130*, 1053.

[102] P. B. Hitchcock, M. F. Lappert and S. J. Smith, *J. Organomet. Chem.* **1987**, *320*, C27.

[103] D. R. Armstrong, D. Barr, W. Clegg, S. M. Hodgson, R. E. Mulvey, D. Reed, R. Snaith and D. S. Wright, *J. Am. Chem. Soc.* **1989**, *111*, 4719.

[104] W. Clegg, M. MacGregor, R. E. Mulvey and P. A. O'Neil, *Angew. Chem.* **1992**, *104*, 74; *Angew. Chem. Int. Ed. Engl.* **1992**, *31*, 93.

[105] K. F. Tesh, B. D. Jones and T. P. Hanusa, *J. Am. Chem. Soc.* **1992**, *114*, 6590.

[106] R. A. Jones, A. L. Stuart and T. C. Wright, *J. Am. Chem. Soc.* **1983**, *105*, 7459.

[107] (a) G. W. Rabe, S. Kheradmandan, L. M. Liable-Sands, I. A. Guzei and A. L. Rheingold, *Angew. Chem.* **1998**, *110*, 1495; *Angew. Chem. Int. Engl.* **1998**, *37*, 1404; (b) J. D. Smith, *Angew. Chem.* **1998**, *110*, 2181; *Angew. Chem. Int. Engl.* **1998**, *37*, 2071.

[108] R. P. Davies, P. R. Raithby and R. Snaith, *Angew. Chem.* **1997**, *109*, 1261; *Angew. Chem. Int. Ed. Engl.* **1997**, *36*, 1215.

[109] N. D. R. Barnett, W. Clegg, L. Horsburgh, D. M. Lindsay, Q.-Y. Liu, F. M. Mackenzie, R. E. Mulvey, *Chem. Commun.* **1996**, 2321.

[110] D. R. Armstrong, D. Barr, W. Clegg, S. R. Drake, R. J. Singer, R. Snaith, D. Stalke and D. S. Wright, *Angew. Chem.* **1991**, *101*, 1702; *Angew. Chem. Int. Ed. Engl.* **1991**, *30*, 1707.

[111] N. P. Lorenzen, J. Kopf, F. Olbrich, U. Schümann and E. Weiss, *Angew. Chem.* **1990**, *102*, 1481; *Angew. Chem. Int. Ed. Engl.* **1990**, *29*, 1441.

[112] E. Hey-Hawkins and E. Sattler, *J. Chem. Soc., Chem. Commun.* **1992**, 775.
[113] (a) C. Lambert, F. Hampel and P. von R. Schleyer, *J. Organomet. Chem.* **1993**, *455*, 29; (b) C. Lambert, P. von R. Schleyer, G. M. Newton, P. Schreiner and P. Otto, *Z. Naturforsch.* **1992**, *47b*, 869.
[114] D. R. Armstrong, R. E. Mulvey, D. Barr, R. W. Porter, P. R. Raithby, T. R. E. Simpson, R. Snaith, D. S. Wright, K. Gregory and P. Mukulcik, *J. Chem. Soc., Dalton Trans.* **1991**, 765.
[115] G. R. Kowach, C. J. Warren, R. C. Haushalter and F. J. DiSalvo, *Inorg. Chem.* **1998**, *37*, 156.
[116] R. Fleischer and D. Stalke, *Chem. Commun.* **1998**, 343.
[117] A. R. Kennedy, R. E. Mulvey and A. Robertson, *Chem. Commun.* **1998**, 89.
[118] J. Eppinger, E. Herdtweck and R. Anwander, *Polyhedron* **1998**, *17*, 1195.
[119] R. A. Bartlett, M. M. Olmstead and P. P. Power, *Inorg. Chem.* **1986**, *25*, 1243.
[120] E. Hey and F. Weller, *J. Chem. Soc., Chem. Commun.* **1988**, 782.
[121] G. W. Rabe, G. P. A. Yap and A. L. Rheingold, *Inorg. Chem.* **1997**, *36*, 1990.
[122] V. Jordan and E. Weiss, unpublished results, cited in ref. [4].
[123] N. Kuhn, G. Henkel and J. Kreutzberg, *Angew. Chem.* **1990**, *102*, 1179; *Angew. Chem. Int. Engl.* **1990**, *29*, 1143.
[124] (a) P. J. Bailey, A. J. Blake, M. Kryszcuk, S. Parsons and D. Reed, *J. Chem. Soc., Chem. Commun.* **1995**, 1647; (b) P. J. Bailey, L. A. Mitchell, P. R. Raithby, M.-A. Rennie, K. Verhorevoort and D. S. Wright, *Chem. Commun.* **1996**, 1351.
[125] D. C. Bradley, R. C. Mehrotra and D. P. Gaur, *Metal Alkoxides*, Academic Press, London, **1978**.
[126] D. Seebach, *Angew. Chem.* **1988**, *100*, 1685; *Angew. Chem. Int. Ed. Engl.* **1988**, *27*, 1624.
[127] K. C. Malhotra and R. L. Martin, *J. Organomet. Chem.* **1982**, *239*, 159.
[128] M. Geissler, *Ph.D. Thesis*, Universität Hamburg, **1987**.
[129] B. Cetinkaya, I. Gümrükçü, M. F. Lappert, J. L. Atwood and R. Shakir, *J. Am. Chem. Soc.* **1980**, *102*, 2086.
[130] B. Bogdanovic, C. Krüger and B. Wermeckes, *Angew. Chem.* **1980**, *92*, 844; *Angew. Chem. Int. Ed. Engl.* **1980**, *19*, 817.
[131] M. B. Hursthouse, M. A. Hossain, M. Motevalli, M. Sanganee and A. C. Sullivan, *J. Organomet. Chem.* **1990**, *381*, 293.
[132] P. B. Hitchcock, M. F. Lappert and I. A. MacKinnon, *J. Chem. Soc., Chem. Commun.* **1993**, 1015.
[133] H.-J. Gais, G. Hellmann and H. J. Lindner, *Angew. Chem.* **1990**, *102*, 96; *Angew. Chem. Int. Ed. Engl.* **1990**, *29*, 100.
[134] M. D. Janssen, E. Rijnberg, C. A. de Wolf, M. P. Hogerheide, D. Kruis, H. Kooijman, A. L. Spek, D. M. Grove and G. van Koten, *Inorg. Chem.* **1996**, *35*, 6735.
[135] S. Brooker, F. T. Edelmann, T. Kottke, H. W. Roesky, G. M. Sheldrick, D. Stalke and K. H. Whitmire, *J. Chem. Soc., Chem. Commun.* **1991**, 144.
[136] M. P. Hogerheide, S. N. Ringelberg, M. D. Janssen, J. Boersma, A. L. Spek and G. van Koten, *Inorg. Chem.* **1996**, *35*, 1195.
[137] I. Cragg-Hine, M. G. Davidson, O. Kocian, T. Kottke, F. S. Mair, R. Snaith and J. F. Stoddart, *J. Chem. Soc., Chem. Commun.* **1993**, 1355.
[138] U. Englich, S. Chadwick and K. Ruhlandt-Senge, *Inorg. Chem.* **1998**, *37*, 283.
[139] (a) P. A. van der Schaaf, M. P. Hogerheide, D. M. Grove, A. L. Spek and G. van Koten, *J. Chem. Soc., Chem. Commun.* **1992**, 1703; (b) P. A. van der Schaaf, J. T. B. H. Jastrzebski, M. P. Hogerheide, W. J. J. Smeets, A. L. Spek, J. Boersma and G. van Koten, *Inorg. Chem.* **1993**, *32*, 4111.
[140] (a) K. Ruhlandt-Senge, U. Englich, M. O. Senge and S. Chadwick, *Inorg. Chem.* **1996**, *35*, 5820; (b) K. Ruhlandt-Senge and P. P. Power, *Inorg. Chem.* **1993**, *32*, 4505.
[141] E. M. Arnett, M. A. Nichols and A. T. McPhail, *J. Am. Chem. Soc.* **1990**, *112*, 7095.
[142] (a) E. Weiss, H. Alsdorf and H. Kühr, *Angew. Chem.* **1967**, *79*, 816; *Angew. Chem. Int. Ed. Engl.* **1967**, *6*, 801; (b) E. Weiss, H. Alsdorf, H. Kühr and H.-F. Grützmacher, *Chem. Ber.* **1968**, *101*, 3777.
[143] J. A. Samuels, E. B. Lobkovsky, W. E. Streib, K. Folting, J. C. Huffman, J. W. Zwanziger and K. G. Caulton, *J. Am. Chem. Soc.* **1993**, *115*, 5093.

[144] F. Pauer and G. M. Sheldrick, *Acta Crystallogr.* **1993**, *C49*, 1283.
[145] G. R. Fuentes, P. S. Coan, W. E. Streib and K. G. Caulton, *Polyhedron* **1991**, *10*, 2371.
[146] G. E. Hartwell and T. L. Brown, *Inorg. Chem.* **1966**, *5*, 1257.
[147] M. H. Chisholm, S. R. Drake, A. A. Naini and W. E. Streib, *Polyhedron* **1991**, *10*, 805.
[148] T. A. Bazhenova, R. M. Lobkovskaya, R. P. Shibaeva, A. E. Shilov and A. K. Shilova, *J. Organomet. Chem.* **1987**, *330*, 9.
[149] R. A. Jones, S. U. Koschmieder, J. L. Atwood and S. G. Bott, *J. Chem. Soc., Chem. Commun,* **1992**, 726.
[150] B. Goldfuss, P. von R. Schleyer and F. Hampel, *Organometallics* **1997**, *16*, 5032.
[151] (a) T. Greiser and E. Weiss, *Chem. Ber.* **1977**, *110*, 3388; (b) J. E. Davies, J. Kopf and E. Weiss, *Acta Crystallogr.* **1982**, *B38*, 2251.
[152] S. Chadwick, U. Englich and K. Ruhlandt-Senge, *Organometallics* **1997**, *16*, 5792.
[153] (a) R. E. Dinnebier, M. Pink, J. Sieler and P. W. Stephens, *J. Am. Chem. Soc.* **1997**, *36*, 3398; (b) R. E. Dinnebier, M. Pink, J. Sieler, P. Norby and P. W. Stephens, *Inorg. Chem.* **1998**, *37*, 4996.
[154] D. W. Cruickshank, H. Lynton and G. A. Barclay, *Acta Crystallogr.* **1962**, *15*, 491.
[155] (a) G. van Koten, *Pure Appl. Chem.* **1990**, *62*, 1155; (b) B. Hogerheide, J. Boersma and G. van Koten, *Coord. Chem. Rev.* **1996**, *155*, 87.
[156] D. E. Pearson and C. A. Buehler, *Chem. Rev.* **1974**, *74*, 45.
[157] P. Beak and D. B. Reiz, *Chem. Rev.* **1978**, *78*, 275.
[158] P. Beak, W. J. Zajdel and D. B. Reiz, *Chem. Rev.* **1984**, *84*, 471.
[159] P. Beak and A. I. Meyers, *Acc. Chem. Res.* **1986**, *19*, 356.
[160] C. H. Heathcock and J. Lampe, *J. Org. Chem.* **1983**, *48*, 4330.
[161] P. G. Williard and M. J. Hintze, *J. Am. Chem. Soc.* **1987**, *109*, 5539.
[162] P. J. Wheatley, *J. Chem. Soc.* **1961**, 4270.
[163] H. Dunken and J. Krause, *Z. Chem.* **1961**, *1*, 27.
[164] E. Weiss, *Z. Anorg. Allg. Chem.* **1964**, *332*, 197.
[165] E. Weiss and W. Büchner, *Angew. Chem.* **1963**, *75*, 1116; *Angew. Chem. Int. Ed. Engl.* **1963**, *3*, 152.
[166] E. Weiss, *Helv. Chim. Acta.* **1963**, *46*, 2051.
[167] E. Weiss and H. Alsdorf, *Z. Anorg. Allg. Chem.* **1970**, *372*, 206.
[168] T. Greiser and E. Weiss, *Chem. Ber.* **1979**, *112*, 844.
[169] E. Weiss and U. Joergens, *Chem. Ber.* **1972**, *105*, 481.
[170] J. Sieler and M. Kunert, *Z. Kristallogr. Suppl.* **1993**, *7*, 188.
[171] A. S. Lindsay and H. Jeskey, *Chem. Rev.* **1957**, *57*, 583.
[172] O. Boullard, H. Leblanc and B. Bresson, In *Ullmanns Encyclopedia of Industrial Chemistry*, 5th ed.; VCH, Weinheim, Germany, **1993**; *Vol. A23*, p. 478.
[173] L. Lochmann, J. Pospisil and D. Lim, *Tetrahedron Lett.* **1966**, *7*, 257.
[174] M. Schlosser, *J. Organomet. Chem.* **1967**, *8*, 9.
[175] M. Schlosser and S. Strunk, *Tetrahedron Lett.* **1984**, *25*, 741.
[176] W. Bauer and L. Lochmann, *J. Am. Chem. Soc.* **1992**, *114*, 7482.
[177] T. Kremer, S. Harder, M. Junge and P. von R. Schleyer, *Organometallics* **1996**, *15*, 585.
[178] S. Harder and A. Streitwieser, *Angew. Chem.* **1993**, *105*, 1108; *Angew. Chem. Int. Ed. Engl.* **1993**, *32*, 1066.
[179] L. Lochmann, J. Pospisil, J. Vodnansky, J. Trekoval and D. Lim, *Coll. Czech. Chem. Commun.* **1965**, *30*, 2187.
[180] J. F. McGarrity and C. A. Ogle, *J. Am. Chem. Soc.* **1985**, *107*, 1805.
[181] V. Halaska and L. Lochmann, *Coll. Czech. Chem. Commun.* **1973**, *38*, 1780.
[182] K. Huml, *Czech. J. Phys.* **1965**, *B15*, 699.
[183] M. Marsch, K. Harms, L. Lochmann and G. Boche, *Angew. Chem.* **1990**, *102*, 334; *Angew. Chem. Int. Ed. Engl.* **1990**, *29*, 308.
[184] B. J. Wakefield in: *Comprehensive Organometallic Chemistry* (Editors: G. Wilkinson, F. G. A. Stone and E. W. Alell), Pergamon Press, **1982**, Vol. 7, 1.
[185] S. S. Al-Juaid, C. Eaborn, P. B. Hitchcock, C. A. McGeary and J. D. Smith, *J. Chem. Soc., Chem. Commun.* **1989**, 273.
[186] E. Weiß, *J. Organomet. Chem.* **1964**, *2*, 314.

[187] E. Weiß, *J. Organomet. Chem.* **1965**, *4*, 101.
[188] A. I. Snow and R. E. Rundle, *Acta Crystallogr.* **1951**, *4*, 348.
[189] E. Weiß, *Chem. Ber.* **1965**, *98*, 2805.
[190] P. R. Markies, G. Schat, O. S. Akkerman, F. Bickelhaupt, W. J. J. Smeets, A. J. M. Duisenberg and A. L. Spek, *J. Organomet. Chem.* **1989**, *375*, 11.
[191] H. Bock, K. Ziemer and C. Näther, *J. Organomet. Chem.* **1996**, *511*, 29.
[192] G. Wittig, F. J. Meyer and G. Lange, *Liebigs Ann. Chem.* **1951**, *571*, 167.
[193] L. M. Seitz and T. L. Brown, *J. Am. Chem. Soc.* **1967**, *89*, 1602.
[194] D. Thoennes, *Ph.D. Thesis*, Universität Hamburg, **1978**.
[195] D. Thoennes and E. Weiß, *Chem. Ber.* **1978**, *111*, 3726.
[196] E. P. Squiller, R. R. Whittle and H. G. Richey, Jr., *J. Am. Chem. Soc.* **1985**, *107*, 432.
[197] E. P. Squiller, R. R. Whittle and H. G. Richey, Jr., *Organometallics* **1985**, *4*, 1154.
[198] A. D. Pajerski, M. Parvez and H. G. Richey, Jr., *J. Am. Chem. Soc.* **1988**, *110*, 2660.
[199] O. N. D. Mackey and C. P. Morley, *J. Organomet. Chem.* **1992**, *426*, 279.
[200] B. Schubert and E. Weiss, *Chem. Ber.* **1984**, *117*, 366.
[201] M. Geissler, J. Kopf and E. Weiss, *Chem. Ber.* **1989**, *122*, 1395.
[202] H. Viehbrock and E. Weiss, *unpublished results*, cited in ref. [4].

7 Supramolecular Self-Assembly as a Result of π-Interactions

7.1 Introduction

Supramolecular self-assembly as a result of π-interactions has been found in many organometallic compounds of main group and transition metals containing for example olefin, acetylide, cyclopentadienyl, or arene ligands. The last two ligands form particularly interesting and fairly large classes of compounds. Often the borderline between supramolecular compounds associated via ionic interactions and those involving π-systems is not clear-cut. Although many compounds discussed in this chapter, especially those of the alkali metals and f-elements, are clearly almost entirely ionic, they are included here if at least part of the π-system is coordinated to the particular metal. This means, for example, that an allyllithium compound with terminal carbon atoms bridging two lithium atoms will be considered purely ionic, whereas a lithium alkyl containing an η^3-allyl ligand will be discussed in this chapter.

7.2 Main Group Elements

7.2.1 π-Interactions with acetylenes and allyl ligands

π-Interactions between carbon–carbon triple bonds and alkali metals have frequently been found in polar alkali metal acetylides. There are several lithium acetylides, however, in which π-interactions do not play in important role. These include the oligomeric compounds [(LiC≡CBu^t)(THF)]$_4$, **1**, [(LiC≡CBu^t)$_{12}$(THF)$_4$], **2** [1], [LiC≡CPh(tmpda)]$_2$, **3**, and [(LiC≡CPh)$_4$(tmhda)$_2$]$_n$, **4** [2]. In these complexes the RC≡C– ligands act as bridging ligands with the formally anionic carbon atom only being coordinated to two or three lithium atoms. On the other hand, for example, the parent compounds MC≡CH, **5** (M = Na, K, Rb) [3] and the propynides MC≡CMe, **6** (M = Na, K, Rb, Cs) [4] form polymeric sheet structures in the solid state, in which π-bonding is clearly evident. As a result of their polymeric sheet structures these compounds are not volatile and they are completely insoluble in non-polar organic solvents; because of this their crystal structures had to be determined by X-ray powder diffraction [5]. In the complex [Me$_3$SiC(C≡CBu^t)$_2$Li]$_2$, **7**, which contains a 'tweezer-like' diacetylene ligand, the lithium cations form strong electrostatic π-interactions with the acetylenic moieties [6].

More recently, two interesting hexameric lithium complexes involving acetylenic π-coordination have been prepared and structurally characterized [7]. One, [LiC≡CSiMe$_2$C$_6$H$_4$OMe]$_6$, **8**, has acetylide anions complexed with lithium whereas the other, [LiOCMe$_2$C≡CH]$_6$, **9**, is a hexameric lithium alkoxide containing non-metalated acetylene groups. Intramolecular coordination of the *o*-anisyl methoxy functions in [LiC≡CSiMe$_2$C$_6$H$_4$OMe]$_6$, **8**, eliminates external solvent effects and facilitates π-interactions between Li and the acetylide units. Similar additional π-coordination was established for the 'drum-shaped' alkoxide derivative [LiOCMe$_2$C≡CH]$_6$, **9** (cf. Chapter 6, Section 1).

π-Bonding to acetylenic moieties has been reported in several organometallic compounds of the alkaline earth metals. For example, short Be–C$_\beta$ distances in dimeric [Be(C≡CMe)$_2$(NMe$_3$)]$_2$, **10**, indicate self-assembly via π-interactions [8].

A series of heterometallic alkynylmagnesates has been prepared and structurally characterized. In these compounds the alkali metal cations connect the acetylenic moieties of the [Mg(C≡CR)$_3$]$^-$ fragments by π-contacts. Typical examples are Li$_2$[Mg(C≡CPh)$_3$(tmeda)]$_2$, **11** [9], Na$_2$[Mg(C≡CBu^t)$_3$(tmeda)]$_2$, **12**, and Na$_2$[Mg(C≡CBu^t)$_3$(pmdta)]$_2$, **13** [10]. π-Coordination has been shown to be present in the organomagnesium acetylide [(MgEt)$_2$(C≡CBu^t)$_3$(tmeda)]$_2$·C$_6$H$_6$, **14** [11].

η^3-Coordination of an allyl system was found in the diphenyl derivative [Li(η^3-1,3-diphenylallyl)(Et$_2$O)]$_n$, **15**, which forms polymeric chains in the solid state [12].

7.2.2 Cyclopentadienyl, indenyl, and fluorenyl complexes

Cyclopentadienyl complexes of the alkali metals are of fundamental importance as reagents in organometallic chemistry [13]. The structural chemistry of these species

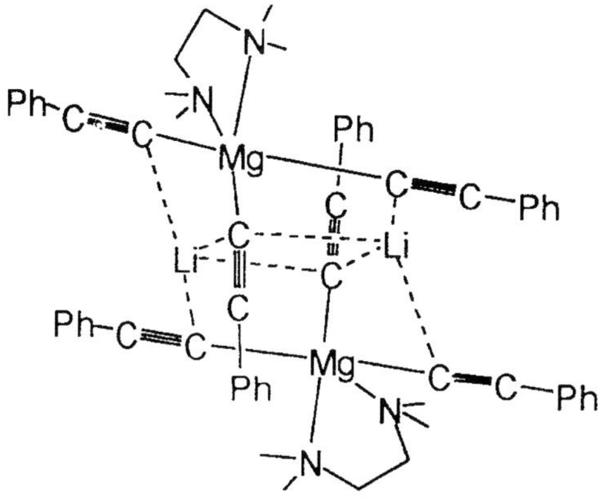

11

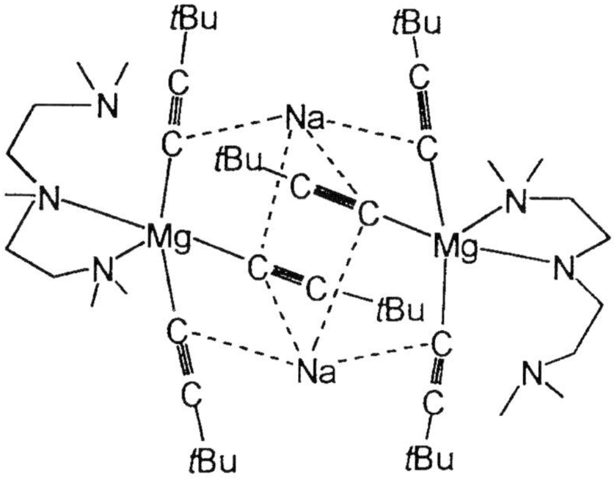

13

is characterized by extensive supramolecular self-assembly, and, especially, the formation of polymers. Although the compounds MC$_5$H$_5$, **16a–e** (M = Li, Na, K, Rb, Cs) have been known for almost one hundred years, it was not until very recently that their solid-state structures were elucidated – by powder diffraction methods using high-resolution synchrotron radiation [14]. Both [LiC$_5$H$_5$]$_n$, **16a**, and [NaC$_5$H$_5$]$_n$, **16b**, form polymeric multidecker structures in which the metal atoms are linearly coordinated by two η^5-bonded cyclopentadienyl rings (Figure 7.1).

The crystal structure of [KC$_5$H$_5$]$_n$, **16c**, differs from the 'string of pearls' structure of the lighter homologs in that the potassium atoms and η^5-coordinated cyclopentadienyl rings form polymeric zigzag chains with a K'–K–K'' angle of 138.0°. The coordination sphere around potassium is filled by additional interactions between neighboring chains. Similar chain-structures have recently been reported for unsolvated [RbC$_5$H$_5$]$_n$, **16d** [15], and [CsC$_5$H$_5$]$_n$, **16e** [16].

428 7 *Supramolecular Self-Assembly as a Result of π-Interactions*

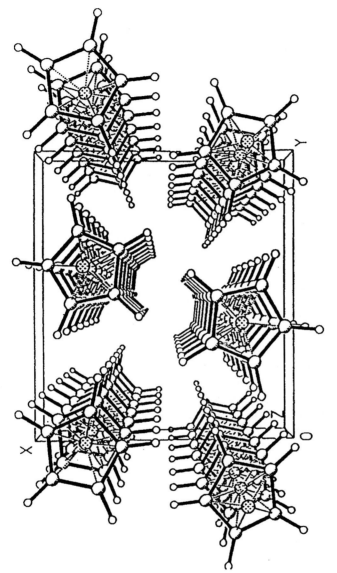

Fig. 7.1. The crystal structure of $[NaC_5H_5]_n$, **16b** (Reproduced from ref. [14]).

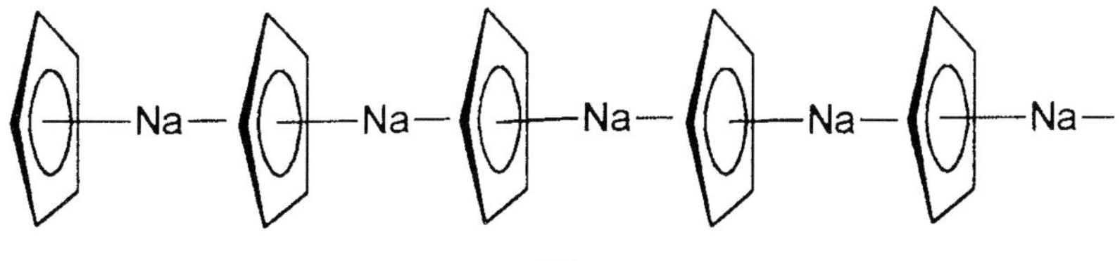

16b

16c

A similar situation was reported for the trimethylsilyl-substituted complexes [Li(η^5-C$_5$H$_4$SiMe$_3$)]$_n$, **17** [17], and [K(η^5-C$_5$H$_4$SiMe$_3$)]$_n$, **18** [18], the latter being the first structurally characterized unsolvated alkali metal cyclopentadienyl compound. Once again the lithium derivative forms a 'supersandwich' or 'string of pearls' structure consisting of stacked metal-ring units in an almost linear arrangement. The structure of the potassium homolog differs significantly from that of **17**. It consists of zigzag chains containing bent metallocene units. In the crystal structure there are additional weak η^2-interactions between adjacent chains, although the presence of bulky trimethylsilyl substituents might also be responsible for the bending of the chains. Apparently in **17** any interchain space is filled by the trimethylsilyl substituents.

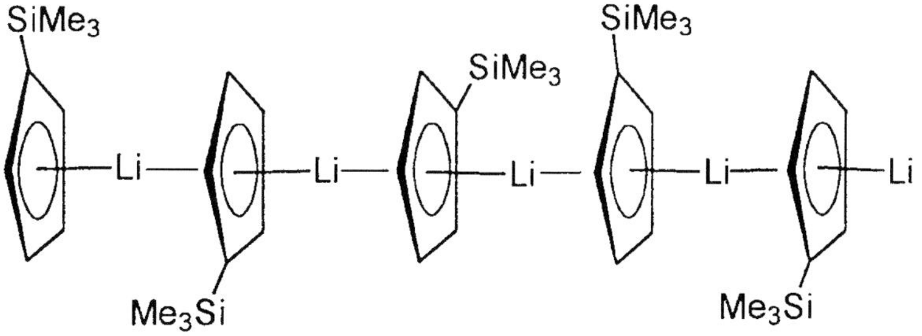

17

Related chain polymers have also been found in several Lewis-base adducts of alkali metal cyclopentadienyl, indenyl, and fluorenyl complexes. Such adducts with ethers or amines have been thoroughly investigated because they form stable, crys-

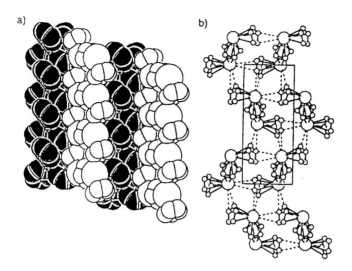

Fig. 7.2. (a) The layered crystal structure of $[PPh_4][Cs_2(C_5H_5)_3]$, **22**; (b) the supramolecular arrangement of the $[Cs_2(C_5H_5)_3]^-$ anions in **22** (Reproduced from ref. [22]).

talline materials which are well-defined and easy to use as reagents in organometallic syntheses. Zigzag chains containing additional diethyl ether or amine ligands have been found in $[Na(\eta^5\text{-}C_5H_5)(tmeda)]_n$, **19** (Na–Na–Na 128° and 119°) [19], $[K(\eta^5\text{-}C_5H_5)(Et_2O)]_n$, **20** (K–K–K 145°) [20], and $[K(\eta^5\text{-}C_5Me_5)(py)_3]_n$, **21** [21]. An interesting example of polymeric self-organization has been reported for the unusual cesium triple-decker complex $[PPh_4][Cs_2(C_5H_5)_3]$, **22** [22]. This orange –red compound was prepared by reacting cesium cyclopentadienide with $[PPh_4]Cl$ in THF solution. The anion consists of a strongly bent ($Cp_{centroid}$–Cs–$Cp_{centroid}$ 115.6(2)°) triple-decker sandwich complex. The coordination of only two cyclopentadienyl ligands leaves room in the coordination sphere of the large cesium ion. This accounts for several types of intermolecular interaction in the crystal structure, which finally lead to a two-dimensional coordination polymer (Figure 7.2b). In addition to the two η^5-coordinated cyclopentadienyl rings (one terminal and one bridging) each cesium has η^5-interactions with two cyclopentadienyl rings from neighboring chains (one 'side-on' and one 'face-on') as well as weak contact with a phenyl group of the phosphonium cation. The overall supramolecular arrangement results in a formal coordination number of 9 around cesium. The crystal structure contains alternating layers of tetraphenylphosphonium cations and $[Cs_2(C_5H_5)_3]^-$ anions which are both stacked along the crystallographic c-axis (Figure 7.2a).

Closely related is the unusual mixed-ligand cesium cyclopentadienyl/pentamethylcyclopentadienyl complex $[\{Cs(18\text{-crown-}6)\}_2(\mu\text{-}C_5Me_5)][\{Cs(18\text{-crown-}6)(C_5Me_5)\}_2\{Cs_2(C_5H_5)_3\}]$, **23** [23]. The anion consists of a central $Cs_2(C_5H_5)_3$ core to which two $\{Cs(18\text{-crown-}6)(C_5Me_5)\}$ units are attached. All interactions between cesium and the cyclopentadienyl ring systems involve pentahapto coordination. Most recently these investigations have been extended to mixed-metal alkali cyclopentadienides, showing that cyclopentadienyl chemistry of the heaviest alkali metals rubidium and cesium differs often greatly from that of the lighter homologues. An interesting example is the novel coordination polymer [Rb(C_5H_5)(18-crown-6)-Cs(C_5H_5)]$_n$, **24**, which has been prepared by reaction of CsC_5H_5, **16e**, with RbC_5H_5(18-crown-6), **25**. The central backbone of the supramolecular structure can be viewed as the polymeric zigzag chain of base-free cesium cyclopentadienide. To this chain RbC_5H_5(18-crown-6) moieties are added in an alternating fashion to give the quasi-one-dimensional structure found in **24**.

Several types of supramolecular self-assembly are found in indenyl and fluorenyl complexes of the alkali metals. It has recently become possible for the first time to elucidate the crystal structures of unsolvated lithium indenide, [LiC_9H_7]$_n$, **26**, and sodium fluorenide, [Na$C_{13}H_9$]$_n$, **27** [24]. The crystal structure of **26** closely resembles that of unsolvated lithium cyclopentadienide because it also consists of a polymeric multidecker array in which the Li atoms are symmetrically coordinated by two η^5-cyclopentadienyl rings of the indenyl ligands. In contrast, unsolvated sodium fluorenide, **27**, forms a two-dimensional supramolecular structure in the solid state (Figure 7.3). In this unusual coordination polymer both the five- and six-membered rings of the fluorene system participate in coordination to sodium via η^5- and η^3- interactions, respectively.

Even more diverse is the structural chemistry of solvated alkali metal indenides and fluorenides. For example, the green potassium complexes [K(η^5-C_7H_9)-(tmeda)]$_9$, **28**, and [K(η^5-C_7H_9)(pmdta)]$_9$, **29**, both form the familiar polymeric zigzag chains in the solid state [25a]. In the tmeda-supported derivative the indenyl anions act as η^5:η^5 bridges between potassium cations, whereas the bridging mode is η^5:η^3 in the complex containing the tridentate amine ligand pmdta.

The type of supporting ligand greatly influences the structure of organoalkali metal compounds containing cyclopentadienyl-type ligands. For example, four different structural types have been reported for solvated sodium fluorenide [25b]. The synthesis of a monomeric molecule can be achieved with the use of the tridentate pmdta ligand, while the combination with tmeda or tmpda affords a polymer or a tetramer, respectively. Finally, complexation of sodium fluorenide with two THF ligands results in polymeric zigzag chain. Fluorenylrubidium has been prepared in the form of dark orange, cube-shaped crystals by transmetalation of fluorenyllithium with RbOBu^t in THF solution. Subsequent addition of tridentate ligands such as pmdta or diglyme leads to the formation of interesting supramolecular aggregates. [Rb(η^5-C_9H_{13})(pmdta)]$_n$, **30**, crystallizes as a 'square-wave-like' polymeric chain, in which each rubidium cation is coordinated to two fluorenyl anions and one tridentate amine ligand (Figure 7.4).

Most unusual are the recently discovered nonameric structures of [Rb(η^5-C_9H_{13})-(diglyme)]$_9$, **31**, and [Cs(η^5-C_9H_{13})(diglyme)]$_9$, **32** [26]. They result from simply changing the ancillary ligand from pmdta to diglyme. In the crystal the two compounds form heart-shaped cyclic nonamers in which two neighboring Rb or Cs cations are bridged by a fluorenide anion. The coordination involves η^5-bonding to the central five-membered ring but also parts of the six-membered rings. The coordination sphere of the alkali metal is completed by the tridentate diglyme ligand. Figure 7.5 depicts the structure of the cesium derivative **32**.

432 7 *Supramolecular Self-Assembly as a Result of π-Interactions*

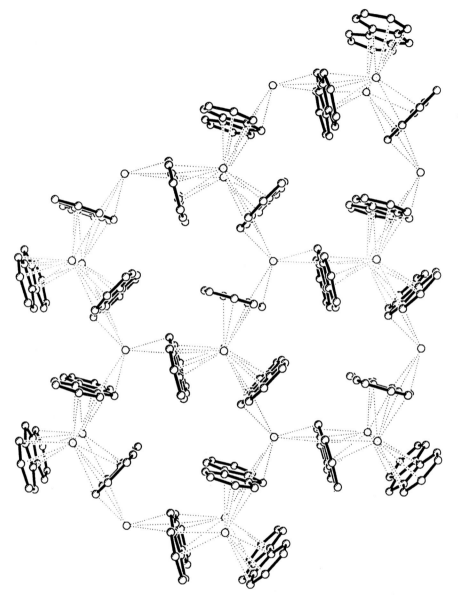

Fig. 7.3. The polymeric structure of unsolvated sodium fluorenide, **27** (Reproduced from ref. [24]).

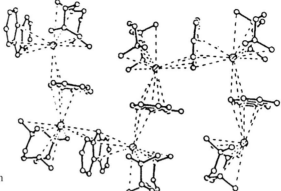

Fig. 7.4. The 'square-wave-like' polymeric chain structure of [Rb(η^5-C$_9$H$_{13}$)(pmdta)]$_n$, **30** (Reproduced from ref. [25]).

The dianion of dihydroacepentalene is of considerable interest as a potential precursor for the elusive hydrocarbon acepentalene, $C_{10}H_6$, which has been predicted to have a triplet diradical ground state. The corresponding dianion, $C_{10}H_6^{2-}$, is significantly more stable and can be generated in solution by treating triquinacene with a superbasic mixture of n-butyllithium and potassium t-pentoxide in the presence of tmeda [27]. More recently a crystalline sample of the dimethoxyethane solvate [{Li(DME)}$_2$C$_{10}$H$_6$]$_2$, **33**, has been successfully prepared according to Eq.(1) by means of a transmetallation reaction between 4,7-bis(trimethylstannyl)-dihydroacepentalene and methyllithium in DME solution at −60 °C [28].

$$\text{Me}_3\text{Sn} \diagup\diagdown \text{SnMe}_3 \quad \xrightarrow[-\ 2\ \text{SnMe}_4]{+\ 2\ \text{MeLi}} \quad 2\ \text{Li}^+ \left[\diagup\diagdown\right]^{2-} \tag{1}$$

In the crystal the compound forms dimers consisting of two bowl-shaped $C_{10}H_6^{2-}$ dianions with the convex surfaces facing each other (Figure 7.6) Two DME-solvated lithium ions are sandwiched between these convex surfaces, whereas other two [Li(DME)]$^+$ cations are located above the concave surfaces of the bowl-shaped acepentalenediide dianions. In all cases the type of interaction between lithium and the hydrocarbon ligand is η^5 coordination.

A recently reported example of a polymeric self-assembled alkali earth metallocene is the THF adduct of bis(indenyl)strontium, [Sr(η^5-C$_7$H$_9$)(THF)]$_n$, **34** [29]. In the solid state this compound forms an infinite coordination polymer in which each strontium atom is coordinated by a terminal indenyl and THF ligand. Approximately tetrahedral geometry around the metal center results from an additional association with two bridging indenyl moieties.

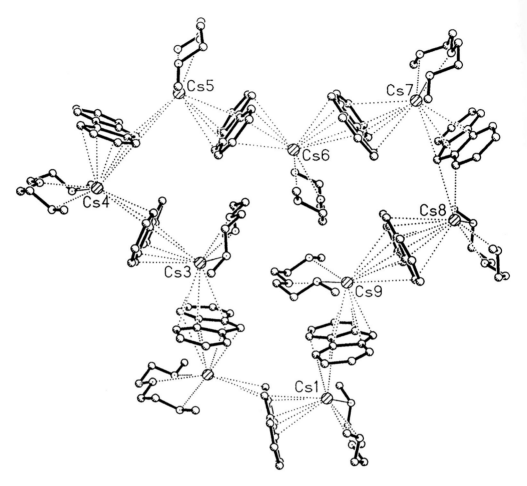

Fig. 7.5. The nonameric structure of $[Cs(\eta^5\text{-}C_9H_{13})(\text{diglyme})]_n$, **32** (Reproduced from ref. [26]).

Interesting varieties of supramolecular self-assembly have been found for some cyclopentadienyl compounds of the heavier Group 13 metals gallium, indium, and thallium. Polymeric dimethyl(cyclopentadienyl)gallium(III), **35**, is a borderline case, because this compound is not associated via η^5-cyclopentadienyl groups [30]. The structure consists of chains of dimethylgallium groups bridged by cyclopentadienyl rings along the b-axis. This results in a distorted tetrahedral coordination environment around each gallium atom. Depending on the steric bulk of the cyclopentadienyl ligands, three fundamentally different types of supramolecular aggregate have been found for cyclopentadienylindium(I) complexes. In the gas phase compounds such as $In(C_5H_5)$, **36a**, or $In(C_5H_4Me)$, **37a**, are monomeric half-sandwich complexes with the indium atom situated above the ring centroid [31, 32]. In the solid state, however, most cyclopentadienylindium(I) derivatives, including

7.2 Main Group Elements   435

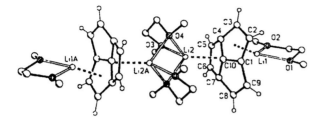

Fig. 7.6. The molecular structure of [{Li(DME)}$_2$C$_{10}$H$_6$]$_2$, **33** (Reproduced from ref. [28]).

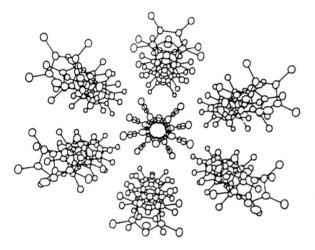

Fig. 7.7. The crystal structure of tetramethylcyclopentadienylindium(I), [In(C$_5$HMe$_4$)]$_n$, **40** (Reproduced from ref. [36]).

[In(C$_5$H$_5$)]$_n$, **36b** [24, 25, 33], [In(C$_5$H$_4$Me)]$_n$, **37b** [25], [In(C$_5$H$_4$SiMe$_3$)]$_n$, **38** [34], [In(C$_5$H$_4$Bu^t)]$_n$, **39** [35], and [In(C$_5$HMe$_4$)]$_n$, **40** [36], are associated in the form of infinite zigzag chains. Both [In(C$_5$H$_5$)]$_n$, **36b**, and [In(C$_5$H$_4$Me)]$_n$, **37b**, have significant In$\cdots$In interactions (3.986 Å in both) between adjacent chains, whereas with the more bulky cyclopentadienyl ligands no interstrand In$\cdots$In interactions have been found (5.869–6.648 Å). The supramolecular structure of tetramethylcyclopentadienylindium(I), **40**, is shown here as an example (Figure 7.7).

The shortest indium–indium distance (3.631(2) Å) in this series of cyclopentadienyl complexes has been found in the pentabenzyl derivative In[C$_5$(CH$_2$Ph)$_5$], **41**, which forms quasi-dimeric units in the crystal structure [37]. Self-assembly of pentamethylcyclopentadienylindium(I) results in a particularly fascinating supramolecular arrangement. [In(C$_5$Me$_5$)]$_6$, **42**, was prepared as a yellow–orange, exceedingly air-sensitive crystalline solid by treatment of InCl with Li(C$_5$Me$_5$) in diethyl ether [38]. In the crystal the hexameric compound forms an octahedral cluster which is not associated through π-bonding. The indium–indium distances in the octahedral In_6 core are 3.943(1) and 3.963(1) Å. The volatility of pentamethylcyclopentadienylindium(I) (sublimation at 55 °C and 0.001 torr) suggests that the octahedral cluster has only marginal stability and that monomeric half-sandwich species are probably formed in the gas phase.

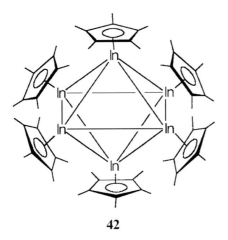

42

Tris(cyclopentadienyl)indium(III), **43**, was first prepared by Fischer and Hofmann, who obtained the material in low yields as a by-product in the preparation of cyclopentadienylindium(I) from InCl and $Na(C_5H_5)$ [26]. Higher yields are obtained by use of lithium cyclopentadienide [39]. The crystal structure of the light yellow solid consists of infinite polymeric chains in which each In atom is linked to two terminal and two bridging cyclopentadienyl ligands. This results in a slightly distorted InC_4 tetrahedral coordination environment around indium. Each bridging cyclopentadienyl is shared between two indium atoms related by a twofold screw axis along b.

Cyclopentadienylthallium(I) compounds are important organometallic reagents which are widely employed as precursors of cyclopentadienyl complexes of other metals, including f-elements [40]. These compounds are synthetically attractive as readily accessible, non-reducing cyclopentadienyl-transfer reagents. In the gas phase both cyclopentadienylthallium(I), **44** [41], and pentamethylcyclopentadienylthallium(I), **45** [42], have been shown to form open-faced half-sandwich complexes of C_{5v} symmetry. In contrast, their solid-state structures have a variety of supramolecular arrangements. The prevailing type of supramolecular self-assembly in cyclopentadienylthallium(I) compounds is that of a polymeric zigzag chain. This structure has been found in the parent compound **44** [43] and in the substituted derivatives $[Tl(C_5HMe_4)]_n$, **46** [29], $[Tl(C_5Me_5)]_n$, **45** [44], $[Tl(C_5H_4SiMe_3)]_n$, **47** [45], $[Tl\{C_5Me_4(SiMe_2Ph)\}]_n$, **48** [46], $[Tl\{C_5H_4(SiMe_2CH_2Ph)\}]_n$, **49** [39], and $[Tl\{C_5H_4(C(CN)C(CN)_2\}]_n$, **50** [47]. The polymeric structure of cyclopentadienylthallium(I), **44**, is depicted here as a typical example. It is interesting to note that despite their very similar structures some physical properties can be dramatically different in this series. For example, **44** is stable towards air and water, and it is completely insoluble in the usual organic solvents except DMSO. In marked contrast, the pentamethylcyclopentadienyl homolog, **45**, readily dissolves in organic solvents, but is exceedingly air-sensitive.

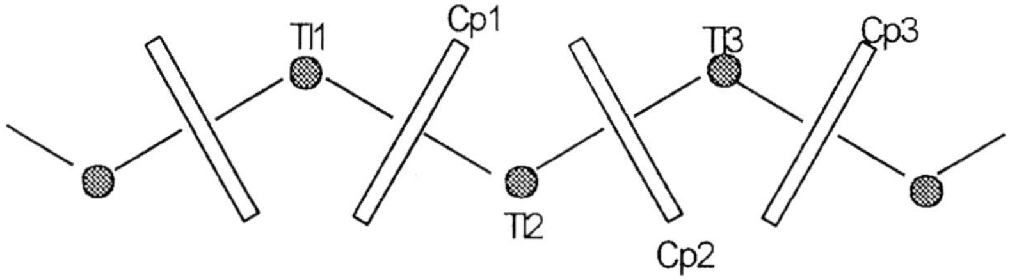

44

Three other varieties of supramolecular self-assembly have been reported for cyclopentadienylthallium(I) compounds. Particularly remarkable is pentabenzylcyclopentadienylthallium(I), **51**, two modifications of which have been found, depending on the crystallization conditions. In the first modification, **51a**, the crystal structure consists of covalent molecules which are associated to dimeric units by thallium–thallium interactions [48]. The second modification, **51b**, forms pale yellow, needle-like crystals. In the crystal structure the molecules are associated as almost linear polymeric chains in which each thallium atom is effectively shielded by five benzyl groups [49].

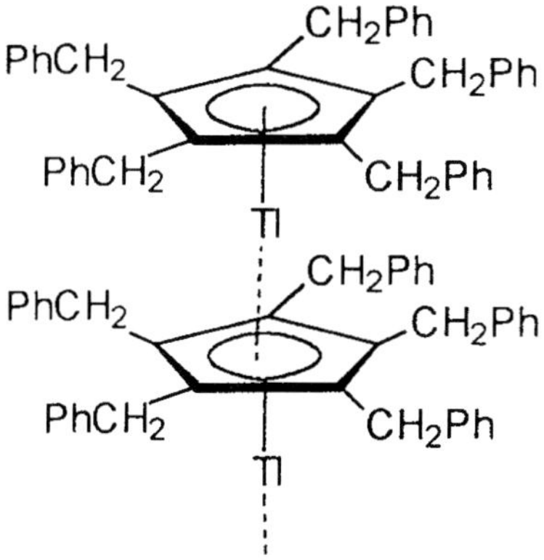

51b

Perhaps the most unusual example of supramolecular self-assembly in a cyclopentadienylthallium(I) compound was reported for the 1,3-bis(trimethylsilyl)-cyclopentadienyl derivative, $[Tl\{C_5H_3(SiMe_3)_2\text{-}1,3\}]_6$, **52** [50]. In contrast with all other substituted derivatives this compound forms a cyclic hexamer ('doughnut

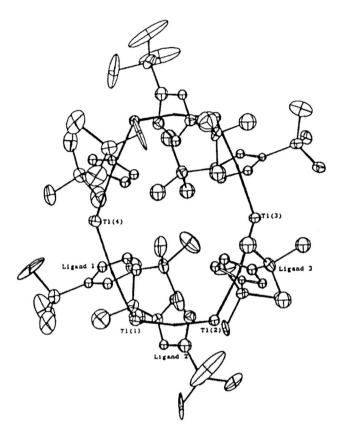

Fig. 7.8. The cyclic hexamer structure of [Tl{C$_5$H$_3$(SiMe$_3$)$_2$-1,3}]$_6$, **52** (Reproduced from ref. [50]).

molecule') (Figure 7.8). The unusual shape of the hexamers has been explained as a consequence of the preference of each thallium center in this essentially covalent molecule to employ approximately sp^2-hybrid σ-orbitals (including one occupied by a stereochemically active non-bonding electron pair).

A particularly fascinating example of supramolecular self-assembly of a main-group metal metallocene derivative is plumbocene (or bis(η^5-cyclopentadienyl)-lead(II), **53**). Originally prepared by Fischer and Grubert [51], it was shown by various methods, including electron diffraction, that Pb(C$_5$H$_5$)$_2$, **53**, has a bent sandwich structure in solution and in the gas phase [52]. The first samples suitable for X-ray diffraction were obtained by vacuum sublimation. Determination of the structure of the thus obtained orthorhombic modification revealed a polymeric zigzag chain, but it was hampered by low crystal quality [53]. The more than forty year story of plumbocene recently reached a new height when it was shown that in the solid state this compound can crystallize in no fewer than three different supramolecularly associated forms (Figure 7.9)! It was found that the originally reported zigzag morphology can also be obtained by crystallization of plumbocene from a toluene–THF mixture (Figure 7.9a) [54]. Moreover, simple recrystallization of or-

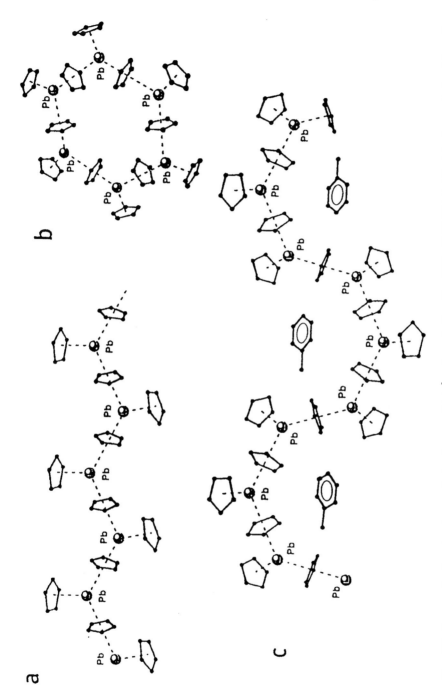

Fig. 7.9. Supramolecular structures of plumbocene (bis(η^5-cyclopentadienyl)lead(II), Pb(C$_5$H$_5$)$_2$), **53**: (a) zigzag chain; (b) cyclic hexamer; (c) sinusoidal chain (Reproduced from ref. [55]).

thorhombic plumbocene from toluene solution yields two other completely different supramolecular arrangements [55]. One is a polymeric toluene solvate (Figure 7.9c) which differs from the simple two-dimensional zigzag structure in the way that the polymer chains adopt a sinusoidal pattern to avoid the interstitial toluene molecules. Molecules of the third form are composed of six identical plumbocene units linked together into a cyclic arrangement (Figure 7.9b). Stacking of the resulting [Pb(η^5-C$_5$H$_5$)$_2$]$_6$ rings results in infinite cylindrical channels through the crystal lattice. The hexameric structure is unique for any unfunctionalized main group metallocene, and the study illustrates for the first time that non-coordinating solvent interactions can have profound consequences on the conformation and supramolecular self-assembly of a main group metallocene.

A small number of self-assembled cyclopentadienyl compounds containing Group 15 elements have been reported in the literature. Thermolabile, light-sensitive (C$_5$H$_5$)SbCl$_2$, **54**, was prepared according to Eqs(2) and (3) [56–58]. The route via trimethylstannylcyclopentadiene gives quantitative yield.

$$SbCl_3 + Na(C_5H_5) \xrightarrow{Et_2O} (C_5H_5)SbCl_2 + NaCl \qquad (2)$$
$$\mathbf{54}$$

$$SbCl_3 + Me_3SnC_5H_5 \xrightarrow{Et_2O} (C_5H_5)SbCl_2 + Me_3SnCl \qquad (3)$$
$$\mathbf{54}$$

The corresponding pentamethylcyclopentadienyl derivative (C$_5$Me$_5$)SbCl$_2$, **55**, was made analogously [59, 60]. The compounds have a very similar polymeric chain

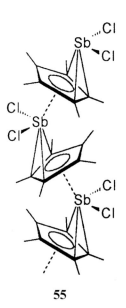

55

structure in the solid state. In a monomeric unit the SbCl$_2$ group is connected to the cyclopentadienyl moiety via an η^3 interaction. The formation of a chain polymer results from additional pentahapto Sb–Cp contacts. With distances of 3.41–3.58 Å the latter are in the range typical of the so-called 'Menshutkin complexes', e.g. C$_6$H$_6$(SbCl$_3$), **56** [61]. It was pointed out that although the intermolecular interactions are quite weak they nevertheless determine the solid-state structure of these compounds.

7.2.3 Five-membered heterocycles

There are a few reports of pyrrole derivatives in which supramolecular self assembly occurs via π-interactions. Among the structurally characterized examples are (C$_4$H$_4$N)InEt$_2$, **57** [62], (C$_4$Me$_4$N)InMe$_2$, **58** [63], (C$_4$Me$_4$N)GaMe$_2$, **59** [64], and (C$_4$H$_4$N)SnMe$_3$, **60** [65]. The distances between the metal atoms and π-bonds of neighboring pyrrolyl rings range from very weak contacts (e.g. 3.18 Å in **57**) to interactions approaching the sum of the covalent radii (e.g. 2.24 Å in **59**). As a typical example the solid-state structure of **60** is depicted here. The crystal structure consists of parallel polymeric zigzag chains. Two types of chain can be distinguished in which the intermolecular distances are slightly different.

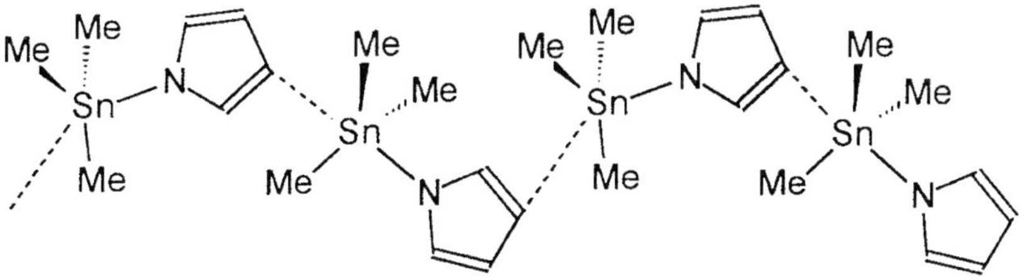

60

7.2.4 π-Interactions with arene rings

In recent years some fascinating supramolecular structures have been uncovered which result from π-interactions between alkali metals and arene rings. Classes of compounds in which this type of self-assembly has been encountered include certain lithium aryls, alkali metal silyls, and alkali metal derivatives of polycyclic hydrocarbons and various mono- and bimetallic alkoxides. As mentioned in Chapter 5, Section 1, base-free organolithium compounds have a strong tendency to associate in the form of oligomers or, more often, polymers. The polymeric species, in particular, are often difficult to characterize. The preparation of well-defined unsolvated lithium aryls suitable for X-ray diffraction was first made possible by the use of very bulky aryl ligands. Some of these ligands were originally developed with the goal of eventually crystallizing a monomeric, unsolvated lithium organometallic derivative. Instead, some interesting cases of supramolecular self-assembly via π-interactions have been uncovered for such compounds. The dimeric compound [LiC$_6$H$_3$-{C$_6$H$_3$Pr$^i_2$-2,6}$_2$]$_2$, **61**, containing a sterically highly demanding *ortho*-substituted terphenyl ligand, was prepared by treatment of the corresponding aryl iodide with *n*-butyllithium. In the unusual dimeric molecule each lithium cation is η^1-bonded to the central phenyl ring of the terphenyl system. Dimerization occurs via an additional η^6-interaction with the 2,6-diisopropylphenyl substituent of a second aryl ligand. The average Li–C(η^1) distance is 2.068(6) Å whereas the Li$\cdots$C(η^6) bond lengths vary between 2.366(7) and 2.544(6) Å [66]. The π-interaction between lithium and the aryl rings is eliminated on addition of a coordinating solvent, such as diethyl ether, which forms a monomeric 1:1 adduct [67].

61

Supramolecular self-assembly via η^6 interactions has also been found for two other structural types of unsolvated lithium aryls [68]. A mixed-ligand organolithium dimer of the composition [(LiBu^n)(LiMes*)]$_2$, **62** (Mes* = C$_6$H$_3$Bu$^t_3$-2,4,6 = 'supermesityl') resulted from treatment of Mes*Br with *n*-butyllithium. The formation of this material indicated that such a metal–halogen exchange reaction can be incomplete under certain conditions. The molecular structure of **62** involves two distinct types of Li coordination. The first coordination mode involves an η^1 bond to the α carbon atom of an *n*-butyl group and an η^6-interaction with the aromatic supermesityl ring. The other two Li atoms are η^1-bonded both to an α carbon of an *n*-butyl unit and the *ipso* carbon of a supermesityl ligand.

Yet another variety of self-association via π-interactions was discovered in the solid-state structure of unsolvated 2,4,6-triisopropylphenyllithium. In the crystal this compound forms the tetramer [Li(2,4,6-Pr$^i_3$C$_6$H$_2$)]$_4$, **63**. This structure has not been observed for other aryllithium species. It consists of an almost planar arrangement of four lithium atoms, each σ-bonded to just one phenyl ring (average Li–C distance 2.12 Å). In addition, each Li interacts with the π-electron system of a

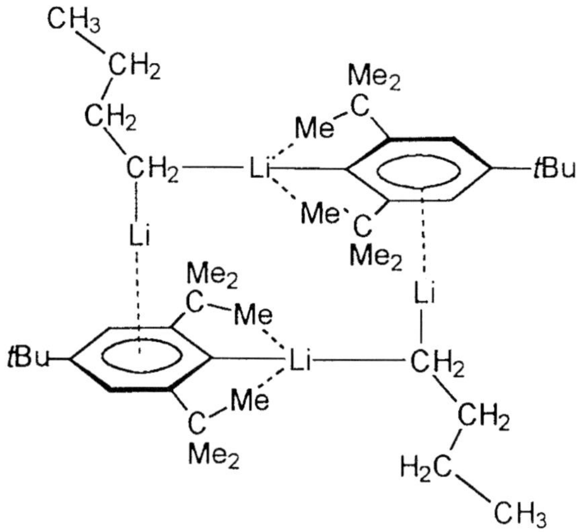

62

second phenyl ring. In some sense these interactions play a solvating role while the Pr^i substituents provide solubility in hydrocarbon solvents. ^{13}C NMR data indicate different amounts of self-assembly in solution. Whereas the tetrameric structure is presumably retained in hydrocarbon solvents, it is easily disrupted on addition of diethyl ether. From these solutions a solvated dimer of formula [Li(2,4,6-Pr$^i_3$C$_6$H$_2$)(Et$_2$O)]$_2$, **64**, can be crystallized [69].

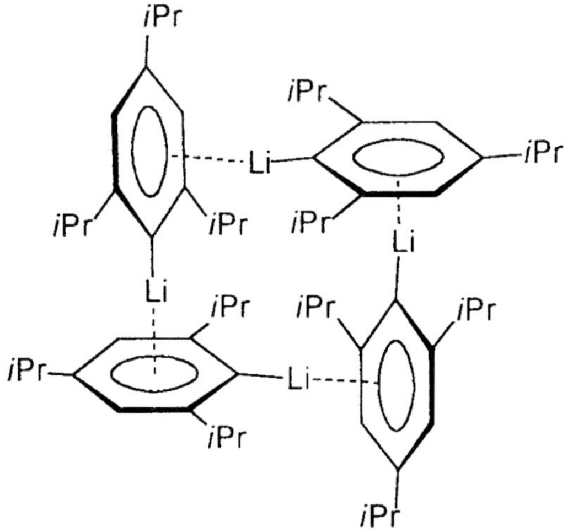

63

One of the most spectacular recent results in this field was the successful structure determination of base-free phenyllithium, **65**, by high-resolution synchrotron X-ray powder diffraction [70]. Samples of highly pure base-free phenyllithium were obtained by rapid crystallization from diethyl ether–cyclopentane mixtures yielding a white, pyrophoric powder. The fundamental unit of the polymeric solid-state structure is a dimeric Li$_2$Ph$_2$ molecule with a central four-membered Li$_2$C$_2$ ring. The phenyl rings are in a perpendicular orientation relative to the Li$_2$C$_2$ units. Self-assembly occurs as a result of strong interaction of the π-electrons of the phenyl rings with the Li atoms of neighboring Li$_2$Ph$_2$ molecules. As a result, base-free phenyllithium crystallizes as a polymeric, infinite zigzag ladder along the crystallographic b-axis, a structure which is unprecedented in organolithium chemistry.

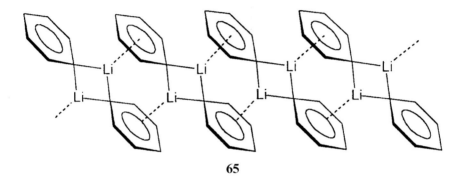

65

Other reports on supramolecular self-assembly via η^6 interactions in alkali metal hydrocarbyls focus on benzyl and triphenylmethyl derivatives of these metals and complexes containing extended anionic π-systems. In the former mostly polymeric solid-state structures have been found. For example, metalation of toluene with the 'superbasic' 1:1 mixture of Bu^{n}Li and MOBu^t (M = K, Rb) at room temperature afforded orange–red powders of benzylpotassium and benzylrubidium [71]. These could be redissolved upon addition of N,N,N′,N′,N′-pentamethyldiethylenetriamine (pmdta) to give burgundy-red solutions, from which needle-like crystals of [PhCH$_2$K(pmdta)(toluene)$_{0.5}$]$_n$, **66**, and [PhCH$_2$Rb(pmdta)]$_n$, **67**, were isolated. In the crystal both compounds have polymeric zigzag arrangements, in which each potassium or rubidium ion bridges two benzyl moieties by η^3 and η^6 interactions. The coordination sphere of the alkali metal cation is completed by the tridentate pmdta ligand. The methylene group in these compounds is essentially in the plane of the phenyl ring, thus enabling maximum charge delocalization. ^{13}C NMR data indicate that this coplanar arrangement of the benzylic carbon atom is retained in solution. The bright red alkali metal derivatives of triphenylmethane have attracted chemists for many years and their structural chemistry has been thoroughly investigated. For example, polymeric supramolecular structures involving extensive π-coordination have been found for the triphenylmethyl (or 'trityl') derivatives of the heaviest alkali metals rubidium and cesium [72]. Whereas several structurally characterized triphenylmethyl compounds of the lighter alkali metals, for example

[Ph$_3$CLi(tmeda)], **68**, [Ph$_3$CLi(Et$_2$O)$_2$], **69**, or [Li(12-crown-4)$_2$][Ph$_3$C], **70**, are monomeric in the solid state, both [Ph$_3$CRb(pmdta)]$_n$, **71**, and [Ph$_3$CCs(pmdta)]$_n$, **72**, form one-dimensional polymers. In the zigzag chain of **71** each rubidium cation bridges two trityl units by η^6-interactions with two phenyl rings; the tridentate amine ligand pmdta completes the coordination sphere of Rb. Interesting structural differences are found in the cesium homolog **72** which also crystallizes as a zigzag chain polymer. The Cs cation interacts with the trityl moieties in a quite unexpected fashion, however. It is η^6-bonded to the phenyl ring of one carbanion, whereas the second triphenylmethyl moiety is coordinated in a propeller-like arrangement with three shorter and four longer Cs–C distances (3.348(4)–3.820(4) Å). This propeller-like coordination involves all three phenyl rings and the deprotonated central carbon.

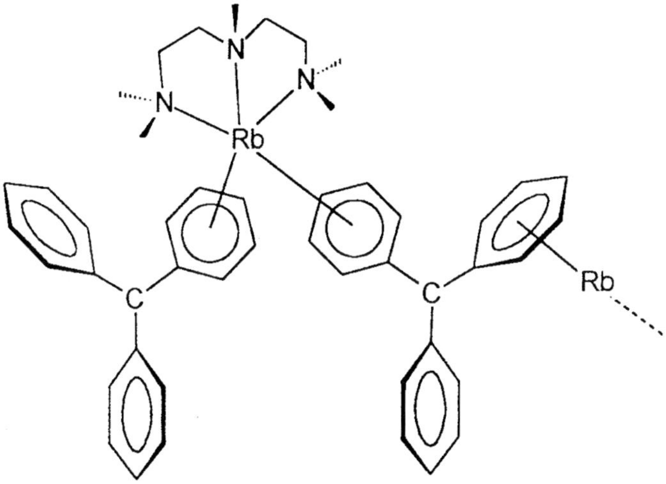

71

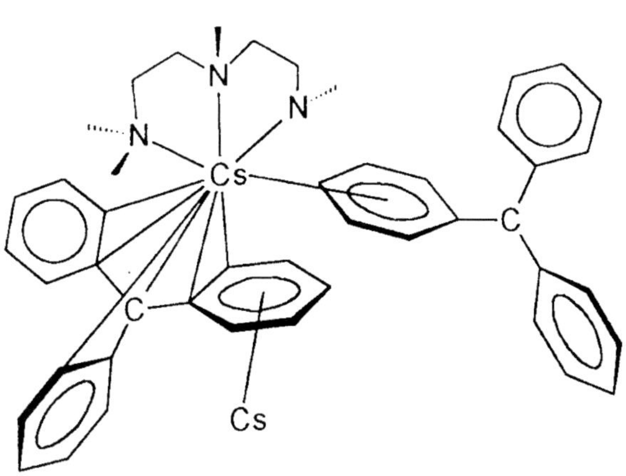

72

π-Interactions between phenyl rings and alkali metals can even occur in some more 'exotic' molecular structures as exemplified by the unusual heterobimetallic lutetium butadiene complex [K(THF)$_2$(μ-Ph$_2$C$_4$H$_4$)$_2$Lu(thf)$_2$]$_n$, **73** [73]. In **73** the central Lu atom is coordinated by two η^4-1,4-diphenyl-1,3-butadiene ligands and two THF molecules. Self-assembly with the formation of one-dimensional polymeric chains occurs because potassium forms η^6-interactions with the π-systems of two phenyl substituents from neighboring molecules.

73

The above-mentioned 1,4-diphenyl-1,3-butadiene ligand immediately leads us to the other area of organoalkali metal chemistry in which self-assembly through π-interactions plays an important role, i.e. the alkali metal derivatives of hydrocarbons containing extended π-electron systems. For example, 1,1,2,2-tetraphenylethane and 1,1,4,4-tetraphenyl-1,3-butadiene react with elemental cesium in the presence of appropriate ethers such as diglyme with the formation of deeply colored, highly air-sensitive organocesium species containing the corresponding hydrocarbon dianions [74]. Surface reactivation of the metal by occasional brief ultrasonic sound irradiation has been found to have a positive effect on these reactions. Black, crystalline [(Ph$_2$CCPh$_2$)Cs(diglyme)$_2$]$_n$, **74**, forms a coordination polymer in which the cesium interacts with four phenyl rings of two different tetraphenylethanediyl dianions in addition to the tridentate coordination of the diglyme ligand (Figure 7.10). One of the π-interactions can be described as η^6-coordination. In the dianion, the Ph$_2$C– halves are twisted by 76° around the central C–C bond, which is elongated from 1.36 Å in the neutral hydrocarbon to 1.51 Å, i.e. by 0.15 Å. The closely related violet–blue compound [(Ph$_2$CCHCHCPh$_2$){Cs(diglyme)}$_2$-(CsOCH$_2$CH$_2$OMe)$_2$]$_n$, **75**, contains a 1,1,4,4-tetraphenylbutadiene-1,4-diyl dianion, with which cesium has η^6- and η^7-interactions. The incorporated cesium 2-methoxyethanolate results from reductive ether cleavage of the diglyme solvent by Cs metal. In the crystal both compounds form one-dimensional supramolecular polymers.

The structural chemistry of solvated alkali metal derivatives of planar polycyclic hydrocarbon dianions has been well investigated. Monomeric complexes have been isolated with hydrocarbons such as naphthalene, anthracene, perylene, pentalene,

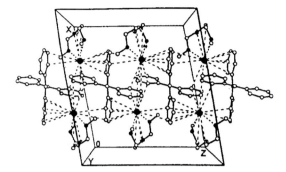

Fig. 7.10. The crystal structure of [(Ph$_2$CCPh$_2$)Cs(diglyme)$_2$]$_n$, **74** (Reproduced from ref. [74]).

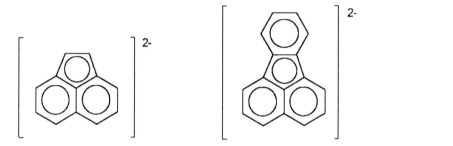

Fig. 7.11. The structures of the acenaphthylene dianion (left) and the fluoranthene dianion (right).

or acenaphthylene [75]. Systematic comparison of the Li and Na derivatives of the hydrocarbons acenaphthylene and fluoranthene revealed that the negative charge in the acenaphthylene dianion is mainly located within the five-membered ring and in the fluoranthene dianion within the six-membered rings of the naphthalene subunit (Figure 7.11) [76].

As a consequence, the (15-crown-5)-solvated sodium salt of acenaphthylene is a monomeric contact ion triple in the solid state, whereas the DME- or diglyme-solvated sodium and lithium salts of the fluoranthene dianion crystallize in polymer strings held together by π-interactions with the six-membered rings. As a typical example involving η^6-coordination the molecular and crystal structure of [{Na(dme)}$_2$(C$_{16}$H$_{10}$)]$_n$, **76**, is depicted here (Figure 7.12) [76]. A similar example of the preference of an alkali metal for coordination to a six-membered ring in a ligand system containing both five- and six-membered rings is the sheet-structure of THF-solvated potassium fluorenide [(μ-thf){K(C$_{13}$H$_9$)}$_2$]$_n$, **77** (Figure 7.13). Surprisingly, the potassium ions are not coordinated primarily to the cyclopentadienyl part of the anion but rather interact in two ways (η^4 and η^6) with the arene rings. Two potassium ions are connected by a bridging THF ligand. The η^6-interactions are responsible for the supramolecular self-assembly into a two-dimensional polymeric network [23].

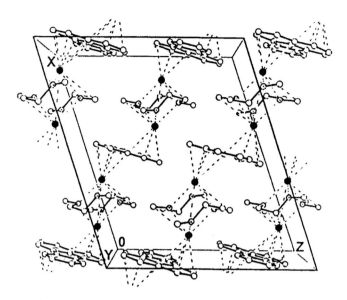

Fig. 7.12. Crystal structure of $[\{Na(DME)\}_2(C_{16}H_{10})]_n$, **76** (Reproduced from ref. [76]).

There is one report in the literature on the supramolecular association of alkali metal silanides via π-interactions. In a very thorough and detailed study Klinkhammer investigated the structural chemistry of alkali metal tris(trimethylsilyl)-silanides (or 'hypersilanides'; cf. Chapter 6, Section 1) and found various kinds and degrees of self-assembly, including one very remarkable case involving π-bonding [77]. Addition of biphenyl to a suspension of cesium hypersilanide produced a pale yellow solution from which pale yellow crystals of composition $[\{CsSi(SiMe_3)_3\}_2$-(biphen)(pentane)$_{0.5}]_n$, **78**, could be isolated at $-60\,°C$. In the complicated crystal structure biphenylene acts as a bridging ligand between $[CsSi(SiMe_3)_3]_2$ dimers. One phenyl ring is nearly η^6-coordinated to one cesium (Cs2) and the second phenyl ring acts as a η^1-donor to the other Cs atom (Cs1) of a neighboring dimer. Then there are additional intra- and intermolecular Cs$\cdots$CH$_3$ interactions with the SiMe$_3$ groups leading to infinite zigzag chains of $[CsSi(SiMe_3)_3]_2$ dimers. Finally, the coordination sphere of one Cs atom is completed by interactions with the terminal C–H bonds of an intercalated n-pentane molecule. All these interactions combined lead to a polymeric layer structure shown in Figure 7.14.

Most of the remaining examples of supramolecular alkali metal compounds involving π-coordination are alkoxides, siloxides, and thiolates. In polymeric sodium phenolate $[\{NaOC_6H_4Me-4\}_2]_n$, **79** [78], the η^6-arene interaction of the phenolate ligands with sodium plays a crucial role, because it restricts the structure of the resulting polymer to a one-dimensional stack of $[NaOC_6H_4Me-4]_2$ dimers and prevents the formation of a fully extended three-dimensional network.

The formation of polymeric chains by η^6-arene coordination to alkali metals has

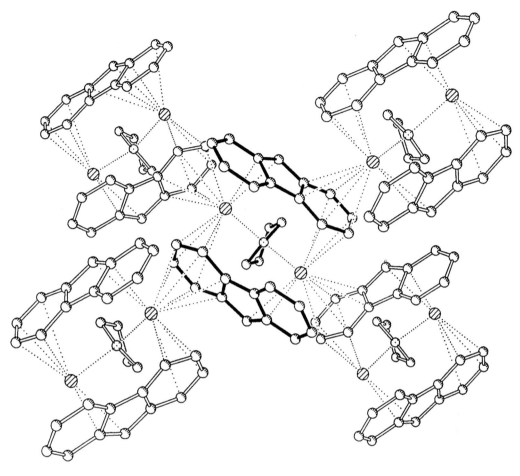

Fig. 7.13. Polymeric network structure of $[(\mu\text{-THF})\{K(C_{13}H_9)\}_2]_n$ **77** (Reproduced from ref. [23]).

been found to occur frequently in 'ate' or 'double alkoxide' complexes of the lanthanide elements. The first example was the pale blue heterobimetallic neodymium alkoxide $[K\{Nd(OC_6H_3Pr^i_2\text{-}2,6)_4\}]_n$, **80**, prepared by reacting anhydrous $NdCl_3$ with three equivalents of potassium 2,6-diisopropylphenoxide [79]. It is noteworthy that even though the reaction was conducted in THF solution the product was isolated as a base-free material. In the solid state the compound has a quasi-one-dimensional infinite chain structure of $[Nd(OC_6H_3Pr^i_2\text{-}2,6)_4]^-$ anions bridged by potassium cations via bis(η^6-arene) interactions between adjacent $[Nd(OC_6H_3Pr^i_2\text{-}2,6)_4]^-$ units. The principal potassium interaction is with the oxygen atom of one of the bulky aryloxide ligands; the remainder of the coordination sphere is filled by close K–C contacts with the arene rings of three separate aryloxide ligands. A

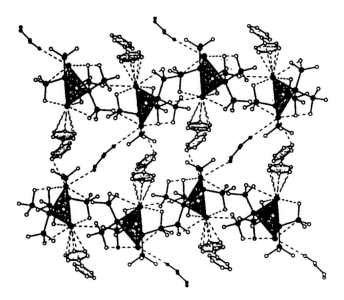

Fig. 7.14. The polymeric network structure of $[\{CsSi(SiMe_3)_3\}_2(biphen)(pentane)_{0.5}]_n$, **78** (Reproduced from ref. [77]).

similar type of supramolecular self-assembly was found in the heterobimetallic samarium(II) alkoxide complex $[K\{Sm(OC_6H_2Bu^t_2\text{-}2,6\text{-}Me\text{-}4)_3\}]_n$, **81** [80]. In this compound the potassium cation is coordinated by two aryloxide oxygen atoms and one phenyl ring from an adjacent monomeric unit, resulting in a polymeric chain structure.

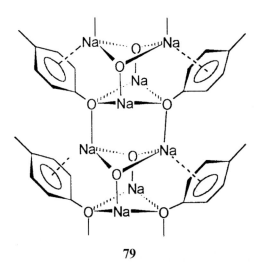

79

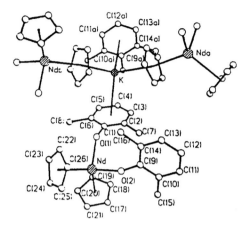

Fig. 7.15. The polymeric structure of [K{(μ-C$_5$H$_5$)$_2$Nd(μ-OC$_6$H$_3$Me$_2$-2,6)$_2$}]$_n$, **82** (Reproduced from ref. [79]).

80

A related compound was subsequently reported in which the potassium cation is coordinated to carbon atoms only. The neodymium 'ate' complex [K{(μ-C$_5$H$_5$)$_2$-Nd(μ-OC$_6$H$_3$Me$_2$-2,6)$_2$}]$_n$, **82**, was prepared according to Eq. (4) [81].

$$1/2[Nd(OC_6H_3Pr^i_2\text{-}2,6)_3(THF)_2]_2 + 2K(C_5H_5)$$

$$\rightarrow 1/n[K\{(\mu\text{-}C_5H_5)_2Nd(\mu\text{-}OC_6H_3Me_2\text{-}2,6)_2\}]_n + K(OC_6H_3Pr^i_2\text{-}2,6)$$
$$\mathbf{82}$$

(4)

The crystal structure of **82** comprises [(μ-C$_5$H$_5$)$_2$Nd(μ-OC$_6$H$_3$Me$_2$-2,6)$_2$]$^-$ anions connected into a two-dimensional layered structure (Figure 7.15). The most unusual structural feature is the coordination around potassium. The metal is coordinated in a distorted pseudo-tetrahedral arrangement by two cyclopentadienyl rings (η^5) and two aryloxide arene rings (η^6) with no close contacts with oxygen donor atoms (shortest K$\cdots$O distance 4.4 Å). An interesting structural difference is noted when cesium as the largest alkali metal ion is combined with an 'ate' complex of a lanthanide element, in this case lanthanum. The complex $[Cs_2\{La(OC_6H_3Pr^i_2\text{-}2,6)_5\}]_n$, **83**, was the first five-coordinate anionic aryloxolanthanide complex. In the solid state **83** forms a three-dimensional extended structure held together exclusively by multiple cesium–π-arene interactions. The overall structure features channels which contain four toluene solvent molecules for every three $[Cs_2\{La(OC_6H_3Pr^i_2\text{-}2,6)_5\}]$ units.

7.3 Transition Elements

Transition metal complexes self-assembled via π-interactions are not large in number, although some very interesting examples have been reported. This area of organometallic chemistry is far from being systematically investigated and it is quite difficult to make any predictions concerning possible supramolecular structures of particular compounds. With the exception of several polymeric olefin and arene complexes, the majority of the complexes described in this section are cyclopentadienyl derivatives.

7.3.1 Olefin complexes

Supramolecular self-assembly has been found in several silver(I) complexes containing bridging olefinic ligands. Typical examples are the silver nitrate olefin adducts $[(\mu\text{-norbornadiene})(AgNO_3)_2]_n$, **84** [82], and $[(\mu\text{-COT})AgNO_3]_n$, **85** (COT = 1,3,5,7-cyclooctatetraene) [83]. The supramolecular structure of **85** is depicted here as an example. The silver cation is coordinated to two non-adjacent double bonds of the neutral cyclooctatetraene ligand with average Ag–C distances ranging from 2.46 to 2.84 Å. Longer Ag–C interactions (3.17, 3.29 Å) join these units into infinite chains along the c-axis of the crystal.

7.3.2 Cyclopentadienyl complexes

Supramolecular self-assembly via bridging cyclopentadienyl ligands occurs, for example, in the metallocenes $[Zn(C_5H_5)_2]_n$, **86** [84], and $[Cd(C_5H_5)_2]_n$, **87**. In the crystalline state these compounds form coordination polymers and dissolve only in strongly coordinating solvents, thereby forming monomeric adducts. From their

85

insolubility and low volatility the substituted derivatives [Zn(C$_5$H$_4$Me)$_2$]$_n$, **88**, and [Zn(C$_5$H$_4$Bu^t)$_2$]$_n$, **89**, are also thought to be coordination polymers in the solid state [85a], whereas pentasubstituted cyclopentadienyl ligands have been reported to yield monomeric zinc complexes with one η^1- and one η^5-bonded ring [85b]. Bis-(cyclopentadienyl)zinc, **86**, was first prepared by Fischer and co-workers in 1969 [84a]. Initially a ferrocene-like structure with predominantly ionic metal-ring bonding was proposed; other authors suggested a structure with only η^1-bonded cyclopentadienyl rings, on the basis that all cyclopentadienyl ring protons are equivalent in the ^{1}H NMR spectrum. Crystal structure determination revealed, however, that in the solid state **86** consists of infinite chains of Zn atoms with bridging cyclopentadienyl groups. In addition, each zinc carries a terminal cyclopentadienyl ligand. Neither the terminal nor the bridging cyclopentadienyls are attached to Zn in a purely η^1-fashion. Clearly the π-electrons participate in the bonding and the overall coordination mode has been described as '$\eta^{2.5}$'.

Also polymeric is the self-assembled structure of (cyclopentadienyl)methylzinc, [Zn(C$_5$H$_5$)Me]$_n$, **90** [85c]. This compound is readily obtained by reacting methylzinc iodide with sodium cyclopentadienide; it forms colorless, needle-shaped crystals. IR and ^{1}H NMR data are consistent with η^5-coordination of the cyclopentadienyl ligand in solution. In the solid state **90** forms polymeric zigzag chains in which methylzinc moieties are bridged by cyclopentadienyl rings. Once again, hapticities are assumed to be in the range 2–3.

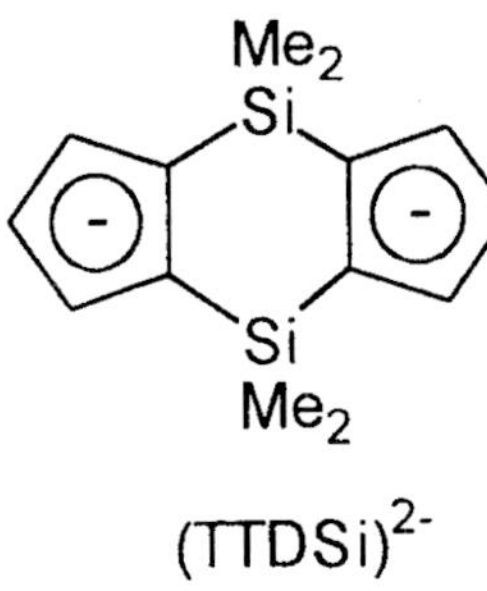

(TTDSi)$^{2-}$

Fig. 7.16. The molecular structure of the bis(cyclopentadienyl)ligand, (TTDSi)$^{2-}$.

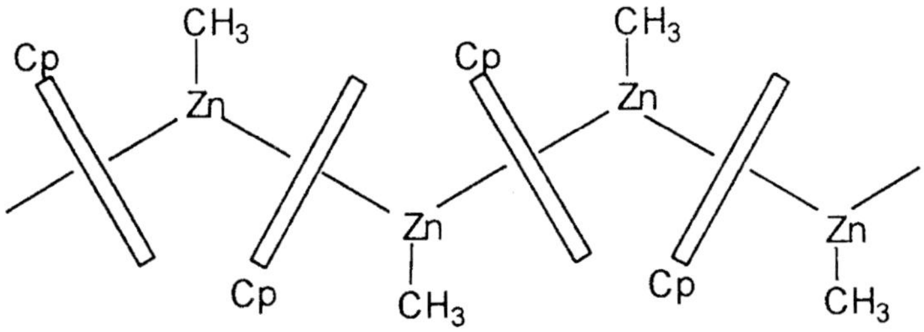

90

Organoiron chemistry includes a beautiful example of a highly unusual and esthetically appealing supramolecular arrangement [86]. Reactions of transition metal halides with the bis(cyclopentadienyl) ligand (TTDSi)$^{2-}$ (Figure 7.16) containing two silyl bridges often yield polymeric materials. For example, treatment of CoCl$_2$ or NiCl$_2$ with Li$_2$(TTDSi) afforded polymeric solids of composition [M(TTDSi)]$_n$ (M = Co, **91**, Ni, **92**) which proved to be insoluble in all common organic solvents. The corresponding reaction with iron(II) chloride, however, gave a crude product which could be partially extracted with hexane. Recrystallization of the soluble portion from benzene yielded orange–red crystals. An X-ray study revealed the presence of a cyclic heptamer of composition [Fe(TTDSi)]$_7$, **93**, in which seven ferrocene units bridged by seven pairs of Me$_2$Si groups form an almost regular ring (Figure 7.17). The outside of the resulting 'paddlewheel' bears fourteen methyl groups. This hydrocarbon character of the surface is responsible for the high solubility of the complex. A cyclovoltammetric study of **93** indicated three reversible oxidation steps involving successive transfer of three, one, and again three electrons.

7.3.3 Arene complexes

Fairly little is known about self-assembly of transition metal complexes via π-arene interactions. A prominent example, [AgClO$_4$(C$_6$H$_6$)]$_n$, **94**, was structurally characterized at a time when X-ray crystallography was still far from being a routine method [87]. Crystals of **94** can be obtained by recrystallizing silver(I) perchlorate

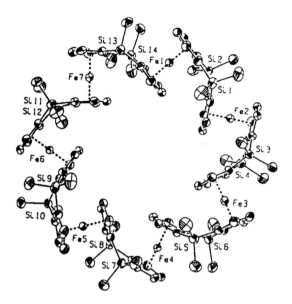

Fig. 7.17. The molecular structure of [Fe(TTDSi)]$_7$, **93**.

from benzene. The crystal structure contains polymeric chains of silver ions bridged by benzene ligands. Each benzene ring bridges two silver ions above and below the ring by η^2-interactions. A most recent example of a polymeric arene-bridged transition metal complex is [Rh$_2$(O$_2$CCF$_3$)$_4$·C$_6$Me$_6$]$_n$ [88].

7.4 *f*-Elements

Several types of supramolecular structures involving π-bonds have been reported for organolanthanide compounds, and, to a much lesser extent, organoactinides. The tendency to undergo self-assembly in the solid state can be attributed to the general tendency of *f*-element ions to achieve high coordination numbers. Usually these are attained by addition of solvent molecules or formation of 'ate' complexes. This can be illustrated by the synthetically important bis(pentamethylcyclopentadienyl)-lanthanide chlorides [89]. Despite the presence of two sterically demanding pentamethylcyclopentadienyl ligands, a monomeric (C$_5$Me$_5$)$_2$LnCl unit seems to be coordinatively unsaturated. This leads to a strong tendency to dimerize or to add either solvent molecules or alkali metal halides. Thus typical products which are obtained from reactions of anhydrous lanthanide trichlorides and two equivalents of alkali metal pentamethylcyclopentadienides are complexes such as (C$_5$Me$_5$)$_2$LnCl(thf), [(C$_5$Me$_5$)$_2$Ln(μ-Cl)]$_2$, and, most often, "ate" complexes of the type (C$_5$Me$_5$)$_2$Ln(μ-Cl)$_2$MS$_2$ where M is an alkali metal and S stands for a coordinated solvent molecule. For example, the compounds (C$_5$Me$_5$)$_2$Ln(μ-Cl)$_2$Li(thf)$_2$ are obtained when the LiC$_5$Me$_5$ is used as pentamethylcyclopentadienyl transfer reagent and the reactions are performed in THF solution. If no alkali metal halides or coordinating solvent are present, self-assembly can occur via participation of other organic ligands. This leads to some interesting examples of supramolecular solid-state structures, which are most prominent in the unsolvated tris(cyclopentadienyl)-lanthanides.

7.4.1 Acetylide complexes

An unusual group of organolanthanide complexes in which self-assembly occurs via both acetylide and pentamethylcyclopentadienyl ligands has been reported by Evans et al [90]. The "ate" complexes $(C_5Me_5)_2Ln(\mu\text{-}Cl)_2K(thf)_2$ (Ln = Ce, Nd, Sm), made from $LnCl_3$ and KC_5Me_5, were found to react cleanly with two equivalents of $KC\equiv CPh$ according to Eq. (5) with formation of the bimetallic acetylide complexes $[(C_5Me_5)_2Ln(C\equiv CPh)_2K]_n$, **95**.

$$(C_5Me_5)_2Ln(\mu\text{--}Cl)_2K(thf)_2 + 2KC\equiv CPh$$

$$\rightarrow (1/n)[(C_5Me_5)_2Ln(C\equiv CPh)_2K]_n + 2KCl + 2THF$$

$$\text{Ln} = \text{Ce}, \mathbf{95a}, \text{Nd}, \mathbf{95b}, \text{Sm}, \mathbf{95c} \tag{5}$$

Despite their polymeric nature and the presence of potassium the complexes **95** are surprisingly soluble in arene solvents. The overall structure of **95a** is similar to that of $(C_5Me_5)_2Y(C\equiv CBu^t)_2Li(thf)$, **96** [91], with potassium being coordinated to the two acetylide units. In contrast with **96** there is no solvent molecule present in **95a** and the formation of a polymeric structure occurs via additional η^5-coordination to a bridging pentamethylcyclopentadienyl ligand.

95a

7.4.2 Cyclopentadienyl complexes

Cyclopentadienyl complexes form the largest and most investigated class of supramolecular organolanthanide complexes [89]. It has been shown that in the solid state di- and trivalent lanthanide cyclopentadienyls have oligomeric or polymeric structures in which aggregation occurs via π-interactions with bridging cyclopentadienyl ligands. One example of the parent lanthanide(II) metallocenes, brick-red [Yb(C$_5$H$_5$)$_2$]$_n$, **97**, has been crystallized by sublimation at 420°C and structurally characterized [92]. The polymeric structure of **97** is isostructural with that of [Ca(C$_5$H$_5$)$_2$]$_n$ [93]. Polymeric zigzag chains are linked by weak interchain η^1-cyclopentadienyl Yb–C interactions. Each ytterbium is bonded to one terminal and two bridging C$_5$H$_5$ ligands. In the presence of coordinating solvents such as THF or DME the polymeric structure of **97** is readily disrupted.

Soluble unsolvated lanthanide(II) metallocenes have been obtained with the use of the bulky 1,3-bis(trimethylsilyl)cyclopentadienyl ligand. Purple [Yb{C$_5$H$_3$(SiMe$_3$)$_2$-1,3}$_2$]$_n$, **98**, and red [Eu{C$_5$H$_3$(SiMe$_3$)$_2$-1,3}$_2$]$_n$, **99**, were prepared by sublimation of the solvated precursors Yb[C$_5$H$_3$(SiMe$_3$)$_2$-1,3]$_2$(Et$_2$O) and Eu[C$_5$H$_3$(SiMe$_3$)$_2$-1,3]$_2$(thf), respectively [94]. The ytterbium complex adopts a bent metallocene conformation with a centroid–Yb–centroid angle of 138°. There is a close contact between the metal of one [Yb{C$_5$H$_3$(SiMe$_3$)$_2$-1,3}$_2$] unit and a methyl group of a neighboring unit (Yb$\cdots$C 2.872(7) Å), which is much shorter than the sum of the van der Waals radii (3.8 Å). The intermolecular interactions results in a 'herring bone weave' pattern of the polymeric structure. Similar short intermolecular contacts are found in the solid-state structure of the europium analog. In addition, complex **99** has an unprecedented conformation with a cyclopentadienyl ring bridging η^3:η^5 between two non-equivalent europium atoms.

Not only unsolvated lanthanide(II) metallocenes, but also certain solvated derivatives, can have supramolecular structures. The THF adduct of bis(methylcyclopentadienyl)ytterbium(II), [(MeC$_5$H$_4$)$_2$Yb(thf)]$_n$, **100**, has been synthesized by several preparative routes, including photolysis or thermolysis of [(MeC$_5$H$_4$)Yb(μ-Me)]$_2$ [95]. In the solid state, yellow crystalline **100** forms polymeric chains oriented along the b-axis. The repeating units, (μ-MeC$_5$H$_4$)(MeC$_5$H$_4$)Yb(thf), are related by a twofold screw axis and are connected by one bridging methylcyclopentadienyl ligand per Yb. Two related heterobimetallic lanthanide(II) cyclopentadienyl complexes, [NaYb(C$_5$H$_5$)$_3$]$_n$, **101** [92], and [NaSm(C$_5$H$_4$Bu^t)$_3$(thf)]$_n$, **102** [96], have also been reported to be self-assembled in the solid state. Green **101** was made according to Eqs. (6) or (7) followed by sublimation at ca 400 °C.

$$Yb(C_5H_5)_3 + Na(C_{10}H_8) \rightarrow (1/n)[NaYb(C_5H_5)_3]_n + C_{10}H_8 \quad (6)$$
$$\mathbf{101}$$

$$NaC_5H_5 + Yb(C_5H_5)_2 \rightarrow (1/n)[NaYb(C_5H_5)_3]_n \quad (7)$$
$$\mathbf{101}$$

The X-ray crystal structure of **101** shows a highly symmetrical three-dimensional array of Na and Yb atoms bridged by μ-η^5:η^5-cyclopentadienyl ligands. The Na–C$_5$H$_5$–Yb arrangement is nearly linear and both sodium and ytterbium are coordinated by three C$_5$H$_5$ ligands in a pseudo-triangular arrangement [92]. The solid state structure of **102** is closely related although in this compound the

[NaSm(C$_5$H$_4$Bu^t)$_3$] units form planar polymer grids between which molecules of THF of crystallization are located [96].

101

By far the most important organolanthanide complexes with supramolecular structures are the tris(cyclopentadienyl)lanthanides(III) and some of their ring-substituted derivatives. Homoleptic complexes of the type Ln(C$_5$H$_5$)$_3$ are the longest known and most thoroughly investigated organolanthanide compounds [89]. They were first described by Birmingham and Wilkinson in 1954 [97]. Some of their ring-substituted derivatives, such as Ln(C$_5$H$_4$Pr^i)$_3$, are currently of interest in materials science [98]. These complexes are among the most volatile lanthanide organometallics and play an important role as precursors for MOCVD processes used for lanthanide doping of semiconductor materials. Structurally the tris-(cyclopentadienyl)lanthanides(III) are perhaps the most diverse group of self-assembled organolanthanide complexes. Although the formal coordination number in a Ln(C$_5$H$_5$)$_3$ complex containing three η^5-coordinated cyclopentadienyl rings is 9, such monomers are coordinatively unsaturated and associate to form oligomers or chain polymers, unless a coordinating solvent is present. Complexes of composition Ln(C$_5$H$_5$)$_3$ have been prepared for the whole lanthanide series including the radioactive promethium. Several synthetic routes leading to these compounds have been developed, with the most straightforward being the original preparation which involves treatment of anhydrous lanthanide trichlorides with three equivalents of sodium cyclopentadienide according to Eq.(8) [89, 97].

$$LnCl_3 + 3NaC_5H_5 \rightarrow Ln(C_5H_5)_3 + 3NaCl \qquad (8)$$

Ln = Sc, Y, La, Ce, Pr, Nd, Sm, Eu, Gd, Tb, Dy, Ho, Er, Tm, Yb, Lu

103a–p,

The preparation outlined in Eq.(8) initially affords the THF adducts $Ln(C_5H_5)_3(THF)$, which can be desolvated by vacuum-sublimation of the crude products at elevated temperatures (up to ca. 400 °C). The resulting base-free tris-(cyclopentadienyl)lanthanides have been found to have three different polymeric solid-state structures. In all of these supramolecular self-assembly occurs via bridging cyclopentadienyl ligands with the bridging mode varying between the three structural types. As expected, the highest formal coordination number of eleven is found in the derivatives containing the largest ions of the lanthanide series, i.e. the early lanthanides. The crystal structures of $La(C_5H_5)_3$, **103c**, and $Pr(C_5H_5)_3$, **103e**, consist of zigzag chain polymers in which each lanthanide ion is bonded to three cyclopentadienyl ligands in an η^5-fashion [98]. Self-assembly occurs through μ-η^5:η^2 bridging cyclopentadienyl rings.

103c, e (Ln = La, Pr)

Zigzag chain polymers are also characteristic of the solid-state structures of **7.103b, g, m, n, o** (Ln = Y, Sm, Er, Tm, Yb) [89]. Because of the smaller ionic radii of the lanthanide ions the formal coordination number is reduced to ten. Once

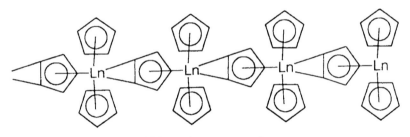

103b, g, m, n, o (Ln = Y, Sm, Er, Tm, Yb)

again each lanthanide ion is surrounded by three η^5-coordinated cyclopentadienyl ligands. In these compounds, however, there is only an additional η^1-interaction of one cyclopentadienyl ring with a neighboring Ln ion.

A further subtle structural change occurs with lutetium, the smallest lanthanide ion. In polymeric Lu(C$_5$H$_5$)$_3$, **103p**, the central metal can accommodate only two pentahapto-coordinated cyclopentadienyl ligands [99]. The third ligand bridges two lutetium ions via two different carbon atoms of the ring. The overall geometry results in a formal coordination number of eight around lutetium. Surprisingly, an X-ray investigation revealed that the scandium analog Sc(C$_5$H$_5$)$_3$, **103a**, is isostructural with **103p** [100]. This can be regarded as a consequence of the well-known lanthanide contraction.

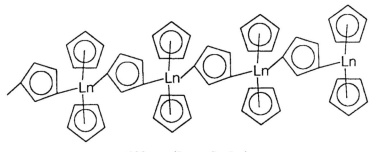

103a, p (Ln = Sc, Lu)

Several ring-substituted derivatives of the tris(cyclopentadienyl)lanthanides(III) have been found to be self-assembled in the crystalline state. In contrast with the polymeric parent compounds tetrameric assemblies prevail in these compounds. Tetramers have been structurally characterized for the methylcyclopentadienyl complexes [La(C$_5$H$_4$Me)$_3$]$_4$, **104a** [101a], and [Ce(C$_5$H$_4$Me)$_3$]$_4$, **104b** [101b], as well as for the *t*-butylcyclopentadienyl complex [Ce(C$_5$H$_4$Bu^t)$_3$]$_4$, **105** [101b]. In all three

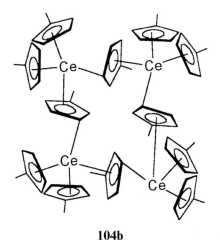

104b

tetrameric compounds each lanthanide ion is surrounded by two terminal substituted cyclopentadienyl ligands and a bridging ring which is η^5-coordinated to one lanthanide ion and η^1-bonded to an adjacent Ln. The formal coordination number around the central lanthanides is 10.

An unprecedented cyclic structure was found in a rare example of an organoactinide complex associated via π-interactions. The reaction of ThBr$_4$(thf)$_4$ in THF with one equivalent of cyclopentadienyl thallium followed by treatment with three equivalents of KOPr^i led to the formation of the highly unusual hexanuclear complex [(C$_5$H$_5$)Th$_2$(OPr^i)$_7$]$_3$, **106** [102]. In the cyclic molecule three dinuclear Th$_2$(OPr^i)$_7$ units are connected by (μ-η^5:η^5) bridging cyclopentadienyl ligands. The formation of such a cyclic structure via μ-η^5:η^5 cyclopentadienyl interactions is without precedent among d- and f-block elements.

106

7.4.3 Five-membered heterocycles

Whereas the structural chemistry of supramolecularly associated cyclopentadienyl lanthanide complexes has been thoroughly investigated, there is only one report of a self-assembled organolanthanide compound containing a five-membered heterocyclic ligand. Treatment of SmCl$_3$ with KTmp (Tmp = 1,2,3,4-tetramethylphospholyl) in a 1:3 molar ratio produced orange [(Tmp)$_6$Sm$_2$(KCl)$_2$(toluene)$_3$]$_n$, **107**, for which X-ray analysis revealed a complicated polymeric network [103]. The supramolecular structure of **107** consists of interconnected eight-membered rings. A representation of an individual eight-membered ring unit is depicted in Figure 7.18. Interestingly the structure contains three differently coordinated Tmp ligands: one is only η^5-bonded to Sm, one is η^1-bonded to Sm and η^5-bonded to potassium, and one is η^5-bonded to Sm and η^1-bonded to potassium.

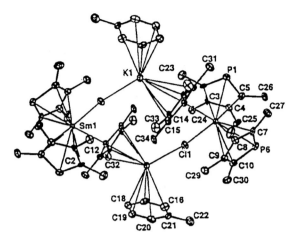

Fig. 7.18. The supramolecular structure of [(Tmp)$_6$Sm$_2$(KCl)$_2$(toluene)$_3$]$_n$, **107** (Reproduced from ref. [103]).

7.4.4 Arene complexes

In recent years significant advances have been made in the chemistry of organo-lanthanides containing neutral π-donors [104]. It is now fully recognized that neutral π-donors are suitable ligands for lanthanides once the correct choice of oxidation state and coordination environment is ensured. Only very rare examples of self-assembled species have, however, been discovered by X-ray crystallography. The unexpected formation of dimeric species via intermolecular π-interaction with an arene ring was found in the lanthanide aryloxide complexes [Ln(OC$_6$H$_3$Pr$^i_2$-2,6)$_3$]$_2$, **108a** (Ln = Nd), **7.108b** (Ln = Sm), **7.108c** (Ln = Er) [105]. The complexes have been prepared by an amide route according to Eq.(9).

$$2Ln[N(SiMe_3)_2]_3 + 6\ 2,6\text{-Pr}^i_2C_6H_3OH$$

$$\rightarrow [Ln(OC_6H_3Pr^i_2\text{-}2,6)_3]_2 + 6HN(SiMe_3)_2 \qquad (9)$$

108a–c, Ln = Nd, Sm, Er

According to X-ray studies on **108a** and **108b** the dimeric molecules are held together by η^6-arene–Ln interactions involving one unique aryloxide ligand of each Ln(OC$_6$H$_3$Pr$^i_2$-2,6)$_3$ unit. The overall coordination geometry results in a formal coordination number of six around each lanthanide ion.

The other known case of an organolanthanide complex self-assembled via π-arene interactions is the unusual bimetallic naphthalene complex [(C$_5$H$_5$)V(μ-η^6:η^2-C$_{10}$H$_8$)Yb(THF)(C$_5$H$_5$)]$_n$, **109** [106]. The compound was made by reaction of vanadocene with excess (C$_{10}$H$_8$)Yb(THF)$_2$ in tetrahydrofuran. In the solid state **109** forms a polymer built up of cyclopentadienylytterbium units, (C$_5$H$_5$Yb)$_n$. Each Yb atom of the chain is further η^2-coordinated to a naphthalene ligand of a $(C_5H_5)V(C_{10}H_8)$ moiety. The Yb–C_5H_5–Yb chain is bent at ytterbium, but linear at the cyclopentadienyl centroids bridging two Yb atoms in a $(\mu\text{-}\eta^5\text{-}\eta^5)$ fashion. The diagram of the structure of **109** illustrates the coordination geometry around ytterbium and vanadium.

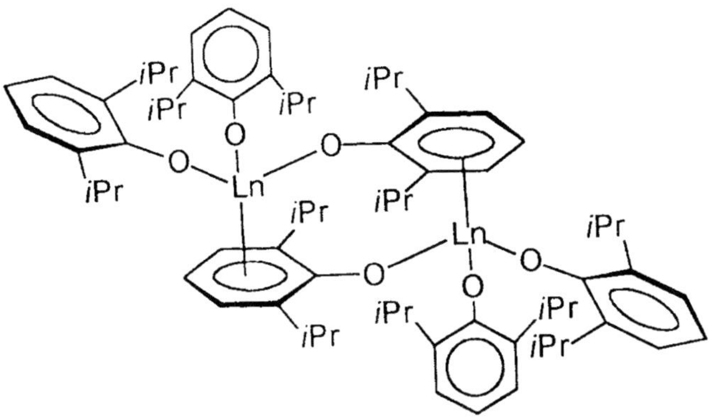

108

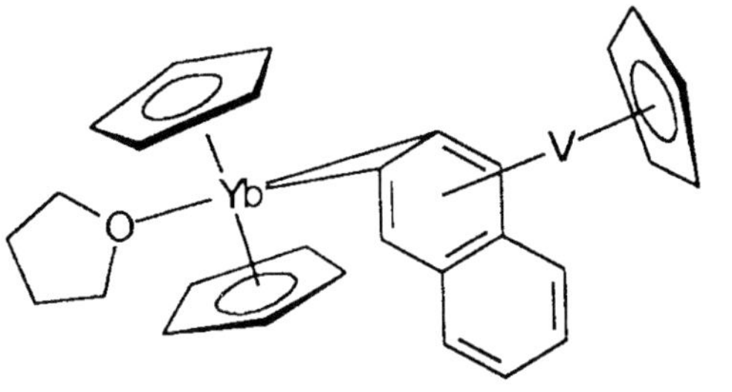

109

References

[1] M. Geissler, J. Kopf, B. Schubert, E. Weiss, W. Neugebauer and P. von R. Schleyer, *Angew. Chem.* **1987**, *99*, 569; *Angew. Chem. Int. Ed. Engl.* **1987**, *26*, 587.
[2] (a) B. Schubert and E. Weiss, *Chem. Ber.* **1983**, *116*, 3212; (b) B. Schubert and E. Weiss, *Angew. Chem.* **1983**, *95*, 499; *Angew. Chem. Int. Ed. Engl.* **1983**, *22*, 496.
[3] (a) E. Weiss and H. Plass, *Chem. Ber.* **1968**, *101*, 2947; (b) M. Atoji, *J. Chem. Phys.* **1972**, *56*, 4947.
[4] (a) R. J. Pulham and D. P. Weston, *J. Chem. Res. (S)* **1995**, 406; (b) R. J. Pulham, D. P. Weston, T. A. Salvesen and J. J. Thatcher, *J. Chem. Res. (S)* **1995**, 254.
[5] C.-C. Chang, B. Srinivas, M.-L. Wu, W.-H. Chiang, M. Y. Chiang and C.-S. Hsiung, *Organometallics* **1995**, *14*, 5150.

[6] W. N. Setzer and P. von R. Schleyer, *Adv. Organomet. Chem.* **1985**, *27*, 169.
[7] B. Goldfuss, P. von R. Schleyer and F. Hampel, *J. Am. Chem. Soc.* **1997**, *119*, 1072.
[8] N. A. Bell, I. W. Nowell and H. M. M. Shearer, *J. Chem. Soc. Chem. Commun.* **1982**, 147.
[9] B. Schubert and E. Weiss, *Chem. Ber.* **1984**, *117*, 366.
[10] M. Geissler, J. Kopf and E. Weiss, *Chem. Ber.* **1989**, *122*, 1395.
[11] H. Viebrock, D. Abeln and E. Weiss, *Z. Naturforsch.* **1994**, *B49*, 89.
[12] (a) G. Boche, H. Etzrodt, M. Marsch, W. Massa, G. Baum, H. Dietrich and W. Mahdi, *Angew. Chem.* **1986**, *98*, 84; *Angew. Chem. Int. Engl.* **1986**, *25*, 194; (b) G. Boche, G. Fraenkel, J. Cabral, K. Harms, N. J. R. van Eikema Hommes, J. Lorenz, M. Marsch and P. von R. Schleyer, *J. Am. Chem. Soc.* **1992**, *114*, 1562.
[13] (a) P. Jutzi, *Adv. Organomet. Chem.* **1986**, *26*, 217; (b) P. Jutzi, *Pure Appl. Chem.* **1989**, *61*, 1731; (c) P. Jutzi, *J. Organomet. Chem.* **1990**, *400*, 1; (d) E. Weiss, *Angew. Chem.* **1993**, *105*, 1565; *Angew. Chem. Int. Ed. Engl.* **1993**, *32*, 1501; (e) D. Stalke, *Angew. Chem.* **1994**, *106*, 2256; *Angew. Chem. Int. Ed. Engl.* **1994**, *33*, 2168.
[14] R. E. Dinnebier, U. Behrens and F. Olbrich, *Organometallics* **1997**, *16*, 3855.
[15] R. E. Dinnebier, F. Olbrich, S. van Smaalen and P. W. Stephens, *Acta Crystallogr.* **1997**, *B53*, 153.
[16] R. E. Dinnebier, F. Olbrich and G. M. Bendele, *Acta Crystallogr.* **1997**, *B53*, 699.
[17] W. J. Evans, T. J. Boyle and J. W. Ziller, *Organometallics* **1992**, *11*, 3903.
[18] P. Jutzi, W. Leffers, B. Hampel, S. Pohl and W. Saak, *Angew. Chem.* **1987**, *99*, 563; *Angew. Chem. Int. Ed. Engl.* **1987**, *26*, 583.
[19] T. Aoyagi, H. M. M. Shearer, K. Wade and G. Whitehead, *J. Organomet. Chem.* **1979**, *175*, 21.
[20] V. Jordan, U. Behrens, F. Olbrich and E. Weiss, *J. Organomet. Chem.* **1996**, *517*, 81.
[21] G. Rabe, H. W. Roesky, D. Stalke, F. Pauer and G. M. Sheldrick, *J. Organomet. Chem.* **1991**, *403*, 11.
[22] S. Harder and M. H. Prosenc, *Angew. Chem.* **1996**, *108*, 101; *Angew. Chem. Int. Ed. Engl.* **1996**, *35*, 97.
[23] F. Olbrich, private communication.
[24] F. Olbrich, R. E. Dinnebier, S. Neander and M. Behrens, *Organometallics* **1999**, in press.
[25] (a) S. Corbelin, J. Kopf and E. Weiss, *Chem. Ber.* **1991**, *124*, 2417; (b) D. Hoffmann, F. Hampel and P. von R. Schleyer, *J. Organomet. Chem.* **1993**, *456*, 13.
[26] F. Olbrich and S. Neander, private communication.
[27] T. Lendvai, T. Friedel, H. Butenschön, T. Clark and A. de Meijere, *Angew. Chem.* **1986**, *98*, 734; *Angew. Chem., Int. Ed. Engl.* **1986**, *25*, 719.
[28] R. Haag, R. Fleischer, D. Stalke and A. de Meijere, *Angew. Chem.* **1995**, *107*, 1642; *Angew. Chem., Int. Ed. Engl.* **1995**, *34*, 1492.
[29] J. S. Overby and T. P. Hanusa, *Organometallics* **1996**, *15*, 2205.
[30] K. Mertz, F. Zettler, H. D. Hausen and J. Weidlein, *J. Organomet. Chem.* **1976**, *122*, 159.
[31] S. Shibata, L. S. Bartell and R. M. Gavin, *J. Chem. Phys.* **1964**, *41*, 717.
[32] O. T. Beachley, Jr., J. C. Pazik, T. E. Glassman, M. R. Churchill, J. C. Fettinger and R. Blom, *Organometallics* **1988**, *7*, 1051.
[33] (a) E. O. Fischer and H. P. Hofmann, *Angew. Chem.* **1957**, *69*, 639; (b) C. Peppe, D. G. Tuck and L. Victoriano, *J. Chem. Soc. Dalton Trans.* **1981**, 2592.
[34] O. T. Beachley, Jr., J. F. Lees, T. E. Glassmann, M. R. Churchill and L. A. Buttrey, *Organometallics* **1990**, *9*, 2488.
[35] O. T. Beachley, Jr., J. F. Lees and R. D. Rogers, *J. Organomet. Chem.* **1991**, *418*, 165.
[36] H. Schumann, H. Kucht, A. Kucht, F. H. Görlitz and A. Dietrich, *Z. Naturforsch.* **1992**, *47b*, 1241.
[37] H. Schumann, C. Janiak, F. Görlitz, J. Loebel and A. Dietrich, *J. Organomet. Chem.* **1989**, *363*, 243.
[38] O. T. Beachley, Jr., M. R. Churchill, J. C. Fettinger, J. C. Pazik and L. Victoriano, *J. Am. Chem. Soc.* **1986**, *108*, 4666.
[39] F. W. B. Einstein, M. M. Gilbert and D. G. Tuck, *Inorg. Chem.* **1972**, *11*, 2832.
[40] W. C. Spink and M. D. Rausch, *J. Organomet. Chem.* **1986**, *308*, C1.
[41] (a) S. Shibata, L. S. Bartell and R. M. Gavin, Jr., *J. Chem. Phys.* **1964**, *41*, 717; (b) J. K. Tyler, A. P. Cox and J. Sheridan, *Nature* **1959**, *183*, 1182.

[42] R. Blom, H. Werner and J. Wolf, *J. Organomet. Chem.* **1988**, *354*, 293.
[43] (a) E. Frasson, F. Menegus and C. Panattoni, *Nature* **1963**, *199*, 1078; (b) J. F. Berar, G. Calvarin, C. Pommier and D. Weigel, *J. Appl. Crystallogr.* **1975**, *8*, 386; (c) F. Olbrich and U. Behrens, *Z. Kristallogr.* **1997**, *212*, 47.
[44] H. Werner, H. Otto and H. J. Kraus, *J. Organomet. Chem.* **1986**, *315*, C57.
[45] S. Harvey, C. L. Raston, B. W. Shelton, A. H. White, M. F. Lappert and G. Srivastava, *J. Organomet. Chem.* **1978**, *328*, C1.
[46] H. Schumann, H. Kucht, A. Dietrich and L. Esser, *Chem. Ber.* **1990**, *123*, 1811.
[47] M. B. Freeman, L. G. Sneddon and J. C. Huffman, *J. Am. Chem. Soc.* **1977**, *99*, 5194.
[48] H. Schumann, C. Janiak, J. Pickardt and U. Börner, *Angew. Chem.* **1987**, *99*, 788; *Angew. Chem. Int. Ed. Engl.* **1987**, *26*, 789.
[49] H. Schumann, C. Janiak, M. A. Khan and J. J. Zuckerman, *J. Organomet. Chem.* **1988**, *354*, 7.
[50] S. Harvey, C. L. Raston, B. W. Skelton, A. H. White, M. F. Lappert and G. Srivastava, *J. Organomet. Chem.* **1987**, *328*, C1.
[51] E. O. Fischer and H. Grubert, *Z. Anorg. Allg. Chem.* **1956**, *286*, 237.
[52] (a) H. P. Fritz and E. O. Fischer, *J. Chem. Soc.* **1961**, 547; (b) L. D. Dave, D. F. Evans and G. Wilkinson, *J. Chem. Soc.* **1959**, 3684; (c) E. Weiss, *Z. Anorg. Allg. Chem.* **1956**, *287*, 236; (d) A. Almenningen, A. Haaland and T. Motzfeldt, *J. Organomet. Chem.* **1967**, *7*, 97.
[53] C. Panattoni, C. Bombieri and U. Croatto, *Acta. Crystallogr.* **1966**, *21*, 823.
[54] J. S. Overby, T. P. Hanusa and V. G. Young, *Inorg. Chem.* **1998**, *37*, 163.
[55] M. A. Beswick, C. Lopez-Casideo, M. A. Paver, P. R. Raithby, C. A. Russell, A. Steiner and D. S. Wright, *J. Chem. Soc. Chem. Commun.* **1997**, 109.
[56] P. Jutzi, M. Kuhn and F. Herzog, *Chem. Ber.* **1975**, *108*, 2439.
[57] T. F. Berlitz, J. Pebler and J. Lorberth, *J. Organomet. Chem.* **1988**, *348*, 175.
[58] W. Frank, *J. Organomet. Chem.* **1991**, *406*, 331.
[59] P. Jutzi, U. Meyer, S. Opiela, M. M. Olmstead and P. P. Power, *Organometallics* **1990**, *9*, 1459.
[60] R. A. Bartlett, A. Cowley, P. Jutzi, M. M. Olmstead and H.-G. Stammler, *Organometallics* **1992**, *11*, 2837.
[61] (a) D. Mootz and V. Händler, *Z. Anorg. Allg. Chem.* **1986**, *533*, 23 (b) P. Jutri and N. Burford, *Chem. Rev.* **1999**, *99*, 969.
[62] M. Porchia, F. Benetollo, N. Brianese, G. Rossetto, P. Zanella and G. Bombieri, *J. Organomet. Chem.* **1992**, *424*, 1.
[63] J. Tödtmann, W. Schwarz, J. Weidlein and A. Haaland, *Z. Naturforsch.* **1993**, *48b*, 1437.
[64] H.-D. Hausen, J. Tödtmann and J. Weidlein, *Z. Naturforsch.* **1994**, *49b*, 430.
[65] J. Hillmann, H.-D. Hausen, W. Schwarz and J. Weidlein, *Z. Anorg. Allg. Chem.* **1995**, *621*, 1785.
[66] B. Schiemenz and P. P. Power, *Angew. Chem.* **1996**, *108*, 2288; *Angew. Chem. Int. Ed. Engl.* **1996**,
[67] B. Schiemenz and P. P. Power, *Organometallics* **1996**, *15*, 958.
[68] K. Ruhlandt-Senge, J. J. Ellison, R. J. Wehmschulte, F. Pauer and P. P. Power, *J. Am. Chem. Soc.* **1993**, *115*, 11353.
[69] R. A. Bartlett, H. V. R. Dias and P. P. Power, *Organometallics*, **1988**, *341*, 1.
[70] R. E. Dinnebier, U. Behrens and F. Olbrich, *J. Am. Chem. Soc.* **1998**, *120*, 1430.
[71] D. Hoffmann, W. Bauer, F. Hampel, N. J. R. van Eikema Hommes, P. von R. Schleyer, P. Otto, U. Pieper, D. Stalke, D. S. Wright and R. Snaith, *J. Am. Chem. Soc.* **1994**, *116*, 528.
[72] D. Hoffmann, W. Bauer, P. von R. Schleyer, U. Pieper and D. Stalke, *Organometallics* **1993**, *12*, 1193.
[73] N. S. Emel'yanova, A. A. Trifonov, L. N. Zakharov, M. N. Bochkarev, A. F. Shestakov and Y. T. Struchkov, *Metalloorg. Khim.* **1993**, *6*, 363.
[74] H. Bock, T. Hauck and C. Näther, *Organometallics* **1996**, *15*, 1527.
[75] (a) J. J. Brooks, W. E. Rhine and G. Stucky, *J. Am. Chem. Soc.* **1972**, *94*, 7346; (b) W. E. Rhine, J. Davis and G. Stucky, *J. Am. Chem. Soc.* **1975**, *97*, 2079; (c) H. Bock, K. Ruppert, C. Näther, Z. Havlas, H.-F. Herrmann, C. Arad, I. Göbel, A. John, J. Meuret, S. Nick, A. Rauschenbach, W. Seitz, T. Vaupel and B. Solouki, *Angew. Chem.* **1992**, *104*, 564; *Angew. Chem. Int. Ed. Engl.* **1992**, *31*, 550.

[76] H. Bock, C. Arad and C. Näther, *J. Organomet. Chem.* **1996**, *520*, 1.
[77] K. W. Klinkhammer, *Chem. Eur. J.* **1997**, *3*, 1418.
[78] (a) W. J. Evans, R. E. Golden and J. W. Ziller, *Inorg. Chem.* **1993**, *32*, 4111; (b) M. P. Hogerheide, S. N. Ringelberg, M. D. Janssen, J. Boersma, A. L. Spek and G. van Koten, *Inorg. Chem.* **1996**, *35*, 1195.
[79] D. L. Clark, J. G. Watkin and J. C. Huffman, *Inorg. Chem.* **1992**, *31*, 1556.
[80] W. J. Evans, R. Anwander, M. A. Ansari and J. W. Ziller, *Inorg. Chem.* **1995**, *34*, 5.
[81] (a) W. J. Evans, M. A. Ansari and S. I. Khan, *Organometallics* **1995**, *14*, 558; (b) D. L. Clark, G. B. Deacon, T. Feng, R. V. Hollis, B. L. Scott, B. W. Skelton, J. G. Watkin and A. H. White, *Chem. Commun.* **1996**, 1729.
[82] (a) N. C. Baenziger, H. L. Haight, R. Alexander and J. R. Doyle, *Inorg. Chem.* **1966**, *5*, 1399; (b) R. E. Marsh, *Acta Crystallogr.* **1983**, *B39*, 280.
[83] (a) F. S. Mathews and W. N. Lipscomb, *J. Am. Chem. Soc.* **1958**, *80*, 4745; (b) F. S. Mathews and W. N. Lipscomb, *J. Am. Chem. Soc.* **1959**, *81*, 845.
[84] (a) E. O. Fischer, H. P. Hoffmann and A. Treiber, *Z. Naturforsch.* **1969**, *B14*, 599; (b) J. Lorberth, *J. Organomet. Chem.* **1969**, *19*, 189; (c) P. H. M. Budzelaar, J. Boersma, G. J. M. van der Kerk, A. L. Spek and A. J. M. Duisenberg, *J. Organomet. Chem.* **1985**, *281*, 123; (d) O. G. Garkusha, B. V. Lokshin and G. K. Borisov, *J. Organomet. Chem.* **1998**, *553*, 59.
[85] (a) B. Fischer, J. Boersma, G. van Koten, W. J. J. Smeets and A. L. Spek, *Organometallics* **1989**, *8*, 667; (b) B. Fischer, P. Wijkens, J. Boersma, G. van Koten, W. J. J. Smeets, A. L. Spek and P. H. M. Budzelaar, *J. Organomet. Chem.* **1989**, *376*, 223; (c) T. Aoyagi, H. M. M. Shearer, K. Wade and G. Whitehead, *J. Organomet. Chem.* **1978**, *146*, C29.
[86] B. Grossmann, J. Heinze, E. Herdtweck, F. H. Köhler, H. Nöth, H. Schwenk, M. Spiegler, W. Wachter and B. Weber, *Angew. Chem.* **1997**, *109*, 384; *Angew. Chem. Int. Ed. Engl.* **1997**, *36*, 387.
[87] (a) R. E. Rundle and J. H. Goring, *J. Am. Chem. Soc.* **1950**, *72*, 5337; (b) H. G. Smith and R. E. Rundle, *J. Am. Chem. Soc.* **1958**, *80*, 5075.
[88] F. A. Cotton, E. V. Dikarev and S.-E. Stiriba, *Organometallics* **1999**, *18*, 2724.
[89] (a) F. T. Edelmann, in: *Comprehensive Organometallic Chemistry II* (Eds.: E. W. Abel, F. G. A. Stone and G. Wilkinson), Vol. 4, Pergamon Press, London, **1995**; (b) H. Schumann, J. A. Meese-Marktscheffel and L. Esser, *Chem. Rev.* **1995**, *95*, 865; (c) M. N. Bochkarev, L. N. Zakharov and G. S. Kalinina, *Organoderivatives of Rare Earth Elements*, Kluwer Academic Publishers, Dordrecht, **1995**; (d) R. D. Köhn, G. Kociok-Köhn and H. Schumann, in: *Encyclopedia of Inorganic Chemistry* (Ed.: R. B. King), Wiley, London, New York, **1994**; (e) C. Schaverien, *Adv. Organomet. Chem.* **1994**, *36*, 283.
[90] W. J. Evans, R. A. Keyer and J. W. Ziller, *Organometallics* **1993**, *12*, 2618.
[91] W. J. Evans, D. K. Drummond, T. P. Hanusa and J. M. Olofson, *J. Organomet. Chem.* **1989**, *376*, 311.
[92] C. Apostolidis, G. B. Deacon, E. Dornberger, F. T. Edelmann, B. Kanellakopoulos, P. MacKinnon and D. Stalke, *Chem. Commun.* **1997**, 1047.
[93] R. Zerger and G. Stucky, *J. Organomet. Chem.* **1974**, *80*, 7.
[94] P. B. Hitchcock, J. A. K. Howard, M. F. Lappert and S. Prashar, *J. Organomet. Chem.* **1992**, *437*, 177.
[95] H. A. Zinnen, J. J. Pluth and W. J. Evans, *J. Chem. Soc., Chem. Commun.* **1980**, 810.
[96] V. K. Bel'sky, Yu. K. Gunko, B. M. Bulychev, A. I. Sizov and G. L. Soloveichik, *J. Organomet. Chem.* **1990**, *390*, 35.
[97] G. Wilkinson and J. M. Birmingham, *J. Am. Chem. Soc.* **1954**, *76*, 6210.
[98] S. H. Eggers, J. Kopf and R. D. Fischer, *Organometallics* **1986**, *5*, 383.
[99] S. H. Eggers, H. Schultze, J. Kopf and R. D. Fischer, *Angew. Chem.* **1986**, *98*, 631; *Angew. Chem. Int. Ed. Engl.* **1986**, *25*, 656.
[100] J. L. Atwood and K. D. Smith, *J. Am. Chem. Soc.* **1973**, *95*, 1488.
[101] (a) Z. Xie, E. Hahn and C. Qian, *J. Organomet. Chem.* **1991**, *414*, C12; (b) S. D. Stults, R. A. Andersen and A. Zalkin, *Organometallics* **1990**, *9*, 115.
[102] D. M. Barnhart, R. J. Butcher, D. L. Clark, J. C. Gordon, J. G. Watkin and B. D. Zwick, *New. J. Chem.* **1995**, *19*, 503.
[103] H.-J. Gosink, F. Nief, L. Ricard and F. Mathey, *Inorg. Chem.* **1995**, *34*, 1306.

[104] G. B. Deacon and Q. Shen, *J. Organomet. Chem.* **1996**, *506*, 1.
[105] D. M. Barnhart, D. L. Clark, J. C. Gordon, J. C. Huffman, R. L. Vincent, J. G. Watkin and B. D. Zwick, *Inorg. Chem.* **1994**, *33*, 3487.
[106] M. N. Bochkarev, I. L. Fedushkin, V. K. Cherkasov, V. I. Nevodchikov, H. Schumann and F. H. Görlitz, *Inorg. Chim. Acta.* **1992**, *201*, 69.

Subject Index

Alkali metal alkoxides 371, 402, 407, 448
Alkali metal amides 371, 386
Alkali metal cyclopentadienyls 426
Alkali metal fluorenides 429, 431
Alkali metal hydrocarbyls 371
Alkali metal thiolates 405, 408
Alkynol complexes 374
Alumoxanes 109, 113
Aminoalanes 117, 118
Arboroles 3
Arene metal complexes 462
Arsolidine 265
Aryllithium ompounds 374
Aurophilic attraction, aurophilicity 18, 201
Azides, organoaluminum 118
Azides, organoantimony 282
Azides, organogallium 130
Azides, organomercury 212
Azides, organotin 251

Benzenechromium tricarbonyl 78
Benzylmercury cloride 205
Bonding, dative 9
Bonding, hydrogen-bonding 17, 18
Bonding, ionic 20
Bonding, normal covalent 9
Bonding, pi-bonding 23
Bonding, secondary 14
Butyllithium 376

Cadmium cyclopentadienyls 452
Cage structures 13
Cages, aluminum-oxygen 113
Cages, gallium-nitrogen 129
Cages, gallium-oxygen 125

Cages, gallium-sulfur 126
Cages, lithium-nitrogen 395
Cages, lithium-sulfur 409
Cages, organoaluminum-nitrogen 119
Cages, organozinc-oxygen 99
Cages, organozinc-sulfur 102
Cages, sodium-oxygen 410
Cages, tin-oxygen 154, 169
Calixarenes 58
Cobaltocene 68
Cobaltocenium cation 67, 361
Crystal engineering 19, 337
Cubane clusters 70
Cubane structures 13
Cubane, aluminum-nitrogen 120
Cubane, aluminum-phosphorus 122
Cubane, gallium-selenium 128
Cubane, indium-nitrogen 138
Cubane, lithium-oxygen 414
Cubane, manganese-oxygen 349
Cubane, organoaluminum phosphonate 114
Cubane, rhenium-oxygen 349
Cubane, tin-oxygen 175
Cubane, zinc-oxygen 95
Cyanometallates, organometallic 258, 265
Cyclodextrins 76
Cyclopentadienylmercury chloride 203
Cyclopentadienylmetal carbonyls 67
Cyclopentadienyltin(II) chloride 232
Cyclosiloxane rings 29-35
Cyclotetrastibanes 198
Cycloveratrylenes 64

Dendrimers 3
Dibenzenechromium 359

Dibenzenechromium guest 66
Dibismuthanes 199
Diethylmagnesium 72
Diethylzinc 73
Dimethylcadmium 104
Dioxastannolanes 235
Diphenylmagnesium 71
Disiloxanediols, hydrogen bond association 323
Distibanes 195, 196
Distiboles 197
Dithiastannolanes 244
Dithiastibolanes 276
Double complementarity 3

Ethylzinc iodide 95

Ferrocene 77
Ferrocene carboxylic acids 338, 341
Ferrocene-thiourea inclusion compound 79

Gallium cyclopentadienyls 434
Grignard reagents 415

Heterostannocanes 239
Horner-Wittig reagents 382
Host-guest complexes, organoaluminum 73
Host-guest complexes, organogallium 74
Host-guest complexes, organothallium 74

Indium cyclopentadienyls 434
Intercalation of organometallic compounds 68

Ladder, alkali metal-sulfur 412
Ladder, aluminum-nitrogen 120
Ladder, lithium-nitrogen 394
Ladder, organostannoxanes 149, 153, 170
Lanthanides, organometallic compounds 455
Lithium aryls 441

Macrocycles, organocyclosiloxane 29, 30
Macrocycles, cyclosiloxanes 29-34
Macrocycles, organocopper 29
Macrocycles, organosilver 29
Macrocycles, organotin 35
Magnetic materials 70
Menschutkin complexes, gallium 213
Menschutkin complexes, antimony 273
Menschutkin complexes, bismuth 285

Metallocenes 69
Metallocenium cations 69
Metallophilic attraction 18
Metallosiloxane sandwich complexes 34
Metal-organic compounds, definition 2
Methylarsenic 195
Methyllithium 374
Methylmercury derivatives 208
Methylsodium 374
Molecular receptors 4
Molecular recognition 3
Molecular solids 7

Non-molecular solids 7

Organoalkali reagents 371
Organoaluminum alkoxides 110
Organoaluminum amides 116
Organoaluminum carboxylates 113
Organoaluminum halides 108
Organoaluminum hydroxides 111, 113
Organoaluminum selenolates 116
Organoaluminum thiolates 114
Organoantimony carboxylates 275
Organoantimony halides 268
Organoantimony, dithiophosphates, dithiophosphinates 280
Organoarsinic acids 333
Organobismuth alkoxides 287
Organobismuth carboxylates 287
Organobismuth halides 282
Organocadmium halides 104
Organodistannoxanes 148
Organogallium amides 128
Organogallium arsinates 125
Organogallium carboxylates 125
Organogallium halides 123, 213
Organogallium hydroxides 124
Organogallium phosphinates 125
Organogallium selenolates 127
Organogallium sulfides 127
Organogallium sulfinates 125
Organogermanium hydroxides 327
Organoindium alkoxides 135
Organoindium amides 137, 217
Organoindium carboxylates 216
Organoindium halides 134, 214
Organoindium phosphides 139
Organoindium selenolates 136
Organoindium thiolates 136

Organolead alkoxides 261
Organolead carboxylates 261
Organolead halides 259
Organolead hydroxides 261
Organomercury carboxylates 205
Organomercury crown derivatives 73
Organomercury halides 107, 203
Organomercury, dithiophosphates, dithiophosphinates 210
Organometallic carboxylic acids 339
Organometallic compounds, definition 2
Organoselenium halides 290
Organosilanols 319
Organotellurium dithiophosphates, dithiophosphinates 299
Organotellurium halides 293
Organothallium carboxylates 142
Organothallium halides 218
Organothallium hydroxides 141
Organothallium phenoxides 141
Organothallium xanthates 221
Organotin alkoxides 155
Organotin alkoxides 236
Organotin carbonates 171
Organotin carboxylates 156, 168
Organotin compounds, hydrogen bond association 328
Organotin halides 146, 227, 328
Organotin hydroxides 150, 328
Organotin nitrates 171
Organotin phosphates, phosphinates, phosphonates 173
Organotin sulfates, sulfinates 177
Organotin tetrazoles 255
Organotin xanthates 156
Organozinc alkoxides 96
Organozinc dithiocarbamates 100
Organozinc halides 95
Organozinc selenolates 102
Organozinc thiolates 100
Oxadithiabismocane 289
Oxathiastannolanes 238

Phenoarsine 266
Phenyllithium 377, 378
Phosphinoalanes 122
Phosphonate host-guest complexes 35
Plumbocene 438
Prismane structures 13
Prismane, aluminum-nitrogen 120
Prismane, tin-oxygen 175

Receptors, cobaltocene-containing 52
Receptors, organomercury 27
Receptors, ruthenocene-containing 57

Second sphere complexes 75
Secondary bonding 14
Self-assembly, definition 4
Self-organization, definition 5
Silanols, hydrogen bond association 319
Structure, primary 7
Structure, secondary 7
Structure, tertiary 7
Substrate, definition 3
Supermolecule, definition 2
Supramolecular array, definition 3
Supramolecular assembly, definition 2
Supramolecular assistance to molecular synthesis 8
Supramolecular chemistry, definition 1
Supramolecular synthesis, definition 8
Synthons, definition 8

Tectons, definition 8
Telluradiazole 302
Tetramethyldiarsane 195
Thallium cyclopentadienyls 434
Trimethylaluminum 73
Trithiastannocanes 246
Trithiastibocanes 277

Zeolites 65
Zinc cyclopentadienyls 452